AF602137

The Institute of Mathematics
and its Applications
Conference Series

Volumes in the previous series were published by
Academic Press to whom all enquiries should be addressed.
The following and all forthcoming titles are published by
Oxford University Press throughout the world.

NEW SERIES

1. *Supercomputers and parallel computation* Edited by D. J. Paddon
2. *The mathematical basis of finite element methods* Edited by David F. Griffiths
3. *Multigrid methods for integral and differential equations* Edited by D. J. Paddon and H. Holstein
4. *Turbulence and diffusion in stable environments* Edited by J. C. R. Hunt
5. *Wave propagation and scattering* Edited by B. J. Uscinski
6. *The mathematics of surfaces* Edited by J. A. Gregory
7. *Numerical methods for fluid dynamics II* Edited by K. W. Morton and M. J. Baines
8. *Analysing conflict and its resolution* Edited by P. G. Bennett
9. *The state of the art in numerical analysis* Edited by A. Iserles and M. J. D. Powell
10. *Algorithms for approximation* Edited by J. C. Mason and M. G. Cox
11. *The mathematics of surfaces II* Edited by R. R. Martin
12. *Mathematics in signal processing* Edited by T. S. Durrani, J. B. Abbiss, J. E. Hudson, R. N. Madan, J. G. McWhirter, and T. A. Moore
13. *Simulation and optimization of large systems* Edited by Andrzej J. Osiadacz
14. *Computers in mathematical research* Edited by N. M. Stephens and M. P. Thorne
15. *Stably stratified flow and dense gas dispersion* Edited by J. S. Puttock
16. *Mathematical modelling in non-destructive testing* Edited by Michael Blakemore and George A. Georgiou
17. *Numerical methods for fluid dynamics III* Edited by K. W. Morton and M. J. Baines
18. *Mathematics in oil production* Edited by Sir Sam Edwards and P. R. King
19. *Mathematics in major accident risk assessment* Edited by R. A. Cox
20. *Cryptography and coding* Edited by Henry J. Beker and F. C. Piper
21. *Mathematics in remote sensing* Edited by S. R. Brooks
22. *Applications of matrix theory* Edited by M. J. C. Gover and S. Barnett
23. *The mathematics of surfaces III* Edited by D. C. Handscomb
24. *The interface of mathematics and particle physics* Edited by D. G. Quillen, G. B. Segal, and Tsou S. T.
25. *Computational methods in aeronautical fluid dynamics* Edited by P. Stow
26. *Mathematics in signal processing II* Edited by J. G. McWhirter
27. *Mathematical structures for software engineering* Edited by Bernard de Neumann, Dan Simpson, and Gil Slater
28. *Computer modelling in the environmental sciences* Edited by D. G. Farmer and M. J. Rycroft
29. *Statistics in medicine* Edited by F. Dunstan and J. Pickles
30. *The mathematical revolution inspired by computing* Edited by J. H. Johnson and M. J. Loomes
31. *The mathematics of oil recovery* Edited by P. R. King
32. *Credit scoring and credit control* Edited by L. C. Thomas, J. N. Crook, and D. B. Edelman
33. *Cryptography and coding II* Edited by Chris Mitchell
34. *The dynamics of numerics and numerics of dynamics* Edited by D. S. Broomhead and A. Iserles
35. *The unified computation laboratory: modelling, specifications, and tools* Edited by Charles Rattray and Robert G. Clark
36. *Mathematics in the automotive industry* Edited by James R. Smith

Continued overleaf

37. *The mathematics of control theory* Edited by N. K. Nichols and D. H. Owens
38. *Mathematics in transport planning and control* Edited by J. D. Griffiths
39. *Computational ordinary differential equations* Edited by J. R. Cash and I. Gladwell
40. *Waves and turbulence in stably stratified flows* Edited by S. D. Mobbs and J. C. King
41. *Robotics: applied mathematics and computational aspects* Edited by Kevin Warwick
42. *Mathematical modelling for materials processing*
Edited by Mark Cross, J. F. T. Pittman, and Richard D. Wood
43. *Wavelets, fractals, and Fourier transforms*
Edited by M. Farge, J. C. R. Hunt, and J. C. Vassilicos
44. *Computational aeronautical fluid dynamics* Edited by J. Fezoui, J. Hunt, and J. Periaux
45. *Cryptography and coding III* Edited by M. J. Ganley
46. *Parallel computation* Edited by A. E. Fincham and B. Ford
47. *Aerospace vehicle dynamics* Edited by M. V. Cook and M. J. Rycroft
48. *Computer-aided surface geometry and design: mathematics of surfaces IV*
Edited by A. Bowyer
49. *Mathematics in signal processing III* Edited by J. G. McWhirter
50. *Design and application of curves and surfaces: mathematics of surfaces V*
Edited by R. B. Fisher
51. *Artificial intelligence in mathematics* Edited by J. H. Johnson, S. McKee, and A. Vella
52. *Stably stratified flows: flow and dispersion over topography*
Edited by I. P. Castro and N. Rockliffe
53. *Computational learning theory: EuroCOLT 93* Edited by J. S. Shawe-Taylor and M. Antony
54. *Complex stochastic systems and engineering* Edited by D. M. Titterington
55. *Mathematics of dependable systems* Edited by C. J. Mitchell and V. Stavridou
56. *The mathematics of deforming surfaces* Edited by D. G. Dritschel and R. J. Perkins
57. *Mathematical education of engineers* Edited by L. R. Mustoe and S. Hibberd
58. *The mathematics of surfaces VI* Edited by G. Mullineux
59. *Applications of finite fields* Edited by Dieter Gollmann
60. *Applications of combinatorial mathematics* Edited by C. J. Mitchell
61. *Image processing: mathematical methods and applications* Edited by J. M. Blackledge
62. *Flow and dispersion through groups of obstacles* Edited by R. J. Perkins and S. E. Belcher
63. *The state of the art in numerical analysis* Edited by I. S. Duff and G. A. Watson

The State of the Art in Numerical Analysis

Based on the proceedings of a conference on the State of the Art in Numerical Analysis. Organized by the Institute of Mathematics and its Applications and held at the University of York in April 1996.

Edited by

I. S. DUFF

Rutherford Appleton Laboratory

and

G. A. WATSON

University of Dundee

CLARENDON PRESS • OXFORD •

This book has been printed digitally and produced in a standard specification in order to ensure its continuing availability

OXFORD
UNIVERSITY PRESS

Great Clarendon Street, Oxford OX2 6DP

Oxford University Press is a department of the University of Oxford. It furthers the University's objective of excellence in research, scholarship, and education by publishing worldwide in

Oxford New York

Auckland Bangkok Buenos Aires Cape Town Chennai Dar es Salaam Delhi Hong Kong Istanbul Karachi Kolkata Kuala Lumpur Madrid Melbourne Mexico City Mumbai Nairobi São Paulo Shanghai Taipei Tokyo Toronto

Oxford is a registered trade mark of Oxford University Press in the UK and in certain other countries

Published in the United States

Reprinted 2003

ISBN 0-19-850014-9

PREFACE

If it is assumed that a mathematical paper is read by an average of 2.3 people, then the preface to a book of mathematical papers may well be read by even fewer. Nevertheless, we feel it is important to set the scene and to say something about the background to this volume, including some justification for its existence.

Those who are closely involved with Numerical Analysis are very much aware of its continuing growth and development. They recognize the way in which it has broadened in scope over the years and how it has found an ever increasing range of applications. Despite this expansion, many numerical analysts continue to maintain a broad interest in, and knowledge of, the subject. Although it is perhaps becoming increasingly difficult, this global appreciation for the subject is at least partially motivated by the fact that so many branches are closely interdependent: numerical linear algebra underpins much of the subject, different strands of the numerical solution of differential and integral equations are closely intertwined, many numerical methods which apply to approximation problems depend heavily on optimization techniques, and so on.

Therefore it has long been regarded as useful to provide the numerical analysis community, and users of numerical methods, with a forum where an account of the important recent developments in the subject can be presented in a coherent and concentrated way in a manner accessible to the non-specialist in the sub-area. The first State of the Art in Numerical Analysis Meeting was held in Birmingham in 1966, sponsored by the Institute of Mathematics and its Applications. The decision to hold such a meeting in 1966 may have reflected a recognition that the subject had by then reached a certain level of maturity, and indeed a certain level of respectability. The next two State of the Art Meetings took place in York in 1976, and in Birmingham in 1986, again organized by the IMA although their American counterpart, SIAM, was a co-sponsor of the meeting in 1986. The choice of a 10-year cycle was perhaps somewhat arbitrary, but certainly a step-length was fixed by the 1976 meeting, and equi-spaced intervals have always had a particular attraction.

Therefore, when the suggestion was put to the IMA that they should continue their previous support for a meeting in 1996, this was enthusiastically agreed, and a Conference Committee was formed. This consisted of Christopher Baker, Iain Duff, Arieh Iserles, Bill Morton, Mike Powell, and Alistair Watson, who agreed to serve as Chairman. It was decided to choose York as a venue, and the Committee selected a range of topics and speakers. A self-imposed constraint was that the talks should fit into 3.5 days. The competition for space for the more traditional subject areas was further increased when it was decided to include some new applications of the subject, a departure from the pattern of previous meetings which concentrated on mainstream topics. In any event, after much discussion, a list of topics and speakers was agreed that was felt to provide a reasonably satisfactory coverage of the subject.

This volume is meant to extend to a much wider audience the opportunity given to those who attended the meeting. It contains full versions of all the papers presented, with one exception on Neural Nets. It is an interesting exercise to try to identify the main changes in the subject as reflected in the topics covered and the emphasis placed on these in this volume, compared with the corresponding volume produced in 1986. The inclusion of some new applications has already been mentioned, and indeed two important areas, numerical tomography and image processing, are included here.

The other sessions might at first sight appear to present a very traditional framework: linear algebra, ordinary differential equations, integral equations, partial differential equations, approximation and optimization. However, while some of these maintain a fairly conventional form, and represent to a greater or lesser extent a continuous and natural development of topics covered in the 1986 meeting, the impact on others of significant departures from conventional trends is very apparent. This development is particularly marked in ordinary differential equations, with a move away from traditional integrators, and from what might be considered to be the classical framework, to a dynamical systems context. The papers on integral equations on the other hand show a continued development of the analysis of integral and integro-differential equations of Volterra type, and further research on the solution of boundary integral equations. In numerical linear algebra, three papers are primarily concerned with large sparse systems as opposed to one in 1986, and a recurring theme is the exploitation of structure and the increased use of parallelism. The idea of the exploitation of structure carries over to optimization and approximation. In both those areas, there is again increased emphasis on the solution of large problems, and the use of interior point methods plays a much more prominent role than previously. In approximation, wavelets and radial basis functions appear for the first time, and both have already made considerable impact. Last, but by no means least, comes the numerical analysis of partial differential equations. Here there are again considerable advances across a broad front, with finite-element methods and their analysis still prominent. The application of finite-volume methods for hyperbolic differential equations, and the solution of equations which arise in modelling various forms of motion have also seen much progress.

Underlying many of the developments is the fact that more and more complicated and sophisticated problems are now amenable to an increasingly powerful range of numerical techniques. This is greatly helped by the advent of excellent computer packages, like MATLAB, and state of the art Fortran software such as LAPACK, so that the fundamental building blocks for much of the armoury of a numerical analyst are now readily accessible. The amount of computing power on our desks continues to increase, and our access to information by electronic mail or via the internet continues to grow. This is something which still fills with wonder those of us who used to have to trudge through the rain with decks of punched cards and queue up for our turn on the local mainframe, or who remember when remote computing often meant waiting for output from a delivery van, a very different type of bus overload to that found in computing systems today!

As far as the Conference itself is concerned, all speakers deserve thanks for their efforts in preparing and delivering their talks, and we are also grateful to those who so ably assisted as Chairmen for the various sessions. A special thanks should go to John Mason for delivering a typically well crafted and extremely entertaining after dinner speech at the Conference Dinner. We are grateful to the IMA for its support, and in particular to Adrian Lepper and Pamela Bye of that organization who were in attendance throughout, and also to the University of York for its hospitality. The session on Linear Algebra was sponsored by IBM, and we gratefully acknowledge that assistance.

Turning to the production of this Volume, we would like to thank all those who contributed for their co-operation, for their patience, and for their willingness to cope with the technicalities of achieving uniformity of style. The Editors sought comments on all the papers from appropriate experts in the subject, and we are grateful to them, and also to the authors for responding so readily to suggestions which were relayed to them. The papers were checked and put into final camera ready copy form at the IMA before their final journey to Oxford University Press. We are grateful, finally, to that organization, for including this volume in their Conference Series.

We think it fair to say that Numerical Analysis is in an extremely healthy condition. Importantly, there still seems to be a satisfactory flow of young people coming into the subject, bringing new ideas and directions, and keeping the rest of us on our toes. It is certain that over the next ten years there will be further major advances, some of these predictable, and others totally unexpected. We can look forward with confidence to another exciting period of further growth and development.

Iain S. Duff
Rutherford Appleton Laboratory, Oxfordshire

G. Alistair Watson
University of Dundee

ACKNOWLEDGEMENTS

The Institute thanks the authors of the papers, the editors Professor Iain S. Duff (Rutherford Appleton Laboratory, Oxfordshire) and Professor G. Alistair Watson (University of Dundee), and from the Institute Debbie Brown and Karen Jenkins for typesetting the papers, and Simon Byles for proof reading them.

CONTENTS

CONTRIBUTORS

K.E. ATKINSON; Department of Mathematics, University of Iowa, Iowa City, IO 52242, USA.

C.T.H. BAKER; Department of Pure and Applied Mathematics, University of Manchester, Oxford Road, Manchester, M13 9PL.

F. BREZZI; Dipartimento di Matematica, Università di Pavia, via Abbiategrasso 209, 27100 Pavia, Italy.

A.R. CONN; IBM T.J. Watson Research Center, P.O. Box 218, Yorktown Heights, NY 10598, USA.

I.S. DUFF; Numerical Analysis Group, Central Computing Laboratory, Rutherford Appleton Laboratory, Atlas Centre, Chilton, Didcot, Oxon, OX11 0QX.

C.M. ELLIOTT; Centre for Mathematical Analysis and its Applications, School of Mathematical Sciences, University of Sussex, Falmer, Brighton, BN1 9GH.

L.P. FRANCA; Department of Mathematics, University of Colorado at Denver, P.O. Box 173364, Campus Box 170, Denver, CO 80217-3364, USA.

G.H. GOLUB; Department of Computer Science, Stanford University, Stanford, CA 94305, USA.

N.I.M. GOULD; Department for Computation and Information, Rutherford Appleton Laboratory, Atlas Centre, Chilton, Didcot, Oxon, OX11 0QX.

F. GUICHARD; CEREMADE, Université Paris IX Dauphine, Place de Lattre-de-Tassigny, 75775 Paris, Cedex 16, France.

N.J. HIGHAM; Department of Pure and Applied Mathematics, University of Manchester, Oxford Road, Manchester, M13 9PL.

P. HOUSTON; Oxford University Computing Laboratory, Wolfson Building, Parks Road, Oxford, OX1 3QD.

T.J.R. HUGHES; Division of Applied Mechanics, Durand Building, Stanford University, Stanford, CA 94305, USA.

A. ISERLES; Department of Applied Mathematics and Theoretical Physics, University of Cambridge, Silver Street, Cambridge, CB3 9EW.

J-M. MOREL; CEREMADE, Université Paris IX Dauphine, Place de Lattre-de-Tassigny, 75775 Paris, Cedex 16, France.

K.W. MORTON; Numerical Analysis Group, Oxford University Computing Laboratory, Wolfson Building, Parks Road, Oxford, OX1 3QD.

F. NATTERER; Institut fur Numerische and Instrumentelle Mathematik, Universität Münster, Einsteinstrasse 62, D-48149 Münster, Germany.

J. NOCEDAL; Department of Electrical and Computer Engineering, Northwestern University, Evanston, IL 60208, USA.

M.J.D. POWELL; Department of Applied Mathematics and Theoretical Physics, University of Cambridge, Silver Street, Cambridge, CB3 9EW.

A. RUSSO; Istituto di Analisi Numerica del CNR, via Abbiategrasso 209, 27100 Pavia, Italy.

J.M. SANZ-SERNA; Departamento de Matemática Aplicada y Computacioń, Universidad de Valladolid, Facultad de Ciencias, Valladolid, Spain.

D.F. SHANNO; Rutgers Center for Operations Research, Rutgers University, New Brunswick, NJ 08903, USA.

E.M. SIMANTIRAKI; Rutgers Center for Operations Research, Rutgers University, New Brunswick, NJ 08903, USA.

A. STUART; Scientific Computing and Computational Mathematics Program, Durand 257, Stanford University, CA 94305-4040, USA.

E. SÜLI; Oxford University Computing Laboratory, Wolfson Building, Parks Road, Oxford, OX1 3QD.

Ph.L. TOINT; Department of Mathematics, Facultés Universitaires ND de la Paix, 61 rue de Bruxelles, B-5000 Namur, Belgium.

H.A. van der VORST; Mathematical Institute, Universiteit Utrecht, P.O. Box 80.010, 3508 TA Utrecht, The Netherlands.

G.A. WATSON; Department of Mathematics and Computer Science, University of Dundee, Dundee, DD1 4HN.

Recent Developments in Dense Numerical Linear Algebra

Nicholas J. Higham

Department of Pure and Applied Mathematics, University of Manchester

Abstract

We survey recent developments in dense numerical linear algebra, covering linear systems, least squares problems and eigenproblems. Topics considered include the design and analysis of block, partitioned and parallel algorithms, condition number estimation, componentwise error analysis, and the computation of practical error bounds. Frequent reference is made to LAPACK, the state of the art package of Fortran software designed to solve linear algebra problems efficiently and accurately on high-performance computers.

1 Introduction

Numerical linear algebra with dense matrices is still today, as it was at the time of the first IMA meeting on the State of the Art in Numerical Analysis in 1965, an extremely active area of research. There are several reasons. First, we still do not fully understand some of the classical algorithms. For example, the behaviour of the growth factors for Gaussian elimination and its variants with partial and complete pivoting is not yet completely understood. Second, novel computer architectures force us to design new algorithms, modify old ones, and reassess algorithms that were once discarded. In particular, parallel computers have made the flop counts traditionally used to measure the cost of an algorithm of dubious relevance, so that some algorithms hitherto considered inefficient are now of interest again. Third, applications stimulate the development of new theory and techniques. For example, real-time signal processing applications have stimulated much work on rank-revealing factorizations and the updating of factorizations after low rank changes.

We survey research in dense numerical linear algebra over the last decade. Limitations of space force our treatment to be brief and selective. We assume that the reader is familiar with the material in Golub and Van Loan's classic text [69]. We give particular attention to the mathematics underlying the LAPACK library of Fortran software [2], because of LAPACK's practical importance. We say relatively little about accuracy and stability issues, because they are comprehensively treated in our recent book [85].

2 Tools, techniques and software

We begin by describing some of the tools, techniques and software that are an integral part of modern numerical linear algebra.

2.1 BLAS

The notion of Basic Linear Algebra Subprograms (BLAS) goes back at least as far as Wilkinson [129] in the early days of digital computers, who suggested the "preparation of standard routines of considerable generality for the more important processes involved in computation." The first standardized set of Fortran BLAS was published in 1979 [99]. It is now referred to as the level 1 BLAS, because it comprises routines operating on vectors, computing the inner product, the sum $\alpha x + y$ (a "saxpy"), a vector 2-norm and so on. The benefits of using the level 1 BLAS (principally modularity, clarity and potential speed improvements by using optimized BLAS) were recognised by the developers of LINPACK [44], who coded the LINPACK routines so as to call the BLAS from the innermost loops whenever possible. A natural extension to matrix-vector operations called the level 2 BLAS, which supports operations such as matrix times vector and the solution of a triangular system, followed some years later [50], [51]. Subsequently, a level 3 BLAS standard was published [45], [46], supporting matrix-matrix operations. The level 3 BLAS have proved to be very useful for obtaining high efficiency on high-performance computers. The basic reason is that the level 3 BLAS carry out an order of magnitude more work per unit of data than the level 1 and 2 BLAS, as shown in Table 1. Hence on machines with a hierarchical memory (e.g., main memory, cache memory, vector registers) the level 3 BLAS involve less data movement per floating point operation, leading to faster execution. For a more detailed explanation see, for example, [52], [65], or [69, Ch. 1].

Work is ongoing to extend the BLAS standards to support parallelism and sparsity; see [48].

It is important to realise that the BLAS comprise subprogram specifications only; there is freedom in the method used to match the specifications. This freedom is most relevant in the case of the level 3 BLAS, where fast matrix multiplication techniques can be applied. There has been interest in using Strassen's method, which has been demonstrated to be viable for practical computation in

Table 1. BLAS data use and arithmetic work

Level of BLAS	Amount of data	Amount of work
1	$O(n)$	$O(n)$
2	$O(n^2)$	$O(n^2)$
3	$O(n^2)$	$O(n^3)$

terms of both speed and accuracy [12], [81]. With the use of level 3 BLAS based on fast matrix multiplication techniques the existing normwise backward error bounds for block and partitioned algorithms remain valid with appropriate increases in the constant terms [40].

2.2 Block and partitioned algorithms

The recognition that on high-performance computers it is generally desirable to operate on large chunks of data, phrasing algorithms in terms of level 3 BLAS, has led to increased development and use of block algorithms. The term "block algorithm" tends to be used with two different meanings. To be precise, we will adopt the following terminology. A *partitioned algorithm* is a scalar (or point) algorithm in which the operations have been grouped and reordered into matrix operations. Thus, for example, Gaussian elimination expressed in terms of level 3 BLAS is a partitioned algorithm. A *block algorithm* is a generalization of a scalar algorithm in which the basic scalar operations become matrix operations ($\alpha+\beta$, $\alpha\beta$, and α/β become $A+B$, AB, and AB^{-1}), and a matrix property based on the nonzero structure becomes the corresponding property blockwise (in particular, the scalars 0 and 1 become the zero matrix and the identity matrix, respectively). A *block factorization* is defined in an analogous way and is usually what a block algorithm computes.

To illustrate, we describe a partitioned Cholesky factorization algorithm. For a symmetric positive definite $A \in \mathbb{R}^{n\times n}$ and a given block size r, write

$$\begin{bmatrix} A_{11} & A_{12} \\ A_{12}^T & A_{22} \end{bmatrix} = \begin{bmatrix} R_{11}^T & 0 \\ R_{12}^T & I_{n-r} \end{bmatrix} \begin{bmatrix} I_r & 0 \\ 0 & S \end{bmatrix} \begin{bmatrix} R_{11} & R_{12} \\ 0 & I_{n-r} \end{bmatrix}, \tag{2.1}$$

where A_{11} is $r\times r$. One step of the algorithm consists of computing the Cholesky factorization $A_{11} = R_{11}^T R_{11}$, solving the multiple right-hand side triangular system $R_{11}^T R_{12} = A_{12}$, and then forming the Schur complement $S = A_{22} - R_{12}^T R_{12}$; this procedure is repeated on S. The block operations defining R_{12} and S are level 3 BLAS operations. This partitioned algorithm does precisely the same arithmetic operations as any other variant of Cholesky factorization, but it does the operations in an order that permits them to be expressed as matrix operations. In contrast, a block $\mathrm{LDL^T}$ factorization (the most useful block factorization for a symmetric positive definite matrix) has the form $A = LDL^T$, where

$$L = \begin{bmatrix} I & & & \\ L_{21} & I & & \\ \vdots & & \ddots & \\ L_{m1} & \dots & L_{m,m-1} & I \end{bmatrix}, \quad D = \mathrm{diag}(D_{ii}),$$

where the diagonal blocks D_{ii} are, in general, full matrices. This factorization is mathematically different from a point Cholesky or $\mathrm{LDL^T}$ factorization (in fact, for an indefinite matrix it may exist when the point factorization does not).

LAPACK uses exclusively partitioned factorization alg s, because they provide the factorizations to which users are accustomed at higher speed. Block factorizations are in use in various applications, for example for block tridiagonal matrices arising in the discretization of partial differential equations.

2.3 Condition number estimation

One of the novel features of LINPACK [44] was the inclusion with the linear equation solvers of a method for estimating the matrix condition number $\kappa(A) = \|A\|\|A^{-1}\|$. Armed with this estimate, the user could estimate the accuracy of a computed solution. LINPACK's condition estimation algorithm estimates $\|A^{-1}\|_1$ by constructing a vector x for which the lower bound $\|A^{-1}x\|_1/\|x\|_1$ is actually a good approximation to $\|A^{-1}\|_1$ [28]. It makes explicit use of an LU factorization of A, but requires only $O(n^2)$ operations beyond the $O(n^3)$ operations required to factorize the $n \times n$ matrix A. The algorithm produces an approximate null vector (namely, $A^{-1}x$), and it is this rather than the $\|A^{-1}\|$ estimate that is required in some applications. Indeed a precursor of the LINPACK algorithm was developed for such an application [71].

The now widespread use of condition estimation relies on two developments. First, Hager [76] devised a method for estimating $\|B\|_1$ that computes a (usually small) number of matrix-vector products involving B and B^T. An immediate advantage over the LINPACK estimator is that Hager's method can be programmed as a black box that requires no knowledge of the details of how the products Bx and B^Tx are computed, making it of general applicability (to mimic the LINPACK estimator, take $B = A^{-1}$ and use an LU factorization to compute the products with B). The key to the widespread use of Hager's method was the observation by Arioli, Demmel and Duff [3] that the method can be used to estimate $\| |A^{-1}|d \|_\infty$ for any given nonnegative vector d, which could, for example, be a matrix-vector product. Writing $D = \text{diag}(d)$ and $e = [1, 1, \ldots, 1]^T$, we have

$$\| |A^{-1}|d \|_\infty = \| |A^{-1}|De \|_\infty = \| |A^{-1}D|e \|_\infty = \| |A^{-1}D| \|_\infty = \|A^{-1}D\|_\infty. \tag{2.2}$$

Since $\|B\|_\infty = \|B^T\|_1$, Hager's method is clearly applicable, just requiring the solution of linear systems with coefficient matrices A and A^T and multiplication of a vector by D. LAPACK makes extensive use of condition estimation to provide both condition estimates and forward error bounds. It implements an algorithm of Higham [80], which incorporates modifications to Hager's algorithm that make it more robust, reliable and efficient. The norm estimates are nearly always within a factor 3 of the true norm.

Counterexamples are known for all existing condition estimators, that is, classes of matrix are known where the estimate is wrong by an arbitrarily large factor. Indeed, it seems likely that estimating $\|A^{-1}\|$ to within a factor depending only on the dimension of A is at least as expensive as computing A^{-1} [35].

2.4 LAPACK and ScaLAPACK

LAPACK, first released in 1992 and regularly updated, is a collection of Fortran 77 programs for solving various linear equation, linear least squares and eigenvalue problems. It can be regarded as a successor to the 1970s packages LINPACK [44] and EISPACK [117], [66]. It has virtually all the capabilities of these two packages and much more besides. LAPACK improves on LINPACK and EISPACK in four main respects: speed, accuracy, robustness and functionality. While LINPACK and EISPACK are based on level 1 BLAS, LAPACK was designed at the outset to use partitioned algorithms wherever possible, so as to exploit the level 3 BLAS. LAPACK can achieve improved accuracy in solving linear equations and certain types of eigenproblems by the use of techniques and methods described later in this paper. A C translation of LAPACK is available, as well as a C++ wrapper for most of the Fortran version of LAPACK [49]. For more about LAPACK, see the users' guide [2].

ScaLAPACK is a subset of LAPACK routines redesigned for distributed memory parallel computers [27], [55]. ScaLAPACK routines make use of BLAS, Parallel BLAS (distributed memory versions of the level 2 and 3 BLAS) and a set of low level communication primitives called the BLACS. For solving square linear systems implementations of partitioned LU and Cholesky factorization are provided that use a block cyclic data distribution.

2.5 MATLAB

MATLAB is an interactive package for matrix computations that was originally developed by Moler for teaching purposes [104]. It provides easy access to software from the EISPACK and LINPACK libraries. The original Fortran version was rewritten in C in the mid 1980s and made into a commercial package [103]. While MATLAB remains widely used for teaching, it has become an almost essential computational tool for researchers in numerical linear algebra, as a glance at current journals reveals. The introduction of sparse matrix handling in MATLAB 4.0 significantly enhanced the usefulness of the package [67].

2.6 IEEE arithmetic

Two IEEE standards for floating point arithmetic have been published, one for base 2 [89] and one that is independent of the base [90]. The base 2 standard is supported by virtually all current workstations and high-performance computers, as well as PCs, via the Intel 80x86/7 and Pentium chips. The benefits of using a well-designed and precisely specified arithmetic such as the IEEE standard are now well understood and documented. They include easier handling or avoidance of underflow and overflow and other arithmetic exceptions. More subtly, a number of algorithms of interest for their speed or accuracy can fail or must run slowly in arithmetics that do not satisfy certain properties possessed by the IEEE standard, such as the correct rounding of floating point operations [38], [85, Chs. 2, 25].

3 Linear systems

Developments in dense linear systems have mainly been motivated by the need to solve large systems efficiently and the desire to estimate or improve the accuracy and stability of computed solutions. We summarize work in these directions, beginning with new results on traditional algorithms.

3.1 LU factorization

Progress has been made on understanding the behaviour of the growth factor for Gaussian elimination with partial and complete pivoting. Recall that the growth factor is defined by

$$\rho_n = \frac{\max_{i,j,k} |a_{ij}^{(k)}|}{\max_{i,j} |a_{ij}|},$$

where $a_{ij}^{(k)}$ $(k = 1\colon n)$ are the intermediate elements occurring during the elimination on $A \in \mathbb{R}^{n \times n}$. The long-standing conjecture that $\rho_n \le n$ for complete pivoting is now known to be false. Gould [70] found a counterexample in floating point arithmetic and Edelman modified it to create a counterexample in exact arithmetic [57], [58]. By how much ρ_n can exceed n for complete pivoting is not known. Examples that can occur in practical applications where partial pivoting yields exponentially large growth factors are identified by Foster [64] and Wright [131], while Higham and Higham [86] identify matrices for which *any* pivoting strategy gives growth factors of order n. An experimental investigation by Trefethen and Schreiber [126] suggests that the behaviour of the growth factor can be explained by statistical means, but no theorems are proved. Yeung and Chan [133] prove probabilistic results for Gaussian elimination without pivoting. The most convincing argument to explain why the growth factor is almost always small for partial pivoting is due to Trefethen [125]: he shows that for a matrix to produce a large growth factor with partial pivoting its column spaces must possess a certain skewness property and he proves that this property holds with low probability for random matrices with elements from a normal distribution.

Componentwise perturbation theory and backward errors for linear systems can be used to produce practical error bounds, but it is interesting to note that the best a posteriori forward error bound for an approximate solution $\widehat{x}$ to $Ax = b$ is perhaps the most trivial. Defining the residual $r = b - A\widehat{x}$ we have $|x - \widehat{x}| \le |A^{-1}||r|$, which is as sharp a bound as can be obtained if we ignore the signs in A^{-1} and r. In floating point arithmetic we obtain not r but $r + \Delta r$, where $|\Delta r| \le f(n,u)(|A||\widehat{x}| + |b|)$, where $f(n,u) \approx (n+1)u$ is a function of the problem size n and the unit roundoff u. Hence our practical bound is

$$\frac{\|x - \widehat{x}\|_\infty}{\|\widehat{x}\|_\infty} \le \frac{\|\,|A^{-1}|(|\widehat{r}| + f(n,u)(|A||\widehat{x}| + |b|)\,\|_\infty}{\|\widehat{x}\|_\infty}.$$

This is the bound estimated and returned to the user by LAPACK's expert driver routine for linear systems. The bound is optimal in a sense explained in [85, Lem. 7.9].

Symmetric indefinite linear systems are of much current interest because they arise in applications such as interior methods in constrained optimization, the least squares problem, and the Stokes problem in PDEs. Dense symmetric indefinite systems are usually solved using a block $\mathrm{LDL^T}$ factorization $PAP^T = LDL^T$, where P is a permutation matrix, D is block diagonal with diagonal blocks of dimension 1 or 2, and L is unit lower triangular. In the 1970s a complete pivoting strategy ($O(n^3)$ searching) was developed by Bunch and Parlett [24] and a partial pivoting strategy ($O(n^2)$ searching) was proposed by Bunch and Kaufman [22]. The Bunch–Kaufman pivoting strategy is used in LINPACK and LAPACK. While an exponential bound for the growth factor was proved in [22], a proof that the strategy is backward stable in the absence of large element growth has only recently been given [84]. For the Bunch–Kaufman pivoting strategy $\|L\|/\|A\|$ is unbounded, which does not affect the stability but is undesirable in certain applications. Ashcraft, Grimes and Lewis [4] develop some new pivoting strategies for the block $\mathrm{LDL^T}$ factorization; in particular, they give a "bounded Bunch–Kaufman" strategy that bounds $\|L\|/\|A\|$ and is usually of similar cost to the Bunch–Kaufman strategy, although it can require $O(n^3)$ searching in the worst case.

Vectorized and partitioned algorithms for the standard matrix factorizations are now well understood, and the algorithms in LAPACK represent the state of the art. Three representative references for LU factorization spanning the last 12 years include Dongarra, Gustavson and Karp [47], Ortega [105] and Dongarra, Duff, Sorensen and van der Vorst [52]. For variants of LU factorization, developing or choosing a partitioned algorithm may not be trivial. For example, developing a partitioned version of the block $\mathrm{LDL^T}$ factorization is somewhat complicated for the Bunch–Kaufman partial pivoting strategy [4], [52], [94], [97]. For matrix inversion there is a bewildering choice of variants of methods based on LU factorization, each with slightly different error bounds, performance properties and storage requirements [56]. The need for care in designing partitioned algorithms is illustrated by the fact that the (stable) partitioned algorithm for inverting a triangular matrix used in LAPACK has an equally plausible variant that is unstable.

Few genuine block algorithms are used for linear system solution, an exception being block LU factorization. How and when to pivot in block LU factorization are key questions. A positive result is that if the matrix is diagonally dominant by columns in either the point or the block sense, then appropriate implementations are perfectly stable [41]. On the other hand, for a symmetric positive definite matrix stability is assured only if the matrix is well conditioned (this being a rare example where symmetric positive definiteness is not the most desirable property a matrix can have).

3.2 Iterative refinement

If y is an approximate solution to a linear system $Ax = b$ then by forming the residual $r = b - Ay$ and solving $Ad = r$ we obtain the correction d such that $x = y+d$. In floating point arithmetic the calculation of r is traditionally done in higher precision and the process iterated. Nowadays it is more common to carry out the whole process at the working precision. This fixed precision iterative refinement was first advocated and shown to be effective in the 1970s [91], [115]. It is used in LAPACK in conjunction with the linear equation solvers based on LU factorization. Current understanding can be summarized as follows [82]. Consider *any* linear equation solver. We require only that the computed solution $\widehat{x}$ satisfies $(A + \Delta A)\widehat{x} = b$ with $\|\Delta A\|_\infty \le f(A,n)u\|A\|_\infty$, where f is a scalar function depending only on A and the dimension n. Provided that $f(A,n)$ is not too large, A is not too ill conditioned, and the vector $|b| + |A||x|$ is not too badly scaled (in particular, it has no zero elements), fixed precision iterative refinement will eventually produce an improved solution $\widetilde{x}$ satisfying

$$(A + \Delta A)\widetilde{x} = b + \Delta b, \qquad |\Delta A| \le \epsilon|A|, \quad |\Delta b| \le \epsilon|b|, \tag{3.1}$$

where $\epsilon \le 4nu$. Technically, this result says that $\widetilde{x}$ has a small componentwise relative backward error. The important feature of (3.1) is that the elements of the perturbation ΔA are bounded relative to the corresponding elements of A, and similarly for b. Thus any scaling or sparsity in the data is preserved in the perturbations. When the solver is Gaussian elimination with partial pivoting, a *single* step of iterative refinement is usually enough to achieve (3.1), as shown by Skeel [115].

3.3 Parallel algorithms

Much research has been devoted to the parallel solution of linear systems. Software in practical use parallelizes Gaussian elimination with partial pivoting in a straightforward way, but other, more unusual, methods exist. While new techniques have been developed, old methods have also attracted renewed interest. For example, pivoting can be incorporated into Gaussian elimination in a way that avoids the sequential search down a column required by partial pivoting: at each element-zeroing step we interchange (if necessary) the pivot row and the row whose first element is to be zeroed to ensure that the multiplier is bounded by 1. In one particular algorithm, the first stage introduces zeros into elements $(2,1)$, $(3,1)$, ..., $(n,1)$, in this order, and the potential row interchanges are $1 \leftrightarrow 2$, $1 \leftrightarrow 3$, ..., $1 \leftrightarrow n$, instead of just one row interchange as for partial pivoting. As well as saving on the pivot searching, this method has the advantage of permitting eliminations to be performed in parallel. For a 6×6 matrix we can represent the elimination as follows, where an integer k denotes elements that can be eliminated in parallel on the kth stage:

$$\begin{bmatrix} \times & \times & \times & \times & \times & \times \\ 1 & \times & \times & \times & \times & \times \\ 2 & 3 & \times & \times & \times & \times \\ 3 & 4 & 5 & \times & \times & \times \\ 4 & 5 & 6 & 7 & \times & \times \\ 5 & 6 & 7 & 8 & 9 & \times \end{bmatrix}.$$

In general, there are $2n - 3$ stages in each of which up to $n/2$ elements are eliminated in parallel. This algorithm is discussed by Wilkinson [130, pp. 236–237] and Gallivan et al. [65]. Sorensen [119], derives an error bound for the factorization that is proportional to 4^n which, roughly, comprises a factor 2^{n-1} bounding the growth factor and a factor 2^{n-1} bounding "L". This method, along with other variants such as pairwise pivoting, in which all row operations are between pairs of adjacent rows only, has not attracted wide use for parallel computing.

Vavasis [127] shows that Gaussian elimination with partial pivoting (GEPP) is P-complete, a complexity result whose implication is that GEPP cannot be efficiently implemented on a highly parallel computer with a large number of processors. This result suggests that we look at methods other than Gaussian elimination for efficient solution of linear systems on massively parallel machines. Other methods are in existence. For example, Csanky [29] gives a method for inverting an $n \times n$ matrix (and thereby solving a linear system) in $O(\log^2 n)$ time on $O(n^4)$ processors. This method has the optimal complexity amongst currently known methods, but it involves the use of the characteristic polynomial and has abysmal numerical stability properties!

A more practical method is Newton's method for A^{-1}, which is known as the Schulz iteration [113]:

$$X_{k+1} = X_k(2I - AX_k) = (2I - X_kA)X_k.$$

It is known that X_k converges to A^{-1} (or, more generally, to the pseudo-inverse if A is rectangular) if $X_0 = \alpha_0 A^T$ and $0 < \alpha_0 < 2/\|A\|_2^2$ [118]. The Schulz iteration is attractive because it involves only matrix multiplication—an operation that can be implemented very efficiently on high-performance computers. The rate of convergence is quadratic, since if $E_k = I - AX_k$ or $E_k = I - X_kA$ then

$$E_{k+1} = E_k^2 = \cdots = E_0^{2^{k+1}}.$$

Like Csanky's method, the Schulz iteration has polylogarithmic complexity, but, unlike Csanky's method, it is numerically stable [118]. Unfortunately, for scalar $a > 0$,

$$0 < x_k \ll a^{-1} \quad \Rightarrow \quad x_{k+1} = 2x_k - x_k a x_k \lesssim 2x_k,$$

and since $x_k \to a^{-1}$, convergence can initially be slow. Overall, about $2\log_2 \kappa_2(A)$ iterations are required for convergence in floating point arithmetic. Pan and

Schreiber [107] show how to accelerate the convergence by at least a factor 2 via scaling parameters and how to use the iteration to carry out rank and projection calculations. For an interesting application of the Schulz iteration to a sparse matrix possessing a sparse inverse, see [1].

4 The least squares problem

In this section we consider the least squares (LS) problem $\min_x \|Ax-b\|_2$, where $A \in \mathbb{R}^{m\times n}$ with $m \geq n$. The most thorough and up-to-date treatment of the LS problem is given by Björck's book [19].

4.1 The seminormal equations

Given a QR factorization $A = QR$, where $Q \in \mathbb{R}^{m\times n}$ has orthonormal columns and $R \in \mathbb{R}^{n\times n}$ is upper triangular, the normal equations $A^TAx = A^Tb$ transform to the triangular system $Rx = Q^Tb$. Intermediate between these two systems are the seminormal equations (SNE) $R^TRx = A^Tb$; since they do not involve Q, they are attractive for multiple right-hand side problems where Q is too large to store. The SNE method has been in use since the early 1970s, and its stability is explained by analysis of Björck [18], who makes the assumption that R is obtained from a stable QR factorization method. Björck obtains a forward error bound proportional to $\kappa_2(A)^2$, just like for the normal equations method that Cholesky factorizes A^TA; hence the SNE method is not backward stable. To improve the stability a step of iterative refinement (in fixed precision) can be added, giving the corrected seminormal equations (CSNE) method:

$$
\begin{aligned}
&R^TRx = A^Tb \text{ (solve for } x) \\
&r = b - Ax \\
&R^TRw = A^Tr \text{ (solve for } w) \\
&y = x + w
\end{aligned}
$$

Björck obtains a forward error bound for the CSNE method that can be smaller or larger than that for a backward stable method, depending on the conditioning of the problem: hence the refinement process can bring a useful improvement in accuracy.

4.2 Modified Gram–Schmidt

Perhaps the oldest method for computing the QR factorization of a matrix is the Gram–Schmidt method. It exists in both "classical" and "modified" forms, which are equivalent mathematically but different numerically, with the modified Gram–Schmidt (MGS) method having the better stability properties. The MGS method is widely used, for example within iterative methods such as the Arnoldi method and GMRES. A remarkable connection between the MGS method and Householder QR factorization has been known since the 1960s but has only recently been fully exploited [20]: the MGS method applied to $A \in \mathbb{R}^{m\times n}$ is equivalent, both mathematically *and numerically*, to Householder QR factorization of the padded matrix $\left[\begin{smallmatrix} 0_n \\ A \end{smallmatrix}\right] \in \mathbb{R}^{(m+n)\times n}$. This connection brings two benefits. First, it leads to shorter and more insightful error analysis for the MGS method. Second, it leads to new stable algorithms. Björck and Paige [20], [21] derive a new backward stable MGS-based algorithm for solving the "augmented system"

$$\begin{bmatrix} I & A \\ A^T & 0 \end{bmatrix} \begin{bmatrix} y \\ x \end{bmatrix} = \begin{bmatrix} b \\ c \end{bmatrix}.$$

This system characterizes the solutions to the following two problems:

$$\min_x \|b - Ax\|_2^2 + 2c^T x,$$
$$\min_y \|y - b\|_2 \quad \text{subject to} \quad A^T y = c.$$

Note that these problems include the LS problem and the minimum 2-norm solution of an underdetermined system as special cases. Björck and Paige's algorithm reduces to the usual method for using MGS to solve the LS problem, but gives a new backward stable method for computing the minimum 2-norm solution to an underdetermined system.

4.3 Block and partitioned QR factorization algorithms

Just as for LU factorization, there are block and partitioned versions of QR factorization. The key to deriving a partitioned QR factorization algorithm is to aggregate a product of Householder transformations so that their application becomes a matrix multiplication. The idea is to represent the product $Q_r = P_r P_{r-1} \dots P_1$ of r Householder transformations $P_i = I - v_i v_i^T \in \mathbb{R}^{m\times m}$ (where $v_i^T v_i = 2$) in the form

$$Q_r = I + W_r Y_r^T, \qquad W_r, Y_r \in \mathbb{R}^{m\times r},$$

as suggested by Bischof and Van Loan [17]. This is achieved using the recurrence

$$W_1 = -v_1, \; Y_1 = v_1, \qquad W_i = [W_{i-1} \;\; -v_i], \; Y_i = [Y_{i-1} \; Q_{i-1}^T v_i]. \tag{4.1}$$

A partitioned QR factorization can be developed as follows. Partition $A \in \mathbb{R}^{m\times n}$ $(m \geq n)$ as

$$A = [A_1 \quad B], \qquad A_1 \in \mathbb{R}^{m\times r}, \tag{4.2}$$

and compute the Householder QR factorization of A_1,

$$P_r P_{r-1} \dots P_1 A_1 = \begin{bmatrix} R_1 \\ 0 \end{bmatrix}.$$

Accumulate the product $P_r P_{r-1} \dots P_1 = I + W_r Y_r^T$ using (4.1), as the P_i are generated, and then update B according to

$$B \leftarrow (I + W_r Y_r^T) B = B + W_r (Y_r^T B),$$

which involves only level 3 BLAS operations. Repeat the process on the last $m-r$ rows of B. Extra work is required to accumulate the WY factors but on a high-performance machine this is outweighed by the gain in speed from applying level 3 BLAS operations.

LAPACK uses this partitioned QR factorization algorithm, though with a more storage efficient variant of the WY form developed by Schreiber and Van Loan [112].

To develop a true block QR factorization a generalization of a Householder matrix called a block reflector can be used. Given $Z \in \mathbb{R}^{m\times n}$ $(m \geq n)$ the "reflector that reverses the range of Z" is

$$H = H(Z) = I_m - ZWZ^T, \qquad W = 2(Z^TZ)^+ \in \mathbb{R}^{n\times n}.$$

Given $E \in \mathbb{R}^{m\times n}$ $(m > n)$ a basic task is to find a block reflector H such that

$$HE = \begin{bmatrix} F \\ 0 \end{bmatrix}, \qquad F \in \mathbb{R}^{n\times n} \tag{4.3}$$

If this task is carried out within the general framework of the partitioned QR factorization described above, then a factorization $A = QR$ is produced with a block triangular R. Schreiber and Parlett [111] develop theory and algorithms for block reflectors. The solution of the task (4.3) is quite expensive and involves computing one or more polar decompositions, though the whole procedure can be implemented in a way that is rich in matrix multiplication [111].

4.4 Backward error

One possible definition of backward error for an approximate solution y to the LS problem $\min_x \|Ax - b\|_2$ is

$$\eta_F(y) := \min\{\, \|[\Delta A,\ \theta\Delta b]\|_F : \|(A+\Delta A)y - (b+\Delta b)\|_2 = \min\,\}, \tag{4.4}$$

where θ is a parameter. It has been known since the 1960s that the computed solution $\widehat{x}$ obtained via a QR factorization has a backward error $\eta_F(\widehat{x})$ of order u, for $\theta = 1$ (say). However, we had no way of evaluating η_F (or any minor variant of it) numerically until recently, when Waldén, Karlson and Sun [128] showed that

$$\eta_F(y) = \begin{cases} \dfrac{\|r\|_2}{\|y\|_2}\sqrt{\mu}, & \lambda_* \geq 0, \\[2ex] \left(\dfrac{\|r\|_2^2}{\|y\|_2^2}\mu + \lambda_*\right)^{1/2}, & \lambda_* < 0, \end{cases}$$

where $r = b - Ay$ and

$$\lambda_* = \lambda_{\min}\left(AA^T - \mu\frac{rr^T}{\|y\|_2^2}\right), \quad \mu = \frac{\theta^2\|y\|_2^2}{1+\theta^2\|y\|_2^2}.$$

Here, we denote by $\lambda_{\min}$ and $\sigma_{\min}$ the smallest eigenvalue of a symmetric matrix and the smallest singular value of a general matrix, respectively. For computation, the expression for η_F is best evaluated as

$$\eta_F(y) = \min\{\, \eta_1,\ \sigma_{\min}([A \ \ \eta_1 C])\,\}, \qquad \eta_1 = \frac{\|r\|_2}{\|y\|_2}\sqrt{\mu}, \quad C = I - \frac{rr^T}{r^Tr}.$$

Sun [123] has found an expression for a generalization of the backward error (4.4) to multiple right-hand side problems, and Sun and Sun [124] have evaluated the backward error for the minimum 2-norm solution to an underdetermined system.

4.5 Rank-revealing factorizations

A topic of much research interest in recent years has been the computation of rank-revealing factorizations. There is no generally agreed definition of what is a rank-revealing factorization (see [26] for an insightful discussion), but a basic property required is that for a rank-deficient matrix the factorization indicates the rank and readily yields information about the range space and the null space. Practical interest focuses on nearly rank-deficient matrices, for which we want a factorization that "displays" the near rank-deficiency. The singular value decomposition (SVD) is the ultimate rank-revealing factorization, and although it is not significantly more expensive than alternative rank-revealing factorizations to compute, it is too expensive to update the SVD when a row is added or removed from the matrix, as happens repeatedly in signal processing applications. Therefore other factorizations involving orthogonal matrices have been investigated.

For QR factorization we try to choose a permutation matrix Π so that $A\Pi = QR$ is a rank-revealing QR factorization. The standard column pivoting strategy [69, §5.4.1] tends to be rank-revealing, but it can completely fail to reveal near rank-deficiency.

Foster [63] and Chan [25] develop iterative algorithms for computing rank-revealing QR factorizations; for both algorithms the bounds that show by how much, in the worst case, the factorizations may fail to reveal the rank contain factors exponential in the rank deficiency. Both methods need a condition estimator, to produce approximate null vectors of certain triangular matrices. Chandrasekaran and Ipsen [26] give a systematic treatment of algorithms for computing rank-revealing QR factorizations. In an important recent development, Gu and Eisenstat [75] derive an algorithm that is guaranteed to compute a strong form of rank-revealing QR factorization whose properties include that it stably provides an approximation to the null space; the algorithm has the same complexity as the column pivoting strategy, though is up to 50% more expensive in practice.

Hong and Pan [87] gave the first proof of the existence of a rank-revealing QR factorization (that is, the existence of a suitable Π) with constants that are not exponential in the dimensions. The proof involves determinant maximization, so does not lead to a practical algorithm.

Stewart [120] considers a URV decomposition of a rank r matrix $A \in \mathbb{R}^{m\times n}$:

$$A = U \begin{bmatrix} R & 0 \\ 0 & 0 \end{bmatrix} V^T,$$

where $R \in \mathbb{R}^{r\times r}$ is upper triangular and $U \in \mathbb{R}^{m\times m}$ and $V \in \mathbb{R}^{n\times n}$ are orthogonal. This is traditionally known as a complete orthogonal decomposition, and was introduced by Faddeev, Kublanovskaja and Faddeeva [59] and by Hanson and Lawson [77]. Denote the ith largest singular value of A by $\sigma_i(A)$. Stewart defines a rank-revealing URV decomposition of a matrix $A \in \mathbb{R}^{m\times n}$ considered to be nearly of rank r by

$$\begin{aligned} A &= U \begin{bmatrix} R & F \\ 0 & G \end{bmatrix} V^T, \qquad R \in \mathbb{R}^{r\times r}, \\ & \sigma_r(R) \approx \sigma_r(A), \quad \|F\|_2^2 + \|G\|_2^2 \approx \sigma_{r+1}^2 + \cdots + \sigma_n^2. \end{aligned}$$

Stewart shows that the URV decomposition is easy to update (when a row is added) using Givens rotations and is suitable for parallel implementation [121]; for downdating of the decomposition see [14], [15], [108]. Initial determination of the URV decomposition can be done by applying the updating algorithm as the rows are brought in one at a time. An analogous ULV decomposition exists in which the middle factor is lower triangular. As shown by the analysis of Fierro and Bunch [62], the rank-revealing ULV decomposition tends to produce better approximations to the numerical null space, while the rank-revealing URV decomposition tends to produce better approximations to the numerical range space (a clue as to why this should be can be obtained by looking at the forms of R^TR and L^TL).

5 The nonsymmetric eigenvalue problem

The QR algorithm, now over thirty years old, remains the method of choice for computing the complete eigensystem of a nonsymmetric matrix. Although the method is widely used and regarded as being extremely reliable, no convergence proof exists. Indeed, matrices are known for which the QR algorithm (with exceptional shifts, as implemented in EISPACK and LAPACK 2.0) fails to converge, both in exact arithmetic and in floating point arithmetic [16], [31]; heuristic remedies are proposed by Day [31].

LAPACK includes an implementation of the QR algorithm that uses a multishift strategy to enhance the performance on high-performance machines [5]. In the Hessenberg QR iteration, instead of using a single or double shift and chasing the resulting bulge of dimension 1 or 2 a column at a time down the matrix, k simultaneous shifts are used and the $k \times k$ bulge is chased p columns at a time. Here, k and p are implementation-dependent parameters. The k shifts are chosen as the eigenvalues of the trailing principal submatrix of order k. This multishift QR algorithm enjoys a high proportion of level 2 and 3 BLAS operations.

The reduction of a general matrix to Hessenberg form that precedes the QR iteration can be organized so that it involves level 3 BLAS operations, making use of the WY Householder aggregation technique described in §4.3 [53]. Extra arithmetic operations and storage are required, but greater efficiency is obtained through the use of block operations.

Error bounds and condition estimation for eigenvalue computations are now well developed, and comprehensive summaries are given in [11], [2, Chap. 4]. A complicating factor is the large number of different error bounds that one can try to estimate: bounds for a simple eigenvalue and its eigenvector or for a cluster and the corresponding invariant subspace, and both asymptotic and strict bounds of each type. The LAPACK eigenvalue and singular value expert driver routines provide selected condition estimates, which make easy the computation of error bounds by the user. Some of the condition estimation procedures require the reordering of the upper triangular Schur form, so that the eigenvalues of interest are in the top left-hand corner. This is a computation that arises in several applications, including that of computing an invariant subspace corresponding to a given group of eigenvalues. LAPACK uses an improved, guaranteed-stable version of an earlier reordering algorithm [8].

For parallel solution of the nonsymmetric eigenvalue problem there is no clear method of choice. The QR algorithm is a fine-grained algorithm that has proven difficult to parallelize, though progress has been made in [79]. Various interesting alternative approaches have been proposed, which we now briefly summarize.

Two new divide and conquer methods work on a upper Hessenberg matrix, and therefore, like the QR algorithm, require an initial unitary reduction to Hessenberg form. Both tear an upper Hessenberg matrix by a subdiagonal element, solve the resulting independent, smaller Hessenberg eigenproblems in parallel, then merge the answers into a solution for the original problem. Dongarra and Sidani [54] use Newton's method for the merging, while Li and his co-workers use homotopy continuation [100]. Drawbacks of both approaches, which include possibly ill conditioned subproblems and convergence difficulties, are described in [92].

An approach called spectral divide and conquer is defined as follows. Let $A \in \mathbb{R}^{n\times n}$, let $Q_1 \in \mathbb{C}^{n\times p}$ have orthonormal columns that span an invariant subspace for A, and choose Q_2 so that $Q = [Q_1, Q_2]$ is unitary. Then

$$Q^*AQ = \begin{bmatrix} B_{11} & B_{12} \\ 0 & B_{22} \end{bmatrix}, \tag{5.1}$$

and the eigenvalues of B_{11} are those of A corresponding to the invariant subspace spanned by Q_1. This process is repeated on B_{11} and B_{22} until the desired eigenvalues are found. The question is how to choose Q. One way is to use the matrix sign function [110], which generalizes the usual sign of a complex number. For $A \in \mathbb{C}^{n\times n}$ with Jordan canonical form

$$A = X \begin{bmatrix} J_+ & 0 \\ 0 & J_- \end{bmatrix} X^{-1},$$

where J_+ has eigenvalues in the open right half-plane and J_- has eigenvalues in the open left half-plane, we define

$$\operatorname{sign}(A) = X \begin{bmatrix} I & 0 \\ 0 & -I \end{bmatrix} X^{-1}.$$

Suppose J_+ is k-by-k. Then the first k columns of Q in the (rank-revealing) QR factorization $\operatorname{sign}(A) + I = QR$ span the invariant subspace corresponding to J_+. This means that (5.1) holds, where B_{11} has eigenvalues in the right half-plane and B_{22} has eigenvalues in the left half-plane; thus the spectrum has been divided along the imaginary axis. By computing the QR factorization of the sign function of $\alpha A + \beta I$ for complex α and β, or of $(A + \beta I)^2 + \alpha I$ for real α and β, we can divide the spectrum along other lines in the complex plane, retaining real arithmetic if A is real [7], [88], [122]. Using this approach, we can determine the eigenvalues lying within quite general regions of the complex plane.

The matrix sign function can be computed using the Newton iteration $A_{i+1} = \frac{1}{2}(A_i + A_i^{-1})$, $A_0 = A$, which converges globally and quadratically to $\operatorname{sign}(A)$ whenever $\operatorname{sign}(A)$ is defined. Other iterations, some with more natural parallelism, are available too [98]. Bai and Demmel [7] develop a toolbox of routines based on the matrix sign function for finding all the eigenvalues in a region of the complex plane and the corresponding invariant subspace. The building blocks are BLAS, QR and LU factorizations and the matrix sign function. They also develop supporting perturbation theory, stability analysis and refinement schemes [9].

Tools other than the sign function can be used to obtain the desired invariant subspace Q_1. One alternative is an algorithm involving no matrix inversions that is explored by Bai, Demmel and Gu [10] and is based on original algorithms of Bulgakov, Godunov and Malyshev. Other schemes have been suggested; see the references in [10].

The spectral divide and conquer approach can be extended to compute deflating subspaces of a matrix pencil $A - \lambda B$ and hence to solve the generalized eigenproblem; see [9], [10]. Condition numbers and error bounds for the generalized eigenproblem and algorithms for reordering the generalized Schur form are given by Kågström and Poromaa [95].

A comprehensive survey of existing parallel eigenroutines and their limitations is given in [39].

6 The symmetric eigenvalue problem

An important recent development in solution of the symmetric eigenproblem concerns a divide and conquer algorithm for tridiagonal matrices. The algorithm writes a symmetric tridiagonal T in the form

$$T = \begin{bmatrix} T_{11} & 0 \\ 0 & T_{22} \end{bmatrix} + \alpha v v^T,$$

where only the trailing diagonal element of T_{11} and the leading diagonal element of T_{22} differ from the corresponding elements of T. The eigensystems of T_{11} and T_{22} are found by applying the algorithm recursively, yielding $T_{11} = Q_1 \Lambda_1 Q_1^T$ and $T_{22} = Q_2 \Lambda_2 Q_2^T$. Then we have

$$\begin{aligned} T &= \begin{bmatrix} Q_1 \Lambda_1 Q_1^T & 0 \\ 0 & Q_2 \Lambda_2 Q_2^T \end{bmatrix} + \alpha v v^T \\ &= \operatorname{diag}(Q_1, Q_2)\big(\operatorname{diag}(\Lambda_1, \Lambda_2) + \alpha \widetilde{v}\widetilde{v}^T\big) \operatorname{diag}(Q_1, Q_2)^T, \end{aligned}$$

where $\widetilde{v} = \operatorname{diag}(Q_1, Q_2)^T v$. The eigensystem of a rank-one perturbed diagonal matrix $D + \rho z z^T$ can be found by solving the *secular equation* obtained by equating the characteristic polynomial to zero:

$$f(\lambda) = 1 + \rho \sum_{j=1}^{n} \frac{z_j^2}{d_{jj} - \lambda} = 0.$$

Hence by solving such an equation we can obtain the eigendecomposition

$$\operatorname{diag}(\Lambda_1, \Lambda_2) + \alpha \widetilde{v}\widetilde{v}^T = \widetilde{Q}\widetilde{\Lambda}\widetilde{Q}^T.$$

Finally, the eigendecomposition of T is given by

$$T = U \widetilde{\Lambda} U^T, \qquad U = \operatorname{diag}(Q_1, Q_2)\widetilde{Q}.$$

The formation of U is a matrix multiplication and dominates the operation count.

This algorithm was originally suggested by Cuppen [30], and how to solve the secular equation efficiently was shown by Bunch, Nielson and Sorensen [23], building on work of Golub [68]. Until recently, it was thought that extended precision arithmetic was needed in the solution of the secular equation to guarantee that sufficiently orthogonal eigenvectors are produced when there are close eigenvalues. However, Gu and Eisenstat [73] have found a new approach that does not require extended precision.

The divide and conquer algorithm has natural parallelism. Even on serial computers it can be many times faster than the QR algorithm, though it needs more workspace, hence LAPACK includes the divide and conquer algorithm.

A novel algorithm for the symmetric eigenvalue problem, intended for parallel computation, is suggested by Yau and Lu [132]. The algorithm involves computing the matrix exponential $\exp(iA)$ and is rich in matrix multiplication, though its practical efficiency is unclear.

When some but not all eigenvalues and eigenvectors of a symmetric tridiagonal matrix T are required, the bisection algorithm followed by inverse iteration is attractive. Recall that if the diagonal entries of T are $a_1, \ldots, a_n$ and the off-diagonal entries are $b_1, \ldots, b_{n-1}$ then we have the Sturm sequence recurrence

$$d_i = (a_i - \sigma) d_{i-1} - b_{i-1}^2 d_{i-2},$$

where d_i is the determinant of the leading i-by-i principal submatrix of $T - \sigma I$. The number of sign changes in the sequence of d_i's is the number of eigenvalues of T less than σ, denoted count(σ), and this fact is the basis for the application of the bisection method. An interesting way to introduce parallelism into the Sturm sequence evaluation is rewrite it as the two-term vector recurrence

$$\begin{bmatrix} d_i \\ d_{i-1} \end{bmatrix} = \begin{bmatrix} a_i - \sigma & -b_{i-1}^2 \\ 1 & 0 \end{bmatrix} \cdot \begin{bmatrix} d_{i-1} \\ d_{i-2} \end{bmatrix} \equiv M_i \begin{bmatrix} d_{i-1} \\ d_{i-2} \end{bmatrix} = M_i \cdot M_{i-1} \cdots M_1 \cdot \begin{bmatrix} d_0 \\ d_{-1} \end{bmatrix}.$$

By using the parallel prefix operation (a generalization of the well known fan-in operation for forming a sum or product of n quantities in $\log_2 n$ time steps), the recurrence can be evaluated in $O(\log_2 n)$ time. The same approach can be used to gain parallelism in other contexts, for example in the solution of triangular systems [83]. Mathias [102] shows that using parallel prefix to evaluate the Sturm sequence is numerically unstable, although instability appears to occur rarely. This is an example of the common phenomenon that increased parallelism brings decreased numerical stability [36].

Although bisection is a simple and robust algorithm, it can give incorrect results if the function count(σ) is not a monotonic increasing function of σ, which can happen for EISPACK's implementation (even in IEEE arithmetic) but not for LAPACK's, or for ScaLAPACK's parallel implementation. Demmel, Dhillon and Ren [37] give a thorough analysis of the correctness of the bisection algorithm for different implementations of the count function and under a variety of assumptions on the arithmetic.

The Jacobi method for diagonalizing a symmetric matrix has attracted renewed interest in recent years, for two reasons. First, the rotations that eliminate pairs of off-diagonal elements can be applied in parallel for suitable orderings; see, e.g., [114]. Second, the Jacobi method has been shown by Demmel and Veselić to be more accurate than was previously thought [43]. Specifically, the Jacobi method determines the eigenvalues of a symmetric positive definite matrix to the accuracy to which they are determined by small componentwise perturbations in the data. This is a much stronger result than holds for the QR algorithm, or any other algorithm that begins by reducing the matrix to tridiagonal form, because this initial reduction can induce large perturbations in small eigenvalues in the presence of roundoff. The proof of the result just described combines rounding error analysis with a new style of perturbation theory that has been developed in a series of papers beginning with one by Barlow and Demmel [13]. Extensions of the perturbation theory and error analysis, including to indefinite matrices, have been made; see Mathias [101] and Slapničar [116]. On most machines the Jacobi method is slower than alternative methods based on tridiagonalizing the matrix, so it is not currently implemented in LAPACK.

7 The singular value decomposition

The standard method for computing the singular value decomposition is the Golub–Reinsch algorithm, which reduces a full matrix to bidiagonal form B and then applies the QR algorithm implicitly to the tridiagonal matrix $B^T B$. In an unpublished technical report in the 1960s, Kahan [96] showed that the singular values of a bidiagonal matrix are determined to approximately the same relative accuracy as the elements of the matrix. Demmel and Kahan [42] use this result, together with rounding error analysis, to show that a zero-shift version of the QR algorithm computes the singular values of a bidiagonal matrix to high relative accuracy; this algorithm obtains the singular vectors to high accuracy, too [34]. Such strong statements do not hold for the previously standard version of the QR algorithm used, for example, in LINPACK. Fernando and Parlett [60], [109] have developed a quotient-difference (qd) algorithm for computing the singular values of a bidiagonal matrix. This algorithm, which goes back to much earlier work of Rutishauser, is more accurate than the algorithm of Demmel and Kahan and, because it allows the incorporation of shifts, several times faster.

Implicit Cholesky algorithms with shifts for computing the SVD of a triangular matrix without reducing it to bidiagonal form are described, and put into historical context, by Fernando and Parlett [61].

Jacobi algorithms for computing the SVD have received attention in recent years, for the same reasons as for the symmetric eigenproblem. Either two-sided or one-sided transformations can be used; in the former case, diagonal form is approached directly, whereas with one-sided transformations the aim is to orthogonalize the columns of the matrix, after which the SVD is readily obtained. Relevant references include Hari and Veselić [78] and de Rijk [33].

Divide and conquer algorithms for finding the SVD of a bidiagonal matrix are developed by Jessup and Sorensen [93] and Gu and Eisenstat [74]; they are related to the divide and conquer algorithms for the symmetric eigenproblem. A new algorithm for computing the SVD of a dense matrix that first reduces to bidiagonal form and then applies divide and conquer is described by Gu, Demmel and Dhillon [72]. Their method incorporates algorithmic refinements that make it faster than the current LAPACK QR algorithm-based SVD code and that enable solution of the least squares problem with a reduction in operation count by a factor of more than 4 compared with the current LAPACK code.

Much progress has been made in the last ten years in theory and computation of generalized singular value decompositions (GSVDs). A GSVD can be defined in various ways, depending on the number of matrices involved (two in the simplest case), and the form of the factorization used; see, for example, [32] and the references therein. LAPACK includes a code for computing a GSVD called the quotient SVD, which uses a Kogbetliantz algorithm of Paige [106] as refined by Bai and Demmel [6].

8 Concluding remarks

Research in numerical linear algebra, and in particular research in dense matrix computations, is as active as ever, with new problems and challenges continually arising from application areas and from the development of new computer architectures. This brief survey has given just a flavour of recent work in the subject. More details can be obtained from the many references and their references.

Acknowledgements

I thank Jesse Barlow and an anonymous referee for comments on draft manuscripts. This work was supported by Engineering and Physical Sciences Research Council grant GR/H/94528.

References

1. B. Alpert, G. Beylkin, R. Coifman, and V. Rokhlin. Wavelet-like bases for the fast solution of second-kind integral equations. *SIAM J. Sci. Comput.*, 14(1):159–184, 1993.

2. E. Anderson, Z. Bai, C. H. Bischof, J. W. Demmel, J. J. Dongarra, J. J. Du Croz, A. Greenbaum, S. J. Hammarling, A. McKenney, S. Ostrouchov, and D. C. Sorensen. *LAPACK Users' Guide, Release 2.0.* Second edition, Society for Industrial and Applied Mathematics, Philadelphia, PA, USA, 1995. xix+325 pp. ISBN 0-89871-345-5.

3. M. Arioli, I. S. Duff, and P. P. M. de Rijk. On the augmented system approach to sparse least-squares problems. *Numer. Math.*, 55:667–684, 1989.

4. Cleve Ashcraft, Roger G. Grimes, and John G. Lewis. Accurate symmetric indefinite linear equation solvers. Manuscript, September 1995. 51 pp.

5. Zhaojun Bai and James W. Demmel. On a block implementation of Hessenberg multishift QR iteration. *Int. J. High Speed Computing*, 1(1):97–112, 1989.

6. Zhaojun Bai and James W. Demmel. Computing the generalized singular value decomposition. *SIAM J. Sci. Comput.*, 14(6):1464–1486, 1993.

7. Zhaojun Bai and James W. Demmel. Design of a parallel nonsymmetric eigenroutine toolbox, Part I. In *Proceedings of the Sixth SIAM Conference on Parallel Processing for Scientific Computing, Volume I*, Richard F. Sincovec, David E. Keyes, Michael R. Leuze, Linda R. Petzold, and Daniel A. Reed, editors, Society for Industrial and Applied Mathematics, Philadelphia, PA, USA, 1993, pages 391–398.

8. Zhaojun Bai and James W. Demmel. On swapping diagonal blocks in real Schur form. *Linear Algebra and Appl.*, 186:73–95, 1993.

9. Zhaojun Bai and James W. Demmel. Design of a parallel nonsymmetric eigenroutine toolbox, part II. Mathematics Research Report 95–11, University of Kentucky, 1995.

10. Zhaojun Bai, James W. Demmel, and Ming Gu. Inverse free parallel spectral divide and conquer algorithms for nonsymmetric eigenproblems. Computer Science Division Report UCB/CSD-94-793, University of California, Berkeley, CA, USA, February 1994. 34 pp.

11. Zhaojun Bai, James W. Demmel, and Alan McKenney. On computing condition numbers for the nonsymmetric eigenproblem. *ACM Trans. Math. Software*, 19(2):202–223, 1993.

12. David H. Bailey, King Lee, and Horst D. Simon. Using Strassen's algorithm to accelerate the solution of linear systems. *J. Supercomputing*, 4:357–371, 1991.

13. Jesse L. Barlow and James W. Demmel. Computing accurate eigensytems of scaled diagonally dominant matrices. *SIAM J. Numer. Anal.*, 27(3):762–791, 1990.

14. Jesse L. Barlow and Peter A. Yoon. An efficient rank detection procedure for modifying the ULV decomposition. 1995. Submitted for publication.

15. Jesse L. Barlow, Peter A. Yoon, and Hongyuan Zha. An algorithm and a stability theory for downdating the ULV decomposition. *BIT*, 36:14–40, 1996.

16. Steve Batterson. Convergence of the shifted QR algorithm on 3 × 3 normal matrices. *Numer. Math.*, 58:341–352, 1990.

17. Christian H. Bischof and Charles F. Van Loan. The WY representation for products of Householder matrices. *SIAM J. Sci. Stat. Comput.*, 8(1):s2–s13, 1987.

18. Åke Björck. Stability analysis of the method of seminormal equations for linear least squares problems. *Linear Algebra and Appl.*, 88/89:31–48, 1987.

19. Åke Björck. *Numerical Methods for Least Squares Problems.* Society for Industrial and Applied Mathematics, Philadelphia, PA, USA, 1996. xvii+408 pp. ISBN 0-89871-360-9.

20. Åke Björck and C. C. Paige. Loss and recapture of orthogonality in the modified Gram–Schmidt algorithm. *SIAM J. Matrix Anal. Appl.*, 13(1):176–190, 1992.

21. Åke Björck and C. C. Paige. Solution of augmented linear systems using orthogonal factorizations. *BIT*, 34:1–24, 1994.

22. James R. Bunch and Linda Kaufman. Some stable methods for calculating inertia and solving symmetric linear systems. *Math. Comp.*, 31(137):163–179, 1977.

23. James R. Bunch, Christopher P. Nielsen, and Danny C. Sorensen. Rank-one modification of the symmetric eigenproblem. *Numer. Math.*, 31:31–48, 1978.

24. James R. Bunch and Beresford N. Parlett. Direct methods for solving symmetric indefinite systems of linear equations. *SIAM J. Numer. Anal.*, 8(4):639–655, 1971.

25. Tony F. Chan. Rank revealing QR factorizations. *Linear Algebra and Appl.*, 88/89:67–82, 1987.

26. Shivkumar Chandrasekaran and Ilse C. F. Ipsen. On rank-revealing factorisations. *SIAM J. Matrix Anal. Appl.*, 15(2):592–622, 1994.

27. Jaeyoung Choi, Jack J. Dongarra, Susan Ostrouchov, Antoine P. Petitet, David W. Walker, and R. Clint Whaley. The design and implementation of the ScaLAPACK LU, QR and Cholesky factorization routines. Report ORNL/TM-12470, Oak Ridge National Laboratory, Oak Ridge, TN, USA, September 1994. 26 pp. LAPACK Working Note 80.

28. A. K. Cline, C. B. Moler, G. W. Stewart, and J. H. Wilkinson. An estimate for the condition number of a matrix. *SIAM J. Numer. Anal.*, 16(2):368–375, 1979.

29. L. Csanky. Fast parallel matrix inversion algorithms. *SIAM J. Comput.*, 5(4):618–623, 1976.

30. J. J. M. Cuppen. A divide and conquer method for the symmetric tridiagonal eigenproblem. *Numer. Math.*, 36:177–195, 1981.

31. David Day. How the QR algorithm fails to converge and how to fix it. Report 96-0913J, Sandia National Laboratory, Albuquerque, New Mexico, April 1996. 22 pp.

32. Bart L. R. De Moor and Paul Van Dooren. Generalizations of the singular value and QR decompositions. *SIAM J. Matrix Anal. Appl.*, 13(4):993–1014, 1992.

33. P. P. M. de Rijk. A one-sided Jacobi algorithm for computing the singular value decomposition on a vector computer. *SIAM J. Sci. Stat. Comput.*, 10(2):359–371, 1989.

34. Percy Deift, James W. Demmel, Luen-Chau Li, and Carlos Tomei. The bidiagonal singular value decomposition and Hamiltonian mechanics. *SIAM J. Numer. Anal.*, 28:1463–1516, 1991.

35. James W. Demmel. Open problems in numerical linear algebra. IMA Preprint Series #961, Institute for Mathematics and its Applications, University of Minnesota, Minneapolis, MN, USA, April 1992. 21 pp. LAPACK Working Note 47.

36. James W. Demmel. Trading off parallelism and numerical stability. In *Linear Algebra for Large Scale and Real-Time Applications*, Marc S. Moonen, Gene H. Golub, and Bart L. De Moor, editors, volume 232 of *NATO ASI Series E*, Kluwer Academic Publishers, Dordrecht, The Netherlands, 1993, pages 49–68.

37. James W. Demmel, Inderjit Dhillon, and Huan Ren. On the correctness of some bisection-like parallel eigenvalue algorithms in floating point arithmetic. *Electronic Transactions on Numerical Analysis*, 3:116–149, 1995.

38. James W. Demmel, J. J. Dongarra, and W. Kahan. On designing portable high performance numerical libraries. In *Numerical Analysis 1991, Proceedings of the 14th Dundee Conference*, D. F. Griffiths and G. A. Watson, editors, volume 260 of *Pitman Research Notes in Mathematics*, Longman Scientific and Technical, Essex, UK, 1992, pages 69–84.

39. James W. Demmel, Michael T. Heath, and Henk A. van der Vorst. Parallel numerical linear algebra. In *Acta Numerica*, Cambridge University Press, 1993, pages 111–198.

40. James W. Demmel and Nicholas J. Higham. Stability of block algorithms with fast level-3 BLAS. *ACM Trans. Math. Software*, 18(3):274–291, September 1992.

41. James W. Demmel, Nicholas J. Higham, and Robert S. Schreiber. Stability of block *LU* factorization. *Numerical Linear Algebra with Applications*, 2(2):173–190, 1995.

42. James W. Demmel and W. Kahan. Accurate singular values of bidiagonal matrices. *SIAM J. Sci. Stat. Comput.*, 11(5):873–912, 1990.

43. James W. Demmel and Krešimir Veselić. Jacobi's method is more accurate than *QR*. *SIAM J. Matrix Anal. Appl.*, 13(4):1204–1245, 1992.

44. J. J. Dongarra, J. R. Bunch, C. B. Moler, and G. W. Stewart. *LINPACK Users' Guide*. Society for Industrial and Applied Mathematics, Philadelphia, PA, USA, 1979. ISBN 0-89871-172-X.

45. J. J. Dongarra, J. J. Du Croz, I. S. Duff, and S. J. Hammarling. A set of Level 3 basic linear algebra subprograms. *ACM Trans. Math. Software*, 16:1–17, 1990.

46. J. J. Dongarra, J. J. Du Croz, I. S. Duff, and S. J. Hammarling. Algorithm 679. A set of Level 3 basic linear algebra subprograms: Model implementation and test programs. *ACM Trans. Math. Software*, 16:18–28, 1990.

47. J. J. Dongarra, F. G. Gustavson, and A. Karp. Implementing linear algebra algorithms for dense matrices on a vector pipeline machine. *SIAM Review*, 26(1):91–112, 1984.

48. Jack Dongarra, Sven Hammarling, and Susan Ostrouchov. BLAS technical workshop. Technical Report CS-95-317, Department of Computer Science, University of Tennessee, Knoxville, November 1995. 21 pp. LAPACK Working Note 109.

49. Jack Dongarra, Roldan Pozo, and David W. Walker. LAPACK++ V1.0: High performance linear algebra users' guide. Technical Report CS-95-290, Department of Computer Science, University of Tennessee, Knoxville, TN, USA, May 1995. 31 pp. LAPACK Working Note 98.

50. Jack J. Dongarra, Jeremy J. Du Croz, Sven J. Hammarling, and Richard J. Hanson. An extended set of Fortran basic linear algebra subprograms. *ACM Trans. Math. Software*, 14(1):1–17, 1988.

51. Jack J. Dongarra, Jeremy J. Du Croz, Sven J. Hammarling, and Richard J. Hanson. Algorithm 656. An extended set of Fortran basic linear algebra subprograms: Model implementation and test programs. *ACM Trans. Math. Software*, 14(1):18–32, 1988.

52. Jack J. Dongarra, Iain S. Duff, Danny C. Sorensen, and Henk A. van der Vorst. *Solving Linear Systems on Vector and Shared Memory Computers.* Society for Industrial and Applied Mathematics, Philadelphia, PA, USA, 1991. x+256 pp. ISBN 0-89871-270-X.

53. Jack J. Dongarra, Sven J. Hammarling, and Danny C. Sorensen. Block reduction of matrices to condensed forms for eigenvalue computations. *J. Comp. Appl. Math.*, 27: 215–227, 1989.

54. Jack J. Dongarra and Majed Sidani. A parallel algorithm for the nonsymmetric eigenvalue problem. *SIAM J. Sci. Comput.*, 14(3):542–569, 1993.

55. Jack J. Dongarra and David W. Walker. Software libraries for linear algebra computations on high performance computers. *SIAM Review*, 37(2):151–180, 1995.

56. Jeremy J. Du Croz and Nicholas J. Higham. Stability of methods for matrix inversion. *IMA J. Numer. Anal.*, 12:1–19, 1992.

57. Alan Edelman. The complete pivoting conjecture for Gaussian elimination is false. *The Mathematica Journal*, 2:58–61, 1992.

58. Editor's note. *SIAM J. Matrix Anal. Appl.*, 12(3), 1991.

59. D. K. Faddeev, V. N. Kublanovskaja, and V. N. Faddeeva. Solution of linear algebraic systems with rectangular matrices. *Proc. Steklov Inst. Math.*, 96:93–111, 1968.

60. K. Vince Fernando and Beresford N. Parlett. Accurate singular values and differential qd algorithms. *Numer. Math.*, 67:191–229, 1994.

61. K. Vince Fernando and Beresford N. Parlett. Implicit Cholesky algorithms for singular values and vectors of triangular matrices. *Numerical Linear Algebra with Applications*, 2 (6):507–531, 1995.

62. Ricardo D. Fierro and James R. Bunch. Bounding the subspaces from rank revealing two-sided orthogonal decompositions. *SIAM J. Matrix Anal. Appl.*, 16(3):743–759, 1995.

63. Leslie V. Foster. Rank and null space calculations using matrix decomposition without column interchanges. *Linear Algebra and Appl.*, 74:47–71, 1986.

64. Leslie V. Foster. Gaussian elimination with partial pivoting can fail in practice. *SIAM J. Matrix Anal. Appl.*, 15(4):1354–1362, 1994.

65. K. A. Gallivan, R. J. Plemmons, and A. H. Sameh. Parallel algorithms for dense linear algebra computations. *SIAM Review*, 32(1):54–135, 1990.

66. B. S. Garbow, J. M. Boyle, J. J. Dongarra, and C. B. Moler. *Matrix Eigensystem Routines—EISPACK Guide Extension*, volume 51 of *Lecture Notes in Computer Science*. Springer-Verlag, Berlin, 1977. viii+343 pp. ISBN 3-540-08254-9.

67. J. R. Gilbert, C. B. Moler, and R. S. Schreiber. Sparse matrices in MATLAB: Design and implementation. *SIAM J. Matrix Anal. Appl.*, 13(1):333–356, 1992.

68. Gene H. Golub. Some modified matrix eigenvalue problems. *SIAM Review*, 15(2):318–334, 1973.

69. Gene H. Golub and Charles F. Van Loan. *Matrix Computations*. Second edition, Johns Hopkins University Press, Baltimore, MD, USA, 1989. xix+642 pp. ISBN 0-8018-3772-3 (hardback), 0-8018-3739-1 (paperback).

70. N. I. M. Gould. On growth in Gaussian elimination with complete pivoting. *SIAM J. Matrix Anal. Appl.*, 12(2):354–361, 1991.

71. W. B. Gragg and G. W. Stewart. A stable variant of the secant method for solving nonlinear equations. *SIAM J. Numer. Anal.*, 13(6):889–903, 1976.

72. Ming Gu, James W. Demmel, and Inderjit Dhillon. Efficient computation of the singular value decomposition with applications to least squares problems. Technical Report CS-94-257, Department of Computer Science, University of Tennessee, Knoxville, TN, USA, October 1994. 19 pp. LAPACK Working Note 88.

73. Ming Gu and Stanley C. Eisenstat. A stable and efficient algorithm for the rank-one modification of the symmetric eigenproblem. *SIAM J. Matrix Anal. Appl.*, 15(4):1266–1276, 1994.

74. Ming Gu and Stanley C. Eisenstat. A divide-and-conquer algorithm for the bidiagonal SVD. *SIAM J. Matrix Anal. Appl.*, 16(1):79–92, 1995.

75. Ming Gu and Stanley C. Eisenstat. Efficient algorithms for computing a strong rank-revealing QR factorization. *SIAM J. Sci. Comput.*, 17(4):848–869, 1996.

76. William W. Hager. Condition estimates. *SIAM J. Sci. Stat. Comput.*, 5(2):311–316, 1984.

77. Richard J. Hanson and Charles L. Lawson. Extensions and applications of the Householder algorithm for solving linear least squares problems. *Math. Comp.*, 23(108):787–812, 1969.

78. Vjeran Hari and Krešimar Veselić. On Jacobi methods for singular value decompositions. *SIAM J. Sci. Stat. Comput.*, 8(5):741–754, 1987.

79. Greg Henry and Robert Van de Geijn. Parallelizing the QR algorithm for the unsymmetric algebraic eigenvalue problem: Myths and reality. *SIAM J. Sci. Comput.*, 17(4):870–883, 1996.

80. Nicholas J. Higham. FORTRAN codes for estimating the one-norm of a real or complex matrix, with applications to condition estimation (Algorithm 674). *ACM Trans. Math. Software*, 14(4):381–396, December 1988.

81. Nicholas J. Higham. Exploiting fast matrix multiplication within the level 3 BLAS. *ACM Trans. Math. Software*, 16(4):352–368, December 1990.

82. Nicholas J. Higham. Iterative refinement and LAPACK. Numerical Analysis Report No. 277, Manchester Centre for Computational Mathematics, Manchester, England, September 1995. 17 pp. To appear in IMA J. Numer. Anal.

83. Nicholas J. Higham. Stability of parallel triangular system solvers. *SIAM J. Sci. Comput.*, 16(2):400–413, March 1995.

84. Nicholas J. Higham. Stability of the diagonal pivoting method with partial pivoting. *SIAM J. Matrix Anal. Appl.*, 18(1):52–65, 1997.

85. Nicholas J. Higham. *Accuracy and Stability of Numerical Algorithms*. Society for Industrial and Applied Mathematics, Philadelphia, PA, USA, 1996. xxviii+688 pp. ISBN 0-89871-355-2.

86. Nicholas J. Higham and Desmond J. Higham. Large growth factors in Gaussian elimination with pivoting. *SIAM J. Matrix Anal. Appl.*, 10(2):155–164, April 1989.

87. Y. P. Hong and C. T. Pan. Rank-revealing QR factorizations and the singular value decomposition. *Math. Comp.*, 197:213–232, 1991.

88. James Lucien Howland. The sign matrix and the separation of matrix eigenvalues. *Linear Algebra and Appl.*, 49:221–232, 1983.

89. *IEEE Standard for Binary Floating-Point Arithmetic, ANSI/IEEE Standard 754-1985.* Institute of Electrical and Electronics Engineers, New York, 1985. Reprinted in SIGPLAN Notices, 22(2):9–25, 1987.

90. *A Radix-Independent Standard for Floating-Point Arithmetic, IEEE Standard 854-1987.* IEEE Computer Society, New York, 1987.

91. M. Jankowski and H. Woźniakowski. Iterative refinement implies numerical stability. *BIT*, 17:303–311, 1977.

92. E. R. Jessup. A case against a divide and conquer approach to the nonsymmetric eigenvalue problem. *Applied Numerical Mathematics*, 12:403–420, 1993.

93. E. R. Jessup and D. C. Sorensen. A parallel algorithm for computing the singular value decomposition of a matrix. *SIAM J. Matrix Anal. Appl.*, 15(2):530–548, 1994.

94. Mark T. Jones and Merrell L. Patrick. Factoring symmetric indefinite matrices on high-performance architectures. *SIAM J. Matrix Anal. Appl.*, 15(1):273–283, 1994.

95. Bo Kågström and Peter Poromaa. Computing eigenspaces with specified eigenvalues of a regular matrix pair (A, B) and condition estimation: Theory, algorithms and software. Report UMINF 94.04, Institute of Information Processing, University of Umeå, Umeå, Sweden, September 1994. 65 pp. LAPACK Working Note 87.

96. W. Kahan. Accurate eigenvalues of a symmetric tri-diagonal matrix. Technical Report No. CS41, Department of Computer Science, Stanford University, 1966. 55 pp. Revised June 1968.

97. Linda Kaufman. Computing the MDM^T decomposition. *ACM Trans. Math. Software*, 21(4):476–489, 1995.

98. Charles S. Kenney and Alan J. Laub. The matrix sign function. *IEEE Trans. Automat. Control*, 40(8):1330–1348, 1995.

99. C. L. Lawson, R. J. Hanson, D. R. Kincaid, and F. T. Krogh. Basic linear algebra subprograms for Fortran usage. *ACM Trans. Math. Software*, 5(3):308–323, 1979.

100. T.-Y. Li, Z. Zeng, and L. Cong. Solving eigenvalue problems of real nonsymmetric matrices with real homotopies. *SIAM J. Numer. Anal.*, 29(1):229–248, 1992.

101. Roy Mathias. Accurate eigensystem computations by Jacobi methods. *SIAM J. Matrix Anal. Appl.*, 16(3):977–1003, 1995.

102. Roy Mathias. The instability of parallel prefix matrix multiplication. *SIAM J. Sci. Comput.*, 16(4):956–973, 1995.

103. *MATLAB User's Guide*. The MathWorks, Inc., Natick, MA, USA, 1992.

104. Cleve B. Moler. Demonstration of a matrix laboratory. In *Numerical Analysis, Mexico 1981*, J. P. Hennart, editor, volume 909 of *Lecture Notes in Mathematics*, Springer-Verlag, Berlin, 1982, pages 84–98.

105. James M. Ortega. *Introduction to Parallel and Vector Solution of Linear Systems*. Plenum Press, New York, 1988. xi+305 pp. ISBN 0-306-42862-8.

106. C. C. Paige. Computing the generalized singular value decomposition. *SIAM J. Sci. Stat. Comput.*, 7(4):1126–1146, 1986.

107. Victor Pan and Robert Schreiber. An improved Newton iteration for the generalized inverse of a matrix, with applications. *SIAM J. Sci. Stat. Comput.*, 12(5):1109–1130, 1991.

108. Haesun Park and Lars Eldén. Downdating the rank-revealing URV decomposition. *SIAM J. Matrix Anal. Appl.*, 16(1):138–155, 1995.

109. Beresford N. Parlett. The new qd algorithms. In *Acta Numerica*, Cambridge University Press, 1995, pages 459–491.

110. J. D. Roberts. Linear model reduction and solution of the algebraic Riccati equation by use of the sign function. *Int. J. Control*, 32(4):677–687, 1980. First issued as report CUED/B-Control/TR13, Department of Engineering, University of Cambridge, 1971.

111. Robert S. Schreiber and Beresford N. Parlett. Block reflectors: Theory and computation. *SIAM J. Numer. Anal.*, 25(1):189–205, 1988.

112. Robert S. Schreiber and Charles F. Van Loan. A storage efficient WY representation for products of Householder transformations. *SIAM J. Sci. Stat. Comput.*, 10:53–57, 1989.

113. Günther Schulz. Iterative Berechnung der reziproken Matrix. *Z. Angew. Math. Mech.*, 13:57–59, 1933.

114. Gautam M. Shroff and Robert S. Schreiber. On the convergence of the cyclic Jacobi method for parallel block orderings. *SIAM J. Matrix Anal. Appl.*, 10(3):326–346, 1989.

115. Robert D. Skeel. Iterative refinement implies numerical stability for Gaussian elimination. *Math. Comp.*, 35(151):817–832, 1980.

116. Ivan Slapničar. *Accurate Symmetric Eigenreduction by a Jacobi Method*. PhD thesis, Fernuniversität Hagen, Germany, 1992. 132 pp.

117. B. T. Smith, J. M. Boyle, J. J. Dongarra, B. S. Garbow, Y. Ikebe, V. C. Klema, and C. B. Moler. *Matrix Eigensystem Routines—EISPACK Guide*, volume 6 of *Lecture Notes in Computer Science*. Second edition, Springer-Verlag, Berlin, 1976. vii+551 pp. ISBN 3-540-06710-8.

118. Torsten Söderström and G. W. Stewart. On the numerical properties of an iterative method for computing the Moore–Penrose generalized inverse. *SIAM J. Numer. Anal.*, 11(1):61–74, 1974.

119. D. C. Sorensen. Analysis of pairwise pivoting in Gaussian elimination. *IEEE Trans. Comput.*, C-34:274–278, 1985.

120. G. W. Stewart. An updating algorithm for subspace tracking. *IEEE Trans. Signal Processing*, 40(6):1535–1541, 1992.

121. G. W. Stewart. Updating a rank-revealing ULV decomposition. *SIAM J. Matrix Anal. Appl.*, 14(2):494–499, 1993.

122. Eberhard U. Stickel. Separating eigenvalues using the matrix sign function. *Linear Algebra and Appl.*, 148:75–88, 1991.

123. Ji-guang Sun. Optimal backward perturbation bounds for the linear least-squares problem with multiple right-hand sides. *IMA J. Numer. Anal.*, 16(1):1–11, 1996.

124. Ji-guang Sun and Zheng Sun. Optimal backward perturbation bounds for underdetermined systems. *SIAM J. Matrix Anal. Appl.*, 1997. To appear.

125. Lloyd N. Trefethen. Why Gaussian elimination is stable for almost all matrices. Manuscript, September 1994. 14 pp.

126. Lloyd N. Trefethen and Robert S. Schreiber. Average-case stability of Gaussian elimination. *SIAM J. Matrix Anal. Appl.*, 11(3):335–360, 1990.

127. S. A. Vavasis. Gaussian elimination with partial pivoting is P-complete. *SIAM J. Disc. Math.*, 2:413–423, 1989.

128. Bertil Waldén, Rune Karlson, and Ji-guang Sun. Optimal backward perturbation bounds for the linear least squares problem. *Numerical Linear Algebra with Applications*, 2(3): 271–286, 1995.

129. J. H. Wilkinson. The Automatic Computing Engine at the National Physical Laboratory. *Proc. Roy. Soc. London Ser. A*, 195:285–286, 1948.

130. J. H. Wilkinson. *The Algebraic Eigenvalue Problem.* Oxford University Press, 1965. xviii+662 pp. ISBN 0-19-853403-5 (hardback), 0-19-853418-3 (paperback).

131. Stephen J. Wright. A collection of problems for which Gaussian elimination with partial pivoting is unstable. *SIAM J. Sci. Stat. Comput.*, 14(1):231–238, 1993.

132. Shing-Tung Yau and Ya Yan Lu. Reducing the symmetric matrix eigenvalue problem to matrix multiplications. *SIAM J. Sci. Comput.*, 14(1):121–136, 1993.

133. Man-Chung Yeung and Tony. F. Chan. Probabilistic analysis of Gaussian elimination without pivoting. CAM Report 95-29, Department of Mathematics, University of California, Los Angeles, USA, September 1995. 21 pp. Revised version.

Sparse Numerical Linear Algebra: Direct Methods and Preconditioning

Iain S. Duff

Rutherford Appleton Laboratory, Oxon

Abstract

Most of the current techniques for the direct solution of linear equations are based on supernodal or multifrontal approaches. An important feature of these methods is that arithmetic is performed on dense submatrices and Level 2 and Level 3 BLAS (matrix-vector and matrix-matrix kernels) can be used. Both sparse LU and QR factorizations can be implemented within this framework.

Partitioning and ordering techniques have seen major activity in recent years. We discuss bisection and multisection techniques, extensions to orderings to block triangular form, and recent improvements and modifications to standard orderings such as minimum degree.

We also study advances in the solution of indefinite systems and sparse least-squares problems.

The desire to exploit parallelism has been responsible for many of the developments in direct methods for sparse matrices over the last ten years. We examine this aspect in some detail, illustrating how current techniques have been developed or extended to accommodate parallel computation.

Preconditioning can be viewed as a way of extending direct methods or of accelerating iterative ones. We will view it in the former way in this talk and leave the other perspective to the talk on iterative methods.

Finally, we will briefly comment on recent attempts to develop tools and platforms towards a sparse problem solving environment.

1 Introduction

In common with many of my co-authors in this volume, my starting point was to read the review on this subject given at the last State of the Art meeting in Birmingham. The article "Sparse Matrices" was authored by John Reid, and it is perhaps significant that it covered both iterative and direct methods, the subject of two talks at this meeting. Indeed, although I have been working in this field for about twenty five years, I find it amazing how many advances have occurred in the last ten, necessitating the dual lectures of this meeting. Although this paper is primarily on direct methods, I also include a section on preconditioning techniques for use with iterative methods. The viewpoint of this discussion will be quite different from that presented in the chapter on Iterative Methods since

I will look at the construction of preconditioners as being an alternative to direct methods when, for one reason or another (usually storage), direct methods are infeasible. Moreover, in common with proponents of iterative methods, I feel strongly that the only way of solving really challenging linear algebra problems is by combining direct and iterative methods through either conventional or novel preconditioning.

Although I intend this paper to be a general State of the Art survey, I wish to emphasize what I consider to be the most important developments of the last ten years. Accordingly, I have structured the text to highlight: implementations that use higher level BLAS kernels, advances in ordering strategies particularly for symmetric problems, the exploitation of parallelism, and the use of direct techniques as preconditioners for iterative methods.

Undoubtedly, the kernel that gets closest to peak performance on modern computers is a **dense** matrix-matrix multiply. A standard version of this kernel is provided by subroutine _GEMM in the Level 3 Basic Linear Algebra Subprograms or BLAS [62] and is supported by many vendors of high performance computers. Most sparse direct codes use this and allied kernels to achieve high performance, and we discuss how they are able to do this in Section 2.

Ten years ago, one might have been forgiven for thinking that the problem of ordering a symmetric system for subsequent Gaussian elimination was resolved in favour of minimum degree, one of the earliest proposed orderings. However, quite recently there has been significant work on developing other classes of orderings, related to dissection methods, which are showing great promise of overhauling this perceived wisdom. Additionally, there has been further understanding of the minimum degree ordering and attempts to make it more efficient and to adapt it for orderings more suited to parallel computation. We discuss orderings for symmetric systems in Section 3.

In 1986, there was little software for unsymmetric sparse problems. Early examples included NSPIV [194], Y12M [212], MA28, MA32, and MA37 from the Harwell Subroutine Library (HSL) [134] and a surrogate MA28 in the NAG Library (F01BRF/F01BSF/F04AXF). The development and production of sparse unsymmetric solvers has now become a veritable growth industry. The multifrontal and supernodal techniques, so powerful in the symmetric case, have been extended to unsymmetric systems. In other approaches, various decompositional and preordering techniques are used to facilitate later factorization, particularly on parallel computers. We consider these issues in Section 4. We discuss the case of sparse symmetric indefinite systems in Section 5 and the solution of sparse least-squares problems in Section 6.

A major impetus for the development of algorithms and software in all linear algebra has been the availability of parallel architectures of various shapes and forms. Although the ICL DAP was originally produced in the 1970s, it was a bit-oriented SIMD architecture, and really the first commercially available parallel machines were the Denelcor HEP in 1984 and the Alliant FX/8 in 1985. Thus any practical implementations of techniques for exploiting parallelism belong strongly to the last decade. Although we have considered issues related to parallelism in previous sections, we study this in more depth in Section 7.

In Section 8, we discuss the use of direct methods to obtain preconditioners for iterative methods. As we mentioned earlier, it is this route that we believe to be the most important for the solution of really large problems from, say, three dimensional partial differential equations.

We briefly review tools for sparse matrix manipulation and computation in Section 9. These include consideration of some tentative first steps towards a problem solving environment for sparse problems.

It is not the intention of this talk or this meeting to perform any crystal-ball gazing, but we indulge in a little of this in our concluding section.

Before continuing with the main paper, it is useful to define a few terms that will be used extensively in the following sections. The solution of a sparse system is usually divided into several phases:

1. Analysis of the sparsity structure to determine a pivot ordering.

2. Symbolic factorization to generate a structure for the factors.

3. Numerical factorization.

4. Solution of set(s) of equations.

In some cases, particularly when it is important to consider numerical values when choosing pivots, the first three phases are combined into an analyse-factorize phase. Additionally, there may be some partitioning scheme prior to all these phases, which are then executed on submatrices from the partition. The relative speeds of the four phases are very dependent on the details of the algorithm and implementation, the problem being solved, and the machine being used. However, one reason for separating the first two phases from the third is that it is usually much faster to perform an analysis and symbolic factorization without reference to numerical values. This does mean that there must be some way of incorporating subsequent numerical pivoting in the numerical factorization if general problems are to be solved.

Most of the algorithms we consider are based on matrix factorization methods like Gaussian elimination, and part-way through the matrix factorization, when we have calculated some of the factors, the remaining matrix (which need not be held explicitly) is composed of original matrix entries and fill-in from the earlier stages. We refer to this matrix as the *reduced* matrix.

Finally, the relationship between graphs and sparse matrices is ubiquitous in sparse matrix work. We do not make heavy use of graphs in the ensuing discussion, but it is important to define the main graphs associated with sparse matrices. With a symmetric sparse matrix of order n, we associate a graph with n vertices, and an edge (i, j) between vertices i and j, if and only if entry $a_{ij} \neq 0$. A clique is a complete subgraph, that is all vertices of the subgraph are pairwise connected by edges in the graph. For an unsymmetric matrix, we associate a directed graph where the edges are now directed. A bipartite graph is sometimes associated with an unsymmetric (even non-square) matrix. This has two sets of disjoint vertices identified with rows and columns of the matrix respectively. An edge (i, j) exists from the row vertices to the column vertices if and only if entry $a_{ij} \neq 0$. The elimination tree is defined by the Cholesky factors of a symmetric matrix and has an edge (i, j) if the first nonzero entry below the diagonal in column j of the lower triangular Cholesky factor is in row i. Elimination trees were used by Duff [66] and Schreiber [191] and are surveyed in depth by Liu [153].

As a postscript to this section, I should stress that, although error analysis is not discussed, I am concerned throughout that the solution methods will give good answers, often in the sense of providing the solution to a nearby problem. It is worth mentioning that this notion of backward error can be developed to include only perturbations that preserve the sparsity structure [15].

Finally, lest I tread on too many toes, let me say that I have not been entirely consistent in quoting the original source of an idea but sometimes favour an overview or more recent report that itself includes further references and original attributions. I did consult widely to avoid the worst abuses of this (see Acknowledgement Section). Of course, in such a rapidly developing field, many ideas are simultaneously "discovered". Although you might think that there are too many references in this paper, I can assure you that I have been quite selective!

2 High performance sparse factorization

It is a truism that really large problems need large computers to solve them, and it is equally true that at the heart of most large-scale computations lies the solution of a large sparse set of linear equations. Currently, high performance computers come in various guises although nearly all are either vector processors or RISC processors, sometimes as many as a few hundred of them but more commonly only about two or four! In this section, we concentrate more on the uniprocessor performance and leave the discussion of parallelism until Section 7.

On some of these machines, the performance of a general code can be much less than the peak performance, usually because of delays in getting data to the arithmetic processors. The kernel that gets closest to the peak is the matrix-matrix multiply routine, _GEMM in the Level 3 Basic Linear Algebra Subprograms [62], which on most machines will achieve over 90% of peak on matrices of order only a few hundred. The Level 3 BLAS kernels can be expressed as calls to _GEMM (and trivial calls to _TRSM) and so all can execute at high speed [57,141]. Many years ago, before the Level 3 BLAS were established, Duff [66] extolled the virtues of using dense matrix kernels within sparse direct codes and gave some examples of techniques that used such kernels. It is now true to say that nearly all sparse direct codes use dense matrix kernels in their inner loops and, by doing so, achieve high performance over a wide range of architectures and problems. Indeed, it is on machines with caches or a hierarchical memory structure that the Level 3 BLAS realize their full potential. The importance of using Level 3 BLAS in sparse factorizations for efficiency on cache-based machines is illustrated by [165,181].

An easy way to use such kernels is to recognize that, as the elimination progresses, the reduced matrix in sparse Gaussian elimination becomes denser and at some point it is more efficient to switch to dense code. The point at which to do this depends on the matrix structure and machine, but recent experience indicates that it can be beneficial to switch when the density is as low as 10%. Switching to dense code was discussed by [60] and was incorporated in an experimental version of the Harwell Subroutine Library code MA28 [65]. However, a switch to dense matrix processing was only included in the analyse-factorize phase. A switch to dense matrix processing in the other phases was introduced in the HSL code MA48 [81,83]. However, although such a switch is clearly beneficial, it only addresses the computations towards the end of the factorization. There are two main approaches that can reap the benefit of higher level BLAS working throughout the entire factorization. These are the frontal methods and the supernodal approach.

Since the basic frontal method and many of its advantages have been known for some time and are described in John Reid's 1986 State of the Art paper, we will not discuss them in detail here but rather concentrate on new developments in these methods over the last decade. The important aspect of these methods is that arithmetic is performed within a dense *frontal* matrix, and pivots can be chosen from anywhere within a submatrix of this frontal matrix. In a frontal scheme, when all possible pivots have been chosen and the elimination operations performed, further data from the problem is assembled or summed into the frontal matrix and more pivots are chosen. This scheme has the merits of fairly simple storage management but can suffer significant fill-in for general problems and has restricted possibilities for exploiting parallelism. These two concerns are addressed by the multifrontal methods [79,154,195], which were again discussed by Reid [176]. Three principal problems with multifrontal methods, as practised in the 1980s, were that

- there could be significant overheads for data movement,
- they assumed structural symmetry, and
- the analyse phase assumed that any diagonal entry could be chosen as pivot.

We now consider how these concerns have been addressed. Other recent work on developing parallel versions of multifrontal methods is considered in Section 7. In the first case, data movement can be reduced by continuing with one frontal matrix rather than stacking it and starting another. Taken to its extreme, this strategy would give a (uni-)frontal scheme. This technique has been incorporated in the HSL code MA38 and is discussed by [55]. Although one of the early multifrontal codes MA37 [80] solved sets of unsymmetric equations, the analyse phase was performed on the pattern of the matrix with entries a_{ij} if either a_{ij} or a_{ji} were an entry in the original matrix. This is clearly fine if the matrix is symmetric in structure but surprisingly can perform quite well for unsymmetric matrices, although it can be inefficient when the matrix is markedly unsymmetric. For unsymmetric systems, a preordering to place nonzeros on the diagonal is often very helpful and is an option in the HSL code MA41, which is designed for shared memory computers and makes much more use of higher level BLAS than its precursor MA37. Davis and Duff [54] have extended the multifrontal scheme to general unsymmetric matrices, using rectangular fronts and directed acyclic graphs [109]. This extension works well on most highly unsymmetric systems but can be poorer than MA41 when the matrix is not highly unsymmetric. We discuss both approaches further in Section 4. The third problem is more difficult and is present for any technique that performs an analyse phase separate from the numerical factorization. Clearly, one way to resolve this is to combine the analyse and factorize phases but then many of the benefits of the fast analyse phase are lost. A certain amount of numerical or additional structural information can be supplied to the analyse phase, and we consider an example of this when we study the use of structured 2×2 pivots in Section 5.

The other main approach to using higher level BLAS in sparse direct solvers is a generalization of a sparse column factorization. These can either be left-looking (or fan-in) algorithms, where updates are performed on each column in turn by all the previous columns that contribute to it, then the pivot is chosen in that column and the multipliers calculated; or a right-looking (or fan-out) algorithm where, as soon as the pivot is selected and multipliers calculated, that column is immediately used to update all future columns that it modifies. Higher level BLAS can be used if columns with a common sparsity pattern are considered together as a single block or supernode and algorithms are termed column-supernode, supernode-column, and supernode-supernode depending on whether target, source, or both are supernodes.

Several authors have experimented with these different algorithms (right-looking, left-looking, and multifrontal) and different blockings. Ng and Peyton [165] favour the left-looking approach and Amestoy and Duff [10] show the benefits of Level 3 BLAS within a multifrontal code on vector processors. Rothberg and Gupta [181] find that on cache-based machines it is the blocking that affects the efficiency (by a factor of 2 to 3) and the algorithm that is used has a much less significant effect. The supernodal concept has been extended to unsymmetric systems by [61] although, for irregular problems, they cannot use regular supernodes for the target columns and so they resort to Level 2.5 BLAS. By doing this, the source supernode can be held in cache and applied to the target columns or blocks of columns of the "irregular" supernode, thus getting a high degree of reuse of data and a performance similar to the Level 3 BLAS.

The multifrontal and supernodal methods form the basis for some of the approaches that exploit parallelism, and we consider this aspect further in Section 7.

Although we have stressed the importance of using higher level BLAS to obtain high performance, we should be aware that, if the matrix is very sparse and the factors are also, there will normally be no benefit in using BLAS kernels. It is, however, quite likely that there will be much parallelism in both factorization and solution. The extreme case of this is a diagonal matrix or a permutation thereof.

3 Orderings for symmetric problems

Although predated by some ten years by the paper of Markowitz [159] on unsymmetric orderings, scheme S2 in the paper by Tinney and Walker [199] established the main ordering for symmetric problems that has remained almost unchallenged until this present day. Scheme S2 is commonly termed the minimum degree ordering because, at each stage, the pivot chosen corresponds to a node of minimum degree in the undirected graph associated with the reduced matrix. In matrix terms, this corresponds to choosing the entry from the diagonal that has the least number of entries in its row within the reduced matrix. This ordering algorithm has proved remarkably resistant to competitors and, although based only on a local criterion, does an excellent job of keeping subsequent work and fill-in low over a wide range of problems. The evolution of the minimum degree ordering is studied by George and Liu [103].

George [97] proposed a different class of orderings based on a non-local strategy of dissection. In his *nested dissection* approach, a set of nodes is selected to partition the graph, and this set is placed at the end of the pivotal sequence. The subgraphs corresponding to the partitions are themselves similarly partitioned and this process is nested with pivots being identified in reverse order. Minimum degree, nested dissection and several other symmetric orderings were included in the SPARSPAK package [102,104]. Many experiments were performed using the orderings in SPARSPAK and elsewhere, and the empirical experience at the beginning of the 90s indicated that minimum degree was the best ordering method for general or unstructured problems.

A major problem with the minimum degree ordering is that it is not susceptible to analysis, in the sense of computing the complexity of the resulting factorization on a regular grid problem. Part of the problem in analysing minimum degree is caused by tie-breaking; that is, choosing which node of minimum degree to use as pivot when, as is usually the case, there are many to choose from. Tie-breaking strategies have become a study in their own right (for example, [45]) and significantly complicate the algorithm while still not guaranteeing a better ordering nor allowing a theoretical analysis. Additionally, various studies have shown that tie-breaking can have a profound effect on the amount of fill-in [34,72]. In contrast, George [97] showed that the nested dissection ordering algorithm could be analysed for regular grid problems and, furthermore, Hoffman, Martin, and Rose [133] proved that the amount of fill-in and work for such an ordering was of the lowest order (in terms of the grid size) that could be obtained by a direct method. This led to the hope that a dissection ordering could be found that was both theoretically and practically superior to minimum degree. Many attempts were made to do this but, although the analysis was extended to planar graphs [149], it was difficult to do for general matrices and practical implementations were superior to minimum degree for only fairly restricted classes of problems, usually arising from regular grids. Analysis has shown that nested dissection is close to optimal on graphs of bounded degree, although the proofs are not constructive [40,108].

We now consider recent advances in ordering strategies: first to the minimum degree orderings and then to methods based on dissection.

Although the minimum degree ordering is simple enough to describe, it is not quite so simple to implement efficiently. There are three main issues:

- selection of pivot,
- update of reduced matrix after selection of pivot, and
- update of degree counts.

For the first, it is easy to keep a list of nodes in order of increasing degree and to choose the node at the head of that list each time. There are two ways in which this task can be made significantly more efficient. The first is to observe that, once a node is selected, all nodes that were in a complete subgraph (or clique) containing that node, have degree one less and so can immediately be eliminated without any extra fill-in, and subsequently all nodes in the clique can be eliminated. This is usually termed *mass node elimination* and was included in some early minimum degree codes. Since there is no fill-in within the clique, a better measure of the "damage" done to the matrix by a potential pivot can be obtained not from its degree but rather from its *external degree*, which corresponds to the number of edges to nodes outside its clique. It is quite natural, in a finite-element application, to perform the minimum degree ordering on a graph where nodes in the same clique are treated as a single node, and this was done by the minimum degree ordering algorithm in the HSL code MA47 [82]. An exhaustive study of this has been performed by Ashcraft [21], who has obtained speed-ups of over 5 for some standard test matrices. The second improvement to pivot selection stems from the observation that, if two nodes of the same degree are not adjacent in the graph, they can be eliminated simultaneously. This clearly has implications for parallelism and for the degree and graph update phases and is termed *multiple elimination* [150]. We will revisit this multiple selection of pivots in Section 7 when we consider parallel computing.

The resolution of the second issue, graph update, was the main reason why minimum degree codes improved by several orders of magnitude over the decade

1976-1986. The principal saving was made by using the clique structure of the reduced matrix and updating this rather than individual edges of the graph. This was discussed in the review [176].

We thus come to the final issue, that of updating the degree counts. There are two main approaches here. One is to have a threshold and compute the new degrees only if they could fall below this threshold. The threshold must, of course, be changed dynamically, at which point some recalculation of degrees is necessary. The second is to replace the minimum degree count by an approximate degree count that is easier to compute. There have been several attempts at this, for example [111], but most give worse orderings than full minimum degree. Recently, however, Amestoy, Daydé, and Duff [9] have designed an approximate minimum degree ordering (AMD) where the bound is equal to the degree in many cases. They have found that their AMD ordering is almost indistinguishable from the minimum degree ordering in quality but is very much faster to compute. An interesting twist to this is given by the work of Rothberg [179] who shows surprising promise with a similarly conceived implementation of an approximate minimum fill-in algorithm.

One problem with the minimum degree ordering is that it tends to give elimination trees that are not well balanced and so not ideal for using as a computational graph for driving a parallel algorithm. Liu [152] has developed a technique for massaging the elimination tree so that it is more suitable for parallel computation but the effect of this is fairly limited for general matrices. Duff, Gould, Lescrenier, and Reid [73] propose modifications to the minimum degree criterion to directly enhance parallelism but have only compared their algorithms using fairly crude models of parallelism. The use of dissection techniques would appear to offer the promise of much better balanced trees, although the inferior performance of the early dissection codes needs to be addressed for them to be viable. We now discuss recent advances in dissection techniques.

It is only within the last year or so that the supremacy of minimum degree has been challenged (by dissection orderings, as may have been expected). The beauty of dissection orderings is that they take a global view of the problem; their difficulty until recently has been the problem of extending them to unstructured problems. Recently, there have been several tools and approaches that make this extension more realistic. The essence of a dissection technique is a bisection algorithm that divides the graph of the matrix into two partitions. If node separators are used, a third set will correspond to the node separators. Such a bisection is then repeated in a nested fashion to obtain an ordering for the matrix. Perhaps the bisection technique that has achieved the most fame has been spectral bisection. In this approach, use is made of the Laplacian matrix that is defined as a symmetric matrix whose diagonal entries are the degrees of the nodes and whose off-diagonals are -1 if and only if the corresponding entry in the matrix is nonzero. This matrix is singular because its row sums are all zero, but if the matrix is irreducible, it is positive semidefinite with only one zero eigenvalue. We can use this matrix to define a bisection by constructing a vector x that has components x_i equal to $+$ or -1, according to which partition node i

lies in. Then the quantity $x^T Ax$ is 4 times the number of edges between the two halves of the bisection. We can thus obtain an "optimal" bisection by minimizing $x^T Ax$ subject to $\sum_i x_i = 0$ with $x_i = \pm 1$. Since the first constraint corresponds to finding a vector orthogonal to the vector of all ones, which is the eigenvector for the zero eigenvalue, in the corresponding continuous problem it is the eigenvector corresponding to the smallest nonzero eigenvalue (called the Fiedler vector) that is of interest. Normally some variant of the Lanczos algorithm is used to compute this [29,30,170,171]. The graph is then bisected according to the components of this eigenvector. If a balanced bisection is desired, commonly all components greater than the median are put in one partition and those lower than the median in the other.

In saying this, of course, one must establish a criterion for "optimality". There are clearly two sometimes conflicting goals: balancing the bisection and minimizing the number of edges joining each set (or if a node separator is used, minimizing the number of nodes in the separator). Rothberg [178] has experimented with various criteria and has found that minimizing the quantity $|S|/(|C| \times |D|)$, where $|S|$ is the number of nodes in the separator set, and $|C|$ and $|D|$ are the number of nodes in each partition, was the best of the criteria he examined in terms of floating-point operations for a Cholesky factorization with a nested dissection ordering. Another measure that has been used by [27] is the quantity $|S|\left(1+\alpha\frac{\max(|C|,|D|)}{\min(|C|,|D|)}\right)$, although it is somewhat sensitive to the choice of the parameter α.

The spectral method requires much computing time, does not always yield optimal bisections, and naturally produces edge separators, requiring some postprocessing to obtain a node separator set. For these reasons, this technique is not now so strongly favoured, and there has been much current research on alternatives that we now describe.

Gilbert, Miller, and Teng [110] have developed a geometric partitioning scheme, and Chan, Gilbert, and Teng [46] have proposed a hybrid of geometric and spectral methods. Ashcraft and Liu [25–28] have explored a different approach to obtaining separators for graph bisection. They base their method on a domain decomposition approach, defining a multisection by the nodes on the boundaries of the domains and using these to dissect or bisect the graph. The resulting vertex separator can then be refined using variants of methods like that of [92]. The other main approach to graph bisection is to perform graph reductions, compute a partition cheaply on the resulting coarse graph, and from this construct a partition of the original graph, using some kind of iterative improvement on the projection of this coarse partition on the finer graph (for example, [92,146]. This approach is nested and is termed a multilevel scheme [42]. Multilevel schemes have been used by [129,131,143], inter alios.

In most of these approaches, the dissection technique is only used for the top levels and the resulting subgraphs are ordered by a minimum degree scheme. This hybrid technique was used many years ago by [106] and is included in many current implementations (for example, [28,131]. As can be seen by the dates on the references, these new schemes are all very recent, but current empirical evidence would suggest that they are at least competitive with minimum degree on some large problems from structural analysis. They also perform far better than a minimum degree ordering on some matrices from financial modelling where Berger, Mulvey, Rothberg and Vanderbei [33] have found practical problems that exhibit similar behaviour to the pathological examples of Rose [177] that give an arbitrarily poor performance for minimum degree. In several studies on problems from structural analysis, Rothberg [178] and Ashcraft and Liu [28] have shown that dissection techniques can outperform minimum degree by on average about 15% in terms of floating-point operations for Cholesky factorization using the resulting ordering, although Ashcraft and Liu [28] report that the cost of these orderings is several times that of minimum degree.

Of course, dissection techniques are important for purposes other than generating an ordering for a Cholesky factorization. They can be used to partition an underlying grid for domain decomposition and are equally useful for the parallel implementation of many iterative methods. Two of the major software efforts for developing graph partitioning based on some of the above techniques are CHACO [130] and METIS [144].

4 Solution of sets of sparse unsymmetric equations

The same revolution that was just discussed for symmetric orderings has not happened in the case of unsymmetric systems, where usually a Markowitz ordering [159] or a column ordering based on minimum degree on the normal equations is used together with a threshold criterion for numerical pivoting. The main recent activities for unsymmetric systems have been the development of methods based on partitioning and the production of efficient codes using higher level BLAS operations.

A decade ago, the only widespread use of partitioning was to preorder the matrix to block triangular form prior to performing the analyse-factorize phase on the blocks on the diagonal of that form. Although there were applications where such matrices arose naturally (for example, chemical engineering), this technique did not apply to most systems, and it was common that the largest irreducible block was close to the order of the original matrix. Arioli and Duff [16] tried to extend block triangularization ideas by using tearing techniques that yield a bordered block triangular form. They found that usually there were too many columns in the border and the amount of arithmetic was much greater than using a standard sparse code on the whole system. Geschiere and Wijshoff [107] have pursued the use of tearing further and developed a package called MCSPARSE. Gallivan, Hansen, Ostromsky, and Zlatev [94] propose a partitioning similar to a block triangular form for exploitation of parallelism and gain some more advantage through allowing the blocks on the diagonal to be rectangular. We consider these techniques further in Section 7. A major problem with this approach is that the partitioning does not guarantee that the diagonal blocks are well conditioned, or even nonsingular. Thus some *a posteriori* measures must be taken to improve the stability of the factorization. This issue of stability was discussed by [18,91] and, more recently, methods for maintaining stability have been developed by [126,204].

A related approach, which I find very appealing, is to expand the matrix, perhaps artificially, in order to obtain a system that is larger and sparser. Normally, there is a choice of pivots for this system that would reduce it to the original system. The logic is that we might be able to do better by using the extra degree of freedom on the expanded system. A simple example of this matrix stretching technique [118,205] is the augmented system discussed in Section 5.

A major tool in the efficient implementation of codes for unsymmetric systems has been the observation of Gilbert and Peierls [113] that partial pivoting can be performed in time proportional to the number of arithmetic operations, so avoiding any potentially costly sorting operations. A nice refinement of their technique was provided by Eisenstat and Liu [89], who suggested ways of pruning a search tree to reduce work in the symbolic phase. Variants of this technique are used in nearly all sparse partial pivoting codes, for example [61,81].

Frontal, multifrontal, and supernodal approaches for the solution of unsymmetric problems have all seen significant recent developments and were discussed in Section 2 with respect to their use of higher level BLAS. The HSL frontal code MA42, which solves unsymmetric systems, was redesigned to use standard Level 2 and Level 3 BLAS and can accommodate entry by both equations and elements [85,86]. If the matrix is close to symmetric in structure in the sense that a_{ij} is usually nonzero when a_{ji} is, then methods adapted from symmetric multifrontal approaches can be used, with the analyse phase performed on the pattern of $A + A^T$ [10,80]. This approach can work well even if the matrix is unsymmetric, although for highly unsymmetric cases, as we discussed in Section 2, the approach of [54] that uses unsymmetric fronts and a directed acyclic graph may be preferable. The main problem with generalizing supernodal techniques to unsymmetric systems is that some regularity in the pattern of the factors is lost and it is not possible to make efficient use of the Level 3 BLAS. Demmel, Eisenstat, Gilbert, Li, and Liu [61] have overcome this difficulty by using Level 2.5 BLAS, as discussed in Section 2. Liu [154] has surveyed the use of multifrontal techniques, and recent surveys of frontal and multifrontal methods that include more discussion of the unsymmetric case can be found in [63,70]. A further strength of the methods considered in this paragraph is that they can be easily developed to exploit parallelism, as we discuss further in Section 7.

5 Solution of indefinite symmetric systems

During the last ten years, the solution of indefinite symmetric systems has been pursued with some vigour. A major impetus for this interest has come from the solution of sparse least-squares problems as a subproblem in the solution of interior-point problems. See [13] for an extensive list of references in this area. The problem with indefinite systems is that numerical pivoting is required and it may not be possible to form a Cholesky factorization. The standard approach is to generalize the 2×2 pivoting technique of [43] and form the factorization LDL^T where L is lower triangular and D is block diagonal with blocks of order 1 or 2. Although the HSL code MA27 [79], which implements such a strategy, has been widely used for some time to solve indefinite problems, it has recently become apparent that it has severe limitations if the problem is significantly indefinite. This is because it uses an ordering based on structure alone on the assumption that any diagonal entry is numerically suitable for a pivot in the subsequent numerical factorization. An extreme example of this is the solution of the ubiquitous augmented system

$$\begin{pmatrix} H & A \\ A^T & 0 \end{pmatrix} \begin{pmatrix} x \\ y \end{pmatrix} = \begin{pmatrix} b \\ c \end{pmatrix}, \tag{5.1}$$

where the (2,2) block of zeros can mean that the initial analyse phase gives a pivot sequence where many prospective pivots are zero. Additionally, as shown by [74], judicious selection of structured pivots of the form $\begin{pmatrix} \times & \times \\ \times & 0 \end{pmatrix}$ or $\begin{pmatrix} 0 & \times \\ \times & 0 \end{pmatrix}$ can preserve much of the structure of the original matrix.

Another common application that gives rise to sparse indefinite systems, usually of more general structure than in Equation 5.1, is the shift and invert strategy for obtaining interior eigenvalues of large sparse matrices.

Duff and Reid [84] have developed a code, called MA47 in the Harwell Subroutine Library, based on the work of [74], but have found that, although it sometimes performs far better than MA27, it can sometimes be significantly worse. This is in part because of the added complications in the code and in part because even the extension to encourage the use of structured pivots can fare badly when numerical considerations cause initial pivot choices to be unsuitable. As in the case of unsymmetric systems, it would appear that to obtain a robust code it is necessary to combine the analyse and factorization phases so that the numerical values of entries are considered when the initial pivot selection is made. This is done by [93], who show some convincing results in their use of structured pivots when solving problems from interior-point methods. Ashcraft, Grimes, and Lewis [24] have studied and analysed various aspects of the extension of threshold pivoting to the case of 2×2 pivots and show that the particular implementation can have quite different numerical properties. They extend their investigation to larger pivot blocks in a attempt to gain performance through the use of higher level BLAS, but the jury is still out on the benefits of using larger blocks.

6 Sparse least-squares

It is possibly fair to say that one of the major growth areas in sparse matrix computations over the last decade has been the solution of sparse least-squares problems. At least part of the reason for this has been the popularity and success of interior-point methods in optimization (see the paper [193] in this volume), where the central and most costly part of the calculation is the solution of a heavily weighted linear least-squares problem. Three main techniques are used to solve these least-squares problems: normal equations, augmented systems, and QR factorization. For a recent excellent review of this area, the book by Björck [39] is strongly recommended. We discuss each approach in turn.

The attraction of normal equations is that there are several efficient and readily available codes for the solution of sparse symmetric positive definite equations. The problem with this approach is that, because of the scaling, the systems can be very ill-conditioned. However, experience has shown [190,210] that the stability of these methods is better than one would expect or deserve, a phenomenon that is discussed in more detail by [193]. A further problem with using the normal equations is that a single dense row in A would result in the normal equations being completely dense. This problem was recognized some time ago [98,101] and is avoided by partitioning the system and treating the dense rows by an updating scheme. The selection of which rows to include in the dense part is still an open question (for example, [198]). There is difficulty in selecting rows so that the remaining problem remains sparse and is not singular or seriously ill-conditioned. Saunders [190] gives a handy list of alternatives to using Cholesky factorization, while Rothberg and Hendrickson [183] have explored the performance of various sparse ordering techniques on normal equations matrices from interior-point methods.

We have already discussed the solution of augmented systems in the previous section. For the solution of least-squares problems, the matrix H is diagonal and the vector c on the right-hand side of Equation 5.1 is zero. The beauty here is that, by suitable choice of 2×2 pivots, small entries on the diagonal of the (1,1) block in Equation 5.1 do not cause stability problems. Arioli, Duff, and Ruiz [17] have used the analysis of [15] and have conducted several experiments to support the viability of using 2×2 pivots. Björck [38] provides a detailed error analysis for this approach. Further computational experience in using a scaling on the (1,1) block is presented in [69].

At first glance, QR methods do not seem very attractive because of the fill-in to the factor Q. Additionally, dense rows in A will cause the factor R to be full. The storage of the denser Q can be avoided, at the cost of possible instability, by using the semi-normal equations (SNE)

$$R^T R = A^T b.$$

Moreover, analysis by [37] has shown that, in most cases, numerically satisfactory results can be obtained by using the corrected semi-normal equations (CSNE), where one step of iterative refinement is used. Sparse QR factorization uses the observation exploited by [98] that the factor R is the same as the Cholesky factor of the normal equations matrix. Although this is true in exact arithmetic, the difficulty in recognizing numerical cancellation means that the computed structure of the Cholesky factor can overestimate the structure of R. However, in many cases, this structure accurately predicts that of the factor R [49], particularly if the original matrix is first permuted to block triangular form [169]. A column ordering of the rectangular matrix A is obtained using a symmetric ordering on the structure of the normal equations, although the ordering can be obtained from A without requiring the formation of $A^T A$ [111]. The column ordering can then be used to construct a computational tree to drive the numerical factorization, which is performed using a QR factorization (for example, [12,161,173]). This can be implemented using a supernodal or multifrontal approach that can borrow extensively from the techniques used for sparse LU factorization. These methods are amenable to parallelization (see Section 7). It is also possible to develop rank revealing factorizations by delaying pivoting on columns that are deemed linearly independent. This can be incorporated economically within a multifrontal factorization scheme [35,168]. For cases that are particularly badly scaled, the factor Q can be stored and used in a conventional QR solution scheme [12]. Lu and Barlow [155] show that, for regular problems, the storage of the factor Q for a multifrontal scheme need be no more than that for the factor R, and Gilbert, Ng and Peyton [112] consider structure prediction techniques for the factor Q.

7 Parallel computing

In spite of the demise of many vendors of parallel machines, there are still several different parallel architectures and the methods for exploiting them differ accordingly. In this review, we will distinguish only the two main classes of machine: shared memory multiprocessors (SMP), including virtual shared memory machines, and distributed memory computers that require some form of message passing to communicate data between processors. The extreme case of the latter, termed network computing, is currently very fashionable but we will consider it only as a form of distributed computing.

There are three levels at which parallelism can be exploited in the solution of sparse linear systems by direct methods. As we discussed in Section 2, many codes use higher level BLAS to perform most or all of the floating-point operations. Many vendors of shared memory computers offer parallel versions of the BLAS and so at this level parallelization is trivial. Given an appropriate distribution of the matrix, parallel versions of the higher level BLAS for distributed memory machines can be constructed, possibly using tools like the BLACS (Basic Linear Algebra Communications Routines) [209].

At the coarsest level, techniques that we introduced in Section 4 for partitioning the matrix are often designed for parallel computing and are especially appropriate for distributed memory computers. Indeed these methods are often only competitive when parallelism is considered. Pothen and Fan [169] have developed a partitioning, based on the Dulmage and Mendelsohn canonical decomposition of a bipartite graph, that generalizes the block triangular form to rectangular systems. This has been used by [12] in the QR factorization of sparse rectangular matrices. Zlatev, Waśniewski, Schaumburg, Hansen, and Ostromsky [211,213] have developed a package, PARASPAR, for parallel solution that uses a preordering to partition the original problem. The MCSPARSE package [95,107] similarly uses a coarse matrix decomposition to obtain an ordering to bordered block triangular form.

At an intermediate level, we can use the sparsity of the matrix to advantage. This could be simply using the ability to choose several pivots simultaneously. Two matrix entries a_{ij} and a_{rs} can be used as pivots simultaneously if a_{is} and a_{rj} are zero. These pivots are termed *compatible*. This observation [44] has been the basis for several algorithms and parallel codes for general matrices. The central theme is to select a number of compatible pivots that would give a diagonal block if ordered to the top left of the matrix. The update from all these pivots is then performed in parallel. The procedure is then repeated on the reduced matrix. The algorithms differ in how the pivots are selected (clearly one must compromise the Markowitz criterion to get a large compatible pivot set) and in how the update is performed. Alaghband [2] uses compatibility tables to assist in the pivot search. She uses a two-stage implementation where first pivots are chosen in parallel from the diagonal and then off-diagonal pivots are chosen sequentially to stabilize the ordering. She sets thresholds for both sparsity and stability when choosing pivots. Her original experiments were performed on a Denelcor HEP. Davis and Yew [56] perform their pivot selection in parallel, which results in the nondeterministic nature of their algorithm because the compatible set will be determined by the order in which potential compatible pivots are found. Their algorithm, D2, was designed for shared-memory machines and was tested extensively on an Alliant FX/8. The Y12M algorithm [211] extends the notion of compatible pivots by permitting the pivot block to be upper triangular rather than diagonal, which gives a larger number of pivots, although the update is more complicated. For distributed memory architectures, van der Stappen, Bisseling, and van de Vorst [201] distribute the matrix over the processors in a grid fashion, perform a parallel search for compatible pivots, choosing entries of low Markowitz cost that satisfy a pivot threshold, and perform a parallel rank-m update of the reduced matrix, where m is the number of compatible pivots chosen. Their code was originally written in OCCAM and run on a network of 400 transputers, but they have since developed a version using PVM [148].

A common structure for both visualizing and implementing parallelism is the elimination tree, or derivatives of it. The main property that we exploit in this tree is that computations corresponding to nodes that are not ancestors or descendants of each other are independent (see, for example, [67,151]). The tree can thus be used to schedule parallel tasks. For shared memory machines, this can be accomplished through a shared pool of work with fairly simple synchronizations that can be controlled using locks protecting critical sections of the code [8,68]. In this way, nearly all the approaches that we have discussed in earlier sections that use frontal, multifrontal, or supernodal techniques to effect LU or QR factorization can be implemented to exploit parallelism. A major issue for an efficient implementation on shared memory machines concerns the management of data, which must be organized so that book-keeping operations such as garbage collection do not cause too much interference with the parallel processing. Johnson and Davis [139] examine aspects of a parallel buddy memory system, while Amestoy and Duff [11] discuss and compare several approaches and recommend a hybrid scheme with different memory management in different subregions of storage.

The original experiments of [114] on the massively parallel SIMD Connection machine were a little disappointing although they did indicate the possibility of using massive parallelism in sparse factorization. The experiments did show that a grid-distributed multifrontal implementation substantially outperformed a "Router Cholesky" algorithm based on a fan-in approach. Conroy, Kratzer, and Lucas [52] used a mapping strategy that can trade work against data movement to design a multifrontal algorithm for the TMC CM-5 that is competitive with codes on vector supercomputers. They tested their code on a MasPar MP-2. Manne and Hafsteinsson [157] have implemented a supernodal fan-out algorithm on the MasPar MP-2 and use a graph colouring algorithm to map the matrix to processors.

The earliest work on parallelizing sparse codes for distributed memory machines was based on column oriented Cholesky factorizations, either the fan-in or the fan-out algorithm. The original codes just used a column-column formulation of the algorithm as in the fan-in algorithm of [99] but it was soon apparent that better efficiency could be obtained as in the supernode-column fan-in approach of [166]. Some of this early work on parallel algorithms for distributed memory computers is reviewed by [127]. For distributed memory machines, processors can be assigned work corresponding to subtrees, but this requires quite balanced trees. Geist and Ng [96] use a breadth-first search strategy to assign work to processors using a heuristic bin packing algorithm to achieve reasonable load balancing. However, the results of runs on L-shape domains on an INTEL iPSC/2 comparing their strategy with wrap mapping and with a nested-dissection ordering and subtree-to-subcube mapping are rather flat, showing only a slight advantage for their heuristic. Pothen and Sun [172] have adapted the heuristic of [96] to a multifrontal scheme and have compared this with a generalized version of a subtree-to-subcube mapping and have found their algorithm to be twice as fast on an INTEL iPSC/2. All approaches to sparse Cholesky factorization have been used to develop parallel factorization routines on hypercubes: a fan-out algorithm [100], a fan-in algorithm [22], and a multifrontal approach [196]. Ashcraft, Eisenstat, Liu and Sherman [23] have compared all three approaches but none of the methods shows very high performance. Sun [197] describes a package of subroutines implementing his parallel multifrontal algorithm [196]. Geist and Ng [105] show that, even in a distributed environment, it is very beneficial if some memory is available as shared memory to hold information such as mapping vectors. The first work to exploit parallelism at all phases of the sparse solution process was by Zmijewski [214], later developed by [115,215]. More recently, the work of Heath and

Raghavan [128] also exploits parallelism in all phases, although they require that the matrix be held in Cartesian form; that is, in a two or three dimensional coordinate system. While this is quite natural in the context of discretized PDEs, it is not a convenient interface in general.

Schreiber [192] presents a clear discussion showing that a one-dimensional mapping of columns or block columns to processors is inherently unscalable and that a two-dimensional mapping is needed to obtain a scalable algorithm. Rothberg [180] compares a block fan-out algorithm using two-dimensional blocking with a panel multifrontal method using one-dimensional blocking and favours the former, obtaining a performance of over 1.7 Gflop/s on 128 nodes of an Intel Paragon. He points out that the benefit of using higher level BLAS kernels, coupled with the then recent increases in local memory and communication speed of parallel processors, had at last made the solution of large sparse systems feasible on such architectures. The 2-D block fan-out algorithm is further investigated by [182], and Rothberg and Schreiber [184] propose some block mapping heuristics to improve the performance to over 3 Gflop/s for a 3-D grid problem on a 196-node Intel Paragon. A similar type of 2-dimensional mapping is used by Gupta, Karypis and Kumar [120] in their implementation of a multifrontal method, where much of the high performance is obtained through balancing the tree near its root and using a careful and regular mapping of the dense matrices near the root to enable a high level of parallelism to be maintained when the trivial parallelism from subtree assignment is exhausted. This work has provided possibly the best performance for distributed memory implementation of a direct sparse Cholesky code. Although the headline figure of nearly 20 Gflop/s on the CRAY T3D was obtained on a fairly artificial and essentially dense problem, large sparse problems from structural analysis were factorized at between 8 and 15 Gflop/s on the same machine. Karypis, Gupta and Kumar [142] have used this parallel multifrontal algorithm in the solution of interior-point problems, and similarly Bisseling, Doup, and Loyens [36] and Lustig and Rothberg [156] have used parallel Cholesky factorizations to speed up this computation (see also [193])

Partly because of the success of fast and parallel methods for performing the numerical factorization, other phases of the solution are now becoming more critical on parallel computers. The package of [128] executes all phases in parallel, and there has been much recent work in finding parallel methods for performing the reordering. This has been another reason for the growth in dissection approaches (for example, see [145,175]). It is possible to parallelize the triangular solve by using a tree, similar or identical to that for the numerical factorization. Anderson and Saad [14] generate a tree given the sparsity structure of the triangular factor, while Amestoy, Duff, and Puglisi [12] use the same elimination tree as for the earlier multifrontal factorization. However, in order to avoid the intrinsically sequential nature of a sparse triangular solve, Alvarado, Yu, and Betancourt [7] have proposed holding L^{-1} or rather a partitioned form of this to avoid some of the fill-in that would be associated with forming the inverse explicitly. Various schemes for this partitioning have been proposed to balance the parallelism (limited by number of partitions) with the fill-in (for example, [5,6,167]). Very recently, Raghavan [174] has proposed the selective inversion of submatrices produced by a multifrontal factorization algorithm.

8 Preconditioning

One of the main problems with sparse LU factorization is that often the number of entries in the factors is substantially greater than in the original matrix so that, even if the original matrix can be stored, the factors cannot. If we assume that we can store the original matrix, then one possibility is to start a sparse LU factorization but drop some fill-in entries so that the partial factors can still be stored. Algorithms for doing this are called incomplete LU (or ILU) factorizations and they differ depending on the criteria for deciding which entries to drop. At one extreme, we could hold all the factors, while at the other we could store no fill-ins. This partial or incomplete factorization is then used to precondition the matrix for iterative solution, normally using a fairly standard Krylov-sequence based iterative technique like conjugate gradients in the symmetric case or GMRES or BiCG when the matrix is unsymmetric.

The main criteria for deciding which entries to include in an incomplete factorization are location and numerical value. The commonest location-based criterion is to allow a set number of levels of fill-in, where original entries have level zero, original zeros have level ∞ and a fill-in in position (i,j) has level $Level_{ij}$ determined by

$$\min_{1\leq k\leq \min(i,j)} \{Level_{ik} + Level_{kj} + 1\}.$$

In the case of simple discretizations of partial differential equations, this gives a simple pattern for incomplete factorizations with different levels of fill-in. For example, if the matrix is from a five-point discretization of the Laplacian in two-dimensions, level 1 fill-in will give the original pattern plus a diagonal inside the outermost band. The other main criterion for deciding which entries to omit is to drop entries less than a predetermined numerical value. For the regular problems just mentioned, it is interesting that the level fill-in and drop strategies give a somewhat similar incomplete factorization, because the numerical value of successive fill-in levels decreases markedly, reflecting the characteristic decay in the entries of the inverse matrix (see, for example, [163]). For general problems, however, the two strategies can be significantly different. Since it is usually not known *a priori* how many entries will be above a selected threshold, the dropping strategy is normally combined with restricting the number of fill-ins allowed to any one column [188]. When using a threshold criterion, it is possible to change it dynamically during the factorization to attempt to achieve a target density of the factors [164]. Tismenetsky [200] has developed a more robust but generally denser incomplete factorization by only excluding entries outside a specified pattern (in which case the diagonal is modified) and those fill-ins that would be caused by the product of two small entries. Although the notation is not yet fully standardized, the nomenclature commonly adopted for incomplete factorizations is ILU(k), when k levels of fill-in are allowed and ILUT(α, f), for the threshold criterion when entries of modulus less than α are dropped and the maximum number of fill-ins allowed in any column is f. There are many variations on these strategies and the criteria are sometimes combined. In some cases, constraining the row sums of the incomplete factorization to match those of the matrix can help [122]. Additionally, in the symmetric positive-definite case, steps are often taken to ensure that the resulting preconditioner is also positive definite by modifying the matrix being factorized [158] or adjusting diagonal entries of the factorization [1,137,164].

The use of incomplete factorizations as preconditioners for symmetric systems has a long pedigree [162] and good results have been obtained for a wide range of problems. An incomplete Cholesky factorization where one level of fill-in is allowed (ICCG(1)) has proven to provide a good balance between reducing the number of iterations and the cost of computing and using the preconditioning. Although it may be thought that a preordering that would result in low fill-in for a complete factorization (for example, minimum degree) might be advantageous for an incomplete factorization, Duff and Meurant [78] and Eijkhout [88] show that it is not true in general and that sometimes the number of iterations of ICCG(0) can double if a minimum degree ordering is used, although this effect is not apparent for ILUT preconditioners. The situation with symmetric systems is quite well analysed and understood. Much recent work for symmetric systems has been to develop preconditioners that can be computed and used on parallel computers. Most of this work has, however, been applicable to highly structured problems from discretizations of elliptic partial differential equations in two and three dimensions, for example [203]. Heroux, Vu, and Yang [132] and Jones and Plassmann [140] have experimented with unstructured matrices, with reasonable speed-ups being achieved in the latter paper.

The situation for unsymmetric systems is, however, much less clear. Although there have been many experiments on using incomplete factorizations and there have been studies of the effect of orderings on the number of iterations [59,87] that show similar behaviour to the symmetric case, there is very little theory governing the behaviour for general systems and indeed the performance of ILU preconditioners is very unpredictable. Allowing high levels of fill-in can help but again there is no guarantee. In fact, a major problem is that the ILU factors can become very much more ill-conditioned than the original system and so the preconditioned system can perform much worse than the original matrix with respect to convergence of the iterative method. This phenomenon is analysed by [90] for the case of an ILU(0) preconditioner for systems from convection-diffusion problems.

The QR or LQ factorizations can also be used in the unsymmetric case to derive an incomplete factorization [136,186]. Here the orthogonal factor need not be kept and the resulting incomplete triangular factor can be used to precondition the normal equations. Preconditioners based on these factorizations are generally more expensive to compute and use than ILU preconditioners but they are usually more robust. Saad [187] has examined the use of incomplete LQ factorizations as preconditioners for nonsymmetric and indefinite systems, and Benzi and Tůma [32] have confirmed that preconditionings based on incomplete orthogonalization methods can succeed where ILU preconditioners fail. One way of computing an incomplete orthogonal factorization is to use incomplete modified Gram-Schmidt (IMGS). The IMGS approach has been explored by [207] and is described in detail in the thesis by [206].

Of course, the LU and LQ factorizations are ways of representing the inverse of a sparse matrix in a way that can be economically used to solve linear systems. The main reason why explicit inverses are not used is that, for irreducible matrices, the inverse will always be dense (because we neglect numerical cancellation, see [71]). However, this need not be a problem if we follow the flavour of ILU factorizations and compute and use a sparse approximation to the inverse. Perhaps the most interesting technique for this is to solve the problem

$$\min_{M} ||I - AM||_F, \tag{8.1}$$

where M has some fully or partially prescribed sparsity structure. One advantage of this is that this problem can be split into n independent least-squares problems for each of the n columns of M. Each of these least-squares problems only involves a few variables (corresponding to the number of entries in the column of M) and, because they are independent, they can be solved in parallel. A further benefit of such techniques is that it is possible to successively increase the density of the approximation to reduce the value of Equation 8.1 and so, in principle, ensure convergence of the preconditioned iterative method [53]. Cosgrove, Diaz, and Griewank [53], Huckle and Grote [135], and Gould and Scott [117] use a (dense) QR factorization to solve the small least-squares problems while Chow and Saad [48] use GMRES. Gould and Scott [117] show that this technique gives almost as good a preconditioner as ILU but is much more expensive to compute both in terms of time and storage, at least if computed sequentially. One problem with these approaches is that, although the residual for the approximation of a column of M can be controlled (albeit perhaps at the cost of a rather dense column in M), the nonsingularity of the matrix M is not guaranteed. Partly to avoid this, Kolotilina and Yeremin [147] have proposed approximating the triangular factors of the inverse and, to this end, Benzi and Tůma [31] generate sparse approximations to an A-biconjugate set of vectors using drop tolerances. In a scalar or vector environment, it is also much cheaper to generate the factors by this means than to solve the least-squares problems for columns of the approximate inverse.

One of the main reasons for the interest in sparse approximate inverse preconditioners is the difficulty of parallelizing ILU preconditioners, not only in their construction but also in their use, which requires a sparse triangular solution. However, although almost every paper on approximate inverse preconditioners states that the authors are working on a parallel implementation, there are relatively few such studies available. Grote and Simon [119] perform limited studies when the matrix is highly structured. Gustafsson and Lindskog [124], using an idea of van der Vorst [202], have implemented a fully parallel preconditioner based on truncated Neumann expansions to approximate the inverse SSOR factors of the matrix. Their experiments on a CM-200 show a worthwhile improvement over a simple diagonal scaling.

Note that, because the inverse of the inverse of a sparse matrix is sparse (a fact not detected because we do not allow numerical cancellation), there are classes of dense matrices for which a sparse approximate inverse might be a very appropriate preconditioner.

There is a midway house between the LU factors and the inverse and that is to use the explicit inverses of L and U. This has been proposed by [6] because such a factorization is easier to use in parallel than an LU factorization. An incomplete form of this factorization for use as a preconditioner has been proposed by [4].

Of course, it is possible to represent the inverse by a polynomial in the matrix and use this polynomial as a preconditioner. One approach, by [64], is to use the low order terms of a Neumann expansion of $(I-B)^{-1}$, where $A = I - B$ and the spectral radius of B is less than 1. They use a matrix splitting $A = M - N$ and a truncated power series for $M^{-1}N$ when the condition on B is not satisfied. More general polynomial preconditioners have also been proposed (see, for example, [20,138,185]). For efficiency, low degree polynomials are normally used. A polynomial preconditioner is simple to use and can be parallelized, but the results are not generally very encouraging and have been particularly disappointing for unsymmetric problems.

In the case of unassembled element problems, it is important to develop preconditioners that do not require assembly of the matrix. This sometimes involves the factorization of the element submatrices, and hence they use a direct method in computing the preconditioner, albeit usually on a small dense matrix [123]. Performing some subassemblies before computing the preconditioning can help [58], but the evidence suggests that any savings in iterations for the iterative method is about balanced by the extra work in using the preconditioner.

This concludes our discussion of incomplete factorizations or other incomplete representations of the inverse. Another whole class of preconditioners that use direct methods are those where the direct method is used to solve a subproblem of the original problem. This is often used in a domain decomposition setting, where problems on subdomains are solved by the direct method but the interaction between the subproblems is handled by an iterative technique. A related example of this is the work on block projection methods like Block Cimmino [19] or Block Kacmarz [41]. Block preconditioning for symmetric systems is discussed by [50], and Concus and Meurant [51] use incomplete factorizations within the diagonal blocks. Attempts have been made to preorder matrices to put large entries into the diagonal blocks so that the inverse of the matrix would be well approximated by the block diagonal matrix whose block entries are the inverses of the diagonal blocks [47]. We do not expand on these techniques here but leave further consideration to the chapter on iterative methods.

Multigrid techniques also often combine aspects of both iterative and direct methods. These methods were originally developed for solving partial differential equations but developments such as algebraic multigrid extend their applicability to more general systems, although the jury is still out on how wide this range is. The basic idea is to use corrections on a sequence of coarser grids to update the required solution on a fine grid. In our context, it is common to use a direct method for the solution on the coarsest grid with one or two iterations of usually a simple iterative method on the other grids. Hackbusch [125] and Wesseling [208] are worth reading for a background to multigrid methods, which are further considered in the chapter on iterative methods [116].

9 Towards a sparse problem solving environment

Often the solution of a set of sparse linear equations is the most costly part of the computation but, to the user of sparse codes, there might be as much cost in organizing and managing the data for the application in preparation for and after the solution step. A major problem is that a sparse data structure can be represented in many different ways. Most are very problem specific and it may not be trivial to organize the data for a call to the linear equation solution routine. This might be even more complicated should parallelism be exploited. There are some tools being developed to assist in the manipulation and management of sparse matrices.

Gilbert, Moler, and Schreiber [111] have introduced a sparse matrix structure and some sparse algorithms into MATLAB. Their aim has been ease of use and functionality rather than efficiency, although increasingly researchers are making codes available to MATLAB users through M-files (for example, [160]). Saad [189] has developed, over many years, a tool kit called SPARSKIT for sparse matrix computations, Gupta and Rothberg [121] have proposed an environment for handling sparse matrices on distributed memory machines, and Alvarado [3] has designed an integrated package as a teaching and development tool called SMMS (Sparse Matrix Manipulation System). We have been keen to stress that the important kernels are those for dense linear algebra. However, in the case of iterative methods and the use of preconditioning matrices, a sparse version of the BLAS is appropriate [77]. The provision of a standard set of sparse matrix test problems, the Harwell-Boeing Collection [75,76], has proven to be very useful for the design and comparison of algorithms, and there is currently an effort underway to update this collection and create a more friendly interface through the World Wide Web (presently at URL `http://math.nist.gov/MatrixMarket`).

We do not feel that a software review is appropriate here and indeed a thorough one would require as much space and effort as this present survey. Suffice it to say that there are a number of sparse direct solvers available through `netlib`, at URL `http://www.netlib.org/`, and possibly the largest collection of Library quality sparse direct codes is included in the Harwell Subroutine Library and in a subset of that Library, the Harwell Sparse Matrix Library (HSML), marketed by NAG. Further information on HSL and HSML can be obtained from the Web page `http://www.dci.clrc.ac.uk/Data/HSL/`.

10 Concluding remarks

Far be it from me to try to steal the thunder of the author of the equivalent talk ten years from now. However, after nine sections of reflections, it might be entertaining to suggest where the excitement may lie in the years to come.

First, I believe that the iterative and direct talks will be combined at the next meeting as they were ten years ago. I believe this because it is already becoming clear that the huge (order greater than 500,000) problems of tomorrow can only be solved by combining direct and iterative techniques, which is why I spent a full section on preconditioning methods. It is not that smaller problems (order 10,000 to 100,000) do not need to be solved but we essentially already have the tools to do this efficiently on serial computers, and this will fairly soon be routine on parallel computers also. Looking into my crystal ball, I think that soon the symmetric ordering problem will be resolved in favour of a class of hybrid methods (that is, methods involving both dissection techniques and minimum degree) parameterized to accommodate any sparse structure; the unsymmetric multifrontal and supernodal approaches will be available on distributed memory machines and will be so efficient that partitioning methods will only be used on huge systems prior to constructing a preconditioner. I think that row projection methods will be developed further and robust direct solvers will be used on the projected problems. Rather than developing a LAPACK approach to providing software for direct solution of sparse equations, a MATLAB-like environment will handle everything from problem formulation to post-analysis of the solution.

One thing I am sure of ... the sparse specialist will still have a job in ten years time ... at least I sincerely hope so!!

Acknowledgements

I asked a large number of people if they would look at a draft of this paper, particularly to check if I had the best reference to their work. I received many replies and am grateful to Patrick Amestoy, Cleve Ashcraft, Michele Benzi, Åke Björck, Randall Bramley, John Gilbert, Jacko Koster, John Reid, Edward Rothberg, Michael Saunders, Jennifer Scott, Henk van der Vorst, and Zahari Zlatev for exceeding their brief by commenting more generally on various aspects of the draft.

References

1. M. A. Ajiz and A. Jennings. A robust incomplete Choleski-conjugate gradient algorithm. *Int. J. Numerical Methods in Engineering*, 20:949–966, 1984.
2. G. Alaghband. Parallel sparse matrix solution and performance. *Parallel Computing*, 21(9):1407–1430, 1995.
3. F. L. Alvarado. Manipulation and visualisation of sparse matrices. *ORSA J. Computing*, 2:186–207, 1989.
4. F. L. Alvarado and H. Dağ. Incomplete partitioned inverse preconditioners. Technical Report (to appear), Department of Electrical and Computer Engineering, University of Wisconsin at Madison, 1994. Submitted to Parallel Computing.
5. F. L. Alvarado, A. Pothen, and R. Schreiber. Highly parallel sparse triangular solution. In Alan George, John R. Gilbert, and Joseph W. H. Liu, editors, *Graph Theory and Sparse Matrix Computation*. Springer-Verlag, 1993.
6. F. L. Alvarado and R. Schreiber. Optimal parallel solution of sparse triangular systems. *SIAM J. Scientific Computing*, 14:446–460, 1993.
7. F. L. Alvarado, D. C. Yu, and R. Betancourt. Partitioned sparse A^{-1} methods. *IEEE Trans. Power Systems*, 3:452–459, 1990.
8. Patrick R. Amestoy. Factorization of large sparse matrices based on a multifrontal approach in a multiprocessor environment. INPT PhD Thesis TH/PA/91/2, CERFACS, Toulouse, France, 1991.
9. Patrick R. Amestoy, Timothy A. Davis, and Iain S. Duff. An approximate minimum degree ordering algorithm. Technical Report TR/PA/95/09, CERFACS, Toulouse, France, 1995. To appear in *SIAM J. Matrix Analysis and Applications*.
10. Patrick R. Amestoy and Iain S. Duff. Vectorization of a multiprocessor multifrontal code. *Int. J. of Supercomputer Applics.*, 3:41–59, 1989.
11. Patrick R. Amestoy and Iain S. Duff. Memory management issues in sparse multifrontal methods on multiprocessors. *Int. J. Supercomputer Applics*, 7:64–82, 1993.
12. Patrick R. Amestoy, Iain S. Duff, and Chiara Puglisi. Multifrontal QR factorization in a multiprocessor environment. *Numerical Linear Algebra with Applications*, 3(4):275–300, 1996.
13. E. D. Andersen, J. Gondzio, C. Mészáros, and X. Xu. Implementation of interior point methods for large scale linear programming. Technical Report 1996.3, Logilab, University of Geneva, Switzerland, 1996.
14. E. C. Anderson and Y. Saad. Solving sparse triangular systems on parallel computers. *Int J. High Speed Computing*, 1:73–95, 1989.
15. Mario Arioli, James W. Demmel, and Iain S. Duff. Solving sparse linear systems with sparse backward error. *SIAM J. Matrix Analysis and Applications*, 10:165–190, 1989.
16. Mario Arioli and Iain S. Duff. Experiments in tearing large sparse systems. In M. G. Cox and S. Hammarling, editors, *Reliable Numerical Computation*, pages 207–226, Oxford, 1990. Oxford University Press.
17. Mario Arioli, Iain S. Duff, and Peter P. M. de Rijk. On the augmented systems approach to sparse least-squares problems. *Numer. Math.*, 55:667–684, 1989.
18. Mario Arioli, Iain S. Duff, Nicholas I. M. Gould, and John K. Reid. Use of the P^4 and P^5 algorithms for in-core factorization of sparse matrices. *SIAM J. Sci. Stat. Comput.*, 11:913–927, 1990.
19. Mario Arioli, Iain S. Duff, Joseph Noailles, and Daniel Ruiz. A block projection method for sparse equations. *SIAM J. Scientific and Statistical Computing*, 13:47–70, 1992.
20. S. F. Ashby. Minimax polynomial preconditioning for Hermitian linear systems. *SIAM J. Matrix Analysis and Applications*, 12(4):766–789, 1991.

21. C. Ashcraft. Compressed graphs and the minimum degree algorithm. *SIAM J. Scientific Computing*, 16:1404–1411, 1995.

22. C. Ashcraft, S. C. Eisenstat, and J. W. H. Liu. A fan-in algorithm for distributed sparse numerical factorization. *SIAM J. Scientific and Statistical Computing*, 11:593–599, 1990.

23. C. Ashcraft, S. C. Eisenstat, J. W. H. Liu, and A. H. Sherman. A comparison on three column-based distributed sparse factorization schemes. Technical Report CS-90-09, Department of Computer Science, York University, York, Ontario, Canada, 1990.

24. C. Ashcraft, R. G. Grimes, and J. G. Lewis. Accurate symmetric indefinite linear equation solvers. Technical Report ISSTECH-95-029, Boeing Computer Services, Seattle, 1995.

25. C. Ashcraft and J. W. H. Liu. Generalized nested dissection: some recent progress. In J. G. Lewis, editor, *Proceedings of the Fifth SIAM Conference on Applied Linear Algebra*, pages 130–139, Philadelphia, 1994. SIAM Press.

26. C. Ashcraft and J. W. H. Liu. A partition improvement algorithm for generalized nested dissection. Technical Report BCSTECH-94-020, Boeing Computer Services, Seattle, 1994.

27. C. Ashcraft and J. W. H. Liu. Using domain decomposition to find graph bisectors. Technical Report ISSTECH-95-024, Boeing Information and Support Services, Seattle, 1995. Also Report CS-95-08, Department of Computer Science, York University, Ontario, Canada.

28. C. Ashcraft and J. W. H. Liu. Robust ordering of sparse matrices using multisection. Technical Report ISSTECH-96-002, Boeing Information and Support Services, Seattle, 1996. Also Report CS-96-01, Department of Computer Science, York University, Ontario, Canada.

29. S. T. Barnard, A. Pothen, and H. Simon. A spectral algorithm for envelope reduction of sparse matrices. *Numerical Linear Algebra with Applications*, 2(4):317–334, 1995.

30. S. T. Barnard and H. Simon. A fast multilevel implementation of recursive spectral bisection for partitioning unstructured problems. In R. F. Sincovec, D. E. Keyes, M. R. Leuze, L. R. Petzold, and D. A. Reed, editors, *Proceedings of the Sixth SIAM Conference on Parallel Processing for Scientific Computing*, pages 711–718. SIAM Press, 1993.

31. M. Benzi and M. Tůma. A sparse approximate inverse preconditioner for nonsymmetric linear systems. Technical Report No. 653, Institute of Computer Science, Academy of Sciences of the Czech Republic, October 1995. To appear in *SIAM J. Scientific Computing*.

32. M. Benzi and M. Tůma. A comparison of some preconditioning techniques for general sparse matrices. In S. D. Margenov and P. S. Vassilevski, editors, *Iterative Methods in Linear Algebra, II*, Volume 3 in the IMACS Series in Computational and Applied Mathematics, pages 191–203. IMACS, 1996.

33. A. Berger, J. Mulvey, E. Rothberg, and R. Vanderbei. Solving multistage stochastic programs using tree dissection. Technical Report SOR-97-07, Programs in Statistics and Operations Research, Princeton University, Princeton, New Jersey, 1995.

34. P. Berman and G. Schnitger. On the performance of the minimum degree ordering for Gaussian elimination. *SIAM J. Matrix Analysis and Applications*, 11(1):83–88, 1990.

35. C. H. Bischof, J. G. Lewis, and D. J. Pierce. Incremental condition estimation for sparse matrices. *SIAM J. Matrix Analysis and Applications*, 11:644–659, 1990.

36. R. H. Bisseling, T. M. Doup, and L. D. J. C. Loyens. A parallel interior point algorithm for linear programming on a network of transputers. *Annals of Operations Research*, 43:51–86, 1993.

37. Å. Björck. Stability analysis of the method of semi-normal equations for least squares problems. *Linear Algebra and its Applications*, 88/89:31–48, 1987.

38. Å. Björck. Pivoting and stability in the augmented system method. In D. F. Griffiths and G. A. Watson, editors, *Numerical Analysis 1991, Proceedings of the 14th Dundee Conference, June 1991*, Pitman Research Notes in Mathematics Series. **260**, pages 1–16, Harlow, England, 1992. Longman Scientific & Technical.

39. Å. Björck. *Numerical Methods for Least Squares Problems.* SIAM Press, Philadelphia, 1996.

40. Hans L. Bodlaender, John R. Gilbert, Hjálmtýr Hafsteinsson, and Ton Kloks. Approximating treewidth, pathwidth, frontsize, and shortest elimination tree. *Journal of Algorithms*, 18:238–255, 1995.

41. R. Bramley and A. Sameh. Row projection methods for large nonsymmetric linear systems. *SIAM J. Scientific and Statistical Computing*, 13:168–193, 1992.

42. T. Bui and C. Jones. A heuristic for reducing fill-in in sparse matrix factorization. In R. F. Sincovec, D. E. Keyes, M. R. Leuze, L. R. Petzold, and D. A. Reed, editors, *Proceedings of the Sixth SIAM Conference on Parallel Processing for Scientific Computing*, pages 445–452. SIAM, 1993.

43. J. R. Bunch and B. N. Parlett. Direct methods for solving symmetric indefinite systems of linear equations. *SIAM J. Numerical Analysis*, 8:639–655, 1971.

44. D. A. Calahan. Parallel solution of sparse simultaneous linear equations. In *Proceedings 11th Annual Allerton Conference on Circuits and System Theory, University of Illinois*, pages 729–735, 1973.

45. I. A. Cavers. Using deficiency measure for tie-breaking the minimum degree algorithm. Technical Report 89-2, University of British Columbia, Canada, 1989.

46. T. Chan, J. Gilbert, and S-H Teng. Geometric spectral partitioning. Technical Report CSL-94-15, Palo Alto Research Center, Xerox Corporation, California, December 1994.

47. H. Choi and D. B. Szyld. Threshold ordering for preconditioning nonsymmetric problems with highly varying coefficients. Technical Report 96-51, Department of Mathematics, Temple University, Philadelphia, 1996.

48. E. Chow and Y. Saad. Approximate inverse preconditioners for general sparse matrices. Technical Report UMSI 94/101, University of Minnesota Supercomputer Institute, 1994.

49. T. F. Coleman, A. Edenbrandt, and J. R. Gilbert. Predicting fill for sparse orthogonal factorization. *J. ACM*, 33:517–532, 1986.

50. P. Concus, G. H. Golub, and G. Meurant. Block preconditioning for the conjugate gradient method. *SIAM J. Scientific and Statistical Computing*, 6:220–252, 1985.

51. P. Concus and G. Meurant. On computing INV block preconditionings for the conjugate gradient method. *BIT*, 26:493–504, 1986.

52. J. M. Conroy, S. G. Kratzer, and R. F. Lucas. Data-parallel sparse matrix factorization. In J. G. Lewis, editor, *Proceedings 5th SIAM Conference on Linear Algebra*, pages 377–381, Philadelphia, 1994. SIAM Press.

53. J. D. F. Cosgrove, J. C. Diaz, and A. Griewank. Approximate inverse preconditionings for sparse linear systems. *Int. J. Computer Mathematics*, 44:91–110, 1992.

54. Timothy A. Davis and Iain S. Duff. An unsymmetric-pattern multifrontal method for sparse LU factorization. Technical Report RAL 93-036, Rutherford Appleton Laboratory, 1993. To appear in *SIAM J. Matrix Analysis and Applications.*

55. Timothy A. Davis and Iain S. Duff. A combined unifrontal/multifrontal method for unsymmetric sparse matrices. Technical Report TR-95-020, Computer and Information Science Department, University of Florida, 1995.

56. Timothy A. Davis and P. C. Yew. A nondeterministic parallel algorithm for general unsymmetric sparse LU factorization. *SIAM J. Matrix Analysis and Applications*, 11:383–402, 1990.

57. Michel J. Daydé, Iain S. Duff, and Antoine Petitet. A parallel block implementation of Level 3 BLAS kernels for MIMD vector processors. *ACM Trans. Math. Softw.*, 20:178–193, 1994.

58. Michel J. Daydé, Jean-Yves L'Excellent, and Nicholas I. M. Gould. On the preprocessing of sparse unassembled linear systems for efficient solution using element-by-element preconditioners. Technical Report RT/APO/96/2, Département Informatique, ENSEEIHT-IRIT, Toulouse, 1996.

59. E. F. D'Azevedo, P. A. Forsyth, and W. P. Tang. Ordering methods for preconditioned conjugate gradient methods applied to unstructured grid problems. *SIAM J. Matrix Analysis and Applications*, 13:944–961, 1992.

60. B. Dembart and A. M. Erisman. Hybrid sparse matrix methods. *IEEE Trans. Circuit Theory*, CT-20:641–649, 1973.

61. J. W. Demmel, S. C. Eisenstat, J. R. Gilbert, X. S. Li, and J. W. H. Liu. A supernodal approach to sparse partial pivoting. Technical Report UCB//CSD-95-883, Computer Science Division, U. C. Berkeley, Berkeley, California, July 1995.

62. Jack J. Dongarra, Jeremy Du Croz, Iain S. Duff, and Sven Hammarling. A set of Level 3 Basic Linear Algebra Subprograms. *ACM Trans. Math. Softw.*, 16:1–17, 1990.

63. Jack J. Dongarra, Iain S. Duff, Danny C. Sorensen, and Henk A. van der Vorst. *Solving Linear Systems on Vector and Shared Memory Computers.* SIAM Press, Philadelphia, 1991. Second edition in preparation.

64. D. F. Dubois, A. Greenbaum, and G. H. Rodrigue. Approximating the inverse of a matrix for use on iterative algorithms on vector processors. *Computing*, 22:257–268, 1979.

65. Iain S. Duff. MA28 – A set of Fortran subroutines for sparse unsymmetric linear equations. Technical Report AERE R8730, Her Majesty's Stationery Office, London, 1977.

66. Iain S. Duff. Full matrix techniques in sparse Gaussian elimination. In G. A. Watson, editor, *Numerical Analysis Proceedings, Dundee 1981*, Lecture Notes in Mathematics 912, pages 71–84, Berlin, 1981. Springer-Verlag.

67. Iain S. Duff. The use of vector and parallel computers in the solution of large sparse linear equations. In P. Deuflhard and B. Engquist, editors, *Large scale scientific computing. Progress in Scientific Computing Volume 7*, pages 331–348, Boston, 1986. Birkhäuser.

68. Iain S. Duff. The influence of vector and parallel computers in the solution of large sparse linear equations. In M J D Powell and A Iserles, editors, *The State of the Art in Numerical Analysis*, pages 359–407, Oxford, 1987. Oxford University Press.

69. Iain S. Duff. The solution of augmented systems. In D. F. Griffiths and G. A. Watson, editors, *Numerical Analysis 1993, Proceedings of the 15th Dundee Conference, June-July 1993*, Pitman Research Notes in Mathematics Series. **303**, pages 40–55, Harlow, England, 1994. Longman Scientific & Technical.

70. Iain S. Duff. A review of frontal methods for solving linear systems. *Computer Physics Communications*, 97:45–52, 1996.

71. Iain S. Duff, A. M. Erisman, C. W. Gear, and John K. Reid. Sparsity structure and Gaussian elimination. *SIGNUM Newsletter*, 23(2):2–8, April 1988.

72. Iain S. Duff, A. M. Erisman, and John K. Reid. On George's nested dissection method. *SIAM J. Numerical Analysis*, 13:686–695, 1976.

73. Iain S. Duff, Nicholas I. M. Gould, Marc Lescrenier, and John K. Reid. The multifrontal method in a parallel environment. In M. G. Cox and S. Hammarling, editors, *Reliable Numerical Computation*, pages 93–111, Oxford, 1990. Oxford University Press.

74. Iain S. Duff, Nicholas I. M. Gould, John K. Reid, Jennifer A. Scott, and Kathryn Turner. Factorization of sparse symmetric indefinite matrices. *IMA J. Numerical Analysis*, 11:181–204, 1991.

75. Iain S. Duff, Roger G. Grimes, and John G. Lewis. Sparse matrix test problems. *ACM Trans. Math. Softw.*, 15(1):1–14, March 1989.

76. Iain S. Duff, Roger G. Grimes, and John G. Lewis. Users' guide for the Harwell-Boeing sparse matrix collection (Release I). Technical Report RAL 92-086, Rutherford Appleton Laboratory, 1992.

77. Iain S. Duff, Michele Marrone, Guiseppe Radicati, and Carlo Vittoli. A set of Level 3 Basic Linear Algebra Subprograms for sparse matrices. Technical Report TR-RAL-95-049, RAL, 1995.

78. Iain S. Duff and Gérard A. Meurant. The effect of ordering on preconditioned conjugate gradients. *BIT*, 29:635–657, 1989.

79. Iain S. Duff and John K. Reid. The multifrontal solution of indefinite sparse symmetric linear systems. *ACM Trans. Math. Softw.*, 9:302–325, 1983.

80. Iain S. Duff and John K. Reid. The multifrontal solution of unsymmetric sets of linear systems. *SIAM J. Scientific and Statistical Computing*, 5:633–641, 1984.

81. Iain S. Duff and John K. Reid. MA48, a Fortran code for direct solution of sparse unsymmetric linear systems of equations. Technical Report RAL 93-072, Rutherford Appleton Laboratory, 1993.

82. Iain S. Duff and John K. Reid. MA47, a Fortran code for direct solution of indefinite sparse symmetric linear systems. Technical Report RAL 95-001, Rutherford Appleton Laboratory, 1995.

83. Iain S. Duff and John K. Reid. The design of MA48, a code for the direct solution of sparse unsymmetric linear systems of equations. *ACM Trans. Math. Softw.*, 22(2):187–226, 1996.

84. Iain S. Duff and John K. Reid. Exploiting zeros on the diagonal in the direct solution of indefinite sparse symmetric linear systems. *ACM Trans. Math. Softw.*, 22(2):227–257, 1996.

85. Iain S. Duff and Jennifer A. Scott. MA42 – a new frontal code for solving sparse unsymmetric systems. Technical Report RAL 93-064, Rutherford Appleton Laboratory, 1993.

86. Iain S. Duff and Jennifer A. Scott. The design of a new frontal code for solving sparse unsymmetric systems. *ACM Trans. Math. Softw.*, 22(1):30–45, 1996.

87. Laura C. Dutto. The effect of ordering on preconditioned GMRES algorithm, for solving the Navier-Stokes equations. *Int. J. Numerical Methods in Engineering*, 36(3):457–497, February 1993.

88. V. Eijkhout. Analysis of parallel incomplete point factorizations. *Linear Algebra and its Applications*, 154-156:723–740, 1991.

89. S. C. Eisenstat and J. W. H. Liu. Exploiting structural symmetry in unsymmetric sparse symbolic factorization. *SIAM J. Matrix Analysis and Applications*, 13:202–211, 1992.

90. H. C. Elman. A stability analysis of incomplete LU factorizations. *Mathematics of Computation*, 47:191–217, 1986.

91. A. M. Erisman, R. G. Grimes, J. G. Lewis, and W. G. Jnr. Poole. A structurally stable modification of Hellerman-Rarick's P^4 algorithm for reordering unsymmetric sparse matrices. *SIAM J. Numerical Analysis*, 22:369–385, 1985.

92. C. M. Fiduccia and R. M. Mattheyses. A linear-time heuristic for improving network partitions. In *Proceedings 19th ACM IEEE Design Automation Conference*, pages 175–181, New York, NY, 1982. IEEE Press.

93. R. Fourer and S. Mehrotra. Solving symmetric indefinite systems in an interior-point method for linear programming. *Mathematical Programming*, 62:15–39, 1993.

94. K. Gallivan, P. C. Hansen, Tz. Ostromsky, and Z. Zlatev. A locally optimized reordering algorithm and its application to a parallel sparse linear system solver. *Computing*, 54:39–67, 1995.

95. K. Gallivan, B. A. Marsolf, and H. A. G. Wijshoff. Solving large nonsymmetric sparse linear systems using MCSPARSE. *Parallel Computing*, 22:1291–1333, 1996.

96. A. Geist and E. Ng. Task scheduling for parallel sparse Cholesky factorization. *Int. J. Parallel Programming*, 18:291–314, 1989.

97. A. George. Nested dissection of a regular finite element mesh. *SIAM J. Numerical Analysis*, 10:345–363, 1973.

98. A. George and M. T. Heath. Solution of sparse linear least squares problems using Givens rotations. *Linear Algebra and its Applications*, 34:69–83, 1980.

99. A. George, M. T. Heath, J. W. H. Liu, and E. Ng. Solution of sparse positive-definite systems on a shared memory multiprocessor. *Int. J. Parallel Programming*, 15:309–325, 1986.

100. A. George, M. T. Heath, J. W. H. Liu, and E. Ng. Solution of sparse positive definite systems on a hypercube. *J. Comput. Appl. Math.*, 27:129–156, 1989.

101. A. George, M. T. Heath, and E. Ng. A comparison of some methods for solving sparse linear least-squares problems. *SIAM J. Scientific and Statistical Computing*, 4:177–187, 1983.

102. A. George and J. W. H. Liu. The design of a user interface for a sparse matrix package. *ACM Trans. Math. Softw.*, 5(2):139–162, 1979.

103. A. George and J. W. H. Liu. The evolution of the minimum degree ordering algorithm. *SIAM Review*, 31(1):1–19, 1989.

104. A. George, J. W. H. Liu, and E. G. Ng. User's guide for SPARSPAK: Waterloo sparse linear equations package. Technical Report CS-78-30 (Revised), University of Waterloo, Canada, 1980.

105. A. George and E. Ng. Shared versus local memory in parallel sparse matrix computations. *SIGNUM Newsletter*, 23(2):9–13, 1988.

106. A. George, J. W. Poole, and R. Voigt. Incomplete nested dissection for solving n by n grid problems. *SIAM J. Numerical Analysis*, 15:663–673, 1978.

107. J. P. Geschiere and H. A. G. Wijshoff. Exploiting large grain parallelism in a sparse direct linear system solver. *Parallel Computing*, 21(8):1339–1364, 1995.

108. John R. Gilbert. Some nested dissection order is nearly optimal. *Information Processing Letters*, 26:325–328, 1988.

109. John R. Gilbert and J. W. H. Liu. Elimination structures for unsymmetric sparse LU factors. *SIAM J. Matrix Analysis and Applications*, 14:334–354, 1993.

110. John R. Gilbert, Gary L. Miller, and Shang-Hua Teng. Geometric mesh partitioning: Implementation and experiments. In *Proceedings of the 9th International Parallel Processing Symposium*, pages 418–427. IEEE, 1995.

111. John R. Gilbert, C. Moler, and R. Schreiber. Sparse matrices in MATLAB: Design and implementation. *SIAM J. Matrix Analysis and Applications*, 13(1):333–356, 1992.

112. John R. Gilbert, E. G. Ng, and B. W. Peyton. Separators and structure prediction in sparse orthogonal factorization. Technical Report CSL-93-15, Palo Alto Research Center, Xerox Corporation, California, November 1993. To appear in *Linear Algebra and Its Applications*.

113. John R. Gilbert and T. Peierls. Sparse partial pivoting in time proportional to arithmetic operations. *SIAM J. Scientific and Statistical Computing*, 9:862–874, 1988.

114. John R. Gilbert and R. Schreiber. Highly parallel sparse Cholesky factorization. *SIAM J. Scientific and Statistical Computing*, 13:1151–1172, 1992.

115. John R. Gilbert and Earl Zmijewski. A parallel graph partitioning algorithm for a message-passing multiprocessor. *Int. J. Parallel Programming*, 16:427–449, 1987.

116. Gene H. Golub and Henk A. van der Vorst. Closer to the solution: Iterative linear solvers. In I. S. Duff and G. A. Watson, editors, *State of the Art in Numerical Analysis - 1996*, Oxford, 1996. Oxford University Press.

117. N. I. M. Gould and J. A. Scott. On approximate-inverse preconditioners. Technical Report RAL-TR-95-026, Rutherford Appleton Laboratory, 1995. To appear in *SIAM J. Scientific Computing*.

118. J. F. Grcar. Matrix stretching for linear equations. Technical Report SAND90-8723, Sandia National Laboratories, Albuquerque, 1990.

119. M. Grote and H. Simon. Parallel preconditioning and approximate inverses on the connection machine. In R. F. Sincovec, D. E. Keyes, M. R. Leuze, L. R. Petzold, and D. A. Reed, editors, *Proceedings of the Sixth SIAM Conference on Parallel Processing for Scientific Computing*, pages 519–523, Philadelphia, PA, 1993. SIAM Press.

120. Anshul Gupta, George Karypis, and Vipin Kumar. Highly scalable parallel algorithms for sparse matrix factorization. Technical Report TR-94-63, Department of Computer Science, University of Minnesota, 1994.

121. Satya Gupta and Edward Rothberg. DME: A distributed matrix environment. In IEEE, editor, *Proceedings of SHPCC '94, Scalable High-Performance Computing Conference. May 23-25, 1994, Knoxville, Tennessee*, pages 629–636, Los Alamitos, California, 1994. IEEE Computer Society Press.

122. I. Gustafsson. *Stability and rate of convergence of modified incomplete Cholesky factorization methods.* PhD thesis, Chalmers University, Göteborg, Sweden, 1979.

123. I. Gustafsson and G. Lindskog. A preconditioning technique based on element matrix factorizations. *Comput. Methods Appl. Mech. Eng.*, 55:201–220, 1986.

124. I. Gustafsson and G. Lindskog. Completely parallelizable preconditioning methods. *Numerical Linear Algebra with Applications*, 2:447–465, 1995.

125. W. Hackbusch. *Multigrid Methods and Applications*, volume 4 of *Computational Mathematics*. Springer–Verlag, Berlin, 1985.

126. P. C. Hansen, Tz. Ostromsky, and Z. Zlatev. Two enhancements in a partitioned sparse solver. In J. J. Dongarra and J. Waśniewski, editors, *Parallel Scientific Computing. Proceedings of the PARA94 Conference, Copenhagen 1994*, Lecture Notes in Mathematics 879, pages 296–303. Springer, Berlin, 1994.

127. M. T. Heath, E. G. Y. Ng, and B. W. Peyton. Parallel algorithms for sparse linear systems. *SIAM Review*, 33:420–460, 1991.

128. M. T. Heath and P. Raghavan. Performance of a fully parallel sparse solver. In IEEE, editor, *Proceedings of SHPCC '94, Scalable High-Performance Computing Conference. May 23-25, 1994, Knoxville, Tennessee*, pages 334–341, Los Alamitos, California, 1994. IEEE Computer Society Press.

129. Bruce Hendrickson and Robert Leland. A multilevel algorithm for partitioning graphs. Technical Report SAND93-1301, Sandia National Laboratories, Albuquerque, 1993.

130. Bruce Hendrickson and Robert Leland. The CHACO User's Guide. Version 2.0. Technical Report SAND94-2692, Sandia National Laboratories, Albuquerque, October 1994.

131. Bruce Hendrickson and Edward Rothberg. Improving the runtime and quality of nested dissection ordering. Technical Report SAND96-0868J, Sandia National Laboratories, Albuquerque, 1996.

132. M. Heroux, P. Vu, and C. Yang. A parallel preconditioned conjugate gradient package for solving sparse linear systems on a CRAY Y-MP. *Applied Numerical Math.*, 8:93–115, 1991.

133. A. J. Hoffman, M. S. Martin, and D. J. Rose. Complexity bounds for regular finite difference and finite element grids. *SIAM J. Numerical Analysis*, 10:364–369, 1973.

134. HSL. *Harwell Subroutine Library. A Catalogue of Subroutines (Release 12).* AEA Technology, Harwell Laboratory, Oxfordshire, England, 1996. For information concerning HSL contact: Dr Scott Roberts, AEA Technology, 552 Harwell, Didcot, Oxon OX11 0RA, England (tel: +44-1235-434714, fax: +44-1235-434136, email: Scott.Roberts@aeat.co.uk).

135. T. Huckle and M. Grote. A new approach to parallel preconditioning with sparse approximate inverses. Technical Report SCCM-94-03, School of Engineering, Stanford University, California, 1994. To appear in *SIAM J. Scientific Computing*.

136. A. Jennings and M. A. Ajiz. Incomplete methods for solving $A'Ax = b$. *SIAM J. Scientific and Statistical Computing*, 5:978–987, 1984.

137. A. Jennings and G. M. Malik. Partial elimination. *J. Institute of Mathematics and its Applications*, 20:307–316, 1977.

138. O. G. Johnson, C. A. Micchelli, and G. Paul. Polynomial preconditioning for conjugate gradient calculations. *SIAM J. Numerical Analysis*, 20:362–375, 1983.

139. T. Johnson and T. A. Davis. Parallel buddy memory management. *Parallel Processing Letters*, 2(4):391–398, 1992.

140. Mark T. Jones and Paul E. Plassmann. The efficient parallel iterative solution of large sparse linear systems. In A. George, J. R. Gilbert, and J. W. H. Liu, editors, *Graph Theory and Sparse Matrix Computations*, IMA Vol 56, pages 229–245. Springer-Verlag, 1994.

141. B. Kågström, P. Ling, and C. Van Loan. Portable high performance GEMM-based Level-3 BLAS. In R. F. Sincovec, D. E. Keyes, M. R. Leuze, L. R. Petzold, and D. A. Reed, editors, *Proceedings of the Sixth SIAM Conference on Parallel Processing for Scientific Computing*, pages 339–345, Philadelphia, PA, 1993. SIAM.

142. G. Karypis, A. Gupta, and V. Kumar. A parallel formulation of interior point algorithms. Technical Report TR-94-020, Department of Computer Science, University of Minnesota, 1994.

143. G. Karypis and V. Kumar. A fast and high quality multilevel scheme for partitioning irregular graphs. Technical Report TR-95-035, Department of Computer Science, University of Minnesota, November 1995.

144. G. Karypis and V. Kumar. METIS: unstructured graph partitioning and sparse matrix ordering system. Technical report, Department of Computer Science, University of Minnesota, 1995.

145. G. Karypis and V. Kumar. Parallel multilevel graph partitioning. Technical Report TR-95-036, Department of Computer Science, University of Minnesota, May 1995.

146. B. Kernighan and S. Lin. An efficient heuristic procedure for partitioning graphs. *Bell System Technical J.*, 49(2):291–307, February 1970.

147. L. Yu. Kolotilina and A. Yu. Yeremin. Factorized sparse approximate inverse preconditioning I: Theory. *SIAM J. Matrix Analysis and Applications*, 14:45–58, 1993.

148. J. Koster and R. H. Bisseling. Parallel sparse LU decomposition on a distributed-memory multiprocessor, 1994. Submitted to *SIAM J. Scientific Computing*.

149. Richard J. Lipton, Donald J. Rose, and Robert Endre Tarjan. Generalized nested dissection. *SIAM J. Numerical Analysis*, 16:346–358, 1979.

150. J. W. H. Liu. Modification of the minimum degree algorithm by multiple elimination. *ACM Trans. Math. Softw.*, 11(2):141–153, 1985.

151. J. W. H. Liu. On the storage requirement in the out-of-core multifrontal method for sparse factorization. *ACM Trans. Math. Softw.*, 12:249–264, 1987.

152. J. W. H. Liu. Reordering sparse matrices for parallel elimination. *Parallel Computing*, 11:73–91, 1989.

153. J. W. H. Liu. The role of elimination trees in sparse factorization. *SIAM J. Matrix Analysis and Applications*, 11:134–172, 1990.

154. J. W. H. Liu. The multifrontal method for sparse matrix solution: Theory and Practice. *SIAM Review*, 34:82–109, 1992.

155. S.-M. Lu and J. L. Barlow. Multifrontal computation with the orthogonal factors of sparse matrices. Technical Report CSE-94-050, Department of Computer Science and Engineering, The Pennsylvania State University, University Park, PA, August 1994.

156. I. J. Lustig and E. Rothberg. Gigaflops in linear programming. *Operations Research Letters*, 18(4):157–165, February 1996.

157. Fredrik Manne and Hjálmtýr Hafsteinsson. Efficient sparse Cholesky factorization on a massively parallel SIMD computer. *SIAM J. Scientific Computing*, 16(4):934–950, July 1995.

158. T. A. Manteuffel. An incomplete factorization technique for positive-definite linear systems. *Mathematics of Computation*, 34:473–498, 1980.

159. H. M. Markowitz. The elimination form of the inverse and its application to linear programming. *Management Science*, 3:255–269, Apr. 1957.

160. P. Matstoms. Sparse QR factorization in MATLAB. *ACM Trans. Math. Softw.*, 20:136–159, 1994.

161. P. Matstoms. Parallel sparse QR factorization on shared memory architectures. *Parallel Computing*, 21:473–486, 1995.

162. J. A. Meijerink and H. A. van der Vorst. An iterative solution method for linear systems of which the coefficient matrix is a symmetric M-matrix. *Mathematics of Computation*, 31(137):148–162, 1977.

163. G. Meurant. A review on the inverse of symmetric tridiagonal and block tridiagonal matrices. *SIAM J. Matrix Analysis and Applications*, 13:707–728, 1992.

164. N. Munksgaard. Solving sparse symmetric sets of linear equations by preconditioned conjugate gradients. *ACM Trans. Math. Softw.*, 6:206–219, 1980.

165. Esmond G. Ng and Barry W. Peyton. Block sparse Cholesky algorithms on advanced uniprocessor computers. *SIAM J. Scientific Computing*, 14:1034–1056, 1993.

166. Esmond G. Ng and Barry W. Peyton. A supernodal Cholesky factorization algorithm for shared-memory multiprocessors. *SIAM J. Scientific Computing*, 14:761–769, 1993.

167. B. W. Peyton, A. Pothen, and X. Yuan. Partitioning a chordal graph into transitive subgraphs for parallel sparse triangular solution. Technical Report ORNL/TM-12270, Engineering Physics and Mathematics Division, Oak Ridge National Laboratory, Tennessee, December 1992.

168. D. J. Pierce and J. G. Lewis. Sparse multifrontal rank revealing QR factorization. Technical Report MEA-TR-193-Revised, Boeing Information and Support Services, Seattle, WA, 1995.

169. A. Pothen and C. Fan. Computing the block triangular form of a sparse matrix. *ACM Trans. Math. Softw.*, 16(4):303–324, 1990.

170. A. Pothen, H. D. Simon, and K. P. Liou. Partitioning sparse matrices with eigenvectors of graphs. *SIAM J. Matrix Analysis and Applications*, 11(3):430–452, 1990.

171. A. Pothen, H. D. Simon, L. Wang, and S. Barnard. Towards a fast implementation of spectral nested dissection. In *Proceedings of Supercomputing '92*, pages 42–51, New York, NY, 1992. ACM Press.

172. A. Pothen and C. Sun. A mapping algorithm for parallel sparse Cholesky factorization. *SIAM J. Scientific Computing*, 14(5):1253–1257, 1993. Timely Communication.

173. P. Raghavan. Distributed sparse Gaussian elimination and orthogonal factorization. *SIAM J. Scientific Computing*, 16(6):1462–1477, 1995.

174. P. Raghavan. Efficient parallel sparse triangular solution with selective inversion. Technical Report CS-95-314, Department of Computer Science, University of Tennessee, Knoxville, Tennessee, 1995.

175. P. Raghavan. Parallel ordering using edge contraction. Technical Report CS-95-293, Department of Computer Science, University of Tennessee, Knoxville, Tennessee, 1995. Submitted to *Parallel Computing*.

176. J. K. Reid. Sparse matrices. In A. Iserles and M. J. D. Powell, editors, *The State of the Art in Numerical Analysis*, pages 59–85, Oxford, 1987. Oxford University Press.

177. D. J. Rose. A graph-theoretic study of the numerical solution of sparse positive definite systems of linear equations. In R. C. Read, editor, *Graph Theory and Computing*, pages 183–217. New York: Academic Press, 1973.

178. Edward Rothberg. Exploring the tradeoff between imbalance and separator size in nested dissection ordering. Technical Report Unnumbered, Silicon Graphics Inc, 1996.

179. Edward Rothberg. Ordering sparse matrices using approximate minimum local fill. Technical Report Unnumbered, Silicon Graphics Inc, 1996.

180. Edward Rothberg. Performance of panel and block approaches to sparse Cholesky factorization on the iPSC/860 and Paragon multicomputers. *SIAM J. Scientific Computing*, 17(3):699–713, 1996.

181. Edward Rothberg and Anoop Gupta. An evaluation of left-looking, right-looking and multifrontal approaches to sparse Cholesky factorization on hierarchical-memory machines. Technical Report STAN-CS-91-1377, Department of Computer Science, Stanford University, 1991.

182. Edward Rothberg and Anoop Gupta. An efficient block-oriented approach to parallel sparse Cholesky factorization. *SIAM J. Scientific Computing*, 15(6):1413–1439, December 1994.

183. Edward Rothberg and Bruce Hendrickson. Sparse matrix ordering methods for interior point linear programming. Technical Report 96-0475J, SANDIA National Laboratory, 1996.

184. Edward Rothberg and Robert Schreiber. Improved load distribution in parallel sparse Cholesky factorization. Technical Report 94-13, Research Institute for Advanced Computer Science, 1994.

185. Y. Saad. Practical use of polynomial preconditionings for the conjugate gradient method. *SIAM J. Scientific and Statistical Computing*, 6:865–881, 1985.

186. Y. Saad. Preconditioning techniques for nonsymmetric and indefinite linear systems. *J. Comput. Appl. Math.*, 24:89–105, 1988.

187. Y. Saad. Preconditioning techniques for nonsymmetric and indefinite linear systems. Technical Report 792, CSRD, University of Illinois, Urbana, Illinois, USA, 1988.

188. Y. Saad. ILUT: a dual threshold incomplete LU factorization. *Numerical Linear Algebra with Applications*, 1(4):387–402, 1994.

189. Y. Saad. SPARSKIT: a basic tool kit for sparse matrix computations. Version 2. Technical report, Computer Science Department, University of Minnesota, 1994.

190. Michael A. Saunders. Major Cholesky would feel proud. *ORSA J. Computing*, 6(1):23–27, 1994.

191. R. Schreiber. A new implementation of sparse Gaussian elimination. *ACM Trans. Math. Softw.*, 8:256–276, 1982.

192. R. Schreiber. Scalability of sparse direct solvers. In A. George, J. R. Gilbert, and J. W. H. Liu, editors, *Graph Theory and Sparse Matrix Computation*, The IMA Volumes in Mathematics and its Applications, Volume 56, pages 191–209, New York, 1993. Springer-Verlag.

193. David F. Shanno and Evangelia M. Simantiraki. Interior point methods for linear and nonlinear programming. In I. S. Duff and G. A. Watson, editors, *State of the Art in Numerical Analysis - 1996*, Oxford, 1996. Oxford University Press.

194. A. H. Sherman. Algorithm 533. NSPIV, A Fortran subroutine for sparse Gaussian elimination with partial pivoting. *ACM Trans. Math. Softw.*, 4:391–398, 1978.

195. B. Speelpenning. The generalized element method. Technical Report Technical Report UIUCDCS-R-78-946, Dept. of Computer Science, Univ. of Illinois, Urbana, IL, 1978.

196. C. Sun. Efficient parallel solutions of large sparse SPD systems on distributed-memory multiprocessors. Technical Report CTC92TR102, Advanced Computing Research Institute, Cornell University, Ithaca, NY, 1992.

197. C. Sun. A package for solving sparse symmetric positive definite systems on distributed-memory multiprocessors. Technical Report CTC92TR114, Advanced Computing Research Institute, Cornell University, Ithaca, NY, November 1992.

198. C. Sun. Dealing with dense rows in the solution of sparse linear least squares problems. Technical Report CTC95TR227, Advanced Computing Research Institute, Cornell University, Ithaca, NY, October 1995.

199. W. F. Tinney and J. W. Walker. Direct solutions of sparse network equations by optimally ordered triangular factorization. *Proc. of the IEEE*, 55:1801–1809, 1967.

200. M. Tismenetsky. A new preconditioning technique for solving large sparse linear systems. *Linear Algebra and its Applications*, 154-156:331–353, 1991.

201. A. F. van der Stappen, R. H. Bisseling, and J. G. G. van de Vorst. Parallel sparse LU decomposition on a mesh network of transputers. *SIAM J. Matrix Analysis and Applications*, 14:853–879, 1993.

202. H. A. van der Vorst. A vectorizable variant of some ICCG methods. *SIAM J. Scientific and Statistical Computing*, 3:350–356, 1982.

203. H. A. van der Vorst. High performance preconditioning. *SIAM J. Scientific and Statistical Computing*, 10:1174–1185, 1989.

204. A. C. N. van Duin, P. C. Hansen, Tz. Ostromsky, H. A. G. Wijshoff, and Z. Zlatev. Improving the numerical stability and the performance of a parallel sparse solver. *Computers Math. Applic.*, 30:81–96, 1995.

205. R. J. Vanderbei. Splitting dense columns in sparse linear systems. *Linear Algebra and its Applications*, 152:107–117, 1991.

206. X. Wang. *Incomplete factorization preconditioning for least squares problems.* PhD thesis, Department of Mathematics, University of Illinois, Urbana, Illinois, USA, 1993.

207. X. Wang, K. A. Gallivan, and R. Bramley. CIMGS: an incomplete orthogonal factorization preconditioner. *SIAM J. Scientific Computing*, 18, 1997.

208. P. Wesseling. *An Introduction to Multigrid Methods.* John Wiley & Sons, Chichester, 1992.

209. R. Clint Whaley. Lapack working note 73 : Basic Linear Algebra Communication Subprograms: analysis and implementation across multiple parallel architectures. Technical Report CS-94-234, Computer Science Department, University of Tennessee, Knoxville, Tennessee, May 1994.

210. Margaret H. Wright. Interior methods for constrained optimization. In A. Iserles, editor, *Acta Numerica 1992*, pages 341–407. Cambridge University Press, 1992.

211. Z. Zlatev, J. Waśniewski, P. C. Hansen, and Tz. Ostromsky. PARASPAR: a package for the solution of large linear algebraic equations on parallel computers with shared memory. Technical Report 95-10, Tech Univ Denmark, Lyngby, 1995.

212. Z. Zlatev, J. Waśniewski, and K. Schaumburg. *Y12M - Solution of Large and Sparse Systems of Linear Algebraic Equations*, volume 121 of *Lecture Notes in Computer Science*. Springer-Verlag, New York, 1981.

213. Z. Zlatev, J. Waśniewski, and K. Schaumburg. Introduction to PARASPAR. solution of large and sparse systems of linear algebraic equations, specialised for parallel computers with shared memory. Technical Report 93-02, Tech Univ Denmark, Lyngby, 1993.

214. Earl Zmijewski. *Sparse Cholesky Factorization on a Multiprocessor*. PhD thesis, Cornell University, 1987.

215. Earl Zmijewski and John R. Gilbert. A parallel algorithm for sparse symbolic Cholesky factorization on a multiprocessor. *Parallel Computing*, 7:199–210, 1988.

Closer to the Solution: Iterative Linear Solvers

Gene H. Golub* and Henk A. van der Vorst**

**Stanford University and ** Utrecht University*

Abstract

Iterative linear solution methods have received much attention in the past ten years and some rather powerful, and now popular, methods have emerged. Until ten years ago, the SOR/SUR method was the method of choice in many large industrial applications with unsymmetric systems, while preconditioned CG was popular for positive definite large sparse systems. In the past ten years, we have observed a growing interest in the 'parameter-free' Krylov subspace methods for unsymmetric systems. We will highlight the important new ideas as embodied in GMRES, CGS, Bi-CGSTAB, and QMR. Much attention has also been devoted in the past period to preconditioning. We will discuss some of these approaches, especially in connection with parallelism.

Our presentation will conclude with a short discussion on some open problems.

1 Introduction

The solution of dense linear systems received much attention after the second world war, and by the end of the sixties, most of the problems associated with it had been solved. For a long time, Wilkinson's "The Algebraic Eigenvalue Problem" [107], other than the title suggests, became also the standard textbook for the solution of linear systems. When it became clear that partial differential equations could be solved numerically, to a level of accuracy that was of interest for application areas (such as reservoir engineering, and reactor diffusion modeling), there was a strong need for the fast solution of the discretized systems, and iterative methods became popular for these problems.

Although, roughly speaking, the successive overrelaxation methods and the first Krylov subspace methods were developed in the same period of time, the class of Successive Overrelaxation methods became the methods of choice, since they required a small amount of computer storage. These methods were quite well understood by the work of Young [108], and the theory was covered in great detail in Varga's book [102]. This book was the source of reference for iterative methods for a long time.

It may seem evident that iterative methods gained in importance as scientific modeling led to larger linear problems, since direct methods are often too expensive in terms of computer memory and CPU-time requirements. However, the convergence behavior of iterative methods depends very much on (often *a priori* unknown) properties of the linear system, which means that one has to consider the origin of the problem. Since large linear systems may arise from various sources, for instance, data fitting problems, discretization of PDE's, Markov chains, or tomography, these properties may also vary widely. This means that one has to study the usefulness of iterative schemes for each class of problems separately and one can hardly rely on expertise gained from other classes. Most often preconditioning is required to obtain or improve convergence. Only for some classes of problems the effects of preconditioning are well understood; for most problems it remains largely a matter of trial and error. In our final Section we will give some hints on preconditioning as well as pointers to literature.
Also the (sparsity) structure of the matrix is important for the decision between algorithms: special algorithms can often be designed for special structures, e.g., Toeplitz matrices, banded matrices, red-black ordering, p-cyclic matrices, etc.
Some important matrix properties can be easily deduced from the originating problem, for instance, symmetry and positive definiteness. Iterative methods that rely on these properties, such as the Lanczos method and the conjugate gradient method, have therefore the most widespread use.

The conjugate gradient method, and the method of Lanczos, were first viewed as exact projection methods, and since their behavior in finite precision arithmetic was not in agreement with that point of view, they were initially considered with suspicion. Through work of Reid [79], it became clear that these methods, if used as iterative techniques, could be used to advantage for many classes of linear systems. With preconditioning [21], these methods could be made even more efficient. Among the first popular preconditioned methods was ICCG [69, 63].

In the period after 1975, we have seen that symmetric positive definite sparse systems were usually solved by preconditioned CG methods, when *very* large, and by sparse direct solvers if moderately large. For indefinite symmetric sparse systems special variants were proposed, like MINRES, and SYMMLQ [74]. However, for very large unsymmetric systems, the SOR methods, and the method of Chebyshev [102, 54, 53], nicely tuned by Manteuffel [68], were still the methods of choice. Due to several poorly understood numerical problems, the two-sided Lanczos method, and a special variant Bi-CG [41], were not so popular at that time, but slowly more robust variants of Krylov subspace methods, with longer recursion formulas, entered the field. We mention GENCG [32], FOM [84], ORTHOMIN [103], ORTHODIR and ORTHORES [62].

In the mid-eighties we see the start of the popularity of Krylov subspace methods for large unsymmetric systems, and many new powerful and robust methods were proposed. The past 10 years have formed a very lively area of research for these methods, but we believe that by now this has come more or less to an end. We are now facing other challenging problems for the solution of very large sparse problems. For large linear systems, associated with 3-dimensional modeling, the iterative methods are often the only means we have for solution.

Of course, for structured and rather regular problems the multigrid methods have usually a better rate of convergence, but even from the CFD community we see an increasing attention to iterative schemes. We believe that it is not fruitful to see the two classes of methods as competitive, they may be combined, where depending on the point of view, multigrid is used as a preconditioner, or the iterative scheme as a smoother. We will not further discuss multigrid in our paper; it deserves separate attention. An excellent introduction to multigrid and the related multilevel methods, in the context of solving linear systems stemming from certain classes of problems, can be found in [61].

In the next Section we will highlight the Krylov subspace methods as the source of one of the most important developments in the period 1985-1995. In subsequent Sections, we will discuss a number of new variants.
For ease of presentation we restrict ourselves to real-valued problems. Matrices will be denoted by capitals (A, B, ...), vectors are represented in lower case (x, y, ...), and scalars are represented by Greek symbols (α, Θ, ...). Unless stated differently, matrices will be of order n: $A \in \mathbb{R}^{n \times n}$, and vectors will have length n.

2 Krylov subspace methods: the basics

The past ten years have led to well-established and popular methods, that all lead to the construction of approximate solutions in the so-called Krylov subspace. Given a linear system $Ax = b$, with a large, usually sparse, unsymmetric nonsingular matrix A, then the standard Richardson iteration

$$x_k = (I - A)x_{k-1} + b$$

generates approximate solutions in shifted Krylov subspaces

$$x_0 + K^k(A; r_0) = x_0 + \{r_0, Ar_0, \ldots, A^{k-1}r_0\},$$

with $r_0 = b - Ax_0$, for some given initial vector x_0.
The Krylov subspace projection methods fall in three different classes:

1. The *Ritz-Galerkin approach*: Construct the x_k for which the residual is orthogonal to the current subspace: $b - Ax_k \perp K^k(A; r_0)$.

2. The *minimum residual approach*: Identify the x_k for which the Euclidean norm $\|b - Ax_k\|_2$ is minimal over $K^k(A; r_0)$.

3. The *Petrov-Galerkin approach*: Find an x_k so that the residual $b - Ax_k$ is orthogonal to some other suitable k-dimensional subspace.

The Ritz-Galerkin approach leads to such popular and well-known methods as Conjugate Gradients, the Lanczos method, FOM, and GENCG. The minimum residual approach leads to methods like GMRES, MINRES, and ORTHODIR. If we select the k-dimensional subspace in the third approach as $K^k(A^T; s_0)$, then we obtain the Bi-CG, and QMR methods. More recently, hybrids of the three approaches have been proposed, like CGS, Bi-CGSTAB, BiCGSTAB(ℓ), TFQMR, FGMRES, and GMRESR.

Most of these methods have been proposed in the last ten years. GMRES, proposed in 1986 by Saad and Schultz [84], is the most robust of them, but, in terms of work per iteration step it is also the most expensive. Bi-CG, which was suggested by Fletcher in 1977 [41], is a relatively inexpensive alternative, but it has problems with respect to convergence: the so-called breakdown situations. This aspect has received much attention in the past period. Parlett et al [75] introduced the notion of look-ahead, in order to overcome breakdowns, and this was further perfected by Freund and Nachtigal [45], and by Brezinski and Redivo-Zaglia [12]. The theory for this look-ahead technique was linked to the theory of Padé approximations by Gutknecht [59]. Other contributions to overcome specific breakdown situations were made by Bank and Chan [9], and Fischer [39]. We will discuss these approaches in our Section on QMR.
The hybrids were developed primarily in the second half of the last 10 years; the first of these was CGS, published in 1989 by Sonneveld [89], and followed by Bi-CGSTAB, by van der Vorst in 1992 [98], and others. The hybrid variants of GMRES: Flexible GMRES and GMRESR, in which GMRES is combined with some other iteration scheme, have only been proposed very recently.

A nice overview of Krylov subspace methods, with focus on Lanczos-based methods, is given in [44]. Simple algorithms and unsophisticated software for some of these methods is provided in [10]. Iterative methods with much attention to various forms of preconditioning have been described in [5]. Recently a book on iterative methods was published also by Saad [83]; it is very algorithm oriented, with, of course, a focus on GMRES and preconditioning techniques, like threshold ILU, ILU with pivoting, and incomplete LQ factorizations.
An annotated entrance to the vast literature on preconditioned iterative methods is given in [14].

2.1 Working with the Krylov subspace

In order to identify optimal approximate solutions in the Krylov subspace we need a suitable basis for this subspace, one that can be extended in a meaningful way for subspaces of increasing dimension. The obvious basis r_0, Ar_0, ..., $A^{i-1}r_0$, for $K^i(A; r_0)$, is not very attractive from a numerical point of view, since the vectors $A^j r_0$ point more and more in the direction of the dominant eigenvector for increasing j (the power method!), and hence the basis vectors become dependent in finite precision arithmetic.

```
v_1 = r_0/||r_0||_2;
for j = 1,..,m-1
    t = Av_j;
    for i = 1,...,j
        h_{i,j} = v_i^T t;
        t = t - h_{i,j} v_i;
    end;
    h_{j+1,j} = ||t||_2;
    v_{j+1} = t/h_{j+1,j};
end
```

Figure 1. Arnoldi's method with modified Gram–Schmidt orthogonalization

Instead of the standard basis one usually prefers an orthonormal basis, and Arnoldi [1] suggested computing this basis as follows. Start with $v_1 \equiv r_0/\|r_0\|_2$. Assume that we have already an orthonormal basis $v_1, \ldots, v_j$ for $K^j(A; r_0)$, then this basis is expanded by computing $t = Av_j$, and by orthonormalizing this vector t with respect to $v_1, \ldots, v_j$. In principle the orthonormalization process can be carried out in different ways, but the most commonly used approach is to do this by a modified Gram-Schmidt procedure [53]. This leads to an algorithm for the creation of an orthonormal basis for $K^m(A; r_0)$, as in Fig 1.

It is easily verified that $v_1, \ldots, v_m$ form an orthonormal basis for $K^m(A; r_0)$ (that is, if the construction does not terminate at a vector $t = 0$). The orthogonalization leads to relations between the v_j, that can be formulated in a compact algebraic form. Let V_j denote the matrix with columns v_1 up to v_j, then it follows that

$$AV_{m-1} = V_m H_{m,m-1}.$$

The m by $m-1$ matrix $H_{m,m-1}$ is upper Hessenberg, and its elements $h_{i,j}$ are defined by the Arnoldi algorithm.

From a computational point of view, this construction is composed from three basic elements: a matrix vector product with A, innerproducts, and updates. We see that this orthogonalization becomes increasingly expensive for increasing dimension of the subspace, since the computation of each $h_{i,j}$ requires an inner product and a vector update.

Note that if A is symmetric, then so is $H_{m-1,m-1} = V_{m-1}^T A V_{m-1}$, so that in this situation $H_{m-1,m-1}$ is tridiagonal. This means that in the orthogonalization process, each new vector has to be orthogonalized with respect to the previous two vectors only, since all other innerproducts vanish. The resulting three term recurrence relation for the basis vectors of $K_m(A; r_0)$ is known as the *Lanczos method* and some very elegant methods are derived from it. In the symmetric case the orthogonalization process involves constant arithmetical costs per iteration step: one matrix vector product, two innerproducts, and two vector updates.

3 The Ritz-Galerkin approach: FOM and CG

Let us assume, for simplicity, that $x_0 = 0$, and hence $r_0 = b$. The Ritz-Galerkin conditions imply that $r_k \perp K^k(A; r_0)$, and this is equivalent to

$$V_k^T(b - Ax_k) = 0.$$

Since $b = r_0 = \|r_0\|_2 v_1$, it follows that $V_k^T b = \|r_0\|_2 e_1$ with e_1 the first canonical unit vector in $\mathbb{R}^k$. With $x_k = V_k y$ we obtain

$$V_k^T A V_k y = \|r_0\|_2 e_1.$$

This system can be interpreted as the system $Ax = b$ projected onto $K^k(A; r_0)$.

Obviously we have to construct the $k \times k$ matrix $V_k^T A V_k$, but this is, as we have seen, readily available from the orthogonalization process:

$$V_k^T A V_k = H_{k,k},$$

so that the x_k for which $r_k \perp K^k(A; r_0)$ can be easily computed by first solving $H_{k,k} y = \|r_0\|_2 e_1$, and forming $x_k = V_k y$. This algorithm is known as FOM or GENCG.

Note that for some $j \leq n - 1$ the construction of the orthogonal basis must terminate. In that case we have that $AV_{j+1} = V_{j+1} H_{j+1,j+1}$. Let y be the solution of the reduced system $H_{j+1,j+1} y = \|r_0\| e_1$, and $x_{j+1} = V_{j+1} y$. Then it follows that $x_{j+1} = x$, i.e., we have arrived at the exact solution, since

$$\begin{aligned} Ax_{j+1} - b &= AV_{j+1} y - b = V_{j+1} H_{j+1,j+1} y - b \\ &= \|r_0\| V_{j+1} e_1 - b = 0 \end{aligned}$$

(we have assumed that $x_0 = 0$).

When A is symmetric, then $H_{k,k}$ reduces to a tridiagonal matrix $T_{k,k}$, and the resulting method is known as the *Lanczos* method [65]. In clever implementations, it is possible to avoid storing all the vectors v_j. When A is in addition positive definite then we obtain, at least formally, the *Conjugate Gradient* method. In commonly used implementations of this method, one implicitly forms an *LU* factorization for $T_{k,k}$ and this leads to very elegant short recurrencies for the x_j and the corresponding r_j. The positive definiteness is necessary to guarantee the existence of the *LU* factorization, but it allows also for another useful interpretation. From the fact that $r_i \perp K^i(A; r_0)$, it follows that $A(x_i - x) \perp K^i(A; r_0)$, or $x_i - x \perp_A K^i(A; r_0)$. The latter observation expresses the fact that the error is A–orthogonal to the Krylov subspace and this is equivalent to the important observation that $\|x_i - x\|_A$ is minimal[1] For an overview of the history of CG and main contributions on this subject, see [51].

The local convergence behavior of CG, and especially the occurrence of superlinear convergence, was first explained in a qualitative sense in [21], and later in a quantitative sense in [94]. In both papers it was linked to the convergence of eigenvalues (Ritz values) of $T_{i,i}$ towards eigenvalues of A, for increasing i. The global convergence can be bounded with expressions that involve condition numbers, for details see for instance [21, 53, 2]. In [2] the situation is analysed where the eigenvalues of A are in disjunct intervals.

4 The minimum residual approach

The creation of an orthogonal basis for the Krylov subspace, with basis vectors $v_1, \ldots, v_{i+1}$, leads to

$$AV_i = V_{i+1} H_{i+1,i}, \tag{4.1}$$

where V_i is the matrix with columns v_1 to v_i. We look for an $x_i \in K^i(A; r_0)$, that is $x_i = V_i y$, for which $\|b - Ax_i\|_2$ is minimal. This norm can be rewritten as

$$\|b - Ax_i\|_2 = \|b - AV_i y\|_2 = \|\, \|r_0\|_2 V_{i+1} e_1 - V_{i+1} H_{i+1,i} y\|_2.$$

Now we exploit the fact that V_{i+1} is an orthonormal transformation with respect to the Krylov subspace $K^{i+1}(A; r_0)$:

$$\|b - Ax_i\|_2 = \|\|r_0\|_2 e_1 - H_{i+1,i} y\|_2,$$

and this final norm can simply be minimized by solving the minimum norm least squares problem for the $i + 1$ by i matrix $H_{i+1,i}$ and right-hand side $\|r_0\|_2 e_1$.

In GMRES [84] this is done efficiently with Givens rotations, that annihilate the subdiagonal elements in the upper Hessenberg matrix $H_{i+1,i}$.
Note that when A is Hermitian (but not necessarily positive definite), the upper Hessenberg matrix $H_{i+1,i}$ reduces to a tridiagonal system. This simplified structure can be exploited in order to avoid storage of all the basis vectors for the Krylov subspace, in a way similar as has been pointed out for CG. The resulting method is known as MINRES [74].

In order to avoid excessive storage requirements and computational costs for the orthogonalization, GMRES is usually restarted after each m iteration steps. This algorithm is referred to as GMRES(m); the not-restarted version is often

[1] The A–norm is defined by $\|y\|_A^2 = (y, y)_A \equiv (y, Ay)$, and we need the positive definiteness of A in order to get a proper innerproduct $(\cdot, \cdot)_A$.

called 'full' GMRES. There is no simple rule to determine a suitable value for m; the speed of convergence may vary drastically for nearby values of m.

There is an interesting and simple relation between the Ritz-Galerkin approach (FOM and CG) and the minimum residual approach (GMRES and MINRES). In GMRES the projected system matrix $H_{i+1,i}$ is transformed by Givens rotations to an upper triangular matrix (with last row equal to zero). So, in fact, the major difference between FOM and GMRES is that in FOM the last $((i+1)$-th row is simply discarded, while in GMRES this row is rotated to a zero vector. Let us characterize the Givens rotation, acting on rows i and $i+1$, in order to zero the element in position $(i+1, i)$, by the sine s_i and the cosine c_i. Let us further denote the residuals for FOM with an superscript F and those for GMRES with superscript G. Then we have the following relation between FOM and GMRES:
If $c_k \neq 0$ then the FOM and the GMRES residuals are related by

$$||r_k^F||_2 = \frac{||r_k^G||_2}{\sqrt{1 - (||r_k^G||_2/||r_{k-1}^G||_2)^2}}, \tag{4.2}$$

([23]: theorem 3.1). From this relation we see that when GMRES has a significant reduction at step k, in the norm of the residual (i.e., s_k is small, and $c_k \approx 1$), then FOM gives about the same result as GMRES. On the other hand when FOM has a breakdown ($c_k = 0$), then GMRES does not lead to an improvement in the same iteration step. Because of these relations we can link the convergence behaviour of GMRES with the convergence of Ritz values (the eigenvalues of the "FOM" part of the upper Hessenberg matrix). This has been exploited in [100], for the analysis and explanation of local effects in the convergence behaviour of GMRES.

There are various different implementations of FOM and GMRES. Among those equivalent to GMRES are: Orthomin [103], Orthodir [62], GENCR [33], and Axelsson's method [3]. These methods are often more expensive than GMRES per iteration step. Orthomin continues to be popular, since this variant can be easily truncated (Orthomin(s)), in contrast to GMRES. The truncated and restarted versions of these algorithms are not necessarily mathematically equivalent.
Methods that are mathematically equivalent to FOM are: Orthores [62] and GENCG [19, 106]. In these methods the approximate solutions are constructed such that they lead to orthogonal residuals (which form a basis for the Krylov subspace; analogously to the CG method). A good overview of all these methods and their relations is given in [83].

The GMRES method and FOM are closely related to vector extrapolation methods, when the latter are applied to linearly generated vector sequences. For a discussion on this, as well as for implementations for these matrix free methods, see [85].

4.1 Inner-outer iteration schemes

Although GMRES is optimal, in the sense that it leads to a minimal residual solution over the Krylov subspace, it has the disadvantage that this goes with increasing computational costs per iteration step. Of course, with a suitable preconditioner one can hope to reduce the dimension of the required Krylov subspace to acceptable values, but in many cases such preconditioners have not been identified.

The preconditioner K^{-1} is generally viewed as an approximation for the inverse of the matrix A of the system $Ax = b$ to be solved. Instead of an optimal update direction $A^{-1}r$, we compute the direction $p = K^{-1}r$. That is, we solve p from $Kp = r$, and K is constructed in such a way that this is easy to do. An attractive idea is to try to get better approximations for $A^{-1}r$, and there are two related approaches to accomplish this. One is to try to improve the preconditioner with updates from the Krylov subspace. This has been suggested first by Eirola and Nevanlinna [31]. Their approach leads to iterative methods that are related to Broyden's method [13], which is a Newton type of method. The Broyden method can be obtained from this update-approach if we do not restrict ourselves to Krylov subspaces. See [104] for a discussion on the relation of these methods.

The updated preconditioners can not be applied immediately to GMRES, since the preconditioned operator now changes from step to step, and we are not forming a regular Krylov subspace, but rather a subset of a higher dimensional Krylov subspace. However, we can still minimize the residual over this subset.

The idea of variable preconditioning has been exploited in this sense, by different authors. Axelsson and Vassilevski [7] have proposed a Generalized Conjugate Gradient method with variable preconditioning, Saad [81, 83] has proposed a scheme very similar to GMRES, called Flexible GMRES (FGMRES), and Van der Vorst and Vuik have published a GMRESR scheme. FGMRES has received most attention, presumably because it is so easy to implement: only the update directions in GMRES have to be preconditioned, and each update may be preconditioned differently. This means that only one line in the GMRES algorithm has to be adapted. The price to be paid is that the method is no longer robust; it may break down. The GENCG and GMRESR schemes are slightly more expensive in terms of memory requirements and in computational overhead per iteration step. The main difference between the two schemes is that GENCG in [7] works with Gram-Schmidt orthogonalization, whereas GMRESR makes it possible to use Modified Gram-Schmidt. This may give GMRESR an advantage in actual computations. In exact arithmetic GMRESR and GENCG should produce the same results for the same preconditioners. Another advantage of GMRESR over Algorithm 1 in [7] is that only one matrix vector product is used per iteration step in GMRESR; GENCG needs two matrix vector products per step.

In GMRESR the residual vectors are preconditioned and if this gives a further reduction then GMRESR does not break down. This gives slightly more control over the method in comparison with FGMRES. In most cases though the results are about the same, but the efficient scheme for FGMRES has an advantage.

We will briefly discuss the GMRESR method. In [101] it has been shown how the GMRES-method, or more precisely, the GCR-method, can be combined with other iterative schemes. The iteration steps of GMRES (or GCR) are called outer iteration steps, while the iteration steps of the preconditioning iterative method are referred to as inner iterations. The combined method is called GMRES$\star$, where $\star$ stands for any given iterative scheme; in the case of GMRES as the inner iteration method, the combined scheme is called GMRESR[101]. It was shown in [26, 25], that GMRESR can be implemented in a way that avoids about 30% of the overhead in the outerloop, which makes the method about as expensive per outer iteration step as FGMRES.

The GMRES$\star$ algorithm can be described as in Fig. 2.

x_0 is an initial guess; $r_0 = b - Ax_0$;
for $i = 0, 1, 2, 3, \ldots$
  Let $z^{(m)}$ be the approximate solution
  of $Az = r_i$, obtained after m steps of
  an iterative method.
  $c = Az^{(m)}$ (often available from the
    iteration method)
  for $k = 0, \ldots, i-1$
    $\alpha = c_k^T c$
    $c = c - \alpha c_k$
    $z^{(m)} = z^{(m)} - \alpha u_k$
  $c_i = c/\|c\|_2$; $u_i = z^{(m)}/\|c\|_2$
  $x_{i+1} = x_i + (c_i^T r_i) u_i$
  $r_{i+1} = r_i - (c_i^T r_i) c_i$
  if x_{i+1} is accurate enough then quit
end

Figure 2. The GMRES$\star$ algorithm

A sufficient condition to avoid breakdown in this method ($||c||_2 = 0$) is that the norm of the residual at the end of an inner iteration is smaller than the norm of the right-hand side residual: $||Az^{(m)} - r_i||_2 < ||r_i||_2$. This can easily be controlled during the inner iteration process. If stagnation occurs, i.e. no progress at all is made in the inner iteration, then it is suggested in [101] to do one (or more) steps of the LSQR method, which guarantees a reduction (but this reduction is often very small).

The idea behind these flexible iteration schemes is that we explore parts of high-dimensional Krylov subspaces, hopefully determining essentially the same approximate solution that full GMRES would find over the entire subspace, but at much lower computational costs. For the inner iteration we may select any appropriate solver, for instance, one cycle of GMRES(m), since then we have also locally an optimal method, or some other iteration scheme, like for instance Bi-CGSTAB.

In [26] it is proposed to keep the Krylov subspace, that is built in the inner iteration, orthogonal with respect to the Krylov basis vectors generated in the outer iteration. Under various circumstances this helps to avoid inspection of already investigated parts of Krylov subspaces in future inner iterations. The procedure works as follows. In the outer iteration process the vectors c_0, ..., c_{i-1} build an orthogonal basis for the Krylov subspace. Let C_i be the n by i matrix with columns c_0, ..., c_{i-1}. Then the inner iteration process at outer iteration i is carried out with the operator A_i instead of A, and A_i is defined as

$$A_i = (I - C_i C_i^T)A. \tag{4.3}$$

Of course, $I - C_i C_i^T$ can be stored as a product of Householder transformations, which makes the updates more efficient. It is easily verified that $A_i z \perp c_0, ..., c_{i-1}$ for all z, so that the inner iteration process takes place in a subspace orthogonal to these vectors. The additional costs, per iteration of the inner iteration process, are i inner products and i vector updates. In order to save on these costs, one should realize that it is not necessary to orthogonalize with respect to all previous c-vectors, and that "less effective" directions may be dropped, or combined with others. In [72, 8, 26] suggestions are made for such strategies. Of course, these strategies are only attractive in cases where we see too little residual reducing effect in the inner iteration process in comparison with the outer iterations of GMRES$\star$.

Note that if we carry out the preconditioning by doing a few iterations of some other iteration process, then we have *inner-outer iteration* schemes; these have been discussed earlier in [52]. It is difficult to analyse their convergence behavior and in a first attempt it is wise to look at the analysis for simplified schemes. We will give the flavor of this analysis and the type of results [52].

while $\frac{\|q_k\|_2}{\|r_k\|_2} < \tau_k$ **do**
  $r_k = b - Ax_k$
  Solve $Mz = r_i$ approximately $\Rightarrow \bar{z}_k$
    $(q_k \equiv M\bar{z}_k - r_k)$
  $x_{k+1} = x_k + \bar{z}_k$
  if x_{k+1} is accurate enough **then** quit
end

Figure 3. A simplified Inner-Outer iteration scheme

We consider a simplified inner-outer iteration scheme associated with the standard iteration for the splitting $A = M - N$, as given in Fig. 3; the inner iterations are carried out with M.
With $e_k \equiv x - x_k$, it follows that

$$e_{k+1} = (I - M^{-1}A)e_k - M^{-1}q_k,$$

and

$$\begin{aligned} \|e_{k+1}\| &\leq \|I - M^{-1}A\|\|e_k\| + \|M^{-1}q_k\| \\ &\leq (\|I - M^{-1}A\| + \tau_k\|M^{-1}\|\|A\|)\|e_k\|. \end{aligned}$$

Hence

$$\frac{\|e_k\|}{\|e_0\|} \leq (\|I - M^{-1}A\| + \eta)^k,$$

with $\eta \equiv \max \tau_k\|M^{-1}\|\|A\|$, so that the convergence is largely determined by the norm of $I - M^{-1}A$, if τ_k is small enough.
It can be shown that for $\tau_k = \Theta^k$, for a fixed $0 < \Theta < 1$:

$$\limsup_{k\to\infty} \left(\frac{\|e_k\|}{\|e_0\|}\right)^{\frac{1}{k}} = \|I - M^{-1}A\|,$$

so that essentially the convergence rate for the splitting $A = M - N$, with exact inversion of M, is restored. A similar analysis can be carried out in relation with Chebyshev acceleration, see [52, 49]. In [49] the question is studied how Θ can be chosen so that the *total* amount of work is minimized.

For an application of inner-outer iterations for Stokes problems, see the inexact Uzawa scheme, proposed in [36]; for Navier-Stokes related problems, see [55].

5 The Petrov-Galerkin approach: Bi-CG

For unsymmetric systems we can, in general, not reduce the matrix A to a symmetric system in a lower-dimensional subspace, by orthogonal projections. The reason is that we can not create an orthogonal basis for the Krylov subspace by a 3-term recurrence relation [38]. We can, however, try to obtain a suitable non-orthogonal basis with a 3-term recurrence, by requiring that this basis is orthogonal with respect to some other basis.
We start by constructing an arbitrary basis for the Krylov subspace:

$$h_{i+1,i}v_{i+1} = Av_i - \sum_{j=1}^{i} h_{j,i}v_j, \tag{5.1}$$

which can be rewritten in matrix notation as $AV_i = V_{i+1}H_{i+1,i}$. The coefficients $h_{i+1,i}$ define the norm of v_{i+1}, and a natural choice would be to select them such that $\|v_{i+1}\|_2 = 1$. In Bi-CG implementations, a popular choice is to select $h_{i+1,i}$ such that $\|v_{i+1}\|_2 = \|r_{i+1}\|_2$.

Clearly, we can not use V_i for the projection, but suppose we have a W_i for which $W_i^T V_i = D_i$ (an i by i diagonal matrix with diagonal entries d_i), and for which $W_i^T v_{i+1} = 0$.
Then

$$W_i^T AV_i = D_i H_{i,i}, \tag{5.2}$$

and now our goal is to find a W_i for which $H_{i,i}$ is tridiagonal. This means that $V_i^T A^T W_i$ should be tridiagonal too. This last expression has a similar structure as the right-hand side in (5.2), with only W_i and V_i reversed. This suggests to generate the w_i with A^T.
We start with an arbitrary $w_1 \neq 0$, such that $w_1^T v_1 \neq 0$. Then we generate v_2 with (5.1), and orthogonalize it with respect to w_1, which means that $h_{1,1} = w_1^T Av_1/(w_1^T v_1)$. Since $w_1^T Av_1 = (A^T w_1)^T v_1$, this implies that w_2, generated with

$$h_{2,1}w_2 = A^T w_1 - h_{1,1}w_1,$$

is also orthogonal to v_1.
This can be continued, and we see that we can create bi-orthogonal basis sets $\{v_j\}$, and $\{w_j\}$, by making the new v_i orthogonal to w_1 up to w_{i-1}, and then by generating w_i with the same recurrence coefficients, but with A^T instead of A.
Now we have that $W_i^T AV_i = D_i H_{i,i}$, and also that $V_i^T A^T W_i = D_i H_{i,i}$. This implies that $D_i H_{i,i}$ is symmetric, and hence $H_{i,i}$ is a tridiagonal matrix, which gives us the desired 3-term recurrence relation for the v_j's, and the w_j's. Note that $v_1, \ldots, v_i$ form a basis for $K^i(A; v_1)$,and $w_1, \ldots, w_i$ form a basis for $K^i(A^T; w_1)$.

We may proceed in a similar way as in the symmetric case:

$$AV_i = V_{i+1}T_{i+1,i}, \tag{5.3}$$

but here we use the matrix $W_i = [w_1, w_2, ..., w_i]$ for the projection of the system

$$W_i^T(b - Ax_i) = 0,$$

or

$$W_i^T AV_i y - W_i^T b = 0.$$

Using (5.3), we find that y_i satisfies

$$T_{i,i} y = \|r_0\|_2 e_1,$$

and $x_i = V_i y$. The resulting method is known as the Bi-Lanczos method [65].

We have assumed that $d_i \neq 0$, that is $w_i^T v_i \neq 0$. The generation of the bi-orthogonal basis breaks down if for some i the value of $w_i^T v_i = 0$, this is referred to in literature as a *serious breakdown*. Likewise, when $w_i^T v_i \approx 0$, we have a near-breakdown. The way to get around this difficulty is the so-called Look-ahead strategy, which comes down to taking a number of successive basis vectors for the Krylov subspace together and to make them blockwise bi-orthogonal. This has been worked out in detail in [75, 45, 46, 47].

Another way to avoid breakdown is to restart as soon as a diagonal element gets small. Of course, this strategy looks surprisingly simple, but one should realise that at a restart the Krylov subspace, that has been built up so far, is thrown away, and this destroys the possibility of faster (i.e., superlinear) convergence.

We can try to construct an LU-decomposition, without pivoting, of $T_{i,i}$. If this decomposition exists, then, similar to CG, it can be updated from iteration to iteration and this leads to a recursive update of the solution vector, which avoids saving all intermediate r and w vectors. This variant of Bi-Lanczos is usually called Bi-Conjugate Gradients, or for short Bi-CG [41]. In Bi-CG, the d_i are chosen such that $v_i = r_{i-1}$, similarly to CG.

Of course one can in general not be certain that an LU decomposition (without pivoting) of the tridiagonal matrix $T_{i,i}$ exists, and this may lead also to breakdown (a breakdown of the *second kind*), of the Bi-CG algorithm. Note that this breakdown can be avoided in the Bi-Lanczos formulation of the iterative solution scheme, e.g., by making an LU-decomposition with 2 by 2 block diagonal elements [9]. It is also avoided in the QMR approach (see Section 5.1).

Note that for symmetric matrices Bi-Lanczos generates the same solution as Lanczos, provided that $w_1 = r_0$, and under the same condition Bi-CG delivers the same iterates as CG for positive definite matrices. However, the Bi-orthogonal variants do so at the cost of two matrix vector operations per iteration step.
For a computational scheme for Bi-CG, without provisions for breakdown, see [10].

5.1 QMR

The QMR method [47] relates to Bi-CG in a similar way as MINRES relates to CG. We start with the recurrence relations for the v_j:

$$AV_i = V_{i+1}T_{i+1,i}.$$

We would like to identify the x_i, with $x_i \in K^i(A; r_0)$, or $x_i = V_i y$, for which

$$||b - Ax_i||_2 = ||b - AV_i y||_2 = ||b - V_{i+1}T_{i+1,i}y||_2$$

is minimal, but the problem is that V_{i+1} is not orthogonal. However, we pretend that the columns of V_{i+1} are orthogonal. Then

$$||b - Ax_i||_2 = ||V_{i+1}(||r_0||_2 e_1 - T_{i+1,i}y)||_2 = ||(||r_0||_2 e_1 - T_{i+1,i}y)||_2,$$

and in [47] it is suggested to solve the projected minimum norm least squares problem $||(||r_0||_2 e_1 - T_{i+1,i}y)||_2$. The minimum value of this norm is called the quasi residual and will be denoted by $||r_i^Q||_2$.
Since, in general, the columns of V_{i+1} are not orthogonal, the computed $x_i = V_i y$ does not solve the minimum residual problem, and therefore this approach is referred to as a Quasi-minimum residual approach [47]. It can be shown that the norm of the residual r_i^{QMR} of QMR can be be bounded in terms of the quasi residual

$$||r_i^{QMR}||_2 \le \sqrt{i+1}\,||r_i^Q||_2.$$

The sketched approach leads to the simplest form of the QMR method. A more general form arises if the least squares problem is replaced by a weighted least squares problem [47]. No strategies are yet known for optimal weights.

In [47] the QMR method is carried out on top of a look-ahead variant of the bi-orthogonal Lanczos method, which makes the method more robust. Experiments indicate that although QMR has a much smoother convergence behaviour than Bi-CG, it is not essentially faster than Bi-CG. This is confirmed explicitly by the following relation for the Bi-CG residual r_k^B and the quasi residual r_k^Q (in exact arithmetic):

$$||r_k^B||_2 = \frac{||r_k^Q||_2}{\sqrt{1 - (||r_k^Q||_2/||r_{k-1}^Q||_2)^2}},$$

see [23]: Theorem 4.1. This relation, which is similar to the relation for GMRES and FOM, shows that when QMR gives a significant reduction at step k, then Bi-CG and QMR have arrived at residuals of about the same norm (provided, of course, that the same set of starting vectors has been used).

It is tempting to compare QMR with GMRES, but this is difficult. GMRES really minimizes the 2-norm of the residual, but at the cost of increasing the work of keeping all residuals orthogonal and increasing demands for memory space.

QMR does not minimize this norm, but often it has a convergence comparable to GMRES, at a cost of twice the amount of matrix vector products per iteration step. However, the generation of the basis vectors in QMR is relatively cheap and the memory requirements are limited and modest. In view of the relation between GMRES and FOM, it will be no surprise that there is a similar relation between QMR and Bi-CG, for details of this see [47]. This relation expresses that at a significant local error reduction of QMR, Bi-CG and QMR have arrived almost at the same residual vector (similar to GMRES and FOM). However, QMR is preferred to Bi-CG in all cases because of its much smoother convergence behaviour, and also because QMR removes one break-down condition (even when implemented without look-ahead). Several variants of QMR, or rather Bi-CG, have been proposed, which increase the effectiveness of this class of methods in certain circumstances.

In a recent paper, Zhou and Walker [110] have shown that the Quasi-Minimum Residual approach can be followed for other methods, such as CGS and Bi-CGSTAB, as well. The main idea is that in these methods the approximate solution is updated as

$$x_{i+1} = x_i + \alpha_i p_i,$$

and the corresponding residual is updated as

$$r_{i+1} = r_i - \alpha_i A p_i.$$

This means that $AP_i = W_i R_{i+1}$, with W_i a lower bidiagonal matrix. The x_i are combinations of the p_i, so that we can try to find the combination $P_i y_i$ for which $\|b - AP_i y_i\|_2$ is minimal. If we insert the expression for AP_i, and ignore the fact that the r_i are not orthogonal, then we can minimize the norm of the residual in a quasi-minimum least squares sense, similar to QMR.

5.2 CGS

It is well known that the bi-conjugate gradient residual vector can be written as $r_j\ (= \rho_j v_j) = P_j(A) r_0$, and, likewise, the so-called shadow residual $\hat{r}_j\ (= \rho_j w_j)$ can be written as $\hat{r}_j = P_j(A^T)\hat{r}_0$. Because of the bi-orthogonality relation we have that

$$\begin{aligned}(r_j, \hat{r}_i) &= (P_j(A) r_0, P_i(A^T)\hat{r}_0) \\ &= (P_i(A) P_j(A) r_0, \hat{r}_0) = 0,\end{aligned}$$

for $i < j$. The iteration parameters for bi-conjugate gradients are computed from innerproducts like the above. Sonneveld [89] observed that we can also construct the vectors $\tilde{r}_j = P_j^2(A) r_0$, using only the latter form of the innerproduct for recovering the bi-conjugate gradients parameters (which implicitly define the polynomial P_j). By doing so, the computation of the vectors $\hat{r}_j$ can be avoided and so can the multiplication by the matrix A^T.

The resulting CGS [89] method works in general very well for many unsymmetric linear problems. It converges often much faster than BI-CG (about twice as fast in some cases) and has the advantage that fewer vectors are stored than in GMRES. These three methods have been compared in many studies (see, e.g., [78, 15, 76, 71]).
CGS, however, usually shows a very irregular convergence behaviour. This behaviour can even lead to cancellation and a "spoiled" solution [98]; see also Section 5.3. Freund [48] suggested a squared variant of QMR, which was called TFQMR. His experiments show that TFQMR is not necessarily faster than CGS, but it has certainly a much smoother convergence behavior.

5.3 How serious is irregular convergence?

By very irregular convergence we refer to the situation where successive residual vectors in the iterative process differ in orders of magnitude in norm, and some of these residuals may even be much larger in norm than the starting residual. We will give an indication why this is a point of concern, even if eventually the (updated) residual satisfies a given tolerance. For more details we refer to Sleijpen et al [86, 87].

We say that an algorithm is *accurate* for a certain problem if the *updated residual* r_j and the *true residual* $b - Ax_j$ are of comparable size for the j's of interest.

In most iteration schemes the approximation for the solution and the corresponding residual are updated independently:

$$\begin{aligned} x_{j+1} &= x_j + w_{j+1} \\ r_{j+1} &= r_j - Aw_{j+1}. \end{aligned}$$

In finite precision arithmetic, there will at least be a discrepancy between the two updated quantities due to the multiplication by A. We can account for that by writing

$$r_{j+1} = r_j - Aw_{j+1} - \Delta_A w_{j+1} \quad \text{for each } j, \tag{5.4}$$

where Δ_A is an $n \times n$ matrix for which $|\Delta_A| \preceq n_A \overline{\xi}\, |A|$: n_A is the maximum number of non-zero matrix entries per row of A, $|B| \equiv (|b_{ij}|)$ if $B = (b_{ij})$, $\overline{\xi}$ is the relative machine precision, the inequality $\preceq$ refers to element-wise $\leq$.

Note that we have ignored the finite precision errors due to the vector additions. In this case (i.e. situation (5.4) whenever we update the approximation), we have that

$$r_k - (b - Ax_k) = \sum_{j=1}^{k} \Delta_A w_j = \sum_{j=1}^{k} \Delta_A (e_{j-1} - e_j), \tag{5.5}$$

where the perturbation matrix Δ_A may depend on j and e_j is the approximation error in the jth approximation: $e_j \equiv x - x_j$. Hence,

$$\begin{aligned} |||r_k|| - ||b - Ax_k||| &\leq 2k\, n_A\, \bar{\xi}\, |||A|||\, \max_j ||e_j|| \\ &\leq 2k\, n_A\, \bar{\xi}\, |||A|||\, ||A^{-1}||\, \max_j ||r_j||. \end{aligned} \tag{5.6}$$

Except for the factor k, the first upper–bound appears to be rather sharp. We see that approximations with large approximation errors may ultimately lead to an inaccurate result. Such large local approximation errors are typical for CGS; Van der Vorst[98] describes an example of the resulting numerical inaccuracy. If there are a number of approximations with comparable large approximation errors, then their multiplicity may replace the factor k; otherwise it will be only the largest approximation error that makes up virtually the bound for the deviation.

A more rigorous analysis, leading to essentially the same results, has been given by Greenbaum [56]. She also includes the errors introduced by the multiplication and subtraction in $r_{i+1} = r_i - \alpha_i A w_i$. However, the qualitative results of this analysis are the same, since in general the errors in the matrix vector product dominate.

5.4 Bi-CGSTAB

Bi-CGSTAB [98] is based on the following observation. Instead of squaring the Bi-CG polynomial, we can construct other iteration methods, by which x_i are generated so that $r_i = \tilde{P}_i(A)P_i(A)r_0$ with other i^{th} degree polynomials $\tilde{P}$. An obvious possibility is to take for $\tilde{P}_j$ a polynomial of the form

$$Q_i(x) = (1 - \omega_1 x)(1 - \omega_2 x)...(1 - \omega_i x), \tag{5.7}$$

and to select suitable constants ω_j. This expression leads to an almost trivial recurrence relation for the Q_i. In Bi-CGSTAB ω_j in the jth iteration step is chosen as to minimize r_j, with respect to ω_j, for residuals that can be written as $r_j = Q_j(A)P_j(A)r_0$.
Bi-CGSTAB needs two innerproducts more per iteration than CGS.

Bi-CGSTAB can be viewed as the product of Bi-CG and GMRES(1), or rather GCR(1). Of course, other product methods can be formulated as well, and this will be the subject of the next subsection.

5.4.1 Variants of Bi-CGSTAB

Because of the local minimization, Bi-CGSTAB displays a much smoother convergence behavior than CGS, and more surprisingly it often also converges (slightly) faster. One weak point in Bi-CGSTAB is that there is a break-down if $\omega_j = 0$. One may also expect negative effects when ω_j is small. As soon as the GCR(1) step in Bi-CGSTAB (nearly) stagnates, then the BiCG part in the next iteration step cannot (or can only poorly) be constructed. Another dubious aspect of Bi-CGSTAB is that the factor Q_k has only real roots by construction. It is well-known that optimal reduction polynomials for matrices with complex eigenvalues may have complex roots as well. If, for instance, the matrix A is real skew-symmetric, then GCR(1) stagnates forever, whereas a method like GCR(2) (or GMRES(2)), in which we minimize over two combined successive search directions, may lead to convergence, and this is mainly due to the fact that then complex eigenvalue components in the error can be effectively reduced.

This point of view was taken in [60] for the construction of the Bi-CGSTAB2 method. In the odd-numbered iteration steps the Q-polynomial is expanded by a linear factor, as in Bi-CGSTAB, but in the even-numbered steps this linear factor is discarded, and the Q-polynomial from the previous even-numbered step is expanded by a quadratic $1-\alpha_k A-\beta_k A^2$. For this construction the information from the odd-numbered step is required. It was anticipated that the introduction of quadratic factors in Q might help to improve convergence for systems with complex eigenvalues, and, indeed, some improvement has been observed in practical situations (see also [77]).

In [88], another and even simpler approach was taken to arrive at the desired even-numbered steps, without the necessity of the construction of the intermediate Bi-CGSTAB-type step in the odd-numbered steps. In this approach the polynomial Q is constructed straight-away as a product of quadratic factors. In fact, it is shown in [88] that the polynomial Q can also be constructed as the product of ℓ-degree factors, without the construction of the intermediate lower degree factors. The main idea is that ℓ successive Bi-CG steps are carried out, where for the sake of an A^T-free construction the already available part of Q is expanded by simple powers of A. This means that after the Bi-CG part of the algorithm, vectors from the Krylov subspace $s, As, A^2s, ..., A^\ell s$, with $s = P_k(A)Q_{k-\ell}(A)r_0$ are available, and it is then relatively easy to minimize the residual over that particular Krylov subspace. There are variants of this approach in which more stable bases for the Krylov subspaces are generated [87], but for low values of ℓ a standard basis satisfies, together with a minimum norm solution obtained through solving the associated normal equations (which requires the solution of an ℓ by ℓ system. In most cases Bi-CGSTAB(2) will already give nice results for problems where Bi-CGSTAB may fail.

5.5 Other product methods

In the paper by Sonneveld [89], it was made clear that we can construct product methods for which $r_k = H_k(A)P_k(A)r_0$, in which P_k denotes the Bi-CG iteration polynomial, and H_k is any other arbitrary monic polynomial of exact degree k, and he showed that this can be done with the same number of matrix vector multiplications as in k steps of Bi-CG. CGS and the Bi-CGSTAB methods are special instances of these hybrid Bi-CG schemes, but other product methods have been proposed as well.

In [42] it was suggested to take for H_k Bi-CG-like polynomials in an attempt to keep attractive convergence aspects of CGS, while avoiding the irregular convergence behavior. One idea is to take for H_k the Bi-CG polynomial obtained with a different starting vector w_1, which can be done for virtually no additional costs in Bi-CG; this leads to a generalized CGS process. Another idea is to take for H_k the product of P_{k-1} and a suitable fixed first degree factor. The idea is that the zeros of P_k and P_{k-1} do not coincide, so that it is likely that a zero of P_{k-1} will be close to a local maximum of P_k. This avoids the squaring effects in CGS, while it maintains the desired squaring effects in converged eigendirections. Both ideas are easily implemented in an existing CGS code.

In [109] it is suggested to generate H_k by a three term recurrence relation. This leads to an expression for r_k of the form

$$r_k = t_{k-1} - \eta_{k-1} y_{k-1} - \zeta_{k-1} A t_{k-1},$$

where t_{k-1} and y_{k-1} are auxiliary vectors in the hybrid Bi-CG iteration process, and η_{k-1} and ζ_{k-1} represent the coefficients in a Bi-CG like three term recurrence for H_k. The idea suggested in [109] is to minimize the Euclidean norm of r_k as a function of η_{k-1} and ζ_{k-1}, and this leads to product methods which, according to given examples, can compete with CGS and Bi-CGSTAB methods. It is not clear whether these generalized product methods offer an advantage over the higher order Bi-CGSTAB(ℓ) methods.

6 Preconditioning

In order to improve the speed of convergence of iterative methods, one applies usually some form of preconditioning. Many different preconditioners have been suggested over the years, each of these preconditioners more or less successful for restricted classes of problems.

Among all these preconditioners the Incomplete LU factorizations [69, 21] are the most popular ones, and attempts have been made to improve them, for instance by including more fill [70], or by modifying the diagonal of the ILU factorization in order to force rowsum constraints [58, 6, 5, 73, 95, 34], or by changing the ordering of the matrix [96, 97]. A collection of experiments with respect to the effects of ordering is contained in [30]. More recently, it was discovered that a multigrid-inspired ordering can be very effective for discretized diffusion-convection equations, leading in some cases to almost grid-independent speeds of convergence [93, 11], see also [24]. In these publications the ordering strategy is combined with a drop-tolerance strategy for discarding small enough fill-in elements.

Red-black ordering is an obvious approach to improve parallel properties for well-structured problems, but it got a bad reputation from experiments, like those reported in [30]. If carefully done though, they can lead to significant gains in efficiency. Elman and Golub [35] suggested such an approach, in which Red-Black ordering was combined with a reduced system technique. The idea is simple, eliminate the red points, and construct an ILU for the reduced system of black points. Recently, DeLong and Ortega [28] suggested carrying out a few steps of red-black ordered SOR as a preconditioner for GMRES and Bi-CGSTAB. The key to success in these cases seems to be combined effect of fast convergence of SOR for red-black ordering, and the ability of the Krylov subspace to remove stagnations in convergence behaviour associated with a few isolated eigenvalues of the preconditioned matrix.

Saad [80] has proposed some interesting variants on the incomplete LU approach for the matrix A, one of which is in fact an incomplete LQ decomposition. In this approach it is not necessary to form the matrix Q explicitly, and it turns out that the lower triangular matrix L can be viewed as the factor of an incomplete Choleski factorization of the matrix A^TA. This can be exploited in the preconditioning step, avoiding the use of Q. The second approach was to introduce partial pivoting in ILU, which appears to have some advantages for convection-dominated problems. This approach was further improved by including a threshold technique for fill-in [82].

Another major step forward, for important classes of problems, was the introduction of block variants of incomplete factorizations [92, 20, 4], and modified variants of them [4, 66]. It was observed, by Meurant, that these block variants were more successful for discretized 2-dimensional problems than for 3-dimensional problems, unless the '2-dimensional' blocks in the latter case were solved accurately. For discussions and analysis on ordering strategies, in relation to modified block incomplete factorizations, see [67].

Domain decomposition methods were motivated by parallel computing, but it appeared that the approach could be used with success also for the construction of preconditioners. Domain decomposition has been used for problems that arise from discretization of a PDE over a given domain. The idea is to split the given domain into subdomains, and to solve the discretized PDE's over each subdomain separately. The main problem is to find proper boundary conditions along the interior boundaries of the subdomains. Domain decomposition is used in an iterative fashion and usually the interior boundary conditions are based upon information on the approximate solution of neighboring subdomains that is available from a previous iteration step.

It was shown by Chan and Goovaerts [17] that domain decomposition can actually lead to improved convergence rates, provided the number of domains is not too large. A splitting of the matrix with overlapping subblocks along the diagonal, which can be viewed as a splitting of the domain, if the matrix is associated with a discretized PDE and has been ordered properly, was suggested by Radicati and Robert [78]. They suggested to construct incomplete factorizations for the sub-blocks. These sub-blocks are then applied to corresponding parts of the vectors involved, and some averaging was applied on the overlapping parts. A more sophisticated domain-oriented splitting was suggested in [105], for SSOR and MILU decompositions, with a special treatment for unknowns associated with interfaces between the subdomains.

The isolation of sub-blocks was done by Tang [91] in such a way that the sub-blocks corresponded to subdomains with proper internal boundary conditions. In this case it is necessary to modify the sub-blocks of the original matrix such that the subblocks could be interpreted as the discretizations for subdomains with Dirichlet and Neumann boundary conditions in order to force some smoothness of the approximated solution across boundaries. In [90] this was further improved by requiring also continuity of cross-derivatives of the approximate solution across boundaries. The local fine-tuning of the resulting interpolation formulae for the discretizations was carried out by local Fourier-analysis. It was shown that this approach could lead to impressive reductions in numbers of iterations for convection dominated problems.

Another approach that received attention is the concept of constructing an explicit approximation for the inverse of a given matrix A. The idea is to find a sparse matrix M such that $||AM - I||$ is small for some convenient norm. Kolotilina and Yeremin [64] presented an algorithm in which the inverse was delivered in factored form, which has the advantage that singularity of M can be easily detected. In [22] an algorithm is presented which uses the 1-norm for the minimization. We also mention Chow and Saad [18], who use GMRES for the minimization of $||AM - I||_F$. This approach has the disadvantage that one does not have easy control over the amount of fill-in in M, and drop-tolerance strategies have to be applied. The approach has the advantage that it can be used to correct explicitly some given implicit approximation, like for instance an ILU decomposition.

An elegant approach was suggested by Grote and Huckle [57]. They also attempt to minimize the F-norm, which is equivalent to the Euclidean norm for the errors in the columns m_i of M:

$$||AM - I||_F^2 = \sum_{i=1}^{n} ||Am_i - e_i||_2^2.$$

Based on this observation they derive an algorithm that produces the sparsity pattern for the most error-reducing elements of M. This is done in steps, starting with a diagonal approximation, each steps adds more non-zero entries to M, and the procedure is stopped when the norms are small enough or when memory requirements are violated.

7 Quo Vadis?

In the previous Sections we have seen a large number of new iterative methods, which have come in addition to older methods like Richardson's method, Gauss-Jacobi, Gauss-Seidel, SOR, SSOR, BSOR, S2LOR, Chebyshev iteration, and many more. From the positive point of view, this huge variety in methods has provided us with a powerful toolbox from which we can select the appropriate tool for a given problem. From the point of view of the poor user however, this toolbox is confusingly large. In many practical situations it is not clear at all what method to select. One is often faced with the question from outside the expert community which method is the best, but there is in general no best method. This is nicely illustrated in [71], where examples are given of different types of linear systems with quite different convergence behavior for a few archetypes of the methods discussed before. It is shown that for each method there is a linear system for which the given method is the clear winner, as well as a linear system for which the same method is the clear looser. For instance, there is an example for which the, in general slowly converging, LSQR method is better than the, in general fast converging, GMRES method by orders of magnitudes. This also puts the efforts in perspective where the methods are compared on the basis of some examples. Hopefully this gives us some insight in the properties of these methods, but what we really need is some insight into the characteristics of the linear systems that discriminate between methods.

At present the situation is even more complicated due to the existence of distributed memory parallel computers. These architectures can be used efficiently only if one succeeds in keeping communication between processors below certain levels, and it turns out that the innerproducts, which are needed for the Krylov subspace methods, become a bottleneck for the efficiency for larger numbers of processors [27]. This has led to a revival of interest in innerproduct-free methods, like Chebyshev's method and SOR. It has been suggested to use these methods in combination with the Krylov subspace methods (as a polynomial preconditioner, see [53, 83] for discussions on this), see also our Section on preconditioning.
An interesting approach to construct an innerproduct-free variant of CG from information obtained after a few steps of CG is hinted in [40], where orthogonal polynomials are constructed based on partial information obtained from CG-iteration polynomials. It has been shown by Golub and Kent [50] that the optimal parameters for the Chebyshev method can also be deduced from the Chebyshev iteration vectors. This has been further worked out by Calvetti et al [16], who show how adaptive Chebyshev methods can be constructed with iteration parameters that are obtained as in [50].

It is well-known that the distribution of eigenvalues has to do with the speed of convergence of Krylov subspace methods, and a way to improve this distribution is known as *Preconditioning.* With a good preconditioner there is not much difference between the various Krylov subspace methods, but this shifts the problem to the identification and construction of good preconditioners. Although much work has been done in this area, there is still a strong need for identifying good preconditioners, in particular for highly non-normal matrices and for indefinite problems. These problems play an important role in *Computational Fluid Dynamics*, and also *Helmholtz* problems lead to indefinite problems. For particular cases, for instance the *Stokes* problem, effective indefinite preconditioners have been proposed [55, 37].

The main difficulty in indefinite preconditioning can be explained as follows. The Krylov subspace methods converge fast when the eigenvalues are clustered, say around 1. This means that the preconditioner, often viewed as an approximation to the inverse of the given matrix, must transfer the eigenvalues to 1. It may easily happen that, due to approximation errors that are intentionally made in order to keep the process efficient, eigenvalues are transferred to values close to 0. In that case the convergence of Krylov subspace methods will be slow, and it may happen that the unpreconditioned iteration process converges faster than the preconditioned iteration.

Complex-valued problems can be treated as the problems discussed in this paper, in particular Hermitian problems can be treated as real symmetric ones, by replacing A^T by A^H, and by the appropriate change of innerproduct. Some problems, for instance in *Electromagnetics*, lead to symmetric complex matrices. Krylov subspace methods that take advantage of this property have been suggested in [43, 99], but preconditioning for this type of problems has remained an open problem.

The approximation error in iterative schemes is not directly available, and has to be estimated or bounded in terms of information obtained in the iteration process. This can be done for special cases, for instance for the Conjugate Gradient method (see [10] and references therein). For most cases, however, the problem of getting realistic error bounds is yet unsolved. In some applications we need information on errors in specified components of the solution, this is also mainly an open problem.

Parallel computing is a very important issue in solving large linear systems, but we have treated this issue only in passing. One might easily dedicate an entire paper for highlighting developments in this area. We restrict ourselves to referring to the overview paper by Demmel et al [29], in which parallel approaches are discussed as well as open problems in this area.

Although the list of interesting open problems can be made arbitrarily longer, space limitations forced us to mention only a few of them.

8 Acknowledgements

The authors wish to thank Andy Wathen for his careful reading of the manuscript, which has helped in improving the presentation.
The work of Gene H. Golub was in part supported by the National Science Foundation, Grant Number CCR-9505393.
The work of Henk A. Van der Vorst was in part supported by the Dutch Research Organization NWO, MPR cluster project 95MPR04.

References

1. W. E. Arnoldi. The principle of minimized iteration in the solution of the matrix eigenproblem. *Quart. Appl. Math.*, 9:17–29, 1951.
2. O. Axelsson. Solution of linear systems of equations: iterative methods. In V. A. Barker, editor, *Sparse Matrix Techniques*, Berlin, 1977. Copenhagen 1976, Springer Verlag.
3. O. Axelsson. Conjugate gradient type methods for unsymmetric and inconsistent systems of equations. *Lin. Alg. Appl.*, 29:1–16, 1980.
4. O. Axelsson. A general incomplete block-factorization method. *Lin. Alg. Appl.*, 74:179–190, 1986.
5. O. Axelsson. *Iterative Solution Methods.* Cambridge University Press, Cambridge, 1994.
6. O. Axelsson and G. Lindskog. On the eigenvalue distribution of a class of preconditioning methods. *Numer. Math.*, 48:479–498, 1986.
7. O. Axelsson and P. S. Vassilevski. A black box generalized conjugate gradient solver with inner iterations and variable-step preconditioning. *SIAM J. Matrix Anal. Appl.*, 12(4):625–644, 1991.
8. Z. Bai, D. Hu, and L. Reichel. A Newton basis GMRES implementation. *IMA J. Numer. Anal.*, 14:563–581, 1991.
9. R. E. Bank and T. F. Chan. An analysis of the composite step biconjugate gradient method. *Num. Math.*, 66:295–319, 1993.
10. R. Barrett, M. Berry, T. Chan, J. Demmel, J. Donato, J. Dongarra, V. Eijkhout, R. Pozo, C. Romine, and H. van der Vorst. *Templates for the Solution of Linear Systems: Building Blocks for Iterative Methods.* SIAM, Philadelphia, PA, 1994.
11. E.F.F. Botta and A. van der Ploeg. Nested grids ILU-decomposition (NGILU). *J. of Comput. and Appl. Math.*, 66:515–526, 1996.
12. C. Brezinski and M. Redivo-Zaglia. Treatment of near breakdown in the CGS algorithm. *Numerical Algorithms*, 7:33–73, 1994.
13. G. C. Broyden. A new method of solving nonlinear simultaneous equations. *Comput. J.*, 12:94–99, 1969.
14. A.M. Bruaset. *A Survey of Preconditioned Iterative Methods.* Longman Scientific & and Technical, Harlow, UK, 1995.
15. G. Brussino and V. Sonnad. A comparison of direct and preconditioned iterative techniques for sparse unsymmetric systems of linear equations. *Int. J. for Num. Methods in Eng.*, 28:801–815, 1989.
16. D. Calvetti, G. H. Golub, and L. Reichel. Adaptive Chebyshev iterative methods for nonsymmetric linear systems based on modified moments. *Numerische Mathematik*, 67:21–40, 1994.
17. T.F. Chan and D.Goovaerts. A note on the efficiency of domain decomposed incomplete factorizations. *SIAM J. Sci. Stat. Comp.*, 11:794–803, 1990.

18. E. Chow and Y. Saad. Approximate inverse preconditioners for general sparse matrices. Technical Report Research Report UMSI 94/101, University of Minnesota Supercomputing Institute, Minneapolis, Minnesota, 1994.

19. P. Concus and G. H. Golub. A generalized Conjugate Gradient method for nonsymmetric systems of linear equations. Technical Report STAN-CS-76-535, Stanford University, Stanford, CA, 1976.

20. P. Concus, G. H. Golub, and G. Meurant. Block preconditioning for the conjugate gradient method. *SIAM J. Sci. Stat. Comput.*, 6:220–252, 1985.

21. P. Concus, G. H. Golub, and D. P. O'Leary. A generalized conjugate gradient method for the numerical solution of elliptic partial differential equations. In J. R. Bunch and D. J. Rose, editors, *Sparse Matrix Computations*. Academic Press, New York, 1976.

22. J. D. F. Cosgrove, J. C. Diaz, and A. Griewank. Approximate inverse preconditionings for sparse linear systems. *Intern. J. Computer Math.*, 44:91–110, 1992.

23. J. Cullum and A. Greenbaum. Relations between Galerkin and norm-minimizing iterative methods for solving linear systems. *SIAM J. Matrix Anal. Appl.*, 17:223–247, 1996.

24. E. F. D'Azevedo, F. A. Forsyth, and W. P. Tang. Ordering methods for preconditioned conjugate gradient methods applied to unstructured grid problems. *SIAM J. Matrix Anal. Appl.*, 13:944–961, 1992.

25. E. De Sturler. Nested Krylov methods based on GCR. *J. of Comput. Appl. Math.*, 67:15–41, 1996.

26. E. De Sturler and D. R. Fokkema. Nested Krylov methods and preserving the orthogonality. In N. Duane Melson, T.A. Manteuffel, and S.F. McCormick, editors, *Sixth Copper Mountain Conference on Multigrid Methods*, volume Part 1 of *NASA Conference Publication 3324*, pages 111–126. NASA, 1993.

27. E. De Sturler and H.A. Van der Vorst. Reducing the effect of global communication in GMRES(m) and CG on parallel distributed memory computers. *J. Appl. Num. Math.*, 18:441–459, 1995.

28. M. A. DeLong and J. M. Ortega. SOR as a preconditioner. *J. Appl. Num. Math.*, 18:431–440, 1995.

29. J. Demmel, M. Heath, and H. Van der Vorst. Parallel numerical linear algebra. In *Acta Numerica 1993*. Cambridge University Press, Cambridge, 1993.

30. I. S. Duff and G. A. Meurant. The effect of ordering on preconditioned conjugate gradient. *BIT*, 29:635–657, 1989.

31. T. Eirola and O. Nevanlinna. Accelerating with rank-one updates. *Lin. Alg. Appl.*, 121:511–520, 1989.

32. S. C. Eisenstat, H. C. Elman, and M. H. Schultz. Variational iterative methods for nonsymmetric systems of linear equations. *SIAM J. Numer. Anal.*, 20:345–357, 1983.

33. H. C. Elman. *Iterative methods for large sparse nonsymmetric systems of linear equations*. PhD thesis, Yale University, New Haven, CT, 1982.

34. H. C. Elman. Relaxed and stabilized incomplete factorizations for non-self-adjoint linear systems. *BIT*, 29:890–915, 1989.

35. H. C. Elman and G. H. Golub. Line iterative methods for cyclically reduced discrete convection-diffusion problems. *SIAM J. Sci. Stat. Comput.*, 13:339–363, 1992.

36. H. C. Elman and G. H. Golub. Inexact and preconditioned Uzawa algorithms for saddle point problems. *SIAM J. Numer. Anal*, 31:1645–1661, 1994.

37. H. C. Elman and D. J. Silvester. Fast nonsymmetric iteration and preconditioning for Navier-Stokes equations. *SIAM J. Sci. Comput.*, 17:33–46, 1996.

38. V. Faber and T. A. Manteuffel. Necessary and sufficient conditions for the existence of a conjugate gradient method. *SIAM J. Numer. Analysis*, 21(2):352–362, 1984.

39. B. Fischer. *Polynomial Based Iteration Methods for Symmetric Linear Systems.* Wiley and Teubner, Chichester/Stuttgart, 1996.

40. B. Fischer and G. H. Golub. How to generate unknown orthogonal polynomials out of known orthogonal polynomials. *J. of Comput. and Appl. Math.*, 43:99–115, 1992.

41. R. Fletcher. *Conjugate gradient methods for indefinite systems*, volume 506 of *Lecture Notes Math.*, pages 73–89. Springer-Verlag, Berlin–Heidelberg–New York, 1976.

42. D.R. Fokkema, G.L.G. Sleijpen, and H.A. van der Vorst. Generalized conjugate gradient squared. Technical Report Preprint 851, Mathematical Institute, Utrecht University, 1994.

43. R. W. Freund. Conjugate gradient-type methods for a linear systems with complex symmetric coefficient matrices. *SIAM J. Sci. Comput.*, 13:425–448, 1992.

44. R. W. Freund, G. H. Golub, and N. M. Nachtigal. Iterative solution of linear systems. In *Acta Numerica 1992*. Cambridge University Press, Cambridge, 1992.

45. R. W. Freund, M. H. Gutknecht, and N. M. Nachtigal. An implementation of the look-ahead Lanczos algorithm for non-Hermitian matrices. *SIAM J. Sci. Comput.*, 14:137–158, 1993.

46. R. W. Freund and N. M. Nachtigal. An implementation of the look-ahead Lanczos algorithm for non-Hermitian matrices, part 2. Technical Report 90.46, RIACS, NASA Ames Research Center, 1990.

47. R. W. Freund and N. M. Nachtigal. QMR: a quasi-minimal residual method for non-Hermitian linear systems. *Numerische Mathematik*, 60:315–339, 1991.

48. Roland Freund. A transpose-free quasi-minimal residual algorithm for non-Hermitian linear systems. *SIAM J. Sci. Comput.*, 14:470–482, 1993.

49. E. Giladi, G.H. Golub, and J.B. Keller. Inner and outer iterations for the Chebyshev algorithm. Technical Report SCCM-95-10, Computer Science Department, Stanford University, Stanford, CA, 1995. (accepted for publication in SINUM)

50. G. H. Golub and M. Kent. Estimates of eigenvalues for iterative methods. *Math. of Comp.*, 53:619–626, 1989.

51. G. H. Golub and D.P. O'Leary. Some history of the conjugate gradient and Lanczos algorithms: 1948-1976. *SIAM Review*, 31:50–102, 1989.

52. G. H. Golub and M. L. Overton. The convergence of inexact Chebyshev and Richardson iterative methods for solving linear systems. *Numerische Mathematik*, 53:571–594, 1988.

53. G. H. Golub and C. F. Van Loan. *Matrix Computations.* The Johns Hopkins University Press, Baltimore, 1989.

54. G. H. Golub and R. S. Varga. Chebyshev semi-iterative methods, Successive Over-Relaxation methods, and second-order Richardson iterative methods, Parts I and II. *Numerische Mathematik*, 3:147–168, 1961.

55. G.H. Golub and A.J. Wathen. An iteration for indefinite systems and its application to the Navier-Stokes equations. Technical Report SCCM-95-07, Computer Science Department, Stanford University, Stanford, CA, 1995. To appear in *SIAM J. Sci. Comput.*

56. A. Greenbaum. Estimating the attainable accuracy of recursively computed residual methods. Technical report, Courant Institute of Mathematical Sciences, New York, NY, 1995.

57. M. Grote and T. Huckle. Effective parallel preconditioning with sparse approximate inverses. In D. H. Bailey, editor, *Proceedings of the Seventh SIAM Conference on Parallel Processing for Scientific Computing*, pages 466–471, Philadelphia, 1995. SIAM.

58. I. Gustafsson. A class of first order factorization methods. *BIT*, 18:142–156, 1978.

59. M. H. Gutknecht. A completed theory of the unsymmetric Lanczos process and related algorithms, Part I. *SIAM J. Matrix Anal. Appl.*, 13:594–639, 1992.

60. M. H. Gutknecht. Variants of BICGSTAB for matrices with complex spectrum. *SIAM J. Sci. Comput.*, 14:1020–1033, 1993.

61. W. Hackbusch. *Iterative solution of large linear systems of equations.* Springer Verlag, New York, 1994.

62. K. C. Jea and D. M. Young. Generalized conjugate-gradient acceleration of nonsymmetrizable iterative methods. *Lin. Alg. Appl.*, 34:159–194, 1980.

63. D. S. Kershaw. The Incomplete Choleski-Conjugate Gradient method for the iterative solution of systems of linear equations. *J. of Comp. Phys.*, 26:43–65, 1978.

64. L. Yu. Kolotilina and A. Yu. Yeremin. Factorized sparse approximate inverse preconditionings. *SIAM J. Matrix Anal. Appl.*, 14:45–58, 1993.

65. C. Lanczos. Solution of systems of linear equations by minimized iterations. *J. Res. Natl. Bur. Stand*, 49:33–53, 1952.

66. M. M. Magolu. Modified block-approximate factorization strategies. *Numerische Mathematik*, 61:91–110, 1992.

67. M. M. Magolu. Ordering strategies for modified block incomplete factorizations. *SIAM J. Sci. Comput.*, 16:378–399, 1995.

68. T. A. Manteuffel. The Tchebychev iteration for nonsymmetric linear systems. *Numerische Mathematik*, 28:307–327, 1977.

69. J. A. Meijerink and H. A. Van der Vorst. An iterative solution method for linear systems of which the coefficient matrix is a symmetric M-matrix. *Math.Comp.*, 31:148–162, 1977.

70. J. A. Meijerink and H. A. Van der Vorst. Guidelines for the usage of incomplete decompositions in solving sets of linear equations as they occur in practical problems. *J.of Comp. Physics*, 44:134–155, 1981.

71. N. M. Nachtigal, S. C. Reddy, and L. N. Trefethen. How fast are nonsymmetric matrix iterations? *SIAM J. Matrix Anal. Appl.*, 13:778–795, 1992.

72. N. M. Nachtigal, L. Reichel, and L. N. Trefethen. A hybrid GMRES algorithm for nonsymmetric matrix iterations. Technical Report 90-7, MIT, Cambridge, MA, 1990.

73. Y. Notay. DRIC: a dynamic version of the RIC method. *Numerical Linear Algebra with Applications*, 1:511–532, 1994.

74. C. C. Paige and M. A. Saunders. Solution of sparse indefinite systems of linear equations. *SIAM J. Numer. Anal.*, 12:617–629, 1975.

75. B. N. Parlett, D. R. Taylor, and Z. A. Liu. A look-ahead Lanczos algorithm for unsymmetric matrices. *Math. Comp.*, 44:105–124, 1985.

76. C. Pommerell and W. Fichtner. PILS: An iterative linear solver package for ill-conditioned systems. In *Supercomputing '91*, pages 588–599, Los Alamitos, CA., 1991. IEEE Computer Society.

77. C. Pommerell. *Solution of large unsymmetric systems of linear equations.* PhD thesis, Swiss Federal Institute of Technology, Zürich, 1992.

78. G. Radicati di Brozolo and Y. Robert. Parallel conjugate gradient-like algorithms for solving sparse non-symmetric systems on a vector multiprocessor. *Parallel Computing*, 11:223–239, 1989.

79. J. K. Reid. The use of conjugate gradients for systems of equations possessing 'Property A'. *SIAM J. Numer. Anal.*, pages 325–332, 1972.

80. Y. Saad. Preconditioning techniques for indefinite and nonsymmetric linear systems. *J. Comput. Appl. Math.*, 24:89–105, 1988.

81. Y. Saad. A flexible inner-outer preconditioned GMRES algorithm. *SIAM J. Sci. Comput.*, 14:461–469, 1993.

82. Y. Saad. ILUT: A dual threshold incomplete LU factorization. *Num. Lin. Alg. with Appl.*, 1:387–402, 1994.

83. Y. Saad. *Iterative methods for sparse linear systems.* PWS Publishing Company, Boston, 1996.

84. Y. Saad and M. H. Schultz. GMRES: a generalized minimal residual algorithm for solving nonsymmetric linear systems. *SIAM J. Sci. Stat. Comput.*, 7:856–869, 1986.

85. A. Sidi. Efficient implementation of minimal polynomial and reduced rank extrapolation methods. *J. of Comp. App. Math.*, 36:305–337, 1991.

86. G. L. G. Sleijpen and H.A. Van der Vorst. Reliable updated residuals in hybrid Bi-CG methods. *Computing*, 56:141–163, 1996.

87. G. L. G. Sleijpen, H.A. Van der Vorst, and D. R. Fokkema. Bi-CGSTAB(ℓ) and other hybrid Bi-CG methods. *Numerical Algorithms*, 7:75–109, 1994.

88. G. L. G. Sleijpen and D. R. Fokkema. BICGSTAB(ℓ) for linear equations involving unsymmetric matrices with complex spectrum. *ETNA*, 1:11–32, 1993.

89. P. Sonneveld. CGS: a fast Lanczos-type solver for nonsymmetric linear systems. *SIAM J. Sci. Stat. Comput.*, 10:36–52, 1989.

90. K.H. Tan. *Local coupling in domain decomposition.* PhD thesis, Utrecht University, Utrecht, 1995.

91. W. P. Tang. Generalized Schwarz splitting. *SIAM J. Sci. Stat. Comput.*, 13:573–595, 1992.

92. R. R. Underwood. An approximate factorization procedure based on the block Cholesky factorization and its use with the conjugate gradient method. Technical Report Tech Report, General Electric, San Jose, CA, 1976.

93. A. van der Ploeg. Preconditioning for sparse matrices with applications. PhD Thesis, University of Groningen, Groningen, 1994.

94. A. Van der Sluis and H. A. Van der Vorst. The rate of convergence of conjugate gradients. *Numerische Mathematik*, 48:543–560, 1986.

95. H. A. Van der Vorst. Iterative solution methods for certain sparse linear systems with a non-symmetric matrix arising from PDE-problems. *J. Comp. Phys.*, 44:1–19, 1981.

96. H. A. Van der Vorst. Large tridiagonal and block tridiagonal linear systems on vector and parallel computers. *Parallel Computing*, 5:45–54, 1987.

97. H. A. Van der Vorst. High performance preconditioning. *SIAM J. Sci. Stat. Comput.*, 10:1174–1185, 1989.

98. H. A. Van der Vorst. Bi-CGSTAB: A fast and smoothly converging variant of Bi-CG for the solution of non-symmetric linear systems. *SIAM J. Sci. Stat. Comput.*, 13:631–644, 1992.

99. H. A. Van der Vorst and J. B. M. Melissen. A Petrov-Galerkin type method for solving $ax = b$, where a is symmetric complex. *IEEE Trans. on Magn.*, pages 706–708, 1990.

100. H. A. Van der Vorst and C. Vuik. The superlinear convergence behaviour of GMRES. *J. of Comp. Appl. Math.*, 48:327–341, 1993.

101. H. A. Van der Vorst and C. Vuik. GMRESR: A family of nested GMRES methods. *Num. Lin. Alg. with Appl.*, 1:369–386, 1994.

102. R. S. Varga. *Matrix Iterative Analysis.* Prentice-Hall, Englewood Cliffs N.J., 1962.

103. P. K. W. Vinsome. ORTOMIN: an iterative method for solving sparse sets of simultaneous linear equations. In *Proc.Fourth Symposium on Reservoir Simulation*, pages 149–159. Society of Petroleum Engineers of AIME, 1976.

104. C. Vuik and H.A. Van der Vorst. A comparison of some GMRES-like methods. *Lin. Alg. Appl.*, 160:131–162, 1992.

105. T. Washio and K. Hayami. Parallel block preconditioning based on SSOR and MILU. *Num. Lin. Alg. with Appl.*, 1:533–553, 1994.

106. O. Widlund. A Lanczos method for a class of nonsymmetric systems of linear equations. *SIAM J. Numer. Anal.*, 15:801–812, 1978.

107. J. H. Wilkinson. *The Algebraic Eigenvalue Problem.* Clarendon Press, Oxford, 1965.

108. D. Young. *Iterative Solution of Large Linear Systems.* Academic Press, New York, 1971.

109. Shao-Liang Zhang. GPBi-CG: Generalized product-type methods based on Bi-CG for solving nonsymmetric linear systems. Technical Report 9301, Nagoya University, Nagoya, Japan, 1993.

110. L. Zhou and H. F. Walker. Residual smoothing techniques for iterative methods. *SIAM J. Sci. Stat. Comp.*, 15:297–312, 1994.

150 Years Old and Still Alive: Eigenproblems[1]

Henk A. van der Vorst* and Gene H. Golub**

** Utrecht University and ** Stanford University*

Abstract

In this presentation, we first sketch the situation about ten years ago with respect to numerical algorithms for eigenproblems of various kinds. This will provide us with a basis for a discussion of the main new developments: Bi-Lanczos and look-ahead, Implicitly Restarted Arnoldi, pseudospectrum, and Jacobi-Davidson methods. We will also give an overview of related developments and improvements to existing algorithms.

Our presentation will conclude with the identification of a number of problems that are still difficult and for which we expect progress in the next ten years.

1 Introduction

Linear eigenproblems continue to be an important and highly relevant area of research in numerical linear algebra. Therefore, it should be no surprise that numerical algorithms for eigenproblems are among the oldest known in the modern literature. In 1846, exactly 150 years ago, Jacobi [33] wrote his famous paper on the computation of solutions for the problem $Ax = \lambda x$, with $A^T = A$. Strictly speaking, this is not correct, since matrix notation was unknown at that time. So Jacobi formulated the problem in terms of systems of equations, written out elementwise. His paper contains computational elements that are still in use, like the plane rotations for making a system more diagonally dominant, and the (Gauss)-Jacobi iteration method for diagonally dominant systems. These elements were combined in his approach for the computation of eigenvalues, together with a clever way of setting up a correction equation for eigenvalue and eigenvector approximations.

Later on other methods became popular, such as the Power iteration method, and, when tools became more sophisticated, eigenproblems other than the standard $Ax = \lambda x$ problem required solution: for instance, the generalized eigenproblems $Ax = \lambda Bx$, higher order polynomial eigenproblems

$$(A_0 + \lambda A_1 + ..\lambda^n A_n)\, x = 0,$$

[1]Dedicated to J.H. Wilkinson on the 10th Anniversary of his death.

and, of course, the singular value decomposition.

Through pioneering work of such people as Arnoldi, Francis, Givens, Householder, Kublanovskaya, Lanczos, Ostrowski, Rutishauser, Wilkinson, and many others, a sophisticated toolbox of algorithms, together with analysis, became available. This work was continued by Kahan, Paige, Parlett, Stewart, to mention only a handful of leaders in a lively and productive research area. Ten years ago we were in the situation that many eigenproblems, of order a few hundreds at most, could be solved routinely. This toolbox contains algorithms like, for instance, Householder reduction, the QR-algorithm, and the numerically stable SVD. Quite essential in the computation of eigenvalues and eigenvectors is the concept of inverse iteration, which leads to quadratic convergence or even cubic convergence (in the diagonalizable case) [52, 60].
Most of these algorithms had been described in the famous book by Wilkinson and Reinsch [77], following their earlier publication in Numerische Mathematik. The Algol-60 codes in these documents formed the basis for the standard libraries EISPACK and LAPACK, and modern codes in NAG, and other libraries, still bear the traces of these 'ancient' Algol-60 procedures. Most of this software is available through *netlib*.

With larger problems coming into the picture, it soon became clear that the matrix transforming techniques could not solve these problems with reasonable computing resources, and, as an alternative, iterative methods were investigated. The Power method was not robust enough, although the method is still hidden in state-of-the-art methods, like the powerful (literally) QR method [30, 57]. Lanczos [34] and Arnoldi [2] started the research on modern iteration methods in the early fifties. After a period of little interest in these methods, mainly because of poor understanding of their numerical properties, Paige [53, 54] showed the potential of the Lanczos method. This marked the start of an entire new area of research. About the same time Davidson, a chemist, designed a popular method that became widely known as Davidson's method [18]. Among numerical analysts the method did not receive much attention, partly because it appeared to be applicable only to diagonal dominant problems, and partly because its effects were poorly understood from a numerical point of view. The Lanczos method, for symmetric matrices, was well understood in 1986, see for instance [57], but the methods for unsymmetric systems have begun to blossom by that time.

It should be noted that in the modern iterative *subspace* methods, like Arnoldi's, Lanczos', and Davidson's method, the given large problem is reduced to a much smaller problem. This smaller problem can then be solved by the, by now standard techniques for dense matrices.

This is a rough sketch of the situation halfway through the eighties, say 1986. In the remaining part of this paper we will highlight what we consider to be the main new developments during the last 10 years.

We are dedicating our paper to J.H. Wilkinson, who died on October 5, 1986. He led the way in numerical linear algebra, and we have all benefited by his vision.

2 Highlights of the period 1986-1996

In the past ten years we have seen many new developments in this exciting area. In our opinion the major steps, from an algorithmic point of view, are:

- Attempts to make the two-sided Lanczos process more robust, by introducing a so-called look-ahead strategy [59, 27, 9, 32].
- The idea of an implicit restart technique for the Arnoldi process, which helps to keep memory requirements reasonable, and which makes the Arnoldi process an attractive algorithm for eigencomputations [68, 38].
- Further improvements on the Davidson method, culminating in the Jacobi-Davidson algorithm [66]. In this algorithm a major deficiency in the original Davidson method has been removed. The Jacobi-Davidson method can be implemented as an accelerated inner-outer iteration scheme.

Of course, much more has happened, and the above reflect only personal impressions. Other important developments and improvements were (the list is still personally colored and incomplete):

- The concept and use of the ε-pseudospectrum [71] as a means to make sensitivities in the spectrum of a nonnormal matrix easily visible. The pseudospectrum does not make classical perturbation theory superfluous, but it helps to detect situations that need further analysis, and it does so in a way that is easily understandable for non-expert users. We will discuss the use of this tool briefly.
- The convergence behavior of Ritz values, for symmetric matrices, has been further analyzed and is now quite well understood. The so-called misconvergence phenomenon, a Ritz value that lingers near some eigenvalue before it converges towards some other eigenvalue [73, 58], and various local effects in the convergence behavior have been analysed and are fairly well understood by now [73]. The notion that rounding errors lead to multiple (spurious) eigenvalues, has been translated into practical strategies also for the unsymmetric Lanczos process [16].
- Strategies have been proposed to improve the efficiency of the various subspace methods. One type of approach amounts to filtering undesired eigenvector components from the starting vector: Chebychev polynomial preconditioning [65]:Chapter VII. The general idea goes back to Flanders and Shortley [25]; Lanczos also suggested polynomial filtering techniques [35]. Other strategies aim for improving the speed of convergence by considering a transformed problem: rational shift-and-invert techniques [62, 63],

inexact shift-invert preconditioning for eigenproblems [50, 45], and [65]: Chapter VII; for a general discussion, see also [46].

- Generalizations of the Davidson method, making the method also suitable for unsymmetric matrices. These methods come down to the incorporation of more general preconditioning, instead of the original proposed diagonal preconditioning [49, 48, 14], or to second order corrections for the current eigenvector approximations [51]. The latter variant is strongly related to the Jacobi-Davidson type of algorithms.

- Eigenproblems can be regarded as nonlinear problems, which means that one can employ homotopy methods for the efficient computation of eigenvalues, or tracking these eigenvalues as function of a parameter in the underlying model [43, 44]. We will sketch this approach and point at some relations with other iterative methods

- Parallelism (Cuppen's Divide and Conquer [17, 22], Restructuring of iterative algorithms in an attempt to combine innerproducts [21, 5, 20]).

- Subspace methods for interior eigenvalues [49, 63], rational Lanczos [62, 63], harmonic Ritz values [28, 47, 55]. We will discuss some of these approaches in this paper.

The remainder of this paper has been organized as follows. We start with an introduction to Krylov subspace methods. Then we highlight the implicit restart technique for Arnoldi's method, as well as some shift-and-invert strategies that help to improve the speed of convergence. Special attention will be given to the Jacobi-Davidson method, which works with different subspaces. A novel extension for this method, which makes it possible to compute several eigenvalues in a part of the spectrum efficiently, will be discussed. Then we will pay some attention to the look-ahead techniques for the two-sided Lanczos method.

Interior eigenvalues are always difficult to compute with subspace methods, if one wants to avoid expensive shift-and-invert operations. Approximate shift-and-invert operations have been suggested for the Jacobi-Davidson method [66], for the Lanczos method [50], and for the Arnoldi method [45]. The notion of harmonic Ritz values offers a helpful tool for restart purposes, since they identify the best approximations with respect to interior eigenvalues.

The concept of homotopy received attention as a means to compute some eigenvalues for 'difficult', or perturbed matrices, starting with available knowledge for a given matrix. We shall briefly examine such techniques. Also, we discuss the concept of ε-pseudospectrum, as a means to study the sensitivity of a spectrum, for not too large matrices. Finally, we will conclude our paper by an outlook on some problems that are still hard to solve, as a motivation for further research.

3 Krylov subspaces

Krylov subspaces play a central role also in iterative methods for eigenvalue computations. To illustrate this, we consider the very well-known Power Method. Assume A is real symmetric, then it has real eigenvalues and a complete set of orthonormal eigenvectors

$$Au_k = \lambda_k u_k \quad , \quad \|u_k\|_2 = 1 \quad (k = 1, 2, \cdots, n).$$

We further assume that the largest eigenvalue in modulus is single and that

$$|\lambda_1| > |\lambda_2| \geq \cdots .$$

Now suppose we are given a vector v_1, which can be expressed in terms of the eigenvectors as $v_1 = \sum_i \gamma_i u_i$, and we assume that $\gamma_1 \neq 0$ (that means that v_1 has a nonzero component in the direction of the largest eigenvector).

Given this v_1, we compute $Av_1, A(Av_1), \ldots$, and it follows that

$$\lim_{j \to \infty} \frac{\|A^j v_1\|}{\|A^{j-1} v_1\|} = |\lambda_1|$$

It is not hard to see why the ratios of these norms approximate the dominant eigenvalue, since

$$\begin{aligned} A^j v_1 &= \sum \gamma_i \lambda_i^j u_i \\ &= \lambda_1^j \left\{ \gamma_1 u_1 + \sum_{i \geq 2} \gamma_i \left(\frac{\lambda_i}{\lambda_1} \right)^j u_i \right\} . \end{aligned}$$

Hence

$$\begin{aligned} \frac{\|A^j v_1\|}{\|A^{j-1} v_1\|} &= |\lambda_1| \frac{\|\gamma_1 u_1 + \sum_{i \geq 2} \gamma_i \left(\frac{\lambda_i}{\lambda_1} \right)^j u_i\|}{\|\gamma_1 u_1 + \sum_{i \geq 2} \gamma_i \left(\frac{\lambda_i}{\lambda_1} \right)^{j-1} u_i\|} \\ &= |\lambda_1| \frac{\|\sum_{i \geq 1} \gamma_i P_j(\lambda_i) u_i\|}{\|\sum_{i \geq 1} \gamma_i P_{j-1}(\lambda_i) u_i\|} , \end{aligned}$$

with $P_j(t) \equiv (t/\lambda_1)^j$.

With the Power method we have built a *Krylov Subspace*

$$K_m(A; v_1) \equiv \text{span}\{v_1, Av_1, \ldots, A^{m-1} v_1\},$$

but note that the method exploits only the last two vectors. The result is that the speed of convergence depends on how fast the polynomial values of $P_j(t)$ decrease for $t = \lambda_k$, for increasing j. The methods of Lanczos and Arnoldi exploit the whole Krylov subspace, and they implicitly construct polynomials P_j that may decrease much faster depending on the eigenvalue distribution.

3.1 Orthogonal basis (Arnoldi, Lanczos)

In order to identify better solutions in the Krylov subspace we need a suitable basis for this subspace, one that can be extended in a meaningful way for subspaces of increasing dimension. The obvious basis r_0, Ar_0, ..., $A^{i-1}r_0$, for $K^i(A; r_0)$, is not very attractive from a numerical point of view, since the vectors $A^j r_0$ point more and more in the direction of the dominant eigenvector for increasing j (the power method !), and hence the basis vectors become dependent in finite precision arithmetic.

Lanczos [34] proposed to use an orthogonal basis for the Krylov subspace, and Arnoldi [2] suggested to compute this basis for unsymmetric matrices as follows. Start with $v_1 \equiv r_0/||r_0||_2$. Assume that we have already an orthonormal basis v_1, ..., v_j for $K^j(A; r_0)$, then this basis is expanded by computing $\widetilde{v} = Av_j$, and by orthonormalizing this vector $\widetilde{v}$ with respect to v_1, ..., v_j. In principle the orthonormalization process can be carried out in different ways, but the most commonly used approach is to do this by a modified Gram-Schmidt procedure [30]. This leads to the following algorithm for the creation of an orthonormal basis for $K^m(A; r_0)$:

$$
\begin{array}{l}
v_1 = r_0/||r_0||_2; \\
\text{for } j = 1, .., m-1 \\
\quad \widetilde{v} = Av_j; \\
\quad \text{for } i = 1, ..., j \\
\quad\quad h_{i,j} = v_i^* \widetilde{v}; \\
\quad\quad \widetilde{v} = \widetilde{v} - h_{i,j} v_i; \\
\quad \text{end}; \\
\quad h_{j+1,j} = ||\widetilde{v}||_2; \\
\quad v_{j+1} = \widetilde{v}/h_{j+1,j}; \\
\text{end}
\end{array}
$$

A more stable implementation, useful for ill-conditioned matrices A, was suggested by Walker [75]; his approach was to use Householder transformations rather than modified Gram-Schmidt.

It is easily verified that v_1, ..., v_m form an orthonormal basis for $K^m(A; r_0)$ (that is, if the construction does not terminate at a value $h_{j+1,j} = 0$). The orthogonalization leads to relations between the v_j, that can be formulated in a compact algebraic form. Let V_j denote the matrix with columns v_1 up to v_j, $V_j \equiv [v_1 \mid v_2 \mid \cdots \mid v_j]$, then it follows that

$$AV_{m-1} = V_m H_{m,m-1}. \tag{3.1}$$

The m by $m-1$ matrix $H_{m,m-1}$ is upper Hessenberg, and its elements $h_{i,j}$ are defined by the Arnoldi orthogonalization algorithm. From a computational point of view, this construction is composed from three basic elements: a matrix vector product with A, innerproducts, and updates. We see that this orthogonalization becomes increasingly expensive for increasing dimension of the subspace, since the computation of each $h_{i,j}$ requires an inner product and a vector update.

Note that if A is symmetric, then so is $H_{m-1,m-1} = V_{m-1}^* A V_{m-1}$, so that in this situation $H_{m-1,m-1}$ is tridiagonal. This means that in the orthogonalization process, each new vector has to be orthogonalized with respect to the previous two vectors only, since all other innerproducts vanish. The resulting three term recurrence relation for the basis vectors of $K_m(A; r_0)$ is known as the *Lanczos method* and some very elegant methods are derived from it. In the symmetric case the orthogonalization process involves constant arithmetical costs per iteration step: one matrix vector product, two innerproducts, and two vector updates. In [13, 74] it has been shown that the Lanczos algorithm can also be applied to matrices for which $A^* = -A$, with a skew symmetric tridiagonal matrix as a result.

3.2 Subspace Iteration Methods

The standard iteration method for eigenproblems is the power method. From it we can derive many well-known iterative methods, almost in a similar way as the iterative solvers for linear systems could be derived from the standard Richardson iteration method.

The main idea is to create some subspace and to compute approximations for the eigenvalues and eigenvectors with respect to this subspace. Starting from the Power method, the most natural choice for this subspace is the Krylov subspace, for which the standard basis vectors are generated by the Power method. A numerically more stable basis is created by generating an orthogonal basis straight away. This leads to the *Lanczos Method* for symmetric matrices, and to the *Arnoldi Method* for unsymmetric matrices.

The orthogonality relation for the basis vectors of $K^m(A; r_0)$ is:

$$AV_m = V_{m+1} H_{m+1,m}. \tag{3.2}$$

and we observe that

$$V_m^* A V_m = H_{m,m}. \tag{3.3}$$

The matrix $H_{m,m}$ is the projection of A onto $K^m(A; r_0)$. If (θ, y) is an eigenpair of $H_{m,m}$ then

$$H_{m,m} y = \theta y$$

$$V_m^* A V_m y - \theta V_m^* V_m y = 0$$

$$V_m^* (A V_m y - \theta V_m y) = 0$$

$$V_m^* (As - \theta s) = 0,$$

where $s = V_m y$.
Hence, the residual for the approximate eigenpair (θ, s) is orthogonal to the current Krylov subspace. The value θ is called a *Ritz value* of A with respect to $K^m(A; r_0)$, and s is the corresponding *Ritz vector.*

If A is unsymmetric then $H_{m,m}$ is an m by m upper Hessenberg matrix and the method is then known as the *Arnoldi Method*; when A is symmetric (or antisymmetric) then $H_{m,m}$ is tridiagonal and symmetric (or antisymmetric), and the method is then known as the *Lanczos Method.* In this case the orthogonality of a new basis vector is obtained with only two innerproducts.

In order to avoid the expensive construction of the upper Hessenberg matrix as in the Arnoldi method, we can follow a dual-basis approach for the construction of a basis that satisfies a three term construction. This leads to the *Bi-Lanczos Method*, which is very similar to the Bi-CG method. Both methods lead to the same projected tridiagonal matrix. Freund [27] and Cullum and Willoughby [16] have published codes for the computation of eigenvalues using this approach.

We can also construct other subspaces for the restriction of the matrix A. One such approach has been suggested by Davidson [18].

In actual situations subspace methods are often applied for $(A - \sigma I)^{-1}$, so that convergence towards eigenvalues close to σ is favored. This is called the *shift-and-invert* approach. This technique is attractive if one can compute the vector $(A - \sigma I)^{-1} z$ at relatively low costs. Note that with a direct solver this can be done with an LU decomposition of $A - \sigma I$, and this LU decomposition can be used for subsequent iteration steps.

For a good overview of subspace methods, see [65].

4 Improvements to Arnoldi's method

The main problem with Arnoldi's method is that it becomes increasingly expensive per iteration step, and in order to restrict memory storage, as well as computational work, restarts are necessary. However, at restart we throw away useful information. Instead of restarting with the most current approximation for the approximated eigenvector (the Ritz vector), corresponding to the desired eigenvalue, it has been suggested to restart with a vector that is a mix of Ritz vectors, corresponding to relevant Ritz values. See [65]:Ch.VII and [61] for such strategies. These strategies are related to an earlier approach suggested in [29]. These strategies may have the advantage that the new starting vector also contains information for nearby eigenvalues. The main problem is that in this kind of restart we try to catch the information for an approximate subspace in one single vector, and apart from this, it is not easy to find the optimal mix.

The restart problem has been solved very elegantly by Sorensen [68]. The idea behind his *Implicitly Restarted Arnoldi* (IRA) method is the following. Suppose we are at step $k+m$, and we want to shrink the current subspace to dimension k again. Obviously we want to maintain the subspace with the best k Ritz vectors, that is, the Ritz vectors corresponding to k Ritz values that we have selected, for instance the k rightmost Ritz values if we are after the k rightmost eigenvalues of A. We compute all $k+m$ Ritz values of H_{k+m} and then we carry out m shifted QR iterations with the m undesired Ritz values as shifts. This has the effect that the k wanted Ritz values are contained in the leading k by k part of the transformed matrix. Also the projection matrix V_{m+k} is transformed correspondingly to the transformations on H_{m+k}, and after the operation the first k columns of the transformed V_{m+k} span the subspace of the Ritz vectors for the wanted Ritz values. This subspace should be expanded again in order to improve the current eigenvalue approximations, and the nice observation is that the leading k by k part of the transformed Hessenberg matrix H_{m+k} is still upper Hessenberg. This means that the first k columns of the transformed V_{m+k} and the leading part H_k can be regarded as representing the first k Arnoldi steps on the transformed basis vector v_1. Sorensen [68] then proposes to continue this new Arnoldi process with another m steps, and to repeat the sketched subspace shrinking procedure. The whole procedure can be viewed as a mechanism that rakes repeatedly new supplementary information with m vectors from the entire space, to filter the desired supplements for the k approximated Ritz values, and to keep the gradually improved k-dimensional subspace in stock.

The attractive property of the IRA process is that we explore effectively subspaces of high-dimensional Krylov subspaces, without the high computational costs that would go with the unrestarted process. The process has been further refined by Lehoucq [37], who considers the many practical aspects involved in a careful implementation. In particular, an analysis and comparison of restarting an Arnoldi iteration, its numerical stability and deflation rules which allow computation of clustered and/or multiple eigenvalues is examined. In [39] existing software for this process, as well as for other methods, is evaluated.

In [68] it has been shown that the implicit restart technique effectively is equivalent with applying a polynomial filter on the starting vector, filtering out undesired eigenvetor components. In [23] it is shown how this polynomial filter, or the multishift QR, can be carried out without actually computing the shifts. The implicit-restart idea has been applied also in combination with the two-sided Lanczos method, for model reduction in electronical applications (pole-zero analysis, stability of CD-players) [31]. Similar ideas have recently been exploited to improve the Jacobi-Davidson method (see Section 6) and the rational Krylov method [64].

The Arnoldi iteration is often carried out with the shift-and-invert approach. For instance, when solving the generalized eigenproblem $Ax = \lambda Bx$, the method is applied to the operator $(A-\sigma B)^{-1}B$. This transformation requires expensive operations with an inverted operator but the advantage is much faster convergence. Meerbergen [45] considers the use of inexact forms of the *Cayley transform* $(A-\sigma B)^{-1}(A-\tau B)$, where the inverse operation is approximated by a few steps of an iterative method, for Arnoldi's method. This technique has a close relation with polynomial preconditioning. Ruhe [63] considers a more general shift-and-invert transform, the so-called *Rational Krylov Sequence (RKS)*:

$$(\delta_j A - \gamma_j B)^{-1}(\sigma_j A - \rho_j B),$$

in which the coefficients may be different for each iteration step j. It has been shown that by generating a subspace with this operator, the given problem can be reduced to a small projected generalized system

$$(\zeta K_{j,j} - \eta L_{j,j})s = 0,$$

where $K_{j,j}$ and $L_{j,j}$ are upper Hessenberg matrices of dimension j. This small system may be solved by the QZ algorithm in order to obtain approximate values for an eigenpair. The parameters in RKS can be chosen to obtain faster convergence to interior eigenvalues. For a comparison of RKS and Arnoldi, see [63, 62].

5 Davidson's method and new variants

The main idea behind Davidson's method is the following one. Suppose we have some subspace K of dimension k, over which the projected matrix A has a Ritz value θ_k (e.g., θ_k is the largest Ritz value) and a corresponding Ritz vector u_k. Let us assume that an orthogonal basis for K is given by the vectors v_1, v_2, ..., v_k.

Now we want to find a successful update for u_k, in order to expand our subspace. To that end we compute the *defect:* $r = Au_k - \theta_k u_k$. Then Davidson, in his original paper [18], suggests to compute $\widetilde{v}$ from $(D_A - \theta_k I)\widetilde{v} = r$, where D_A is the diagonal of the matrix A. The vector $\widetilde{v}$ is made orthogonal to the basis vectors v_1, ..., v_k, and the resulting vector is chosen as the new v_{k+1}, by which K is expanded.

It has been reported that this method can be quite successful in finding dominant eigenvalues of (strongly) diagonally dominant matrices. The matrix $(D_A - \theta_k I)^{-1}$ can be viewed as a preconditioner for the vector r. Davidson [19] suggests that his algorithm (more precisely: the Davidson-Liu variant of it) may be interpreted as a Newton-Raphson scheme, and this has been used as an argument to explain its fast convergence. It is tempting to see the preconditioner also as an approximation for $(A - \theta_k I)^{-1}$, and, indeed, this approach has been followed for the construction of more complicated preconditioners (see, e.g., [14, 47, 50]). However, note that $(A - \theta_k I)^{-1}$ would map r onto u_k, and hence it would not lead to an expansion of our search space. Clearly this is a wrong interpretation for the preconditioner. Originally, the Davidson method had been proposed for symmetric matrices, but in many of the cited publications it has been mentioned that the method can be used successfully for some unsymmetric matrices as well.

5.1 The Jacobi-Davidson iteration method

In this Section the matrix A may be unsymmetric and complex, and in order to express this we use the notation v^* for the complex conjugate of a vector (if complex), or the transpose (if real), and likewise for matrices.
For the construction of effective subspaces we observe that for a given approximate *Ritz pair* (θ, s), the residual is given by

$$r = As - \theta s.$$

Following an old and forgotten technique of Jacobi [33] (for strongly diagonally dominant matrices), it was suggested in [66] to compute a correction Δs for s in the subspace orthogonal to s, such that the residual vanishes in that subspace. That is, we want to solve

$$(I - ss^*)(A - \theta I)(I - ss^*)\Delta s = -r, \tag{5.1}$$

for $\Delta s \perp s$.
It can be shown that for $\theta = \lambda$ (an eigenvalue of A), this correction Δs leads immediately to the corresponding eigenvector $y = s + \Delta s$: $Ay = \lambda y$.

For the expansion of the subspace we solve equation (5.1), for a given Ritz value θ, and we expand the subspace with Δs. We compute a new Ritz pair with respect to the expanded subspace and we repeat the above procedure. This is the basis for the *Jacobi-Davidson Method.* The sketched procedure leads to quadratic convergence of the Ritz value to an eigenvalue if A is unsymmetric, and to cubic convergence if A is symmetric.

Of course, solving (5.1) may be an expensive affair, and for that reason we discuss what happens if this is equation is only solved approximately. This question has been addressed in [66], and has led to the following observations:

1. If we take the very crude approximation $\Delta s = -r$, then this method becomes equivalent with the Arnoldi method (and with Lanczos if A is symmetric).
2. If we approximate the projected operator by $D_A - \theta I$, and skip the condition that $\Delta s \perp s$ then we obtain Davidson's method. More recent suggestions made in [14, 47, 48, 50, 19] come down to better approximations for the inverse of $A - \theta_k I$, e.g., incomplete decompositions for this operator. However, as is well-known, this is a risky approach (see [65, 14]), since the exact inverse of this operator leads to failure of the method[2], and therefore the approximation should not be too accurate [65].

[2] Any progress in this case may be attributed to the effects of rounding errors

3. If we take full account of the restriction to the subspace orthogonal to s the we obtain the Jacobi - Davidson methods.

The algorithm for the improved Davidson method then becomes as follows (in the style of [65], in particular we have skipped indices for variables that overwrite old values in an iteration step, e.g., u instead of u_k).

1. **Start:**
   - Compute $v_1 = v/||v||$, $w_1 = Av_1$,
     set $V_1 = [v_1]$, $W_1 = [w_1]$, $H_1 = [h_{11}]$,
     $u = v_1$, $\theta = h_{11}$, compute $r = w_1 - \theta u$.
2. **Iterate:** Until convergence do:
3. **Inner Loop:** For $k = 1, ..., m$ do:
   - Solve (approximately)
     $(I - u u^*)(A - \theta I)(I - u u^*)\Delta s = -r$.
   - Orthogonalize Δs against V_k via Modified Gram-Schmidt, and expand V_k with this vector to V_{k+1}.
   - Compute $w_{k+1} := Av_{k+1}$ and expand W_k with this vector to W_{k+1}.
   - Compute $V_{k+1}^* w_{k+1}$, the last column of $H_{k+1} := V_{k+1}^* A V_{k+1}$, and $v_{k+1}^* W_k$, the last row of H_{k+1} (only if $A \neq A^*$).
   - Compute the largest eigenpair (θ, s) of H_{k+1} (with $||s|| = 1$).
   - Compute the Ritz vector $u := V_{k+1}s$, compute $\widehat{u} := Au$ $(= W_{k+1}s)$, and the associated residual vector
   - Test for convergence. Stop if satisfied.
4. **Restart:** Set $v_1 := u$ and goto 3.

Now we will discuss convenient ways for the approximate solution of

$$(I - u_k u_k^*)(A - \theta_k I)(I - u_k u_k^*)\Delta s = -r \quad \text{and} \quad \Delta s \perp u_k. \tag{5.2}$$

Since $\Delta s \perp u_k$, it follows from (5.2) that

$$(A - \theta_k I)\Delta s - \varepsilon u_k = -r \tag{5.3}$$

or

$$(A - \theta_k I)\Delta s = \varepsilon u_k - r.$$

When we have a suitable preconditioner M, for which $M^{-1} \approx (A - \theta_k I)^{-1}$, then we can compute an approximation $\widetilde{\Delta s}$ for Δs:

$$\widetilde{\Delta s} = \varepsilon M^{-1} u_k - M^{-1} r. \tag{5.4}$$

The value of ε is determined by the requirement that $\widetilde{\Delta s}$ should be orthogonal with respect to u_k:

$$\varepsilon = \frac{u_k^* M^{-1} r}{u_k^* M^{-1} u_k}. \tag{5.5}$$

Equation (5.4) leads to several interesting observations:

1. If we choose $\varepsilon = 0$ then we obtain the Davidson method (with preconditioner M). In this case $\widetilde{\Delta s}$ will not be orthogonal to u_k.

2. If we choose ε as in (5.5) then we have a Jacobi-Davidson method. Note that this method requires two operations with the preconditioning matrix per iteration.

3. If $M = A - \theta_k I$, then (5.4) reduces to

$$\Delta s = \varepsilon (A - \theta_k I)^{-1} u_k - u_k .$$

   Since Δs is made orthogonal to u_k afterwards, this choice is equivalent with $\Delta s = (A - \theta_k I)^{-1} u_k$. In this case the method is mathematically equivalent with (accelerated) shift and invert iteration (with optimal shift).

If we solve (5.2) approximately with a preconditioned iterative method, like Bi-CGSTAB or GMRES, then we do not need two preconditioning operations per iteration step (as is necessary if we do only unaccelerated preconditioning), for details see [66].

Successful implementations largely depend on how well an effective preconditioner can be identified. Note that the operator $A - \theta I$ will be indefinite in general, so that one has to be careful with incomplete decomposition techniques. The operator restricted to the subspace orthogonal to the Ritz vector corresponding to θ, however, is not indefinite, and in [67] it is shown how available preconditioners for $A - \theta I$ can be restricted to that subspace.

6 A novel extension for the Jacobi-Davidson method

In some circumstances the Jacobi-Davidson method has apparent disadvantages with respect to Arnoldi's method. For instance, in many cases we see rapid convergence to one single eigenvalue, and what to do if we want more eigenvalues? For Arnoldi this is not a big problem, since the usually slower convergence towards a particular eigenvalue goes hand in hand with simultaneous convergence towards other eigenvalues. So after a number of steps Arnoldi produces approximations for several eigenvalues.
For Jacobi-Davidson the obvious approach would be to restart with a differently selected Ritz pair, with no guarantee that this leads to a new eigenpair. Also the detection of multiple eigenvalues is a problem, but this problem is shared with the other subspace methods.

A well-known way out of this problem is to use a technique, known as *deflation*. If an eigenvector has converged, then we continue in a subspace spanned by the remaining eigenvectors. A problem is then how to re-use information obtained in a previous Jacobi-Davidson cycle. In [26] an algorithm is proposed by which several eigenpairs can be computed. The algorithm is based on the computation of a partial Schur form of A:

$$AQ_k = Q_k R_k,$$

where Q_k is an $n \times k$ orthonormal matrix, and R_k is a $k \times k$ upper triangular matrix, with $k \ll n$. Note that if (x, λ) is an eigenpair of R_k, then $(Q_k x, \lambda)$ is an eigenpair of A.

We now proceed in the following way in order to obtain this partial Schur form for eigenvalues close to a target value τ.

Step I: Given an orthonormal subspace basis $v_1, \ldots, v_i$, with matrix V_i, compute the projected matrix $M = V_i^* A V_i$. For the $i \times i$ matrix M we compute the complete Schur form $MU = US$, with $U^*U = I$, and S upper triangular. This can be done with the standard QR algorithm [30].
Then we order S such that the $|s_{i,i} - \tau|$ form a nondecreasing row for increasing i. The first few diagonal elements of S then represent the eigenapproximations closest to τ, and the first few of the correspondingly reordered columns of V_i represent the subspace of best eigenvector approximations. If memory is limited then this subset can be used for restart, that is the other columns are simply discarded. The remaining subspace is expanded according to the Jacobi-Davidson method. After convergence of this procedure we have arrived at an eigenpair (q, λ) of A: $Aq = \lambda q$. The question is how to expand this partial Schur form of dimension 1. This will be shown in step II.

Step II: Suppose we have already a partial Schur form of dimension k, and we want to expand this by a convenient new column q:

$$A\,[Q_k, q] = [Q_k, q] \begin{bmatrix} R_k & s \\ & \lambda \end{bmatrix}$$

with $Q^* q = 0$.
After some standard linear algebra manipulations it follows that

$$(I - Q_k Q_k^*)(A - \lambda I)(I - Q_k Q_k^*)q = 0,$$

which expresses that the new pair (q, λ) is an eigenpair of

$$\widetilde{A} = (I - Q_k Q_k^*)A(I - Q_k Q_k^*).$$

This pair can be computed by applying the Jacobi-Davidson algorithm (with Schur form reduction, as in step I) for $\widetilde{A}$.

Some notes are appropriate:

1. Although we see that after each converged eigenpair the explicitly

deflated matrix $\tilde{A}$ leads to more expensive computations, it is shown in [26], by numerical experiments, that the entire procedure leads to a very efficient computational process. An explanation for this is that after convergence of some eigenvectors, the matrix $\tilde{A}$ will be better conditioned, so that the correction equation in the Jacobi-Davidson step is more easily solved.

2. The correction equation may be solved by a preconditioned iterative solver, and it is shown in [26] that the same preconditioner can be used with great efficiency for different eigenpairs. Hence, it pays to construct better preconditioners.

3. In [26] a similar algorithm for generalized eigenproblems $Ax = \lambda Bx$ is proposed, based on partial QZ reduction [30].

7 Bi-Lanczos and look-ahead techniques

In the unsymmmetric Lanczos method [34] dual bases $\{r_j\}$ and $\{s_j\}$ are generated for the Krylov subspace $K^i(A; r_0)$ and its adjoint $K^i(A^*; s_0)$. The r_j are generated with a three term recurrence relation, with A:

$$\gamma_j r_{j+1} = Ar_j - \alpha_j r_j - \beta_j r_{j-1},$$

and the s_j with a similar recurrence for A^*:

$$\gamma_j s_{j+1} = A^* s_j - \alpha_j s_j - \beta_j s_{j-1}.$$

The constants γ_j, α_j, and β_j, are chosen so that $s_k^* r_i = 0$ for $k \neq i$, and $s_j^* r_j = 1$ (this requirement can not always be fulfilled). In algebraic form these recurrencies read as

$$AR_j = R_{j+1} T_{j+1,j},$$

and

$$A^* S_j = S_{j+1} T_{j+1,j}.$$

The matrix S_j is now used for the projection of the first equation:

$$S_j^* A R_j = T_{j,j}.$$

The eigenvalues of $T_{j,j}$ are taken as approximations for those of A, and the eigenvector approximations are taken as $R_j y_k$, where y_k is an eigenvector of $T_{j,j}$.

The breakdown occurs when $s_j^* r_j = 0$, and for numerical stability reasons one also wants to avoid the situation that

$$\frac{s_j^* r_j}{\|s_j\|_2 \|r_j\|_2} \approx 0.$$

See [76]:Ch6.36 for a discussion on the failures in the Lanczos process. Despite the bad reputation of the standard unsymmetric Lanczos process, software produced by Cullum and Willoughby [16] has been quite successful. In their approach the inaccurate eigenapproximations, due to numerical instabilities, were identified from comparison of results for submatrices of $T_{j,j}$. The main idea is that the starting vector has nonzero components in directions of eigenvectors corresponding to eigenvectors that we are interested in. With this starting vector the Krylov subspaces are generated. When we compare the eigenvalues of $T_{j,j}$ with those of the tridiagonal matrix of order $j-1$ that is obtained if we skip the first row and column from $T_{j,j}$, then we compare Ritz values of the current Krylov subspace with Ritz values for a subspace from which the starting vector has been removed. Since this starting vector contains essential information of the desired eigenvector directions, no Ritz value can have converged *unless* all information for the corresponding eigenvector has entered through rounding errors. This kind of heuristics help to identify the so-called *spurious* eigenvalues.

In the early eighties Taylor [70] and Parlett et al [59], suggested to expand the Krylov subspace with sets of basis vectors that were block-dual. The idea is that certain dimensions, namely those for which breakdown occurs, can not be used for projection, and one has to postpone the inspection of the projected system until a block R_i of sufficient large dimension was discovered for which $S_i^* R_i$ is not too small. This process of delaying the actual projection step was called *look-ahead.* In the period '85–'95 the look-ahead technique was generally accepted as a necessary element for the Lanczos method. Gutknecht gave a detailed theoretic basis for the look-ahead mechanism [32]. Freund and Nachtigal incorporated this technique in their QMR method [27], and it was shown later that their codes could also be used for effective solution of large sparse unsymmetric eigenproblems, see for instance [24] for an example of this.

Brezinski and co-workers, see e.g., [9], considered the breakdown problem from a polynomial point of view. The vectors in the Krylov subspaces can be seen as the results of matrix polynomials acting on the starting vectors, and likewise, the bi-orthogonality relations can be viewed as orthogonality relations for polynomials, with respect to a finite innerproduct (in which the weights are defined by the starting vectors). They showed that for some degrees the polynomials do not exist, and they also showed how higher degree orthogonal polynomials could be defined, by temporarily using polynomial factors of degree larger than 1 in the recursion formulas. This is equivalent with the block-wise expansion for the dual bases in the look-ahead approach.

8 Related issues

8.1 Approximations for interior eigenvalues

It is well-known that the subspace methods lead to eigenvalue approximations that tend to converge towards exterior eigenvalues. It may happen that an eigenvalue approximation is close to an interior eigenvalue of A, but in the next iteration step this needs not to be the case. In that situation we say that the eigenvalue approximation was on its way towards some exterior eigenvalue.

It is easy to reverse the direction of convergence of the eigenvalue approximations to the interior eigenvalues closest to the origin, by working with A^{-1}, but this is expensive. It is also possible to obtain eigenvalue approximations that converge (slowly) to the eigenvalues of A closest to the origin, from the subspaces generated with A. We will explain this for the Arnoldi process.

The Arnoldi process leads to

$$AV_m = V_{m+1}H_{m+1,m},$$

with $H_{m+1,m}$ a upper Hessenberg matrix with $m+1$ rows and m columns. The upper m by m part of this matrix will be denoted as H_m.

With $H^*_{m+1,m}V^*_{m+1} = V^*_mA^*$ it follows that:

$$V^*_mA^*AV_m = H^*_{m+1,m}V^*_{m+1}V_{m+1}H_{m+1,m}$$

or

$$V^*_mA^*AV_m = H^*_{m+1,m}H_{m+1,m} \equiv M^2_m.$$

From the equation

$$M^{-1}_mV^*_mA^*AV_mM^{-1}_m = I$$

it follows that the columns of $AV_mM^{-1}_m$ form a set of m orthonormal vectors.

Note that these vectors form an orthonormal basis for $AK_m(A;v_1)$. The matrix A^{-1} maps $AK_m(A;v_1)$ onto $K_m(A;v_1)$, and we may try to find a suitable form for the projection of the operator A^{-1}, that is the orthogonal restriction of A^{-1} with respect to $AK_m(A;v_1)$. To that end we use the orthonormal basis, and find that the projection can be given as:

$$M^{-1}_m(AV_m)^*A^{-1}AV_mM^{-1}_m = M^{-1}_m(AV_m)^*V_mM^{-1}_m = M^{-1}_mH^*_mM^{-1}_m.$$

So approximations for the eigenvalues of A^{-1} follow from

$$M^{-1}_mH^*_mM^{-1}_it = \theta t,$$

or

$$M^{-2}_mH^*_ms = \theta s\,.$$

The vector s can be represented in terms of the basis vectors for $AK_m(A;v_1)$, and can then be viewed as an approximate eigenvector of A^{-1}:

$$y = AV_ms.$$

Since we are actually looking for eigenvalue approximations for A, we may wish to solve

$$H^{-*}_mM^2_ms = \theta^{-1}s.$$

The idea behind this approach is the following. The Arnoldi projection process for A leads to approximations that tend to converge to exterior eigenvalues of A. Likewise, we might hope that the exterior eigenvalues of the projected A^{-1} converge (slowly) to the exterior eigenvalues of A^{-1}. Note that these happen to be the eigenvalues of A closest to the origin. Note that we then can force (slow) convergence to eigenvalues close to any point in the spectrum of A.

In [55] these eigenvalue approximations, in connection with the related Lanczos process, were called Harmonic Ritz values, and some nice relations for Harmonic Ritz values for symmetric indefinite matrices are given in that paper. Harmonic Ritz values had already been studied from a different viewpoint by other authors. Freund [28] has studied them as the zeros of the GMRES and MINRES iteration polynomials. Morgan [47] had observed that the Harmonic Ritz values and vectors are very suitable for restarting purposes if one wants to compute interior eigenvalues with subspaces of restricted dimension. In [66, 26] the Harmonic Ritz values are considered in connection with the Jacobi-Davidson process.

8.2 Sensitivity of Eigenproblems

Eigenvalues are used as a source of information on stability or convergence problems, and the question arises how valid the information of the mere values is. Many authors have studied the problem of sensitivity of the eigenvalues with respect to perturbations, see for instance [76, 69, 12]. These studies are usually related to perturbations caused by rounding errors, and not so much by the relevance of the eigenvalues due to the particular representation of a given problem, for instance the choice of basis.

Around 1987 Trefethen [72] started to emphasize this aspect of eigencomputations, and he propagated the idea of inspecting the pseudospectrum of a matrix as a relatively simple means for getting an idea of the significance of a particular part of the spectrum, without getting involved in complicated matters such as angles between eigenvectors or eigenspaces.
The definition of pseudospectrum $\Lambda_\varepsilon(A)$ for a matrix A is directly related to perturbations:

$$\Lambda_\varepsilon(A) \equiv \{z \in \mathbb{C} : z \in \Lambda(A+E) \text{ for some } E \text{ with } ||E|| \leq \varepsilon\}.$$

The pseudospectrum is usually graphically shown as a set of level curves for various values of ε. The level curves, or contour integrals, are more apparent from the original definition for the ε-pseudospectrum, in terms of the norm of the resolvent $(zI-A)^{-1}$:

$$\Lambda_\varepsilon(A) \equiv \{z \in \mathbb{C} : ||(zI-A)^{-1}|| \geq \varepsilon^{-1}\},$$

with the convention $||(zI-A)^{-1}|| = \infty$ for $z \in \Lambda(A)$.

For symmetric matrices, the pseudospectrum of A is a collection of discs around the eigenvalues of A (note that the perturbation E needs not be symmetric). For unsymmetric matrices the pseudospectrum can be any collection of closed curves, containing the set of eigenvalues of A. These level curves may give information that is hidden by the information provided by the eigenvalues themselves. For instance, when studying stability of integration methods for systems of ODE's, or in bifurcation problems, the eigenvalues may be in a proper region, for instance, in the left-half plane, while the level curves even for rather small values of ε may intersect with the right-half plane. In such cases it may be time to ask further questions about the problem. On the other hand the pseudospectrum may not tell the full story. For instance, the sensitivity problems may be due to a single pair of ill-conditioned eigenvectors for which the more global level curves are too pessimistic. It may be the case that it is not realistic to assume equal perturbations for all matrix entries, but nevertheless the pseudospectrum points the attention to critical places in the spectrum. A nice introduction to the relevance of pseudospectra is given in [71], where for a number of matrices pseudospectra are actually computed and discussed.

Due to the nature of pseudospectra, this useful tool is often restricted to matrices of moderate size. This poses another problem: if we study the pseudospectrum for a discretized PDE with rather course meshsize (since we have to compute the smallest singular value of $zI - A$, for various values of z), what does this tell us on the effects to be expected for finer meshsizes?

We have carried out some experiments, not reported here, with a simple discretized convection-diffusion equation:

$$-u_{xx} - \beta u_x = f.$$

If we discretize this equation over the interval $[0, 1]$, with central difference approximation, then for large values of β and relatively large meshsize, we see that the spectrum is on a line parallel to the imaginary axis. If we decrease the meshsize, then the spectrum gradually shrinks to one single point on the real axis (a highly defective case), and then spreads along the real axis. It turns out that amazingly soon, after the spectrum starts to spread along the real axis, the pseudospectrum gives quite accurate information on the sensitivity of the given problem, information that is not essentially different from the information that we would have obtained for very fine meshsizes. On the other hand, the information for rather crude meshsizes can be quite misleading. This illustrates that pseudospectra do not necessarily represent more reliable information than the standard spectrum, when the information is obtained for lower-dimensional problems than the problems that are used in the actual large-scale scientific computations.

More recently, tools have become available for computing the pseudospectrum of large sparse matrices. Carpraux et al [10] propose an algorithm for computing the smallest singular value of $zI - A$, that is based on Davidson's method with ILU preconditioning. Lui [41] (see also [8]) suggests to use Lanczos' method in combination with continuation techniques. This is a plausible approach, since we need to do the computation for many values of z, well-distributed over the region of interest, in order to obtain a complete picture of the pseudospectrum. We know that currently such tools are being used for the analysis of instability problems of large sets of ODE's, related to climate modelling, but results have not yet been published.

8.3 Homotopy methods

The subspace methods that we have discussed before, are often applied in combination with shift and invert operations. That means that if one wants to have eigenvalues close to a value σ, then the methods are applied to the inverse $(A - \sigma I)^{-1}$ of the shifted matrix. As we have seen, the Jacobi-Davidson method can be interpreted as an inexact shift-invert method, since the invert step is usually approximated by a few steps of some convenient preconditioned inner iteration method.

Related to these inexact shift-invert approaches is the homotopy approach, that has received attention in the past five years. The idea is to compute some of the eigenvalues of a perturbed matrix $A + E$, when eigenvalues of A are known, or can be relatively easily computed. In order to this we use the homotopy $H(t) = A + tE$, $0 \le t \le 1$. If eigenpairs of $H(t_0)$ are known, then they are used as approximations for those of $H(t_0 + \Delta t)$. These approximations are improved by a convenient subspace iteration, for instance in [43] Rayleigh quotient iterations are used for symmetric A and E (see references in [43] for earlier work on homotopy for eigenproblems). For the Rayleigh quotient iteration one needs to solve systems like $(H(t_0 + \Delta t) - \lambda I)y = x$, where (λ, x) represents the current approximation for an eigenpair of $H(t_0 + \Delta t)$. In the context of large sparse matrices, it may be undesirable to do this with a direct solver, and in [43] the system is solved with SYMMLQ [56]. Of course, one could restrict oneself to only a few steps with SYMMLQ, and then try to accelerate the inexact Rayleigh quotient steps, as is done in the Jacobi-Davidson method. This indicates relations between these different approaches, but as far as we know, these relations have not yet been explored. There seem to be more relations. In [43] it is observed that SYMMLQ may find difficulty in converging for the nearly singular system $H(t_0 + \Delta t) - \lambda I)y = x$, and it is suggested to improve the situation by applying the Rayleigh quotient iteration to the approximately deflated matrix $H(t_0 + \Delta t) + xx^T$, where the term approximately deflated is used to indicate that x is only an approximation to the desired eigenvector. Note that similar deflation procedures are incorporated in the Jacobi-Davidson process.

The whole procedure is repeated for successive increments Δt, until the final value $t = 1$ is reached. In [43] an elegant approach is followed for the selection of the step size Δt.

The homotopy approach lends itself quite naturally for situations where the matrix A varies in time, or where it varies as a linearization of a nonlinear operator, as in bifurcation problems. Other situations that are obvious are for instance the Schrödinger eigenvalue problem [43]:

$$-\Delta u + fu = \lambda u$$

in the unit square in two space dimensions with homogeneous Dirichlet boundary condition. With the usual finite difference approximations on a uniform grid, this leads to the discrete Laplacian for $-\Delta u$, of which we know the eigensystem.

In [78] the homotopy approach is used for symmetric generalized eigenproblems, very much along the same lines as sketched above. The application for real unsymmetric eigenproblems is considered in [44].

8.4 Implementation aspects

In the past ten years sophisticated software has been produced for the computation of eigenvalues and eigenvectors. We have mentioned already the Implicitly Restarted Arnoldi method, for which Lehoucq et al have written the package ARPACK [40]. We also mention the code SRRIT [6] for the identification of a dominant invariant subspace of a nonsymmetric matrix. We have discussed the approaches by Cullum and Willougby, and by Freund and Nachtigal, but we also want to mention a recent approach to design robust software for the unsymmetric Lanczos process, the so-called ABLE package [4].

In [39] an overview of relevant software is given as well as numerical results for representative test problems. In [3] a project is announced for the production of software for relevant algorithm for large eigenproblems.

Most of the iterative methods reduce the given large problem to a much smaller problem. The smaller problem can then be handled by standard software for dense or banded matrices. Excellent software is available in LAPACK [1], the modern successor of the famous packages EISPACK and LINPACK.

Parallelism in the subspace methods is, except for the (approximate) shift-and-invert steps, usually no big problem. For some modern computers the required innerproducts may form a bottleneck with respect to scalable performance. For an overview of techniques to improve parallel behavior of algorithms for eigenproblems, see [21].

9 Some open problems

We have highlighted some of the progress that has been made in the past period. A novice in this area might easily have got the impression that most of the relevant problems, associated with the computation of eigenvalues and eigenvectors have been solved. Fortunately, this is not true. Like in most other areas of science there are many challenging problems that wait for solution and for further analysis.

We mention only a few of these open problems. Despite all our knowledge and experience with the Lanczos process, its simple basic three-term recurrence still contains some mysteries, when computing in finite precision arithmetic. We know that duplicate eigenvalues enter the process, due to rounding errors, and we see that they enter at more or less regular intervals during the iteration process, but this phenomenon is still not well understood.
For unsymmetric problems the picture is still rather obscure. Although the two-sided Lanczos process often leads to good approximations for some eigenvalues, its convergence behavior is not well understood. Even the convergence behavior of the Arnoldi method is not well understood, notwithstanding the fact that it can be viewed as an accelerated power iteration. In a recent paper, Cullum [15] shows that the convergence behavior of the two-sided Lanczos process, for eigenvalues, can be mimicked by Arnoldi's process and vice versa. For the unsymmetric eigenproblem we also face the problem that there are no efficient and reliable algorithms for computing the eigenvalues and eigenvectors of unsymmetric tridiagonal matrices, in contrast with the symmetric situation.

Although serious attempts have been undertaken for the computation of the Kronecker canonical form, by for instance Kågström and Van Dooren, this still needs much further research. Also the computation of invariant subspaces of highly nonnormal matrices is still in its infancy, notwithstanding useful contributions by, for instance, Chaitin-Chatelin et al [11, 7] and Lee [36]. It is also necessary that we get efficient tools for checking the condition of (partial) eigensystems, angles between invariant subspaces, etcetera. There is a need for efficient algorithms for the computation of a (partial) SVD for very large sparse matrices.

Several special types of eigenproblems require special algorithms, for instance large sparse polynomial eigenproblems, of the form as has been mentioned in our Introduction. Such problems occur in acoustics and in structural engineering.

For very large sparse matrices we need efficient approximate inverses to replace shift-and-invert operations. This leads to the problem of approximating the inverse of (highly) indefinite matrices, and these inverses should be effective in the eigendirections close to the one associated with the shift. Currently available approximation techniques are not very good in this respect, most often they do not work at all for interior eigenvalues.

Recently, it has been shown by Lui [42] that domain decomposition techniques can be used for the computation of eigenvalues of partial differential operators. This is certainly a promising direction in view of parallel computation.

Some problems lead to eigenproblems with constraints, for instance Stokes problems, but also semi-indefinite optimization problems. For general cases, the required tools are missing at the moment.

Finally we mention the well-known QR iteration for eigenproblems, which still needs attention when it comes to parallel implementation.

We hope that the reader is convinced that there are still many open problems in this exciting field and these problems involve interesting work to be done by theorists as well as more practical oriented computational scientists.

Acknowledgements

We would like to thank Rich Lehoucq, Karl Meerbergen, Nick Trefethen and Gerard Sleijpen, for specific comments that helped to improve our presentation. The work of Gene H. Golub was in part supported by the National Science Foundation, Grant Number DMS-9403899.
The work of Henk van der Vorst was in part supported by the Dutch Research Organization NWO, MPR cluster project 95MPR04.

References

1. E. Anderson, Z. Bai, C. Bischof, J. Demmel, J. Dongarra, J. DuCroz, A. Greenbaum, S. Hammarling, A. McKenney, S. Ostrouchov, and D. So ren sen. *LAPACK User's Guide*. SIAM, Philadelphia, PA, 1992.

2. W. E. Arnoldi. The principle of minimized iteration in the solution of the matrix eigenproblem. *Quart. Appl. Math.*, 9:17–29, 1951.

3. Z. Bai, D. Day, J. Demmel, J. Dongarra, M. Gu, A. Ruhe, and H. van der Vorst. Templates for linear algebra problems. In J. van Leeuwen, editor, *Computer Science Today*. Springer Verlag, Berlin, 1995.

4. Z. Bai, D. Day, and Q. Ye. ABLE: An adaptive block Lanczos method for non-Hermitian eigenvalue problems. Technical Report Research Report 95-04, University of Kentucky, Lexington, KY, 1995.

5. Z. Bai, D. Hu, and L. Reichel. A Newton basis GMRES implementation. *IMA J. Numer. Anal.*, 14:563–581, 1991.

6. Z. Bai and G.W. Stewart. SRRIT - A FORTRAN subroutine to calculate the dominant invariant subspace of a nonsymmetric matrix. Technical Report 2908, Department of Computer Science, University of Maryland, 1992.

7. T. Braconnier, F. Chatelin, and V. Frayssé. The influence of large nonnormality on the quality of convergence of iterative methods in linear algebra. Technical Report TR-PA-94/07, CERFACS, Toulouse, 1994.

8. T. Braconnier and N. J. Higham. Computing the field of values and pseudospectra using the Lanczos method with continuation. Technical Report 279, Manchester centre for computational mathematics, Manchester, UK, 1995.

9. C. Brezinski and M. Redivo-Zaglia. Treatment of near breakdown in the CGS algorithm. *Numerical Algorithms*, 1994.
10. J. F. Carpraux, J. Erhel, and M. Sadkane. Spectral portrait for non-Hermitian large sparse matrices. *Computing*, 53:301–310, 1994.
11. F. Chaitin-Chatelin. Is nonnormality a serious difficulty? Technical Report TR-PA-94/18, CERFACS, Toulouse, 1994.
12. F. Chatelin. *Valeurs propres de matrices*. Collection mathématiques appliquées pour la maîtrise. Masson, 1988.
13. P. Concus and G. H. Golub. A generalized Conjugate Gradient method for nonsymmetric systems of linear equations. Technical Report STAN-CS-76-535, Stanford University, Stanford, CA, 1976.
14. M. Crouzeix, B. Philippe, and M. Sadkane. The Davidson method. *SIAM J. Sci. Comp.*, 15:62–76, 1994.
15. J. Cullum. Arnoldi versus nonsymmetric Lanczos algorithms for solving matrix eigenvalue problems. *BIT*, 36(3):470–493, 1996.
16. J. Cullum and R.A. Willoughby. A practical procedure for computing eigenvalues of large sparse nonsymmetric matrices. In J. Cullum and R.A. Willoughby, editors, *Large Scale Eigenvalue Problems*, pages 193–240, Amsterdam, 1986. North-Holland.
17. J. J. M. Cuppen. A divide and conquer method for the symmetric eigenproblem. *Numerische Mathematik*, 36:177–195, 1981.
18. E.R. Davidson. The iterative calculation of a few of the lowest eigenvalues and corresponding eigenvectors of large real symmetric matrices. *J. Comp. Phys.*, 17:87–94, 1975.
19. E.R. Davidson. Monster matrices: their eigenvalues and eigenvectors. *Computers in Physics*, 7:519–522, 1993.
20. E. De Sturler and H.A. van der Vorst. Reducing the effect of global communication in GMRES(m) and CG on parallel distributed memory computers. *J. Appl. Num. Math.*, 18:441–459, 1995.
21. J. Demmel, M. Heath, and H. van der Vorst. Parallel numerical linear algebra. In *Acta Numerica 1993*. Cambridge University Press, Cambridge, 1993.
22. J. J. Dongarra and D. C. Sorensen. A fully parallel algorithm for the symmetric eigenvalue problem. *SIAM J. Scientific and Statistical Computing*, 8:139–154, 1987.
23. A. A. Dubrulle and G. H. Golub. A Multishift QR iteration without computation of the shifts. *Numerical Algorithms*, 7:173–181, 1994.
24. P. Feldmann and R.W. Freund. A linear circuit analysis by Padé approximation via the Lanczos process. *IEEE Trans. Computer-Aided Design*, 14, 1995.
25. D. A. Flanders and G. Shortley. Numerical determination of fundamental modes. *J. Appl. Phys.*, 21:1322–1328, 1950.
26. D.R. Fokkema, G.L.G. Sleijpen, and H.A. van der Vorst. Jacobi-Davidson style QR and QZ algorithms for the partial reduction of matrix pencils. Accepted for publication in *SIAM J. Sci. Comp.*
27. R. W. Freund and N. M. Nachtigal. QMR: a quasi-minimal residual method for non-Hermitian linear systems. *Num. Math.*, 60:315–339, 1991.
28. R.W. Freund. Quasi-kernel polynomials and their use in non-Hermitian matrix iterations. *J. Comp. Appl. Math.*, 43:135–158, 1992.
29. G. H. Golub and R. Underwood. The block Lanczos method for computing eigenvalues. In J.R. Rice, editor, *Mathematical Software III*, pages 361–377. Academic Press, New York, 1977.
30. G. H. Golub and C. F. van Loan. *Matrix Computations*. The Johns Hopkins University Press, Baltimore, 1989.

31. E. J. Grimme, D. C. Sorensen, and P. Van Dooren. Model reduction of state space systems via an Implicitly Restarted Lanczos method. Technical Report Report TR94-21, Dept. of Comp. and Appl. Math., Rice University, Houston, TX, 1994.

32. M. H. Gutknecht. A completed theory of the unsymmetric Lanczos process and related algorithms, Part I. *SIAM J. Matrix Anal. Appl.*, 13:594–639, 1992.

33. C.G.J. Jacobi. Ueber ein leichtes Verfahren, die in der Theorie der Säcularstörungen vorkommenden Gleichungen numerisch aufzulösen. *Journal für die reine und angewandte Mathematik*, pages 51–94, 1846.

34. C. Lanczos. An iteration method for the solution of the eigenvalue problem of linear differential and integral operators. *J. Res. Natl. Bur. Stand*, 45:225–280, 1950.

35. C. Lanczos. *Applied Analysis*. Prentice Hall, Englewood Cliffs, NJ, 1956.

36. S. L. Lee. A practical upper bound for departure from normality. *SIAM J. Matrix Anal. Appl.*, 16:462–468, 1995.

37. R. B. Lehoucq. *Analysis and Implementation of an Implicitly Restarted Iteration*. PhD thesis, Rice University, Houston, TX, 1995.

38. R. B. Lehoucq. Restarting an Arnoldi reduction. Technical Report Preprint MCS-P591-0496, Math. and Comp. Science Div., Argonne National Laboratory, Argonne, IL 60439, 1996.

39. R. B. Lehoucq and J. A. Scott. An evaluation of software for computing eigenvalues of sparse nonsymmetric matrices. Technical Report Preprint MCS-P547-1195, Math. and Comp. Science Div., Argonne National Laboratory, Argonne, IL 60439, 1996.

40. R.B. Lehoucq, D.C. Sorensen, P. Vu, and C. Yang. ARPACK: An implementation of the Implicitly Re-started Arnoldi Iteration that computes some of the eigenvalues and eigenvectors of a large sparse matrix. Available from `netlib@ornl.gov` under the directory scalapack, 1996.

41. S. H. Lui. Computation of pseudospectra by continuation. *SIAM J. Scient. Comput.*, 1996.

42. S. H. Lui. Domain decomposition methods for eigenvalue problems. Technical report, Hong Kong University of Science and Technology, Hong Kong, 1996.

43. S. H. Lui and G. H. Golub. Homotopy method for the numerical solution of the eigenvalue problem of self-adjoint partial differential operators. *Numerical Algorithms*, 10:363–378, 1995.

44. S. H. Lui, H. B. Keller, and W. C. Kwok. Homotopy method for the large sparse real nonsymmetric eigenvalue problem. Technical Report, Department of Mathematics, Hong Kong University of Science and Technology, Kowloon, Hong Kong, 1996.

45. K. Meerbergen. *Robust methods for the calculation of rightmost eigenvalues of non-symmetric eigenvalue problems*. PhD thesis, Katholieke Universiteit Leuven, Leuven, Belgium, 1996.

46. K. Meerbergen and D. Roose. Matrix transformations for computing rightmost eigenvalues of real nonsymmetric matrices. *IMA J. Numer. Anal.*, 16:297–346, 1996.

47. R.B. Morgan. Computing interior eigenvalues of large matrices. *Lin. Alg. and its Appl.*, 154–156:289–309, 1991.

48. R.B. Morgan. Generalizations of Davidson's method for computing eigenvalues of large nonsymmetric matrices. *J. Comp. Phys.*, 101:287–291, 1992.

49. R.B. Morgan and D.S. Scott. Generalizations of Davidson's method for computing eigenvalues of sparse symmetric matrices. *SIAM J. Sci. Stat. Comput.*, 7(3):817–825, 1986.

50. R.B. Morgan and D.S. Scott. Preconditioning the Lanczos algorithm for sparse symmetric eigenvalue problems. *SIAM J. Sci. Comput.*, 14:585–593, 1993.

51. J. Olsen, P. Jørgensen, and J. Simons. Passing the one-billion limit in full configuration-interaction (FCI) calculations. *Chemical Physics Letters*, 169:463–472, 1990.

52. A.M. Ostrowski. On the convergence of the Rayleigh quotient iteration for the computation of characteristic roots and vectors. V. *Arch. Rational. Mesh. Anal.*, 3:472–481, 1959.

53. C. C. Paige. Computational variants of the Lanczos method for the eigenproblem. *J. Inst. Math. Appl.*, 10:373–381, 1972.

54. C. C. Paige. Error analysis of the Lanczos algorithm for tridiagonalizing a symmetric matrix. *J. Inst. Math. Appl.*, 18:341–349, 1976.

55. C. C. Paige, B. N. Parlett, and H. A. van der Vorst. Approximate solutions and eigenvalue bounds from Krylov subspaces. *Num. Lin. Alg with Appl.*, 2(2):115–134, 1995.

56. C. C. Paige and M. A. Saunders. Solution of sparse indefinite systems of linear equations. *SIAM J. Numer. Anal.*, 12:617–629, 1975.

57. B. N. Parlett. *The Symmetric Eigenvalue Problem*. Prentice-Hall, Englewood Cliffs, N.J., 1980.

58. B. N. Parlett. Misconvergence in the Lanczos algorithm. *Reliable Numerical Computation*, Editors: M.G. Cox and S. Hammarling, 7–24, Oxford University Press, Oxford, 1990.

59. B. N. Parlett, D. R. Taylor, and Z. A. Liu. A look-ahead Lanczos algorithm for unsymmetric matrices. *Math. Comp.*, 44:105–124, 1985.

60. B.N. Parlett. The Rayleigh Quotient iteration and some generalizations. *Math. Comp.*, 28:679–693, 1974.

61. S. G. Petiton. Parallel subspace method for non-Hermitian eigenproblems on the Connection Machine (CM2). *Appl. Num. Math.*, 10:19–36, 1992.

62. A. Ruhe. The rational Krylov algorithm for nonsymmetric eigenvalue problems. iii: Complex shifts for real matrices. *BIT*, 34:165–176, 1994.

63. A. Ruhe. Rational Krylov algorithms for nonsymmetric eigenvalue problems. ii. matrix pairs. *Lin. Alg. and its Appl.*, 197, 198:283–295, 1994.

64. A. Ruhe. Rational Krylov, a practical algorithm for large sparse nonsymmetric matrix pencils. Technical Report UCB/CSD-95-871, Computer Science Division, University of California, Berkeley, CA, 1995.

65. Y. Saad. *Numerical methods for large eigenvalue problems.* Manchester University Press, Manchester, UK, 1992.

66. G. L. G. Sleijpen and H.A. Van der Vorst. A Jacobi-Davidson iteration method for linear eigenvalue problems. *SIAM J. Matrix Anal. Appl.*, 17:401–425, 1996.

67. G.L.G. Sleijpen, J.G.L. Booten, D.R. Fokkema, and H.A. van der Vorst. Jacobi-Davidson type methods for generalized eigenproblems and polynomial eigenproblems. *BIT*, 36(3):595–633, 1996.

68. D. C. Sorensen. Implicit application of polynomial filters in a k-step Arnoldi method. *SIAM J. Mat. Anal. Appl.*, 13(1):357–385, 1992.

69. G. W. Stewart and Ji-Guang Sun. *Matrix Perturbation Theory.* Academic Press, San Diego, CA, 1990.

70. D.R. Taylor. *Analysis of the look ahead Lanczos algorithm.* PhD thesis, University of California, Berkeley, CA, 1982.

71. L. N. Trefethen. Pseudospectra of matrices. In D. F. Griffiths and G. A. Watson, editors, *Numerical Analysis 1991*, pages 234–266. Longman, 1992.

72. L. N. Trefethen and M. R. Trummer. An instability phenomenon in spectral methods. *SIAM J. Numer. Anal.*, 24:1008–1023, 1987.

73. A. van der Sluis and H. A. van der Vorst. The convergence behavior of Ritz values in the presence of close eigenvalues. *Lin. Alg. and its Appl.*, 88/89:651–694, 1987.

74. H. A. van der Vorst. A generalized Lanczos scheme. *Math. Comp.*, 39(160):559–561, 1982.

75. H. F. Walker. Implementation of the GMRES method using Householder transformations. *SIAM J. Sci. Stat. Comp.*, 9:152–163, 1988.

76. J. H. Wilkinson. *The Algebraic Eigenvalue Problem*. Clarendon Press, Oxford, 1965.

77. J.H. Wilkinson and C. Reinsch, editors. *Handbook for Automatic Computation, Vol 2, Linear Algebra*. Springer Verlag, Heidelberg - Berlin - New York, 1971.

78. T. Zhang, K. H. Law, and G. H. Golub. On the homotopy method for symmetric modified generalized eigenvalue problems. Technical Report, Stanford University, Stanford, CA, 1996.

Geometric Integration

J.M. Sanz-Serna

Departamento de Matemática Aplicada y Computación, Universidad de Valladolid, Spain

1 Introduction

"The last decade has seen such a deluge of papers on the numerical solution of the initial value problems for ordinary differential equations that it is quite impracticable to list, far less to summarize, all the contributions." These are the opening words of J. D. Lambert's paper [54] in the State of the Art 1976 Conference Proceedings. In spite of such a disclaimer, the paper succeeds in presenting, in less than fifty large-print pages, a unified view of all that, at the time of its writing, was known on numerical initial-value problems (IVP) for ordinary differential equations (ODEs). Ten years later, the subject was just too big and although the State of the Art 1986 Conference [50] featured three ODE speakers (J.D. Lambert, A.R. Curtis and G. Wanner), their combined contributions are very far from surveying all the numerical ODE field. After another decade of growth in the subject, the present paper must limit itself to the presentation of an individual topic, geometric integration, without any ambition of being exhaustive. We have tried to convey to the reader a feeling for what we see as a new way of doing numerical ODEs. We have also tried to direct him or her to the relevant literature, but we have not considered it possible to present detailed mathematical arguments.

Before defining geometric integration let us place ourselves in the classical, 1976 State of the Art Conference point of view. Considered then were two situations, "general" and "stiff". For general problems, Lambert [54] perceived a "consensus of opinion on what are the 'best' methods" and reported that "some highly tuned and thoroughly tested packages" were available. The stiff field was not nearly as mature with "new methods ... continually being proposed", and, as a consequence, we were not "yet at the stage of being able firmly to recommend 'best' packages."

Given an initial value problem

$$\dot{y} = f(y), \quad y(0) = \alpha \in \mathcal{R}^D \tag{1.1}$$

(a dot represents differentiation with respect to time t), in 1976 we had

- *A well defined goal.* This was to find as cheaply as possible within the desired accuracy the vectors $y(t_i)$ at some prescribed output locations.
- *A tool to achieve the goal.* This was a package, or rather a couple of them, general/stiff, in which to plug the subroutine evaluating f. The general package dealt with all general f's and the stiff package dealt with all stiff f's.
- *A theoretical framework to design/understand the tool.* This included two main groups of ideas (i) consistency, local error, error constants, (ii) stability, error propagation, stability region.

The second of the items above deserves some comments. Surely, twenty years ago people were aware of the limitations of solving everything with two packages. Lambert mentioned "special classes of IVPs ... such as problems with periodic solutions" and regretted that they "received relatively little attention." (Was "relatively" a slight understatement when only one paper on special classes was quoted in [54]?) While Lambert aptly concluded that "where there is structure, we ought to be able to use it in the numerical method," for the numerical analyst, the field of special problems was a large, completely uncharted land.

The classical approach outlined above has clearly made an outstanding contribution to the solution of scientific problems. But, as all human things, it has limitations. These will be illustrated in two examples.

In July 1992, the pretigious journal Science featured in its Research News section [52] the announcement that "From Mercury to Pluto, Chaos pervades the solar system", as borne out by a numerical integration by Sussman and Wisdom [102] of the planetary equations of motion over a time span of nearly 100 million years. Does this widely publicized numerical ODE integration fall within the classical paradigm?

- The aim was not to compute accurately the state of the solar system after a long time; it was rather a matter of deciding whether the motion is regular or chaotic. When comparing results of different integration techniques in this sort of study, it was observed [115] that "the plots are remarkably similar", a clear indication that the authors had no illusions of having achieved any accuracy in a conventional sense, i.e. of having achieved small global errors.
- The tool was a special purpose splitting method, tailored to the problem at hand.
- The tool was designed through considerations that went a long way beyond the classical consistency/stability approach.

The second example concerns the simulation of the dynamics of biomolecules [10], a matter of integrating Newton's second law for the motion of the atoms in the molecule (the number of atoms could be as high as 10,000 or 100,000 and there are six differential equations per atom).

- The aim is to obtain information on things like average energies, conformational distributions, large-scale protein bending, etc. There is no hope of computing the solution with any accuracy: perturbations to the system being integrated typically double in size every picosecond and current simulations may cover time intervals of 1000 picoseconds. Furthermore the initial velocities are unknown and ***assigned randomly.*** Also the expressions currently used for the interatomic forces are only approximations that involve fitting parameters semiemperically.
- The current method of choice, leapfrog (known in molecular dynamics as Verlet [3]), is far away from the notion of package of the classical paradigm. Even if conventional packages could be applied to such a large problem, it is unlikely that they would outperform the simple leapfrog method.
- The success of the leapfrog method cannot be explained via the traditional ideas of stability and consistency. The scheme is only second order accurate. Being only marginally stable for linear forces and small time-steps, one would fear that the smallest nonlinearity could make it unstable (see [87,95] for a discussion).

2 Mathematical preliminaries

In this section we present the notation that will be used throughout the paper. We also include some background material that will later simplify the discussions.

2.1 Vector fields, flows and Lie operators

Each system of differential equations $\dot{y} = f(y)$, $y \in \mathcal{R}^D$ is defined by a *vector field* f [2,79]. With each system/vector field we associate its *flow* $\phi_{t,f}$ [2,6,79]. For each value of the real parameter t, $\phi_{t,f}$ maps $\mathcal{R}^D$ in $\mathcal{R}^D$ in such a way that $\phi_{t,f}(\alpha)$ is the value at time t of the solution of the system with initial value α at time 0. Thus, for fixed α and varying t, $\phi_{t,f}(\alpha)$ is the solution of the initial value problem Equation 1.1.

Also associated with the field f is the *Lie operator* L_f [2,6,79]. This maps each real-valued function F defined in $\mathcal{R}^D$ into the real-valued function $L_f \cdot F$ such that, for $y \in \mathcal{R}^D$,

$$(L_f \cdot F)(y) = f_1(y)\frac{\partial F}{\partial y_1}(y) + \cdots + f_D(y)\frac{\partial F}{\partial y_D}(y).$$

(Subscripts denote components.) Clearly, for each $\alpha \in \mathcal{R}^D$ and each real t,

$$(L_f \cdot F)(\phi_{t,f}(\alpha)) = \frac{d}{dt}F(\phi_{t,f}(\alpha)). \tag{2.1}$$

By recursively applying Equation 2.1 with $L_f^{k-1} \cdot F$, $k = 2, 3, \ldots$, in lieu of F, we conclude

$$(L_f^k \cdot F)(\phi_{t,f}(\alpha)) = \frac{d^k}{dt^k}F(\phi_{t,f}(\alpha)),$$

and evaluating these derivatives at time $t = 0$ we arrive at the following formula [79]

$$F(\phi_{t,f}(\alpha)) = \left(\left(\sum_{k=0}^{\infty} \frac{t^k L_f^k}{k!}\right) \cdot F\right)(\alpha) = \left(\exp(tL_f) \cdot F\right)(\alpha), \tag{2.2}$$

for the Taylor expansion of F along the solution of the IVP Equation 1.1. Particularization of Equation 2.2 to the case where F is the mapping that associates with each y its ith component y_i, $i = 1, \ldots, D$, provides the Taylor expansion of each of the components of the solution $\phi_{t,f}(\alpha)$.

Given two Lie operators L_f and L_g, their *commutator* $[L_f, L_g] = L_f L_g - L_g L_f$ turns out to be the Lie operator associated with a third vector field called the Lie or Lie-Poisson bracket [2,6,79] of f and g and denoted by $[f, g]$. The jth component of $[f, g]$ is given by

$$\sum_i f_i \frac{\partial g_j}{\partial y_i} - g_i \frac{\partial f_j}{\partial y_i}.$$

2.2 The Baker-Campbell-Haussdorff formula

The Baker-Campbell-Haussdorff (BCH) formula [113] has recently played an important role in the analysis of numerical methods.

Let X and Y be "symbols". (We later think of these as Lie operators, but this is not necessary for the present.) We form the exponentials

$$\begin{aligned} \exp(X) &= I + X + \frac{1}{2}X^2 + \frac{1}{6}X^3 + \cdots \\ \exp(Y) &= I + Y + \frac{1}{2}Y^2 + \frac{1}{6}Y^3 + \cdots \end{aligned}$$

and we multiply them out

$$\begin{aligned} \exp(X)\exp(Y) &= I + X + Y + \frac{1}{2}X^2 + XY + \frac{1}{2}Y^2 \\ &\quad + \frac{1}{6}X^3 + \frac{1}{2}X^2Y + \frac{1}{2}XY^2 + \frac{1}{6}Y^3 + \cdots \end{aligned}$$

According to the BCH formula, the product $\exp(X)\exp(Y)$ can be written as the exponential $\exp(Z)$ of a new symbol

$$\begin{aligned} Z &= X + Y + \frac{1}{2}[X,Y] + \frac{1}{12}([X,X,Y] + [Y,Y,X]) \\ &\quad + \frac{1}{24}[X,Y,Y,X] - \frac{1}{720}([Y,Y,Y,Y,X] + [X,X,X,X,Y]) \\ &\quad + \frac{1}{360}([Y,X,X,X,Y] + [X,Y,Y,Y,X]) \\ &\quad + \frac{1}{120}([X,X,Y,Y,X] + [Y,Y,X,X,Y]) + \cdots \end{aligned} \tag{2.3}$$

Here $[X,Y]$ is the commutator $[X,Y] = XY - YX$ and we use iterated commutators $[X,X,Y] = [X,[X,Y]]$, $[X,Y,Y,X] = [X,[Y,[Y,X]]]$, etc. It is remarkable that Z only consists of X, Y and commutators. In particular, if X and Y are Lie operators L_f, L_g, then Z is a new Lie operator: the operator associated with the vector field $f + g + (1/2)[f,g] + \cdots$, where now the brackets represent the Lie-Poisson bracket of vector fields as discussed above.

2.3 Numerical methods

Throughout the paper we ignore multistep numerical methods. This is due to the fact that the literature we try to survey has almost exclusively dealt with one-step integrators (see [30] however). Each one-step numerical method induces [92] a one-parameter family $\psi_{h,f}$ of maps in $\mathcal{R}^D$ in such a way that $\psi_{h,f}(\alpha)$ is the numerical solution after one step of length h starting from the initial condition α; for instance Euler's method has $\psi_{h,f}(y) = y + hf(y)$. For a numerical method to make sense it is necessary that $\psi_{h,f}$ approximates $\phi_{h,f}$ for small h; the order of the method is defined as the (highest) positive integer r such that $\psi_{h,f}(y) - \phi_{h,f}(y) = O(h^{r+1})$ as $h \to 0$ for each $y \in \mathcal{R}^D$ and each smooth f. A method is consistent if $r \geq 1$.

Given an IVP Equation 1.1 and a step-length $h > 0$, the numerical method obtains approximations y^n to the true solution values $y(t_n)$ corresponding to the gridpoints $t_n = nh$, $n = 0, 1, \ldots$. The numerical solution is recursively defined by $y^0 = \alpha$, $y^{n+1} = \psi_{h,f}(y^n)$, $n = 0, 1, \ldots$, (for the true solution $y(t_{n+1}) = \phi_{h,f}(y(t_n))$). For a method of order r, the global errors $y_n - y(t_n)$ are $O(h^r)$ uniformly in bounded time-intervals. The extension of these considerations to variable steplengths h_n is straightforward.

For Taylor, Runge-Kutta (RK) and multiderivative Runge-Kutta methods, the expansion of $\psi_{h,f}$ in powers of h is a B-series [11,46]

$$y + \sum_{n=1}^{\infty} h^n \sum_{\tau \in RT_n} \frac{1}{\sigma(\tau)} c(\tau) F(\tau)(y). \tag{2.4}$$

Here RT_n denotes the set of rooted trees with n-vertices and, for each rooted tree τ, $\sigma(\tau)$ is the number of symmetries of τ, $F(\tau)(y)$ denotes the corresponding elementary differential evaluated at y and $c(\tau)$ is a real coefficient depending only on the numerical method. For an RK method with weights b_i and coefficients a_{ij}, the c's corresponding to the trees of orders one and two are $\sum_i b_i$, $\sum_{ij} b_i a_{ij}$. (The B-series Equation 2.4 has been normalized as in [11] or [72]; a different normalization is used in [46], cf. [12].)

The expansion Equation 2.4 is the key to writing the conditions for the corresponding method to have order r. Indeed the true solution has an expansion of the form Equation 2.4 with the coefficients $c(\tau)$ replaced by $1/\gamma(\tau)$, where $\gamma(\tau)$ is an integer called the density of τ. By imposing that the expansions of the numerical method and the true solution coincide except for $O(h^{r+1})$, one concludes that a method is of order r if and only if

$$c(\tau) = \frac{1}{\gamma(\tau)}, \quad \tau \in RT_n, \quad n = 1, \ldots, r. \tag{2.5}$$

For an RK method these conditions read $\sum_i b_i = 1$, $\sum_{ij} b_i a_{ij} = 1/2$, etc.

For methods that partition the components of y and treat differently different components, including Partitioned Runge-Kutta (PRK) methods [92] and Runge-Kutta-Nyström (RKN) methods, B-series have to be generalized to P-series [46,72]. For methods based on decompositions of f into N parts, the relevant generalization is called NB-series [5].

3 New error analyses

In recent years it has become increasingly clear that there is a need for making sense of numerical results that have little or no accuracy in a conventional sense, i.e. whose global errors are large. As pointed out in the introduction, there are many instances where scientists have benefited from the use of numerical ODE solvers and yet it is impossible to get numerical results with small global errors. Similar situations have appeared before in numerical analysis and backward error analysis has been most useful. In a backward error analysis the question is not how big is the difference between the exact and computed solutions; one rather shows that the computed solution solves exactly a problem $\tilde{\mathcal{P}}$ that is a perturbation of the problem $\mathcal{P}$ one wants to solve and then tries to estimate the difference between $\tilde{\mathcal{P}}$ and $\mathcal{P}$. In our context, backward error analysis could show that the effect of using a numerical method is to change slightly the IVP being solved. Such a conclusion would be particularly useful in applications, as those mentioned in the introduction, where there is an inherent uncertaintity in the model, i.e. in the exact values of f and α.

Many recent papers in numerical ODEs have compared the numerical solution y^n of the IVP Equation 1.1 with the values $\tilde{y}(t_n)$ of the solution of a neighbouring IVP. One succeeds in obtaining bounds for $y^n - \tilde{y}(t_n)$ that are much smaller than those that can be derived for the conventional global error $y^n - y(t_n)$. While several of those papers use the terminology "backward error analysis", one is in fact dealing with a mixed forward/backward error analysis strategy; there is a backward component because a perturbed problem is introduced and there is a forward component because the numerical solution and the exact solution of the perturbed problem are not quite the same and one has to estimate their difference.

Since the IVP Equation 1.1 is specified by two items f and α, one may perturb it by changing either the vector field f or the initial condition α. The two possible perturbations respectively lead to the ideas of modified equations and shadowing.

An early important reference on the "backward" approach is Eirola [29].

3.1 Modified equations

For partial differential equations, the method of modified equations has been known for a long time as a valuable tool to investigate the behaviour of numerical solutions. A classical reference is [114] and a more rigorous approach may be seen in [42].

In our context, a modified system (of order N) $\dot{y} = \widetilde{f}_N(y)$ for a method $\psi_{h,f}$ is a system for which $\psi_{h,f} - \phi_{h,\widetilde{f}_N} = O(h^{N+1})$. For the solution flow of the system $\dot{y} = f(y)$ to which the method is being applied, $\psi_{h,f} - \phi_{h,f} = O(h^{r+1})$, with r the order of the method and therefore, if $r < N$, the flow of the modified system provides a better description of $\psi_{h,f}$ than the true flow does. In fact, if $\widetilde{y}(t)$ denotes the solution of the modified system with initial condition α (see Equation 1.1), then the numerical solution satisfies $y^n - \widetilde{y}(t_n) = O(h^N)$.

It is obvious that the modified vector field $\widetilde{f}_N$ has to depend on the parameter h, but this dependence has not been incorporated to the notation. It is easy to construct, for any given N and method $\psi_{h,f}$, a modified vector field $\widetilde{f}_N$ of order N. The easiest possibility is to look for $\widetilde{f}_N(y)$ as a polynomial in h of degree N

$$\widetilde{f}_N(y) = f^0(y) + hf^1(y) + \cdots + h^N f^N(y),$$

where the f^n do not depend on h. It is clear that a method is consistent if and only if f^0 coincides with the true f. In a similar manner, a consistent method is of order $r > 1$ if and only if $f^1, \ldots, f^{r-1}$ vanish. It is also easy to guess that the f^n do not change with the order of the approximation being sought. Then it is possible to construct a formal power series

$$f^0(y) + hf^1(y) + h^2 f^2(y) + \cdots \tag{3.1}$$

whose truncations coincide with the $\widetilde{f}_N$'s.

When the method can be expanded in a B-series Equation 2.4, Hairer [43] shows that the f^n can be written of terms of the elementary differentials of f. Indeed Hairer shows that the formal power series Equation 3.1 is a B-series

$$\sum_{n=1}^{\infty} h^{n-1} \sum_{\tau \in RT_n} \frac{1}{\sigma(\tau)} b(\tau) F(\tau)(y). \tag{3.2}$$

and provides a systematic way to construct the coefficients $b(\tau)$ from the B-series coefficient $c(\tau)$ of the method Equation 2.4. An alternative derivation of Hairer's result was given in Murua's thesis [72] (see [15,90]). Note that the consistency of the method translates into the requirement that, for the tree with one vertex, $b(\tau) = 1$, and that, for consistent methods, order $r > 1$ is equivalent to $b(\tau) = 0$ for trees of orders $2, \ldots, r$. This provides an alternative to Equation 2.5 when writing the order conditions for the method.

Ideally one would like to have an "exact" modified system $\dot{y} = \widetilde{f}(y)$, for which $\psi_{h,f} \equiv \phi_{h,\widetilde{f}_\infty}$. For linear problems $\widetilde{f}_\infty$ is easily constructed, see Beyn [8].

However for nonlinear situations, the solutions generated by the mapping $\psi_{h,f}$ may possess features that cannot be present in the solutions generated by flows of differential systems [87]. For instance, in two dimensions, a mapping ψ may generate chaotic orbits, while no solution flow is chaotic. For this reason it is impossible to exactly interpolate the numerical solution by a suitable ϕ_{h,f_∞}.

The problem of how close can a smooth invertible mapping $\psi_{h,f}$ be approximated by flows of differential equations is well known in dynamical systems [65]. We saw above that $O(h^{N+1})$ approximations are easily constructed for any N. Neishtadt [74] was the first in rigorously proving that an "optimal" h-dependent modified vector field $\widetilde{f}$ for which the discrepancy between $\psi_{h,f}$ and $\phi_{h,\widetilde{f}}$ is exponentially small (see also [28]). This optimal $\widetilde{f}$ can be obtained by truncating the series Equation 3.1 (which in general diverges) after a suitably chosen number of terms that increases as $h \to 0$. Two recent papers [7,44] provide new proofs of the exponential smallness of the error with respect to the flow of the optimal modified vector field. These papers also present a number of valuable applications.

3.2 Shadowing

In shadowing, the computed solution y^n of the IVP Equation 1.1 is compared to the exact solution values $\tilde{y}(t_n)$ of the IVP given by the differential equation being solved $\dot{y} = f(y)$ along with a perturbed initial condition $\tilde{y}(0) = \tilde{\alpha}$. The idea of shadowing is well known in dynamical systems; its applications to numerical IVP in ODEs prior to 1993 are surveyed in the paper [93] and lack of space prevents us from repeating that material here. Recent references, not covered in [93], are [4,24–27,55,56,64,71,112].

4 Composition methods

Many of the methods used in geometric integration are *composition methods*, i.e. their associated mappings $\psi_{h,f}$ are a composition of simpler mappings.

- Sometimes one is given a method $\psi^{[B]}_{h,f}$ (the basic method) and constructs a new method

$$\psi_{h,f} = \psi^{[B]}_{b_s h,f} \circ \psi^{[B]}_{b_{s-1} h,f} \circ \cdots \circ \psi^{[B]}_{b_1 h,f}, \qquad (4.1)$$

  where the b_i are suitable real constants chosen in such a way that $\psi_{h,f}$ is of higher order than $\psi^{[B]}_{h,f}$. In practice, the basic method has a favourable geometric property and low order. If the geometric property is preserved by composition, the new $\psi_{h,f}$ will share it.

- Often (see [5] for a discussion) the right-hand side function f can be written in a natural way as a sum of several, say two, contributions $f = f^{[1]} + f^{[2]}$, and one may construct an integrator $\psi_{h,f}$ by combining steps of a method $\chi_{h,f^{[1]}}$ for the system $\dot{y} = f^{[1]}(y)$ and steps of a method $\pi_{h,f^{[2]}}$ for the system $\dot{y} = f^{[2]}(y)$,

$$\psi_{h,f} = \pi_{b_s h,f^{[2]}} \circ \chi_{a_s h,f^{[1]}} \circ \cdots \circ \pi_{b_1 h,f^{[2]}} \circ \chi_{a_1 h,f^{[1]}}. \tag{4.2}$$

An important particular case is that where the systems $\dot{y} = f^{[i]}(y)$ can be integrated in closed form and χ and π are taken to be the corresponding exact flows. Then the method reads

$$\psi_{h,f} = \phi_{b_s h,f^{[2]}} \circ \phi_{a_s h,f^{[1]}} \circ \cdots \circ \phi_{b_1 h,f^{[2]}} \circ \phi_{a_1 h,f^{[1]}}. \tag{4.3}$$

These splitting integrators are of course well known; the different parts of f often correspond to physically different contributions, say different reactions in chemistry or different forces in mechanics.

When $\psi_{h,f}$ is a composition of simpler mappings, its properties may be investigated through the BCH formula Equation 2.3. The methodology is as follows.

- Construct modified systems $\widetilde{f}_{N,i}$ of order N for the individual mappings being composed in Equation 4.1 or 4.2 (here the index i labels the mappings ψ_i being composed). Then (ignoring $O(h^{N+1})$ error terms), each ψ_i is the h-flow of the corresponding $\widetilde{f}_{N,i}$.
- Write, by means of Equation 2.2, the flows of the $\widetilde{f}_{N,i}$'s as exponentials.
- Use the BCH formula to combine all the exponentials into a single exponential $\exp(hL_{\widetilde{f}_N})$; in doing so, care must be taken of the right order of the factors, see [92]. From the exponent $hL_{\widetilde{f}_N}$, one recovers an (order N) modified system $\widetilde{f}_N$ for the overall method. The properties of the method, including the order of consistency r, are then retrieved from $\widetilde{f}_N$.

It can be shown [116] that, if in Equation 4.1 the basic method is of order two, then, for any prescribed r, it is possible to choose the number of stages s and the weights b_i so that the composition $\psi_{h,f}$ achieves order r.

Suzuki [103–111] was among the first in studying composition methods. Further useful references in connection with this approach are [41,60,68,69,81].

The discussion of composition methods will be continued later.

5 Symplectic integration

Symplectic methods for Hamiltonian problems have been the most studied family of geometric integrators. In the opening chapter of their 1987 collection of papers on Hamiltonian problems [66], MacKay and Meiss wrote: "Another neglected problem is the development of computer algorithms which respect the symplectic nature of the Hamiltonian". That situation changed quickly. Only a few years later, the field of symplectic integration had grown large enough to deserve a section in the second edition of the treatise by Hairer, Nørsett and Wanner [46] and even a monograph [92].

5.1 Hamiltonian systems

Suppose that the dimension D of Equation 1.1 is even $D = 2d$ and write $y = (p, q)$ with $p, q \in \mathcal{R}^d$. Then the system in Equation 1.1 is a Hamiltonian problem [6,66] if and only if f is of the form

$$f(y) = J^{-1}\nabla H \tag{5.1}$$

for a suitable real-valued function $H = H(p, q)$ (the Hamiltonian function). In Equation 5.1 ∇ is the operator

$$\left(\frac{\partial}{\partial p_1}, \frac{\partial}{\partial p_2}, \ldots, \frac{\partial}{\partial p_d}, \frac{\partial}{\partial q_1}, \frac{\partial}{\partial q_2}, \ldots, \frac{\partial}{\partial q_d}\right)^T,$$

and J is the skewsymmetric matrix

$$J = \begin{bmatrix} 0_d & I_d \\ -I_d & 0_d \end{bmatrix}. \tag{5.2}$$

The quantity H is conserved along the solutions of the corresponding Hamiltonian system [92]. This often corresponds to the principle of conservation of energy.

If the vector fields f and g in $\mathcal{R}^{2d}$ are both Hamiltonian, i.e. $f = J^{-1}\nabla H$ and $g = J^{-1}\nabla G$, then the Lie-Poisson bracket $[f, g]$ is also a Hamiltonian vector field. The corresponding Hamiltonian function is given by $-\{H, G\}$ (note the sign!), where [6,92]

$$\{H, G\} = \sum_i \frac{\partial H}{\partial q_i}\frac{\partial G}{\partial p_i} - \frac{\partial H}{\partial p_i}\frac{\partial G}{\partial q_i}$$

is the Poisson bracket of Hamiltonian functions. There is a correspondence between fields+Lie-Poisson bracket, Lie operators+commutator and Hamiltonian functions+*negative* Poisson bracket. The reader should be warned that there is no agreement in the literature: sometimes, to avoid the *negative* Poisson bracket in the correspondence above, things are defined with signs that disagree with our choice here. For instance, Arnold [6] reverses the signs of our $[f, g]$ and $[L_f, L_g]$. In [92] commutators and Poisson brackets are defined as in the present paper but the sign of L_f is reversed.

Hamiltonian systems appear very frequently in the applications; virtually all phenomena where dissipation is absent or can be ignored may be modelled by a Hamitonian system. In particular, Hamiltonian systems play a key role in classical, statistical and quantum mechanics, in optics and in plasma physics.

The property of $\dot{y} = f(y)$ being Hamiltonian, which refers to the vector field of the system differential equations, can be translated into a corresponding property of the flow $\phi_{t,f}$. In fact a system in $\mathcal{R}^{2d}$ is Hamiltonian if and only if its flow is, for each t, a *symplectic* transformation. By definition, a transformation Ψ in $\mathcal{R}^{2d}$ is symplectic if its Jacobian matrix $\Psi'(y)$ satisfies [92]

$$\Psi'^T(y)J\Psi'(y) \equiv J,$$

where J is the matrix Equation 5.2. An equivalent definition in terms of differential forms exists [6]. Differential forms lead to manipulations that are easier than those required when using Jacobians. More importantly, differential forms provide a geometric interpretation of symplecticness in terms of conservation of areas.

Hamiltonian systems possess many features not shared by other systems of differential equations. Many properties that "general" systems possess only under exceptional circumstances appear generically in Hamiltonian systems. All these features and properties can be traced back to the symplecticness of Hamiltonian flows.

5.2 What is a symplectic integrator?

The failure of virtually all well-known methods in mimicking Hamiltonian dynamics suggested the consideration of schemes such that the mapping $\psi_{h,f}$ is symplectic whenever the field f is Hamiltonian. Such methods are called *symplectic* or *canonical*. Early references on symplectic integration are Ruth [83], Channel [22], Menyuk [70], Feng [31–33], even though the idea of symplectic integration apparently goes back to DeVogelaere in 1956 (cf. [23]).

Since a flow is symplectic if and only if the corresponding vector field is Hamiltonian, it is not difficult to see that a method is symplectic if and only if the corresponding formal modified vector field Equation 3.1 is Hamiltonian, or equivalently all the $\widetilde{f}_N$ are Hamiltonian. Therefore, ignoring the $O(h^{N+1})$ remainder term, a symplectic discretization of a Hamiltonian problem changes the Hamiltonian system being solved into a nearby Hamiltonian problem. The discretization provided by a nonsymplectic method changes the Hamiltonian system into a nearby non-Hamiltonian perturbation. Thus for a symplectic method, the Hamiltonian vector field Equation 3.1, has an associated Hamiltonian function that we denote by

$$H^0(y) + hH^1(y) + h^2H^2(y) + \cdots . \tag{5.3}$$

As discussed above the properties of a method are encapsulated in the vector series Equation 3.1. In the Hamiltonian context we can rather use the simpler, scalar series Equation 5.3. More details will be given later.

For extensions of the idea of symplectic integration, including constrained systems, partial differential equations and nonstandard symplectic structures, see [37,38,58,59,67,89,92].

5.3 Families of symplectic integrators

There have been three main approaches to the construction of symplectic integrators.

5.3.1 Symplectic methods based on generating functions

Symplectic transformations can also be characterized in terms of so-called generating functions [6,92] and the early papers on symplectic integration resorted to this characterization. Most of the resulting methods were too involved for practical use. Those generating function methods that are practical are best analyzed as members of the classes of symplectic composition methods or symplectic Runge-Kutta methods. For instance, Ruth first derived his methods [83] through the generating function formalism, but later reinterpreted them as composition methods [36]. For these reasons this methodology will not be discussed further here and the interested reader is referred to [92].

5.3.2 Symplectic methods using compositions

Assume that the individual mappings being composed in Equation 4.1–4.3 are symplectic. This will automatically be the case for splitting methods Equation 4.3 if the parts $f^{[i]}$ are Hamiltonian. (Note that splitting f into *Hamiltonian* parts $f^{[1]}+f^{[2]}$ corresponds to splitting the Hamiltonian function H in *arbitrary* pieces $H = H^{[1]}+H^{[2]}$.) Under this assumption of symplecticness of the parts being composed, the resulting overall method $\psi_{h,f}$ will also be symplectic.

The methodology presented in Section 4 to analyze $\psi_{h,f}$ is particularly suitable in the symplectic case, because rather than working with vector fields and their Lie-Poisson brackets one works with Hamiltonian functions and their Poisson brackets, see [35,92,94].

Perhaps separable Hamiltonians

$$H = T(p) + V(q) \tag{5.4}$$

have provided the most common application of the composition approach. These Hamiltonians appear often in practice with T and V respectively giving the kinetic and potential energies. After the splitting $H = H^{[1]} + H^{[2]}$, $H^{[1]} = T(p)$, $H^{[2]} = V(q)$, the individual pieces are integrable in closed form; the solution flow of for $T(p)$ is $(p^0, q^0) \to (p^0, q^0 + t\nabla T(p^0))$ (sometimes referred to as a "drift") and the solution flow of for $V(q)$ is $(p^0, q^0) \to (p^0 - t\nabla V(q^0), q^0)$ (sometimes referred to as a "kick"). Then a step of the method Equation 4.3 is given by

$$Q_i = Q_{i-1} + a_i h\nabla T(P_i), \quad P_{i+1} = P_i - b_i h\nabla V(Q_i), \quad i = 1, \ldots, s, \tag{5.5}$$

where (P_1, Q_0) (respectively (Q_s, P_{s+1})) is the numerical solution at the beginning (respectively the end) of the step.

It is also possible [63] to modify the kicks by modifying the force $-\nabla V$ with a term involving the Hessian matrix of V. A pioneering paper is due to Rowlands

[82]. Rowlands method is very efficient and achieves fourth order by a change of variables. The idea of changing variables to increase the order goes back to Butcher, see the discussion in [61–63].

Splittings of the Hamiltonian also arise when using multiple time-steps [9], so that different forces are sampled at different rates.

5.3.3 Symplectic Runge-Kutta methods

The standard class of implicit RK methods happens to contain symplectic integrators. This was shown independently by Lasagni [57], Sanz-Serna [86] and Suris [99]. For a method with coefficients (a_{ij}) and weights (b_i) the condition

$$\forall i, j, \quad b_i a_{ij} + b_j a_{ji} - b_i b_j = 0, \tag{5.6}$$

guarantees symplecticness. The Gauss methods (order $2s$ with s stages) are symplectic. There are also diagonally implicit, symplectic methods; with s stages they include s free parameters. These diagonally implicit methods are compositions of the midpoint rule (the lowest order Gauss method) and can be analyzed via the BCH formula. Other references on the construction of methods satisfying Equation 5.6 are [48,49,85].

The condition Equation 5.6 is also essentially necessary for symplecticness [92]. The first proof of the necessity of Equation 5.6 was given by Lasagni in an unpublished manuscript and is very delicate; a similar proof was published in [1].

A more modern approach to the symplecticness condition Equation 5.6 is via B-series [19]. First, a necessary and sufficient condition is derived on the coefficients $c(\tau)$ of a B-series Equation 2.4 for this series to define a symplectic transformation whenever the underlying vector field is Hamiltonian. In a second stage, the B-series coefficients are expressed in terms of the tableau elements (a_{ij}) and (b_i) and the symplecticness condition for the $c(\tau)$'s is shown to be equivalent to Equation 5.6. The proof of the necessity of Equation 5.6 via B-series is easier and more powerful than the original Lasagni proof. Furthermore the B-series technique is also more general, because it allows to deal with methods, other than RK methods, expressible as B-series. Hairer, Murua and Sanz-Serna, building on the B-series characterization of symplecticness, showed that there are not any nontrivial examples of symplectic multiderivative RK methods [45].

Another important contribution based on B-series is due to Hairer [43]. For Hamiltonian f, Hairer investigates the modified vector field Equation 3.2 and writes a necessary and sufficient condition on the coefficients $b(\tau)$ for this series to be symplectic, i.e. for the underlying method to be symplectic. He then shows that, when the $b(\tau)$'s are written as functions of the method coefficients $c(\tau)$ Equation 2.4, his symplecticness condition on the $b(\tau)$'s becomes the Calvo and Sanz-Serna [19] symplecticness condition on the $c(\tau)$'s. Furthermore when

Equation 3.2 is symplectic, Hairer gives an explicit formula for the modified Hamiltonian Equation 5.3. This is of the form

$$\sum_{n=1}^{\infty} h^{n-1} \sum_{\omega \in T_n^*} \frac{1}{\sigma(\omega)} d(\omega) H(\omega)(y), \tag{5.7}$$

where the meaning all the various symbols will be discussed presently. The inner summation extends to all *trees* with n-vertices, while in conventional B-series Equation 2.4, Equation 3.2 one deals with *rooted* trees. A rooted tree is a tree in which a vertex has been highlighted to play the role of the root; conversely a tree can be seen as an equivalence class of rooted tree obtained by grouping all rooted trees that only differ in the location of the root. Thus Equation 5.7 has *fewer* terms than Equation 3.2. The star in T_n^* means that not all trees are present in the summation: so-called superfluous trees [91] are not to be included. For each tree ω, $\sigma(\omega)$ represents the number of symmetries and $H(\omega)(y)$ is the corresponding elementary Hamiltonian, a real-valued function of y obtained by combining H and its partial derivatives. Elementary Hamiltonians were first introduced in [91] with the name canonical elementary differentials. Finally $d(\omega)$ is a real coefficient that can be computed in terms of the coefficients $b(\tau)$ in Equation 3.2. Murua [72] calls expressions like Equation 5.7 H-series and shows how to manipulate them.

It was first proved in [91] (see also [84]) that, for an RK method, the symplecticness condition Equation 5.6 acts as a simplifying assumption, i.e. when Equation 5.6 holds not all the order conditions Equation 2.5 are independent. Indeed when Equation 5.6 holds, to achieve order r it is enough to impose $c(\tau) = 1/\gamma(\tau)$ for *one* rooted tree τ in each nonsuperfluous tree of order $\leq r$. This is a consequence of two facts:

1. order r can be imposed by demanding that Equation 5.7 agrees with the true Hamiltonian H except for a remainder of order $O(h^r)$ and

2. in Equation 5.7 there is a term for each nonsuperfluous tree.

(In [91] generating functions were used instead of the modified Hamiltonians, but the argument is essentially the same.)

We finally discuss other classes of Runge-Kutta-like methods that contain symplectic integrators.

- Partitioned Runge-Kutta (PRK) methods for separable Hamiltonian systems Equation 5.4. These are Runge-Kutta methods with two tableaux; one is used for the p variables and the other for the q variables. A symplecticness condition analogous to Equation 5.6 was first presented by Sanz-Serna at the 1989 London ODE meeting [88], and discovered independently by Suris [101]. The order conditions were studied in [1]. This class of methods is useful because it contains *explicit* schemes that are symplectic. These turn out to be [77] of the form Equation 5.5 and may therefore be studied through the BCH formula, see [14]. Specific methods are constructed in [1,20,77,88].

- Runge-Kutta-Nyström (RKN) methods for Hamiltonian problems of the form $\dot{p} = -\partial V/\partial q$, $\partial q = M^{-1}p$, where $V = V(q)$ is the potential and M the mass matrix. The corresponding symplecticness conditions were first derived by Suris [99,100]. For the order conditions see [16]. Specific methods are constructed in [17,18,77,78]. See also [76].

- Partitioned Runge-Kutta methods for general (not necessarily separable) Hamiltonians. This family is methodologically useful because it includes the families of RK methods, PRK methods for separable Hamiltonians and RKN methods. It is the class considered in important recent theoretical papers such as [43]. Interesting references are [51,73].

- Additive Runge-Kutta methods [5]. This class applies to splittings of the Hamiltonian functions and contains all the others as particular cases.

5.4 Properties of symplectic integrators

Hamiltonian flows exactly preserve the symplectic structure *and* the value of the Hamiltonian function (energy). It was proved by Ge and Marsden [40] that it is in general impossible for a symplectic integrator to exactly preserve energy. However in practice symplectic integrators do a very good job of preserving the value of the energy. This conservation is linked to the existence of a modified Hamiltonian function, as first noted by Lasagni in a set of unpublished notes mentioned in [57]. See also [7,44,65] and the discussion in [92].

Sometimes symplectic integrators have more favourable error-growth properties than their general counterparts. See [13,18,21,39]. It is interesting that those favourable properties are not shared [80] by schemes obtained by fitting the classical stability region of the method; this shows that we are dealing with methods that cannot be analyzed or obtained within the classical paradigm.

Unfortunately the advantages of symplecticness cannot be combined with variable stepsizes [17]. This is due to fact that with variable stepsizes the backward error interpretation breaks down, see [17,92].

6 Reversible integration

In this section the presentation emphasizes the analogies with symplectic integration.

6.1 Reversible systems

Let ρ be a linear involution, i.e., a linear mapping in $\mathcal{R}^D$ such that $\rho^2 = Id$. An example often found in mechanics has $y = (p, q)$ and $\rho(p, q) = (-p, q)$, where p is the collective vector of velocities of the system being studied and

q the corresponding collective vector of positions or coordinates. A system of differential equations $\dot{y} = f(y)$ is ρ-reversible if

$$f(\rho y) \equiv -\rho f(y).$$

This property, that operates at the vector field level, is equivalent to the requirement that the corresponding flow $\phi_{t,f}$ be, for each t, a ρ-reversible mapping. By definition, a mapping Ψ in $\mathcal{R}^D$ is ρ-reversible if

$$\rho^{-1} \circ \Psi \circ \rho = \Psi^{-1}. \tag{6.1}$$

Thus a mapping Ψ is ρ-reversible if it is transformed in its inverse Ψ^{-1} by the change of variables $y \to \rho y$. In the particular case of a flow, the reversibility condition becomes

$$\rho^{-1} \circ \phi_{t,f} \circ \rho = \phi_{t,f}^{-1} = \phi_{-t,f};$$

for a reversible system changing variables $y \to \rho y$ just reverses the sense of the arrow of time.

6.2 What is a reversible integrator

Methods for which $\psi_{h,f}$ is a ρ-reversible mapping for each h and each ρ-reversible vector field are called ρ-reversible. D. Stoffer has been one of the leading contributors to the area of reversible integration, starting with his thesis [96].

6.3 Families of reversible integrators

6.3.1 Reversible composition methods

By using Equation 6.1 it is easy to derive sufficient conditions for a composition method to be reversible. For instance a symmetric composition of reversible mapp $\Psi^{[1]}\Psi^{[2]}\Psi^{[1]}$ is reversible.

6.3.2 Reversible Runge-Kutta methods

For RK, PRK, RKN and other standard families of numerical methods, ρ-reversibility turns out to be equivalent to time-reversibility, i.e. to the requirement

$$\psi_{-h,f} = (\psi_{h,f})^{-1}. \tag{6.2}$$

This is because such methods are equivariant with respect to linear changes of variables

$$\rho^{-1} \circ \psi_{h,f} \circ \rho = \psi_{h,\rho f}$$

(changing variables by ρ in the computed solution is the same as applying the method to the differential equation resulting from changing variables).

Note that Equation 6.2 is independent of ρ. Time-reversible (also called selfadjoint [92]) methods were of course well known before the introduction of the notion reversible integration. Therefore there has been no need to study the classes of ρ-reversible RK, PRK and RKN method, as distinct from the situation for symplectic RK, PRK and RKN methods.

6.4 Properties of reversible integrators

Reversible integrators share many of the favourable properties of symplectic methods [13,21], with the advantage that they retain those properties even when applied with variable stepsizes, provided that the stepsize selection is carried out so as to preserve reversibility [47,97].

7 Volume preserving flows

A divergence-free $\nabla \cdot f = 0$ vector field gives rise to a volume-preserving flow $\phi_{t,f}$, i.e., to a flow whose Jacobian determinant is $\equiv 1$, leading to the property that, for any domain Ω in phase-space, Ω and $\phi_{t,f}(\Omega)$ possess the same volume. Volume-preserving integration has received some attention [34,65,75,98], but not nearly as much as symplectic or reversible integration.

8 Epilogue

We shall finish as we started: quoting Lambert. In his (1973) textbook [53], he writes: "Remarkably little is required by way of prerequisites for the study of computational methods for ordinary differential equations." In fact in the black-box subroutine mode of operation, the numerical analyst was screened from the specific structure of the problem being solved, from its mathematical singularities and from the application fields. Things have now changed and it is our feeling that they will change even more in the future. Contributors to *numerical* ordinary differential equations will need to be familiar with relevant developments in the *theory* of differential equations. Furthermore, many important developments are likely to arise when considering specific problems from the different application fields and therefore some degree of interdisciplinary work will be essential.

Acknowledgement

The author has been supported by grant DGICYT PB92-0254.

References

1. Abia, L. and Sanz-Serna, J.M. (1993). Partitioned Runge-Kutta methods for separable Hamiltonian problems. *Math. Comput.*, **60**, 617–634.

2. Abraham, R., Marsden, J.E. and Ratiu, T. (1988). *Manifolds, Tensor Analysis and Applications*, 2nd Edition, Springer, New York.

3. Allen, M.P. and Tildesley, D.J. (1989). *Computer Simulation of Liquids*, Clarendon Press, Oxford.

4. Alouges, F. and Debussche, A. (1993). On the discretization of a partial differential equation in the neighborhood of a periodic orbit. *Numerische Mathematik*, **65**, 143–175.

5. Araújo, A.L., Murua, A. and Sanz-Serna, J.M. (1995). Symplectic methods based on decompositions. *Applied Maths. Computer Report 1995/10*, Valladolid. Also to appear in *SIAM J. Numerical Analysis.*

6. Arnold, V.I. (1989). *Mathematical Methods of Classical Mechanics*, 2nd Edition, Springer Verlag, New York.

7. Benettin, G. and Giorgilli, A. (1994). On the Hamiltonian interpolation of near to the identity symplectic mappings with applications to symplectic integration algorithms. *J. Statistical Physics*, **74**, 1117–1143.

8. Beyn, W-J. (1991). Numerical methods for dynamical systems. *Advances in Numerical Analysis*, Editor: W. Light, **I**, Clarendon Press, Oxford, 175–236.

9. Biesiadecki, J.J. and Skeel, R.D. (1993). Dangers of multiple time step methods. *J. Comput. Physics*, **109**, 318–328.

10. Board, J.A., Kalé, L.V., Schulten, K., Skeel, R.D. and Schlick, T. (1994). Modeling biomolecules: Larger scales, longer durations. *IEEE Computer Science Engineering*, **1–4**, 19–30.

11. Butcher, J.C. (1987). *The Numerical Analysis of Ordinary Differential Equations*, John Wiley, Chichester.

12. Butcher, J.C. and Sanz-Serna, J.M. (1996). The number of order conditions for a Runge-Kutta method to have effective order *p*. *Appl. Numerical Maths.*, **22**, 103–111.

13. Calvo, M.P. and Hairer, E. (1995). Accurate long-term integration of dynamical systems. *App. Numerical Maths.*, **18**, 95–105.

14. Calvo, M.P. and Hairer, E. (1995). Further reduction in the number of independent order conditions for symplectic, explicit Partitioned Runge-Kutta and Runge-Kutta-Nyström methods. *Appl. Numerical Maths.*, **18**, 107–114.

15. Calvo, M.P., Murua, A. and Sanz-Serna, J.M. (1994). Modified equations for ODEs. *Contemporary Maths.*, **172**, 63–74.

16. Calvo, M.P. and Sanz-Serna, J.M. (1992). Order conditions for canonical Runge-Kutta-Nyström methods. *BIT*, **32**, 131–142.

17. Calvo, M.P. and Sanz-Serna, J.M. (1993). The development of variable step integrators, with application to the two-body problem. *SIAM J. Sci. Comput.*, **14**, 936–952.

18. Calvo, M.P. and Sanz-Serna, J.M. (1993). High-order symplectic Runge-Kutta-Nyström methods. *SIAM J. Sci. Comput.*, **14**, 1237–1252.

19. Calvo, M.P. and Sanz-Serna, J.M. (1994). Canonical B-series. *Numer. Math.*, **67**, 161–175.

20. Candy, J. and Rozmus, W. (1991). A symplectic integration algorithm for separable Hamiltonian functions. *J. Comput. Physics*, **92**, 230–256.

21. Cano, B. and Sanz-Serna, J.M. (1995). Error growth in the numerical integration of periodic orbits, with application to Hamiltonian and reversible systems. *Appl. Maths. Comput. Report 1995/1*, Valladolid. Also to appear in *SIAM J. Numerical Analysis.*

22. Channell, P.J. (1983). Symplectic integration algorithms. *Los Alamos National Laboratory Report, Report AT-6:ATN83-9.*

23. Channell, P.J. and Scovel, C. (1990). Symplectic integration of Hamiltonian systems. *Nonlinearity*, **3**, 231–259.

24. Chow, S-N. and Van Vleck, E.S. (1994). A shadowing lemma approach to global error analysis for initial-value ODEs. *SIAM J. Sci. Comput.*, **15**, 959–976.

25. Coomes, B.A., Koçak, H. and Palmer, K.J. (1994). Shadowing orbits of ordinary differential equations. *J. Comput. Appl. Maths.*, **52**, 35–43.

26. Coomes, B.A., Koçak, H. and Palmer, K.J. (1995). Rigorous computational shadowing of orbits of ordinary differential equations. *Numerische Mathematik*, **69**, 401–421.

27. Dawson, S., Grebogi, C., Sauer, T. and Yorke, J.A. (1994). Obstructions to shadowing when a Lyapunov exponent fluctuates about zero. *Phys. Review Letters*, **73**, 1927–1930.

28. Dragt, A.J. and Finn, J.M. (1976). Lie series and invariant functions for analytic symplectic maps. *J. Math. Physics*, **17**, 2215–2227.

29. Eirola, T. (1993). Aspects of backward error analysis of numerical ODEs. *J. Comput. Appl. Maths.*, **45**, 65–73.

30. Eirola, T. and Sanz-Serna, J.M. (1992). Conservation of integrals and symplectic structure in the integration of differential equations by multistep methods. *Numerische Mathematik*, **61**, 281–290.

31. Feng, K. (1985). On difference schemes and symplectic geometry. *Proc. of the 1984 Beijing Sym. on Differential Geometry and Differential Equations*, Editor: K. Feng, Science Press, Beijing, 42–58.

32. Feng, K. (1986). Difference schemes for Hamiltonian formalism and symplectic geometry. *J. Comput. Maths.*, **4**, 279–289.

33. Feng, K. (1986). Symplectic geometry and numerical methods in fluid dynamics. *Tenth Int. Conf. on Numerical Methods in Fluid Dynamics*, Editors: F.G. Zhuang and Y.L. Zhu, *Lecture Notes in Physics*, **264**, Springer Verlag, Berlin, 1–7.

34. Feng, K. and Shang, Z. (1995). Volume-preserving algorithms for source-free dynamical systems. *Numerische Mathematik*, **71**, 451–463.

35. Forest, E. (1992). Sixth-order Lie-group integrators. *J. Comput. Phys.*, **99**, 209–213.

36. Forest, E. and Ruth, R.D. (1990). Fourth-order symplectic integration. *Physica D*, **43**, 105–117.

37. Frutos, J. de, Ortega, T. and Sanz-Serna, J.M. (1990). A Hamiltonian explicit algorithm with spectral accuracy for the "good" Boussinesq equation. *Comput. Methods Appl. Mechanical Engineering*, **80**, 417–423.

38. Frutos, J. de and Sanz-Serna, J.M. (1992). An easily implementable fourth-order method for the time-integration of wave problems. *J. Comput. Phys.*, **103**, 160–168.

39. Frutos, J. de and Sanz-Serna, J.M. (1994). Accuracy and conservation properties in numerical integration: The case of the Korteweg-de Vries equation. *Appl. Maths. Comput. Report 1994/4*, Valladolid. Also to appear in *Numerische Mathematik*.

40. Ge, Z. and Marsden, J.E. (1988). Lie-Poisson Hamilton-Jacobi theory and Lie-Poisson integrators. *Phys. Letters A*, **133**, 134–139.

41. Goldman, D. and Kaper, T.J. (1996). N-th order operator splitting schemes and nonreversible systems. *SIAM J. Numerical Analysis*, **33**, 349–367.

42. Griffiths, D.F. and Sanz-Serna, J.M. (1986). On the scope of the method of modified equations. *SIAM J. Sci. Comput.*, **7**, 994–1008.

43. Hairer, E. (1994). Backward analysis of numerical integrators and symplectic methods. *Annals of Numerical Maths.*, **1**, 107–132.

44. Hairer, E. and Lubich, Ch. (1995). The life-span of backward error analysis for numerical integrators. (Preprint.)

45. Hairer, E., Murua, A. and Sanz-Serna, J.M. (1994). The nonexistence of symplectic multiderivative Runge-Kutta methods. *BIT*, **34**, 80–87.

46. Hairer, E., Nørsett, S.P. and Wanner, G. (1991). *Solving Ordinary Differential Equations I: Nonstiff Problems*, 2nd Edition, Springer Verlag, Berlin.

47. Hairer, E. and Stoffer, D. (1996). Reversible long-term integration with variable step sizes. *SIAM J. Sci. Comput.* (To appear.)

48. Hairer, E. and Wanner, G. (1994). Symplectic Runge-Kutta methods with real eigenvalues. *BIT*, **34**, 310–312.

49. Iserles, A. (1991). Efficient Runge-Kutta methods for Hamiltonian equations. *Bulletin Hellenic Math. Society*, **32**, 3–20.

50. Iserles, A. and Powell, M.J.D. (Editors). (1987). *The State of the Art in Numerical Analysis*, Clarendon Press.

51. Jay, L. (1996). Symplectic partitioned Runge-Kutta methods for constrained Hamiltonian systems. *SIAM J. Numerical Analysis*, **33**, 368–387.

52. Kerr, R.A. (1992). From Mercury to Pluto, chaos pervades the solar system. *Science*, **257**, 33.

53. Lambert, J.D. (1973). *Computational Methods in Ordinary Differential Equations*, John Wiley.

54. Lambert, J.D. (1976). The initial value problem for ordinary differential equations. *The State of the Art in Numerical Analysis*, Editor: D.A.H. Jacobs, Academic Press, 451–500.

55. Larsson, S. and Sanz-Serna, J.M. (1994). The behavior of finite element solutions of semilinear parabolic problems near stationary points. *SIAM J. Numerical Analysis*, **31**, 1000–1018.

56. Larsson, S. and Sanz-Serna, J.M. (1994). A shadowing result with applications to finite element approximation of reaction-diffusion equations. *Preprint No. 1996–05*, Department of Mathematics, Chalmers University of Technology, Göteborg.

57. Lasagni, F.M. (1988). Canonical Runge-Kutta methods. *ZAMP*, **39**, 952–953.

58. Leimkuhler, B.J. and Reich, S. (1994). Symplectic integration of constrained Hamiltonian systems. *Math. Comput.*, **63**, 589–605.

59. Leimkuhler, B.J. and Skeel, R.D. (1994). Symplectic numerical integrators in constrained Hamiltonian systems. *J. Comput. Phys.*, **112**, 117–125.

60. Li, R-C. (1995). Raising the orders of unconventional schemes for ordinary differential equations. *Ph.D. Dissertation*, University of California at Berkeley, USA.

61. López-Marcos, M.A., Sanz-Serna, J.M. and Skeel, R.D. (1996). Cheap enhancement of symplectic integrators. *Numerical Analysis 1995*, Editors: D.F. Griffiths and G.A. Watson, Longman Scientific Press, 107–122.

62. López-Marcos, M.A., Sanz-Serna, J.M. and Skeel, R.D. (1996). An explicit symplectic integrator with maximal stability interval. *Numerical Analysis, A.R. Mitchell 75th Birthday Volume*, Editors: D.F. Griffiths and G.A. Watson, World Scientific, 163–176.

63. López-Marcos, M.A., Sanz-Serna, J.M. and Skeel, R.D. (1995). Explicit symplectic integrators using Hessian-vector products. *Appl. Maths. Comput. Report 1995/8*, Valladolid. Also to appear in *SIAM J. Sci. Comput.*

64. Lubich, Ch., Nipp, K. and Stoffer, D. (1993). Runge-Kutta solutions of stiff differential equations near stationary points. *Research Report Number 93-01, Seminar für Angewandte Mathematik*, ETH, Zürich.

65. MacKay, R.S. (1992). Some aspects of the dynamics and numerics of Hamiltonian systems. *The Dynamics of Numerics and the Numerics of Dynamics*, Editors: D.S. Broomhead and A. Iserles, Clarendon Press, 137–193.

66. MacKay, R.S. and Meiss J.D. (Editors). (1987). *Hamiltonian Dynamical Systems*, Adam Hilger Publishers.

67. McLachlan, R. (1994). Symplectic integration of Hamiltonian wave equations. *Numerische Mathematik*, **66**, 465–492.

68. McLachlan, R.I. (1995). Composition methods in the presence of small parameters. *BIT*, **35**, 258–268.

69. McLachlan, R.I. (1995). On the numerical integration of ordinary differential equations by symmetric composition methods. *SIAM J. Sci. Comput.*, **16**, 151–168.

70. Menyuk, C.R. (1984). Some properties of the discrete Hamiltonian method. *Physica D*, **11**, 109–129.

71. Murdock, J. (1995). Shadowing multiple elbow orbits: An application of dynamical systems to perturbation theory. *J. Differential Equations*, **119**, 224–247.

72. Murua, A. (1995). Métodos simplécticos desarrollables en P-series. *Tesis Doctoral*, Valladolid.

73. Murua, A. (1995). Order conditions for partitioned symplectic methods. *SIAM J. Numerical Analysis*. (To appear.)

74. Neishtadt, A.I. (1984). The separation of motions in systems with rapidly rotating phase. *J. Appl. Math. Mech.*, **48**, 133–139.

75. Okunbor, D.I. (1995). Energy conserving, Liouville and symplectic integrators. *J. Comput. Phys.*, **120**, 375–378.

76. Okunbor, D. and Skeel, R.D. (1992). An explicit Runge-Kutta-Nyström method is symplectic if and only if its adjoint is explicit. *SIAM J. Numerical Analysis*, **29**, 521–527.

77. Okunbor, D. and Skeel, R.D. (1992). Explicit canonical methods for Hamiltonian systems. *Math. Comput.*, **59**, 439–455.

78. Okunbor, D. and Skeel, R.D. (1994). Canonical Runge-Kutta-Nyström methods of orders five and six. *J. Comput. Appl. Maths.*, **51**, 375–382.

79. Olver, P.J. (1986). *Applications of Lie Groups to Differential Equations*, Springer Verlag, New York.

80. Portillo, A. and Sanz-Serna, J.M. (1995). Lack of dissipativity is not symplecticness. *BIT*, **35**, 269–276.

81. Reich, S. (1996). Symplectic integration of constrained Hamiltonian systems by composition methods. *SIAM J. Numerical Analysis*, **33**, 475–491.

82. Rowlands, G. (1991). A numerical algorithm for Hamiltonian systems. *J. Comput. Phys.*, **97**, 235–239.

83. Ruth, R.D. (1983). A canonical integration technique. *IEEE Trans. Nuclear Science*, **30**, 2669–2671.

84. Saito, S., Sugiura, H. and Mitsui, T. (1992). Butcher's simplifying assumption for symplectic integrators. *BIT*, **32**, 345–349.

85. Saito, S., Sugiura, H. and Mitsui, T. (1992). Family of symplectic implicit Runge-Kutta formulae. *BIT*, **32**, 539–543.

86. Sanz-Serna, J.M. (1988). Runge-Kutta schemes for Hamiltonian systems. *BIT*, **28**, 877–883.

87. Sanz-Serna, J.M. (1991). Two topics in nonlinear stability. *Advances in Numerical Analysis*, Editor: W. Light, **I**, Clarendon Press, 147–174.

88. Sanz-Serna, J.M. (1992). The numerical integration of Hamiltonian systems. *Computational Ordinary Differential Equations*, Editors: J.R. Cash and I. Gladwell, Clarendon Press, 437–449.

89. Sanz-Serna, J.M. (1994). An unconventional symplectic integrator of W. Kahan. *Appl. Numerical Maths.*, **16**, 245–250.

90. Sanz-Serna, J.M. (1996). Backward error analysis of symplectic integrators. *Integration Algorithms and Classical Mechanics*, Editors: J.E. Marsden, G.W. Patrick and W.T. Shadwick. Fields Institute Communications, **10**, American Mathematical Society, 193–205.

91. Sanz-Serna, J.M. and Abia, L. (1991). Order conditions for canonical Runge-Kutta schemes. *SIAM J. Numerical Analysis*, **28**, 1081–1096.

92. Sanz-Serna, J.M. and Calvo, M.P. (1994). *Numerical Hamiltonian Problems*, Chapman and Hall.

93. Sanz-Serna, J.M. and Larsson, S. (1993). Shadows, chaos and saddles. *Appl. Numer. Maths.*, **13**, 181–190.

94. Sanz-Serna J.M. and Portillo, A. (1996). Classical numerical integrators for wave-packet dynamics. *J. Chem. Phys.*, **104**, 2349–2355.

95. Sanz-Serna, J.M. and Vadillo, F. (1987). Studies in numerical nonlinear instability III: Augmented Hamiltonian systems. *SIAM J. Applied Maths.*, **47**, 92–108.

96. Stoffer, D.M. (1988). Some geometric and numerical methods for perturbed integrable systems. *Ph.D. Thesis*, ETH Zürich.

97. Stoffer, D.M. (1995). Variable steps for reversible integration methods. *Computing*, **55**, 1–22.

98. Suris, Y.B. (1987). Some properties of methods for the numerical integration of systems of the form $\ddot{x} = f(x)$. *USSR Comput. Maths. Math. Phys.*, **27**, 149–156.

99. Suris, Y.B. (1988). Preservation of symplectic structure in the numerical solution of Hamiltonian systems. *Numerical Solution of Differential Equations*, Editor: S.S. Filippov, Akad. Nauk. SSSR, Inst. Prikl. Mat., Moscow, 148–160. (In Russian.)

100. Suris, Y.B. (1989). The canonicity of mappings generated by Runge-Kutta type methods when integrating the systems $\ddot{x} = -\partial U/\partial x$. *USSR Comput. Maths. Math. Phys.*, **29**, 138–144.

101. Suris, Y.B. (1990). Hamiltonian methods of Runge-Kutta type and their variational interpretation. *Math. Model.*, **2**, 78–87. (In Russian.)

102. Sussman, G.J. and Wisdom, J. (1992). Chaotic evolution of the solar system. *Science*, **257**, 56–62.

103. Suzuki, M. (1976). Generalized Trotter's formula and systematic approximants of exponential operators and inner derivations with applications to many body problems. *Commun. Math. Phys.*, **51**, 183–190.

104. Suzuki, M. (1977). On the convergence of exponential operators—the Zassenhaus formula, BCH formula and systematic approximants. *Commun. Math. Phys.*, **57**, 193–200.

105. Suzuki, M. (1990). Fractal decomposition of exponential operators with applications to many-body theories and Monte Carlo simulations. *Phys. Letters*, **146**, 319–323.

106. Suzuki, M. (1991). General theory of fractal path integrals with applications to many-body theories and statistical physics. *J. Math. Phys.*, **32**, 400–406.

107. Suzuki, M. (1992). General theory of higher-order decomposition of exponential operators and symplectic integrators. *Phys. Letters*, **165**, 387–395.

108. Suzuki, M. (1992). Fractal path integrals with applications to quantum many-body systems. *Physica A*, **191**, 501–515.

109. Suzuki, M. (1992). General nonsymmetric higher-order decompositions of exponential operators and symplectic integrators. *J. Phys. Society Japan*, **61**, 3015–3019.

110. Suzuki, M. (1993). General decomposition theory of ordered exponentials. *Proc. Japan Acad.*, **69B**, 161–166.

111. Suzuki, M. (1993). Improved Trotter-like formula. *Phys. Letters A*, **180**, 232–234.

112. Van Vleck, E.S. (1995). Numerical shadowing near hyperbolic trajectories. *SIAM J. Sci. Comput.*, **16**, 1177–1189.

113. Varadarajan, V.S. (1974). *Lie groups, Lie algebras and their Representations*, Prentice-Hall, Englewood Cliffs, USA.

114. Warming, R.F. and Hyett, B.J. (1974). The modified equation approach to the stability and accuracy analysis of finite difference methods. *J. Comput. Phys.*, **14**, 159–179.

115. Wisdom, J. and Holman, M. (1991). Symplectic maps for the N-body problem. *Astron. J.*, **102**, 1528–1538.

116. Yoshida, H. (1990). Construction of higher order symplectic integrators. *Phys. Letters A*, **150**, 262–268.

Convergence and Stability in the Numerical Approximation of Dynamical Systems

Andrew Stuart

Scientific Computing and Computational Mathematics Program, Stanford University, California, USA

1 Outline

In this article we give an overview of the application of theories from dynamical systems to the analysis of numerical methods for initial-value problems. We start by describing the classical viewpoints of numerical analysis and of dynamical systems and then indicate how the two viewpoints can be merged to provide a framework for both the interpretation of data obtained from numerical simulations and the design of efficient numerical methods. This is done in Section 2.

In addressing the question of how to interpret data, we will show in Section 3 how the concept of convergence can be generalized to the numerical approximation of dynamical systems. The main theory is developed for one-step methods for ordinary differential equations and extensions to the study of multistep methods, adaptive time-stepping algorithms and partial differential equations are then outlined.

In addressing the question of designing efficient schemes we will show in Section 4 how the concept of stability can be generalized to dynamical systems. Stability theory is developed for both one-step and multistep methods for ordinary differential equations; extensions to adaptive time-stepping and to partial differential equations are also outlined.

A variety of surveys of this field already exist: see [13] for a complete study, see [9] for a discussion of convergence in the dynamical systems context, [12] for a discussion of stability in the dynamical systems context and see [5] for a discussion of both issues in relation to discretization of partial differential equations. Since these surveys contain fairly exhaustive bibliographys we refer to them for detailed references.

The impetus for the work described here is the fact that the classical convergence and stability theories do not apply to many dynamical systems of interest in science and engineering. Whilst the classical theories can be usefully developed at a very general level, new theories of convergence and stability are tied in a very strong way to particular classes of problems. This article concerns equations arising from physical systems where an energy loss mechanism, or dissipation, is present. For a discussion of similar issues in the context of conservative physical systems see [6].

2 Background

The problem which we wish to solve is

$$\frac{du}{dt} = f(u), \quad u(0) = U. \tag{2.1}$$

Here $f : \mathbb{R}^p \to \mathbb{R}^p$ satisfies sufficient smoothness and structural assumptions to ensure the existence of a dynamical system on $\mathbb{R}^p$: that is, a unique solution $u(t) \in C^1([0,\infty), \mathbb{R}^p)$ of Equation 2.1 exists for all $U \in \mathbb{R}^p$. Thus we define the one-parameter semigroup $S(t) : \mathbb{R}^p \mapsto \mathbb{R}^p$ for each $t \in \mathbb{R}^+$ by

$$u(t) = S(t)U.$$

For one-step methods generating a sequence $\{U_n\}_{n=0}^N \approx \{u(n\Delta t)\}_{n=0}^N$ we define, for Δt sufficiently small, $S_{\Delta t}^1 : \mathbb{R}^p \mapsto \mathbb{R}^p$ to be the solution operator for one step of the numerical method. Thus

$$U_{n+1} = S_{\Delta t}^1 U_n, \quad U_0 = U. \tag{2.2}$$

We may then define $S_{\Delta t}^n : \mathbb{R}^p \mapsto \mathbb{R}^p$ by

$$U_n = S_{\Delta t}^n U := S_{\Delta t}^1 \circ \ldots \circ S_{\Delta t}^1 U.$$

The primary objectives in solving Equation 2.1 numerically are:

1. to get quantitative and qualitative information about solutions to the equation; and
2. to obtain the information as quickly or cheaply as possible.

The concepts of *convergence* and *stability* have arisen in the theory of numerical analysis to enable these issues to be addressed. See [2,3] for example.

Classical convergence theory in numerical analysis is concerned with the following question. Fix $T > 0$ and $U \in B(0,R)$ and find the maximal real number r such that

$$\limsup_{n\Delta t = T, \Delta t \to 0} \frac{||S(n\Delta t)U - S_{\Delta t}^n U||}{\Delta t^r} \leq C, \tag{2.3}$$

for some constant $C = C(R,T)$. Typically $C \to \infty$ as $T \to \infty$. Thus a convergent scheme yields quantitative information about Equation 2.1 on time-intervals sufficiently small that $C\Delta t^r$ is small. Roughly speaking, the important concepts required to analyse the convergence of numerical methods are those of *truncation error* and *well-posedness.*

Classical (practical) stability theory in numerical analysis is concerned with the following question. Fix $\Delta t > 0$ and $U \in \mathbb{R}^p$ and find conditions which ensure that

$$\lim_{n\to\infty} S^n_{\Delta t} U = 0, \tag{2.4}$$

for the problems:

$$f(u) = \lambda u, \quad Re(\lambda) < 0, \tag{2.5}$$

and

$$\exists \mu > 0 : \langle f(u) - f(v), u - v\rangle \leq -\mu \|u - v\|^2, \quad \forall u, v. \tag{2.6}$$

Note that, for Equation 2.6 we may assume without loss of generality that 0 is the unique equilibrium point. A non-autonomous version of Equation 2.5 is also important, namely:

$$f(u) = \lambda(t)u, \quad Re(\lambda(t)) \leq -\mu < 0 \quad \forall t > 0. \tag{2.7}$$

(Note that, for this class of problems a semigroup does not exist since the equation is not time-translation invariant). For all the problems Equations 2.5–2.7, the true solution converges exponentially to a unique equilibrium point which, without loss of generality, we may assume to be 0. Useful concepts in addressing stability for Equations 2.5–2.7 are *A−stability* (for Equation 2.5 and both Runge-Kutta methods (RKM) and linear multistep methods (LMM)), *AN−stability* (for Equation 2.7 and both Runge-Kutta and multistep methods), *B−stability* (for Runge-Kutta methods applied to Equation 2.6) and *G−stability* (for multistep methods applied to Equation 2.6). See [1,3] for example. These stability theories catergorize methods which preserve property Equation 2.4 for all values of $\Delta t > 0$ and all initial data U. It is interesting to note that, roughly speaking, $AN-$ and $B-$stability are equivalent for Runge-Kutta methods and that $A-$ and $G-$stability are equivalent for multistep methods. Thus, intuition from linear problems is useful in the understanding of nonlinear problems. However, Equations 2.5–2.7 all have the same, essentially trivial, asymptotic behaviour. Nonetheless, the attendant stability concepts developed for them will turn out to be invaluable in many of the more recent stability theories for numerical approximation of dynamical systems with decidedly nontrivial asymptotic behaviour.

In the theory of dynamical systems, interest is focussed on the limit $t \to \infty$ and on the evolution of sets of initial data. Useful concepts include:

1. Invariant sets I :

$$S(t)I \equiv I \quad \forall t \geq 0.$$

2. Positively invariant sets I :

$$S(t)I \subseteq I \quad \forall t \geq 0.$$

3. ω–limit sets ω : for points $U \in \mathbb{R}^p$,

$$\omega(U) = \{x | \exists t_i \to \infty : S(t_i)U \to x\},$$

or, for sets $B \in \mathbb{R}^p$,

$$\omega(B) = \{x | \exists y_i \in B, t_i \to \infty : S(t_i)y_i \to x\}.$$

Here $t_i \in \mathbb{R}^+$ for Equation 2.1 or $t_i \in \mathbb{Z}^+$ for Equation 2.2 (with $S(n) = S^n_{\Delta t}$).

4. Basin of attraction $\mathcal{B}$ of I :

$$\mathcal{B} = \{y | \ \omega(y) \subseteq I\}.$$

5. Attractor $\mathcal{A}$: a compact invariant set $\mathcal{A}$ whose basin of attraction contains an open neighbourhood of $\mathcal{A}$ itself.

6. Global attractor: an attractor $\mathcal{A}$ for which every bounded set B is in its basin of attraction. Thus for all bounded $B \subset \mathbb{R}^p$:

$$\text{dist}(S(t)B, \mathcal{A}) \to 0 \quad \text{as} \quad t \to \infty.$$

Here

$$\text{dist}(A, B) < \epsilon \Leftrightarrow \overline{A} \subseteq \mathcal{N}(\overline{B}, \epsilon),$$

so that

$$\text{dist}(A, B) = 0 \Leftrightarrow \overline{A} \subseteq \overline{B}.$$

To illustrate these concepts we consider some examples. Figure 1 shows the solution of the scalar equation

$$\frac{du}{dt} = u - u^3, \quad u(0) = 10.$$

It is clear that $u \to 1$ as $t \to \infty$ and hence that $\omega(10) = 1$. Thus the ω–limit set is the point $u = 1$, representing a steady state.

Figure 2 shows the solution of the pair of coupled equations

$$\begin{aligned}\frac{dx}{dt} &= x + y - x(x^2 + y^2), \qquad x(0) = 3,\\ \frac{dy}{dt} &= -x + y - y(x^2 + y^2), \quad y(0) = -3.\end{aligned}$$

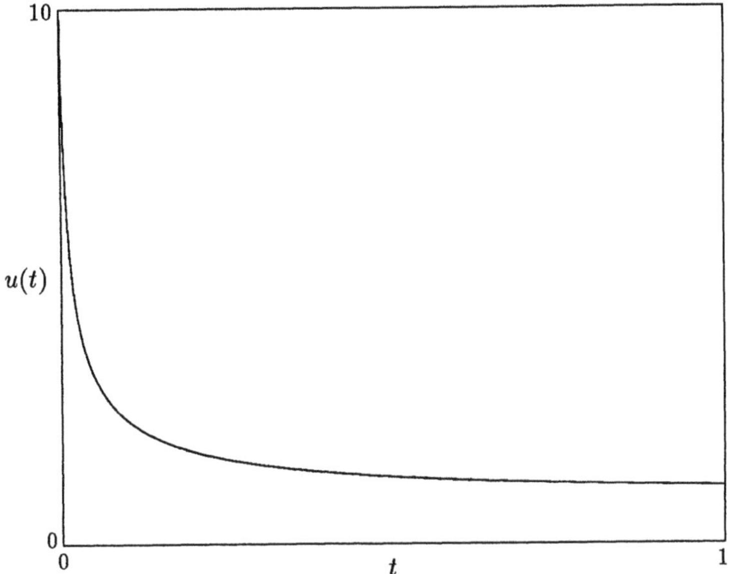

Figure 1. Approach to equilibrium point

In Figure 2 (c) we define $R(t)^2 = x(t)^2 + y(t)^2$. Clearly the solution converges to a limit cycle as $t \to \infty$. Thus the omega limit set in this case is the set $\{(x, y) : x^2 + y^2 = 1\}$, representing a periodic solution.

To illustrate a more complicated limit set, consider Equation 2.1 in $\mathbb{R}^p$, $p > 2$, with solution $u(t) = (u_1(t), \ldots, u_p(t))^T$. Assume that the first two solution components satisfy

$$u_1(t) = e^{-t} + \sin(t), \quad u_2(t) = e^{-t} + \cos(at),$$

and that the remaining solution components approach 0 as $t \to \infty$. Then, if a is irrational, as $t \to \infty$ the limiting solution is known as a *quasi-periodic solution.* Figure 3 (a) illustrates this type of behaviour in the case $a = 10\pi/31$; the notation $x(t) = u_1(t)$, $y(t) = u_2(t)$ is used. The omega limit set here is the set

$$\{u = (u_1, \ldots, u_p) \in \mathbb{R}^p : -1 \le u_i \le 1,\ i = 1, 2 \text{ and } u_i = 0, i \ge 2\}.$$

In Figure 3 (b) a similar phenomenon is illustrated: the function $u(t) = e^{-t} + \sin(t) + \cos(\delta t)$ is shown with $\delta = \sqrt{2}$.

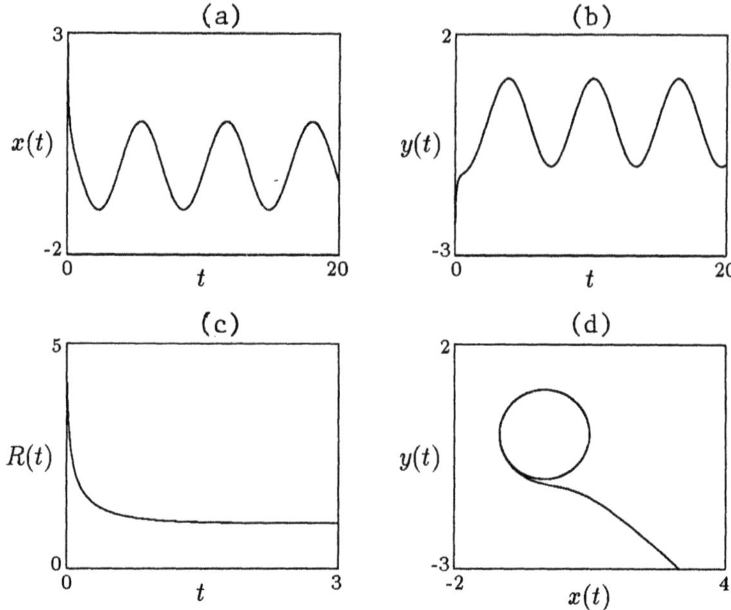

Figure 2. Periodic solution

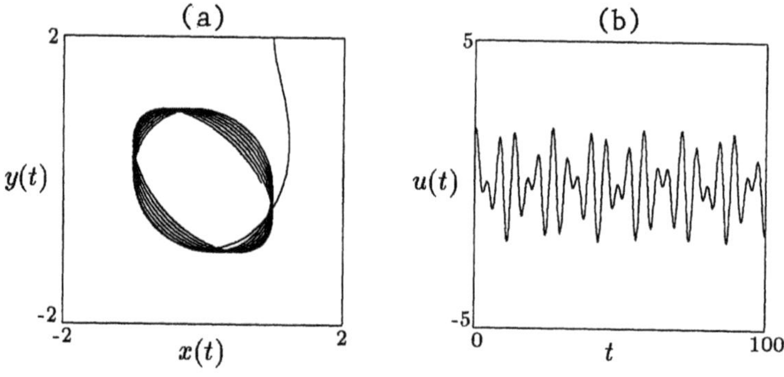

Figure 3. Quasi-periodic behaviour

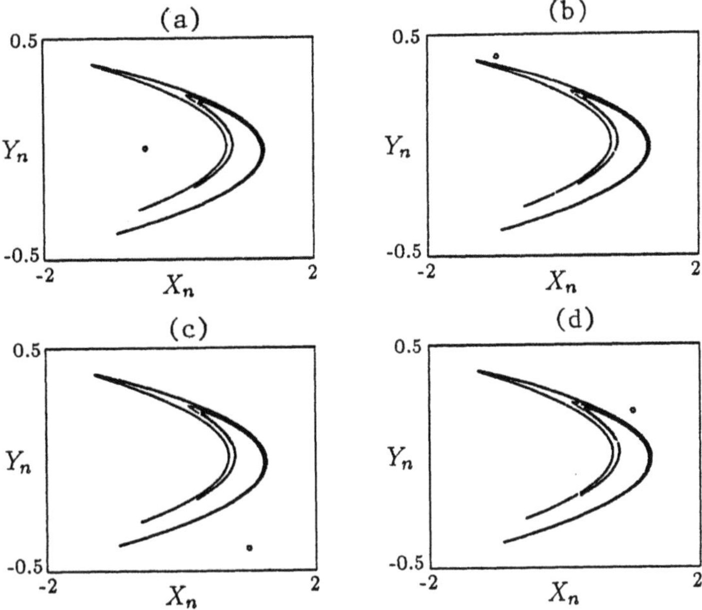

Figure 4. Strange attractor of modified Hénon map. Note that o marks the initial value, (X_0, Y_0)

Figure 4 shows an omega limit set for the modified Hénon map

$$\begin{aligned} X_{n+1} &= 1 + Y_{n+1} - \tfrac{7}{5}X_n^2, \\ Y_{n+1} &= \tfrac{3}{10}X_n. \end{aligned}$$

To produce this figure, four different starting points are chosen (marked with circles) and the map iterated. The transients are not shown so that what remains represents the omega-limit set. The important point illustrated by this figure is that the omega-limit set has a very complicated structure which is the same for all four initial conditions. This is an example of a *strange attractor*.

Strange attractors are also found in differential equations. Figure 5 shows various plots of the strange attractor for the *Lorenz equations*

$$\begin{aligned} \frac{dx}{dt} &= 10(y - x), \\ \frac{dy}{dt} &= 28x - y - xz, \\ \frac{dz}{dt} &= xy - 8z/3. \end{aligned}$$

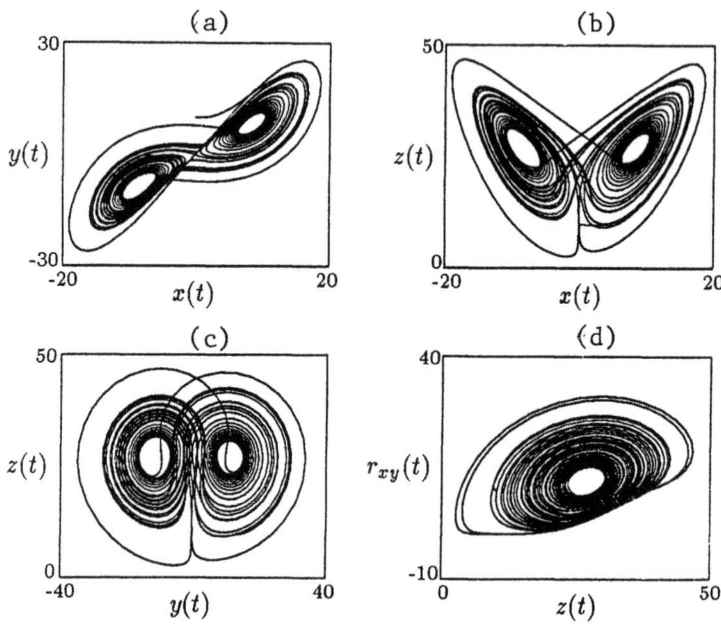

Figure 5. Lorenz attractor

In Figure 5 (d) we have defined $r_{xy}(t)^2 = x^2(t) + y^2(t)$. For these equations the attractor is in fact a global attractor. It is contained inside an ellipsoid which is a positively invariant set for the semigroup and which $S(t)U$ enters at some time $t = T(U)$ for every $U \in \mathbb{R}^p$.

The Figures 1–5 illustrate the wealth of dynamical behaviour that is present in differential equations and mappings when studied over long time intervals. It is clearly desirable that any theory of numerical analysis should encompass this variety. Classical numerical analysis however is somewhat defective in this regard. The focus is on *individual solutions* — $S(t)$ is applied to a single initial data point U. Convergence on finite time intervals is studied and stability theory is developed for problems with trivial asymptotic behaviour — exponential convergence to a unique equilibrium point. In contrast, dynamical systems is focussed on *families* of solutions: $S(t)$ is applied to sets B of initial data points. Much of the interest is in infinite time intervals and non-trivial omega limit sets (as seen in Figures 1–5.) Note that, in many applications of interest, the exact initial condition is not known in practice. In such a situation it is natural to study all possible limit sets and their basins of attraction.

Whilst the classical theories of convergence and stability have been invaluable in the development of accurate and efficient algorithms for initial-value problems, they are far from being the last word in the subject. As an example, consider again the Lorenz equations. These equations have interesting long-time dynamics

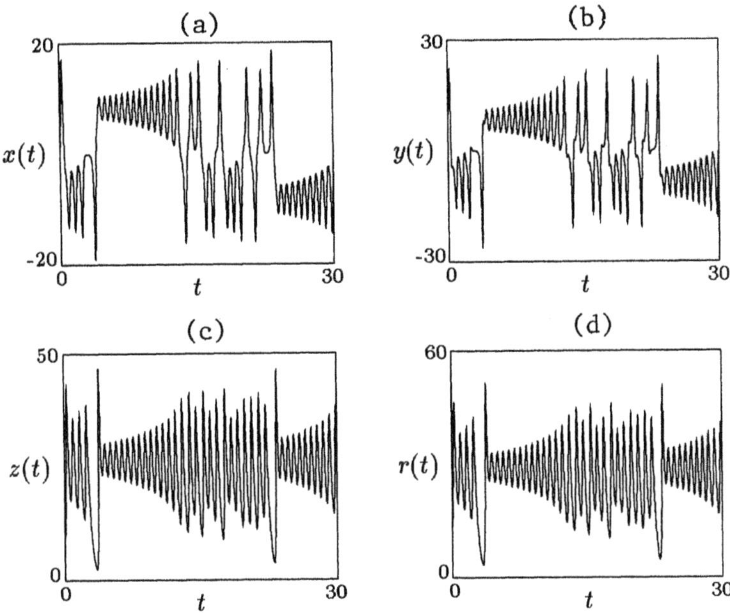

Figure 6. "Unpredictable" nature of the Lorenz equations

(see Figures 5 and 6; in Figure 6 (d) we have defined $r(t)^2 = x(t)^2 + y(t)^2 + z(t)^2$) and, to fully investigate this phenomenon, it is necessary to integrate the equations past the time at which the classical error bound Equation 2.3 is small. (Improved approximate error estimates can be obtained by estimating the constant multiplying Δt^r numerically, rather than majorizing it by an exponential, but nonetheless, the fact remains that for any given Δt there is a time after which the error in the approximation of an individual trajectory will be $\mathcal{O}(1)$). Furthermore, the equations do not satisfy Equation 2.6 and so the stability theories alluded to are of no direct use here — this is manifest in the fact that the asymptotic behaviour is far from trivial.

In this paper we outline some of the theories which help in the design and analysis of appropriate algorithms for equations with complicated long-time dynamics. For instance, the Lorenz equations have unstable periodic solutions and hyperbolic equilibria with stable and unstable manifolds; we will quote theories which show that these periodic solutions persist as unstable invariant curves in any convergent one-step numerical method and that similar results hold for the stable and unstable manifolds. The Lorenz equations have a strange (global) attractor $\mathcal{A}$; we quote a theory which indicates that any convergent numerical method also has a local attractor $\mathcal{A}_{\Delta t}$ and that $\text{dist}(\mathcal{A}_{\Delta t}, \mathcal{A}) \to 0$ as $\Delta t \to 0$.

Finally we will quote stability theories showing that, for certain particular Runge-Kutta and multistep methods, the numerical attractor is actually a global attractor as for the differential equation.

Taking this discussion of the Lorenz equation as prototypical, we now outline how the theories of convergence and stability might be usefully generalized to dynamical systems.

Note that a solution $\{S(t)U\}_{t\geq 0}$ of Equation 2.1 is a positively invariant set. Thus standard convergence of trajectories corresponds to studying the existence and closeness of a nearby positively invariant set (a single trajectory) for $S^n_{\Delta t}$ as $\Delta t \to 0$. This suggests that, in the context of dynamical systems, convergence should correspond to studying the existence of nearby (positively) invariant sets of $S^n_{\Delta t}$ as $\Delta t \to 0$ given the existence of an invariant set for $S(t)$. We pursue this approach in Section 3. (Actually more than this is shown when standard convergence of trajectories is considered, namely that the dynamics on the sets agree; in general it is not possible to extend such results to more complicated invariant sets. For example, phase error prevents this for periodic solutions).

For the problems Equations 2.5–2.7, the global attractor is the origin 0 (without loss of generality). Thus 0 has basin of attraction $\mathbb{R}^p$. Stability is concerned with finding conditions under which the basin of attraction is preserved for all $\Delta t > 0$, or for Δt independent of stiffness. This suggests that, in the context of dynamical systems, stability might usefully be defined to encompass the study of preservation of basins of attraction of (positively) invariant sets of $S(t)$ under discretization, for all $\Delta t > 0$, or for Δt independent of initial data and/or stiffness. This approach to numerical stability is pursued in Section 4.

3 Convergence

We will make the following basic assumption for all the one-step methods considered in this paper: given any bounded $B \subset \mathbb{R}^p$, there is $\Delta t_c = \Delta t_c(B)$ and $K = K(B)$ such that, for all $(\Delta t, U, V) \in [0, \Delta t_c) \times B \times B$,

$$\begin{aligned}
||S(\Delta t)U - S^1_{\Delta t}U|| &\leq K(B)\Delta t^{r+1}, \\
||D_U[S(\Delta t)U - S^1_{\Delta t}U]|| &\leq K(B)\Delta t^{r+1}, \\
||S^1_{\Delta t}U - S^1_{\Delta t}V|| &\leq [1 + K(B)\Delta t]||U - V||, \\
||D_U[S^1_{\Delta t}U] - D_V[S^1_{\Delta t}V]|| &\leq K(B)\Delta t||U - V||.
\end{aligned}$$

Here D_U denotes the derivative (Jacobian) with respect to U. If this assumption holds we will say that $S(\Delta t)$ and $S^1_{\Delta t}$ are C^1–close. All Runge-Kutta methods, for example, satisfy this assumption under sufficient smoothness on f.

Using this it is straightforward to prove the following result:

Result 1. There are constants $C_i = C_i(B), i = 1, \ldots, 4$ such that, for all $U \in B$ and $\Delta t \in [0, \Delta t_c)$ with $n\Delta t = T$

$$\begin{aligned} \|S(T)U - S^n_{\Delta t}U\| &\leq C_1 e^{C_2 T} \Delta t^r, \\ \|D_U[S(T)U - S^n_{\Delta t}U]\| &\leq C_3 e^{C_4 T} \Delta t^r. \end{aligned}$$

□

Clearly such a result is of no direct use in the dynamical systems context where long time-intervals are considered. Thus we consider instead the effect of numerical approximation on invariant sets. The unifying feature of the convergence theory we describe is that the existence theory for both the invariant sets of Equation 2.1 and of Equation 2.2 is formulated in the same way. To achieve this we need to consider solutions of Equation 2.1 as iterates of a map over a time interval of length Δt.

3.1 The ODE as a map

Near an equilibrium point ($u = 0$ without loss of generality) we may write:

$$\frac{du}{dt} = f(u) := -Au + g(u), \quad u(0) = U.$$

Here $A := D_u f(u)|_{u=0}$ and $g(u)$ is $\mathcal{O}(\|u\|^2)$ as $\|u\| \to 0$. Then we have the *variation of constants formula*

$$u(t) = e^{-At}U + \int_0^t e^{-A(t-\tau)} g(S(\tau)U)d\tau.$$

If $u_n = u(n\Delta t)$ then

$$u_{n+1} = S(\Delta t)u_n := Lu_n + N(u_n), \tag{3.1}$$

where

$$L = e^{-A\Delta t} \quad \text{and} \quad N(\bullet) = \int_0^{\Delta t} e^{-A(\Delta t-\tau)} g(S(\tau)\bullet)d\tau.$$

We now introduce $N_{\Delta t}(\bullet) : \mathbb{R}^p \mapsto \mathbb{R}^p$ so that the numerical method may be written

$$U_{n+1} = S^1_{\Delta t}U_n := LU_n + N_{\Delta t}(U_n). \tag{3.2}$$

By C^1-closeness of $S(\Delta t)$ and $S^1_{\Delta t}$ we have

$$\|N(\bullet) - N_{\Delta t}(\bullet)\|_{C^1} = \mathcal{O}(\Delta t^{r+1}).$$

Hence any results proving the existence of an invariant set for Equation 3.1 (and hence for Equation 2.1) which relies on the smallness of $N(u)$ and $D_u N(u)$ relative to properties of L can also be translated into results for Equation 3.2. (A seperate argument is needed to show that the invariant sets of the mapping Equation 3.1 are also invariant under Equation 2.1.) In this manner equilibria, local phase portraits near equilibria and local unstable and stable manifolds of equilibria can all be shown to persist, under numerical approximations satisfying our basic assumptions, provided the equilibrium point is hyperbolic. Similar ideas work in the vicinity of periodic solutions and quasi-periodic solutions. We now outline the general framework in which all these objects may be studied.

3.2 Hyperbolic invariant sets

In this subsection we study equilibria, stable and unstable manifolds, local phase portraits, periodic solutions and quasi-periodic solutions all of which we denote by $\mathcal{M}$. We will make suitable hyperbolicity assumptions but do not detail these here: see [9] or [12] for details. Roughly speaking it is these conditions which place us in the realm of problems where an energy loss mechanism is present and outside that of the conservative physical systems treated in [6]. In all these cases $\mathcal{M}$ can be constructed as a fixed point of a contraction mapping $\mathcal{T}$:

$$\mathcal{M}^{k+1} = \mathcal{T}\mathcal{M}^{k},$$

in an appropriate closed subset X of a Banach space B. For equilibria the Banach space is the set of points in $\mathbb{R}^p$; for unstable and stable manifolds B is a set of Lipschitz graphs $C(P\mathbb{R}^p, (I-P)\mathbb{R}^p)$ where P is an appropriate projection; for construction of the phase portrait B is a set of sequences in $\mathbb{R}^p$.

The mapping $\mathcal{T}$ is constructed from the mapping Equation 3.1. The derivation of $\mathcal{T}$ from Equation 3.1 is different for each of the invariant sets under consideration and will not be elaborated here. The abstract formulation is chosen to show a unifying framework for the analysis. Given $\mathcal{T}$ we may construct a mapping $\mathcal{T}_{\Delta t}$ by following, for each of the invariant sets mentioned, the derivation of $\mathcal{T}$ but using Equation 3.2 instead of Equation 3.1. This leads to the mapping

$$\mathcal{M}^{k+1}_{\Delta t} = \mathcal{T}_{\Delta t}\mathcal{M}^{k}_{\Delta t}.$$

For all the invariant sets described two important facts follow from C^1–closeness, namely

$$\sup_{\Phi \in X} \|\mathcal{T}\Phi - \mathcal{T}_{\Delta t}\Phi\|_B = \mathcal{O}(\Delta t^{r+1}), \tag{3.3}$$

$$\mathrm{Lip}\{\mathcal{T}_{\Delta t}\} = \mathrm{Lip}\{\mathcal{T}\} + \mathcal{O}(\Delta t^{r+1}); \tag{3.4}$$

thus $\mathcal{T}$ is $\mathcal{O}(\Delta t^{r+1})$ close to $\mathcal{T}_{\Delta t}$ as an operator from X into X and the Lipschitz constants are similarly close. The following type of result may be proved using this fact:

Result 2. Consider all the invariant sets of Equation 3.1, and hence Equation 2.1, mentioned at the start of this subsection. Under suitable hyperbolicity assumptions on them there exists $\mathcal{M}_{\Delta t}$, invariant for Equation 3.2 and hence Equation 2.2, satisfying

$$\|\mathcal{M}_{\Delta t} - \mathcal{M}\|_B = \mathcal{O}(\Delta t^r).$$

□

Sketch Proof Existence is proved as follows: $\mathcal{T}$ is set up to be a contraction and, because it is constructed from Equation 3.1, it follows that $\mathcal{T}$ approaches the identity as $\Delta t \to 0$. This is because $L \equiv I$ and $N \equiv 0$ if $\Delta t = 0$. Hence it turns out that

$$\text{Lip}\{\mathcal{T}\} = \lambda = 1 - C\Delta t < 1.$$

By Equation 3.4 we have

$$\text{Lip}\{\mathcal{T}_{\Delta t}\} = \text{Lip}\{\mathcal{T}\} + \mathcal{O}(\Delta t^{r+1}) \leq \lambda_{\Delta t} = 1 - \frac{C\Delta t}{2} < 1,$$

for Δt sufficiently small. Thus by Equation 3.3 we have

$$\begin{aligned}
\|\mathcal{M} - \mathcal{M}_{\Delta t}\|_B &= \|\mathcal{T}\mathcal{M} - \mathcal{T}_{\Delta t}\mathcal{M}_{\Delta t}\|_B, \\
&\leq \|\mathcal{T}\mathcal{M} - \mathcal{T}_{\Delta t}\mathcal{M}\|_B + \|\mathcal{T}_{\Delta t}\mathcal{M} - \mathcal{T}_{\Delta t}\mathcal{M}_{\Delta t}\|_B, \\
&\leq K\Delta t^{r+1} + \lambda_{\Delta t}\|\mathcal{M} - \mathcal{M}_{\Delta t}\|_B, \\
&\Rightarrow \|\mathcal{M} - \mathcal{M}_{\Delta t}\|_B \leq \frac{2K\Delta t^r}{C}.
\end{aligned}$$

□

Thus we have outlined a very general approach to proving convergence of hyperbolic invariant sets. The general technique shown in the proof is known as the ***uniform contraction principle***. See [9] and Chapter 6 of [13], for detailed references to the use of this idea in the context of numerical analysis and dynamical systems. The central individual in the development of the approach outlined is W.-J. Beyn.

3.3 Non-hyperbolic invariant sets: Attractors

For non-hyperbolic objects results as strong as those just quoted for hyperbolic objects do not hold in general. Nonetheless, for attractors it is still possible to say something about the effect of numerical approximation. To illustrate this, consider the *Rossler equations*

$$\begin{aligned}
\frac{dx}{dt} &= -y - z, & x(0) &= X, \\
\frac{dy}{dt} &= x + y/5, & y(0) &= Y, \\
\frac{dz}{dt} &= 1/5 + z(x - 5), & z(0) &= Z.
\end{aligned}$$

Figure 7 shows numerically computed attractors for this system with initial data $(X, Y, Z) = (-8.0578, 0.6288, 0.0154)$. This point is chosen to be close to the attractor so that transients are not present; the numerical attractors are thus computed simply by running the numerical method from this point over a long time interval and plotting a projection onto the $x - y$ plane. The step size is $\Delta t = 0.00005$ in (a) and is halved succesively for each subsequent figure. It is clear that some form of convergence of the attractor is occuring as Δt is refined.

Nonetheless, the trajectories calculated in each figure are not close. This is illustrated in Figure 8 where the solution leading to Figure 7 (d) is taken as "exact" and the errors in trajectories from Figures 7 (a) – (c) calculated. Note that the errors are $\mathcal{O}(1)$ by the end of the interval. Thus Figures 7 and 8 indicate that there is something more to interpreting data from numerical simulations of chaotic systems than simply asking that trajectories remain close. We now try to make this precise.

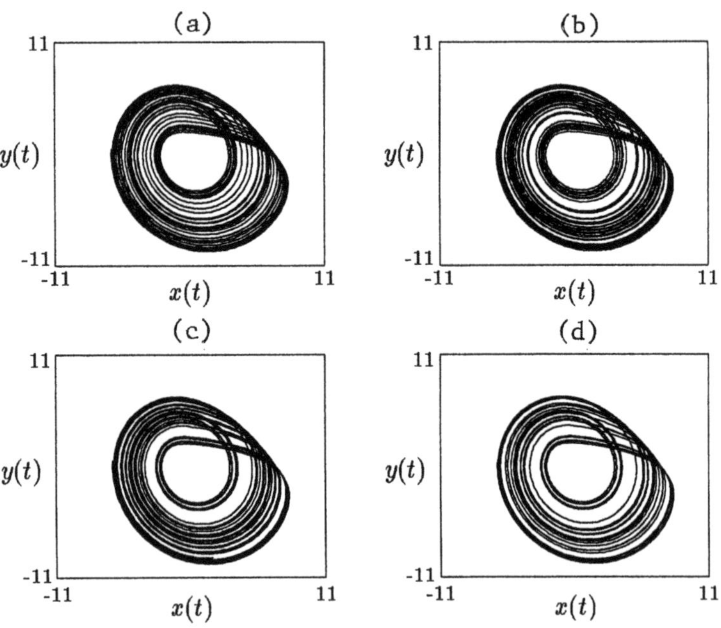

Figure 7. The Rossler attractor with four different Δt

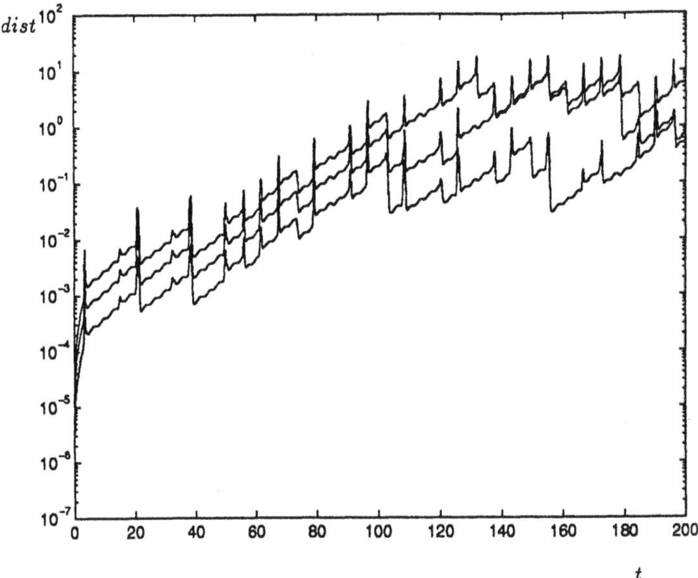

Figure 8. Errors in trajectories for the Rossler system

Result 3. Let Equation 2.1 have an attractor $\mathcal{A}$. For any ϵ there is $\Delta t_c(\epsilon)$ such that Equation 2.2 has an attractor $\mathcal{A}_{\Delta t}$ and

$$\mathcal{A}_{\Delta t} \subseteq \mathcal{N}(\mathcal{A}, \epsilon) \quad \forall \Delta t \leq \Delta t_c.$$

Thus

$$\text{dist}(\mathcal{A}_{\Delta t}, \mathcal{A}) \to 0 \quad \text{as} \quad \Delta t \to 0.$$

□

This shows that every computed point on the numerical attractor is close to a point on the true attractor, for Δt sufficiently small. However $\mathcal{A}$ is not contained in $\mathcal{N}(\mathcal{A}_{\Delta t}, \epsilon)$ in general — it is necessary to make further assumptions about the dynamics on $\mathcal{A}$ to obtain this implication. The most general of these is contained in the following result, where $W^u(\bullet)$ denotes the unstable manifold of an equilibrium point.

Result 4. Let Equation 2.1 have an attractor $\mathcal{A}$ and assume that

$$\mathcal{A} = Cl\{\bigcup_{v \in \mathcal{E}} W^u(v)\},$$

where $\mathcal{E}$ is a set of hyperbolic equilibria. For any ϵ there is $\Delta t_c(\epsilon)$ such that Equation 2.2 has an attractor $\mathcal{A}_{\Delta t}$ and

$$\mathcal{A} \subseteq \mathcal{N}(\mathcal{A}_{\Delta t}, \epsilon) \quad \forall \Delta t \leq \Delta t_c.$$

Thus

$$\operatorname{dist}(\mathcal{A}, \mathcal{A}_{\Delta t}) \to 0 \quad \text{as} \quad \Delta t \to 0.$$

□

See [9] and Chapter 7 of [13] for further elaboration of these results and for a complete bibliography. The study of attractors under numerical perturbation has been led by P.E. Kloeden and J. Lorenz and by J.K. Hale and co-workers; important further developments along the lines of Result 4 have been made by A.R. Humphries and I. Kostin.

3.4 Multistep methods as one-step methods

Multistep methods generate a dynamical system on $\mathbb{R}^{kp}$ where k is the number of steps of the method. One way of doing this is as follows: define

$$\widetilde{U}_n = (U_n^T, \ldots, U_{n+k-1}^T)^T.$$

Then, for Δt sufficiently small, we may define $\widetilde{S}^1_{\Delta t} : \mathbb{R}^{kp} \mapsto \mathbb{R}^{kp}$ by

$$\widetilde{U}_{n+1} = \widetilde{S}^1_{\Delta t}\widetilde{U}_n, \quad \widetilde{U}_0 = \widetilde{U}. \tag{3.5}$$

Composing this map over several steps yields

$$\widetilde{U}_n = \widetilde{S}^n_{\Delta t}\widetilde{U} := \widetilde{S}^1_{\Delta t} \circ \ldots \circ \widetilde{S}^1_{\Delta t}\widetilde{U}.$$

To apply the ideas we have developed for one-step methods we need to extract a map on $\mathbb{R}^p$. For *strictly stable* multistep methods this can be done, using invariant manifold theory.

We illustrate this by considering the second order backward differentiation formula (BDF):

$$U_{n+2} - \frac{4}{3}U_{n+1} + \frac{1}{3}U_n = \frac{2\Delta t f(U_{n+2})}{3}.$$

We can write this as a one-step method as just described. However, an alternative choice of variables for the one-step formulation is particularly helpful here: we introduce the new variables W_n and V_n by

$$\frac{2}{3}W_n = U_{n+1} - \frac{1}{3}U_n, \quad V_n = U_{n+1} - U_n.$$

Then a short calculation reveals that

$$\left.\begin{aligned} V_{n+1} &= \tfrac{1}{3}V_n + \tfrac{2}{3}\Delta t f(W_{n+1} - \tfrac{1}{2}V_{n+1}) \\ W_{n+1} &= W_n + \Delta t f(W_{n+1} - \tfrac{1}{2}V_{n+1}). \end{aligned}\right\}$$

The existence of an exponentially attractive invariant manifold, which is representible as a graph $h : \mathbb{R}^p \mapsto \mathbb{R}^p$ on which $v = \Delta t\, h(w)$, may be proved. To be precise, if f is sufficiently smooth and with compact support then there is a function $h : \mathbb{R}^p \mapsto \mathbb{R}^p$ such that

$$V_{n+1} = \Delta t\, h(W_{n+1}) \Leftrightarrow V_n = \Delta t\, h(W_n),$$

$$\|V_n - \Delta t\, h(W_n)\| \leq C(V_0, W_0)(1/6)^n.$$

On the manifold we have

$$W_{n+1} = W_n + \Delta t f(W_{n+1} - \frac{\Delta t}{2} h(W_{n+1}))$$

and so we have defined a one-step method on $\mathbb{R}^p$, namely:

$$W_{n+1} = S^1_{\Delta t} W_n.$$

The map $S^1_{\Delta t}$ on $\mathbb{R}^p$ is C^1 close to $S(\Delta t)$. Hence the foregoing one-step theories can be applied to the multistep method. (Note that, provided we are considering compact invariant sets, smooth modification of f outside a bounded set can always be made to yield a function with compact support; in this way the invariant manifold ideas can be used).

The ideas we have outlined were first introduced by U. Kirchgraber. They may be generalized to all strictly stable multistep methods and to an appropriate class of general linear methods. See Chapter 4 of [13] for further details and for references to the literature.

3.5 Sectorial evolution equations

We now indicate how the foregoing theories can be adapted to partial differential equations. We consider the equation

$$\frac{du}{dt} + Au = F(u), \quad u(0) = U,$$

where A is a *sectorial operator* in a Hilbert space H — the precise definition is given in [10] as are many references to appropriate literature. For the purposes of this survey it is sufficient to think of the example where A is $-\Delta$ with Dirichlet boundary conditions so that $D(A) = H^2(\Omega) \cap H^1_0(\Omega)$. We equip H with the inner-product $\langle \bullet, \bullet \rangle$ and induced norm $| \bullet |$. If we assume, for some $\gamma \in (0, 1)$,

$$|F(u) - F(v)| \leq K(R)|A^\gamma(u - v)| \quad \forall u, v \in B_{A^\gamma}(0, R),$$

then local existence and uniqueness follow for $U \in D(A^\gamma)$. For our illustrative example of A we have $D(A^{\frac{1}{2}}) = H^1_0(\Omega)$.

A dissipativity condition, such as

$$-|A^{1/2}u|^2 + \langle F(u), u \rangle \leq \alpha^2 - \beta^2 |u|^2, \tag{3.6}$$

will ensure global existence. In this case we define the semigroup $S(t) : D(A^\gamma) \mapsto D(A^\gamma)$ so that

$$u(t) = S(t)U,$$

and consider perturbations to the problem (numerical or otherwise) yielding an approximate semigroup $S^h(t) : D(A^\gamma) \mapsto D(A^\gamma)$. If the discretization is in space only then $t \in \mathbb{R}^+$ whilst if the discretization is in time alone or in space and time then $t \in \Delta t\mathbb{Z}^+$. We let $\mathcal{S}$ denote $\mathbb{R}^+$ or $\Delta t\mathbb{Z}^+$ as appropriate.

There is no concept of truncation error in this abstract setting since the space on which $S(t)$ yields a dynamical system may not possess sufficient smoothness. Hence the approach to hyperbolic invariant sets outlined before cannot be used without modification. However, one typically has finite-time error estimates of the form

$$|A^\gamma(S(t)U - S^h(t)U)| \leq \frac{C(T,R)h}{t^\alpha},$$

$$|A^\gamma(D_U[S(t)U - S^h(t)U)])| \leq \frac{C(T,R)\kappa(h)}{t^\alpha},$$

$$\forall (t,U) \in (0,T]\bigcap \mathcal{S} \times B_{A^\gamma}(0,R).$$

Here $\kappa(h) \to 0$ as $h \to 0$.

For hyperbolic objects contraction arguments can still be set up for the semigroups over fixed time-intervals $[0,T]$ rather than $[0,\Delta t]$. Again it is necessary to use a variation of constants formula to write

$$u(t) = e^{-At}U + \int_0^t e^{-A(t-\tau)}F(S(\tau)U)d\tau$$

and hence to formulate a map on $D(A^\gamma)$ for $u_n = u(nT)$. In fact

$$u_{n+1} = Lu_n + N(u_n)$$

with

$$L = e^{-AT} \quad \text{and} \quad N(\bullet) = \int_0^T e^{-A(T-\tau)}F(S(\tau)\bullet)d\tau.$$

The perturbed problem generates a map which is $\mathcal{O}(h)$ close to $L\bullet + N(\bullet)$ in $C^0(D(A^\gamma), D(A^\gamma))$ with the sup-norm topology and $\mathcal{O}(\kappa(h))$ close in $C^1(D(A^\gamma), D(A^\gamma))$ with the induced sup-norm topology. For fixed T the time singularity integrates out of the error estimates between the two maps.

The basic method for studying hyperbolic invariant sets is similar to those used before but with fixed $T > 0$. Now, since T is fixed independent of the discretization constants, the contraction constant is $\mathcal{O}(1)$. Similar arguments to those used for ODEs, employing the uniform contraction principle and now using the finite time C^1 error estimates instead of truncation error bounds, yield convergence of hyperbolic invariant sets. To conclude the proofs it is simply necessary to establish that the invariant sets of $L\bullet + N(\bullet)$ are also invariant under $S(t)$, and similarly for $S^h(t)$, for $t \neq T$. For attractors the same proofs may be used as in finite dimensions.

References to the literature concerning perturbation of sectorial evolution equations may be found in [10]. Central in the development of this subject area has been the work of S. Larsson and co-workers.

3.6 Adaptive time-stepping algorithms

In real software codes the time-step is varied adaptively as part of the integration procedure. Such methods, when based on one-step integration schemes, can often be formulated as mappings on $\mathbb{R}^p \times \mathbb{R}^+$ of the form

$$U_{n+1} = S^1(U_n, \Delta t_n), \quad U_0 = U,$$

$$\Delta t_{n+1} = \Gamma(U_n, \Delta t_n; \tau), \quad \Delta t_0 = \Delta t_{init}.$$

Here τ is a parameter, input by the user, which controls estimates of the local error commited at each step. The function $S^1(U,t) = S_t^1 U$ defined before. We will not go into the details of realistic models for Γ here. It suffices to say that Γ is chosen to ensure that an estimate of the local error committed at each step is $\mathcal{O}(\tau \Delta t_n)$. Note that in practice Γ is discontinuous due to step-rejections. Further non-smoothness in derivatives of Γ is introduced by maximum step-size and step-size ratio bounds.

If $t_n = \sum_{j=0}^{n-1} \Delta t_j$ then the heuristics underlying the algorithm are such that U_n should approximate $u(t_n)$ to $\mathcal{O}(\tau)$. However, for a fairly realistic model Γ of what is used in practice, the best available result concerning the convergence of such schemes is probabilistic in nature:

Result 5. [11] Let f satisfy certain genericity and smoothness conditions, assume that it has compact support and consider a certain choice of Γ which incorporates step-rejection, maximum step-size ratio and maximum step-size. If U is chosen at random uniformly in $B(0,R)$ with respect to Lebesque measure then, with probability one, there is Δt_c sufficiently small such that, for all $\Delta t_{init} \in (0, \Delta t_c)$,

$$||U_n - u(t_n)|| \leq C(T,U,R)\tau, \quad 0 \leq t_n \leq T.$$

□

Thus there are initial data (of zero measure) in $B(0,R)$ for which convergence may not occur. These are initial data points for which the time-step may become large, because the local error estimate is small, even when the true local error is not small. Furthermore, the error constant $C(T,U,R)$ is not uniform across the bounded set $B(0,R)$ — compare this with the classical case of Equation 2.3. Hence it is not at all clear how to progress further with analysis of this problem to encompass dynamical systems and long time intervals where sets of initial data must be considered, possibly including points for which no error bound exists.

A variety of other authors have considered finite-time convergence and dynamical systems analysis for adaptive time-stepping algorithms prior to Result 5. Stetter [7] avoided the problems leading to the probablistic nature of Result 5 simply by assuming that f and/or U are chosen so that the undesirable initial data points are avoided. This implies that, for some constant $K = K(U_n, \Delta t_n)$,

$$\Delta t_{n+1}^{r} \leq K\tau. \tag{3.7}$$

Stoffer and Nipp in [8] avoided the issue by modifying Γ to enforce Equation 3.7; they were then able to prove some very strong results about the effect of adaptive time-stepping algorithms on periodic solutions. The assumption Equation 3.7 implies that, for some constant $C = C(U_{n+1})$,

$$||S(U_{n+1}, \Delta t_{n+1}) - S^1(U_{n+1}, \Delta t_{n+1})|| \leq C\tau\Delta t_{n+1}. \tag{3.8}$$

In [4] the assumption Equation 3.8 was made and the effect of discretization studied on the long-time behaviour of dissipative and contractive dynamical systems.

In summary the situation is this: for realistic models of Γ only very weak probabilistic finite-time convergence results exist. If certain assumptions (such as Equations 3.7 and 3.8) are made then stronger results can be proved. However it is unclear how to justify these assumptions in general. Thus a coherent analysis of variable time-stepping has yet to emerge.

4 Stability

We discuss three classes of nonlinear problem, starting with the classical theory of contractive systems which is surveyed in [1] and progressing to the dissipative and gradient theories which are surveyed in [12] and Chapter 5 of [13]. The results concerning contractive problems are straightforward modifications of the work of G. Dahlquist and of K. Burrage and J. Butcher from the 1970's. The results concerning dissipative and gradient systems may be found in the work of C.M. Elliott, of A.T. Hill, and of A.R. Humphries and A.M. Stuart. It is often desirable that numerical methods mimic the gross asymptotic features of Equation 2.1 (for example, ultimate boundedness) for intervals $\Delta t \in [0, \Delta t_c]$ with Δt_c independent of initial data. The fact that this does not occur for most methods may be understood by studying ***spurious solutions***; see Chapter 5 of [13]. Active research into spurious solutions has died down, but the existing body of papers provides important motivation for the construction of methods, such as those detailed in Results 7,8,10,11,13 and 14, which avoid spurious behaviour.

The results now stated are all followed by the expression (ODE), (RKM) or (LMM) depending upon whether they apply to the original problem Equation 2.1, its Runge-Kutta approximation or its linear multistep approximation.

4.1 Contractive problems

Let Equation 2.6 hold:

$$\exists \mu > 0 : \langle f(u) - f(v), u - v \rangle \leq -\mu \|u - v\|^2, \quad \forall u, v \in \mathbb{R}^p.$$

Without loss of generality we assume that the unique equilibrium point of f is 0.

Result 6. (ODE) $\omega(U) = \{0\}$ for all $U \in \mathbb{R}^p$.

□

Result 7. (RKM) If the method is B–stable then, for any $\Delta t > 0$, $\omega(U) = \{0\}$ for all $U \in \mathbb{R}^p$.

□

Result 8. (LMM) If the method is A–stable then, for any $\Delta t > 0$, $\omega(\widetilde{U}) = \{0\}$ for all $\widetilde{U} \in \mathbb{R}^{kp}$.

□

Note that B–stability is essentially equivalent to AN–stability (for non-confluent methods) [1] and thus that the preceeding three results show the remarkable fact that linear stability theories give a satisfactory understanding of a certain class of contractive nonlinear problems.

However, class Equation 2.6 admits only trivial limiting behaviour in Equation 2.1. To generalize Equation 2.6 it is natural to consider:

$$\exists c > 0 : \langle f(u) - f(v), u - v \rangle \leq c \|u - v\|^2, \quad \forall u, v \in \mathbb{R}^p. \tag{4.1}$$

But this alone is too broad a class to work in — exponentially growing solutions with unbounded limiting behaviour are admitted. Hence we consider other possibilities.

4.2 Dissipative problems

The first such possibility is to consider dissipative problems where

$$\exists \alpha, \beta > 0 : \langle f(u), u \rangle \leq \alpha^2 - \beta^2 \|u\|^2, \quad \forall u \in \mathbb{R}^p. \tag{4.2}$$

The Lorenz equations satisfy this condition showing that complicated limiting behaviour is certainly present within this class — see Figures 5 and 6. The Navier-Stokes equations also satisfy an infinite-dimensional analogue of this condition. The following result holds for Equation 2.1 under Equation 4.2.

Result 9. (ODE) The set $\mathcal{B} := B(0; \frac{\alpha}{\beta} + \epsilon)$ is positively invariant and has basin of attraction $\mathbb{R}^p$. Hence $\omega(\mathcal{B})$ is a global attractor.

□

Remarkably, the numerical stability theories appropriate for contractive nonlinear problems also turn out to be appropriate for this class of dissipative problems.

Result 10. (RKM) If the method is B–stable then there is a constant $C > 0$ such that, for any $\Delta t > 0$, the set $\mathcal{B}_{\Delta t} := B(0; \frac{\alpha}{\beta} + C\Delta t)$ is positively invariant and has basin of attraction $\mathbb{R}^p$. Thus $\omega(\mathcal{B}_{\Delta t})$ is a global attractor.

□

Result 11. (LMM) If the method is A–stable then there are constants $C, K > 0$ such that, for any $\Delta t > 0$, the set $\mathcal{B}_{\Delta t,k} := B(0; K\frac{\alpha}{\beta} + C\Delta t)$ is positively invariant and has basin of attraction $\mathbb{R}^{kp}$. Thus $\omega(\mathcal{B}_{\Delta t,k})$ is a global attractor.

□

4.3 Gradient problems

Another important class of nonlinear problems are gradient systems. These arise in many physical processes where dynamic energy minimization is present and are also important in the theory of dynamical systems. We consider gradient systems under the condition that all equilibria are hyperbolic and under the well-posedness assumption Equation 4.1. Specifically let the following conditions hold:

$$\left\{ \begin{array}{c} \exists F \in C^1(\mathbb{R}^p, \mathbb{R}) \quad \text{with} \quad F(u) \geq 0, \\ \lim_{\|u\| \to \infty} F(u) = +\infty \quad \text{and} \quad f(u) = -\nabla F(u) \\ \text{The set} \quad \mathcal{E} \quad \text{of equilibria is hyperbolic} \\ \text{Equation 4.1} \quad \text{holds} \end{array} \right\}. \tag{4.3}$$

Note that, under Equation 4.3,

$$\frac{d}{dt} F(u(t)) = -\|f(u(t))\|^2.$$

Using this fact, the following result follows:

Result 12. (ODE) For any $U \in \mathbb{R}^p$ there is $v \in \mathcal{E}$ such that $\omega(U) = \{v\}$.

□

The stability theory for gradient systems is not as well-developed as for dissipative problems. We consider only the one and two stage *theta* methods (which are $A-$stable for $\theta \in [1/2, 1]$) and the BDF methods of order ≤ 3 (which are $A(\alpha)-$stable.

Result 13. (RKM) If the one or two-stage method has $\theta \in [1/2, 1]$ ($A-$stable) then, for $\Delta t \in (0, \frac{1}{c})$ and for any $U \in \mathbb{R}^p$, there is $v \in \mathcal{E}$ such that $\omega(U) = \{v\}$.

□

Result 14. (LMM) If the BDF methods of order ≤ 3 are used ($A(\alpha)-$ stable) then for $\Delta t \in (0, \frac{1}{c})$ and for any $\widetilde{U} \in \mathbb{R}^{kp}$, there is $v \in \mathcal{E}$ such that $\omega(\widetilde{U}) = \{v \otimes e\}$, where e is the unit vector in $\mathbb{R}^k$.

□

Once again the preceeding results show important connections with the linear theory. Note that gradient systems necessarily have symmetric linearization so that it is not entirely surprising that $A(\alpha)$ stability plays an important role. Further work to generalize Results 13 and 14 to other schemes would be interesting.

4.4 Sectorial evolution equations

For dissipative PDEs such as the Kuramoto-Sivashinksy equation, the Navier-Stokes equation, various reaction-diffusion equations and the Cahn-Hilliard equation, the effect of a variety of space and time discretizations on dissipativity have been investigated. For gradient PDEs such as the Cahn-Hilliard equation and Allen-Cahn equation, the effect of space and time discretization on the gradient structure have been investigated. Much of these investigations have centred around the work of C. Foais, R. Temam and co-workers and C.M. Elliott and co-workers.

4.5 Adaptive time-stepping algorithms

An appropriate generalization of stability in this context is to seek numerical methods which replicate basins of attraction for Equation 2.1 under Equation 2.6, Equation 4.2 or Equation 4.3 for an interval of the tolerance τ which is independent of initial data and/or stiffness. Thus τ plays the role of Δt in the fixed time-step theories.

Such stability results do exist for some simple adaptive time-stepping schemes although a complete theory has not emerged. It is an interesting fact that for contractive, dissipative and gradient problems some schemes are stable in adaptive implementations which are not stable in fixed-step mode. See [12] and [14].

5 Summary

In summary, we have highlighted the differences in philosophy between the classical theories of numerical analysis and theories of dynamical systems. We have shown that natural generalizations of the classical theories of numerical analysis can be made to encompass the dynamical systems context, especially for equations with an energy loss mechanism. These generalizations build heavily on the classical theories, especially in the case of stability. A very general approach to perturbation theory for hyperbolic invariant sets and for attractors exists. This theory applies to one-step and multistep methods for ordinary differential equations and has been extended to a broad class of partial differential equations.

The primary limitation of the convergence theory is for non-hyperbolic attractors. However, this limitation exists even in the context of differential equations where the effect of smooth perturbation to the vector field f is not completely understood. The stability theory is fairly complete for most of the systems mentioned, with the exception of gradient systems where much remains to be done.

A major hole in the theory of numerical analysis for dynamical systems is for adaptive time-stepping algorithms, both for convergence and stability. Some isolated papers exist in this area but a coherent and general view has yet to emerge. This area certainly presents hard mathematical challenges; it might potentially influence software development, but this is difficult to predict.

Another area where interesting questions concerning the dynamics of numerical methods remains is that of conservative and Hamiltonian systems where, typically, hyperbolicity does not hold for invariant sets. No satisfactory and complete theory exists which enables us to interpret data from long-time simulations of these problems. In contrast to the convergence theory outlined in this paper for dynamical systems of energy-loss type, that for conservative and Hamiltonian systems will necessarily involve an interaction between structural properties of the method — such as conservation of energy or symplectic two-form — and its long-time dynamics.

A final area where further active or important research may be anticipated is that of *backward error analysis* and *shadowing*. A variety of scattered, but important, results already exist in this field — see [6] and [13] for further references — but a wider range of results, and consequently a more complete picture, is likely to emerge over the next few years.

Thus the formulation, and study, of new and interesting questions concerning numerical approximation of initial-value problems has been an active research area over the last decade and it seems likely that there are sufficient open, interesting and important questions that it will continue to be an active field over the next decade.

6 Acknowledgements

The author is grateful to the participants of the Geiranger, Norway "ODE to NODE" workshop of June 1995, where a preliminary version of this material was discussed and subsequently improved, and to Tony Humphries who made several valuable suggestions. The figures are taken from [13] and used with the permission of Cambridge University Press.

This work was supported by the Office of Naval Research, grant number N00014-92-J-1876, and by the National Science Foundation, grant number DMS-9201727.

References

1. K. Dekker and J.G Verwer, *Stability of Runge-Kutta Methods for Stiff Nonlinear Differential Equations.* North-Holland, Amsterdam, 1984.
2. E. Hairer, S.P. Nørsett and G. Wanner, *Solving Ordinary Differential Equations I: Nonstiff Problems*, Springer-Verlag, New York, 1987.
3. E. Hairer, S.P. Nørsett and G. Wanner, *Solving Ordinary Differential Equations II: Stiff Problems and Differential-Algebraic Equations*, Springer-Verlag, New York, 1991.
4. D.J. Higham and A.M. Stuart, *Analysis of the Dynamics of Local Error Control Via a Piecewise Continuous Residual.* Submitted to BIT.
5. A.R. Humphries, D.A. Jones and A.M. Stuart, *Approximation of dissipative partial differential equations over long time intervals*, in "Numerical Analysis", editors D.F. Griffiths and G.A. Watson, Longman, 1994.
6. J.M. Sanz-Serna, *Geometric Integration, Article in this volume.*
7. H.J. Stetter, *Tolerance proportionality in ODE-codes.* Appears in "Proc. Second Conf. on Numerical Treatment of Ordinary Differential Equations", Ed. R. März, Seminarberichte 32, Humboldt University, Berlin, 1980.
8. D. Stoffer and K. Nipp, *Invariant curves for variable step-size integrators.* BIT **31**(1991), 169–180 and (*Erratum*) BIT **32**(1992), 367–368.
9. A.M. Stuart, *Numerical analysis of dynamical systems.* Acta Numerica 1994, 467–572, Cambridge University Press.
10. A.M. Stuart, *Perturbation theory for infinite dimensional dynamical systems.* Appears in "Theory and numerics of ordinary and partial differential equation", edited by M. Ainsworth, J. Levesley, W.A. Light and M. Marletta.
11. A.M. Stuart *Probabilistic and Deterministic Convergence Proofs for Software for Initial Value Problems.* To appear in Ann. Num. Math.
12. A.M. Stuart and A.R. Humphries, *Model problems in numerical stability theory for initial value problems.* SIAM Review **36**(1994), 226–257.
13. A.M. Stuart and A.R. Humphries, *Dynamical Systems and Numerical Analysis.* Cambridge University Press, 1996.
14. A.M. Stuart and A.R. Humphries, *The essential stability of local error control for dynamical systems.* SIAM J. Num. Anal. **32**(1995), 1940–1971.

Beyond the Classical Theory of Computational Ordinary Differential Equations

Arieh Iserles

Department of Applied Mathematics and Theoretical Physics, University of Cambridge

1 The classical problem and beyond

It is appropriate to commence this brief survey of the numerical analysis of 'strange' ordinary differential equations by quoting verbatim from Jack Lambert's informative survey in the last *State of the Art in Numerical Analysis* proceedings [22]: "By a *problem* we shall mean the initial value problem

$$\dot{\boldsymbol{y}} = \boldsymbol{f}(t, \boldsymbol{y}), \quad \boldsymbol{y}(t_0) = \boldsymbol{y}_0, \quad t_0 \le t \le T, \quad \boldsymbol{f} : \mathbb{R} \times \mathbb{R}^m \to \mathbb{R}^m$$

and by a *method* we shall mean a numerical method of the class defined by

$$\begin{aligned} \boldsymbol{y}_\mu &= \boldsymbol{\eta}_\mu(h), \quad \mu = 0, 1, \ldots, k-1 \\ \sum_{j=0}^{k} \alpha_j \boldsymbol{y}_{n+j} &= h\boldsymbol{\phi}_f(t_n, \boldsymbol{y}_{n+k}, \boldsymbol{y}_{n+k-1}, \ldots, \boldsymbol{y}_n) \end{aligned}$$

where h is the steplength and k the stepnumber." Everything is here, in this single sentence, both the object of study and the tools for algorithmic development and mathematical analysis. Indeed, in many respects the computational work on this 'plain-vanilla ODE problem' was one of the great success stories of modern numerical mathematics. Guided by the seminal work of Germund Dahlquist, a whole generation of first-class numerical analysts have established a profound understanding of the subject on theoretical and practical levels alike. Arguably, computational ODE theory became the gold standard of the numerical analysis of differential equations and a subject for emulation by the more elusive theory of computational PDEs.

Thus, a decade ago matters appeared to settle down to the leisurely course of incremental development: speeding up a programme, tightening an error estimate.... As often, however, once a subject seemingly reaches a plateau, something is bound to happen. Specifically, building on the aforementioned understanding, the numerical community becomes equipped to pose new questions in a meaningful fashion. And, as soon as new answers arrive, it transpires that, beyond their mathematical or æsthetical virtues, they are of relevance to a range of applications. The plateau is abandoned for a climb up new and uncharted summits.

The critical step in abandoning the old paradigm is that by a 'problem' we no longer mean a single, standard ordinary differential system with an anonymous function $\boldsymbol{f}$. Different families of ODEs share certain qualitative attributes and they deserve special numerical treatment that does not trade these attributes away. Moreover, many problems that seemingly look like ODEs are in reality completely different mathematical constructs. Although their algorithmic analysis draws heavily on existing theory of computational ODEs, it poses different questions and seeks different answers.

Actually, our designation of the classical theory as dealing with just one kind of an object, a 'general ODE', is somewhat unfair, because of the long-standing distinction between *stiff* and *nonstiff* ODEs. Yet, it is a measure of the great advances of the last decade that stiff ODEs, once considered to be the most difficult computational nut to crack, are recognised at present as, basically, quite simple. The seeming difficulty of stiff ODEs, namely that trajectories corresponding to different initial values approach each other at an exponential speed, is ultimately also their redeeming grace. As long as the underlying equations are discretized by time-stepping algorithms with decent stability properties, bunching of trajectories means that numerical perturbations are bound to go away.

Two of the themes of this survey can be visualised as limiting cases of stiff ODEs. Firstly, in a well-defined sense, allowing the stiffness ratio to become infinite results in *differential-algebraic equations (DAEs)*, the subject of Section 2. A good starting point for DAEs is the system

$$\begin{aligned} \boldsymbol{x}' &= \boldsymbol{f}(\boldsymbol{x},\boldsymbol{y}), \\ \boldsymbol{0} &= \boldsymbol{g}(\boldsymbol{x},\boldsymbol{y}), \end{aligned} \qquad t \ge t_0, \tag{1.1}$$

which, as its name implies, combines elements of differential and algebraic equations.

Another limiting case of stiff ODEs occurs when, instead of decaying modes, one contemplates solutions that neither die out nor necessarily tend to a fixed point but evolve forever, whilst conserving certain invariants. Since such invariants often possess important physical interpretation or correspond to the outcome of qualitative mathematical analysis, there are obvious advantages to their retention under discretization. A familiar (and perhaps the most important) example is presented by Hamiltonian systems

$$\begin{aligned} \boldsymbol{p}' &= -\frac{\partial H(\boldsymbol{p},\boldsymbol{q})}{\partial \boldsymbol{q}}, \\ \boldsymbol{q}' &= \frac{\partial H(\boldsymbol{p},\boldsymbol{q})}{\partial \boldsymbol{p}}, \end{aligned} \qquad t \ge t_0, \tag{1.2}$$

which are surveyed in great depth elsewhere in this volume, in Chus Sanz-Serna's paper [27]. Another example (about which more later) is the ODE

$$Y' = B(Y)Y, \quad t \ge t_0, \quad Y(t_0) = Y_0, \tag{1.3}$$

where the unknown Y is a $d \times d$ real matrix, the initial value Y_0 is an orthogonal matrix and the function B maps orthogonal matrices to skew-symmetric matrices. It is easy to prove that the invariant is *orthogonality:* the matrices $Y(t)$ are orthogonal for all $t \geq t_0$ [11,19].

It is useful to express the retention of invariants in the language of differential manifolds. In Section 3 we review numerical methods on manifolds, a theme that has already led to significant new results and which is a subject of intensive and exciting current research.

Section 4 is devoted to yet other constructs which, while resembling standard ODEs, are quite different: *functional differential equations (FDEs).* A convenient starting point for their discussion is the equation

$$\boldsymbol{y}'(t) = \boldsymbol{f}(t, \boldsymbol{y}(t), \boldsymbol{y}(t-\tau)), \quad t \geq 0, \tag{1.4}$$

given in tandem with the initial condition $\boldsymbol{y}(t) = \boldsymbol{\phi}(t)$, $-\tau < t \leq 0$.

A common thread running through the numerical analysis of 'nonclassical ODEs' (a misnomer, since many are not ODEs at all!) is that a convenient algorithmic point of departure is the familiar arsenal of standard numerical methods for the ODE

$$\boldsymbol{y}' = \boldsymbol{f}(t, \boldsymbol{y}), \quad t \geq t_0, \quad \boldsymbol{y}(t_0) = \boldsymbol{y}_0,$$

namely *multistep* methods

$$\sum_{k=0}^{s} \rho_k \boldsymbol{y}_{n+k} = h \sum_{k=0}^{s} \sigma_k \boldsymbol{f}(t_{n+k}, \boldsymbol{y}_{n+k}), \quad n \in \mathbb{Z}^+, \tag{1.5}$$

and *Runge–Kutta* schemes

$$\left.\begin{array}{ll} \left.\begin{array}{ll} \boldsymbol{\varphi}_\ell &= \boldsymbol{y}_n + h \sum_{j=1}^{\nu} a_{\ell,j} \boldsymbol{k}_j, \\ \boldsymbol{k}_\ell &= \boldsymbol{f}(t_n + c_\ell h, \boldsymbol{\varphi}_\ell), \end{array}\right\} & \ell = 1, 2, \ldots, \nu, \\ \boldsymbol{y}_{n+1} = \boldsymbol{y}_n + h \sum_{\ell=1}^{\nu} b_\ell \boldsymbol{k}_\ell, & \end{array}\right. \quad n \in \mathbb{Z}^+ \tag{1.6}$$

Classical methods can be occasionally extended fairly easily, for example by the provision of continuous output, although sometimes considerably more effort is required. Their analysis, however, is difficult and often calls for unfamiliar tools: differential topology, Lie group theory, symplectic geometry, harmonic analysis....

Having quoted from Jack Lambert's survey in a spirit of mild criticism (an easy commodity when aided by hindsight), this introduction will not be complete without another quote from an earlier *State of the Art in Numerical Analysis* survey by the same author [21], which admirably (and presciently) sums up the new paradigm: "The initial value problem is a lively mathematical creature (compared with, for example, the linear algebraic problem); where there is structure, we ought to be able to use it in the numerical method".

2 Differential-algebraic equations

DAEs occur naturally in a surprisingly large number of applications: optimal control, the behaviour of constrained mechanical systems (for example in robotics) and singular perturbation problems. Given their ubiquity, it is surprising that they have not received comprehensive attention in the numerical community until the last decade.

As often in an exposition of mathematical material, it is preferable to commence by considering the simplest possible model which is rich enough to expose the underlying structure. Thus, let us consider the equation (1.1), where $\boldsymbol{x} \in \mathbb{R}^{m_1}$, $\boldsymbol{y} \in \mathbb{R}^{m_2}$, the functions $\boldsymbol{f} : \mathbb{R}^{m_1+m_2} \to \mathbb{R}^{m_1}$ and $\boldsymbol{g} : \mathbb{R}^{m_1+m_2} \to \mathbb{R}^{m_2}$ are sufficiently smooth and the initial condition $(\boldsymbol{x}_0, \boldsymbol{y}_0)$ obeys the equation $\boldsymbol{g}(\boldsymbol{x}_0, \boldsymbol{y}_0) = \mathbf{0}$. In addition, we stipulate that the Jacobian matrix

$$\frac{\partial \boldsymbol{g}(\boldsymbol{x}_0, \boldsymbol{y}_0)}{\partial \boldsymbol{y}}$$

is nonsingular. It is then easy to prove, invoking the implicit function theorem, that there exists a function $\boldsymbol{q} : \mathbb{R}^{m_1} \to \mathbb{R}^{m_2}$, defined in a non-empty neighbourhood of $\boldsymbol{x}_0$, such that $\boldsymbol{y}(t) = \boldsymbol{q}(\boldsymbol{x}(t))$ for all $t \in [t_0, t^*]$. Substituting $\boldsymbol{q}$ into (1.1) we derive the *state space form* of the DAE,

$$\boldsymbol{x}' = \boldsymbol{f}(\boldsymbol{x}, \boldsymbol{q}(\boldsymbol{x})), \quad t \geq t_0. \tag{2.1}$$

A very fruitful interpretation of the DAE (1.1) is as a limiting case of a *singular perturbation problem*

$$\begin{aligned} \boldsymbol{x}' &= \boldsymbol{f}(\boldsymbol{x}, \boldsymbol{y}), \\ \varepsilon \boldsymbol{y}' &= \boldsymbol{g}(\boldsymbol{x}, \boldsymbol{y}), \end{aligned} \quad t \geq t_0, \tag{2.2}$$

as $\varepsilon \downarrow 0$. In particular, it allows us to extend standard ODE discretization methods to this framework. Matters are simple with regard to multistep schemes: we apply the method (1.5) to the singularly-perturbed problem (2.2), let $\varepsilon \downarrow 0$ and the outcome is

$$\begin{aligned} \sum_{k=0}^{s} \rho_k \boldsymbol{x}_{n+k} &= h \sum_{k=0}^{s} \sigma_k \boldsymbol{f}(\boldsymbol{x}_{n+k}, \boldsymbol{y}_{n+k}), \\ 0 &= \sum_{k=0}^{s} \sigma_k \boldsymbol{g}(\boldsymbol{x}_{n+k}, \boldsymbol{y}_{n+k}), \end{aligned} \quad n \in \mathbb{Z}^+. \tag{2.3}$$

We hasten to remark that this simplicity is misleading. As soon as we add stability considerations, simple reasoning singles out one family of multistep methods as a natural choice for the task in hand.

The generalization of a Runge–Kutta scheme (1.6) to DAEs requires more effort. For reasons that will become apparent soon, we commence by stipulating that the $\nu \times \nu$ *Runge–Kutta matrix* $A = (a_{i,j})$ is nonsingular and denote by $(\omega_{i,j})$ the components of A^{-1}. We let $\varepsilon \downarrow 0$ in the Runge–Kutta scheme

$$\left.\begin{aligned} \boldsymbol{\varphi}_\ell &= \boldsymbol{x}_n + h\sum_{j=1}^{\nu} a_{\ell,j}\boldsymbol{k}_j, \\ \varepsilon\boldsymbol{\psi}_\ell &= \varepsilon\boldsymbol{y}_n + h\sum_{j=1}^{\nu} a_{\ell,j}\boldsymbol{p}_j, \\ \boldsymbol{k}_\ell &= \boldsymbol{f}(\boldsymbol{\varphi}_\ell, \boldsymbol{\psi}_\ell) \\ \boldsymbol{p}_\ell &= \boldsymbol{g}(\boldsymbol{\varphi}_\ell, \boldsymbol{\psi}_\ell) \end{aligned}\right\} \quad \ell = 1, 2, \ldots, \nu$$

$$\begin{aligned} \boldsymbol{x}_{n+1} &= \boldsymbol{x}_n + h\sum_{\ell=1}^{\nu} b_\ell \boldsymbol{k}_\ell, \\ \varepsilon\boldsymbol{y}_{n+1} &= \varepsilon\boldsymbol{y}_n + h\sum_{\ell=1}^{\nu} b_\ell \boldsymbol{p}_\ell, \end{aligned}$$

whereby nonsingularity of A implies that

$$\boldsymbol{g}(\boldsymbol{\varphi}_\ell, \boldsymbol{\psi}_\ell) = \boldsymbol{0}, \quad \ell = 1, 2, \ldots, \nu.$$

This, of course, is an eminently sensible outcome: the algebraic condition is satisfied exactly at each Runge–Kutta node. It also simplifies the scheme a great deal. Easy calculation affirms that it assumes the form

$$\begin{aligned} &\left.\begin{aligned} \boldsymbol{\varphi}_\ell &= \boldsymbol{x}_n + h\sum_{j=1}^{\nu} a_{\ell,j}\boldsymbol{f}(\boldsymbol{\varphi}_j, \boldsymbol{\psi}_j), \\ \boldsymbol{0} &= \boldsymbol{g}(\boldsymbol{\varphi}_\ell, \boldsymbol{\psi}_\ell), \end{aligned}\right\} \qquad \ell = 1, 2, \ldots, \nu, \\ &\boldsymbol{x}_{n+1} = \boldsymbol{x}_n + h\sum_{\ell=1}^{\nu} b_\ell \boldsymbol{k}_\ell, \\ &\boldsymbol{y}_{n+1} = \left(1 - \sum_{j,\ell=1}^{\nu} b_j\omega_{j,\ell}\right)\boldsymbol{y}_n + h\sum_{j,\ell=1}^{\nu} b_j\omega_{j,\ell}\boldsymbol{\psi}_\ell. \end{aligned} \tag{2.4}$$

Had this been all, numerical treatment of DAEs would have not been much more than a mildly singular perturbation of the standard computational ODE theory. That this is not the case becomes apparent upon the consideration of a scalar linear model of (2.2),

$$\begin{aligned} x' &= \alpha_{1,1}x + \alpha_{1,2}y, \\ \varepsilon y' &= \alpha_{2,1}x + \alpha_{2,2}y. \end{aligned} \tag{2.5}$$

This is also the place to mention another helpful interpretation of DAEs, namely as ODEs with *infinite stiffness.* Writing (2.5) in a vector form, $\boldsymbol{z}' = B\boldsymbol{z}$, we obtain the matrix

$$B = \begin{bmatrix} \alpha_{1,1} & \alpha_{1,2} \\ \varepsilon^{-1}\alpha_{2,1} & \varepsilon^{-1}\alpha_{2,2} \end{bmatrix}.$$

It is easy to verify that the *stiffness ratio* (the quotient between the eigenvalues of the largest and the least magnitude, the usual, albeit sometimes unreliable, measuring rod of stiffness [13,18]) becomes infinite when $\varepsilon \downarrow 0$.

Consider again multistep methods. The standard condition for the ODE time-stepping scheme (1.5) to be convergent and of order $p \geq 1$ can be formulated in the language of the polynomials

$$\rho(w) = \sum_{k=0}^{s} \rho_k w^k \qquad \text{and} \qquad \sigma(w) = \sum_{k=0}^{s} \sigma_k w^k.$$

Thus, ρ must obey the *root condition* (all its zeros must reside in $|w| \leq 1$ and the zeros on $|w| = 1$ must be simple), whereby

$$\rho(w) - \log w\sigma(w) = c(w-1)^{p+1} + \mathcal{O}(|w-1|^{p+2}), \quad w \to 1, \quad c \neq 0,$$

means that the method is of order p [12,18]. However, to make any sense, we need to restrict the step size $h > 0$ so that $h\lambda$ lies in the *linear stability domain* $\mathcal{D}$ of the method for every $\lambda \in \sigma(B)$. Letting $\varepsilon \downarrow 0$ we conclude an additional necessary condition, specific to DAEs, namely that $\infty \in \mathcal{D}$. In that case it is possible to prove that

$$\boldsymbol{x}_n = \boldsymbol{x}(t_n) + \mathcal{O}(h^p), \quad \boldsymbol{y}_n = \boldsymbol{y}(t_n) + \mathcal{O}(h^p), \quad h \downarrow 0,$$

uniformly for $t_n = t_0 + nh$ in a compact interval, provided that the starting values $\boldsymbol{x}_n, \boldsymbol{y}_n$, $n = 0, 1, \ldots, s$, are adequately chosen [13,24].

The condition $\infty \in \mathcal{D}$ is tantamount to the requirement that all the zeros of σ reside in $|w| < 1$. The natural choice is placing all its zeros at the origin: the outcome is the familiar *backward differentiation formula (BDF).*

Order conditions for Runge–Kutta methods are, as always, more intricate. Insofar as the ODE method (1.6) is concerned, the reader is referred to [12] for graph-theoretical techniques which proved themselves useful in error analysis. In the case of DAEs there are two further considerations, both implied by the aforementioned interpretation of such systems as 'infinitely stiff'. We say that the method (1.6) has *stage order* $q \geq 1$ if each $\boldsymbol{\varphi}_\ell$ approximates $\boldsymbol{y}$ at $t_n + c_\ell h$ to order q. Moreover, we recall from the ODE theory that the linear stability domain of the Runge–Kutta method (1.6) is

$$\mathcal{D} = \{z \in \mathbb{C} : |r(z)| < 1\},$$

where

$$r(z) = 1 + z\boldsymbol{b}^T(I - zA)^{-1}\mathbf{1}, \quad z \in \mathbb{C},$$

and $\boldsymbol{b}$ is the vector of the Runge–Kutta weights. Suppose that (1.6) is of order p (as an ODE method). Then the DAE method (2.4) is of order $\tilde{p}$, where

$$\begin{array}{ll} \tilde{p} = p & \text{if } r(\infty) = 0, \\ \tilde{p} = \min\{p, q+1\} & \text{if } r(\infty) \in [-1, 1) \setminus \{0\}, \\ \tilde{p} = \min\{p-1, q\} & \text{if } r(\infty) = 1. \end{array}$$

If $|r(\infty)| > 1$ then the method diverges.

Simple algebra readily verifies that $a_{\nu,j} = b_j$, $j = 1, 2, \ldots, \nu$, implies $r(\infty) = 0$. Such *stiffly accurate* Runge–Kutta schemes possess obvious appeal insofar as DAEs (or very stiff ODEs) are concerned [13].

The analysis of general DAEs is more intricate and, as will be soon apparent, it imposes further restrictions on numerical methods. A general algebraic-differential equation can be written as an implicit ODE

$$\boldsymbol{p}(\boldsymbol{u}, \boldsymbol{u}') = \mathbf{0}. \tag{2.6}$$

We say that (2.6) is of *differential index* m if $m \geq 0$ is the least integer such that, perhaps after some algebraic manipulation,

$$\begin{aligned} \boldsymbol{p}(\boldsymbol{u}, \boldsymbol{u}') &= \mathbf{0}, \\ \frac{\mathrm{d}}{\mathrm{d}t}\boldsymbol{p}(\boldsymbol{u}, \boldsymbol{u}') &= \mathbf{0}, \\ &\vdots \\ \frac{\mathrm{d}^m}{\mathrm{d}t^m}\boldsymbol{p}(\boldsymbol{u}, \boldsymbol{u}') &= \mathbf{0} \end{aligned}$$

is a conventional ODE system.

To illustrate this important concept let us reexamine the DAE (1.1). Differentiating once the second equation, we have

$$\mathbf{0} = \frac{\partial \boldsymbol{g}}{\partial \boldsymbol{x}}\boldsymbol{x}' + \frac{\partial \boldsymbol{g}}{\partial \boldsymbol{y}}\boldsymbol{y}',$$

therefore

$$\boldsymbol{y}' = -\left[\frac{\partial \boldsymbol{g}}{\partial \boldsymbol{y}}\right]^{-1} \frac{\partial \boldsymbol{g}}{\partial \boldsymbol{x}}\boldsymbol{f}$$

(recall that we have stipulated nonsingularity of the Jacobian matrix) and substitution into the derivative of the first equation yields

$$\boldsymbol{x}'' = \left(\frac{\partial \boldsymbol{f}}{\partial \boldsymbol{x}} - \frac{\partial \boldsymbol{f}}{\partial \boldsymbol{y}}\left[\frac{\partial \boldsymbol{g}}{\partial \boldsymbol{y}}\right]^{-1} \frac{\partial \boldsymbol{g}}{\partial \boldsymbol{x}}\right)\boldsymbol{f}.$$

Hence the DAE (1.1) is of index $m = 1$.

An example of an index-two DAE is provided by

$$\begin{aligned} \boldsymbol{x}' &= \boldsymbol{f}(\boldsymbol{x}, \boldsymbol{y}), \\ 0 &= \boldsymbol{g}(\boldsymbol{x}), \end{aligned} \tag{2.7}$$

where we require that both Jacobian matrices

$$\frac{\partial \boldsymbol{f}}{\partial \boldsymbol{y}} \quad \text{and} \quad \frac{\partial \boldsymbol{g}}{\partial \boldsymbol{x}}$$

are nonsingular. Differentiating once the second equation in (2.7), we obtain

$$\begin{aligned} \boldsymbol{x}' &= \boldsymbol{f}(\boldsymbol{x}, \boldsymbol{y}), \\ 0 &= \frac{\partial \boldsymbol{g}(\boldsymbol{x}, \boldsymbol{y})}{\partial \boldsymbol{x}} \boldsymbol{f}(\boldsymbol{x}, \boldsymbol{y}), \end{aligned}$$

an index-one system. Hence (2.7) is of index two. This reduction to a lower-index system by differentiation is typical to index analysis.

It is left to the reader to verify that the DAE

$$\begin{aligned} \boldsymbol{x}' &= \boldsymbol{f}(\boldsymbol{x}, \boldsymbol{y}), \\ \boldsymbol{y}' &= \boldsymbol{q}(\boldsymbol{x}, \boldsymbol{y}, \boldsymbol{z}), \\ 0 &= \boldsymbol{g}(\boldsymbol{x}), \end{aligned} \tag{2.8}$$

where the Jacobian matrices

$$\frac{\partial \boldsymbol{f}}{\partial \boldsymbol{y}}, \quad \frac{\partial \boldsymbol{q}}{\partial \boldsymbol{z}} \quad \text{and} \quad \frac{\partial \boldsymbol{g}}{\partial \boldsymbol{x}}$$

are all nonsingular, is of index three.

Numerical solution of high-index DAE systems (that is, with $m \geq 2$) is associated with a reduction in the order of the underlying ODE method. Although an in-depth consideration of this important phenomenon is well outside the scope of this survey, we present a simple illustration. The reader is referred to [2,13,24] for a more comprehensive exposition.

Let us consider the solution of the innocently-looking scalar, linear index-two system

$$x' + y = s_1(t), \quad x = s_2(t)$$

by the BDF method

$$\sum_{k=0}^{s} \rho_k \boldsymbol{y}_{n+k} = h\sigma_s \boldsymbol{f}(t_{n+s}, \boldsymbol{y}_{n+s}).$$

The outcome is

$$\begin{aligned}\sum_{k=0}^{s} \rho_k x_{n+k} + h\sigma_s y_{n+s} &= hs_1(t_{n+s}),\\ x_{n+s} &= s_2(t_{n+s}).\end{aligned}$$

Substituting the second equation into the first, we solve explicitly for y_{n+s},

$$y_{n+s} = \frac{1}{\sigma_s}\left[s_1(t_{n+s}) - \frac{1}{h}\sum_{k=0}^{s}\rho_k s_2(t_{n+k})\right].$$

Therefore the magnitude of the error is multiplied by $\mathcal{O}(h^{-1})$ and the order is $s-1$, rather than the nominal order s of the s-step BDF method for standard ODEs. This order reduction should cause no surprise since it originates in numerical differentiation.

The general rule is that the order is degraded by $m-1$ in numerical discretization of index-m DAEs.

3 Differential equations on manifolds

Numerical analysis is perhaps the most important tool in the investigation of differential equations. Having said this, many other techniques – functional analysis, nonlinear dynamical systems, perturbation theory, differential geometry, degree theory, Lie group theory, special functions' theory... – can be used to a highly beneficial effect. A growing concensus is that there is no conflict between computational and 'qualitative' approaches and that all means are fair. Indeed, harnessing computation in tandem with mathematical analysis is bound to lead to synergy and transform both for the better [31].

Qualitative information about a differential system being known, it is often worthwhile to incorporate it into the discretized solution. This has not just the obvious virtue of retaining important features of the solution under discretization but it also frequently leads to better behaviour of the global error [28]. Qualitative information, though, comes in two basic varieties: asymptotic information and invariants. Insofar as the recovery of correct ODE asymptotics is concerned, the important developments of the last decade are covered elsewhere in this volume by Andrew Stuart [29] and we shall not dwell upon them in this article.

Let us suppose that the initial-value ODE system

$$\boldsymbol{y}' = \boldsymbol{f}(t, \boldsymbol{y}), \quad t \ge t_0, \quad \boldsymbol{y}(t_0) = \boldsymbol{y}_0 \in \mathbb{R}^d, \tag{3.1}$$

possesses an invariant. In other words, there exists a function $\rho : \mathbb{R}^d \to \mathbb{R}$ such that $\rho(\boldsymbol{y}(t)) \equiv \rho(\boldsymbol{y}_0)$, $t \ge t_0$. (The problem (3.1) might make sense only for initial values in a subset of $\mathbb{R}^d$, requiring a minor and obvious amendment to our argument.) In the language of differential topology we might say that the ODE (3.1) induces a foliation of $\mathbb{R}^d$ into *differentiable manifolds* $\mathcal{M}(\boldsymbol{v})$, $\boldsymbol{v} \in \mathbb{R}^d$, such that $\boldsymbol{y}(t) \in \mathcal{M}(\boldsymbol{y}_0)$, $t \geq t_0$. Letting

$$\mathcal{M} := \{\mathcal{M}(\boldsymbol{v}) : \boldsymbol{v} \in \mathbb{R}^d\},$$

we say that (3.1) is $\mathcal{M}$-*invariant*. Without any danger of confusion, we adopt similar terminology with regard to numerical methods: a discretization of the $\mathcal{M}$-invariant ODE (3.1) is said to be $\mathcal{M}$-invariant if $\boldsymbol{y}_n \in \mathcal{M}(\boldsymbol{y}_0)$ for all $n \in \mathbb{Z}^+$ and *all* $\mathcal{M}$-invariant ODEs (3.1).

Classical ODE methods perform very poorly indeed when it comes to $\mathcal{M}$-invariance. Before formulating this statement in a more precise form, we remind the reader that each differentiable manifold is a level set of some C^1 function [14] and that the Runge–Kutta method (1.6) is said to be *nonconfluent* if $c_k \neq c_\ell$ for every $k \neq \ell$.

Theorem 1. (The cheesecutter theorem) *Suppose that there exists a two-dimensional 'slice' $\mathcal{K}(\boldsymbol{v})$ through the manifold $\mathcal{M}(\boldsymbol{v})$. Then $\mathcal{K}(\boldsymbol{v})$ is itself a manifold, embedded in $\mathbb{R}^2$. Let it be the level set of a function $\varrho : C^3[\mathbb{R}^2 \to \mathbb{R}]$ and suppose that there exists $\boldsymbol{v} \in \mathbb{R}^d$ such that ϱ does* not *obey the partial differential equation*

$$\varrho_y^3 \varrho_{xxx} - 3\varrho_x \varrho_y^2 \varrho_{xxy} + 3\varrho_x^2 \varrho_y \varrho_{xyy} - \varrho_x^3 \varrho_{yyy} = 0. \tag{3.2}$$

Then no nonconfluent Runge–Kutta method may be $\mathcal{M}$-invariant [7].

The situation is, if at all, even worse with regard to multistep methods. In greater generality, suppose that the function $\boldsymbol{f}$ is analytic. In that case we can repeatedly differentiate (3.1) to obtain expressions for progressively higher derivatives of $\boldsymbol{y}$, for example

$$\boldsymbol{y}'' = \frac{\partial \boldsymbol{f}(t,\boldsymbol{y})}{\partial t} + \frac{\partial \boldsymbol{f}(t,\boldsymbol{y})}{\partial \boldsymbol{y}} \boldsymbol{f}(t,\boldsymbol{y}).$$

In general, we have $\boldsymbol{y}^{(r)} = \boldsymbol{g}_r(t,\boldsymbol{y})$, $r \in \mathbb{Z}^+$. Many ODE methods, inclusive of multistep and multiderivative schemes (but, in general, not Runge–Kutta!), can be expanded (at least for some starting values) in the form

$$\boldsymbol{y}_{n+1} = \sum_{r=0}^{\infty} c_r h^r \boldsymbol{g}_r(t_n, \boldsymbol{y}_n), \tag{3.3}$$

where the coefficients $\{c_r\}$ depend only on the underlying method.

Theorem 2. *Let $\mathcal{M}(\boldsymbol{v})$ be an arbitrary analytic nonlinear manifold. The only $\mathcal{M}$-invariant choice of the coefficients $\{c_r\}$ in (3.3) is $c_r = 1/r!$, $r \in \mathbb{Z}^+$, i.e. the exact solution of the ODE (3.1) [7].*

The two theorems indicate that classical numerical methods are unlikely to retain invariants. The one exception is when the equation (3.2) is obeyed by every two-dimensional section of every manifold in the foliation. This takes place in the very important case of *quadratic* manifolds,

$$\mathcal{M}(\boldsymbol{v}) = \{\boldsymbol{x} \in \mathbb{R}^d : \boldsymbol{\beta}^T\boldsymbol{x} + \boldsymbol{x}^T\Gamma\boldsymbol{x} = \boldsymbol{\beta}^T\boldsymbol{v} + \boldsymbol{v}^T\Gamma\boldsymbol{v}\}$$

where $\boldsymbol{\beta} \in \mathbb{R}^d$ and Γ is a $d \times d$ real matrix.

Theorem 3. *The Runge–Kutta method (1.6) is $\mathcal{M}$-invariant for all quadratic manifolds $\mathcal{M}$ subject to the identity*

$$m_{k,\ell} := b_k a_{k,\ell} + b_\ell a_{\ell,k} - b_k b_\ell = 0, \quad k,\ell = 1,2,\ldots,\nu$$

[9]. Moreover, if the above condition fails and the method is nonconfluent, it is possible to find an ODE which is invariant on such a manifold but for which invariance fails under discretization [5].

Specific cases of quadratic invariance are the retention of mechanical energy, conservation of orthogonality [11] and invariance on Stiefel and Grassmann manifolds. Yet, perhaps the most important instance of a quadratic manifold is the symplectic group, which represents the main invariant of Hamiltonian systems (1.2) [27,28].

Special manifolds require special algorithmic remedies and this is true even for quadratic manifolds. Hamiltonian equations are a case in point but they are surveyed elsewhere in this volume [27]. Instead, let us examine briefly orthogonal flows (1.3). Any method (1.6) with $m_{k,\ell} \equiv 0$ retains invariance, but all such methods are implicit, hence expensive. A little bit of algorithmic ingenuity, however, can do wonders! Thus, suppose that (1.3) is solved by an arbitrary Runge–Kutta method of order $p \geq 1$. In that case the *orthogonality error* $e_n := \|Y_n^T Y_n - I\|$ will be also of order p. Together with the equation (1.3), we consider the *adjoint equation*

$$Z' = -ZB(Z^T), \quad t \geq t_0, \quad Z(t_0) = Y_0^{-1}. \tag{3.4}$$

Clearly, the exact solution of (3.4) is $Z(t) = Y^T(t)$, because B maps orthogonal matrices to skew-symmetric matrices. As long as we are employing an *orthogonal* Runge–Kutta method, there is no difference between using (1.3) or (3.4). This, however, is no longer the case for arbitrary methods. Thus, suppose that in each *even* step we solve (1.3) and in every *odd* step we use the adjoint equation (3.4) (of course, taking appropriate initial conditions). In that case, although the order of the combined method remains p, it is possible to prove that $e_n = \mathcal{O}(h^{p+2})$, hence the orthogonality error is of order $p+1$ [6].

We can, as a matter of fact, do much better! Thus, let us designate each step with (1.3) as "0" and each step of (3.4) as "1". No matter what the combination of zeros and ones, the order of the method remains p. Yet, the orthogonality error is

$$
\begin{array}{ll}
00000000\cdots: & p, \\
01010101\cdots: & p+1, \\
01100110\cdots: & p+2, \\
01101001\cdots: & p+3.
\end{array}
$$

In general, suppose that, to advance from t_n to t_{n+1} (with constant step sizes) we use "0" if the number of ones in the binary expansion of n is even, "1" otherwise. Then (in exact arithmetic) $e_{2^r n} = \mathcal{O}(h^{p+r+1})$, $r, n \in \mathbb{Z}^+$, and we can make the orthogonality error arbitrarily small for very little extra cost.

Figure 1 displays the orthogonality error in solving an *orthogonal flow* (1.3) with forward Euler for different sequences of "0"s and "1"s. The attenuation of orthogonality error is self-evident.

Orthogonal flows (1.3) are interesting in their own right but perhaps their most fascinating application is implicit in their connection with *isospectral flows*

$$X' = F(X)X - XF(X), \quad t \geq t_0, \quad X(t_0) = X_0. \tag{3.5}$$

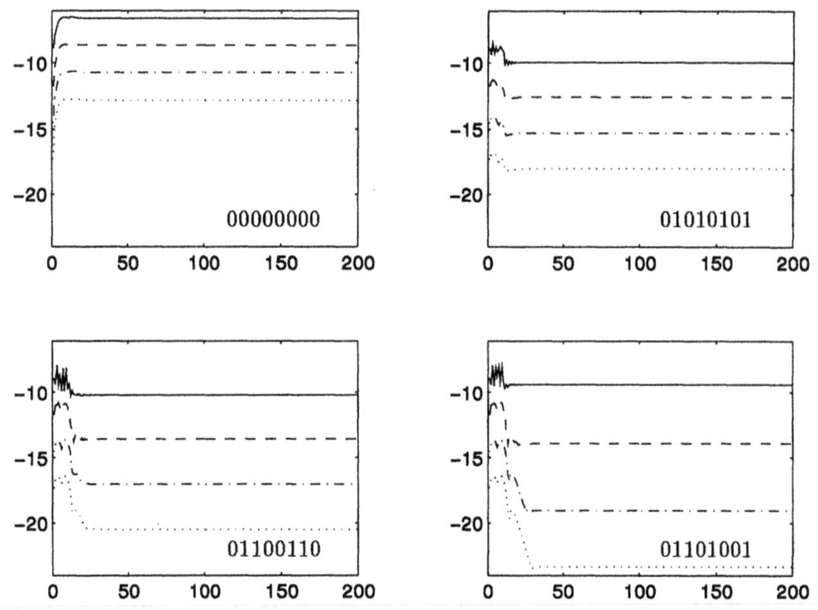

Figure 1. Orthogonality error (on a logarithmic scale) for the orthogonal flow, solved by the forward Euler method. Consecutive lines correspond to step-size refinement

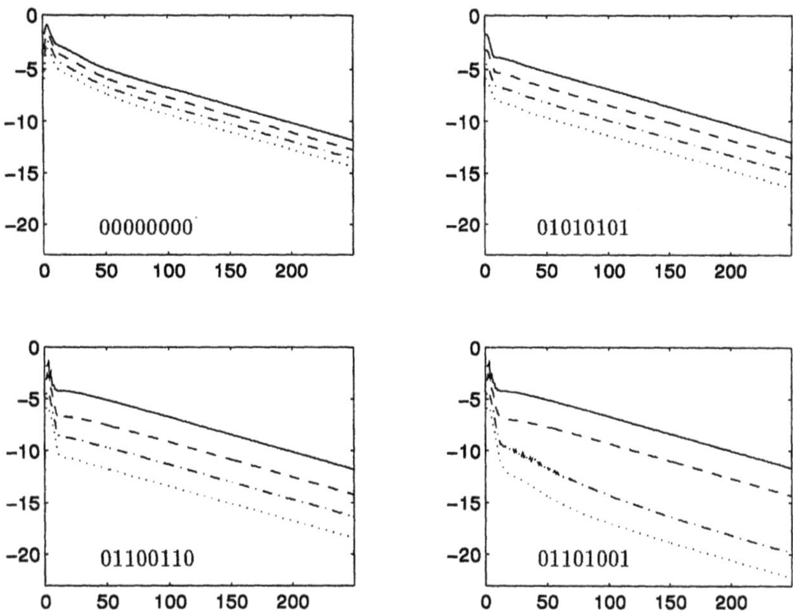

Figure 2. Symmetry error (on a logarithmic scale) for the isospectral flow, solved by the forward Euler method

Again, the unknown $X(t)$ is a $d \times d$ matrix, while the function F maps $d \times d$ matrices to themselves. The invariants of (3.5) are the eigenvalues of the matrix X_0 [30]. Isospectral flows are ubiquitous in many applications, for example in mechanics, quantum chemistry, micromagnetics and molecular dynamics, because of their connection with Toda lattice equations [30]. However, applications nearest to a numerical analyst's heart are in numerical algebra, *inter alia* in the determination of eigenvalues and in inverse eigenvalue problems [8]. In most such applications the matrix X_0 is symmetric and F maps $d \times d$ matrices into skew-symmetric matrices – this ensures that $X(t)$ is symmetric for all $t \geq t_0$. We henceforth stipulate this to be the case.

It is possible to prove, along the lines of Theorem 1, that no nonconfluent Runge–Kutta method may conserve eigenvalues for $d \geq 3$ [5]. However, retention of eigenvalues is possible by exploiting a link between (3.5) and orthogonal flows (1.3). Thus, letting

$$B(Y) := YX_0Y^{-1}, \quad Y_0 = I,$$

it is easy to prove that $X(t) = Y(t)X_0Y^{-1}(t)$, where Y is the solution of (1.3) (incidentally, this proves that the eigenvalues are retained by X!). Subject to this reformulation, isospectral flows can be solved by quadratically-invariant Runge–Kutta methods [5].

Note that eigenvalues are retained no matter how (1.3) is solved. However, unless Y_n is orthogonal, X_n will not be symmetric. Similarly to the use of the adjoint equation to decrease the orthogonality error, we can resort to a similar device to solve isospectral flows while minimizing symmetry breakdown. This is illustrated in Figure 2.

Simultaneously with the work on special families of ODEs that live on 'interesting' manifolds, there is a growing interest in new types of methods that are crafted to retain invariants of a general type. Although many of them are of a tentative and preliminary nature, it is perhaps worthwhile to mention emerging trends that are likely to inform future research. Four approaches have been considered in the last few years:

1. Perhaps the simplest and most natural technique is to project the numerical solution (originating in an arbitrary time-stepping solver) on the underlying manifold. Such *projective methods* have obvious connections with DAEs – indeed, one can interpret differential-algebraic equations as ODEs on manifolds. A conversion to DAEs, followed by, say, the use of BDF or of stiffly-accurate Runge–Kutta, constitutes a powerful approach to many ODEs with invariants [1]. Another projective approach, the *discrete gradient method,* is due to Reinout Quispel and his coworkers [26]. Projective methods, however, are more problematic when the underlying manifold cannot be easily described in a constructive fashion, for example in the case of isospectral flows.

2. Classical Runge–Kutta methods are formulated within the framework of Euclidean spaces. Peter Crouch and Robert Grossman have put forward an alternative formulation, whereby derivatives are restricted to the *tangent space* of the underlying manifold [10]. Although this leads to exceedingly complicated order analysis (assisted a great deal by the recent interpretation of Butcher's graph theory-based order analysis in the terminology of Lie brackets by Hans Munthe-Kaas [25]), third and fourth-order methods have already been developed. The tangent space approach is very powerful because of its generality, although the latter is also its Achilles heel: in general, it is far from easy to produce numerically a good local basis of the tangent space. A special case whereby the tangent space is given for free in a constructive fashion, thereby rendering this approach very attractive, is presented by *Lie-type equations*

$$Y' = F(Y)Y, \quad t \geq t_0, \quad Y(t_0) = Y_0, \tag{3.6}$$

where Y_0 lies in a Lie group $\mathcal{U}$, while F maps $\mathcal{U}$ into its Lie algebra $\boldsymbol{u}$. In this case it is true that $Y(t) \in \mathcal{U}$ for all $t \geq t_0$. (Of course, orthogonal matrices form the Lie group $\mathrm{O}(d, \mathbb{R})$, whose Lie algebra consists of $d \times d$ skew-symmetric matrices – orthogonal flows (1.3) are but an example of Lie-type equations (3.6).) In that case the tangent space is $\boldsymbol{u}$ and the above methods can be formulated with greater ease.

3. In a recent paper, Antonella Zanna has presented very interesting new methods that automatically stay on the Lie group $\mathcal{U}$ whenever applied to the Lie-type equations (3.6) [31]. These methods are explicit and they can attain arbitrarily high orders (although high-order methods are expensive). They extensively exploit the closure of a Lie algebra $\boldsymbol{u}$ under commutation and the fact that the exponential map takes $\boldsymbol{u}$ to $\mathcal{U}$.

4. Suppose that the ODE (3.1) is $\mathcal{M}$-invariant. The topology of the manifold $\mathcal{M}(\boldsymbol{y}_0)$ is often known. In particular, $\mathcal{M}(\boldsymbol{y}_0)$ can be often mapped diffeomorphically into another manifold $\mathcal{N}(\boldsymbol{y}_0)$ with simpler geometry. Provided that we know how to discretize invariantly on $\mathcal{N}(\boldsymbol{y}_0)$, we can subject the ODE to the same diffeomorphism, whereby it becomes $\mathcal{N}$-invariant, discretize it while retaining $\mathcal{N}$-invariance and map back. This simple idea has been exploited in [7] in a systematic fashion but many outstanding questions remain in its implementation.

4 Functional differential equations

A general functional differential equation is of the form

$$\boldsymbol{y}'(t) = \boldsymbol{f}(t, \boldsymbol{y}(t), \boldsymbol{y}(\vartheta(t))), \quad t \geq 0, \tag{4.1}$$

with an initial condition given for $t \in (\inf_{\xi \geq 0} \vartheta(\xi), 0]$. Equations of this kind feature in a multitude of applications and, in the author's opinion, would have probably been much more ubiquitous were their theory and computation based on a firmer footing.

It is deceptively simple to extend standard ODE methods, like (1.5) and (1.6), to an FDE setting. As long as the output is continuous (and this is always attainable, for example by polynomial interpolation), it can be exploited in an obvious fashion and *presto* an ODE method becomes an FDE method. We hasten to add that, needless to say, the actual algorithmic practice is *much* more complicated. The two obvious problems in a naive conversion of ODE to FDE methods are how to deal with derivative discontinuities (which are often present in the *exact* solution of (4.1)) and how to keep storage requirements at bay. However, the most demanding and least explored theme in the numerical analysis of FDEs is the question of stability.

Important strides have been made in the last decade with regard to two aspects of the stability theory of time-stepping methods for (4.1). Firstly, the linear theory for the constant-delay FDE (1.4) is at last reasonably well understood.

Much work has been published in the last few decades on the linear, scalar model equation

$$y'(t) = ay(t) + by(t - \tau), \quad t \geq 0. \tag{4.2}$$

Unfortunately, all this has very little bearing on the vector equation

$$\boldsymbol{y}'(t) = P\boldsymbol{y}(t) + Q\boldsymbol{y}(t - \tau), \quad t \geq 0. \tag{4.3}$$

In the ODE case one can generalize from scalar to vector linear equations since just a single matrix is present and standard Jordan factorization does the trick. However, unless A and B commute and the matrix pencil (A, B) can be factorized in unison, stability results for equation (4.2) (whether exact or discretized) have little bearing on (4.3).

The understanding of the vector equation (4.3) in a numerical setting owes much to the work of Karel in 't Hout. First, however, we must state stability conditions for the exact equation. Standard FDE theory readily provides a condition for the asymptotic stability of the zero solution. Thus, let

$$\gamma(z, w) := \det[zI - (P + wQ)], \quad z, w \in \mathbb{C}.$$

Then $\lim_{t\to\infty} y(t) = 0$ for every initial value if and only if for every $z \in \mathbb{C}$

$$\gamma(z, e^{-\tau z}) = 0 \Rightarrow \operatorname{Re} z < 0.$$

This, however, falls short of being a transparent asymptotic stability requirement, verifiable with minimal fuss from the matrices P and Q. Fortunately, Karel in 't Hout formulated asymptotic stability conditions in a more user-friendly fashion [15].

Theorem 4. *The zero solution of (4.3) is asymptotically stable for all initial conditions and for all delays $\tau > 0$ if and only if the following three conditions are satisfied.*

1. $\operatorname{Re} \lambda < 0$ *for all* $\lambda \in \sigma(P)$;
2. $\rho[(itI - P)^{-1}Q] < 1$ *for all* $t \in \mathbb{R} \setminus \{0\}$; *and*
3. $-1 \notin \sigma(P^{-1}Q)$.

We consider the solution of (4.3) by a *θ-method.* Applied to the standard ODE (3.1), such a method reads

$$y_{n+1} = y_n + h[\theta f(t_{n+1}, y_{n+1}) + (1 - \theta) f(t_n, y_n)],$$

where $\theta \in [0, 1]$ is a given constant. An extension of the method to the FDE (4.1) is

$$y_{n+1} = y_n + h[\theta f(t_{n+1}, y_{n+1}, v_{n+1}) + (1 - \theta) f(t_n, y_n, v_n)],$$

where v_r is a linear interpolation to the numerical solution at $\vartheta(t_r)$. In the case of the linear system (4.3), letting the step h be commensurate with the delay τ, $h = \tau/m$, say, we obtain

$$(I - \theta hP)y_{n+1} = [I + (1 - \theta)hP]y_n + hQ[\theta y_{n-m+1} + (1 - \theta)y_{n-m}]. \quad (4.4)$$

The pencil (hP, hQ) is said to be *stable* if the matrix $(I - \theta hP)$ is invertible and $\lim_{n\to\infty} y_n = 0$ in (4.4). Note that the linear stability domain of the θ-method (in the classical ODE context) is

$$\mathcal{D} = \left\{ z \in \mathbb{C} : \left| \frac{1 + (1-\theta)z}{1 - \theta z} \right| < 1 \right\}.$$

Theorem 5. *If*

$$h\sigma(P) \subset \mathcal{D} \qquad \text{and} \qquad h \sup_{z \in \partial\mathcal{D}} \rho[(zI - hP)^{-1}Q] < 1$$

then (hP, hQ) is stable. Moreover, if this pencil is stable then necessarily

$$h\sigma(P) \subset \mathcal{D} \qquad \text{and} \qquad h \sup_{z \in \partial\mathcal{D}} \rho[(zI - hP)^{-1}Q] \le 1.$$

Therefore, as in the ODE case, numerical stability follows by replacing the complex left half-plane with the linear stability domain $\mathcal{D}$ [15].

A θ-method is a simple instance of a two-stage Runge–Kutta, specifically,

$$c = \begin{bmatrix} 0 \\ 1 \end{bmatrix}, \quad A = \begin{bmatrix} 0 & 0 \\ \theta & 1-\theta \end{bmatrix}, \quad b = \begin{bmatrix} \theta \\ 1-\theta \end{bmatrix}.$$

Little wonder that in 't Hout's analysis has been already extended to more elaborate Runge–Kutta schemes [20].

A second development on the FDE front in the last decade is an initial foray into stability analysis of discretized FDEs with variable delays. The stability model is the *pantograph equation*

$$y'(t) = Py(t) + Qy(qt), \quad t \ge 0, \quad y(0) = y_0 \in \mathbb{C}^d, \tag{4.5}$$

where $q \in (0, 1)$ is given. This equation is also of an independent interest in a number of applications [16].

The difference between constant and proportional delay – to wit, between (4.3) and (4.5) – cannot be greater. The pantograph equation (4.5) requires just a single initial condition and its solution resides in $C^\infty[\mathbb{C}^d]$. Its asymptotic behaviour is well-understood, yet highly nontrivial.

Theorem 6. *The zero solution of (4.5) is asymptotically stable if*

$$\operatorname{Re}\lambda < 0 \quad \forall \lambda \in \sigma(P) \qquad \text{and} \qquad \rho(P^{-1}Q) < 1,$$

where $\rho(\cdot)$ is the spectral radius. The boundary of the asymptotic stability domain decomposes into three portions:

1. $\mathrm{Re}\,\lambda < 0$ *for all* $\lambda \in \sigma(P)$, $\rho(P^{-1}Q) = 1$: *Provided that the eigenvalues of* $P^{-1}Q$ *with unit modulus have the same algebraic and geometric multiplicity, there exists a* periodic *function* ϕ *and for every* $\varepsilon > 0$ *there exists* $t_\varepsilon > 0$ *such that*
$$\|y(t) - \phi(q^{-t})\| < \varepsilon, \quad t \geq t_\varepsilon.$$
Otherwise
$$\limsup_{t\to\infty} \|y(t)\| = \infty \tag{4.6}$$
for some initial values.

2. $\max\{\mathrm{Re}\,\lambda : \lambda \in \sigma(P)\} = 0$, P *is nonsingular and* $\rho(P^{-1}Q) < 1$: *The solution* y *tends to an* almost-periodic *function* ψ: *for every* $\varepsilon > 0$ *there exists* $s_\varepsilon > 0$ *such that*
$$\|\psi(t + s_\varepsilon) - \psi(t)\| < \varepsilon, \quad t \geq 0,$$
and $\liminf_{\varepsilon\downarrow 0} s_\varepsilon > 0$.

3. P *is singular: In that case (4.6) holds for some initial values and the zero solution is unstable [16,23].*

Although stability of the discretized pantograph can be analysed by traditional techniques [4], this is inadequate. All such techniques are based, often implicitly, on Liapunov functions or similar estimates. Therefore, they are at their best if, at least in an asymptotic sense, the norm of the solution decays monotonically. This, however, is not always the case with regard to the pantograph and the equation calls for a more subtle approach.

Although the question of discretized pantograph stability for general methods remains wide-open, there already exists a body of interesting results about scalar equations and $q = 1/r$, where $r \geq 2$ is an integer. Let us restrict this brief debate to $q = \frac{1}{2}$, whereby, without loss of generality, the scalar pantograph becomes
$$y'(t) = ay(t) + by(\tfrac{1}{2}t), \quad t \geq 0, \quad y(0) = 1. \tag{4.7}$$
It is easy to specialize Theorem 6 to the equation (4.7): the zero solution is asymptotically stable if and only if
$$\mathrm{Re}\,a < 0, \quad |b| < |a|, \tag{4.8}$$
the function y tends (under a change of variable) to a periodic curve when $\mathrm{Re}\,a < 0$ and $|b| = |a|$ and it is almost-periodic when $\mathrm{Re}\,a = 0$, $a \neq 0$ and $|b| < |a|$. It is not difficult to verify that, as long as
$$\mathrm{Re}\,a + |b| < 0, \tag{4.9}$$
the function $|y(t)|$ tends to zero monotonically. This opens up the way to Liapunov function-based techniques and discretized stability analysis is relatively

straightforward [4]. The set (4.9), however, is in general much smaller than the set (4.8).

Solving (4.7) with the familiar *trapezoidal rule* (a θ-method with $\theta = \frac{1}{2}$) results in the nonstationary recurrence

$$\begin{aligned} y_0 &= 1, \\ y_{2n+1} &= ry_{2n} + \tfrac{1}{4}s(3y_n + y_{n+1}), \\ y_{2n+2} &= ry_{2n+1} + \tfrac{1}{4}s(y_n + 3y_{n+1}) \end{aligned} \Bigg\} \quad n \in \mathbb{Z}^+, \tag{4.10}$$

where

$$r = \frac{1 + \frac{1}{2}ha}{1 - \frac{1}{2}ha} \quad \text{and} \quad s = \frac{hb}{1 - \frac{1}{2}ha}.$$

To analyse the stability of (4.10), Martin Buhmann and the author proved that the Fourier transform $\hat{x}$ of the sequence $x_n = y_{n+1} - y_n$, $n \in \mathbb{Z}^+$, obeys the equation

$$\hat{x}(\xi) = \frac{1}{1 - re^{i\xi}} \left[(r + s - 1) + \tfrac{1}{4}s(1 + e^{i\xi})^2 \hat{x}(2\xi)\right] \pmod{2\pi}, \quad \xi \in [0, 2\pi].$$

This, together with an application of the *mean ergodic theorem,* is the main step in proving that $\hat{x}$ is an L_2 function, therefore $\{x_n\}_{n\in\mathbb{Z}^+}$ is an ℓ_2 sequence, provided that (4.8) holds. A simple technical argument can be used to prove that also $\{y_n\}_{n\in\mathbb{Z}^+}$ is ℓ_2, whence $\lim_{n\to\infty} y_n = 0$ and the zero solution is asymptotically stable. Therefore the trapezoidal rule recovers faithfully the asymptotic stability domain of the exact pantograph equation (4.7) [3].

The behaviour of (4.10) on the stability boundary is, if at all, even more complicated. Figure 3 displays the sequence $\{y_n\}_{n\in\mathbb{Z}^+}$ for two different choices of $a \in i\mathbb{R} \setminus \{0\}$ and $|b| < |a|$ in the 'phase plane' $(\mathrm{Re}\, y, \mathrm{Im}\, y)$. Both figures are suggestive of fractal sets – rightly so! It has been proved in [17] that, at least in specific simplified cases, the attractor of the underlying dynamical system is equivalent to a *probabilistic mixture of Julia sets* and is of a fractal Hausdorff dimension.

The pantograph equation (4.5) and its generalizations into neutral and integro-differential settings provide an enduring analytic and numerical challenge. And this is, in a sense, the simplest FDE with an unbounded delay! Having said this, the research of last decade has laid here, like elsewhere for different themes of this paper, firm mathematical foundations. It is up to the numerical community to take up the challenge. Many new and exciting research problems are presented by nonclassical breeds of ODEs: differential-algebraic equations, differential equations on manifolds, functional-differential equations.... The underlying problems are difficult and they often call for new tools and profound mathematics. Yet, the lesson of the last decade is that, difficulties notwithstanding, these problems are within the reach of modern numerical analysis.

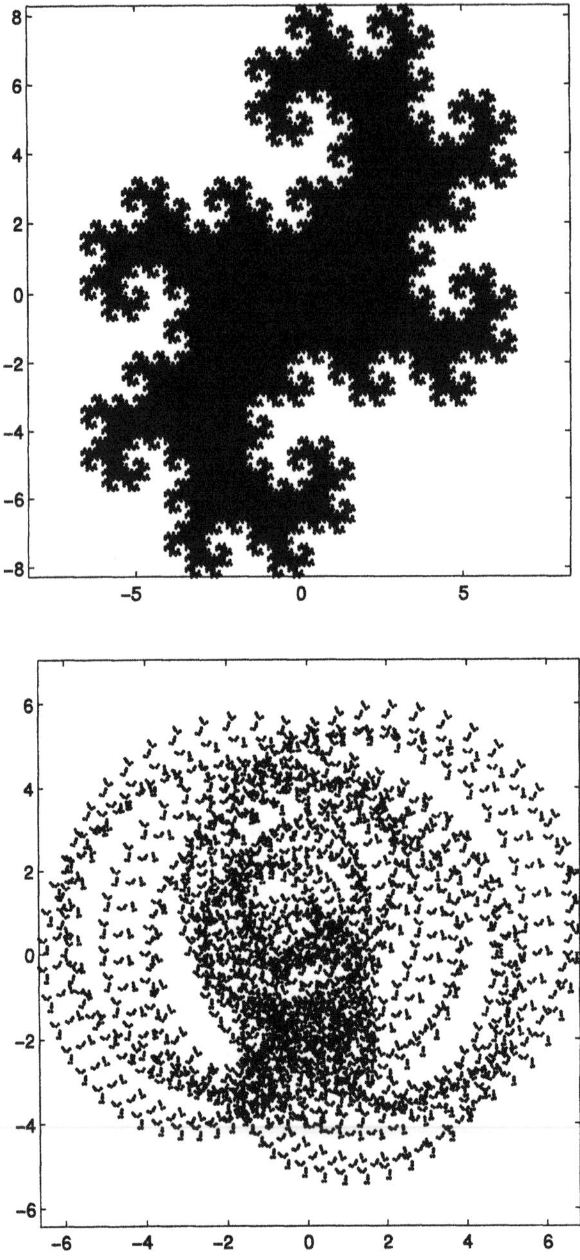

Figure 3. Examples of fractal attractors on the stability boundary of the discretized pantograph equation

Acknowledgements

The author is grateful to Yunkang Liu and Andrew Stuart, who diligently read an earlier version of this paper and offered many illuminating and useful comments.

References

1. Ascher, U.M. (1996). Stabilization of invariants of discretized differential systems. *Tech. Rep.*, University of British Columbia.

2. Brennan, K.E., Campbell, S.L. and Petzold, L.R. (1989). *Numerical Solution of Initial-Value Problems in Differential-Algebraic Equations*, North Holland, New York.

3. Buhmann, M.D. and Iserles, A. (1992). On the dynamics of a discretized neutral equation. *IMA J. Num. Anal.*, **12**, 339–363.

4. Buhmann, M.D. and Iserles, A. (1993). Stability of the discretized pantograph differential equation. *Maths Comp.*, **60**, 575–589.

5. Calvo, M.P., Iserles, A. and Zanna, A. (1995). Numerical solution of isospectral flows. *Maths Comp.* (To Appear.)

6. Calvo, M.P., Iserles, A. and Zanna, A. (1996). Runge–Kutta methods for orthogonal and isospectral flows. *Appl. Num. Maths.*, **22**, 153–163.

7. Calvo, M.P., Iserles, A. and Zanna, A. (1996). Numerical methods on differentiable manifolds. (In Preparation.)

8. Chu, M.T. (1994). A list of matrix flows with applications. *Hamiltonian and Gradients Flows, Algorithms and Control*, Editor: A. Bloch, *American Math. Society*, 87–97.

9. Cooper, G. (1987). Stability of Runge–Kutta methods for trajectory problems. *IMA J. Num. Anal.*, **7**, 1–13.

10. Crouch, P. and Grossman, R. (1993). Numerical integration of ordinary differential equations on manifolds. *J. Nonlinear Sci.*, **3**, 1–33.

11. Dieci, L., Russell, R.D. and van Vleck, E.S. (1994). Unitary integrators and applications to continuous orthonormalization techniques. *SIAM J. Num. Anal.*, **31**, 261–281.

12. Hairer, E., Nørsett, S.P. and Wanner, G. (1991). *Solving Ordinary Differential Equations I: Nonstiff Problems*, 2nd Edition, Springer-Verlag, Berlin.

13. Hairer, E. and Wanner, G. (1991). *Solving Ordinary Differential Equations II. Stiff and Differential-Algebraic Problems*, Springer-Verlag, Berlin.

14. Guillemin, V. and Pollack, A. (1974). *Differential Topology*, Prentice-Hall, Englewood Cliffs.

15. in 't Hout, K.J. (1994). The stability of θ-methods for systems of delay differential equations. *Annals of Num. Maths.*, **1**, 323–334.

16. Iserles, A. (1993). On the generalized pantograph functional-differential equation. *Euro. J. Appl. Maths.*, **4**, 1–38.

17. Iserles, A. (1994). The asymptotic behaviour of certain difference equations with proportional delay. *Comp. Math. App.*, **28**, 141–152.

18. Iserles, A. (1996). *A First Course in the Numerical Analysis of Differential Equations*, Cambridge University Press.

19. Iserles, A. and Zanna, A. (1996). Qualitative numerical analysis of ordinary differential equations. *Lectures in Appl. Maths*, Editors: J. Renegar, M. Shub and S. Smale, American Mathematical Society, Providence, RI, 421–442.

20. Koto, T. (1994). A stability property of A-stable natural Runge–Kutta methods for systems of delay differential equations. *BIT*, **34**, 262–267.

21. Lambert, J.D. (1977). The initial value problem for ordinary differential equations. *State of the Art in Num. Anal.*, Editor: D.A.H. Jacobs, Academic Press, London, 451–500.

22. Lambert, J.D. (1987). Developments in stability theory for ordinary differential equations. *State of the Art in Num. Anal.*, Editors: A. Iserles and M.J.D. Powell, Oxford University Press, 409–431.

23. Liu, Y. (1996). On functional differential equations with proportional delays. *Ph.D. Dissertation*, University of Cambridge.

24. März, R. (1992). Numerical methods for differential algebraic equations. *Acta Numerica*, **1**, 141–198.

25. Munthe-Kaas, H. (1995). Lie–Butcher theory for Runge–Kutta methods. *BIT*, **35**, 572–587.

26. Quispel, G.R.W. and Capel, H.W. (1996). Solving ODE's numerically while preserving a first integral. *Tech. Rep.*, University of La Trobe.

27. Sanz-Serna, J.M. (1997). Geometric integration. *IMA Conf. Proc. on State of the Art in Num. Anal.*, Oxford University Press.

28. Sanz-Serna, J.M. and Calvo, M.P. (1994). *Numerical Hamiltonian Problems*, Chapman and Hall, London.

29. Stuart, A.M. (1997). Convergence and stability in the numerical approximation of dynamical systems. *IMA Conf. Proc. on State of the Art in Num. Anal.*, Oxford University Press.

30. Toda, M. (1981). *Theory of Nonlinear Lattices*, Springer-Verlag, Berlin.

31. Zanna, A. (1996). The method of iterated commutators for ordinary differential equations on Lie groups. *DAMTP Tech. Rep. 1996/NA12*, University of Cambridge.

Numerical Analysis of Volterra Functional and Integral Equations

Christopher T.H. Baker

Department of Pure and Applied Mathematics,
University of Manchester

Dedicated to Donald Kershaw, Reader Emeritus, Lancaster University.

Abstract

Volterra equations are, broadly speaking, evolutionary equations incorporating memory. We indicate where they arise in practice, discuss their significant features (notably smoothness properties of the solutions), describe numerical methods for their approximate solution, and indicate the mathematical foundations of numerical procedures.

1 Mise en scène

The type of problem that we are interested in here has arisen in the study of such diverse topics as electric locomotives, economic growth, and cardiorespiratory control. Our starting point is the solution of such a problem and the first step is to marry the theoretical understanding to a *realistic* mathematical model. The solution of the mathematical problem relies upon insight based on analysis and (numerical or symbolic) computation informed by an understanding of the real-life problem. The transition between analytical and computational techniques and the choice of numerical procedures require an understanding of the underlying mathematics and of what may be feasible, computationally. The efficient implementation of the numerical procedures requires an approach that exploits mathematical insight but borders on computational science. Our perspective requires a broad-based mathematical approach and a scientific understanding of the underlying realities, and is reflected in the structure of this paper. The first section introduces various Volterra equations and their rôle as mathematical models; we then give analytical insight and only subsequently (numerical analysis is driven by the need to model reality accurately) proceed to the numerical aspects indicated by our title.

This survey may be read in conjunction with [13] and [107], which give overviews for delay equations, and [12, 34]. The book [34] on Volterra integral equations is encyclopaedic; the book of Bellen and Zennaro [24] is awaited at the time of writing. The contribution by Brunner in the last State of the Art meeting [69, pp. 563–600] was devoted to collocation techniques, which therefore gain little emphasis in the present paper. In May 1996, two Volterra Centennial meetings in the USA marked the mathematical influence of Volterra (abstracts of the present author's invited talks are given in [45, 50]).

2 Modelling memory in evolutionary problems

In very many real-life situations, the present state and the manner in which it changes are both dependent on the past. In population dynamics, the birth-rate is clearly a function of the state of the population on some previous date, as well as its current state, because of the significance of the gestation period. In this sense, the population has a "memory". In cell kinetics, the division of an individual cell follows its transition through various other stages preliminary to division, during which the cell may suffer environmental changes or death but – if it survives – there is a delay before new cells actually appear. The analogue with gestation is clear. Memory effects also occur in mechanical systems.

We first present illustrative examples of the types of equation we will consider. Since we are concerned with 'evolutionary' equations, we shall associate the variable t with 'time'. In the classical *logistic equation*

$$y'(t) = \lambda y(t)\{1 - y(t)\}, \quad \lambda \in \mathbb{R} \tag{2.1}$$

(valid for $t \geq t_0$, $t_0 \geq 0$ say), the derivative[1] of y at time t is determined by the current value $y(t)$; however, in the *delayed logistic equation*

$$y'(t) = \lambda y(t)\{1 - y(t - \tau_\star)\}, \quad \lambda \in \mathbb{R}, \quad \tau_\star \in \mathbb{R}_+ \tag{2.2}$$

(for $t \geq t_0 \geq 0$), the derivative depends on $y(t)$ and (since $\tau_\star > 0$) the 'earlier' state $y(t - \tau_\star)$; in the *integro-differential equation*

$$y'(t) = \lambda y(t)\left\{1 - \int_{t_0}^{t} k(t-s)y(s)ds\right\}, \quad \lambda \in \mathbb{R} \tag{2.3}$$

($t \geq t_0 \geq 0$), the derivative depends on $y(t)$ and (in a manner determined by the *kernel* k) on all previous states $y(s)$ after the initial moment t_0.

The preceding equations serve to introduce the mathematical concept of "memory". Response in the *ordinary differential equation* (ODE) in eqn (2.1) is instantaneous; in eqn (2.2), which is a *delay differential equation* (DDE), it depends upon a memory of the state at an instant $\tau_\star$ earlier in time; in (2.3) (a *Volterra integro-differential equation* or VIDE) it depends on possibly total recall since timing began. In practice, memory fades and the term $\int_{t_0}^{t} k(t-s)y(s)ds$ in the latter equation is an example of a fading memory integral when $\int_{t_0}^{\infty} |k(s)|ds < \infty$. Thus, fading memory will arise when k has compact

[1] Following the classical literature, derivatives are taken to be right-hand derivatives wherever the distinction is necessary.

support, as in

$$y'(t) = \lambda y(t) \left\{1 - \int_{t-\tau_\star}^{t} k(t-s)y(s)ds\right\}, \quad \lambda \in \mathbb{R},\ \tau_\star \in \mathbb{R}_+.$$

These functional differential equations cannot be solved without specifying appropriate initial data: (2.1) and (2.3) require us to specify $y(t_0) := y_0$ and (2.2) requires us to specify $y(t) := \psi(t)$, for $t \in [t_0 - \tau_\star, t_0]$ in order to find $y(t)$ for $t > t_0$. The exact nature of the initial data will vary from problem to problem, and may depend on t_0. For slightly more complicated equations see [54]. Realistic models often involve *systems* of equations and we concentrate upon scalar equations solely for simplicity in our presentation.

3 Types of functional equations

In this section we shall review various categories of equations.

• *Volterra integral equations*: Classical Volterra integral equations (VIEs) are familiar in the form

$$y(t) = \int_{t_0}^{t} K\big(t, s, y(s)\big)ds + g(t), \quad t \geq t_0, \tag{3.1}$$

(a Volterra equation *of the second kind*) and $g(t) = \int_{t_0}^{t} K\big(t,s,y(s)\big)ds$ $(t \geq t_0)$, a Volterra equation *of the first kind*[2]. Here, g, K are prescribed functions and we seek y. *Convolution equations* arise when $K(t,s,u)$ has the form $k(t-s)\Phi(u)$. The term *Abel equation* (or Abel–Volterra equation) is used when $K(t,s,u)$ has the form $\mathfrak{h}(t,s,u)/(t-s)^\nu$ with $\nu \in (0,1)$ and the function $\mathfrak{h}$ is smooth (i.e., has continuous derivatives of modest order), in particular, for smooth $\mathfrak{K}$ and Φ,

$$K(t,s,u) = (t-s)^{-\nu}\mathfrak{K}(t-s)\Phi(u). \tag{3.2}$$

First kind equations can be ill-posed. As an example, for $\nu \in (0,1)$

$$y(t) = \frac{\sin(\nu\pi)}{\pi}\frac{d}{dt}\int_0^t (t-s)^{\nu-1}g(s)ds$$

solves

$$\int_0^t (t-s)^{-\nu}y(s)ds = g(t) \tag{3.3}$$

$(t \geq t_0 = 0)$ given appropriate restrictions on g. If $\nu = 0$, $y(t) = g'(t)$ (when the derivative exists and $g(0) = 0$) but the determination of g' is ill-posed for $g \in C[0,\infty)$. Informally, the problem becomes progressively 'less ill-posed' as ν increases in $(0,1)$. The problem (3.3) can be expressed in terms of $\mathcal{I}^{1-\nu}$ where

$$\mathcal{I}^\mu\phi(t) \equiv \frac{1}{\Gamma(\mu)}\int_0^t (t-s)^{\mu-1}\phi(s)ds, \qquad \mu \in (0,1]; \tag{3.4}$$

---

[2]At Tempe [50], Brunner reviewed 100 years of Volterra equations of the first kind.

which introduces (see [34, p.8 *et seq.*]) integrals and derivatives of fractional order: For $\nu \in (0,1)$, a solution of (3.3) (if it exists) is expressible in terms of a fractional derivative $\mathcal{D}^\nu g(t)$ (where $\mathcal{D}g(t) = g'(t)$, $\mathcal{D}^\mu = \mathcal{D}\mathcal{I}^{1-\mu}$, $\mu \in (0,1]$).

• *Volterra integro-differential equations*: We encountered an example of a Volterra integro-differential equation (VIDE) with a convolution kernel in (2.3). A fairly general form of VIDE can be written (for an appropriate $\mathcal{F}$) as

$$y'(t) = \mathcal{F}\Big(t, y(t), \int_{t_0}^{t} \mathcal{K}(t,s,y(s))ds\Big) \tag{3.5}$$

or, more generally, $y'(t) = \mathcal{F}\Big(t, y(t), \int_{t_0}^{t} \mathcal{H}(t,s,y(t),y(s),y'(t),y'(s))ds\Big)$, which is a *neutral* VIDE (NVIDE), termed 'neutral' because y' occurs in the right-hand side.

With suitable smoothness assumptions, if we differentiate eqn (3.1) we obtain

$$y'(t) = K(t,t,y(t)) + \int_{t_0}^{t} \frac{\partial}{\partial t} K(t,s,y(s))ds + g'(t) \quad (t \geq t_0); \tag{3.6}$$

with $y(t_0) = g(t_0)$. This is another example of a VIDE. However, let us suppose that, in eqn (3.1), $K(t,s,u)$ is smooth for $s \in [t-\tau_\star, t]$ but vanishes if $s < t - \tau_\star$. Thus, eqn (3.1), with $y(t) = \psi(t)$, $t \in [t_0 - \tau_\star, t_0]$, now reads (for $t \geq t_0$)

$$y(t) = \int_{t-\tau_\star}^{t} K(t,s,y(s))ds + g(t). \tag{3.7}$$

We now obtain, in place of (3.6),

$$y'(t) = K(t,t,y(t)) - K(t,t-\tau_\star,y(t-\tau_\star)) + \int_{t-\tau_\star}^{t} \frac{\partial}{\partial t} K(t,s,y(s))ds + g'(t). \tag{3.8}$$

Note that just as one can sometimes obtain a VIDE by differentiating a VIE of the second kind, one might obtain a VIE by integrating a VIDE.

• *Delay differential equations*: If $K(t,s,u)$ is independent of t then $(\partial/\partial t)K(t,s,u) \equiv 0$ and, if $K(t,t-\tau_\star,u)$ does not vanish identically, then (3.8) is another example of a DDE. Under the heading of DDEs are equations of the form

$$y'(t) = f\Big(t, y(t), y\big(\alpha(t,y(t))\big)\Big) \quad (t \geq t_0) \quad \text{with} \quad y(t) = \psi(t) \quad (t \leq t_0). \tag{3.9}$$

We also encounter systems of DDEs or of neutral delay equations (NDEs), e.g.

$$y'(t) = f(t,\ y(t),\ y(\beta(t,y(t),y'(t))),\ y'(t), y'(\gamma(t,y(t),y'(t)))) \quad (t \geq t_0),$$

with $y(t) = \psi(t)$ $(t \leq t_0)$. The preceding equations are *differential equations with deviating arguments*; they have *retarded* or *delayed* arguments (and are DDEs or NDEs) when $\alpha(t,y(t)) \leq t$, $\beta(t,y(t),y'(t)) \leq t$, $\gamma(t,y(t),y'(t)) \leq t$. As there is no commonly accepted terminology, we proposed (in [13]) to fix on

Delayed argument	Associated terminology
$\alpha \equiv t - \tau(t, y(t))$	state-dependent lag
$\alpha \equiv t_0 + \varrho(t - t_0)$	proportionate lag
$\alpha \equiv t - \tau(t)$	state-independent lag
$\alpha \equiv t - \tau_\star$	constant (or fixed) lag
$\alpha \in [t - \tau^*, t]$ (τ^* finite)	finite-memory (bounded lag)
$\alpha \to \infty$ as $t \to \infty$	fading memory
$\alpha \to t^\star$ as $t \to t^\star$	vanishing lag at $t = t^\star$.

Table 1. Nomenclature for discrete delays

one in which the term $\alpha(t, y(t))$ in eqn (3.9) is called a *delayed argument* and $t - \alpha(t, y(t))$ is called the *lag*: if $\alpha(t, y(t)) = t - \tau_\star$, then $\tau_\star$ is the constant lag. The term 'differential equations with deviating arguments' tends to suggest that these are simple modifications of ODEs. Whilst some delays are 'harmless', delay terms can have a profound effect on the qualitative behaviour of solutions. *Scalar DDEs are in essence infinite-dimensional.* One can encounter chaotic behaviour in a scalar DDE, whilst such behaviour only arises in *systems* of ODEs. The solutions of (2.1) are monotonic whilst solutions of (2.2) may oscillate, and can do so chaotically. The zero solution of the equation $y'(t) = \lambda y(t) + v y(t - \tau_\star)$ $(t \geq t_0)$ is *unstable* with $\lambda > 0$, and $v = 0$ but one can find values $v < -\lambda$ and $\tau > 0$ for which it is *stable*[3]. On the other hand, if $\lambda < 0$ the zero solution of the equation with $v = 0$ is stable but with $v > |\lambda|$ it is unstable if $\tau \geq 0$. The NDE $y'(t) + 2y'(t - \tau_\star) + y(t) = 0$ is often cited: the trivial solution is unstable for $\tau_\star > 0$, but the trivial solution for $\tau_\star = 0$ (the ODE case) is stable. Thus there is a sense in which *delay terms can stabilize* or *destabilize.*

• *Discrete and distributed delays*: The theory of many functional differential equations can be unified if one introduces the Stieltjes integral, as in $y'(t) = \int_{t_0}^{t} y(t - s)[d\mathsf{K}(s)]$ with a suitable choice of K (see [81, p. 48]). VIDEs are sometimes called equations with distributed delays whilst DDEs are called equations with discrete delays. A DDE of the form $y'(t) = f\big(t, y(t), y(t-\tau_\star)\big)$ is replaced, in some analytical studies, by an approximation $y'(t) = f\big(t, y(t), \int_{t_0}^{t} k(t-s)y(s)ds\big)$ where $k(s)$ peaks at $s = \tau_\star$ and is small elsewhere.

• *Volterra- and delay-algebraic equations*: A class of problems that arise in applications is represented by

$$y'(t) = f(t, y(t), y(t - \tau_\star)), \quad 0 = g(t, y(t), y(t - \tau_\star)) \quad (t \geq t_0)$$

with $y(t) = \psi(t)$ $(t < t_0)$; $y \in \mathbb{R}^m, m \geq 1$. This delay differential algebraic

[3] Stability terms are made precise in Definition 1 below. The *stability function* introduced later can be used to verify the statements here.

equation (DDAE), which is potentially much more complex than an ordinary differential algebraic equation (*cf.* [70] and the references therein), has been considered by Ascher and Petzold [1], Campbell [42], and recently by Hauber [58]. Similar equations involving Volterra integral operators also arise, as, for example, with $y = [y_1, y_2]^{\mathrm{T}} \in \mathbb{R}^2$,

$$\begin{aligned} y_1'(t) &= f(t, y_1(t), y_2(t), \textstyle\int_{t_0}^t k_1(t, s, y_1(s), y_2(s))\, ds) \quad (t \geq t_0), \\ 0 &= g(t, y_1(t), y_2(t), \textstyle\int_{t_0}^t k_1(t, s, y_1(s), y_2(s))\, ds) \quad (t \geq t_0), \quad (3.10) \\ y_1(t_0) = \psi_1, &\quad y_2(t_0) = \psi_2, \quad 0 = g(t_0, \psi_1, \psi_2, 0). \end{aligned}$$

Such problems can arise in connection with real-life *singular perturbation problems* of the form (say)

$$\begin{aligned} y_1'(\epsilon; t) &= f(t, y_1(\epsilon; t), y_2(\epsilon; t), \textstyle\int_{t_0}^t k_1(t, s, y_1(\epsilon; s), y_2(\epsilon; s))\, ds) \quad (t \geq t_0) \\ \epsilon y_2'(\epsilon; t) &= g(t, y_1(\epsilon; t), y_2(\epsilon; t), \textstyle\int_{t_0}^t k_1(t, s, y_1(\epsilon; s), y_2(\epsilon; s))\, ds) \quad (t \geq t_0), \end{aligned}$$

with $y_1(t_0)$, $y_2(t_0)$ specified (Kauthen [50, 74, 75]).

• *General classes of equations*: Our equations involve concrete realisations of the concept of an abstract Volterra operator[4]: *An operator V defined on a function space $\mathcal{S}$ of functions is* Volterra *if, for any $\phi_1, \phi_2 \in \mathcal{S}$, $V\phi_1(t) = V\phi_2(t)$ when $\phi_1(s) = \phi_2(s)$ for $s \leq t$.* One may define Volterra equations as equations expressible in terms of Volterra operators.

A fairly general Volterra functional differential equation (VFDE) is represented by

$$y'(t) = \mathfrak{F}\left(t, y(t), y\big(\alpha(t, y(t))\big), \int_{-\infty}^{t} K\big(t, s, y(s)\big) ds\right) \quad (t \geq t_0) \qquad (3.11)$$

subject to the requirement that $\alpha(t, y(t)) \leq t$. (This is a Volterra delay-integro-differential equation or VDIDE.) For (3.11), $y(t)$ is determined by specifying $y(t) = \psi(t)$ for $t \leq t_0$. It is of *practical* concern for numerical analysts to distinguish (3.11) from

$$y'(t) = F\left(t, y(t), y\big(\alpha(t, y(t))\big), \int_{a(t)}^{b(t)} K\big(t, s, y(s)\big) ds\right) \quad (t \geq t_0) \qquad (3.12)$$

wherein $a(t) < b(t) \leq t$, and related equations [19]. Particular attention is given in the literature to the case $a(t) = t - \tau_\star$, $b(t) = t$ ($\tau_\star > 0$ fixed).

• *A canonical type*: To facilitate unified definitions given later, we introduce a canonical problem to represent the problems above:

$$\mathcal{G}(t, y) = 0 \quad (t \geq t_0), \qquad y(t) = \psi(t) \quad (t \leq t_0), \qquad (3.13)$$

[4] The identity operator satisfies this definition and more restrictive definitions, that invoke asymptotic (quasi-) nilpotency for example, are to be found in the literature.

where $\mathcal{G}(t,y) = \begin{cases} y(t) - \mathcal{V}(t,y) \\ \text{or} \\ y'(t) - \mathcal{V}(t,y) \end{cases}$ and $\mathcal{V}$ is a Volterra operator.

We have excluded a discussion of problems involving *partial derivatives*, that involve memory incorporated through delays or Volterra integral terms (either in the differential equation or in the boundary conditions). A review of a selection of such problems is incorporated, with further references, in [95].

4 Analytical behaviour of the solution

A study of the analytical problem is a prerequisite for attempts at its solution.

4.1 Smoothness

Simple examples of convolution VIEs and constant-coefficent DDEs with a constant lag give a feeling for the likely behaviour of solutions.

• *Volterra equations of the second kind* (3.1):

∘ Suppose K, g are smooth in eqn (3.1); then y is smooth. [*Remark:* Differentiate the equation (3.1) to obtain expressions for successive derivatives of y, using an 'inductive' ('recursive') technique: y' is smooth because g' and the function $\frac{\partial}{\partial t}K(t,s,y(s))$ (for $t_0 \le s \le t$) are smooth, etc.]

∘ Suppose $K(t,s,u) = (t-s)^{-\frac{1}{2}}\mathfrak{h}(t,s,u)$ in eqn (3.1) where $\mathfrak{h}$ and g are smooth; then $y(t)$ may be expected to have the form

$$y(t) = t^{\frac{1}{2}}y_1(t) + ty_2(t) + t^{\frac{3}{2}}y_3(t) + t^2y_4(t) + t^{\frac{5}{2}}y_5(t) + \cdots, \tag{4.1}$$

where each function y_ℓ is smooth. (Conversely, if the functions $\mathfrak{h}$ and y are smooth then $g(t)$ has a similar expansion in powers of $t^{\frac{1}{2}}$.) [In the case $y(t) - \int_0^t (t-s)^{-\frac{1}{2}}y(s)ds = g(t)$, we find $y(t) - \pi\int_0^t y(s)ds = g(t) + \int_0^t (t-s)^{-\frac{1}{2}}g(s)ds$ and can deduce y when $g(t) = 1$; we can also find g if $y(t) = 1$.]

∘ In an interesting class of nonlinear VIEs, such as those that arise when modelling explosions in a diffusive medium (*cf.* [91] and its references), one has for the solution $y(t)$ a property ('*blow-up*') that $y(t) \to \infty$ as $t \to t_c$ (or a property $y'(t) \to \infty$ as $t \to t_c$) for some *unknown* critical value t_c. In specific examples, mathematical analysis has given the asymptotic behaviour of $y(t)$ as $t \to t_c$. The numerics of such problems, involving locating t_c numerically, present an as yet open challenge that has attracted the attention of Brunner and of others[5].

• *Volterra integro-differential equations* (3.5):

∘ If y' satisfies (3.5), and $\mathcal{F}$ and $\mathcal{K}$ are smooth, then y is smooth.

• *Delay differential equations* (3.9):

∘ Suppose eqn (3.9) is a scalar equation with smooth f, with $\alpha(t,u) \equiv t - \tau_\star$

[5]Conventional wisdom has it that robust numerical procedures, exploiting known asymptotic behaviour and using error controls, might signal an approach to t_c and could be used to find the value t_c.

($\tau_\star > 0$ fixed) and smooth $\psi(t)$ defined on $[t_0 - \tau_\star, t_0]$. In the case that $\psi'(t_0) \neq f(t_0, \psi(t_0), \psi(t_0 - \tau_\star))$ we may well find that

$$\lim_{t \nearrow t_0+\tau_\star} y''(t) \neq y''(t_0 + \tau_\star), \text{ and } \lim_{t \nearrow t_0+(r-1)\tau_\star} y^{(r)}(t) \neq y^{(r)}(t_0 + (r-1)\tau_\star).$$

[Consider the case $y'(t) = y(t - \tau_\star)$.] Now suppose that $\psi(t)$ has a bounded jump discontinuity at $t = \sigma_{-1}$ where $\sigma_{-1} = t_0 - \mu\tau_\star$, $\mu \in (0,1)$. Then $y'(t)$ may have a bounded jump discontinuity at $t = t_0 + (1-\mu)\tau_\star$, $y''(t)$ may have a bounded jump discontinuity at $t = t_0 + (2-\mu)\tau_\star$, etc.

∘ For a general delay function α in (3.9), suppose $t = \sigma_r$ is (for $r = 1, 2, \ldots$) a zero of odd multiplicity of the function $\alpha(t, y(t)) = \sigma_s$ with $\sigma_s < \sigma_r$ and $\sigma_0 = t_0$. If $\psi'(t_0) \neq f\Big(t_0, y(t_0), y(\alpha(t_0, y(t_0)))\Big)$ then we may well have $\lim_{t \nearrow \sigma_r} y^{(r+1)}(t) \neq y^{(r+1)}(\sigma_r)$ $(r = 0, 1, 2, \ldots)$.

∘ Suppose $\alpha_\star \in (0,1)$. For the *pantograph* equation of the form

$$y'(t) = \lambda y(t) + \nu y(\alpha_\star t), \; (t \geq t_0 \geq 0), \; y(t) = \psi(t), \; (t \leq t_0), \tag{4.2}$$

we have two possibilities: if $t_0 = 0$ the solution is smooth and we only need $\psi(0)$ to determine $y(t)$, but if $t_0 > 0$ and $\lim_{t \nearrow t_0} \psi'(t) \neq y'(t_0)$ then y has jumps in its second, third, fourth, ... derivatives at $t = t_0/\alpha_\star, t_0/\alpha_\star^2, t_0/\alpha_\star^3, \ldots$ respectively. (For generalisations of (4.2), see, for example, [84].)

• *Volterra delay integro-differential equations* (3.11):

∘ For an equation of the form (3.11), the difficulties associated with DDEs arise in the case K is smooth, but are compounded if the integral term is of Abel type (K weakly singular): Consider, for example, the equation $y'(t) = y(t - \tau_\star) + \int_0^t (t-s)^{-\frac{1}{2}} y(s) ds$, $t \geq 0$, with $y(t) = 1$ for $t \in [-\tau_\star, 0]$. Then $y''(t)$ is unbounded at $t = 0$, $y'''(t)$ becomes unbounded at $t = \tau_\star$, etc.

It is apparent from the above that the points $\{\sigma_r\}$ such that

$$\alpha(\sigma_r, y(\sigma_r)) - \sigma_s = 0 \text{ for some } \sigma_r > \sigma_s \text{ with } \sigma_0 = t_0 \tag{4.3}$$

(and where the derivatives of ψ are discontinuous at points $\cdots < \sigma_{-3} < \sigma_{-2} < \sigma_{-1} < t_0$) are potentially significant in the discussion of the smoothness of the solution of a VFDE having a lagging argument. The solution y is smooth on each interval (σ_r, σ_{r+1}), $r \geq 0$. This observation can be extended to *systems* of DDEs with multiple lags, and a network dependency graph can be constructed to see how discontinuities propagate (see [102, 103]).

4.2 Dynamic behaviour

There is a growing literature on oscillation [54], bifurcation, and chaos [38] in DDEs, and on the stability [81] of certain types of solution. The numerical approximations should emulate important features of the analytical solutions.

In the literature, there is frequently a blurring of the distinction between behaviour of a solution as $t \to \infty$ and stability. Bounded solutions of non-linear equations need not be stable with respect to perturbations. Stability is concerned with response to (possibly small) perturbations. Frequently, stability definitions are couched in terms of the *stability of the zero solution*, but they can be rephrased to define the stability of any particular solution.

To illustrate, let us examine the delayed logistic equation (2.2). Suppose that the solution y corresponds to an initial function ψ, and consider the effect δy of a perturbation $\delta\psi$, subject to appropriate (*admissibility*) conditions, in $\delta\psi$. Since y and $y + \delta y$ are both solutions of (2.2), we find $\delta y(t) = \delta\psi(t)$ $(t \leq t_0)$ and

$$\delta y'(t) = \lambda \delta y(t)\{1 - y(t - \tau_\star)\} - \lambda y(t)\delta y(t - \tau_\star) - \lambda \delta y(t)\delta y(t - \tau_\star) \tag{4.4}$$

(for $t \geq t_0$). This is a DDE for δy; we have $\delta y(t) \equiv \delta y(\delta\psi; t)$, given a particular y. If $\delta\psi(t) \equiv 0$, one has the zero solution $\delta y(t) \equiv 0$, and one needs to know how this zero solution responds to changes that take $\delta\psi(t)$ away from zero. Thus, stability of y (with $y(t) \equiv y(\psi; t)$) is reduced to the question of stability of the zero solution, of a related equation, that may be defined as follows:

Definition 1. *Suppose the solution of the evolutionary problem (3.13) is defined by an initial (bounded continuous) function on* $(-\infty, t_0]$ *or some subset thereof, and is zero when the initial function is zero. The zero solution is* (*i*) stable *(or* 'stable under perturbed initial conditions'*) if for each* $\epsilon > 0$ *there exists a* $\delta \equiv \delta(\epsilon, t_0) > 0$ *such that (for any initial function* $\delta\psi$ *with* $\sup_{t\in(-\infty,t_0]} |\delta\psi(t)| < \delta$*), the corresponding solution* δy *satisfies* $\sup_{t\geq t_0} |\delta y(t)| < \epsilon$; (*ii*) uniformly stable *if the number* δ *in definition* (*i*) *is independent of* t_0; (*iii*) asymptotically stable *if it is stable and there exists a constant* δ_0 *such that* $|\delta y(t)| \to 0$ *as* $t \to \infty$, *for any bounded continuous initial function* $\delta\psi$ *with* $\sup_{t\in(-\infty,t_0]} |\delta\psi(t)| < \delta_0$.

For an m-dimensional system of equations $|\ |$ is taken to denote a norm on $\mathbb{R}^m$. The condition $\sup_{t\in(-\infty,t_0]}|\delta\psi(t)| < \delta$ can be replaced by some other measure of smallness. For an *initial–value* problem (ODE or DDE), the solution is defined by $y(t_0)$ and one is concerned only with perturbations $\delta y(t_0) = \delta\psi(t_0)$. There is a well-established theory for linear DDEs with constant coefficients and a constant lag and VIDEs with constant coefficients and $\mathcal{L}_1$-convolution kernels, based upon Laplace transform theory and the requirement that the zeros of a *stability function*, lie in the right half-plane $\Re(\zeta) < 0$. Thus, for

$$y'(t) = \lambda y(t) + \upsilon y(t - \tau_\star) + \omega \int_0^t k(t - s)y(s)ds \tag{4.5}$$

where $k \in \mathcal{L}_1[0, \infty)$, the reader may show that the stability function is

$$\mathcal{S}([\lambda, \upsilon, \omega, \tau_\star]; \zeta) := \zeta - \lambda - \upsilon \exp(-\tau_\star\zeta) - \omega L(k; \zeta) \tag{4.6}$$

where $L(k; \zeta)$ is the Laplace transform of k.

Stability in the first approximation is a concept that permits insight into nonlinear equations that are well-approximated, as $t \to \infty$ and on the solution concerned, by a linear equation. For example, if, in eqn(3.9),

$$f\Big(t, y(t), y\big(\alpha(t, y(t))\big)\Big) = \lambda(t)y(t) + \upsilon(t)y(t - \tau_\star) + \mathcal{O}\left(\{\sup_{t-\tau_\star\leq s\leq t} |y(s)|\}^2\right);$$

(the 'order' term being as $\sup_{t-\tau_\star \le s \le t} |y(s)| \to 0$) and $\lambda(t) \to \lambda_\star$ and $\upsilon(t) \to \upsilon_\star$, as $t \to \infty$, we claim that one has asymptotic stability of y if the limiting equation $u'(t) = \lambda_\star u(t) + \upsilon_\star u(t-\tau_\star)$ has an asymptotically stable solution $u = 0$.

If we return to our example (2.2) (in which $t_0 = 0$), there are two *steady state*[6] solutions $y(t) \equiv 1$ and $y(t) \equiv 0$. Definition 1 applies for $y(t) \equiv 0$ and

$$\delta y'(t) = \lambda\{\delta y(t) - \delta y(t)\delta y(t-\tau_\star)\} \quad (t \ge 0); \quad \delta y(t) = \delta\psi(t) \quad (t \le 0), \tag{4.7}$$

whilst for $y(t) \equiv 1$ we need to consider the stability of the zero solution of

$$\delta y'(t) = -\lambda \delta y(t-\tau_\star)\{1 + \delta y(t)\} \quad (t \ge 0); \qquad \delta y(t) = \delta\psi(t) \quad (t \le 0). \tag{4.8}$$

We derive conclusions by neglecting the second-order terms (stability in the first approximation). For $y(t) \equiv 0$, (4.7) gives $u'(t) = \lambda u(t)$, whose solution $u = 0$ is uniformly asymptotically stable if $\lambda < 0$, implying $y(t) \equiv 0$ is also asymptotically stable. For $y(t) \equiv 1$, (4.8) gives $u'(t) = -\lambda u(t-\tau_\star)$. Using the stability function $\mathcal{S}([-\lambda, 0, 0, \tau_\star]; \zeta)$ of (4.6), one can show that the zero solution $u(t) \equiv 0$ is (along with $y(t) \equiv 1$) asymptotically stable if $\lambda \in (0, \pi/\{2\tau_\star\})$.

A standard mathematical approach to stability involves the construction of suitable Lyapunov functions or Lyapunov functionals. Whilst much discussion focuses on the stability of steady state solutions, it is equally valid to discuss the stability of non-constant solutions. Oscillatory solutions (solutions that have zeros or change sign beyond any arbitrary $t = T$) are of interest, being observed in biological phenomena. Setting $y(t) = \exp(\varpi(t))$ in (2.2) we find $\varpi'(t) = -\lambda\{1 - \exp[\varpi(t-\tau_\star)]\}$. Thus $y(t)$ oscillates about 1 if $\varpi(t)$ oscillates about 0. Neglecting quadratic terms, $\varpi'(t) \approx \lambda\varpi(t-\tau_\star)$ and the stability function (4.6) to determine stability in the first approximation for $\varpi = 0$ is $\mathcal{S}([\lambda, 0, 0, \tau_\star]; \zeta)$. If this function has no real zeros ($\Im(\zeta) \ne 0$ for all zeros ζ), then [54, p. 67], all the nontrivial solutions $u(t)$ are oscillatory. Periodic solutions frequently arrive via Hopf bifurcation, where the parameters in the stability function of the linearized equation cross critical values at which a zero ζ crosses the imaginary axis. For a procedure to analyse this, see [87].

Definition 1 above was concerned with stability under perturbed initial conditions; one can also define stability under persistent disturbances. In particular, for the integral equation (3.1) there is no explicit 'starting value' and such a definition is in fact appropriate. We give one such form, next. (Other definitionss, where stability might automatically imply asymptotic stability, or where the class of admissible perturbations is constrained differently, are possible. One can also define the stability of objects other than solutions, such as the stability of an attracting set associated with a VFDE when the equation is perturbed.)

Definition 2. *Let $\|\ \|$ denote a norm on the normed linear space of (respectively, bounded, decaying, or $\mathcal{L}_1$) continuous functions on $[t_0, \infty)$. Suppose y is*

[6] For a DDE with fading memory delay α, e.g., with constant lags, the steady states are those for the corresponding ODE (where the lags are set to zero).

a continuous solution of (3.13). Then this solution is stable under continuous (bounded, decaying, or $\mathcal{L}_1$) persistent perturbations *if, for* $\epsilon > 0$, *there exists a* $\Delta \equiv \Delta(\epsilon, t_0)$ *such that when* $y(t) = \psi(t)$ $(t \leq t_0)$ *and*

$$\mathfrak{G}\ (t, (y + \delta y)) = \eta(t)\ (t \geq t_0); \tag{4.9}$$

and η *is respectively, a bounded, decaying, or* $\mathcal{L}_1$ *perturbation, satisfying* $||\eta|| \leq \Delta$ *the solution suffers a perturbation* δy *with* $||\delta y|| \leq \epsilon$; *it is* uniformly stable *if* $\Delta(\epsilon, t_0)$ *is independent of* t_0, *it is* asymptotically stable *if it is stable and, for* Δ *sufficiently small,* $\lim_{t\to\infty} \delta y(t) = 0$.

For a *linear* Volterra integral equation, questions of stability can generally be resolved by investigating the so-called resolvent kernel; for the convolution case one can invoke Laplace transforms, and stability can be linked to qualitative behaviour of solutions (are they bounded, do they decay, are they in $\mathcal{L}_1[t_0, \infty)$?). To apply Definition 2 to a convolution equation, consider, for $g \in C[0, \infty)$,

$$y(t) - \lambda \int_0^t k(t-s)y(s)ds = g(t), \quad (t \geq 0); \tag{4.10}$$

suppose $k \in C[0, \infty) \cap \mathcal{L}[0, \infty)$ is positive definite and $\Re(\lambda) > 0$. The perturbations to be considered are continuous perturbations $\eta = \delta g$. Thus

$$\delta y(t) - \lambda \int_0^t k(t-s)\delta y(s)ds = \delta g(t) \quad (t \geq 0). \tag{4.11}$$

If δg is additionally constrained to decay to zero as $t \to \infty$ then with the above conditions (by a well-known theory of Paley and Wiener) δy has the same properties. Hence y is stable, and indeed asymptotically stable, under persistent perturbations that are continuous and decay to zero.

It is important, in the discussion of stability and of numerical stability, to observe that our definitions are related to the *asymptotic effect* of perturbations – so that large *transient* changes in a solution (resulting from possibly small perturbations) are not excluded by stability as defined here.

5 Numerical tools

Our numerical tools comprise (a) approximation with polynomials, piecewise polynomials, and splines, (b) formulae for indefinite integration, and (c) adaptation of collocation, Runge-Kutta (RK) and linear multistep formulae for integral and differential equations. We review essentials as we proceed.

5.1 Discrete and continuous methods

General *discrete methods* are associated with a grid or *mesh*

$$\mathcal{T}_N = \{t_0 < t_1 < t_2 < \cdots < t_N\} \subset \mathbb{R}, \quad h_n := t_{n+1} - t_n, \ \ t_N \leq T \tag{5.1}$$

(with a possibly unequal mesh-spacing $\{h_n\}$, $0 < \check{h} \le h_n \le H$; $n = 0, 1, 2, \ldots$), on which the approximation is defined as a *mesh function*. For theoretical purposes we may also consider $T = \infty$, $\mathcal{T}_\infty = \{t_0 < t_1 < t_2 < t_3 < \cdots\}$. We shall term the value $H \equiv H(\mathcal{T}) := \sup\{h_n\}$ the *width* of the mesh; a mesh for which $h_n = h$ is constant will be called *uniform* or *regular*, and a mesh for which $\{h_n\}$ is determined, possibly in advance, to match the smoothness properties of the solution will be termed *graded* (or *constrained*). It is the aim in a discrete method to compute approximations $\widetilde{y}_n \approx y(t_n)$, $t_n \in \mathcal{T}_N$, and this may suffice. In contrast, general *continuous methods* produce a continuous approximation $\widetilde{y}(t) \approx y(t)$ for all $t \in [t_0, T]$; however, since these approximations are frequently piecewise smooth, they too can be associated with a mesh $\mathcal{T}_N$, or, indeed, can be generated from a discrete approximation defined upon the mesh.

• *Mesh functions and continuous extensions* will be required. Consider, for definiteness, a DDE with one lag $\tau_\star$, such as (2.2). Amongst the meshes $\mathcal{T}_N$ with uniform mesh (of width $H \equiv h$), we can distinguish, given the value $\tau_\star$, those *uniform* meshes for which

$$\tau_\star = Mh,\ M \in \mathbb{Z}_+ \quad \ldots \text{ a } \tau_\star\text{-}\textit{commensurable} \text{ mesh } \mathcal{T}_N;$$
$$\tau_\star = (M - \vartheta)h,\ M \in \mathbb{Z}_+, \vartheta \in (0,1) \quad \ldots \text{ a } \tau_\star\text{-}\textit{incommensurable} \text{ mesh } \mathcal{T}_N.$$

For most methods that generate a mesh function, the computation of $\widetilde{y}$ on $\mathcal{T}_N$ proceeds in an evolutionary fashion. In a typical step, the approximate solution values $\{\widetilde{y}_\ell \approx y(t_\ell)\}_{\ell=0}^{n}$ are used to compute $\widetilde{y}_{n+1} \approx y(t_{n+1})$. (For some methods, there may be subsidiary values defined on a sub-mesh comprising points of the form $t_{n,i} = t_n + c_i h_n$; we illustrate this point later.) Much as in the numerical solution of an ODE, one may require certain initial (or starting) values $\widetilde{y}_0, \widetilde{y}_1, \ldots, \widetilde{y}_m$, in addition to any initial function data supplied as part of the problem. For integral and integro-differential equations or delay equations with one fixed lag $\tau_\star$, one has only to construct equations for $\widetilde{y}_{n+1}$ in terms of $\{\widetilde{y}_\ell\}_0^n$ (and ψ). Thus, for $\tau_\star$-commensurable meshes, no continuous extension is necessary when solving (2.2). But for *general* functional equations of the type considered here, we must seek a companion continuous approximation

$$\widetilde{y}(t) \approx y(t),\ t \in [t_0, t_N] \quad \text{with } \widetilde{y}(t_n) \equiv \widetilde{y}_n \text{ for } t_n \in \mathcal{T}_N. \tag{5.2}$$

Since $\widetilde{y}(t_n) \equiv \widetilde{y}_n$, we refer to $\widetilde{y}$ as a *continuous extension* (or a *prolongation*) of the values $\widetilde{y}_n$. If there is a delay $\alpha(t_n, \widetilde{y}(t_n))$ that assumes a value not amongst previous points $\{t_j : j \le n\}$ then[7] one needs the continuous extension in order to compute $\widetilde{y}(\alpha(t_n, \widetilde{y}(t_n)))$.

Now the standard tools of approximation and interpolation can be employed to derive the continuous extension, and Hermite or Hermite–Birkhoff interpolants

[7] For a true solution of a DDE, one has $\alpha(t, y(t)) \le t$. However, it is possible, particularly with lags that are small relative to h_n, to find that $\alpha(t_n, \widetilde{y}(t_n)) \in [t_n, t_{n+1}]$, or $\alpha(t_{n,i}, \widetilde{y}(t_{n,i})) \in [t_n, t_{n+1}]$. This is a complication associated with a small or vanishing lag, or a poor approximation $\widetilde{y}$.

can provide these tools. But there are more *natural* tools associated with dense output mechanisms for ODEs, that are in an obvious sense extensions of the methods that generate mesh functions. These exist for LMFs and for RK methods. We shall return to continuous RK formulae in a later section. The effect of delay terms has to be represented through the formulae for computing the solution at delayed arguments, and it is not surprising to discover that the choice of the extension formula can affect:

∘ accuracy, and asymptotic behaviour, of the error as a function of the width $H \equiv h$ of the mesh $\mathcal{T}_N$ or $\mathcal{T}_\infty$;

∘ stability of the numerical scheme on $\mathcal{T}_\infty$.

• *Collocation techniques* provide an example of continuous methods. They too can be associated with a mesh $\mathcal{T}_N$; typically, for spline collocation the approximation $\widetilde{y}(t)$ is taken as a spline that reduces to a polynomial on each interval $[t_\ell, t_{\ell+1}]$ and is globally in some prescribed continuity class. Associated with the approximation is a *defect* or residual $\eta(t)$; in the case of (3.11), we have

$$\eta(t) := \widetilde{y}'(t) - \mathfrak{F}\left(t, \widetilde{y}(t), \widetilde{y}(\alpha(t, \widetilde{y}(t))), \int_{-\infty}^{t} K(t, s, \widetilde{y}(s))ds\right) \quad (t \geq t_0). \quad (5.3)$$

(More generally, modify (5.3) using (4.9).) In each interval $[t_\ell, t_{\ell+1}]$ collocation points $\{t_{\ell,i} = t_\ell + c_i h_\ell\}$ are selected and the collocation method consists of determining $\widetilde{y}$ in the given continuity class such that $\eta(t)$ vanishes at the collocation points $\{t_{n,i}\}$ (for $n = 0, n = 1, n = 2, \ldots$, in turn). In practice, the integrals $\int_{-\infty}^{t} K(t, s, \widetilde{y}(s))ds$ have to be approximated, and this results in a perturbed collocation method. There are links between collocation and certain RK processes.

• *Defect:* Any method that produces a continuous approximation $\widetilde{y}$ with a right-hand derivative generates a corresponding residual or *defect* η as in (5.3). The size of this function, combined with a knowledge of the stability properties under global perturbation, can give confidence in the size of $y(t) - \widetilde{y}(t)$, $t \in [t_0, t_N]$.

6 Insight based upon uniform meshes

We present some methods, based upon the use of uniform meshes, for equations where arguments can be found for choosing to work with a fixed stepsize.

6.1 Discretization of Volterra integral equations by quadrature

Volterra integral equations may be treated directly or (sometimes) transformed by differentiation into integro-differential equations. Consider the problem of computing an approximating mesh function $\widetilde{y}(t)$ to the solution $y(t)$ $(t \in \mathcal{T})$ of eqn (3.1), based upon quadrature. To this end, given $\mathcal{T}$ (which, for the moment, may be a nonuniform mesh), set $t = t_n$ in eqn (3.1) and, assuming that K and g are at least continuous, seek a suitable quadrature $\int_{t_0}^{t_n} \psi(s)ds \approx \sum_{j=0}^{n} W_{n,j}\psi(t_j)$

to discretize the integral term in the integral equation. This yields the system

$$\widetilde{y}(t_n) - \sum_{j=0}^{n} W_{n,j} K(t_n, t_j, \widetilde{y}(t_j)) = g(t_n) \quad (n = n_0, n_0+1, n_0+2, \ldots). \tag{6.1}$$

For $n = 0, 1, \ldots, n_0 - 1$ one may decide on special starting formulae. This approach is well-documented (see [34]), though a good procedure for the automatic selection of points $\{t_\ell\}$ is elusive; one does not know how refined a rule one will require at the later stage n'' when computing the solution at stage $n' < n''$. The basic quadrature approach assumes a degree of smoothness of the integrand over intervals $[t_0, t_n]$, $n \in \mathbb{N}$, and has to be modified for Abel equations, for example.
• *Convolution integral equations* have a number of features that can be preserved if $\mathcal{T}_\infty$ is a uniform mesh. We shall therefore set $t_0 = 0$ and use

$$y(t) - \int_0^t k(t-s)\Phi(y(s))ds = g(t) \quad (t \geq 0) \tag{6.2}$$

to develop our discussion and modifications of the method, assuming initially that k, g are smooth and Φ has properties ensuring that *there is a unique solution.*

For (6.2), the repeated trapezium rule with step h gives at the n-th stage ($n > 0$; for $n = 0$, $\widetilde{y}_0 = g(0)$) the equation

$$\widetilde{y}_n - \tfrac{1}{2}hk(0)\Phi(\widetilde{y}_n) = \gamma_n \quad (n \in \mathbb{N}), \tag{6.3}$$

$$\gamma_n \equiv \left\{ g(nh) - h\textstyle\sum_{j=0}^{\prime\,(n-1)} k((n-j)h)\Phi(\widetilde{y}_j) \right\}, \tag{6.4}$$

to be solved for $\widetilde{y}_n =: \widetilde{y}(t_n)$ where $t_n = nh$. (The notation $\sum'$ denotes that the term $j = 0$ is halved.) This raises the question: 'Is the sequence $\{\widetilde{y}_\ell = \widetilde{y}(t_\ell)\}$ uniquely defined?' The answer is affirmative if Φ satisfies a uniform Lipschitz condition[8] and the width $H \equiv h$ is sufficiently small. When eqns (6.3) have more than one solution, one may nevertheless be able to specify a unique recipe for defining $\{y_\ell\}$. If existence and uniqueness are thus assured, one can address *numerical convergence as* $h \to 0$ and ask 'Is $\lim_{h\to 0} \sup_{t_\ell \in \mathcal{T}_\infty, t_\ell \leq T} |\widetilde{y}(t_\ell) - y(t_\ell)| = 0$?' For finite T, this is convergence on $(0, T)$ and for $T = \infty$ it is convergence on $\mathbb{R}_+$. The other theoretical questions involve numerical stability and preserving qualitative behaviour (see [34, 51, 85]). The choice of a uniform mesh has theoretical attractions, since there are classes of kernels k (such as completely monotone functions, or positive definite functions) for which one can *guarantee* that certain qualitative behaviour is correctly modelled if one uses quadrature with appropriate properties.

Turning to computational aspects, we have in (6.3) an implicit equation for $\widetilde{y}_n$ that must in general be solved by an iteration. The right-hand side γ_n is evaluated once only, at the start of the iteration. A constant stepsize h ($t_n = nh$) has

[8] This is easily shown; the case where Φ satisfies a suitable one-sided Lipschitz condition [56, p. 58] that (for appropriate k) defines a contractivity condition is more interesting.

some practical attractions in the case of convolution equations, because (*a*) the derivative used to solve (6.3) by Newton's iteration is independent of n, and (*b*) the terms γ_n in (6.4) can be computed efficiently. In this respect, the discretized equations preserve the convolution structure and fast Fourier transforms (FFTs) can be used in the computation, in a batch, of a number of terms of the form, say,

$$h \sum_{j=0}^{m-1}{}' k((n-j)h)\Phi(y_j) \text{ for } n = m, m+1, \ldots, 2m-1; \quad m = 2^r \quad (r \in \mathbb{Z}_+).$$

Similar techniques can be used if we find alternative discretizations

$$\int_0^{nh} k(nh-s)\Phi\big(y(s)\big)ds \approx \sum_{j=0}^{n} \Omega_{n-j}\Phi\big(\widetilde{y}(jh)\big) + \text{ starting terms} \tag{6.5}$$

(or similar forms, employing RK parameters [7], not discussed here). We outline such alternatives, that are fully described in the literature [34]. First, define

$$\omega^{[\mu]}(\zeta) := \sum_{j=0}^{\infty} \omega_j^{[\mu]}\zeta^j = \left\{\frac{\sigma(\zeta^{-1})}{\rho(\zeta^{-1})}\right\}^{\mu}, \qquad \mu \in (0,1], \tag{6.6}$$

and refer to (3.4) for the meaning of $\mathcal{I}^{\mu}\phi(t)$. Application of a general zero-stable $\{\rho,\sigma\}$-LMF to the problem of integration supplies quadrature formulae

$$\mathcal{I}\phi(nh) \approx h\sum_{j=0}^{n} \omega_{n-j}^{[1]}\phi(jh) + \text{ starting terms} \tag{6.7}$$

which produces the required form on setting $\Omega_j = h\omega_j^{[1]}k(jh)$. Secondly, in analogy with fractional integration, we can (with suitable LMFs) obtain formulae

$$\mathcal{I}^{\mu}\phi(nh) \approx h^{\mu}\textstyle\sum_{j=0}^{n} \omega_{n-j}^{[\mu]}\phi(jh) + \text{ starting terms}, \tag{6.8}$$

where, in the light of the behaviour of the solution $y(t)$ in the neighbourhood of $t=0$, the starting terms involving $\phi(0), \ldots, \phi(kh)$ are constructed to ensure that integrands of the type $s^{\lambda}, \lambda \in \Lambda \subset \mathbb{R}_+$ are integrated exactly for an appropriate set Λ. Thus, if $\mu = \frac{1}{2}$, we might take $\Lambda = \{\frac{1}{2}, 1, \frac{3}{2}, 2, \frac{5}{2}, 3\}$. If $k(t) = t^{\nu}\mathfrak{K}(t)$ we set $\Omega_j = h^{1-\nu}\Gamma(1-\nu)\omega_j^{[1-\nu]}\mathfrak{K}(jh)$. The BDFs of order no more than 6 are satisfactory candidates for generating weights $\{\omega_\ell^{[\mu]}\}$, $\mu \in (0,1]$. These weights can be obtained using FFTs and Newton schemes [6, 8].

An alternative is to employ product integration formulae $\int_0^{t_n}(t_n-s)^{-\nu}u(s)ds \approx \sum_{j=0}^{n} w_{n-j}^{[\nu]}u(t_j)$ that are constructed (say) to be exact whenever u is a piecewise linear "hat function" (shaped like a peaked Welsh hat[9]) φ_ℓ for the mesh $\mathcal{T}$.

[9]The continuous function φ_ℓ has support $(t_{\ell-1}, t_{\ell+1})$, is linear on $[t_{\ell-1}, t_\ell]$, and on $[t_\ell, t_{\ell+1}]$, with $\varphi_\ell(t_\ell) = 1$. We can amend the formulae if $y(t)$ behaves like $(t-t_0)^{\lambda}$, $\lambda \notin \mathbb{N}$.

If $k(t) = t^{-\nu}\mathfrak{K}(t)$ and $\mathcal{T}$ is uniform and of width $H \equiv h$, this gives equations

$$\widetilde{y}_n - \sum_{j=0}^{n} w_{n-j}^{[\nu]}\mathfrak{K}((n-j)h)\Phi(\widetilde{y}_j) = g(nh) + \text{ starting terms} \tag{6.9}$$

$(n = n_0, n_0+1, n_0+2\ldots)$. If $\mathfrak{K}(s) \equiv 1$, this is also a collocation technique.

• *Convolution integro-differential equations* can be treated with modified LMFs. The use of a uniform mesh is prompted by the convolution structure. Each of the quadrature techniques we have described provides discretizations

$$\int_0^{t_n} k(t_n - s)\Phi\big(y(s)\big)ds \approx \sum_{\ell=0}^{n} \Omega_{n-\ell}\Phi\big(\widetilde{y}(t_\ell)\big) + \text{ starting terms.}$$

For the VIDE

$$y'(t) = F(t, y(t), \int_0^t k(t-s)\Phi\big(y(s)\big)ds), \quad (t \geq 0), \quad y(0) = y_0, \tag{6.10}$$

a k-step LMF with first and second characteristic polynomials $\rho(\mu) := \sum \alpha_{k-\ell}\mu^\ell$, $\sigma(\mu) := \sum \beta_{k-\ell}\mu^\ell$ now gives us the formulae

$$\alpha_0\widetilde{y}_n - h\beta_0\widetilde{f}(t_n, \widetilde{y}_n, \widetilde{z}_n) = \sum_{\ell=1}^{k}\left\{\alpha_\ell\widetilde{y}_{n-\ell} + h\beta_\ell\widetilde{f}_{n-\ell}\right\} \qquad (n \geq n_0), \tag{6.11}$$

$$\widetilde{z}_j = \sum_{\ell=0}^{j} \Omega_{j-\ell}\Phi\big(\widetilde{y}(t_\ell)\big) + \text{ starting terms;} \quad \widetilde{f}_\ell = f(t_\ell, \widetilde{y}_\ell, \widetilde{z}_\ell). \tag{6.12}$$

To anyone familiar with *practical* multistep methods for ODEs, the use of a uniform mesh is clearly a limitation imposed by a wish to preserve the convolution structure and qualitative behaviour.

6.2 LMFs in the discretization of delay differential equations

For a DDE with a fixed lag, of the form

$$y'(t) = f\big(t, y(t), y(t-\tau_\star)\big) \quad (t \geq 0), \qquad y(t) = \psi(t) \quad (t \in [-\tau_\star, 0]), \tag{6.13}$$

it is not difficult to see how one may adapt a $\{\rho, \sigma\}$-LMF using a uniform mesh with width $H \equiv h$ along lines similar to those for the integro-differential equation (6.10), *provided* $\mathcal{T}$ is $\tau_\star$-commensurable; in this case no continuous extension is needed. With $\tau_\star = mh$, one arrives at the recurrence

$$\alpha_0\widetilde{y}_n - h\beta_0\widetilde{f}(t_n, \widetilde{y}_n, \widetilde{y}_{n-m}) = \sum_{\ell=1}^{k}\left\{\alpha_\ell\widetilde{y}_{n-\ell} + h\beta_\ell\widetilde{f}_{n-\ell}\right\}, \ \widetilde{f}_\ell = f(t_\ell, \widetilde{y}_\ell, \widetilde{y}_{\ell-m}). \tag{6.14}$$

One has to recall, however, that the classical truncation error analysis for an LMF relies upon smoothness assumptions that are not necessarily satisfied for a number of DDEs (in a transient phase for the cited DDE, and in the longer term for many NDEs). In the case $\mathcal{T}_N$ is $\tau_\star$-incommensurable, one requires a suitable continuous extension to redefine the notation $\widetilde{f}_\ell$ as $f\left(t_\ell, \widetilde{y}_\ell, \widetilde{y}(\alpha(t_\ell, \widetilde{y}(t_\ell)))\right)$. With the aid of an extension (prolongation), the formulae can be modified for the case of a DDE with two different retarded terms, having incommensurate lags, say

$$y'(t) = f(t, y(t), y(t-\tau_1), y(t-\tau_2)) \ (t \geq 0), \quad y(t) = \psi(t) \quad t \in [-\tau_1, 0] \quad (\tau_1 > \tau_2),$$

where the ratio $\tau_1 : \tau_2$ is non-integer. This pinpoints the need to be able to generate an approximation at any previous point $t_n - \tau_\star$ (not merely at previous mesh points). It brings us to look at continuous formulae of the type used in dense output for ODEs. We shall examine continuous RK methods, but it is possible to make considerable headway with adaptive dense-output LMFs (see Willé [101], who built on ideas of Feldstein, of Bader and of Bock and Schlöder). The LMF-based formulae may have some advantages if the solution suffers few discontinuities in low-order derivatives.

7 Some continuous methods for FDEs

7.1 Continuous RK formulae and RK quadrature

For general problems (where there may be derivative discontinuities) the use of one-step methods based upon suitable continuous Runge-Kutta formulae is flexible and attractive (for a DDE code, see Paul [94]). It is convenient to display the parameters of a continuous RK (CRK) formula in the form of a *continuous* Butcher tableau, each row of which defines a quadrature rule. The *conventional* Butcher tableau has θ replaced by the (implied) value 1. The following indicates the form of *CRK parameters and associated quadrature:*

$$\begin{array}{c|c} \mathsf{c} & \mathsf{A} \\ \hline \theta & \mathsf{b}^{\mathrm{T}}(\theta) \end{array} := \begin{array}{c|cccc} c_1 & a_{1,1} & a_{1,2} & \cdots & a_{1,m} \\ c_2 & a_{2,1} & a_{2,2} & \cdots & a_{2,m} \\ \vdots & \vdots & \vdots & & \vdots \\ c_m & a_{m,1} & a_{m,2} & \cdots & a_{m,m} \\ \hline \theta & b_1(\theta) & \cdots & \cdots & b_m(\theta) \end{array} \qquad \begin{array}{l} \int_0^{c_1} \phi(s)ds \approx \sum_{j=1}^m a_{1,j}\phi(c_j) \\ \int_0^{c_2} \phi(s)ds \approx \sum_{j=1}^m a_{2,j}\phi(c_j) \\ \vdots \qquad\qquad \vdots \\ \int_0^{c_m} \phi(s)ds \approx \sum_{j=1}^m a_{m,j}\phi(c_j) \\ \\ \int_0^{\theta} \phi(s)ds \approx \sum_{j=1}^m b_j(\theta)\phi(c_j). \end{array}$$

The *RK abscissae* $\{c_i;\ i = 1, 2, \ldots, m\}$ in a RK tableau define auxiliary mesh points $t_{n,i} := t_n + c_i h_n$. By suitable changes of variables, the quadrature rules implicit in a CRK tableau define the quadrature approximations in the table labelled *'Examples of Quadrature from CRK parameters'*. The quadrature rule given for arbitrary $\theta \in (0, 1)$ often has a lower degree of precision than that achieved for $\theta = 1$, though by the addition of further stages this can be rectified.

For integral terms in integral and integro-differential equations, we can obtain (for example) discretization formulae:

$$\widetilde{z}_{n,i} := \sum_{s=0}^{n-1}\sum_{r=1}^{m} h_s b_r K(t_{n,i},\ t_{s,r}\ , \widetilde{y}(t_{s,r})) + \\ + h_n \sum_{r=1}^{m} c_i b_r K(t_{n,i},\ t_n + c_i c_r h_n, \widetilde{y}(t_n + c_i c_r h_n)), \tag{7.1}$$

where

$$\widetilde{z}_{n,i} \approx \int_0^{t_n} K(t_{n,i}, s, y(s))ds + \int_{t_n}^{t_{n,i}} K(t_{n,i}, s, y(s))ds.$$

Clearly these formulae *rely* on $\widetilde{y}$ being densely defined.

Table 2. *Examples of Quadrature from CRK parameters:*

$$\begin{aligned}
\int_0^{t_n} \varphi(s)ds &\approx \sum_{s=0}^{n-1}\sum_{r=1}^{m} h_s b_r \varphi(t_{sr}) \\
\int_{t_n}^{t_n+\theta h_n} \varphi(s)ds &\approx h_n \sum_{r=1}^{m} b_r(\theta)\varphi(t_n + c_r h_n) \\
\int_{t_n}^{t_n+c_i h_n} \varphi(s)ds &\approx h_n \sum_{r=1}^{m} a_{i,r}\varphi(t_n + c_r h_n) \quad \text{or} \\
\int_{t_n}^{t_n+c_i h_n} \varphi(s)ds &\approx h_n \sum_{r=1}^{m} c_i b_r \varphi(t_n + (c_i c_r) h_n)
\end{aligned}$$

Table 3. *Examples of CRK parameters of natural formulae (Tang [96]):*

m	Components of c	Components of $\mathsf{b}(\theta)$	m	Components of c	Components of $\mathsf{b}(\theta)$
	Rule	Gauss methods		Rule	Radau IIA methods
1	$\frac{1}{2}$	θ	1	$\frac{1}{2}$	θ
2	$\frac{3-\sqrt{3}}{6}$ $\frac{3+\sqrt{3}}{6}$	$\frac{\theta(3-3\theta+\sqrt{3})}{2\sqrt{3}}$ $\frac{\theta(-3+3\theta+\sqrt{3})}{2\sqrt{3}}$	2	$\frac{1}{3}$ 1	$\frac{3\theta(2-\theta^2)}{4}$ $\frac{\theta(3\theta-2)}{4}$

7.2 The CRK approximation formulae

The CRK approximation $\widetilde{y}$ to the solution y of (3.11), with

$$y'(t) = \mathfrak{F}(t, y(t), y(\alpha(t, y(t))), \int_{-\infty}^{t} K(t, s, y(s))ds)\ (t \geq t_0),\quad y(t) = \psi(t)\ (t \leq t_0),$$

is defined by $\psi(t)$ for $t \le t_0$ and is sought on (t_0, ∞) by computation of its values $\{\widetilde{y}(t_n)\}$ on $\mathcal{T}_\infty$. To set down the formulae, suppose that the numerical solution has advanced to $t = t_n$ and the continuous extension $\widetilde{y}(t)$ is available for $t \in [t_0, t_n]$. With $\widetilde{z}_{n,i}$ defined by (7.1), the CRK formulae are

$$\widetilde{y}(t_n + \theta h_n) = \widetilde{y}(t_n) + h_n \sum_{i=1}^{m} b_i(\theta)\widetilde{\mathfrak{F}}_{n,i}, \quad \widetilde{\mathfrak{F}}_{n,i} := \mathfrak{F}(t_{n,i}, \widetilde{Y}_{n,i}, \widetilde{y}(\widetilde{\alpha}_{n,i}), \widetilde{z}_{n,i}), \tag{7.2}$$

with

$$\widetilde{Y}_{n,i} := \widetilde{y}(t_n) + h_n \sum_{j=1}^{m} a_{i,j}\widetilde{\mathfrak{F}}_{n,j}, \quad \widetilde{\alpha}_{n,i} := \alpha(t_{n,i}, \widetilde{y}(t_{n,i})), \tag{7.3}$$

and $\widetilde{y}(\widetilde{\alpha}_{n,i})$ can be computed from (7.2); if $\widetilde{\alpha}_{n,i} = t_k + \vartheta_{n,i}h_k$, with $\vartheta_{n,i} \in [0,1)$, then we set $n = k$ and $\theta = \vartheta_{n,i}$ in eqn (7.2). The CRK formula is called *natural* when (a simplifying but not an essential assumption) $b_j(c_i) = a_{i,j}$ for $i, j \in \{1, 2, \ldots, m\}$; in this case $\widetilde{y}(t_{n,j}) = \widetilde{Y}_{n,j}$. To construct $\widetilde{y}(t)$, it is sufficient to compute either the values $\widetilde{Y}_{n,i}$ or the values $\widetilde{\mathfrak{F}}_{n,i}$. Note that $\widetilde{z}_{n,i}$, and possibly $\widetilde{y}(\widetilde{\alpha}_{n,i})$, depend on unknown values $\widetilde{y}(t) = \widetilde{y}(t_n) + h_n \sum_{j=1}^{m} b_j(\theta(t))\widetilde{\mathfrak{F}}_{n,j}$, where $t = t_n + \theta(t)h_n$, $\theta(t) \in [0,1)$. The equations to be solved at each step are in general implicit. They can be implicit even if A is lower triangular. Existence and uniqueness of the RK solution can be shown by a contraction mapping argument, under suitable assumptions (notably that $\mathcal{T}_N$ is sufficiently fine).

7.3 Event location and tracking discontinuities

From our discussion of smoothness, we know the significance (for both discrete and continuous methods) of the points $\{\sigma_r\}$ in (4.3). In some algorithmic approaches it is considered desirable to compute each point σ_r in order to include those that are 'significant' in the mesh points $\mathcal{T}_\infty$. With fixed lags, these points can be computed in advance, and one might specify a pre-graded mesh; this is highly unlikely with state-dependent delays (state-dependent delays raise interesting questions [102]). Finding a point $\sigma > \sigma_s$ at which $\sigma_s - \alpha(\sigma, \widetilde{y}(\sigma))$ changes sign is an example of *event location*. However, points $\{\sigma_r\}$ are significant only if there is a discontinuity in a derivative of modest order; higher-order derivative discontinuities may be irrelevant. With a one-step method, we would hope that computing significant points σ_r would allow us to ensure that one never takes a step across a point at which a significant derivative has a discontinuity. Unfortunately, the points $\{\sigma_r\}$ can cluster: $\lim_{r\to\infty} \sigma_r = \sigma^\star$. In this case, the point $\sigma^\star$ could be a natural barrier beyond which the step-by-step method cannot proceed if one is not allowed to cross a discontinuity. Whether or not this is the case depends upon the degree of smoothness of the solution in a neighbourhood of $\sigma^\star$. There will be instances where one has to cross a point of discontinuity, and what is required is that in doing so the error incurred remains small and is not subsequently magnified to an unacceptable degree. See [101, 102, 103], etc.

7.4 Error control and practical codes

Error control mechanisms used in the numerical solution of ODEs are sometimes adapted to methods for DDEs or VIDEs. The main strategies are based on:

∘ comparison formulae to estimate truncation error (for example, using embedded formulae);

∘ step-adjustment to estimate truncation error (for example, using two short steps to advance one previous longer step);

∘ estimation of the defect;

∘ estimation of stiffness (for example, monitoring convergence rates of nonlinear equation solvers).

The first three of these generally rely for their theoretical foundation upon *order* arguments that are valid for small stepsizes given sufficient continuity. (Attempts have been made to consider the effects of discontinuous derivatives.)

Having estimated either the local truncation error, the defect, or the stiffness, one has to respond to these estimates (to ensure that one has control over the algorithm) by adjusting stepsize, formula, or method of implementation of a formula, and the mathematical justification again relies upon certain premises. Most methods of controlling error are based on arguments that can fail in practice for some problems. This is strong motivation for *tracking discontinuity points* σ_r and monitoring [102] the degree of smoothness of the solution, at least for nontrivial problems.

Enright and co-workers are advocates of the control of their DDE solvers through estimates of the defect. This is highly appealing, but estimating the norm of the defect accurately is expensive. Asymptotic arguments suggest that sampling the defect at one point gives adequate control, but more work appears to be necessary on robustness of the procedures. Concerning stability, model problems (see below) are often employed to gain insight, but, for ODEs and DDEs, practitioners often rely upon error control 'somehow' to ensure that stability constraints are also satisfied.

Codes for integral and functional differential equations have recently been written by a number of research groups, in particular those associated with the author, with Brunner, with Enright, with Thompson, and with Jackiewicz. At Manchester, Paul [94] has devised a code for DDEs and FDEs that is based on an explicit CRK tableau (A is lower-triangular) that has the following features.

∘ It uses parameters obtained by Shampine to provide a fifth–order Hermite extension for the RK formulae of Dormand and Prince;

∘ it tracks discontinuities $\{\sigma_r\}$ by solving the equations (4.3);

∘ it uses embedded formulae to obtain local truncation error estimates;

∘ Volterra integral terms are dealt with using adaptive integration;

∘ the code permits treatment of neutral terms, vanishing lags, and initial value problems, in NDEs and DDEs.

8 Numerical analysis

8.1 Convergence and order of convergence

Convergence proofs have to take account of the varying continuity of exact solutions, and should therefore incorporate assumptions about the associated mesh $\mathcal{T}$, and the degree of smoothness of the solution on each interval $[t_\ell, t_{\ell+1}]$ or (say) $[t_{\ell-k}, t_{\ell+1}]$. With such assumptions, it is generally possible to establish, for some $\varrho^\star \geq \varrho_\star > 0$, results of the form

$$\sup_{t_\ell \in \mathcal{T}_\infty, t_\ell \leq T} |y(t_\ell) - \widetilde{y}(t_\ell)| = \mathcal{O}(H^{\varrho^\star}); \quad \text{or} \quad \sup_{t_0 \leq t \leq T} |y(t) - \widetilde{y}(t)| = \mathcal{O}(H^{\varrho_\star}) \qquad (8.1)$$

as $H \to 0$, where $H = H(\mathcal{T})$ is the width of $\mathcal{T}$. The significance of such results (though clear) is often misunderstood: it is not, in general, reasonable to expect that the maximum error is asymptotically proportional to H^ϱ for some ϱ. Thus, using a uniform mesh of width $H \equiv h$ to solve a DDE with fixed lag $\tau_\star$, if $h \to 0$ in such a way that $\mathcal{T}$ is $\tau_\star$-commensurable one may well (*cf.* [10]) obtain a higher rate of convergence than if one allows $\tau_\star$-incommensurable meshes: consider an error of the form $e_1(t)(\frac{\tau}{h} - [\frac{\tau}{h}])h^2 + e_2(t)h^3$ where $[z]$ is the integer part of z. The situation is exacerbated when nonuniform meshes are used. The reader will also be familiar with problems in ODEs (see *stiff order* in Butcher [39]) where, for a very stiff problem, there is an effective reduction in the order of a method; it is clear that a corresponding phenomenon could arise for functional differential equations. This said, one can sometimes obtain valid 'proportionality' results for suitable approximations to smooth solutions of integral and integro-differential equations using uniform meshes.

• *Integral equations*: Convergence theory for classical integral equations is well understood; recent developments have centred on *variant forms* of eqn (3.1), such as

$$y(t) = \int_{a(t)}^{b(t)} K(t, s, y(s))ds + g(t) \qquad (t \geq t_0), \qquad (8.2)$$

with (i) equations with $a(t) = -\infty$, $b(t) \leq t$; (ii) equations with $a(t) = t - \tau_\star, b(t) = t$; (iii) equations with $a(t) = t_0$, $b(t) = \max\{t_0, t - \tau_\star\}$; and (iv) equations with more than one integral term each taken from the above. These and other variants appear in the literature (in particular in work of Baker, of Brunner, etc., each with their respective co-workers).

•*Functional differential equations*: The need to consider possibly small or vanishing state-dependent lags in FDEs and varying continuity assumptions led Baker and Paul [10] to produce a rigorous convergence proof for certain RK methods, with orders of convergence. Their approach applies to various formulae.

8.2 Stability with uniform meshes

We commence with a definition of stability of a numerical solution of a FDE, computed by some numerical process using a mesh $\mathcal{T}_\infty$. In many of the discussions to be found in the literature, $\mathcal{T}_\infty$ is taken to be a uniform mesh, and the class of FDEs is restricted (often quite severely). The basic definitions are analogues of those found in the continuous case; stability in the first approximation and Lyapunov theory have analogues in numerical stability.

Suppose the numerical solution of the evolutionary problem is defined by application of some numerical scheme using a mesh $\mathcal{T}_\infty$, in terms of a *starting function* on an interval $(-\infty, t_\star]$. The stability definition takes a slightly different form depending on whether it refers to mesh functions or continuous extensions. The starting data is based upon the given initial bounded continuous function ψ on $(-\infty, t_0]$ or some subset thereof, and – if required by the numerical scheme – a (bounded and continuous) auxiliary starting function on $[t_0, t_\star]$. Suppose the numerical solution is zero when the initial function and auxiliary starting function (if any) are both zero. Then the following applies:

Definition 3. *The zero numerical solution is* (*i*) stable with respect to initial data *if for each* $\epsilon > 0$ *there exists* $\Delta = \Delta(\epsilon, t_0) > 0$ *such that (for any starting function* $\delta\psi$ *that satisfies* $\sup_{t\in(-\infty,t_\star]} |\delta\psi(t)| < \Delta$*), the corresponding solution* $\delta\widetilde{y}(t) \equiv \delta y(\delta\psi; t)$ *satisfies* $\sup_{t\geq t_\star} |\delta\widetilde{y}(t)| < \epsilon$. *It is* uniformly asymptotically stable with respect to initial data *if the number* Δ *in definition* (*i*) *is independent of* t_0 *and there exists a constant* Δ_0 *such that* $|\delta\widetilde{y}(t)| \to 0$ *as* $t \to \infty$, *for any bounded continuous initial function* $\delta\psi$ *with* $\sup_{t\in(-\infty,t_\star]} |\delta\psi(t)| < \Delta_0$.

There are also analogues of Definition 3 which we leave to the reader. The standard equation in DDEs that plays the role of $y'(t) = \lambda y(t)$ in ODEs is

$$y'(t) = \lambda y(t) + \upsilon y(t - \tau_\star), \quad t \geq t_0, \quad y(t) = \psi(t), \quad t \leq t_0 \qquad (8.3)$$

(cited earlier); $\tau_\star$ is assumed constant. The zero function solves this DDE with $\psi(t) \equiv 0$; if a consistent linear numerical method is applied (e.g., a consistent CRK method) then the discretized problem also has zero as a solution. If $\Re(\lambda) + |\upsilon| < 0$ the zero solution of (8.3) is asymptotically stable *whatever the value of* $\tau_\star > 0$, but it may be stable for a given $\tau_\star > 0$ when the condition is not satisfied.

The test equation (8.3) with constant delay leads to the concepts of *P-stability* and *GP-stability* as at least partial generalizations of the concept of *A-stability* for ODEs. In ODEs, it is common to consider λ to be a complex value; systems of the form $y'(t) = Ay(t)$ with diagonalisable $A \in \mathbb{R}^m$ can then be addressed by suitable corollaries. This tradition has been carried over to the study of DDEs $y'(t) = My(t) + Ny(t - \tau_\star)$, with M, N simultaneously diagonalisable. Unfortunately, systems of DDEs may be expected to have multiple lags, as in

$$\begin{aligned} y_1'(t) &= \lambda_{1,1}y_1(t) + \lambda_{1,2}y_2(t) + \upsilon_{1,1}y_1(t - \tau_{1,1}) + \upsilon_{1,2}y_1(t - \tau_{1,2}), \\ y_2'(t) &= \lambda_{2,1}y_1(t) + \lambda_{2,2}y_2(t) + \upsilon_{2,1}y_1(t - \tau_{2,1}) + \upsilon_{2,2}y_1(t - \tau_{2,2}). \end{aligned}$$

Definition 4. *(Barwell) A numerical method for DDEs is* P-stable *if* for every $\tau_\star$–commensurable *mesh* $\mathcal{T}_\infty$, *the zero numerical solution for (8.3) is asymptotically stable whenever* $\Re(\lambda) + |\upsilon| < 0$. *It is* GP-stable *if this solution is asymptotically stable for general* $\tau_\star$–incommensurable *meshes* $\mathcal{T}_\infty$ *of arbitrary width.*

The above concepts can be extended to (nonautonomous) equations with variable coefficients, $y'(t) = \lambda(t)y(t) + \upsilon(t)y(t-\tau_\star)$, $t \geq t_0$, $y(t) = \psi(t)$, $t \leq t_0$, giving rise to concepts called PN-*stability* and GPN-*stability* [107]. The discussion has been extended to nonlinear systems with constant delay, in terms of concepts called R-*stability* and GRN-*stability* (generalizations of the concept of BN-*stability* for ODEs), and there is an extensive theory. (See [107] for an admirable review. We refer to references therein, in particular the interesting work of Bellen, in 't Hout, Torelli, Zennaro, etc., and also to the papers of Koto [77, 78, 79].)

Much work has been performed on finding stability regions for specific combinations of parameters $\lambda, \upsilon, \tau_\star$ in (8.3) when using a fixed step h, see [9, 55]. Care should be taken not to misinterpret such results: A typical formula applied to (8.3) gives a recurrence $\mathbf{u}_{n+1} = \mathbf{M}\mathbf{u}_n + \gamma_n$ (with amplification matrix $\mathbf{M} \equiv \mathbf{M}([\lambda, \mu, \tau_\star]; h)$) which is asymptotically stable with respect to perturbed initial conditions if the spectral radius $\rho(\mathbf{M})$ is less than unity. However, as Spijker also noted [50], the dimension of $\mathbf{M}$ depends upon $\tau_\star$ and h, so that although the condition for stability implies decay of the effect of perturbations, the transient behaviour and the rate of decay can be different for different h. We note the trend towards relating numerical stability to pseudospectra [100].

An example of DDEs with non-constant delay, for which the concept of $\tau_\star$-commensurable meshes is not pertinent, is (4.2) where λ and υ are complex numbers and $0 < \alpha_\star < 1$. We refer to [84] and the work cited therein.

We now turn to the stability behaviour of numerical methods for convolution equations. For linear equations, where convolution structure is preserved in the discretized equations, the basic tool for analysis is (in analogy with the Laplace transform for the continuous case) the z-transform. One can seek the analogue of A-stability by searching for those formulae that preserve stability of approximate solutions of eqn (6.2) when k is positive-definite or one can seek stability regions. See Blank [26]. For non-linear equations, Ford and Baker [51] consider discrete analogues on uniform meshes of a theory of Corduneanu that holds for the case $k, k' \in \mathcal{L}[0, \infty)$ and $u\phi(u) > 0$ for $u \neq 0$. Their work is related to the property that, with analogous conditions, the discrete solutions tend to zero; this implies a form of stability under persistent perturbations.

8.3 Stability for VDIDEs with nonuniform meshes

In the work of Baker and Tang [14] the problems covered are those with realistic *fading memory properties*, the most general being of the type

$$\begin{aligned} u'(t) &= \mathcal{M}(t)u(t) + \mathcal{N}(t)u(t-\tau(t)) + \int_{-\infty}^{t} \mathcal{K}(t,s)u(s)\,ds + \\ &\quad + g\Big(t, u(t), u(t-\tau(t)), \int_{-\infty}^{t} K(t,s,u(s))\,ds\Big), \end{aligned}$$

with "small" nonlinear perturbations g to linear equations. Using a discrete delay difference inequality and discrete Lyapunov functionals, Baker and Tang considered stability of solutions obtained by CRK methods with *nonuniform meshes*. Most of the stability analysis referred to earlier relates to uniform meshes; insights obtained by various approaches often complement each other.

8.4 Qualitative behaviour and implicit formulae

There has been considerable progress of late in the study of the dynamics of discretized ODEs and rather less (but see [2, 4, 59]) in studying the dynamics of discretized DDEs. A complicating feature for nonlinear problems is that the discretization based on an implicit formula (such as an adaptation of the implicit Euler method) can produce a multivalued map: If $y'(t)$ is determined in the functional equation by $\{y(t)\}^2$ and history terms $y(s)$ $(s < t)$ (take as example $y'(t) = \lambda_0\{y(t)\}^2\{\lambda_1 - y(t-1)\}$), then for each n one has to solve a quadratic to determine $\widetilde{y}_{n+1}$ in terms of $\{\widetilde{y}_\ell\}_{\ell=0}^n$. Thus from each historical mesh function there can be two solutions, and we can construct sequences $\{\widetilde{y}_\ell^{[++\cdots+]}\}_{\ell=0}^{n+1}$ and $\{\widetilde{y}_\ell^{[--\cdots-]}\}_{\ell=0}^{n+1}$, for example, by consistently taking positive or negative square roots in the formula for the roots of a quadratic, assuming the discriminant is positive. (Other solution sequences may be $\{\widetilde{y}^{[\sigma_\ell]}\}_{\ell=0}^{n+1}$, where σ_ℓ is any sequence of ℓ 'plus' or 'minus' signs such that $\{\sigma_{\ell+1}\} = \{\sigma_\ell, +\}$ or $\{\sigma_\ell, -\}$ $(\ell \geq 0)$; σ_0 is empty. The application of implicit formulae to general nonlinear problems gives rise to nonlinear equations that cannot be solved exactly and may have no solutions.) In the most trivial dynamics (the case of stationary points), one sequence $\{y_\ell^{[\sigma_\ell]}\}_{\ell=0}^{\infty}$ may have *one* stationary point and another sequence may have another, whilst other sequences may have no stationary point but may cycle (for example), or may terminate because some discriminant becomes negative.

The preceding remarks establish that we are required to distinguish between a *formula* and an *algorithm* when discussing numerics. An algorithm requires us to determine a unique sequence defining the mesh function $\widetilde{y}$. The interesting questions arise (*i*) with algorithms based on nonuniform adaptive meshes, (*ii*) with problems with non-trivial dynamics (much remains to be done to study the numerics of oscillatory solutions of DDEs and FDEs, and the onset of chaos).

9 Concluding remarks

Whilst differential equations are often used to model scientific phenomena, equations of Volterra type can frequently provide better mathematical models (though, for fast-reacting systems, the history may be insignificant). Some attempts have been made to determine appropriate models through parameter fitting to data, using delay–differential equations [16], etc., and Volterra integro-differential equations [73]. This approach can yield some interesting difficulties [11].

No comprehensive and completely practical theory exists (or is likely to be found) for convergence, error control, or stability; theory produces only guidance and insight – and I have occasionally cautioned, above, against misinterpretation of theory. Too much practical significance is sometimes claimed for theoretical numerical analysis by virtue of its mathematical elegance; complicated questions sometimes require complicated and inelegant answers. As we seek better insight and hence better numerical techniques, the subject continues to develop.

Space constraints required me to omit from [5] much literature of interest, but I have ensured that some recent theses and reports that might otherwise be overlooked have been retained.

In concluding, I thank Drs. Luise Blank, Neville Ford, Arieh Iserles, Yunkang Liu, Donald Kershaw, and Chris Paul, for most helpful comments.

Bibliography

1. Ascher, U.M. & Petzold, L.R. (1995). The numerical solution of delay–differential–algebraic equations of retarded and neutral type. *SIAM J. Numer. Anal.*, *32*, 1635-1657.
2. Aves, M.A., Davies, P.J. & Higham, D.J. (1995). The effects of quadrature on the dynamics of a discretized nonlinear integro-differential equation. *Report NA/166*, Dept. of Math. & Comp. Sci., Univ. of Dundee.
3. Bader, G. (1983). *Ein Mehrschrittverfahren mit variabler Schrittweite und variabler Ordnung zur Integration von Systemen retardierter Differentialgleichungen mit zustandsabhangiger Verzögerung*, Diplomarbeit, Univ. Heidelberg.
4. Baker, C.T.H. (1994). Dynamics of discretized equations for DDEs. *Comp. Math. & its Appl.*, *1*, 61–67.
5. Baker, C.T.H. (1996) Numerical analysis of Volterra functional and integral equations (state of the art) *MCCM Tech. Rep. 292*, Manchester Univ.
6. Baker, C.T.H. & Derakhshan, M.S. (1987). Computational approximations to some power series. *ISNM*, *81*, 11–20.
7. Baker, C.T.H. & Derakhshan, M.S. (1987). FFT Techniques in the numerical solution of convolution equations. *JCAM.*, *20*, 5–24.
8. Baker, C.T.H. & Derakhshan, M.S. (1988) Stability barriers to the construction of $\{\rho, \sigma\}$-reducible and fractional quadrature rules. *ISNM*, *85*, 1–15.
9. Baker, C.T.H. & Paul, C.A.H. (1994). Computing stability regions – Runge-Kutta methods for delay differential equations. *IMA J. Numer. Anal.*, *14*, 347–362.
10. Baker, C.T.H. & Paul, C.A.H. (1996). A global convergence theorem for a class of PCERK methods and vanishing lag DDEs. *SIAM J. Numer. Anal. 33* (in press).
11. Baker, C.T.H. & Paul, C.A.H. (1997). Pitfalls in parameter estimation for delay differential equations. *SIAM J. Sci. Comp.* (to appear).
12. Baker, C.T.H., Paul, C.A.H. & Willé, D.R. (1995). A bibliography on the numerical solution of delay differential equations. *MCCM Tech. Rep. 269*, Manchester Univ.
13. Baker, C.T.H., Paul, C.A.H. & Willé, D.R. (1995). Issues in the numerical solution of evolutionary delay differential equations. *Adv. Comp. Math.*, *3*, 171–196.
14. Baker, C.T.H. & Tang, A. (1995). Stability analysis of CIRK methods for VIDEs with unbounded delays. *MCCM Tech. Rep. 270*, Manchester Univ.
15. Baker, C.T.H. & Tang, A. (1995). Stability of variable step CRK methods for scalar DDEs. *MCCM Tech Rep. 278*, Manchester Univ.
16. Banks, H.T. & Lamm, P.K.D. (1983) Estimation of delays and other parameters in nonlinear functional differential equations, *SIAM J. Control Optim.*, *21*, 895–915.

17. Belair, J. 1991. Population Models with State-Dependent Delays. *In:* Arino. O. Axelrod. D., & Kimmel, M. (eds), *Lecture Notes in Pure and Applied Mathematics*, **131**, Academic, New York.

18. Bellen, A. (1984). One-step collocation for delay differential equations. *JCAM.*, *10*, 275–283.

19. Bellen, A. (1985). Constrained mesh methods for functional differential equations. *ISNM*, *74*, 52–70.

20. Bellen, A., Jackiewicz, J. & Zennaro, M. (1988) Stability analysis of one-step methods for neutral delay-differential equations. *Numer. Math.*, *52*, 605–619.

21. Bellen, A., Kolmanovskii, V.B, Torelli, L, & Vermiglio, R. (1995). About stability of some functional-differential equations of neutral type, *J. Math. Anal. Appl.*, *189*, 59-84.

22. Bellen, A. & Zennaro, M. (1985). Numerical solution of delay differential equations by uniform corrections to an implicit Runge-Kutta method. *Numer. Math.*, *47*, 301–316.

23. Bellen, A. & Zennaro, M. (1992). Strong contractivity properties of numerical methods for ordinary and delay differential equations. *Appl. Numer. Math.*, *9*, 321–346.

24. Bellen, A. & Zennaro, M. (book, in preparation). *Numerical methods for delay differential equations.*

25. Blank, L. (1991). *Stabilitätsanalyse der Kollokationsmethode für Volterra-Integral-Gleichungen mit schwach singulaerem Kern.* Ph.D., Univ. Bonn.

26. Blank, L. (1995). Stability of collocation for weakly singular Volterra equations. *IMA J. Numer. Anal.*, *15*, 357-375.

27. Blank, L. (1995). Stability-type results for collocation methods for Volterra integral equations. *MCCM Tech. Rep. 282*, Manchester Univ.

28. Bocharov, G. A, & Romanyukha, A. A. (1994) Numerical treatment of the parameter-identification problem for delay-differential systems arising in immune-response modeling *Appl. Numer. Math.*, *15*, 307 – 326.

29. Bocharov, G. A, Marchuk, G. I. & Romanyukha, A. A. (1996) Numerical-solution by LMMs of stiff delay-differential systems modeling an immune-response. *Numer. Math.*, *73*, 131 – 148.

30. Bock, H. G. & Schlöder, J. P. (1981). Numerical solution of retarded differential equations with state-dependent time lags. *ZAMM*, *61*, 269–271.

31. Brunner, H. (1991). Direct quadrature methods for nonlinear Volterra integro-differential equations with infinite delay. *Utilitas Mathem.*, *40*, 237–250.

32. Brunner, H. (1992). Implicitly linear collocation methods for nonlinear Volterra equations. *Appl. Numer. Math.*, *9*, 235–247.

33. Brunner, H. (1994). Iterated collocation methods for Volterra integral equations with delay arguments. *Math. Comp.*, *62*, 581–599.

34. Brunner, H. & van der Houwen P.J. (1986). *The numerical solution of Volterra equations.* North-Holland, Amsterdam.

35. Buckwar, E. (1992). Anwendung der Methode der gewichteten Metrik auf nichtlineare Volterrasche Integralgleichungen vom Faltungstyp. Diplomarbeit, Mathematics Faculty, Free Univ. Berlin.

36. Buhmann, M.D. & Iserles, A. (1992). On the dynamics of a discretized neutral equation. *IMA J. Numer. Anal.*, *12*, 339–363.

37. Buhmann, M.D. & Iserles, A. (1993). Stability of the discretized pantograph differential-equation. *Math. Comp.*, *60*, 575–589.

38. Busenberg, S. & Martelli, M. (editors) *Delay differential equations and dynamical systems.* Lect. Notes in Math. 1475, Springer Verlag, Berlin.

39. Butcher, J.C. (1987) *The numerical analysis of ordinary differential equations*, John Wiley, Chichester.

40. Cahlon, B. (1995). On the stability of Volterra integral-equations with a lagging argument. *BIT*, 35, 19-29.

41. Campbell, S. L. (1980). Singular linear systems of differential equations with delays. *Applicable Analysis, 11*, 129–136.

42. Campbell, S. L. (1991). 2-D differential-delay implicit systems. *Procs. 1991 IMACS Congress, Dublin.*

43. Claus, H. (1990). Singly-implicit Runge-Kutta methods for retarded and ordinary differential equations. *Computing*, *43*, 209-222.

44. Corduneanu, C. (1991). *Integral equations and applications*, Cambridge Univ. Press, Cambridge.

45. Corduneanu, C. & Kannan, R. (editors) (1996). *Abstracts, UTA Volterra Centennial Symposium.* University of Texas at Arlington.

46. Corwin, S.C. & Thompson, S. (1996) DRKLAG6: Solution of systems of functional differential equations with state dependent delays. *Dept. Math. Stats., Rept. on software package DRKLAG6*, Radford Univ., Va.

47. Crisci, M.R., Kolmanovskii, V. B., Russo, E. & Vecchio, A. (1995) Stability of continuous & discrete Volterra integro–differential equations by Liapunov approach. *J. Integral. Eqns.*, *7*, 393 – 412.

48. Crisci, M.R., Ferraro, N. & Russo, E. (1996) Convergence results for continuous–time waveform methods for Volterra integral equations. *JCAM.*, *72*, *in press.*

49. Feldstein, A., Iserles, A., & Levin, D. (1995) Embedding of delay equations into an infinite-dimensional ODE system. *J. Diff. Eqns.*, *117*, 127–150.

50. Feldstein, A. & Jackiewicz, Z. (editors) (1996). *Abstracts, ASU Volterra Centennial Symposium.* Arizona State University, Tempe.

51. Ford, N. J. & Baker, C. T. H. (1996). Qualitative behaviour and stability of solutions of discretised non-linear Volterra integral equations of convolution type. *JCAM.*, *66*, 213–225.

52. de Gee, M. (1985). Smoothness of solutions of functional differential equations. *J. Math. Anal. & Appl.*, *107*, 103–121.

53. Gorenflo, R. & Vessella, S. O. (1990). *Abel integral equations: analysis and applications.* Springer-Verlag, New York.

54. Gopalsamy, K. (1992). *Stability and oscillations in delay differential equations of population dynamics.* Kluwer Academic Publishers, London.

55. Guglielmi, N. (1995). *Delay dependent stability regions of numerical methods for delay differential equations.* Doctoral thesis, Trieste Univ.

56. Hairer, E., Norsett, S.P., & Wanner, G. (1987) *Solving ordinary differential equations -I*, Springer-Verlag.

57. Hayashi, H. (1995). *Numerical solution of delay differential equations*, Ph.D., Univ. of Toronto.

58. Hauber, R. (1994) *Numerische Behandlung von retardierten differential-algebraischen Gleichungen.* Doctoral thesis, Univ. München.

59. Higham, D. J. (1994). The dynamics of a discretized nonlinear delay differential equation. *In:* Griffiths, D.F. & Watson, G.A. (editors) *Numerical Analysis 1993*, Longman, 167-179.

60. Higham, D. J. (1993). Error control for initial-value problems with discontinuities and delays. *Appl. Numer. Math.*, *12*, 315–330.

61. Higham D. J. & Famelis, I. T. (1995). Equilibrium states of adaptive algorithms for delay-differential equations. *JCAM.*, *58*, 151–169.

62. Hong-Jiong, T. & Jiao-Xun, K. (1996) The numerical stability of linear multistep methods for delay differential equations with many delays. *SIAM J. Numer. Anal.*, *33*, in press.

63. in 't Hout, K.J. (1992) The stability of a class of Runge-Kutta methods for delay differential equations, *Appl. Numer. Math.*, *9* , 347–355.

64. in 't Hout, K. J. (1992) *Runge-Kutta methods in the numerical solution of delay differential equations*, Ph.D., Univ. of Leiden.

65. in 't Hout, K.J. (1994). The stability of θ-methods for systems of delay differential equations. *Annals of Numer. Maths.*, *1*, 323–334.

66. in 't Hout, K.J., Spijker, M.N. (1991). Stability analysis of numerical methods for delay differential equations, *Numer. Math.*, *59*, 807-814

67. Hu, G.D. & Mitsui, T. (1995). Stability analysis of numerical methods for systems of neutral delay differential equations. *BIT*, *35*, 504–515.

68. Hu, Q.Y. 1996. Stieltjes Derivatives and Beta-Polynomial Spline Collocation for Volterra Integrodifferential Equations with Singularities. *SIAM Journal on Numerical Analysis*, **33**, 208-220.

69. Iserles, A. & Powell, M. J. D. (editors) (1987). *The State of the Art in Numerical Analysis.* Clarendon Press, Oxford.

70. Iserles, A. (1996) Beyond the classical theory of computational ordinary differential equations, The State of the Art in Numerical Analysis, York (*this volume*).

71. Jackiewicz, Z. (1987). Variable-step variable-order algorithm for the numerical solution of neutral functional differential equations. *Appl. Numer. Math.*, *3*, 317–329.

72. Jackiewicz, Z. & Lo, E. (1995). The numerical-integration of neutral functional-differential equations by fully implicit one-step methods. *ZAMM.*, *75*, 207-221.

73. Kappel, F., Kunisch, K. & Moyshewitz, G. (1984). An Approximation Scheme for Parameter Estimation in Infinite Delay Equations of Volterra Type. Rep. 51-1984, Institut für Math. Technische U. –Universität Graz.

74. Kauthen, J.–P. (1993) Implicit Runge–Kutta methods for singularly perturbed integro–differential equations. Tech. Rep. 93–1, Inst. de Math., Université de Friburg.

75. Kauthen, J.–P. (1993) Implicit Runge–Kutta methods for some integro–differential algebraic equations. *Appl. Numer. Math.*, *13*, 125–134.

76. Kauthen, J.–P. (1995) Implicit Runge–Kutta methods for singularly perturbed integro–differential systems. *Appl. Numer. Math.*, *18*, 201–210.

77. Koto, T. (1995). Stability properties of Runge-Kutta methods for delay differential equations. *Lect. Notes in Num. Appl. Anal.*, *14*, 107–114.

78. Koto, T. (1995). The stability of natural Runge-Kutta methods for nonlinear delay differential equations. Rep. CSIM 95-04, Univ. of Electro-Communications, Chofo, Tokyo.

79. Koto, T. (1996). NP-stability of Runge-Kutta methods based on classical quadrature. Rep. CSIM 96-02, Univ. of Electro-Communications, Chofo, Tokyo.

80. Kuang, Y. (1993). *Delay differential equations with applications in population dynamics.* Academic Press, New York.

81. Kolmanovskii, V. B. & Nosov, V. R. (1986). *Stability of functional differential equations.* Academic Press, New York.

82. Kolmanovskii, V.B. & Myskhis, A. (1992). *Applied theory of functional differential equations.* Kluwer Academic, Dordrecht.

83. Kolmanovskii, V.B., Torelli, L. & Vermiglio, R. (1994). Stability of some test equations with delay, *SIAM J. Math. Anal.*,*25*, 948-961.

84. Liu, Y. (1996). *On functional equations with proportional delays.* Ph.D., DAMTP, Cambridge Univ.

85. Lubich, C. (1983). On the stability of linear multistep methods for Volterra convolution equations. *IMA J. of Numer. Anal.*, *3*, 439–465.

86. Lubich, C. (1985). Fractional linear multistep methods for Abel–Volterra integral equations of the second kind. *Math. Comp.*, *45*, 463–469.

87. Luzyanina, T. & Roose, D. (1996). Numerical stability analysis and computation of Hopf bifurcation points for delay differential equations. *JCAM.*, *72*, 379-392.

88. Morris, H.C., Ryan, E.E. & Dodd, R.K. (1983). Periodic solutions and chaos in a delay-differential equation modelling haematopoiesis. *Nonlinear Analysis, Theory, Methods & Applications*, *7*, 623-660.

89. Neves, K.W. & Thompson, S. (1992). Software for the numerical-solution of systems of functional-differential equations with state-dependent delay. *Appl. Num. Math.*, *9*, 385–401.

90. Neves, K.W. & Thompson, S. (1992). Solution of systems of functional differential equations with state dependent delays. Tech. Rep. TR–92–003, Comp. Sci., Radford Univ.

91. Olmstead, W.E., Roberts, C.A. & Deng, K. (1995) Coupled Volterra equations with blow–up solutions. *J Integral Eqns. and Applicns.*, *7*, 499–516.

92. Paul, C.A.H. (1992). Developing a delay differential equation solver, *Appl. Num. Math.*, *9*, 403–414.

93. Paul, C.A.H. (1992). *Runge-Kutta methods for functional differential equations.* Ph.D., Manchester Univ.

94. Paul, C.A.H. (1995). A User Guide to ARCHI, *MCCM Tech. Rep. 283*, Manchester Univ.

95. Shaw, S. & Whiteman, J. R. (1997) Toward adaptive finite element schemes for partial differential Volterra equation solvers. *submitted.*

96. Tang, A. (1996). *Analytical and numerical treatment of delay Volterra integro-differential equations.* Ph.D., Manchester Univ.

97. Torelli, L. (1989). Stability of numerical methods for delay differential equations. *J. Comput. Appl. Math.*, *25*, 15–26.

98. Torelli, L. (1991). A sufficient condition for GPN-stability for delay differential equations. *Numer. Math.*, *59*, 311–320.

99. Torelli, L. & Vermiglio, R. (1993). On the stability of continuous quadrature-rules for differential-equations with several constant delays, *Numer. Anal.* *13*, 291-302.

100. Trefethen, L.N. (1992) Pseudospectra of matrices. *In:* Griffiths, D.F. & Watson, G.A. (editors) *Numerical Analysis 1991*, Longman.

101. Willé, D.R. (1989). *The numerical solution of delay-differential equations.* Ph.D., Manchester Univ.

102. Willé, D.R. & Baker, C.T.H. (1992). The tracking of derivative discontinuities in systems of delay differential equations. *Appl. Num. Math.*, *9*, 209–222.

103. Willé, D.R. & Baker, C.T.H. (1992). DELSOL – A numerical code for the solution of systems of delay differential equations. *Appl. Num. Math.*, *9*, 223–234.

104. Zennaro, M. (1986). Natural continuous extensions of Runge-Kutta methods. *Math. Comp.*, *46*, 119–133.

105. Zennaro, M. (1986). P–stability properties of Runge–Kutta methods for delay differential equations. *Numer. Math.*, *49*, 305–318.

106. Zennaro, M. (1993). Contractivity of Runge-Kutta methods with respect to forcing terms, *Appl. Numer. Math.*, *10*, 321–345.

107. Zennaro, M. (1995). Delay differential equations: theory and numerics *in* Ainsworth, M., Levesley, L., Light, W.A. & Marletta, M. (editors) (1995) *Theory and numerics of ordinary and partial differential equations*, Clarendon Press, Oxford.

108. Zhu,W. & Petzold, L.R. (1996) Asymptotic stability of linear delay differential algebraic equations and numerical methods. *Preprint, submitted to Elsevier Science.*

The Numerical Solution of Boundary Integral Equations[1]

Kendall E. Atkinson

Department of Mathematics, University of Iowa, USA

Abstract

Much of the research on the numerical analysis of Fredholm type integral equations during the past ten years has centered on the solution of boundary integral equations (BIE). A great deal of this research has been on the numerical solution of BIE on simple closed boundary curves S for planar regions. When a BIE is defined on a smooth curve S, there are many numerical methods for solving the equation. The numerical analysis of most such problems is now well-understood, for both BIE of the first and second kind, with many people having contributed to the area. For the case with the BIE defined on a curve S which is only piecewise smooth, new numerical methods have been developed during the past decade. Such methods for BIE of the second kind were developed in the mid to late 80s; and more recently, high order collocation methods have been given and analyzed for BIE of the first kind. The numerical analysis of BIE on surfaces S in $\mathbb{R}^3$ has become more active during the past decade, and we review some of the important results. The convergence theory for Galerkin methods for BIE is well-understood in the case that S is a smooth surface, for BIE of both the first and second kind. For BIE of the second kind on piecewise smooth surfaces, important analyses have been given more recently for both Galerkin and collocation methods. In contrast, almost nothing is understood about collocation methods for solving BIE of the first kind, regardless of the smoothness of S. Numerical methods for BIE on surfaces S in $\mathbb{R}^3$ lead to computationally expensive procedures, and a great deal of the research for such BIE has looked at the efficient numerical evaluation of integrals, the use of iterative methods for solving the associated linear systems, and the use of "fast matrix-vector calculations" for use in iteration procedures.

1 Introduction

Boundary integral equations (BIE) are reformulations of boundary value problems for partial differential equations. They have been present in the mathematics literature for almost two centuries, and engineers have made much use of

[1] Dedicated to Professor Ben Noble [59].

them for solving a wide variety of real-world problems. As a few examples of the use of BIE, we cite Hess [43] and Newman [58] for potential flow calculations, Rizzo et al. [70] and Rudolphi et al. [72] for crack problems in elasticity, and Jaswon and Symm [44] for electrostatic and elastostatic calculations.

In spite of the widespread use of BIE, much of their numerical analysis has not been well-understood until much more recently, say from 1975 onwards. One of the major contributions to the subject was the recognition that most BIE can be regarded as strongly elliptic pseudo-differential operator equations on suitably chosen Sobolev spaces, and further, that finite element methods can then be applied to their solution. This theory is well-surveyed in the article [87] of W. Wendland, given during the last meeting in 1986 on the *State of the Art in Numerical Analysis*, and in the papers of Costabel [23], Costabel and Wendland [26], and Wendland [88]. For this reason, we omit further mention of it in this survey.

The theory of finite element methods for strongly elliptic pseudo-differential operator equations is most useful when the boundary S is a smooth curve or surface and when Galerkin methods are being used. During the past decade, there has been much research on numerical methods for BIE on piecewise smooth boundaries and on the study of collocation methods for all types of BIE. In addition, more attention has been given to the practical problems associated with solving discretizations of BIE, including numerical integration, iteration methods for solving the associated linear systems, and the use of "fast matrix-vector calculations".

In this survey, we concentrate on the topics mentioned in the last paragraph. We also restrict the survey to BIE for Laplace's equation $\Delta u = 0$. In part, this this is due to space limitations; but also, most of the major aspects of the theory are adequately understood by considering this case alone.

2 Boundary integral equations

To begin, we summarize the most popular BIE reformulations of Laplace's equation; and most such BIE are obtained as straightforward applications of the divergence theorem. We omit BIE for planar problems that are derived as a consequence of applications of Cauchy's formula from functions of a complex variable.

2.1 Green's representation formulas

Let D be a planar simply-connected bounded region with a boundary S to which the divergence theorem can be applied. For example, let S be piecewise smooth, meaning that S is composed of a finite number of smooth sections, each of which has a C^∞ parameterization. Assume $u \in C^1(\overline{D}) \cap C^2(D)$ and $\Delta u \equiv 0$ on D.

Then

$$\int_S \left\{ \frac{\partial u(Q)}{\partial \mathbf{n}_Q} \log|P-Q| - u(Q)\frac{\partial}{\partial \mathbf{n}_Q}\left[\log|P-Q|\right] \right\} dS_Q$$

$$= \begin{cases} 2\pi u(P), & P \in D \\ \Omega(P)u(P), & P \in S \\ 0, & P \in D_e. \end{cases} \tag{2.1}$$

In this, $\mathbf{n}_Q$ is the inner normal to S at Q, $\Omega(P)$ is the interior boundary angle at P, and $D_e = \mathbb{R}^2 \backslash \overline{D}$.

An exterior variant is obtained by considering $u \in C^1(\overline{D}_e) \cap C^2(D_e)$ with $\Delta u \equiv 0$ on D_e and

$$\sup_{P \in \overline{D}_e} |u(P)| < \infty.$$

Then

$$\int_S \left\{ \frac{\partial u(Q)}{\partial \mathbf{n}_Q} \log|P-Q| - u(Q)\frac{\partial}{\partial \mathbf{n}_Q}\left[\log|P-Q|\right] \right\} dS_Q$$

$$= \begin{cases} 2\pi\left[u(\infty) - u(P)\right], & P \in D_e \\ 2\pi u(\infty) - \left[2\pi - \Omega(P)\right]u(P), & P \in S \\ 2\pi u(\infty), & P \in D. \end{cases} \tag{2.2}$$

2.1.1 Formulae in three dimensions

Let $D \subset \mathbb{R}^3$ be a bounded simply-connected region with a piecewise smooth boundary S. Assume $u \in C^1(\overline{D}) \cap C^2(D)$ and $\Delta u \equiv 0$ on D. Then

$$\int_S \left\{ u(Q)\frac{\partial}{\partial \mathbf{n}_Q}\left[\frac{1}{|P-Q|}\right] - \frac{\partial u(Q)}{\partial \mathbf{n}_Q}\frac{1}{|P-Q|} \right\} dS_Q$$

$$= \begin{cases} 4\pi u(P), & P \in D \\ \Omega(P)u(P), & P \in S \\ 0, & P \in D_e. \end{cases} \tag{2.3}$$

In this, $\mathbf{n}_Q$ is the inner normal to S at Q, $\Omega(P)$ is the interior solid angle at P, and $D_e = \mathbb{R}^3 \backslash \overline{D}$.

For the exterior variant of this formula, assume $u \in C^1(\overline{D}_e) \cap C^2(D_e)$, $\Delta u \equiv 0$ on D_e, and

$$|u(P)| = O\left(|P|^{-1}\right), \quad |\nabla u(P)| = O\left(|P|^{-2}\right) \text{ as } |P| \to \infty.$$

Then

$$\int_S \left\{ u(Q) \frac{\partial}{\partial \mathbf{n}_Q} \left[\frac{1}{|P-Q|} \right] - \frac{\partial u(Q)}{\partial \mathbf{n}_Q} \frac{1}{|P-Q|} \right\} dS_Q$$

$$= \begin{cases} -4\pi u(P), & P \in D_e \\ -[4\pi - \Omega(P)]\, u(P), & P \in S \\ 0, & P \in D. \end{cases} \tag{2.4}$$

2.2 Direct boundary integral equations

Direct BIE are obtained directly from the above formulae, either by using one of the given formulae or by using some derivative of them. The unknown function is then the function u or its normal derivative $\partial u/\partial \mathbf{n}$. One of the reasons that direct BIE are sometimes preferred is because the unknown function has immediate physical significance, as compared to the unknown in an indirect BIE. We give some examples of direct BIE.

Solve the *exterior planar Neumann problem*

$$\begin{aligned} \Delta u(P) &= 0, \qquad P \in D_e \\ \frac{\partial u(p)}{\partial \mathbf{n}_P} &= f(P), \quad P \in S, \end{aligned} \tag{2.5}$$

and assume S is smooth. For uniqueness, we impose

$$|u(P)| = O\left(|P|^{-1}\right), \quad |\nabla u(P)| = O\left(|P|^{-2}\right),$$

as $|P| \to \infty$. Then solve the BIE

$$-\pi u(P) + \int_S u(Q) \frac{\partial}{\partial \mathbf{n}_Q} [\log |P-Q|]\, dS_Q$$

$$= \int_S f(Q) \log |P-Q|\, dS_Q, \quad P \in S. \tag{2.6}$$

This is a classical BIE of the second kind; and it is well known that it is uniquely solvable within $C(S)$ for any given right hand function in $C(S)$, not just those written in the form of the given integral on the right side of (2.6).

Solve the *interior planar Dirichlet problem*

$$\begin{aligned} \Delta u(P) &= 0, \qquad P \in D \\ u(P) &= f(P), \quad P \in S. \end{aligned}$$

Do this by solving

$$\int_S \psi(Q) \log |P - Q| \, dS_Q = \varphi(P), \quad P \in S \tag{2.7}$$

$$\varphi(P) = \int_S f(Q) \frac{\partial}{\partial \mathbf{n}_Q} [\log |P - Q|] \, dS_Q + \Omega(P) f(P),$$

with the unknown

$$\psi(Q) \equiv \frac{\partial u(Q)}{\partial \mathbf{n}_Q}.$$

The above BIE of the first kind has been studied intensively in the past decade; and today it is very well understood for the case that S is a smooth curve.

For other direct BIE reformulations of Laplace's equation, see Blue [16] and Jaswon and Symm [44].

2.3 Indirect boundary integral equations

We will consider some *indirect BIE* in $\mathbb{R}^3$, and analogous formulae can be given in $\mathbb{R}^2$. We begin by deriving some consequences of (2.3) and (2.4).

Let $u_i \in C^1(\overline{D}) \cap C^2(D)$ and $\Delta u_i \equiv 0$ on D; and let $u_e \in C^1(\overline{D}_e) \cap C^2(D_e)$, $\Delta u_e \equiv 0$ on D_e, with

$$|u_e(P)| = O\left(|P|^{-1}\right), \quad |\nabla u_e(P)| = O\left(|P|^{-2}\right) \quad \text{as } |P| \to \infty.$$

For points $P \in S$, introduce

$$\begin{aligned} [u(P)] &= u_i(p) - u_e(P) \\ \left[\frac{\partial u(P)}{\partial \mathbf{n}_P}\right] &= \frac{\partial u_i(P)}{\partial \mathbf{n}_P} - \frac{\partial u_e(P)}{\partial \mathbf{n}_P}. \end{aligned} \tag{2.8}$$

Subtract the exterior version of Green's representation formula (2.4) from the interior version (2.3). This yields

$$\int_S \left\{ [u(Q)] \frac{\partial}{\partial \mathbf{n}_Q} \left[\frac{1}{|P-Q|}\right] - \left[\frac{\partial u(Q)}{\partial \mathbf{n}_Q}\right] \frac{1}{|P-Q|} \right\} dS_Q$$

$$= \begin{cases} 4\pi u_i(P), & P \in D \\ \Omega(P) u_i(P) + [4\pi - \Omega(P)] \, u_e(P), & P \in S \\ 4\pi u_e(P), & P \in D_e. \end{cases} \tag{2.9}$$

We illustrate the use of (2.9) by deriving a *single layer representation* for the solution of Laplace's equation. Given a function $u_i \in C^1(\overline{D}) \cap C^2(D)$ and $\Delta u_i \equiv 0$ on D, choose a function $u_e \in C^1(\overline{D}_e) \cap C^2(D_e)$, $\Delta u_e \equiv 0$ on D_e, with

$$|u_e(P)| = O\left(|P|^{-1}\right), \quad |\nabla u_e(P)| = O\left(|P|^{-2}\right),$$

and with

$$u_e(P) = u_i(P), \quad P \in S. \tag{2.10}$$

Then u_i has the representation

$$u_i(P) = \int_S \frac{\psi(Q)}{|P-Q|} dS_Q, \quad P \in D, \tag{2.11}$$

with

$$\psi(Q) = -\frac{1}{4\pi}\left[\frac{\partial u_i(Q)}{\partial \mathbf{n}_Q} - \frac{\partial u_e(Q)}{\partial \mathbf{n}_Q}\right], \quad Q \in S. \tag{2.12}$$

This in turn leads to both BIE of the first kind and second kind, depending on the type of boundary condition being given.

We can repeat the above type of argument to obtain the *double layer representation*

$$u_i(P) = \int_S \rho(Q) \frac{\partial}{\partial \mathbf{n}_Q}\left[\frac{1}{|P-Q|}\right] dS_Q, \quad P \in D. \tag{2.13}$$

In this case, we replace (2.10) with the matching of the normal derivatives on S. The function ρ has the interpretation

$$\rho(Q) = \frac{1}{4\pi}\left[u_i(Q) - u_e(Q)\right]. \tag{2.14}$$

The formulae (2.12) and (2.14) can be used to obtain regularity results for the respective functions ψ and ρ, by using regularity results on the solutions of Laplace's equation. For an example of this, see Petersdorf and Stephan [62].

As an example of the use of (2.13), consider solving the interior Dirichlet problem

$$\begin{aligned} \Delta u(P) &= 0, & P \in D \\ u(P) &= f(P), & P \in S, \end{aligned} \tag{2.15}$$

as the double layer potential of (2.13). This leads to the classic BIE of the second kind

$$(2\pi + \mathcal{K})\rho = f, \tag{2.16}$$

$$\mathcal{K}\rho(P) = \int_S \rho(Q) \frac{\partial}{\partial \mathbf{n}_Q}\left[\frac{1}{|P-Q|}\right] dS_Q + [2\pi - \Omega(P)]\rho(P), \quad P \in S. \tag{2.17}$$

In the case that S is smooth, this is one of the equations studied by Ivar Fredholm in order to show the existence of solutions to the Dirichlet problem for $\Delta u = 0$; and it is also the classic BIE example given in most textbooks on partial differential equations. When the double layer representation is used to solve the interior Neumann problem, one obtains a ***hypersingular*** BIE. For examples of the solution of this equation in $\mathbb{R}^3$, see Giroire and Nedelec [36] and Petersdorf and Stephan [63].

3 The numerical solution of planar problems

Until the early 1970s, the most commonly studied BIE in numerical analysis was the classic second kind equation (2.16) with a smooth planar boundary S in $\mathbb{R}^2$. Solving the interior Dirichlet problem using a double layer representation for $u(P)$ leads to a second kind equation $(-\pi + \mathcal{K})\,\rho = f$

$$-\pi\rho(t) + \int_0^{2\pi} K(t,s)\rho(s)\,ds = f(t),$$

in which K is smooth and periodic. Essentially any numerical method will work well, and Nyström methods with the trapezoidal rule probably work best.

$$-\pi\rho_n(t) + h\sum_{j=1}^{n} K(t,jh)\rho_n(jh) = f(t), \quad 0 \le t \le 2\pi$$

$$\|\rho - \rho_n\|_\infty = O\left(\|\mathcal{K}\rho - \mathcal{K}_n\rho\|_\infty\right).$$

For a complete discussion of this method, see Section 7.2 of [10]. Most of the work of the past two decades has looked at other planar BIE problems, especially those of the first kind on smooth boundaries and those of both the first and second kind for piecewise smooth boundaries.

3.1 BIE of the first kind on a smooth boundary

A large literature has developed in the past ten to fifteen years for BIE of the first kind. The most studied of such BIE has been

$$\mathcal{L}x(P) \equiv \int_S x(P)\log|P-Q|\,dS_Q = f(P), \quad P \in S, \tag{3.1}$$

often called *Symm's equation*. The curve S is finite, and it can be either open or closed. Unique solvability for all $f \in C(S)$ follows if S is not a Γ-contour; and this is guaranteed by assuming that S has a diameter less than 1 (which can always be attained when solving Laplace's equation by a re-scaling of the variables). For a complete development of the theory of $\mathcal{L}x = f$, see Yan and Sloan [90]. In particular, for S simple, closed, and C^∞

$$\mathcal{L} : H^r(S) \underset{onto}{\overset{1-1}{\to}} H^{r+1}(S), \quad r \geq 0. \tag{3.2}$$

The operator $\mathcal{L}$ can also be extended to $H^r(S)$ for $r < 0$, and then (3.2) will be true for all real r.

The key to understanding the behaviour of $\mathcal{L}$ and to analyzing numerical methods for solving (3.1) is to decompose $\mathcal{L}$ as follows. Let S have a parameterization $\mathbf{r}(t)$, $0 \leq t \leq 2\pi$; and then write

$$\mathcal{L}x(\mathbf{r}(t)) = \int_0^{2\pi} [\log |\mathbf{r}(t) - \mathbf{r}(s)|] \, |\mathbf{r}'(s)| \, x(\mathbf{r}(s)) ds, \quad 0 \leq t \leq 2\pi.$$

Introduce the equivalent operator

$$\mathcal{K}\varphi(t) = \int_0^{2\pi} \varphi(s) \log |\mathbf{r}(t) - \mathbf{r}(s)| \, ds, \quad 0 \leq t \leq 2\pi. \tag{3.3}$$

The equation $\mathcal{L}x = f$ becomes

$$\mathcal{K}\varphi = g,$$

with

$$g(t) = f(\mathbf{r}(t)), \quad \varphi(s) = |\mathbf{r}'(s)| \, x(\mathbf{r}(s)).$$

We can write

$$\mathcal{K}\varphi = -\pi \mathcal{A}\varphi + \mathcal{B}\varphi, \tag{3.4}$$

$$\mathcal{A}\varphi(t) = \int_0^{2\pi} \varphi(s) \log \left| 2e^{-\frac{1}{2}} \sin\left(\frac{t-s}{2}\right) \right| ds \tag{3.5}$$

$$\mathcal{B}\varphi(t) = \int_0^{2\pi} B(t,x)\varphi(s) \, ds$$

$$B(t,s) = \begin{cases} \log \dfrac{|\sqrt{e}\,[\mathbf{r}(t) - \mathbf{r}(s)]|}{\left|2 \sin\left(\dfrac{t-s}{2}\right)\right|}, & t - s \neq 2m\pi \\ \log |\sqrt{e}\mathbf{r}'(t)|, & t - s = 2m\pi. \end{cases}$$

If $\mathbf{r} \in C_p^m(2\pi)$, then $B(t,s)$ is in $C_p^{m-1}(2\pi)$ with respect to both s and t. Consequently, $\mathcal{B}$ is a compact operator on most spaces. For example, $\mathcal{B}$ is compact from H^r to H^{r+1} for $r \geq 0$, provided $m > 2$. Then $\mathcal{K}\varphi = g$ is equivalent to

$$-\pi\varphi + \mathcal{A}^{-1}\mathcal{B}\varphi = \mathcal{A}^{-1}g. \tag{3.6}$$

This is a second kind equation, with a compact integral operator; and as noted earlier, much is known about such equations.

It can be shown that

$$\mathcal{A}\varphi(t) = \frac{1}{\sqrt{2\pi}} \left[\widehat{\varphi}(0) + \sum_{|m|>0} \frac{\widehat{\varphi}(m)}{|m|} e^{ims} \right], \tag{3.7}$$

where

$$\varphi(s) = \frac{1}{\sqrt{2\pi}} \sum_{m=-\infty}^{\infty} \widehat{\varphi}(m) e^{ims}.$$

In essence, $\mathcal{A}$ is a slight modification of $\mathcal{K}$ with S the unit circle; and (3.7) shows

$$\mathcal{A} : H^r \underset{onto}{\overset{1-1}{\longrightarrow}} H^{r+1}, \quad r \geq 0.$$

The decomposition in (3.4) is the major tool used in analyzing many numerical methods for solving $\mathcal{K}\varphi = g$.

3.1.1 Numerical methods

For an excellent survey of numerical methods for solving $\mathcal{L}x = f$, see Sloan [79]. To derive Galerkin methods, regard $\mathcal{L}$ as a strongly elliptic operator from $H^{-\frac{1}{2}}(S)$ to $H^{\frac{1}{2}}(S)$

$$(\mathcal{L}x, x) \geq c \|x\|_{-\frac{1}{2}}^2, \quad x \in H^{-\frac{1}{2}}, \tag{3.8}$$

and then extend finite element methods and their analysis to this case. Define suitable finite element subspaces $\mathcal{X}_n$, and then find x_n from the Galerkin conditions

$$(\mathcal{L}x_n, v) = (f, v), \quad v \in \mathcal{X}_n.$$

The left side uses the extension to $H^{\frac{1}{2}} \times H^{-\frac{1}{2}}$ of the standard inner product on $H^0 = L^2$. Then it follows from Cea's Lemma that x_n exists and

$$\|x - x_n\|_{-\frac{1}{2}} \leq c \inf_{v \in \mathcal{X}_n} \|x - v\|_{-\frac{1}{2}}. \tag{3.9}$$

With x_n piecewise polynomial of degree $r \geq 0$

$$\|x - x_n\|_0 \leq ch^{r+1} \|x\|_{r+1}, \quad x \in H^{r+1}. \tag{3.10}$$

Collocation methods for $\mathcal{L}x = f$ divide into two classes, depending on the form of the approximating functions. With piecewise polynomial approximants, we obtain *boundary element methods;* and with trigonometric polynomial approximants, we obtain *spectral methods.* For boundary element methods, we cite Arnold and Wendland [4] and Saranen [73]. Again, the entire literature is well-surveyed in [79].

3.1.2 Spectral methods

The spectral methods define $\mathcal{X}_n$ as the trig polynomials of degree $\leq n$. For a *Galerkin method* to solve $\mathcal{K}\varphi = g$, let $\mathcal{P}_n$ denote the orthogonal projection of $L^2(0, 2\pi)$ onto $\mathcal{X}_n$. Define $\varphi_n \in \mathcal{X}_n$ as the solution of

$$\mathcal{P}_n \mathcal{K} \varphi_n = \mathcal{P}_n g.$$

Recalling (3.4) and using the special properties of $\mathcal{A}$, especially

$$\mathcal{P}_n \mathcal{A} \psi = \mathcal{A} \mathcal{P}_n \psi, \quad \psi \in \mathcal{X}_n,$$

we have the equivalent equation

$$\varphi_n + \mathcal{P}_n \mathcal{A}^{-1} \mathcal{B} \varphi_n = \mathcal{P}_n \mathcal{A}^{-1} g.$$

A complete error analysis can be based on comparing this with

$$-\pi\varphi + \mathcal{A}^{-1}\mathcal{B}\varphi = \mathcal{A}^{-1} g.$$

For example, see McLean [51]. A spectral discrete Galerkin-collocation method is given in [7] for Laplace's equation with S a smooth simple closed curve, and this is extended in [15] to smooth open arcs. Extensions to a more general framework and to other BIE are given in McLean [52], McLean et al. [53], and Saranen and Vainikko [74].

3.2 Hypersingular BIE on smooth boundaries

Use a double layer potential

$$u(P) = \int_S \rho(Q) \frac{\partial}{\partial \mathbf{n}_Q} [\log |P - Q|] \, dS_Q, \quad P \in D,$$

to solve the interior Neumann problem for $\Delta u = 0$ on D. We find ρ by solving

$$\mathcal{H}\rho = f \tag{3.11}$$

$$\begin{aligned} \mathcal{H}\rho(P) &\equiv \lim_{A \to P} \mathbf{n}_P \cdot \nabla_A \int_S \rho(Q) \frac{\partial}{\partial \mathbf{n}_Q} [\log |A - Q|] \, dS_Q \\ &\equiv \frac{\partial}{\partial \mathbf{n}_P} \int_S \rho(Q) \frac{\partial}{\partial \mathbf{n}_Q} [\log |P - Q|] \, dS_Q. \end{aligned}$$

For $S =$ unit circle, let $\mathcal{H}$ be denoted by $\mathcal{H}_u$. Then

$$\mathcal{H}_u : e^{int} \to \pi \, |n| \, e^{int}, \quad n = 0, \pm 1, , \ldots \; . \tag{3.12}$$

Let $\mathcal{D}$ denote the derivative operator, and let $\mathcal{C}_u$ denote the Cauchy singular integral operator on the unit circle

$$\mathcal{C}_u\varphi(z) = \frac{1}{\pi i} \int_{|\zeta|=1} \frac{\varphi(\zeta)\,d\zeta}{\zeta - z}.$$

Then

$$\mathcal{H}_u = -\pi i \mathcal{D}\mathcal{C}_u = -\pi i \mathcal{C}_u \mathcal{D} \tag{3.13}$$

$$\mathcal{H}_u : H^r \to H^{r-1}, \quad r \geq 1.$$

For the general equation $\mathcal{H}\rho = f$, with a parameterization of S proportional to arc-length and L the length of S

$$\mathcal{H}\rho(\theta) = -\frac{2\pi}{L}\mathcal{H}_u\rho(\theta) + \mathcal{B}\mathcal{D}\rho(\theta), \tag{3.14}$$

with $\mathcal{B}$ a smoothing operator. This can be used to develop spectral methods of solution. Numerical methods for solving $\mathcal{H}\rho = f$ are given in Rathsfeld et al. [69], Kress [50], Chien and Atkinson [22], and Amini and Maines [2].

3.3 Second kind BIE on piecewise smooth boundaries

This is a topic on which enormous progress has been made since the early 1980s, essentially solving the problem. As a particular example, consider solving the interior Dirichlet problem with boundary data f by using a double layer potential representation

$$u(P) = \int_S \rho(Q) \frac{\partial}{\partial \mathbf{n}_Q} [\log |P - Q|]\, dS_Q, \quad P \in D. \tag{3.15}$$

Then ρ satisfies

$$(-\pi + \mathcal{K})\rho = f, \tag{3.16}$$

$$\mathcal{K}\rho(P) = \int_S \rho(Q) \frac{\partial}{\partial \mathbf{n}_Q} [\log |P - Q|]\, dS_Q - [\pi - \Omega(P)]\,\rho(P), \quad P \in S. \tag{3.17}$$

The quantity $\Omega(P)$ denotes the interior angle at $P \in S$. It can be shown that

$$\mathcal{K} : C(S) \to C(S),$$

is bounded. The behaviour of a solution ρ can be studied by using

$$\rho(P) = u(P) - u_e(P),$$

with $u_e(P)$ a solution of an exterior Neumann problem with normal derivative matching that of $u(P)$ on S, as in (2.13) and (2.14).

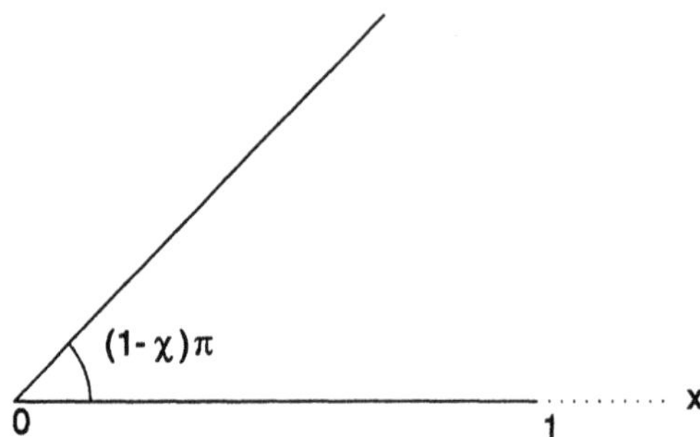

Figure 1. A wedge boundary

Let S have a corner at the origin with the interior angle

$$\phi = (1-\chi)\,\pi, \quad -1 < \chi < 1.$$

It follows from regularity results for Laplace's equation that

$$\rho(P) = O\left(|P|^{\beta}\right), \quad \beta = \frac{1}{1+|\chi|}. \tag{3.18}$$

Therefore, ρ has an algebraic singularity with

$$\frac{1}{2} < \beta < 1.$$

In addition to the lack of smoothness in ρ, the operator $\mathcal{K}$ is no longer compact on $C(S)$. To gain some insight, look at the special case of a wedge boundary, as in Figure 1. Solve the Dirichlet problem with boundary data f as the double layer potential of (3.15), which means solving

$$(-\pi + \mathcal{K})\,\rho = f. \tag{3.19}$$

This is equivalent to the system

$$\begin{aligned} -\pi\rho_1(x) - \int_0^1 \frac{x \sin(\chi\pi)\,\rho_2(y)\,dy}{x^2 + 2xy\cos(\chi\pi) + y^2} &= f_1(x) \\ -\pi\rho_2(x) - \int_0^1 \frac{x \sin(\chi\pi)\,\rho_1(y)\,dy}{x^2 + 2xy\cos(\chi\pi) + y^2} &= f_2(x), \end{aligned}$$

for $0 < x \leq 1$. The function ρ_1 and ρ_2 are the restrictions of ρ to the bottom branch and upper branch of S, respectively. This system can be uncoupled as

$$-\,\pi\psi_{\pm}(x) \pm \int_0^1 \frac{x \sin(\chi\pi)\,\psi_{\pm}(y)\,dy}{x^2 + 2xy\cos(\chi\pi) + y^2} = F_{\pm}(x), \tag{3.20}$$

and each equation can be reduced to a more well-known type of equation.

One approach is to use the change of variables $x = e^{-t}$, $y = e^{-\tau}$. Then (3.20) becomes the Wiener-Hopf equation

$$-\pi\widehat{\psi}(t) \pm \int_0^\infty \frac{\sin(\chi\pi)\,\widehat{\psi}(\tau)}{e^{t-\tau} - 2\cos(\chi\pi) + e^{\tau-t}} = \widehat{F}(t), \quad 0 \le t < \infty. \tag{3.21}$$

From this perspective, we obtain that the spectrum of $\mathcal{K}$ equals $[-\chi\pi, \chi\pi]$, showing clearly $\mathcal{K}$ is not compact.

A second approach is to rewrite (3.20) as the *Mellin convolution equation*

$$-\pi\psi_\pm(x) \pm \int_0^1 \kappa\left(\frac{x}{y}\right)\psi_\pm(y)\frac{dy}{y} = F_\pm(x) \tag{3.22}$$

$$\kappa(u) = \frac{u\sin(\chi\pi)}{u^2 + 2u\cos(\chi\pi) + 1}, \quad 0 \le u < \infty.$$

This has become the preferred way to study the wedge equation and its numerical solution.

3.3.1 Galerkin methods

Regard the wedge operator

$$\mathcal{L}\psi(x) \equiv \int_0^1 \frac{x\sin(\chi\pi)\,\psi(y)\,dy}{x^2 + 2xy\cos(\chi\pi) + y^2}, \quad 0 < x \le 1,$$

from (3.20) as an operator on $L^2(0,1)$. Then

$$\|\mathcal{L}\| \le \pi\left|\sin\left(\frac{\chi\pi}{2}\right)\right| < \pi,$$

and therefore

$$(-\pi \pm \mathcal{L})\,\psi = F, \tag{3.23}$$

is uniquely solvable. Let $\mathcal{X}_n$ be a subspace of piecewise polynomial functions, and let $\mathcal{P}_n$ be the orthogonal projection of $L^2(0,1)$ onto $\mathcal{X}_n$. Then

$$(-\pi \pm \mathcal{P}_n\mathcal{L})\,\psi_n = \mathcal{P}_n F,$$

is uniquely solvable for $\psi_n \in \mathcal{X}_n$, the method is stable, and

$$\|\psi - \psi_n\|_2 \le \pi\left\|(-\pi \pm \mathcal{P}_n\mathcal{L})^{-1}\right\|\|\psi - \mathcal{P}_n\psi\|_2\,. \tag{3.24}$$

By using a graded mesh that compensates for the behaviour in (3.18), we can obtain an optimal order of convergence. Define a graded mesh using

$$x_i = \left(\frac{i}{n}\right)^q, \quad i = 0, 1, ..., n, \quad q \geq 1. \tag{3.25}$$

Let $\mathcal{X}_n$ consist of all piecewise polynomial functions of degree $\leq r$. Recall the order of convergence of the solution about the origin is $O(x^\beta)$ with $\beta = (1 + |\chi|)^{-1}$. Then choose

$$q > \frac{r+1}{\beta + \frac{1}{2}}.$$

It follows that

$$\|\psi - \psi_n\|_2 \leq O\left(n^{-(r+1)}\right). \tag{3.26}$$

Choosing $q \geq r + 1$ will ensure this for all angles. These generalize to results using $\|\cdot\|_\infty$, with a suitably larger choice for q. The above theory was first given in Chandler [17].

3.3.2 Collocation methods

Let $\mathcal{X}_n$ denote all piecewise polynomials of degree $\leq r$ on our graded mesh, with no continuity restrictions. Introduce

$$0 \leq \xi_0 < \xi_1 < \cdots < \xi_r \leq 1,$$

and define collocation nodes

$$x_{i,j} = x_{i-1} + \xi_j \left(x_i - x_{i-1}\right), \quad j = 0, ..., r,$$

for $i = 1, ..., n$. (With $\xi_0 = 0$ and $\xi_r = 1$, there are continuity restrictions on $\mathcal{X}_n$.) Define $\mathcal{P}_n$ as the piecewise interpolatory projection of $C[0, 1]$ onto $\mathcal{X}_n$ (actually we must use a larger space than $C[0, 1]$ to have a well-defined theory, as in [13]).

For the wedge (3.23), the resulting collocation method can be written abstractly as

$$\left(-\pi \pm \mathcal{P}_n \mathcal{L}\right) \psi_n = \mathcal{P}_n F. \tag{3.27}$$

It is more difficult to analyze than Galerkin's method; and just having $\mathcal{P}_n x \to x$ for all $x \in C[0, 1]$, with a suitably graded mesh, is no longer sufficient for convergence (due to an example in [18]). For the numerical method the approximating subspace $\mathcal{X}_n$ must be modified about the origin, with a corresponding change in $\mathcal{P}_n$. On some neighborhood $[0, x_{i_*}]$, we must define the approximations as piecewise constant, with corresponding collocation nodes the centroids of the subintervals.

Assume

$$\int_0^1 \left(\prod_{i=0}^{r} (\xi - \xi_i)\, g(\xi) d\xi\right) = 0, \quad \deg(g) \leq r'.$$

Then for some $i_* \geq 0$ chosen sufficiently large, the inverses $(-\pi \pm \mathcal{P}_n\mathcal{L})^{-1}$ exist and are uniformly bounded for $n \geq N$. Moreover, if

$$q > \frac{r + r' + 2}{\beta},$$

then

$$\max_{i,j} |\psi(x_{i,j}) - \psi_n(x_{i,j})| \leq O\left(n^{-(r+r'+2)}\right). \tag{3.28}$$

This is uniform in χ if

$$q \geq 2\,(r + r' + 2)\,.$$

A crucial role in the analysis is played by the problem of inverting "finite section" equations. In the Mellin convolution formulation, we need to show the stability over ϵ of the approximation

$$-\pi\psi_\epsilon(x) \pm \int_\epsilon^1 \kappa\left(\frac{x}{y}\right)\psi_\epsilon(y)\frac{dy}{y} = F(x), \quad 0 < x \leq 1, \tag{3.29}$$

and its relation to the full equation

$$-\pi\psi(x) \pm \int_0^1 \kappa\left(\frac{x}{y}\right)\psi(y)\,\frac{dy}{y} = F(x), \quad 0 < x \leq 1.$$

This is also the problem of analyzing the approximation of the Wiener-Hopf equation

$$-\pi\widehat{\psi}(t) \pm \int_0^\infty k\,(t-\tau)\,\widehat{\psi}(\tau)\,dy = \widehat{F}(t), \quad t \geq 0,$$

by the finite section equation

$$-\pi\widehat{\psi}_\gamma(t) \pm \int_0^\gamma k\,(t-\tau)\,\widehat{\psi}_\gamma(\tau)\,dy = \widehat{F}(t), \quad t \geq 0, \tag{3.30}$$

and in this case, the problem was first solved by Anselone and Sloan [3].

The general solvability theory for $(-\pi + \mathcal{K})\,\rho = f$ goes back to Radon [65] in 1919. His methods are akin to the present methods: the problem is divided into two parts, one on a small "wedge shaped" section about the corner point and the other on the remaining part of the boundary. This is also the basis for studying the corresponding problems on surfaces in $\mathbb{R}^3$.

For Nyström methods for solving $(-\pi + \mathcal{K})\,\rho = f$, see Graham and Chandler [35]. For a general approach to all of these methods and generalizations of them, see Elschner [29]. For some other references to the approximate solution of $(-\pi + \mathcal{K})\,\rho = f$, see Costabel and Stephan [24], Atkinson and de Hoog [14], Kress [49], Rathsfeld [67], Atkinson and Graham [12], and Jeon [45].

3.4 First kind BIE on piecewise smooth boundaries

For S a piecewise smooth boundary, the choice of study for first kind BIE is again

$$\mathcal{L}x(P) \equiv \int_S x(P) \log |P - Q| \, dS_Q = f(P), \quad P \in S.$$

Preliminary results on collocation methods for this equation were obtained by Yan [89] and Costabel and Stephan [25]. When compared with the earlier work in Section 3.1 for S a smooth boundary, the difficulty is that the earlier split

$$\mathcal{L} = -\pi\mathcal{A} + \mathcal{B},$$

is no longer of much use, in that $\mathcal{B}$ is no longer a compact operator.

The first general collocation method was obtained only recently, by Elschner and Graham [32]. The main idea in their work is to use a carefully constructed parameterization of the polygonal boundary, one which will improve the behaviour of the equation in the vicinity of corners of S . For simplicity, consider only a single corner, at the origin. Then introduce a parameterization

$$\gamma : [-\pi, \pi] \to S,$$

with $\gamma(0) = \mathbf{0}$, and consider here only the corner at the origin. Then $\mathcal{L}x = f$ becomes

$$\mathcal{K}w(\sigma) \equiv \int_{-\pi}^{\pi} \log |\gamma(s) - \gamma(\sigma)| \, w(\sigma) = g(s) \tag{3.31}$$

$$w(\sigma) = |\gamma'(\sigma)| \, x(\gamma(\sigma)), \quad g(s) = f(\gamma(s)).$$

The function γ is so chosen that all low order derivatives vanish around $s = 0$. The form of γ is

$$\gamma(s) = O\left(s^q\right),$$

for s near to 0, and with q chosen sufficiently large.

The resulting solution w is relatively smooth around 0; and the operator $\mathcal{A}$ of (3.5) can again be used in approximating $\mathcal{K}$. We transform $\mathcal{K}w = g$ to

$$\left[I + \mathcal{A}^{-1}\left(\mathcal{K} - \mathcal{A}\right)\right] w = \mathcal{A}^{-1}g, \tag{3.32}$$

It can be shown that

$$\mathcal{A}^{-1}\left(\mathcal{K} - \mathcal{A}\right) = -\mathcal{H}\mathcal{D}\left(\mathcal{K} - \mathcal{A}\right) + \mathcal{E},$$

with $\mathcal{E}$ compact, $\mathcal{D}\varphi = \varphi'$, and $\mathcal{H}$ the Hilbert singular integral operator. The operator

$$\mathcal{B} = -\mathcal{H}\mathcal{D}(\mathcal{K} - \mathcal{A}),$$

can be shown to be equivalent about the corner at the origin to a Mellin convolution operator plus a compact perturbation. Earlier types of results for such operators can then be brought to bear to analyze the second kind reformulation (3.32) of $\mathcal{K}w = g$.

A collocation method with splines is defined for solving (3.31), although modifications depending on some unknown index $i^* \geq 0$ must be introduced. For the error in solving for w, it can be shown that if q is chosen sufficiently large, and if i_* is chosen sufficiently large, then a collocation method with splines of degree k and a uniform mesh h will have an error in L^2 of $O(h^{k+1})$. Generalizations of this work are being given; for example, see Elschner and Stephan [33].

4 BIE problems on smooth surfaces

The study of BIE on smooth surfaces in $\mathbb{R}^3$ is a well-developed area of classical mathematics, and there are many references to it; for example, see Kress [48]. The general theory of numerical methods for solving BIE of the second kind with a compact operator on smooth surfaces is basically well-understood. But significant practical problems remain, especially concerning numerical integration. For BIE of the first kind, only Galerkin methods are well-understood; and virtually nothing is understood about collocation methods for such equations.

Recall the single layer and double layer potential operators introduced in Section 2.3

$$\mathcal{S}\psi(P) = \int_S \frac{\psi(Q)}{|P-Q|}dS_Q, \quad P \in S, \tag{4.1}$$

$$\mathcal{K}\rho(P) = \int_S \rho(Q)\frac{\partial}{\partial \mathbf{n}_Q}\left[\frac{1}{|P-Q|}\right] dS_Q, \quad P \in S, \tag{4.2}$$

For S the smooth boundary of a closed bounded simply connected region in $\mathbb{R}^3$

$$\mathcal{S}, \mathcal{K} : H^r(S) \to H^{r+1}(S),$$

are bounded operators. For $S = U$ the unit sphere, $\mathcal{S} = 2\mathcal{K}$. Using a representation of the function $\rho \in L^2(U)$ in terms of its Laplace series expansion in spherical harmonics, the mapping properties of $\mathcal{S}$ are transparent from the result

$$\mathcal{S} : Y_n^m \to \frac{4\pi}{2n+1} Y_n^m,$$

for spherical harmonics Y_n^m of degree n. An extensive development of the mapping properties of $\mathcal{S}$ and $\mathcal{K}$ is given in [39]. In particular, both $\mathcal{K}$ and $\mathcal{S}$ are compact on $\mathcal{X}$ to $\mathcal{X}$, with $\mathcal{X}$ equal to either $C(S)$ or $L^2(S)$.

We begin with BIE of the second kind, returning later with some brief remarks on BIE of the first kind. For the interior Dirichlet problem (2.15), we represent the solution u as a double layer potential

$$u(P) = \int_S \rho(Q) \frac{\partial}{\partial \mathbf{n}_Q} \left[\frac{1}{|P-Q|} \right] dS_Q, \quad P \in D,$$

and we find ρ by solving the second kind BIE of (2.16)

$$(2\pi + \mathcal{K}) \rho = f. \tag{4.3}$$

We can also obtain other BIE of the second kind, by other applications of the various identities of Section 2.

4.1 Numerical methods

It is straightforward to speak of both Galerkin and collocation methods for solving (4.3). The convergence theory follows easily from the compactness of $\mathcal{K}$ and $\mathcal{S}$. The Galerkin and collocation methods are simply

$$(2\pi + \mathcal{P}_n \mathcal{K}) \rho_n = \mathcal{P}_n f, \tag{4.4}$$

with appropriate choices of projection operators $\mathcal{P}_n$. If we assume $\mathcal{P}_n x \to x$ for all $x \in \mathcal{X}$, then

$$\|(I - \mathcal{P}_n) \mathcal{K}\| \to 0. \tag{4.5}$$

In turn, this implies the existence and uniform boundedness of $(2\pi + \mathcal{P}_n \mathcal{K})^{-1}$ for all $n \geq N$; and moreover

$$\|\rho - \rho_n\| \leq 2\pi \left\| (2\pi + \mathcal{P}_n \mathcal{K})^{-1} \right\| \|(I - \mathcal{P}_n) \rho\|. \tag{4.6}$$

The methods of most interest have been those of *boundary element* type defined using collocation, for which we now describe a framework.

Let $\mathcal{T}_n = \left\{ \Delta_k^{(n)} \mid k = 1, \dots, n \right\}$ denote a sequence of triangulations of S; and let

$$h \equiv h_n = \max_k \operatorname{diam} \left(\Delta_k^{(n)} \right).$$

To refine a triangulation $\mathcal{T}_n$, we usually subdivide each element of $\mathcal{T}_n$ by connecting the midpoints of the sides of $\Delta_k^{(n)}$, in order to get the next triangulation $\mathcal{T}_{4n}$. We usually have a parameterization of each element $\Delta_k^{(n)}$, say

$$m_k : \widehat{\Delta}_k^{(n)} \subset \mathbb{R}^2 \underset{onto}{\overset{1-1}{\longrightarrow}} \Delta_k^{(n)}.$$

Let $p(x,y)$ be a polynomial over $\widehat{\Delta}_k^{(n)}$. Then consider

$$P(m_k(x,y)) = p(x,y),$$

as a "polynomial" over $\Delta_k^{(n)}$.

For some degree $r \geq 0$, we define $\mathcal{X}_n$ as the set of all functions that are piecewise polynomial over the triangulation $\mathcal{T}_n$ of S. In our own work, we often restrict the functions of $\mathcal{X}_n$ to be continuous; but this is sometimes too restrictive. As a particular example for which I have created a public domain package BIEPACK [8], we take $r = 2$ and we base the collocation method on interpolation at the vertices and midpoints of the sides of the elements of $\mathcal{T}_n$. On the boundary of bounded simply connected region, this leads to

$$n_v = 2n + 2,$$

interpolation nodes. These are denoted collectively by $\mathcal{V}_n = \left\{v_i^{(n)}\right\}$, and in terms of the element $\Delta_k^{(n)}$ to which they belong by

$$\{v_{k,1}, ..., v_{k,6}\}.$$

For interpolation over $\Delta_k^{(n)}$, introduce "Lagrange basis polynomials" $\{L_{k,1}, ..., L_{k,6}\}$ and write

$$\mathcal{P}_n f(Q) = \sum_{j=1}^{6} f(v_{k,j})\, L_{k,j}(Q), \quad Q \in \Delta_k^{(n)},$$

with $\mathcal{P}_n$ the projection of $C(S)$ onto $\mathcal{X}_n$. Formulas (4.4)–(4.6) are valid with this definition of $\mathcal{P}_n$, and it is discussed at greater length in [11].

Also, we often approximate the surface S by interpolation, of the same or greater degree. For a "piecewise polynomial surface" approximation of degree $q \geq 1$, interpolate the parameterization function $m_k(x,y)$ by a polynomial of degree q over $\widehat{\Delta}_k^{(n)}$. In BIEPACK, we use quadratic interpolation. The *panel method* uses piecewise constant interpolation of functions combined with piecewise planar approximations of the surface; and it is widely used in the aircraft industry.

With piecewise quadratic interpolation defining the quadrature, and with the refinement of $\mathcal{T}_n$ to $\mathcal{T}_{4n}$ based on subdividing each element into 4 new elements based on connecting the midpoints of the sides, we have

$$\max_{v_i} |\rho(v_i) - \rho_n(v_i)| = O\left(h^4 \log h\right),$$

whereas

$$\|\rho - \rho_n\|_\infty = O\left(h^3\right).$$

This appears to generalize to collocation with $\mathcal{X}_n$ composed of piecewise polynomials of degree $\le r$ for some even r.

When approximating the surface S with a polynomial of degree q, with piecewise polynomial collocation of degree r, the best we have been able to prove is

$$\|\rho - \rho_n\|_\infty = O\left(h^{\min\{r+1,q\}}\right),$$

although empirical results at the node points are better than this. For $r = q = 2$, we obtain something consistent with

$$\max_{v_i} |\rho(v_i) - \rho_n(v_i)| = O\left(h^4\right).$$

As references, see Wendland [86], Chien [20,21], Atkinson [8], and Atkinson and Chien [11]. For extensions of the package BIEPACK to a distributed processor parallel computer, see Natarajan and Krishnaswamy [56].

4.1.1 The linear system

Consider the linear system associated with

$$(2\pi + \mathcal{P}_n\mathcal{K})\,\rho_n = \mathcal{P}_n f.$$

We write

$$\rho_n(Q) = \sum_{j=1}^{6} \rho_n\left(v_{k,j}\right) L_{k,j}(Q), \quad Q \in \Delta_k^{(n)}, \quad k = 1, \ldots, n,$$

and solve

$$2\pi\rho_n\left(v_i^{(n)}\right) + \sum_{k=1}^{n}\sum_{j=1}^{6} \rho_n\left(v_{k,j}\right) \int_{\Delta_k} L_{k,j}(Q) \frac{\partial}{\partial \mathbf{n}_Q}\left[\frac{1}{\left|v_i^{(n)} - Q\right|}\right] dS_Q$$

$$= f\left(v_i^{(n)}\right), \quad i = 1, \ldots, n_v. \tag{4.7}$$

The most time-consuming part of this method is the setup of the coefficient matrix, which involves the accurate and efficient numerical integration of the integrals over the elements. The cost of this is often an order of magnitude more than that of the solution of the linear system. We consider this further in Section 6.

4.2 Spectral methods

For the equation $(2\pi + \mathcal{K})\,\rho = f$

$$2\pi\rho(P) + \int_S K(P,Q)\rho(Q)\,dS_Q = f(P), \quad P \in S, \tag{4.8}$$

transform the integration region to the unit sphere U. Let S be given by a smooth mapping

$$M : U \underset{onto}{\overset{1-1}{\to}} S,$$

with a smooth inverse M^{-1}. Then our BIE becomes

$$2\pi\widehat{\rho}(P) + \int_S \widehat{K}(P,Q)\widehat{\rho}(Q)\,dS_Q = \widehat{f}(P), \quad P \in U$$

$$\widehat{\rho}(P) \equiv \rho(M(P)), \quad \widehat{f}(P) \equiv f(M(P))$$

$$\widehat{K}(P,Q) = K(M(P), M(Q))J_M(Q), \quad Q \in U,$$

with Jacobian $J_M(Q)$. Denote the new integral equation by

$$\left(2\pi + \widehat{\mathcal{K}}\right)\widehat{\rho} = \widehat{f}. \tag{4.9}$$

Let $\mathcal{X}_N$ denote all spherical polynomials of degree$\leq N$, which has dimension $d_N \equiv (N+1)^2$. The standard orthogonal basis is the set of spherical harmonics

$$\{Y_n^m(\theta,\phi) \mid 0 \leq n \leq N,\ -n \leq m \leq n\},$$

which we assume here are of norm 1. Also denote these by

$$\{\Phi_k \mid 1 \leq k \leq d_N\}.$$

Let $\mathcal{P}_N$ denote the orthogonal projection of $L^2(U)$ onto $\mathcal{X}_N$. The Galerkin method is simply

$$\left(2\pi + \mathcal{P}_N\widehat{\mathcal{K}}\right)\widehat{\rho}_N = \mathcal{P}_N\widehat{f}.$$

Write the numerical solution as

$$\widehat{\rho}_N(P) = \sum_{k=1}^{d_N} \alpha_k \Phi_k.$$

It is obtained by solving

$$2\pi\alpha_i + \sum_{k=1}^{d_N} \alpha_k \left(\widehat{\mathcal{K}}\Phi_k, \Phi_i\right) = (f, \Phi_i), \quad i = 1, ..., d_N.$$

The coefficients $\left(\widehat{\mathcal{K}}\Phi_k, \Phi_i\right)$ are double surface integrals; and as such, they must be evaluated carefully to have a Galerkin method of acceptable cost.

For the speed of convergence, say on $C(U)$

$$\|\rho - \rho_N\|_\infty \leq 2\pi \left\|(2\pi + \mathcal{P}_N \mathcal{K})^{-1}\right\| \|(I - \mathcal{P}_N)\rho\|_\infty .$$

For $\rho \in C^{[\ell,\gamma]}(U)$, a Hölder space of ℓ-times differentiable functions with index γ, and for S sufficiently smooth, we have

$$\|\rho - \rho_N\|_\infty \leq O\left(N^{-(\ell+\gamma-0.5)}\right). \tag{4.10}$$

For this, see [5,6]. For discrete versions of this Galerkin method, see Ganesh et al. [34].

4.3 First kind BIE

There is a well-developed theory of Galerkin methods for those BIE which can be regarded as strongly elliptic pseudo-differential equations on suitable Sobolev spaces. The most studied of such equations is

$$\mathcal{S}\psi(P) \equiv \int_S \frac{\psi(Q)}{|P-Q|} dS_Q = f(P), \quad P \in S.$$

In this, we usually assume S is the boundary of a bounded simply-connected $D \subset \mathbb{R}^3$, and we assume S is as smooth as needed. The basic paper in the study of this equation is that of Nedelec [57].

Nedelec shows that $\mathcal{S} : H^{-\frac{1}{2}} \underset{onto}{\overset{1-1}{\longrightarrow}} H^{\frac{1}{2}}$ and

$$(\mathcal{S}\varphi, \varphi) \geq c\|\varphi\|^2_{-\frac{1}{2}}, \quad \varphi \in H^{-\frac{1}{2}}, \tag{4.11}$$

with $(\cdot,\cdot)$ denoting the extension of the usual inner product on $L^2(S)$ to $H^{\frac{1}{2}} \times H^{-\frac{1}{2}}$. Then he applies the standard theory of finite element methods to obtain a convergent finite element Galerkin method.

Using a triangulation $\mathcal{T}_n = \left\{\Delta_k^{(n)}\right\}$ of S, let $\mathcal{X}_n$ denote all functions which are piecewise polynomial of degree $\leq r$ in the parameterization variables. The Galerkin method is defined by

$$(\mathcal{S}\psi_n, \varphi) = (f, \varphi), \quad \varphi \in \mathcal{X}_n.$$

Then using Cea's Lemma, the existence and uniqueness of ψ_n can be shown, together with

$$\|\psi - \psi_n\|_{-\frac{1}{2}} \leq c \inf_{\varphi \in \mathcal{X}_n} \|\psi - \varphi\|_{-\frac{1}{2}} .$$

By standard arguments, this can then be extended to the more standard error bound

$$\|\psi - \psi_n\|_0 \le ch^{r+1} \|\psi\|_{r+1}, \quad \psi \in H^{r+1}. \tag{4.12}$$

Nedelec also considers the approximation of the boundary by interpolation, of some degree $k \ge 1$. Denoting the resulting solution by $\widehat{\psi}_n$, the convergence results become

$$\left\|\psi - \widehat{\psi}_n \circ \Phi_n^{-1}\right\|_0 \le c\left[h^{r+1}\|\psi\|_{r+1} + h^k \|\psi\|_0\right], \tag{4.13}$$

for $\psi \in H^{r+1}(S)$. In this, Φ_n is a special map from the interpolating surface to the original surface S.

These results generalize to more general strongly elliptic pseudo-differential operator equations. For example, see Wendland [87,88]. For Galerkin methods based on using spherical polynomials as the approximants, see [19].

Unfortunately, almost nothing is understood of collocation methods for solving BIE of the first kind on smooth surfaces. In Section 9.2.3 of [10], we give some numerical experiments for some collocation methods for such BIE.

5 BIE problems on piecewise smooth surfaces

When S is only piecewise smooth, the most studied BIE is $(2\pi + \mathcal{K})\rho = f$,

$$\mathcal{K}\rho(P) = \int_S \rho(Q)\frac{\partial}{\partial \mathbf{n}_Q}\left[\frac{1}{|P-Q|}\right] dS_Q + [2\pi - \Omega(P)]\,\rho(P), \quad P \in S, \tag{5.1}$$

which can arise from either the interior Dirichlet problem or the exterior Neumann problem. The first important paper in the study of the numerical analysis of this equation is Wendland [85]. In it, he gives detailed analyses on the mapping properties of single and double layer operators, including showing that $\mathcal{K}$ is a bounded map of $C(S)$ to $C(S)$. The collocation methods studied were those based on piecewise constant and piecewise linear interpolation. More recently, major extensions have been given by J. Elschner, for Galerkin methods, and A. Rathsfeld, for collocation methods. A related discussion is given in Vainikko [83]. We begin with a description of the ideas of Wendland, and then we describe briefly some of the ideas of Elschner and Rathsfeld.

The properties Wendland assumed for the surface are still those used for most error analyses. In essence, S is assumed to be piecewise smooth

$$S = S_1 \cup \cdots \cup S_J,$$

with each S_j the smooth image of a planar polygonal region. In all cases, the interior solid angle $\Omega(P)$ is assumed to satisfy

$$0 < \Omega(P) < 4\pi. \tag{5.2}$$

Also, each face S_j is assumed to have a piecewise smooth boundary with no cusps. A more precise description of the properties of each S_j is given in [85]. The surface S is triangulated as in the case of a smooth surface S

$$S = \bigcup_{k=1}^{n} \Delta_k^{(n)},$$

with the triangulation $\mathcal{T}_n = \left\{\Delta_k^{(n)}\right\}$ "respecting the edges and corners" of S.

We let $\mathcal{X}_n$ be a space of piecewise polynomial functions of some degree $\leq r$, as in Section 4. These functions may be restricted to be continuous; but if they are only piecewise continuous, the space for the error analysis must be enlarged from $C(S)$, much as is discussed in [13]. Let $\mathcal{P}_n$ denote the interpolatory projection of $C(S)$ onto $\mathcal{X}_n$, and then the collocation method for solving $(2\pi + \mathcal{K})\,\rho = f$ is given by

$$(2\pi + \mathcal{P}_n\mathcal{K})\,\rho_n = \mathcal{P}_n f. \tag{5.3}$$

The analysis of Wendland [85] is based on a critical pair of assumptions

1. at each point $P \in S$, the tangent cone is convex or its complement is convex;

2. the surface angles and interpolating projections must satisfy

$$\sup_{P\in S} |2\pi - \Omega(P)| \sup_{n\geq 1} \|\mathcal{P}_n\| < 2\pi.$$

With piecewise constant interpolation and piecewise linear interpolation (at the vertices of a triangular element), this last assumption is always true. A convergence theory can then be based on the geometric series theorem for $(2\pi + \mathcal{P}_n\mathcal{K})^{-1}$, localized to corners and edges of S. This work also assumes well-behaved solution functions, so that uniform meshes work well in creating accurate approximations of the unknown function ρ. For some extensions of this work to higher order methods, see [11].

5.1 Galerkin methods

When S is a polyhedral Lipschitz boundary, Elschner [30] gives additional theory on the properties of $\mathcal{K}$. He localizes the operator $\mathcal{K}$ to the vertices of S and then shows the localized operator can be regarded as a Mellin convolution operator with an operator valued kernel function. He then uses results for such operators to look at the solvability properties of $(2\pi + \mathcal{K})\,\rho = f$, including obtaining regularity results for the solution ρ. In [31], he gives a general theory for Galerkin boundary element methods. This includes using a graded mesh to compensate for singular behaviour in the solution ρ in the vicinity of edges and corners of S.

Elschner requires that the localized version of $\mathcal{K}$ on each infinite cone generated at each corner of the surface S satisfy a *finite section assumption*. This amounts to something akin to the invertibility of the finite section equation in the planar case, as discussed earlier in Section 3.1 for (3.29) and (3.30). This finite section assumption is known to be true in either of two cases (assuming the surface is Lipschitz), and these are cases also covered by the analyses of Wendland and his co-authors.

1. Either the interior or the exterior solid tangent cone is convex at each point of the surface; or
2. The surface S is made up of faces that are parallel to one of the three coordinate planes in $\mathbb{R}^3$.

With the finite section assumption, the author constructs a mesh which is graded towards each edge of S, and it is doubly-graded towards the vertices of S. The subspace $\mathcal{X}_n$ is defined as the set of piecewise polynomial functions degree $\leq r$; and there is no continuity restriction on the approximants in $\mathcal{X}_n$. With a properly graded mesh, Elschner is able to prove stability of a Galerkin method and to show an optimal order of convergence

$$\|\rho - \rho_n\|_{L^2(S)} = O\left(m^{-(r+1)}\right), \quad n = O\left(m^2\right). \tag{5.4}$$

The grading is of the same algebraic type as used for planar problems and described by (3.25).

5.2 Collocation methods

When S is a polyhedral Lipschitz boundary, A. Rathsfeld [66] gives a general theory for collocation boundary element methods for solving $(2\pi + \mathcal{K})\rho = f$. This also includes using a graded mesh to compensate for singular behaviour in the solution ρ in the vicinity of edges and corners of S. As with the analysis of Elschner for Galerkin methods, Rathsfeld makes a "finite section" assumption on the localizations of $\mathcal{K}$.

Rathsfeld uses a graded mesh, similar to that in Elschner's Galerkin methods. For some $r \geq 0$, define $\mathcal{X}_n$ as the functions of degree $\leq r$ in the parameterization variables on each face of S. But as in the planar case, this is too general to allow a successful error analysis for collocation methods. In the vicinity of edges and corners of S, special modifications of the approximating space $\mathcal{X}_n$ are needed; and the collocation nodes are chosen with some care as well. With this modified approximating subspace $\mathcal{X}_n$, Rathsfeld proves convergence and stability of the collocation scheme, with

$$\|\rho - \rho_n\|_\infty = O\left(m^{-(r+1)}\right), \quad n = O(m^2). \tag{5.5}$$

In the work of Elschner and Rathsfeld, the authors study the double layer integral operator $\mathcal{K}$ of (5.1) by localizing it to each corner of the polyhedral surface S. Denote by Γ the infinite cone determined by a generic corner P_0, and consider $\mathcal{K} \equiv \mathcal{W}$ defined over Γ. Let Γ intersect a unit sphere centered at P_0, and call this intersection γ. Parameterize γ by $\omega(s)$, $0 \le s \le L$. Write a typical point on Γ as $P = r\omega(s)$. For $g \in C_0(\Gamma)$

$$\mathcal{W}g(r\omega(s)) = \int_0^\infty \left(\mathcal{B}\left(\frac{r}{\tau}\right) g\left(\tau\omega(\cdot)\right)\right) (\omega(s))\frac{d\tau}{\tau}, \tag{5.6}$$

with $\mathcal{B}(t)$ an operator on $C(\gamma)$ to $C(\gamma)$

$$\left(\mathcal{B}(t)h\right)(\omega(s)) = \int_0^L h(\omega(\sigma))\frac{t\mathbf{n}_{\omega(\sigma)} \cdot \omega(s)}{|t\omega(s) - \omega(\sigma)|}d\sigma, \quad t \ge 0.$$

The equation $(2\pi + \mathcal{W})\,\rho = f$ is a Mellin convolution equation using the operator-valued Mellin convolution kernel $\mathcal{B}$. The authors use properties for such operators to obtain convergence and stability results for the numerical solution of boundary element methods for solving $(2\pi + \mathcal{W})\,\rho = f$. This is then used in obtaining a solvability theory for the original integral equation $(2\pi + \mathcal{K})\,\rho = f$, including regularity results for the solution ρ.

5.2.1 An alternative framework

Using results on the solvability of the associated boundary value problems for $\Delta u = 0$, combined with the Green's representation formulae, one can obtain results on solvability of some BIE, including regularity results on their solutions. For such an approach, see von Petersdorff and Stephan [62]. For an introduction to other graded mesh methods and to *hp*-versions of BIE methods, see Stephan [80] and Stephan and Suri [81].

6 BIE on surfaces – other aspects

There are aspects of the numerical solution of BIE on surfaces which become a source of difficulty because of the rapid increase in cost associated with the surface S being two dimensional. These difficulties are related primarily to the iterative solution of the associated linear system and the calculation of the surface integrals in that system.

The linear system associated with the quadratic collocation method for (5.3) is given by

$$\begin{aligned}2\pi\rho_n\left(v_i^{(n)}\right) + \sum_{k=1}^{n}\sum_{j=1}^{6}\rho_n\left(v_{k,j}\right)\int_{\Delta_k} L_{k,j}(Q)\frac{\partial}{\partial \mathbf{n}_Q}\left[\frac{1}{\left|v_i^{(n)} - Q\right|}\right] dS_Q \\ + \left[2\pi - \Omega(v_i^{(n)})\right]\rho_n\left(v_i^{(n)}\right) = f\left(v_i^{(n)}\right), \quad i = 1, \dots, n_v,\end{aligned} \tag{6.1}$$

and we denote it by

$$(2\pi + K_n)\,\mathbf{v}_n = \mathbf{y}_n. \tag{6.2}$$

There remains the problem of estimating $\Omega(v_i^{(n)})$. Note that when $\rho \equiv 1$, it follows that $(2\pi + \mathcal{K})\,\rho = 4\pi$. Using this, the solid angle can be approximated by

$$\Omega(v_i^{(n)}) \approx \sum_{k=1}^{n}\sum_{j=1}^{6}\int_{\Delta_k} L_{k,j}(Q)\frac{\partial}{\partial \mathbf{n}_Q}\left[\frac{1}{\left|v_i^{(n)} - Q\right|}\right] dS_Q. \tag{6.3}$$

Some discussion of this is given in [11].

Two-grid iteration methods for solving the system in (6.2) are examined in [9], and multigrid methods are examined in Hackbusch [40]. A general examination and comparison of these and other iteration methods is given in Rathsfeld [68]. For problems over smooth surfaces S, there is no difficulty in applying any of these methods; and their error analysis is relatively straightforward. For piecewise smooth surfaces, one must "precondition" the system. To do this, we can "solve exactly" the portion of the linear system corresponding to the nodes on the surface that are at an edge, a corner, or are nearby to such. Examples of this are given in [9] and [68].

For other discussions of iteration methods for solving discretizations of BIE, see Schippers [76], Hebeker [42], Atkinson and Graham [12], Vavasis [84], and Petersdorf and Stephan [63,64], Kelley [46], Kelley and Xue [47]. An interesting discussion is given in Edelman [28] of the significance of BIE within the numerical linear algebra of dense linear systems.

6.1 Numerical integration

In (6.1), the integrals

$$\int_{\Delta_k} L_{k,j}(Q)\frac{\partial}{\partial \mathbf{n}_Q}\left[\frac{1}{\left|v_i^{(n)} - Q\right|}\right] dS_Q, \tag{6.4}$$

must be evaluated numerically in all but the simplest cases. The functions $L_{k,j}(Q)$ are "Lagrange basis functions" for quadratic interpolation over Δ_k. [In the case of piecewise constant interpolation over polyhedral surfaces S, these integrals can be done analytically.] These are the most time consuming part of most boundary element codes. We divide the evaluation of them into two cases.

Case (a). Let $v_i^{(n)} \in \Delta_k$. Then the integrand in (6.4) is singular. This appears to be a difficult integral, but there is a simple fix, called the "Duffy transformation". Consider the integral

$$I = \int_\sigma g(s,t)\,d\sigma,$$

with σ the unit simplex in the plane

$$\sigma = \{(s,t) \mid 0 \le s, t, s+t \le 1\}.$$

Further assume the function has a point singularity at the origin. Introduce the change of variables *wh*.

$$s = (1-y)\,x, \quad t = yx, \quad 0 \le x, y \le 1$$

$$I = \int_0^1 \int_0^1 xg\left((1-y)\,x, yx\right)\,dx\,dy.$$

When the original integral arises from a collocation integral of either single or double layer type, then the new integrand contains no singularity, including in its derivatives. Then simply apply Gaussian quadrature to both integrals, usually with only a few nodes (for example ≈ 5) in each direction. For a complete discussion and extension to other singular integrals, see Schwab and Wendland [78].

Case (b). Let $v_i^{(n)} \notin \Delta_k$. Then the integrals (6.4) are nonsingular. However, they vary from being almost singular, when $v_i^{(n)}$ is near Δ_k, to having very well-behaved integrands, when $v_i^{(n)}$ is distant from Δ_k. Note that of the $O(n^2)$ integrals in the collocation matrix only $O(n)$ are of singular type; and therefore, most of the cost of setting up the matrix for Equation 6.1 is in these nonsingular integrals.

An efficient quadrature method with an exponential rate of convergence is defined and analyzed in Schwab [77]. We have used the following somewhat simpler schema in [8], and it is reasonably efficient in most cases. Introduce a parameter μ to indicate a number of level of subdivisions of Δ_k for a composite quadrature scheme. For the basic quadrature scheme, we use the 7-point rule T2:5-1, of degree of precision 5, from Stroud [82].

1. If $0 < \text{dist}\,(v_i, \Delta_k) \le h$, use μ levels of subdivision of Δ_k and apply a given quadrature scheme to each of the resulting 4^μ subelements.

2. If $h < \text{dist}\,(v_i, \Delta_k) \le 2h$, use $\mu - 1$ levels of subdivision of Δ_k and apply a given quadrature scheme to each of the resulting $4^{\mu-1}$ subelements.

3. Continue this until no subdivision of Δ_k takes place, so that the quadrature scheme is applied directly to Δ_k.

As n increases to $4n$ to $16n$, it seems best to increase μ to $\mu + 1$.

We give an example of solving $(2\pi + \mathcal{K})\,\rho = f$ for an exterior Neumann problem on an ellipsoid. Table 1 contains the errors for varying μ, Table 2 contains timings for the composite scheme described above, and Table 3 contains timings when μ levels of subdivisions are used over all triangular elements. The errors in Table 1 are valid for both Tables 2 and 3; but the timings are far larger for Table 3. The timings are for a Hewlett-Packard HP-720 workstation.

For other discussions of numerical integration over boundary elements for solving integral equations on surfaces, see Chien [21] and Sauter and Schwab [75].

Table 1. Max errors in ρ_n

n	$\mu = 0$	$\mu = 1$	$\mu = 2$	$\mu = 3$
8	3.63E−2	3.30E−2	3.31E−2	3.31E−2
32	3.78E−3	3.79E−3	3.78E−3	3.78E−3
128	5.49E−4	2.92E−4	2.91E−4	2.91E−4
512	2.62E−4	1.88E−5	1.87E−5	1.87E−5

Table 2. Timings for "graded" mesh quadrature (sec)

n	$\mu = 0$	$\mu = 1$	$\mu = 2$	$\mu = 3$	Iteration
8	0.85	0.85	1.9	4.0	
32	3.9	6.1	22.0	59.5	< 0.01
128	31.2	42.7	123	439	0.13
512	388	424	754	2240	2.05

Table 3. Timings using a uniform μ levels of subdivision (sec)

n	$\mu = 0$	$\mu = 1$	$\mu = 2$	$\mu = 3$
8	0.85	0.90	1.6	4.3
32	3.9	7.2	20.8	76.0
128	31.2	89.3	317	1230
512	388	1350	5030	19800

6.2 Fast matrix-vector calculations

Consider a discretization of the single layer operator

$$\mathcal{S}\rho\left(v_i^{(n)}\right) \approx \sum_{k=1}^{n}\sum_{j=1}^{6} \rho\left(v_{k,j}^{(n)}\right) \int_{\Delta_k} \frac{L_{k,j}(Q)}{\left|v_i^{(n)} - Q\right|} dS_Q, \tag{6.5}$$

for $i = 1, ..., n_v$, using a piecewise quadratic interpolation of ρ. Denote the vector of values $\rho\left(v_i^{(n)}\right)$ by ρ_n, and let $A\rho_n$ denote the matrix multiplication inherent in the above approximation. Then all iteration methods for solving a linear system involving A, say $A\mathbf{v} = \mathbf{y}$, will involve repeated multiplications $A\mathbf{w}$ for various vectors $\mathbf{w}$. To evaluate A explicitly requires $O\left(n^2\right)$ operations or more, and to evaluate $A\mathbf{w}$ requires $2n^2$ operations. Reducing the cost of computing $A\mathbf{w}$ is of little consequence if we must still evaluate A explicitly.

In the past ten years, several approaches have been used to speed the calculation of $A\mathbf{w}$, reducing it from $O(n^2)$ to $O\left(n \log^d n\right)$ operations, for some small integer d. The three main types of methods are as follows.

- Fast multipole methods;
- Clustering methods;
- Wavelet compression methods.

With all of these methods, we produce a vector

$$\mathbf{u} \approx A\mathbf{w}, \tag{6.6}$$

within an acceptably small error and at a cost of $O\left(n \log^d n\right)$ operations.

6.2.1 Fast multipole methods

These were developed by L. Greengard and V. Rokhlin in a series of papers, and they have become widely used in the past few years. For papers of Greengard and Rokhlin, see [37,38,71]; and for a paper dealing more explicitly with BIE approximations, see Nabors et al. [55].

The problem as originally posed by Greengard and Rokhlin is to calculate the potential at points $\Upsilon_1, ..., \Upsilon_p$ as determined by charges of strengths $q_1, ..., q_k$ at k points $Q_1, ..., Q_k$

$$\Phi(\Upsilon_i) = \sum_{j=1}^{k} \frac{q_j}{|\Upsilon_i - Q_j|}, \quad i = 1, ..., p. \tag{6.7}$$

Often $\{\Upsilon_i\} = \{Q_j\}$, with the proviso of omitting the term in the sum for $j = i$. The size of k can be quite large, say $k = 10,000$. The methods used depend on the use of spherical harmonics expansions and on clusterings of the points $\{Q_j\}$ according to their distance from Υ_i.

Writing points using their spherical coordinates, let $\Upsilon = (r, \theta, \phi)$ and $Q = (\rho, \alpha, \beta)$. Then for $\rho < r$

$$\frac{1}{|\Upsilon - Q|} = \sum_{n=0}^{\infty} P_n(\cos\gamma)\left(\frac{\rho}{r}\right)^n, \tag{6.8}$$

with γ the angle between Υ and Q, and with $P_n(u)$ the Legendre polynomial of degree n. To give a flavor to the method's techniques, we give one theorem from the development of the method.

Theorem 1. *Suppose that k charges of strengths $q_1, ..., q_k$ are located at the points $Q_i = (\rho_i, \alpha_i, \beta_i)$, $i = 1, ..., k$, and assume that $\rho_i < a$, a given number. Then for any $\Upsilon = (r, \theta, \phi)$ with $r > a$, the potential $\Phi(\Upsilon)$ is given by*

$$\Phi(\Upsilon) = \sum_{n=0}^{\infty} \sum_{m=-n}^{n} \frac{M_n^m}{r^{n+1}} Y_n^m(\theta, \phi) \tag{6.9}$$

$$M_n^m = \sum_{i=1}^{k} q_i \rho_i^n Y_n^{-m}(\alpha_i, \beta_i).$$

For any $\ell \geq 1$

$$\left|\Phi(\Upsilon) - \sum_{n=0}^{\ell} \sum_{m=-n}^{n} \frac{M_n^m}{r^{n+1}} Y_n^m(\theta, \phi)\right| \leq \frac{A}{r-a}\left(\frac{a}{r}\right)^{\ell+1} \tag{6.10}$$

$$A = \sum_{i=1}^{k} |q_i|.$$

The functions $\{Y_n^m(\theta, \phi)\}$ are spherical harmonics of degree n and order m. The formula (6.9) is called the *multipole expansion* of $\Phi(\Upsilon)$. The *fast multipole method* is based on creating graded subdivisions of space and to combine this with approximations as in (6.10) so as to calculate an estimate of $\Phi(\Upsilon)$ at p points Υ in much fewer than $O(pk)$ operations. An excellent introduction to this is Greengard [37], Chapter 3.

6.2.2 Clustering methods

Assume the creation of a sequence of triangulations of S, much as in Sections 4 and 5; and let each triangulation be a refinement of the preceding one. Call these $\mathcal{T}_0$, $\mathcal{T}_1$, $\mathcal{T}_2$, ...,$\mathcal{T}_I$, with $\mathcal{T}_0 = \{\Delta_k \mid 1 \leq k \leq n\}$ the finest mesh, in keeping with the notation in Hackbusch and Nowak [41]. For simplicity here, also assume S is polyhedral. We restrict our interest to the single layer potential approximation of (6.5).

Note that as the distance from v_i is increased, the integrand becomes smoother. For each element τ in some $\mathcal{T}_l$, with $P \notin \tau$, we expand the simple layer in a Taylor series about the "centroid" Q_τ of τ

$$\frac{1}{|P-Q|} \approx \sum_{l \in I_m} \kappa_l(P; Q_\tau)\Phi_l(Q).$$

The degree of the Taylor series is m, and I_m denotes an index set for the expansion. The accuracy of this expansion will depend on the distance of P from Q_τ, the size of m, and the size of τ. We call τ a "cluster" of elements from the finest level $\mathcal{T}_0$.

Note that $\Phi_l(Q)$ is independent of P and Q_τ, which is critically important in obtaining a low operations count. The functions $\Phi_l(Q)$ are to be simple monomials in the components of Q, thus making them simple to integrate over elements Δ_k of $\mathcal{T}_0$ and thence over clusters from the coarser triangulations $\mathcal{T}_l$, some $l > 0$.

Let $\rho_n(Q)$ denote the piecewise quadratic

$$\rho_n(Q) = \sum_{j=1}^{6} \rho_n(v_{k,j}) L_{k,j}(Q), \quad Q \in \Delta_k, \quad 1 \le k \le n.$$

Let the cluster τ consist of a set of elements from $\mathcal{T}_0$, say $\Delta_1, \ldots, \Delta_p$. Then

$$\int_\tau \frac{\rho_n(Q)\, dS_Q}{|P-Q|} \approx \sum_{j=1}^{p} \kappa_l(P; Q_\tau) \sum_{l \in I_m} \int_{\Delta_j} \rho_n(Q)\, \Phi_l(Q)\, dS_Q.$$

With preprocessing, we can evaluate all of the integrals

$$\int_{\Delta_j} \rho_n(Q)\, \Phi_l(Q)\, dS_Q,$$

in $O(n)$ operations, allowing both l and j to vary. Because both $\rho_n(Q)$ and $\Phi_l(Q)$ are polynomials in Q, these integrals are straightforward to evaluate.

For each point $P = v_i$, we do an initial preprocessing to produce the needed clustering of elements of $\mathcal{T}_0$, to have

$$S = [\tau_1 \cup \cdots \cup \tau_c] \cup [\Delta_{i_1} \cup \cdots \cup \Delta_{i_q}],$$

with $\{\tau_j\}$ elements from various $\mathcal{T}_k$ with $k > 0$. Then we have

$$\int_S \frac{\rho_n(Q)\, dS_Q}{|P-Q|} = \sum_{j=1}^{c} \int_{\tau_j} \frac{\rho_n(Q)\, dS_Q}{|P-Q|} + \sum_{j=1}^{q} \int_{\Delta_{i_j}} \frac{\rho_n(Q)\, dS_Q}{|P-Q|}.$$

With a careful attention to detail and with preprocessing where possible, it is possible to evaluate an approximation of $[\mathcal{S}\rho_n(v_i)]$ in $O(n \log^3 n)$ operations. The main reference to such *clustering methods* is Hackbusch and Nowak [41].

6.2.3 Wavelet compression methods

The main idea is to use a wavelet basis such that the most of the elements of the collocation and Galerkin matrix will be small enough to be neglected. Then the cost of evaluating $[S\rho_n(v_i)]$ will be only slightly larger than $O(n)$ operations. For some papers in this very recent topic, see Alpert et al. [1], Dahmen et al. [27], Petersdorff and Schwab [60,61] and Micchelli and Xu [54].

References

1. Alpert, B., Beylkin, G., Coifman, R. and Rokhlin, V. (1993). Wavelets for the fast solution of second-kind integral equations. *SIAM J. Sci. Comp.*, **14**, 159–184.
2. Amini, S. and Maines, N. (1995). Regularisation of strongly singular integrals in boundary integral equations. *Tech. Rep. MCS-95-7*, University of Salford.
3. Anselone, P. and Sloan, I. (1985). Integral equations on the half line. *J. Integral Equations*, **9** (Suppl.), 3–23.
4. Arnold, D. and Wendland, W. (1983). On the asymptotic convergence of collocation methods. *Num. Math.*, **41**, 349–381.
5. Atkinson, K. (1982). The numerical solution of Laplace's equation in three dimensions. *SIAM J. Num. Anal.*, **19**, 263-274.
6. Atkinson, K. (1985). Algorithm 629: An integral equation program for Laplace's equation in three dimensions. *ACM Trans. on Math. Soft.*, **11**, 85–96.
7. Atkinson, K. (1988). A discrete Galerkin method for first kind integral equations with a logarithmic kernel. *J. Integral Equations and Applic.*, **1**, 343–363.
8. Atkinson, K. (1993). User's guide to a boundary element package for solving integral equations on piecewise smooth surfaces. *Reps. on Computational Maths*, **43**, University of Iowa.
9. Atkinson, K. (1994). Two-grid iteration methods for linear integral equations of the second kind on piecewise smooth surfaces in $\mathbf{R}^3$. *SIAM J. Sci. Comp.*, **15**, 1083–1104.
10. Atkinson, K. (1997). *The Numerical Solution of Integral Equations of the Second Kind.* Cambridge University Press. (To appear.)
11. Atkinson, K. and Chien, D. (1995). Piecewise polynomial collocation for boundary integral equations. *SIAM J. Sci. Comp.*, **16**, 651–681.
12. Atkinson, K. and Graham, I. (1990). Iterative variants of the Nyström method for second kind boundary integral equations. *SIAM J. Sci. and Stat. Comp.*, **13**, 694–722.
13. Atkinson, K., Graham, I. and Sloan, I. (1983). Piecewise continuous collocation for integral equations. *SIAM J. Num. Anal.*, **20**, 172–186.
14. Atkinson, K. and de Hoog, F. (1984). The numerical solution of Laplace's equation on a wedge. *IMA J. Num. Anal.*, **4**, 19–41.
15. Atkinson, K. and Sloan, I. (1991). The numerical solution of first kind logarithmic-kernel integral equations on smooth open arcs. *Math. of Comp.*, **56**, 119–139.
16. Blue, J. (1978). Boundary integral solutions of Laplace's equation. *The Bell System Technical J.*, **57**, 2797–2822.
17. Chandler, G. (1984). Galerkin's method for boundary integral equations on polygonal domains. *J. Australian Math. Soc., Series B*, **26**, 1–13.
18. Chandler, G. and Graham, I. (1988). Product integration-collocation methods for noncompact integral operator equations. *Math. Comp.*, **50**, 125–138.

19. Chen, Y. and Atkinson, K. (1994). Solving a single layer integral equation on surfaces in $\mathbf{R}^3$. *SIAM J. Num. Anal., Reps. on Computational Mathematics*, **51**, University of Iowa.

20. Chien, D. (1992). Piecewise polynomial collocation for integral equations with a smooth kernel on surfaces in three dimensions. *J. of Int. Eqns. and Applic.*, **5**, 315–344.

21. Chien, D. (1995). Numerical evaluation of surface integrals in three dimensions. *Maths. of Comp.*, **64**, 727–743.

22. Chien, D. and Atkinson, K. (1995). The numerical solution of a hypersingular planar boundary integral equation. *IMA J. Num. Anal.* (To appear.)

23. Costabel, M. (1987). Principles of boundary element methods. *Computer Physics Reps.*, **6**, 243–274.

24. Costabel, M. and Stephan, E. (1985). Boundary integral equations for mixed boundary value problems in polygonal domains and Galerkin approximation. *Mathematical Models and Methods in Mechanics*, **15**, Banach Center Publications, 175–251.

25. Costabel, M. and Stephan, E. (1987). On the convergence of collocation methods for boundary integral equations on polygons. *Math. Comp.*, **49**, 461–478.

26. Costabel, M. and Wendland, W. (1986). Strong ellipticity of boundary integral operators. *J. Reine Angewandte Math.*, **372**, 34–63.

27. Dahmen, W., Kleemann, B., Prößdorf, S. and Schneider, R. (1994). A multiscale method for the double layer potential equation on a polyhedron. *Advances in Comp. Maths.*, Editors: H. Djkshit and C. Micchelli, World Scientific, 15–57.

28. Edelman, A. (1993). Large dense numerical linear algebra in 1993: The parallel computing influence. *J. Supercomputing Applic.*, **7**, 113–128.

29. Elschner, J. (1990). On spline approximation for a class of non-compact integral equations. *Math. Nachr.*, **146**, 271–321.

30. Elschner, J. (1992). The double layer potential operator over polyhedral domains I: Solvability in weighted Sobolev spaces. *Applic. Anal.*, **45**, 117–134.

31. Elschner, J. (1992). The double layer potential operator over polyhedral domains II: Spline Galerkin methods. *Math. Methods Appl. Sci.*, **15**, 23–37.

32. Elschner, J. and Graham, I. (1995). An optimal order collocation method for first kind boundary integral equations on polygons. *Numerische Math.*, **70**, 1–31.

33. Elschner, J. and Stephan, E. (1995). A discrete collocation method for Symm's integral equation on curves with corners. *J. Comp. Appl. Math.* (To appear.)

34. Ganesh, M., Graham, I. and Sivaloganathan, J. (1993). A pseudospectral 3D boundary integral method applied to a nonlinear model problem from finite elasticity. *SIAM J. Num. Anal.*, **31**, 1378–1414.

35. Graham, I. and Chandler, G. (1988). High order methods for linear functionals of solutions of second kind integral equations. *SIAM J. Num. Anal.*, **25**, 1118–1137.

36. Giroire, J. and Nedelec, J. (1978). Numerical solution of an exterior Neumann problem using a double layer potential. *Math. Comp.*, **32**, 973–990.

37. Greengard, L. (1988). *The Rapid Evaluation of Potential Fields in Particle Systems*, MIT Press, Cambridge, Massachusetts.

38. Greengard, L. and Rokhlin, V. (1987). A fast algorithm for particle simulation. *J. Comput. Phys.*, **73**, 325–348.

39. Günter, N. (1967). *Potential Theory*, Ungar Pub., New York.

40. Hackbusch, W. (1994). *Integral Equations: Theory and Numerical Treatment*, Birkhäuser Verlag, Basel.

41. Hackbusch, W. and Nowak, Z. (1989). On the fast matrix multiplication in the boundary element method by panel clustering. *Num. Math.*, **54**, 463–491.

42. Hebeker, F. (1988). On the numerical treatment of viscous flows against bodies with corners and edges by boundary element and multi-grid methods. *Num. Math.*, **52**, 81–99.

43. Hess, J. (1990). Panel methods in computational fluid dynamics. *Annual Rev. Fluid Mechanics*, **22**, 255–274.

44. Jaswon, M. and Symm, G. (1977). *Integral Equation Methods in Potential Theory and Elastostatics*, Academic Press.

45. Jeon, Y-M. (1993). A Nyström method for boundary integral equations on domains with a piecewise smooth boundary. *J. Int. Eqns. and Applic.*, **5**, 221–242.

46. Kelley, C.T. (1995). A fast multilevel algorithm for integral equations. *SIAM J. on Num. Anal.*, **32**, 501–513.

47. Kelley, C.T. and Xue, Z. (1996). GMRES and integral operators. *SIAM J. Sci. Comp.*, **17**, 217–226.

48. Kress, R. (1989). *Linear Integral Equations*, Springer-Verlag.

49. Kress, R. (1990). A Nyström method for boundary integral equations in domains with corners. *Num. Math.*, **58**, 145–161.

50. Kress, R. (1995). On the numerical solution of a hypersingular integral equation in scattering theory. *J. Comp. and Appl. Maths.*, **61**, 345–360.

51. McLean, W. (1986). A spectral Galerkin method for boundary integral equations. *Math. Comp.*, **47**, 597–607.

52. McLean, W. (1995). Fully-discrete collocation methods for an integral equation of the first kind. *J. Int. Eqns. and Applic.*, **6**, 537–572.

53. McLean, W., Prößdorf, S. and Wendland, W. (1993). A fully-discrete trigonometric collocation method. *J. Int. Eqns. and Applic*, **5**, 103–129.

54. Micchelli, C. and Xu, Y. (1995). Weakly singular Fredholm integral equations I: Singularity preserving wavelet-Galerkin methods. *Approximation Theory VIII*, Editors: C. Chui and L. Schumaker, World Scientific.

55. Nabors, K., Korsmeyer, F., Leighton, F. and White, J. (1994). Preconditioned, adaptive, multipole-accelerated iterative methods for three-dimensional first-kind integral equations of potential theory. *SIAM J. Sci. Comp.*, **15**, 713–735.

56. Natarajan, R. and Krishnaswamy, D. (1995). A case study in parallel scientific computing: The boundary element method on a distributed memory multicomputer. *Proc. on Supercomputing 95*, Assoc. of Comp. Mach.

57. Nedelec, J. (1976). Curved finite element methods for the solution of singular integral equations on surfaces in $\mathbf{R}^3$. *Computer Methods in Appl. Mechanics and Engin.*, **8**, 61–80.

58. Newman, J. (1992). Panel methods in marine hydrodynamics, *Proc. Conf. on Eleventh Australasian Fluid Mechanics.*

59. Noble, B. (1977). The numerical solution of integral equations. *The State of the Art in Num. Anal.*, Editor: D. Jacobs, Academic Press, 915–966.

60. von Petersdorff, T. and Schwab, C. (1994). Wavelet approximations for first kind boundary integral equations on polygons. *Numer. Math.* (To appear.)

61. von Petersdorff, T. and Schwab, C. (1995). Fully discrete multiscale Galerkin BEM. *Multiresolution Analysis and Partial Differential Equations*, Editors: W. Dahmen, P. Kurdila, and P. Oswald, Academic Press, New York.

62. von Petersdorff, T. and Stephan, E. (1990). Regularity of mixed boundary value problems in $\mathbf{R}^3$ and boundary element methods on graded meshes. *Math. Methods in the Appl. Sci.*, **12**, 229–249.

63. von Petersdorff, T. and Stephan, E. (1990). On the convergence of the multigrid method for a hypersingular integral equation of the first kind. *Num. Math.*, **57**, 379–391.

64. von Petersdorff, T. and Stephan, E. (1992). Multigrid solvers and preconditioners for first kind integral equations. *Num. Methods for Partial Diff. Eqns.*, **8**, 443–450.

65. Radon, J. (1919). Über die Randwertaufgaben beim logarithmischen Potential. *Sitzungsberichte der Akademie der Wissenschafter Wien*, **128 Abt.IIa**, 1123–1167.

66. Rathsfeld, A. (1992). The invertibility of the double layer potential operator in the space of continuous functions defined on a polyhedron: The panel method. *Appl. Anal.*, **45**, 135–177.

67. Rathsfeld, A. (1992). Iterative solution of linear systems arising from the Nyström method for the double layer potential equation over curves with corners. *Math. Methods Appl. Sci.*, **15**, 443–455.

68. Rathsfeld, A. (1995). Nyström's method and iterative solvers for the solution of the double layer potential equation over polyhedral boundaries. *SIAM J. Num. Anal.*, **32**, 924–951.

69. Rathsfeld, A., Kieser, R. and Kleemann, B. (1992). On a full discretization scheme for a hypersingular boundary integral equation over smooth curves. *Z. Anal. Anwendungen*, **11**, 385–396.

70. Rizzo, F., Shippy, D. and Rezayat, M. (1985). A boundary integral equation method for radiation and scattering of elastic waves in three dimensions. *Int. J. for Num. Methods in Engineering*, **21**, 115-129.

71. Rokhlin, V. (1983). Rapid solution of integral equations of classical potential theory. *J. Comp. Phys.*, **60**, 187–207.

72. Rudolphi, T., Krishnasamy, G., Schmerr, L. and Rizzo, F. (1988). On the use of strongly singular integral equations for crack problems. *Proc. 10th Int. Conf. on Boundary Elements.*

73. Saranen, J. (1988). The convergence of even degree spline collocation solution for potential problems in the plane. *Num. Math.*, **53**, 499–512.

74. Saranen, J. and Vainikko, G. (1994). Trigonometric collocation methods with product-integration for boundary integral equations on closed curves, *SIAM J. Num. Anal.*, **33**, 1577-1596.

75. Sauter, S. and Schwab, C. (1996). Quadrature for *hp*-Galerkin BEM in $\mathbf{R}^3$. Res. Rep., **96-02**, Seminar für Angew. Math. Eid. Tech. Hoch., Zurich.

76. Schippers, H. (1985). Multi-grid methods for boundary integral equations. *Num. Math.*, **46**, 351–363.

77. Schwab, C. (1994). Variable order composite quadrature of singular and nearly singular integrals. *Computing*, **53**, 173–194.

78. Schwab, C. and Wendland, W. (1992). On numerical cubatures of singular surface integrals in boundary element methods. *Num. Math.*, **62**, 343–369.

79. Sloan, I. (1992). Error analysis of boundary integral methods. *Acta Numerica*, **1**, 287–339.

80. Stephan, E. (1994). The $h-p$ boundary element method for solving 2- and 3-dimensional problems – especially in the presence of singularities and adaptive approaches. (Preprint.)

81. Stephan, E. and Suri, M. (1989). On the convergence of the p-version of the boundary element Galerkin method. *Math. Comp.*, **52**, 32–48.

82. Stroud, A. (1971). *Approximate Calculation of Multiple Integrals*. Prentice-Hall, New Jersey.

83. Vainikko, G. (1993). *Multidimensional Weakly Singular Integral Equations*. Lecture Notes in Mathematics, **1549**, Springer-Verlag, Berlin.

84. Vavasis, S. (1992). Preconditioning for boundary integral equations. *SIAM J. Matrix Anal.*, **13**, 905–925.

85. Wendland, W. (1968). Die Behandlung von Randwertaufgaben im $\mathbf{R}^3$ mit Hilfe von Einfach und Doppelschichtpotentialen. *Numerische Math.*, **11**, 380–404.

86. Wendland, W. (1985). Asymptotic accuracy and convergence for point collocation methods. *Topics in Boundary Element Research: Time-Dependent and Vibration Problems*, Springer-Verlag, Berlin, 230–257.

87. Wendland, W. (1987). Strongly elliptic boundary integral equations. *The State of the Art in Num. Anal.*, Editors: A. Iserles and J. Powell, Clarendon Press, 511–562.

88. Wendland, W. (1990). Boundary element methods for elliptic problems. *Mathematical Theory of Finite and Boundary Element Methods*, Editors: A. Schatz, V. Thomée, and W. Wendland, Birkhäuser, Boston, 219–276.

89. Yan, Y. (1990). The collocation method for first kind boundary integral equations on polygonal domains. *Math. of Comp.*, **54**, 139–154.

90. Yan, Y. and Sloan, I. (1988). On integral equations of the first kind with logarithmic kernels. *J. Integral Eqns and Applics*, **1**, 517–548.

Aspects of Approximation with Emphasis on the Univariate Case

G.A. Watson

Department of Mathematics and Computer Science, University of Dundee

Abstract

A development which has made a huge impact in recent years is that of wavelets, and an introduction is given to this important topic. We also consider some progress in approximation to data by models which contain free parameters, with particular emphasis on the use of the l_1, l_2 and l_∞ norms. Traditionally the favoured methods for linear l_1 and l_∞ problems have had a close relationship with the simplex method of linear programming. However, more recently, other approaches have been proposed, in particular methods which try to smooth the problems. We look in detail at two such methods for l_1 problems, one based on an interior point philosophy, the other based on solving a sequence of Huber M-estimation problems. We also consider some classes of problems, both linear and nonlinear, where there is structure which can be exploited. These arise mainly from a relaxation of conventional assumptions about errors.

1 Introduction

This paper focusses on some aspects of approximation which are seen to be important from the numerical analysis point of view, and which have seen activity in the last 10 years. Its emphasis is on univariate approximation, which is an extensive area, and many topics could be covered. Perhaps the development which has stimulated most interest over this period is that of wavelets. This subject has expanded extremely rapidly, there have been many theoretical advances, and it has been the focus of many potential and actual applications. It is clearly not possible to do justice to all of this here, particularly for a non-specialist audience: we content ourselves by providing an introduction to the topic which we hope will stimulate further interest. Of course, wavelets can also be multivariate, although we will not consider those here. For some approximation problems, the number of independent variables is actually not very important, and this is certainly true of some of the topics considered in the rest of the paper. However, what they all have in common is that they are not inherently multivariate, and so do not fit naturally into that category.

Wavelets provide a particular set of basis functions, but there are many others which may also be used, depending on the specific application. A fundamental problem in approximation is that of approximating a given function by a finite linear combination of given functions, over a given finite region. There are many different ways in which the coefficients can be calculated, depending on the appropriate criterion. For example, if all functions are continuous, a classical approach is to use the Chebyshev norm, a situation when it is important whether or not the functions are univariate.

Problems set on regions may be discretised (so that the region is sampled at a finite set of points), and this process gives rise to another fundamental problem in approximation, typified by the solution of an overdetermined system of linear equations. Such problems also arise naturally in data fitting or regression, where particular mathematical models are to be fitted to observed discrete data, and we will be particularly concerned with this situation. Of course, if the number of unknown coefficients is exactly equal to the number of data points, the system of equations is no longer overdetermined, and the resulting problem is an interpolation problem. It is frequently the case, however, that much more data are available, these data contain errors, and there is no obvious reason to choose a particular subset. It is then appropriate to make use of all the data, and obtain values of the unknowns by solving a best approximation problem, for example, where a norm is to be minimized. The nature of such problems is largely unaffected by the number of independent variables, and the sort of fitting criterion to be used depends very much on the kind of assumptions which can be made about errors in the data. The use of the least squares norm has always been popular, but other norms, in particular the l_∞ and l_1 norms, have attracted much interest.

Increasing attention has been devoted to so-called robust estimators, which can give less weight to gross errors or wild points in the data. These estimators are in fact not always norms, and for example the Huber M-estimator is a valuable tool. However, an important role is played by the l_1 norm, and we consider here some recent developments for linear problems. Of particular interest is the relationship between the Huber M-estimator and the l_1 norm: this provides a smoothing technique which it is claimed can increase computational efficiency over more conventional methods. We also point out a duality property of the Huber problem.

An underlying assumption in all of this is that the errors are only in one variable (the dependent variable), and any other variable values are exact. There are many situations when this is a perfectly valid assumption, but there are also situations when it is not appropriate, and then novel fitting techniques have to be used. Over the last ten years there has been increasing interest in such situations, typified by the idea of total least squares. As Gene Golub says in the foreword to [51], this "represents a technique that synthesizes statistical and numerical methodologies for solving problems arising in many application areas". The efficient solution of the problem as originally posed involves the singular value decomposition, and is essentially a problem in numerical linear algebra.

Recent developments have involved the use of other norms, and the treatment of problems having special structure, and these require techniques from approximation and optimization. It is quite natural that many developments in numerical methods for approximation problems continue to be greatly influenced by methods from optimization, and this is reflected in some of the material of this paper.

Of course mathematical models need not be linear in the free parameters, and considerable attention has also been paid to nonlinear approximation problems. Good methods for small dense problems have been available for some time, and recent interest has mostly been in the treatment of larger problems, possibly with special structure. One class of problems having these features is the nonlinear analogue of the problems mentioned in the previous paragraph, the so-called errors-in-variables problems. Then there is inherent structure which can be exploited, and some recent developments are also included here.

Among the many topics not included is interpolation, and there is no treatment of splines. It could perhaps be argued that most recent attention here has been paid to the multivariate case. However, the recent volume by Späth [46] deals comprehensively with the computation of one-dimensional spline functions which are determined by the requirement of smooth and shape preserving interpolation, and includes algorithms and Fortran subroutines. There is also a large volume of work on orthogonal polynomials and other special functions not included here; see for example [2]. Also not included is complex approximation: in a sense that is a naturally bi-variate problem.

2 Wavelets

Consider the problem of providing an orthogonal basis for $L_2(R)$, the space of square integrable functions defined on the real line R. A wavelet can be thought of as a function, defined on the real line R, which, when acted on in a certain way, generates a sequence of functions which provide an orthogonal basis for $L_2(R)$. The resulting representation is analogous to the Fourier series

$$f(x) = a_0/2 + \sum_{k=1}^{\infty} (a_k \cos kx + b_k \sin kx). \tag{2.1}$$

This has long been recognised as an extremely valuable way of representing a continuous function. However, a disadvantage is that it is not local: evaluation of the right hand side (or a finite number of terms) at a particular point involves all terms of the sum, or put another way, the Fourier matrix is a full matrix. This disadvantage would disappear if there was decay (exponential or polynomial) of the basis functions, or if they had compact support, with their support confined to a small interval of the real line, in other words if the basis was local. The resulting matrix would of course then be sparse. The use of a wavelet generator normally gives an orthonormal basis of functions which has compact support, or decays exponentially.

The term "wavelet" (in French "ondelette") was introduced by J Morlet. Wavelets may be defined in more than one dimension, but here we will confine attention to the univariate case. In common with most elementary introductions to the topic, we will start from the simplest set of such functions, the Haar wavelets. It is difficult to do otherwise, and still make the material digestible to a non-specialist audience. The properties of Haar wavelets are relatively easy to understand, and they point the way to the development of more sophisticated and general objects.

2.1 Haar wavelets

Define $w(x)$ by

$$w(x) = \begin{cases} 0 & \text{when } x < 0 \\ 1 & \text{when } 0 \leq x < 1/2 \\ -1 & \text{when } 1/2 \leq x < 1 \\ 0 & \text{when } 1 \leq x. \end{cases}$$

This is just a particular Haar function. Then it is easy to see that if we define the inner product on functions u, v in $L_2(R)$

$$< u(x), v(x) > = \int_R u(x)v(x)dx, \tag{2.2}$$

then for example

$$< w(x), w(2x) > = < w(x), w(2x-1) > = < w(2x), w(2x-1) > = 0, \tag{2.3}$$

using compact support and symmetry. Extending this result in a straightforward way, we have that for all integers j, k, the function $w(2^j x - k)$ has support on the interval $[k2^{-j},\ (k+1)2^{-j}]$, and further for any integers r, s with $j \neq r$, or $k \neq s$,

$$< w(2^j x - k),\ w(2^r x - s) > = 0.$$

If we define

$$w_{jk}(x) = 2^{j/2} w(2^j x - k), \text{for all } j, k \in Z, \tag{2.4}$$

where Z denotes the set of integers, then it follows that

$$< w_{jk}(x),\ w_{rs}(x) > = \delta_{jr}\delta_{ks}, \tag{2.5}$$

where δ denotes the Kronecker delta; in other words the functions form an orthonormal system. The multiplicative factor on the right hand side of (2.4) just adds normality to orthogonality. Thus we have produced an orthonormal system of compactly supported functions. The question remains: is this a basis for $L_2(R)$? The answer is in fact "yes", and we will give an indication of the proof, since it points to how more general situations may be handled. Indeed, the above process encapsulates the spirit of more general orthogonal wavelets: they are functions $w(x)$, which, when subjected to the fundamental operations which appear in (2.4) (that is, multiplication of the argument by powers of 2, and integer shifts), produce an orthogonal basis for $L_2(R)$.

Consider the function $\phi(x)$, defined by

$$\phi(x) = \begin{cases} 0 & \text{when } x < 0 \\ 1 & \text{when } 0 \le x < 1 \\ 0 & \text{when } 1 \le x. \end{cases}$$

Then $w(x)$ can readily be expressed in terms of $\phi(x)$ by the relation

$$w(x) = \phi(2x) - \phi(2x-1).$$

$\phi(x)$ has two further simple and immediate properties:

(i) $\{\phi(x-k),\ k \in Z\}$ is an orthonormal system (or $\phi(x)$ has orthonormal shifts)

(ii) $\phi(x) = \phi(2x) + \phi(2x-1)$ (or $\phi(x)$ is a scaling function).

Now define the sequence of spaces

$$S_0 = \text{span}\{\phi(x-k),\ k \in Z\}$$
$$S_1 = \text{span}\{\phi(2x-k),\ k \in Z\}$$
$$S_2 = \text{span}\{\phi(2^2 x-k),\ k \in Z\}$$

......

Then clearly S_j has "jumps" at $\frac{1}{j+1}Z$, and further

$$\ldots \subset S_0 \subset S_1 \subset S_2 \ldots$$

Two further properties are readily seen,

(i) S_j has an orthonormal basis $\{\phi_{jk}(x) = 2^{j/2}\phi(2^j x - k),\ k \in Z\}$,

(ii) $f(x) \in S_j$ if and only if $f(2^{k-j}x) \in S_k$.

This sequence of spaces is an example of a *multiresolution analysis.* There are two other significant properties which may be established:

$$\cap_{j \in Z} S_j = \{0\}, \quad \overline{\cup_{j \in Z} S_j} = L_2(R), \tag{2.6}$$

where we have used standard notation for intersection, union and closure. It might be supposed that a basis for $L_2(R)$ could easily be obtained by putting together all the functions $\{\phi_{jk}(x),\ j,k \in Z\}$: it can readily be shown that this does not work. Consider, however, the following question: given that we have an orthonormal basis for S_j, what must we add to complete an orthonormal basis for S_{j+1}? Clearly, we need a set of functions in S_{j+1}, orthogonal to those in S_j, which form a basis for W_j say, where

$$W_j \oplus S_j = S_{j+1},$$

and the operation on the left hand side is the orthogonal sum. Thus W_j is just the orthogonal complement of S_j in S_{j+1}. It is not difficult to show that a suitable basis for W_j is given by $\{w_{jk}(x),\ k \in Z\}$: they are elements of S_{j+1}, they are orthogonal to elements of S_j, and they span W_j. W_j is referred to as a wavelet space. Further for any j, $i < j$,

$$S_{j+1} = S_i \oplus W_i \oplus W_{i+1} \oplus \ldots \oplus W_j.$$

Thus from (2.6), $L_2(R)$ is just the orthogonal sum of all the wavelet spaces, and so the totality of the basis functions for all these spaces forms a (mutually orthonormal) basis for $L_2(R)$. In other words, $\{w_{jk}(x), \quad j,k \in Z\}$ is an orthonormal basis for $L_2(R)$.

2.2 More general wavelets

The steps outlined above are those which can be used to generate a basis of orthogonal wavelets. The original idea is due to Mallat [38] and we can summarise as follows

(1) Choose a scaling function $\phi(x)$ with orthonormal shifts.

(2) Generate a multiresolution analysis $\ldots \subset S_j \subset S_{j+1} \ldots$ such that (2.6) is satisfied.

(3) Form the wavelet spaces $W_j : W_j \oplus S_j = S_{j+1}$, with wavelet generator $w(x)$ such that $\{w_{jk}(x), \quad k \in Z\}$ formed as in (2.4) is a basis for W_j.

Then $\{w_{jk}(x), \quad j,k \in Z\}$ is a wavelet basis for $L_2(R)$.

The smoothness properties of $\phi(x)$ are passed on to the wavelet generator $w(x)$, and of course the Haar wavelet is not continuous. Thus it is of limited practical value, and the question arises as to whether smooth compactly supported wavelets can be constructed. The key is clearly to determine a suitable scaling function $\phi(x)$, after which an associated wavelet function can be obtained. Such functions are rather special, and their discovery has been an important research area. The above question was in fact answered in the affirmative by Daubechies [16] who showed that for any given order of smoothness, scaling functions leading to orthogonal wavelets of finite support can be constructed.

The Haar wavelet is in fact the only wavelet that is compactly supported, orthogonal and also symmetric. Further, the difficulty in constructing convenient scaling functions has lead to various weakenings of the orthogonality condition (2.5). For example wavelets are said to be semi-orthogonal if

$$< w_{jk}(x), \; w_{rs}(x) > = 0, \; j \neq r. \tag{2.7}$$

The resulting wavelet is referred to as a pre-wavelet. If wavelets do not satisfy (2.7), they are said to be non-orthogonal. These weaker conditions are useful in constructing compactly supported spline wavelets, based on spline scaling functions. The method outlined above for constructing wavelets may still be followed, even if the scaling function does not have orthonormal shifts.

An example of a specific scaling function $\phi(x)$ is given by taking the rth order cardinal B-spline (that is, the rth order B-spline with simple knots at consecutive integers). This is in fact a generalization of the function $\phi(x)$ introduced earlier, which is the special case $r = 1$. The support of the B-spline is of length r, and

Chui and Wang [11] construct compactly supported pre-wavelets (or B-wavelets) with support of length $(2r-1)$. The support is minimal among all functions in the wavelet space. Taking the orthonormalized shifts of the cardinal B-spline gives orthogonal wavelets which have smoothness C^{r-2} [6]. Unfortunately, the support of $w(x)$ is the whole of R, although there is exponential decay.

The popularity of splines as scaling functions has resulted in various generalizations. For example, Goodman [25] constructs compactly supported spline wavelets with knots of multiplicity r (analogous to B-splines with knots of multilplicity r) by further weakening the orthogonality requirement. Another general class of scaling functions is due to Meyer [39], which are C^{∞}, with noncompact support and polynomial decay.

2.3 Fast wavelet transforms

Suppose that $w_{jk}(x)$, $j,k \in Z$ is an orthonormal wavelet system for $L_2(R)$, and consider the representation of $f(x) \in L_2(R)$:

$$f(x) = \sum_{j,k \in Z} < f, w_{jk}(x) > w_{jk}(x).$$

In practice we would restrict attention to a finite number of terms of the series, appropriate for the accuracy required.

Let us return for a moment to the Fourier series (2.1). The Fast Fourier Transform is a powerful technique which enables 2^n coefficients of the Fourier series to be computed in $O(n \log_2 n)$ operations. The key is to obtain relationships between the coefficients. There is a precisely analogous process in the wavelet expansion case. We do not need formulae for $\phi(x)$ or $w(x)$, but can work directly with the coefficients

$$< f,\ w_{jk}(x) > = \alpha_{jk}, \ \text{say}.$$

Let

$$< f,\ \phi_{jk}(x) > = \beta_{jk}, \ \text{say},$$

and suppose that we have knowledge of the level n coefficients β_{jn}, $j \in Z$. Then simple formulae exist which connect β_{jn} to both α_{jn} and $\beta_{j,n-1}$. Thus we can use the β values as a means of computing the α values in a systematic way. A total of 2^n coefficients can be computed in $O(2^n)$ operations. For the Haar wavelets each new calculation requires only one multiplication and one addition.

The extent to which wavelets can be used in conjunction with efficient numerical calculation has led to their use in lots of areas, for example signal processing, and data and image compression. It is beyond the scope of this article to pursue this here.

Before leaving the topic of wavelets, we mention some references [10], [17], [20], [47], [49], [28], [32], [48], [27] which are intended not to be too demanding for the non-specialist.

3 Approximation by linear models

The previous Section was concerned with the provision of a particular class of basis functions suitable for approximation. In this Section the approximating function will still be a linear combination of functions, but we will not be particularly concerned with what the basis is; rather we will focus on numerical methods for calculating approximations, or equivalently calculating suitable values of the unknown coefficients. As already indicated, an important class of approximation problems arises when the problem data are only available as observations or readings obtained experimentally. The approximating function can then be thought of as an appropriate mathematical model: the choice of that model in a given situation is obviously an extremely important one, although we will not deal here with that aspect.

The basic problem to be considered in this Section is therefore as follows: given (x_i, y_i), $i = 1, \ldots, n$, and a linear model

$$y = \sum_{j=1}^{p} a_j \phi_j(x), \tag{3.1}$$

where $\phi_j(x)$, $j = 1, \ldots, p < n$ are given functions, find values of the parameters $\mathbf{a} \in R^p$ to give a satisfactory fit. The nature of the errors which are invariably present in such a process plays an important role in deciding the criterion to be used for estimating the parameters, and we will consider first the case when the errors are only present in the values of y_i, $i = 1, \ldots, n$. Letting these errors form the vector $\mathbf{r} \in R^n$, then the components satisfy

$$r_i = \sum_{j=1}^{n} a_j \phi_j(x_i) - y_i, \; i = 1, \ldots .n,$$

and it is appropriate to minimise some function of $\mathbf{r}$.

There is wide recognition of the importance of the l_1 norm in the data fitting context. In contrast to the use of the l_2 or l_∞ norms, it has a property of robustness with respect to wild points or gross errors in the data. It is widely believed that the most effective algorithms for computing best l_1 (and l_∞) approximations are based on exploiting the fact that the problems can be expressed as linear programming problems, and using variants of the simplex method, or essentially equivalent methods. For example the method of Barrodale and Roberts [3] is universally available for l_1 problems, and is frequently cited as the algorithm of choice, although the method of [5] is competitive. For some relevant observations on l_∞ problems, see [4].

We give here an indication of two other approaches which have been proposed recently for l_1 problems. One of these [14] is motivated by the fact that in recent times the simplex method has been shifted somewhat from its pedestal by polynomial time interior point methods which can achieve actual efficiency gains. The question arises: is there a way in which this idea can be used in the l_1 case?

We consider here a recent attempt to answer this question, with a procedure which is in the spirit of such methods (although not actually an interior point method, since the problem has no interior). It is based on the principle of "affine scaling", and attempts to tailor a method of this kind to the l_1 case. We then examine another approach based on solving a sequence of Huber problems.

A method for the linear discrete l_∞ problem which is in the spirit of [14] is given in [15]. Other recent developments in methods for such problems are row relaxation or proximal point methods (for example [18], [19]), which may have potential for large sparse problems.

3.1 An affine scaling method for the l_1 problem

Consider the usual discrete linear l_1 problem: given a matrix $A \in R^{n\times p}$ and a vector $\mathbf{b} \in R^n$, find $\mathbf{x} \in R^p$ to minimise $\sum_{i=1}^n |r_i|$ where

$$\mathbf{r} = A\mathbf{x} - \mathbf{b}. \tag{3.2}$$

We have reverted here to a standard notation for such problems (hopefully without confusion) where $\mathbf{x}$ takes on the role of $\mathbf{a}$ as the vector of unknowns, and $\mathbf{b}$ is the vector of values of y_i, $i = 1, \ldots, n$. This will last throughout the rest of this Section. We will assume that $p < n$ and (for convenience) that rank$(A) = p$. For any $\mathbf{x} \in R^p$, let I denote the set of indices where $r_i = 0$, and let I^C denote its complement. Then it is well known [52] that $\mathbf{x}$ is a solution if and only if there exists $\lambda \in R^n$ with $\lambda_i = \text{sign}(r_i)$, $i \in I^C$, $|\lambda_i| \leq 1$, $i \in I$ such that

$$A^T\lambda = 0. \tag{3.3}$$

Let the QR decomposition of A be given by

$$A = YR,$$

where R is $p \times p$ upper triangular, and $[Y : Z]$ is a $n \times n$ orthogonal matrix. Then (3.3) is equivalent to

$$\lambda = Z\mathbf{w}, \tag{3.4}$$

where $\mathbf{w} \in R^{n-p}$, and (3.2) is equivalent to

$$Z^T(\mathbf{r} + \mathbf{b}) = 0. \tag{3.5}$$

Define D_r to be the diagonal matrix with (i,i) element r_i, and let $\mathbf{g} \in R^n$ be the vector defined by

$$g_i = \text{sign}(r_i), \quad i = 1, \ldots, n.$$

It follows that $\mathbf{x}$ solves the l_1 problem if and only if there exists $\mathbf{r} \in R^n$, $\mathbf{w} \in R^{n-p}$ such that

$$D_r(\mathbf{g} - Z\mathbf{w}) = 0, \tag{3.6}$$

$$Z^T(\mathbf{r}+\mathbf{b}) = 0, \tag{3.7}$$

and in addition

$$-1 \le (Z\mathbf{w})_i \le 1, \quad i = 1,\ldots,n. \tag{3.8}$$

An affine scaling method attempts to solve (3.6),(3.7), (3.8) by an iterative method which computes a direction of progress from the current vector $\mathbf{r}$ by solving the following subproblem:

$$\text{minimize}_{\mathbf{d}\in R^n} \ \mathbf{g}^T\mathbf{d}$$

$$\text{subject to } Z^T\mathbf{d} = 0,$$

$$||D^{-1}\mathbf{d}||_2 \le \tau, \tag{3.9}$$

where D is a given positive definite diagonal matrix, and τ is a given positive number which restricts the size of $\mathbf{d}$: (3.9) can be thought of as scaling the solution. It is a straightforward exercise to show that $\mathbf{d}^* = \alpha\mathbf{d}$ is the solution, where α is a suitably chosen scalar and $\mathbf{d}$ is given by

$$\mathbf{d} = -A(A^TD^{-2}A)^{-1}A^T\mathbf{g}. \tag{3.10}$$

Assume that at the current $\mathbf{r}$, (3.7) is satisfied (and so remains satisfied for subsequent $\mathbf{r}$), and also $I^C = \phi$. Then $\mathbf{g}$ is just the gradient of $||\mathbf{r}||_1$ at the current point, and it follows that the solution $\mathbf{d}$ is a descent direction for the l_1 norm (since $\mathbf{d} = 0$ implies that $A^T\mathbf{g} = 0$). A linesearch to minimize the piecewise linear function $||\mathbf{r} + \alpha\mathbf{d}||$ with respect to α can readily be incorporated, and if we stop short of the optimal step length, then we can start again from a point with $I^C = \phi$. Note that no explicit value of τ is actually required.

Now consider the alternative of applying Newton's method to (3.6) and (3.7). The Newton step in $\mathbf{r}$ and $\mathbf{w}$ is given by solving the system

$$\begin{bmatrix} D_\lambda & -D_rZ \\ Z^T & 0 \end{bmatrix}\begin{bmatrix} \delta\mathbf{r} \\ \delta\mathbf{w} \end{bmatrix}\begin{bmatrix} -D_r(\mathbf{g}-Z\mathbf{w}) \\ 0 \end{bmatrix}. \tag{3.11}$$

It follows from this system that

$$\delta\mathbf{r} = -A(A^TD_r^{-1}D_\lambda A)^{-1}A^T\mathbf{g}, \tag{3.12}$$

where $D_\lambda = \text{diag}\{g_i - (Z\mathbf{w})_i, \ i = 1,\ldots,n\}$. Note that (3.10) and (3.12) have the same general form. Let D be chosen by

$$D = \text{diag}\{|r_i|^{1/2}, \ i = 1,\ldots,n\},$$

and write

$$\delta_\theta\mathbf{r} = -A(A^TW_\theta A)^{-1}A^T\mathbf{g}, \tag{3.13}$$

where

$$W_\theta = \text{diag}\{|r_i^{-1}(g_i - (1-\theta)(Z\mathbf{w})_i)|, \ i = 1,\ldots,n\}.$$

Then the choice $\theta = 1$ in (3.13) gives (3.10), and the choice $\theta = 0$ gives (3.12), provided that $D_r^{-1} D_\lambda$ is positive definite. That this is true for a set of points $\mathbf{r}$ arbitrarily close to a solution but having no component of $\mathbf{r}$ zero is a key observation in the hybrid method of Coleman and Li [14]. By suitable choice of θ at each iteration, and working with (3.13), they develop a method which is globally convergent, with a quadratic convergence rate in non-degenerate problems (essentially unique λ in (3.3)). Note that (3.13) may be solved by first calculating the l_2 solution of the system

$$W_\theta^{1/2} A \, \mathbf{d}_1 \; = \; W_\theta^{-1/2} \mathbf{g},$$

followed by setting

$$\delta_\theta \mathbf{r} \; = \; -A \mathbf{d}_1. \tag{3.14}$$

The new value of $\mathbf{r}$ is then obtained by a linesearch in the direction $\delta_\theta \mathbf{r}$. To obtain a new value of λ, it helps to observe that use of the hybrid method corresponds to solving the system analogous to (3.11), viz

$$\begin{bmatrix} W_\theta & -Z \\ Z^T & 0 \end{bmatrix} \begin{bmatrix} \delta_\theta \mathbf{r} \\ \delta \mathbf{w} \end{bmatrix} \begin{bmatrix} -(\mathbf{g} - Z\mathbf{w}) \\ 0 \end{bmatrix}. \tag{3.15}$$

It follows from this that the current value of λ can be updated to the value

$$Z(\mathbf{w} \, + \, \delta \mathbf{w}) \; = \; \mathbf{g} \, + \, W_\theta \delta_\theta \mathbf{r},$$

with $\delta_\theta \mathbf{r}$ given by (3.14). Of course, (3.3) remains satisfied.

Although Z appears as an aid to the theoretical development of the method, note that its actual computation is unnecessary, and we need only work with $\mathbf{r}$ and λ. An initial approximation may be obtained by choosing the $\mathbf{r}$ satisfying (3.2) of minimum l_2 norm, and taking λ to be a multiple of that $\mathbf{r}$.

It may appear that difficulties are inevitable in practice as components of $\mathbf{r}$ tend to zero. However, it is shown in [14] that theoretically this is not a problem, and neither is it in practice if the method is implemented carefully. Numerical results are given which show improvement over an earlier attempt to provide a method based on an interior point approach. Although comparisons do not seem to be available with other methods, the affine scaling method seems promising for large problems, since it appears to be insensitive to problem size.

3.2 A method for l_1 problems based on the Huber M-estimator

Another method which attempts to smooth the l_1 problem has been developed recently in [36]. It is based on the use of the Huber M-estimator, defined by

$$\psi_\gamma \; \equiv \; \psi_\gamma(\mathbf{r}) = \sum_{i=1}^{n} \rho(r_i), \tag{3.16}$$

where

$$\rho(t) = \begin{cases} t^2/2, & |t| \le \gamma \\ \gamma(|t| - \gamma/2), & |t| > \gamma, \end{cases} \tag{3.17}$$

$\mathbf{r}$ is given by (3.2) and γ is a scale factor or tuning constant. The function (3.16) is convex and once continuously differentiable, but has discontinuous second derivatives at points where $|r_i| = \gamma$. Clearly if γ is chosen large enough, then ψ_γ is just the least squares function; in addition if γ tends to zero, then limit points of the set of solutions minimize the l_1 norm [12]. It is the latter property which concerns us here. Let a partition be defined by an index set σ and its complement σ^c as follows:

$$\sigma = \{i : |r_i| \leq \gamma\}, \quad \sigma \cup \sigma^c = \{1, 2, \cdots, n\}. \tag{3.18}$$

Then the system of equations determined by the necessary conditions for $\mathbf{x}$ to be a solution is

$$A_\sigma^T A_\sigma \mathbf{x} = A_\sigma^T \mathbf{b}_\sigma - \gamma \sum_{i \in \sigma^c} g_i \mathbf{a}_i, \tag{3.19}$$

where $\mathbf{a}_i^T$ denotes the ith row of A, $g_i = \text{sign}(r_i)$ as before, A_σ is obtained from A by deleting rows corresponding to indices $i \in \sigma^c$, and $\mathbf{b}_\sigma$ is defined similarly. It has been suggested ([36], [35]) that the preferred method for solving the l_1 problem is via a sequence of Huber problems for a sequence of scale values $\gamma \to 0$. This algorithmic development has lead to increased interest in the relationship between the Huber M-estimator and the l_1 problem ([37], [35]).

For given γ, the Huber problem can be solved by Newton's method, or a variant, using a line search. For given $\mathbf{x} \in R^p$, define W as a diagonal matrix with elements 1 if $|r_i| \leq \gamma$ and 0 otherwise. Then, assuming that no value of $|r_i|$ is equal to γ, and letting $\mathbf{s} \in R^n$ be defined by

$$s_i = \begin{cases} 0, & i \in \sigma \\ \text{sign}(r_i), & i \in \sigma^c, \end{cases}$$

it is easily seen by differentiating the Huber function that ψ is minimized if and only if

$$A^T \left[\frac{1}{\gamma} W\mathbf{r} + \mathbf{s}\right] = 0.$$

The formal Newton step $\mathbf{d}$ for solving this system of equations ignores the discontinuity in derivative. It satisfies

$$\frac{1}{\gamma} A^T W A \mathbf{d} = -A^T \left[\frac{1}{\gamma} W\mathbf{r} + \mathbf{s}\right], \tag{3.20}$$

or

$$A^T W A \mathbf{d} \;=\; -A^T(W\mathbf{r} \;+\; \gamma\mathbf{s}), \tag{3.21}$$

If A has full rank p then the rank of W can always be taken as $\geq p$ at the solution of the M–estimation problem [42]. However, this does not ensure that W has rank $\geq p$ in a step of the Newton iteration. Problems with singularity of the linear system can be avoided by inserting additional 1's into the diagonal positions of W or indeed the unit matrix can be used. The solution to (3.21) is most efficiently obtained through LU factorization of the matrix on the left hand side; one step of iterative refinement is recommended in [36]. A line search may be needed to ensure descent.

As the iteration proceeds, the partition σ will change, until the partition valid at the solution is obtained. Then the iteration terminates in one further Newton step (because a quadratic is being minimised). Changes in the partition translate into changes in W (and $\mathbf{s}$), and corresponding changes to the LU factors of $A^T WA$. It is here that the efficiency of the algorithm is achieved, because changes of a single index in the partition mean a rank 1 change so that updating of the factors is all that is required, a typical iteration costing $O(n^2)$ operations. Changes of more than one index need refactorization, at a cost of $O(n^3)$ operations. Once a solution has been obtained for a particular γ, the value of γ is reduced and the process repeated, using warm starts. Again only a rank 1 change to $A^T WA$ may be needed.

A key observation here is that it is not necessary to let γ go to zero, but the method can be terminated at a non-zero value. The relevant result is as follows (see [13], [36]). Let $\mathbf{x}_\gamma$ minimize ψ_γ, and let $\mathbf{v}$ be defined by

$$\gamma A^T WA\mathbf{v} = -A^T W\mathbf{r}, \tag{3.22}$$

with variable quantities evaluated at $\mathbf{x}_\gamma$.

Theorem 1. *Suppose that there is no change to* $\mathbf{s}$ *(or* W*) for* $0 < \delta < \gamma$*. Then* $\mathbf{x}_\gamma + \gamma\mathbf{v}$ *solves the* l_1 *problem.*

Note that the matrix on the left hand side of (3.22) is such that no new factorization is needed to compute $\mathbf{v}$. Numerical results given in [36] show that careful implementation of the method can make it superior to the algorithm of [3]. However, it is argued in [43] that wider implications of the comparisons should be treated with some caution:

- Firstly, more efficient implementations of the simplex based methods, with careful attention to line search performance and scaling, are available for solving the l_1 problem [7].
- Secondly, the larger scale calculations using randomly generated data reported in [36] are carried out only for $n/p = 2$. This does not seem appropriate for many problems of practical interest.
- Thirdly, it is legitimate to question conclusions as to which of the two approaches to solving the l_1 problem is superior based on the results of random simulations: such problems are easy to generate, but they are unlikely to have a great deal of relevance to actual practice.

Thus although the approach seems to be a promising one, further investigation would be valuable.

3.3 Huber M-estimator duality

Another recent development of interest which has links with the material in the previous Section is the recognition that the Huber M-estimator problem has a simple dual. This means that a single Huber problem (3.16), (3.17) can easily be solved using standard software. The fact that (3.16), (3.17) gives a measure which can be interpreted as a compromise between the least squares and the l_1 norm has helped to make it popular in its own right, not only in the case where γ is given, but also where γ has to be obtained as part of the calculation.

The duality result seems to be due to Michelot and Bougeard [40]. There it is shown that there is a simple interval bound quadratic program dual to the M–estimation problem, namely:

$$\text{minimize } \frac{\gamma}{2}\mathbf{y}^T\mathbf{y} - \mathbf{b}^T\mathbf{y} \quad \text{subject to}$$

$$\begin{aligned} A^T\mathbf{y} &= 0, \\ -\mathbf{e} \le \mathbf{y} &\le \mathbf{e}, \end{aligned} \tag{3.23}$$

where $\mathbf{e} = (1, 1, \ldots, 1)^T$. This result is readily verified by computing the Kuhn–Tucker conditions. These give

$$\begin{aligned} \gamma\mathbf{y} - \mathbf{b} &= A\mathbf{u} + \mathbf{v} - \mathbf{w}, \quad \mathbf{v},\ \mathbf{w} \ge 0, \\ \mathbf{v}^T(\mathbf{y} + \mathbf{e}) &= 0 \\ \mathbf{w}^T(\mathbf{e} - \mathbf{y}) &= 0, \end{aligned}$$

where $\mathbf{u}$, $\mathbf{v}$, $\mathbf{w}$ are Lagrange multipliers. Let $\mathbf{x} = -\mathbf{u}$. Then we have

$$\mathbf{r} = A\mathbf{x} - \mathbf{b} = \mathbf{v} - \mathbf{w} - \gamma\mathbf{y}.$$

There are 3 cases to consider.

1) $i : v_i = w_i = 0$, so that $r_i = -\gamma y_i$, or $|r_i| \le \gamma$,
2) $i : v_i > 0,\ w_i = 0$, so that $r_i = v_i + \gamma$, or $r_i > \gamma$,
3) $i : v_i = 0,\ w_i > 0$, so that $r_i = -w_i - \gamma$, or $r_i < -\gamma$.

With σ defined by (3.18), we immediately have from (3.23) that (3.19) is satisfied, so that $\mathbf{x}$ solves (3.16), (3.17). Further, a solution of the M–estimation problem $\mathbf{x}$ and corresponding partition σ permits a solution of the dual quadratic program to be written down directly. Note that if $\gamma = 0$, then the above quadratic programming problem is just the dual of the l_1 problem. There may be no advantage in using duality if a sequence of solutions is required (as above), but for a particular solution with a particular value of γ, the idea may be attractive.

An extension of this result is to bounded variable Huber M-estimation problems. For example consider the problem

$$\text{minimize } \psi_\gamma(\mathbf{r})$$

$$\text{subject to } \|\mathbf{x}\|_\infty \le \tau,$$

where $\tau > 0$ is a given constant. This problem can arise as a subproblem to be solved in a trust region method for solving nonlinear Huber M-estimation problems. An additional $2p$ variables $\xi \in R^p$, $\eta \in R^p$ are required, and it is readily seen that this problem has the dual:

$$\text{minimize } \frac{\gamma}{2}\mathbf{y}^T\mathbf{y} - \mathbf{b}^T\mathbf{y} + \tau\mathbf{e}^T\xi + \tau\mathbf{e}^T\eta \text{ subject to}$$

$$A^T\mathbf{y} + \xi - \eta = 0$$
$$-\mathbf{e} \le \mathbf{y} \le \mathbf{e},$$
$$\xi \ge 0, \quad \eta \ge 0.$$

3.4 Total least norm problems

The assumption in methods based on the model equations (3.2) that errors are only present in the observed dependent variable values forming the vector $\mathbf{b}$ and that the independent variable values (which generate the matrix A) can be assumed to be error free is often perfectly reasonable. However, for many problems, it is an over-simplification of the situation and can lead to bias in the estimator. A technique which has been suggested for dealing with this is to allow for errors in A as well as in $\mathbf{b}$ so that the model equations become

$$\mathbf{r} = (A + B)\mathbf{x} - \mathbf{b}, \tag{3.24}$$

and it is appropriate to minimise some norm of the matrix $[B : \mathbf{r}]$. If the norm is the Frobenius norm, then we have the total least squares problem, a name originating in the paper by Golub and Van Loan [24]. The multiplication of $\mathbf{x}$ by B actually makes this a nonlinear problem, but it is included here because it originates from a linear problem.

Applications arise in systems identification, frequency estimation, superresolution, control theory etc, and since the early work mentioned above, contributions dealing with techniques, analysis have been made by many authors. The book [51] is an excellent source to 1991. A follow-on book (which will also include nonlinear problems - see Section 4.1) will appear, based on papers presented at the "Second International Workshop on Total Least Squares and Errors-in-Variables Modelling", held in Leuven in August 1996. Recent developments relevant to this paper have been in the use of other norms, and in the maintenance of structure of A in $A + B$ (for example Hankel, Toeplitz, Vandermonde, or sparse matrices, see for example [1], [41], [50], [44], [45], [34]). The second and third last of these papers are concerned with retaining given structure, and permitting the use of other norms, in particular the l_1 and l_∞ norms. We will illustrate this is the important special case when A has Toeplitz structure, which must be preserved, as occurs in system identification problems [41], and frequency estimation [1]. Then

$$A_{ij} = \alpha_{p+i-j}, \; i = 1, \ldots, n, \;\; j = 1, \ldots, p,$$

so that we can define A entirely by its first row

$$\rho_1(A) = [\alpha_p, \alpha_{p-1}, \ldots, \alpha_1],$$

and its first column

$$\kappa_1(A) = [\alpha_p, \alpha_{p+1}, \ldots, \alpha_{p+n-1}]^T.$$

Thus in (3.24) we can define B by the same form, with unique unknown elements say β_i, $i = 1, \ldots, p+n-1$. Assume that the matrix norm is an l_p norm defined for any matrix M by

$$\|M\| = (\sum |m_{ij}|^p)^{1/p}, \quad 1 \leq p \leq \infty.$$

Then

$$\|B : \mathbf{r}\| = \left\| \begin{array}{c} \mathbf{r}(\beta, \mathbf{x}) \\ W\beta \end{array} \right\| \tag{3.25}$$

where the vector norm is the l_p norm, where

$$\mathbf{r}(\beta, \mathbf{x}) = A\mathbf{x} + B\mathbf{x} - \mathbf{b},$$

and where W is a $(n+p-1) \times (n+p-1)$ diagonal weighting matrix which accounts for repetitions of elements in B. The problem (3.25) is of course nonlinear in β and $\mathbf{x}$, and methods for the l_1, l_2 and l_∞ norms based on linearization are given in [44]. Extensions to problems where A depends nonlinearly on parameters which have to be estimated are given in [45]. The techniques are appropriate for problems which arise in fitting by models which are nonlinear in the free parameters, and so we are lead naturally on to a consideration of that class of problems.

4 Approximation by nonlinear models

Generally speaking, the univariate data fitting problem is as follows: given observations y_i at points x_i, for $i = 1, \ldots, n$, and a model

$$y = f(x, \mathbf{a}), \tag{4.1}$$

which depends on p parameters forming the vector $\mathbf{a} \in R^p$, find values of the parameters to give a satisfactory fit. If the parameters occur nonlinearly in f, then the problem is a nonlinear one. Letting $\mathbf{r} \in R^n$ be defined by

$$r_i = f(x_i, \mathbf{a}) - y_i, \; i = 1, \ldots, n, \tag{4.2}$$

then typically we would minimise a norm of $\mathbf{r}$. Most interest has centred around the use of the l_1, l_2 and l_∞ norms, although other criteria have also been considered. The idea of linearizing r_i, $i = 1, \ldots, n$ about the current approximation to the solution is a natural way to proceed, and leads (normally) to a first order method. Letting A be the $n \times p$ matrix with (i,j) element $\dfrac{\partial f(x_i, \mathbf{a})}{\partial a_j}$, a typical subproblem defined at $\mathbf{a} \in R^p$ is

$$\text{minimise } \|\mathbf{r} + A\delta\mathbf{a}\| \text{ subject to}$$

$$\|\delta\mathbf{a}\|_A \leq \tau,$$

where the next approximation is $\mathbf{a} + \alpha\delta\mathbf{a}$, and $\tau > 0$ is an adaptively chosen restriction on the size of $\|\delta\mathbf{a}\|_A$, for some $\|.\|_A$, which makes the linearised approximation a reasonable one. An analogous approach for the nonlinear Huber M-estimator problem has been considered: for some developments, see [33], [22], [21].

Although there are circumstances when second order convergence is possible, such methods can be slowly convergent. Quadratic or superlinear convergence can be obtained by introducing second derivative information, using or approximating the Hessian matrix of the Lagrange function for the problem. Such methods are inherently more complicated, and even for problems with sparse Jacobian matrices of $\mathbf{r}$, produce full matrices in the subproblems, which are typically quadratic programming problems. Most algorithmic development has assumed that the problems are small and dense, and good methods for most of these problems were available 10 years ago.

However, of interest is the possiblity of tackling large, sparse problems by methods which only use first derivative information, yet can be modified in a simple and computationally efficient way to improve convergence rates. Such methods have linear subproblems for which sparsity can often be expoited. We will not pursue this in detail here, but just mention that for the special case of the l_∞ norm, this problem has been addressed in some work of [31], [30]. In this case, first order methods can involve the solution of linear programming problems, and sparse codes are available. Improved convergence is achieved by incorporating a special correction step and some preliminary numerical evidence (although mainly for small dense problems) is that these methods are competitive with the more complicated second order methods and improve on other first order methods. It may be that similar ideas could be used for other norms, but this remains to be established.

For other large structured problems, only first order methods have been developed. We consider finally a particular class of such problems, which are the nonlinear analogue of the problems considered in Section 3.4.

4.1 Errors-in-variables problems

The model equations (3.24) are only valid for linear models. In addition the elements of B are assumed unrelated, unless further conditions are subsequently imposed. An alternative way to proceed is to recognise explicitly the errors in the x_i values. This means that the model equations become from (4.2)

$$r_i = f(x_i + \epsilon_i, \mathbf{a}) - y_i, \; i = 1, \ldots, n, \tag{4.3}$$

and it is appropriate to minimise some function of the components of both $\mathbf{r}$ and ϵ. In this form, the problem is referred to as an errors-in-variables problem. For example, we could minimise the l_2 norm, or equivalently

$$\mathbf{r}^T\mathbf{r} + \epsilon^T\epsilon = \sum_{i=1}^{n}(r_i^2 + \epsilon_i^2), \tag{4.4}$$

subject to (4.3). This problem is referred to as orthogonal distance regression. The reason for this name is that we are minimizing the sum of squares of the distances from the data points (x_i, y_i) to the model curve. Note that if all the errors are in the y_i values, then we are minimizing the sum of squares of the vertical distances from the points to the curve; on the other hand, if all the errors are in the x_i values, then we are minimizing the sum of squares of the horizontal distances from the points to the curve. An efficient method for minimising (4.4) is given in [8], with software available in [9], where the structure of the problem is exploited. Define the $n \times p$ matrix A and the $n \times n$ diagonal matrix D by

$$A_{ij} = \frac{\partial f(x_i + \epsilon_i, \mathbf{a})}{\partial a_j}, i = 1, \ldots, n, \quad j = 1, \ldots, p,$$

$$D = \text{diag}\{\frac{\partial f(x_i + \epsilon_i, \mathbf{a})}{\partial \epsilon_i}, \quad i = 1, \ldots, n\}.$$

Then setting

$$J = \begin{bmatrix} A & D \\ 0 & I \end{bmatrix},$$

and also $\mathbf{p}^T = [\mathbf{r}^T, \epsilon^T]$, we can expand $\mathbf{p}$ in a Taylor series in the direction $\mathbf{d}^T = [\delta\mathbf{a}^T, \delta\epsilon^T]$, giving the expansion

$$\mathbf{p} + J\mathbf{d} + O(||\mathbf{d}||^2).$$

The vector $\mathbf{d}$ is sought so that replacing $\mathbf{a}$ and ϵ by $\mathbf{a} + \delta\mathbf{a}$ and $\epsilon + \delta\epsilon$ gives an improved value of (4.4). Restricting attention to the first two terms of the expansion and minimizing this in the least squares sense gives a Gauss-Newton step, and the corresponding Levenberg-Marquardt step satisfies

$$(J^T J + \lambda I)\mathbf{d} = -J^T\mathbf{p}, \tag{4.5}$$

where λ is a suitably chosen positive constant. Assuming that D is nonsingular, it is easy to eliminate $\delta\epsilon$ from (4.5), and then we obtain a system of equations to be solved for $\delta\mathbf{a}$ only. The form of these is such that they can be solved as an l_2 problem: find the l_2 solution of the overdetermined linear system:

$$\begin{bmatrix} W^{1/2}A \\ \lambda^{1/2}I \end{bmatrix} \delta\mathbf{a} = \begin{bmatrix} W^{-1/2} \\ 0 \end{bmatrix} \mathbf{c},$$

where the matrix W is given by

$$W = I - D(D^2 + I + \lambda I)^{-1}D,$$

and $\mathbf{c}$ is given by

$$\mathbf{c} = D(D^2 + I + \lambda I)^{-1}(D\mathbf{r} + \epsilon) - \mathbf{r}.$$

For given λ, this system can be solved for $\delta\mathbf{a}$, then $\delta\epsilon$ can be obtained using

$$\delta\epsilon = -(D^2 + I + \lambda I)^{-1}(DA\delta\mathbf{a} + D\mathbf{r} + \epsilon).$$

Thus the full Levenberg-Marquardt step can be obtained in an amount of work comparable with that required for the conventional problem with errors only in the values of y. Notice that moving to multivariate problems means that the matrix D would no longer be diagonal, and some of the efficiencies in exploiting the structure would be lost. A method similar to this but using the l_1 norm is given in [54].

An application area which gives rise to problems having errors in all variables is metrology, or the science of measurement: see for example Forbes [23]. Problems in this area (and others) also arise where there is no variable which can be considered the dependent variable, and these are called *implicit problems*; problems based on the model equations (4.3) are then referred to as *explicit*. The model equations analogous to (4.3) for implicit problems are

$$f(x_i + \epsilon_i, \mathbf{a}) = 0, \; i = 1, \ldots, n, \tag{4.6}$$

where $x_i \in R^k$. Notice that because the k variables are not distinguishable, the present setting is the right one. A trust region method for implicit orthogonal distance regression is given in [26]. The solution of the implicit problem based on (4.6) for some other smooth criteria is considered in [29]. A method which uses the l_1 norm is given in [53]. There is still considerable scope for the application of robust fitting methods to many of these problems.

References

1. Abatzoglou, T. J., Mendel, J. M. and Harada, G. A. (1991). The constrained total least squares technique and its application to harmonic superresolution. *IEEE Transactions on Signal Processing*, *39*, 1070–1087.
2. Allasia, G. (1995) (ed) *Special Functions: Annals of Numerical Mathematics*, *Volume 2*.
3. Barrodale, I. and Roberts, F. D. K. (1973). An improved algorithm for discrete l_1 linear approximation. *SIAM J. Num. Anal.*, *10*, 839–848.
4. Bartels, R. , Conn, A. R. and Li, Y. (1989). Primal methods are better than dual methods for solving overdetermined linear systems in the l_∞ sense? *SIAM J. Num. Anal.*, *26*, 693–726.
5. Bartels, R. , Conn, A. R. and Sinclair, J, W. (1978). Minimisation techniques for piecewise differentible functions: the l_1 solution to an overdetermined linear system. *SIAM J. Num. Anal.*, *15*, 224–241.
6. Battle, G. (1987). A block spin construction of ondelettes, Part I: Lemarie functions. *Comm. Math. Phys.*, *10*, 601–615.
7. Bloomfield, P. and Steiger, W. L. (1983). *Least Absolute Deviations.* Birkhaüser.

8. Boggs, P. T., Byrd, R. H. and Schnabel, R. B. (1987). A stable and efficient algorithm for nonlinear orthogonal distance regression. *SIAMJ Sci. Stat. Comp.*, *8*, 1052–1078.

9. Boggs, P. T., Byrd, R. H., Donaldson, J. R. and Schnabel, R. B. (1989). ODRPACK, Software for weighted orthogonal distance regression. *ACM Trans. Math. Soft.*, *15*, 348–364.

10. Chui, C. K. (1992). *An Introduction to wavelets*, Academic Press, Boston.

11. Chui, C. K. and Wang, J, Z. (1991). On compactly supported spline wavelets and a duality principle. *Trans. Amer. Math. Soc.*

12. Clark, D. I. (1985). The mathematical structure of Huber's M–estimator. *SIAM J. Sci. Stat. Comp.*, *6*, 209–219.

13. Clark, D. I. and Osborne, M. R. (1986). Finite algorithms for Huber's M–estimator. *SIAM J. Sci. Stat. Comp.*, *7*, 72–85.

14. Coleman, T. and Li, Y. (1992). A globally and quadratically convergent affine scaling algorithm for l_1 problems. *Math. Prog.*, *56*, 189–222.

15. Coleman, T. and Li, Y. (1992). A globally and quadratically convergent method for linear l_∞ problems. *SIAM J. Num. Anal.*, *29*, 1166–1186.

16. Daubechies, I. (1988). Orthonormal bases of compactly supported wavelets. *Comm. Pure Appl. Math.*, *XLI*, 909–996.

17. Daubechies, I. (1992). *Ten Lectures on Wavelets*, CBMS/NSF Series in Applied Mathematics No 61, SIAM, Philadelphia.

18. Dax, A. (1989). The minimax solution of linear equations subject to linear constraints. *IMA J Num. Anal.*, *9*, 95–109.

19. Dax, A. (1993). A row relaxation method for large minimax problems. *BIT*, *33*, 262–276.

20. DeVore, R. A. and Lucier, B. J. (1992). Wavelets, *Acta Numerica*, *1*, 1–56.

21. Ekblom, H. and Madsen, K. (1989). Algorithms for nonlinear Huber estimation. *BIT*, *29*, 60–76.

22. Ekblom, H., Li, G. and Madsen, K. (1990). Algorithms for non-linear Huber regression with variable scale. *Report NI-90-10, Inst. for Numer. Anal., Technical University of Denmark.*

23. Forbes, A. B. (1993). Generalised regression problems in metrology, *Numerical Algorithms*, *5*, 523–533.

24. Golub, G. H. and Van Loan, C. F. (1980). An analysis of the total least squares problem. *SIAM J. Num. Anal.*, *17*, 883–893.

25. Goodman, T. N. T. (1994). Interpolating Hermite spline wavelets, *J. Approximation Theory*, *78*, 174–189.

26. Helfrich, H.-P. and Zwick, D. (1993). A trust region method for implicit orthogonal distance regression, *Numerical Algorithms*, *5*, 535–545.

27. Hubbard, B. B. (1996). *The World According to Wavelets: The Story of a Mathematical Technique in the Making*, A. K. Peters.

28. Jawerth, B. and Sweldens, W. (1994). An overview of wavelet based multiresolution analyses, *SIAM Review*, *36*, 377–412.

29. Jefferys, W. H. (1990). Robust estimation when more than one variable per equation of condition has error, *Biometrika*, *77*, 597–607.

30. Jonasson, K. (1993). A projected conjugate gradient method for sparse minimax problems. *Numerical Algorithms*, *5*, 309–323.

31. Jonasson, K. and Madsen, K. (1994). Corrected sequential linear programming for sparse minimax optimization. *BIT*, *34*, 372–387.

32. Kaiser, G. (1994). *A Friendly Guide to Wavelets*, Birkhauser, Boston.

33. Li, G. and Madsen, K. (1988). Robust nonlinear data fitting. *Numerical Analysis, 1987*, eds Griffiths, D. F. and Watson, G. A. Longman, Essex.

34. Lemmerling, P., Van Huffel, S. and De Moor, B. (1997). Structured total least squares problems: formulations, algorithms and applications, *Recent Advances in Total Least Squares Techniques and Errors-in-Variables Modeling*, ed Van Huffel, S., SIAM (to appear).

35. Li, W. and Swetits, J. J. (1995). Linear l_1 estimator and Huber M–estimator. *preprint.*

36. Madsen, K. and Nielsen, H. B. (1993). A finite smoothing algorithm for linear l_1 estimation. *SIAM J. Opt.*, *3*, 223–235.

37. Madsen, K., Nielsen, H. B. and Pinar, M. C. (1994). New characterizations of l_1 solutions of overdetermined linear systems. *Operations Research Letters*, *16*, 159–166.

38. Mallat, S. G. (1989). Multi-resolution approximations and wavelet orthonormal bases of $L^2(R)$. *Trans. Amer. Math. Soc.*, *315*, 69–97.

39. Meyer, Y. (1990). *Ondelettes et Operateurs.* Hermann, Paris.

40. Michelot, C. and Bougeard, M.L. (1994). Duality results and proximal solutions of the Huber M–estimator problem. *Appl. Math. Optim.*, *30*, 203–221, 1994.

41. De Moor, B. (1993). Structured total least squares and L_2 approximation problems, *Lin. Alg. Appl.*, *188/9*, 163–207.

42. Osborne, M. R. (1985). *Finite Algorithms in Optimisation and Data Analysis.* John Wiley, Chichester.

43. Osborne, M. R. and Watson, G. A. (1996). Aspects of M-estimation and l_1 fitting problems. *Numerical Analysis: A. R. Mitchell 75th Birthday Volume*, eds Griffiths, D. F. and Watson, G. A., World Scientific, Singapore.

44. Rosen, J. B., Park, H. and Glick, J. (1996). Total least norm formulation and solution for structured problems. *SIAM Jour. on Matrix Anal. and Appl.*, *17*, 110–126.

45. Rosen, J. B., Park, H. and Glick, J. (1997). Structured total least norm techniques for linearly and nonlinearly structured problems, *Recent Advances in Total Least Squares Techniques and Errors-in-Variables Modeling*, ed Van Huffel, S., SIAM (to appear).

46. Späth, H. (1995). *One Dimensional Spline Interpolation Algorithms.* A K Peters, Wellesley, Massachusetts.

47. Strang, G. (1993). Wavelet transforms versus Fourier transforms, *Bull Amer Math Soc*, *28*, 288–305.

48. Strang, G. and Nguyen, T. (1996). *Wavelets and Filter Banks.* Wellesley-Cambridge Press.

49. Strichartz, R. S. (1993). How to make wavelets, *Amer Math Monthly*, *100*, 539–556.

50. Van Huffel, S., De Moor, B. and Chen, H. (1994). Relationships between constrained and structured TLS with applications to signal enhancement. *Proc. Int. Symp. on the Math. Theory for Networks and Systems.* Regensberg, Germany, August 1993.

51. Van Huffel, S. and Vandewalle, J. (1991). *The Total Least Squares Problem, Computational Aspects and Analysis.* SIAM, Philadelphia.

52. Watson, G. A. (1980). *Approximation Theory and Numerical Methods.* John Wiley, Chichester.

53. Watson, G. A. (1997). The use of the L_1 norm in nonlinear errors-in-variables problems, *Recent Advances in Total Least Squares Techniques and Errors-in-Variables Modeling*, ed Van Huffel, S., SIAM (to appear).

54. Watson, G. A. and Yiu, K. F. C. (1991). On the solution of the errors in variables problem using the l_1 norm. *BIT*, *31*, 697–710.

A Review of Methods for Multivariable Interpolation at Scattered Data Points

M.J.D. Powell

Department of Applied Mathematics and Theoretical Physics, University of Cambridge

Abstract

In many applications, a function s from $\mathcal{R}^d$ to $\mathcal{R}$ is required that satisfies the interpolation conditions $s(\underline{x}_i)=f_i$, $i=1,2,\ldots,n$, where the points $\underline{x}_i \in \mathcal{R}^d$ and the right hand sides $f_i \in \mathcal{R}$, $i=1,2,\ldots,n$, are known. We study three different kinds of methods for constructing the interpolant when the positions of the data points are general. Firstly, s may be formed from polynomial pieces. Secondly, some algorithms are designed for the case when $s(\underline{x})$ is required for only a finite number of values of $\underline{x}$ that can be treated individually, so the form of s as a function is less important than the ability to generate each $s(\underline{x})$ efficiently from the data. Thirdly, we consider the radial basis function method, because it provides some highly successful ways of picking an n-dimensional linear space of functions that is independent of the right hand sides f_i, $i=1,2,\ldots,n$, and then s is the unique element of the space that interpolates the data. Particular attention is given to algorithms that are derived from the Dirichlet tessellation of the points $\underline{x}_i$, $i=1,2,\ldots,n$, and to the calculation of coefficients in the radial basis function method. Some theoretical properties of the algorithms are mentioned too.

1 Introduction

We review some useful algorithms that construct a function s from $\mathcal{R}^d$ to $\mathcal{R}$ that satisfies the interpolation equations

$$s(\underline{x}_i) = f_i, \quad i=1,2,\ldots,n, \tag{1.1}$$

where the interpolation points $\underline{x}_i$, $i = 1,2,\ldots,n$, and the right hand sides f_i, $i = 1,2,\ldots,n$, are given, each $\underline{x}_i$ and f_i being in $\mathcal{R}^d$ and $\mathcal{R}$, respectively, for some fixed positive integer d. If s is required to be an element of a prescribed linear space, $\mathcal{S}$ say, that is spanned by the functions s_j, $j = 1,2,\ldots,n$, then s has the form $\sum_{j=1}^{n} \lambda_j s_j$, so the conditions (1.1) provide the linear constraints

$$\sum_{j=1}^{n} \lambda_j s_j(\underline{x}_i) = f_i, \quad i=1,2,\ldots,n, \tag{1.2}$$

on the coefficients λ_j, $j=1,2,\ldots,n$. Thus the interpolation problem can sometimes be reduced to the solution of an $n\times n$ system of linear equations.

This simple approach raises some important questions. In particular it is well-known that, if the functions s_j, $j=1,2,\ldots,n$, are continuous and independent of the interpolation points, then, for $d\geq 2$, the matrix of the system is singular for infinitely many sets of distinct interpolation points [19]. Therefore, if suitable properties of $\underline{x}_i$, $i=1,2,\ldots,n$, are not guaranteed, it is dangerous to let $\mathcal{S}$ be an n-dimensional space of algebraic or trigonometric polynomials, for example, or of piecewise functions whose joins are independent of the positions of the data points. We restrict attention to algorithms that always provide an interpolant when the only conditions on the points $\underline{x}_i$, $i=1,2,\ldots,n$, are that they are all different and the volume of their convex hull in $\mathcal{R}^d$ is nonzero.

Further, except for a way of choosing a triangulation, all of the algorithms that we consider have the property that the interpolant $s(\underline{x})$, $\underline{x}\in\mathcal{R}^d$, depends linearly on the right hand sides f_i, $i=1,2,\ldots,n$. In other words, if we regard the interpolation points as fixed, and if we let $\ell_j(\underline{x})$, $\underline{x}\in\mathcal{R}^d$, be the interpolant that is defined by the algorithm when the right hand sides have the values $f_i=\delta_{ij}$, $i=1,2,\ldots,n$, where δ_{ij} is the Kronecker delta and j is any integer from $[1,n]$, then the interpolant to the general right hand sides of the equations (1.1) is the expression

$$s(\underline{x}) = \sum_{i=1}^{n} \ell_i(\underline{x})\, f_i, \quad \underline{x}\in\mathcal{R}^d. \tag{1.3}$$

The functions ℓ_i, $i=1,2,\ldots,n$, are called the "Lagrange functions of interpolation". Studying their properties is usually instructive, but they are hardly ever formed explicitly in practice. Expression (1.3) is valid if and only if the equations (1.2) define the coefficients λ_j, $j=1,2,\ldots,n$, uniquely, provided that the basis functions s_j, $j=1,2,\ldots,n$, are independent of the right hand sides.

In many algorithms, the right hand side f_i makes a nonzero contribution to $s(\underline{x})$, $\underline{x}\in\mathcal{R}^d$, only if $\underline{x}$ is relatively close to $\underline{x}_i$. Then we deduce from formula (1.3) that, for each $\underline{x}$, most of the numbers $\ell_i(\underline{x})$, $i=1,2,\ldots,n$, have to vanish when n is large. The methods that are studied in Sections 2 and 3 have this property. Specifically, we consider interpolation by piecewise polynomials of low degree in Section 2. Then Section 3 addresses local interpolation schemes that have been developed from the method of [27]. Some of these schemes are highly useful, because they are very fast and they provide enough smoothness and accuracy for many applications. They also have some interesting theoretical properties.

These algorithms become more difficult to apply successfully when the number of components of $\underline{x}$, namely d, is increased. Indeed, more nonzero terms are usually required in formula (1.3), and more complicated data structures are needed to record the division of $\mathcal{R}^d$ into pieces. For example, one can compare the division of an interval in $\mathcal{R}$ into subintervals with the division of a polygon in $\mathcal{R}^2$ into triangles. On the other hand, if we pick a suitable function $\phi(r)$, $r\geq 0$, and if we interpolate from the n-dimensional linear space that is spanned by the functions

$$s_j(\underline{x}) = \phi(\|\underline{x}-\underline{x}_j\|_2), \quad \underline{x}\in\mathcal{R}^d, \quad j=1,2,\ldots,n, \tag{1.4}$$

then it is straightforward to allow any value of the dimension d. This approach is called the "radial basis function" method, and several choices of ϕ ensure that the interpolation equations (1.2) have a unique solution [21]. For example, the "multiquadric function" $\phi(r)=(r^2+c^2)^{1/2}$, $r\geq 0$, where c is a positive constant, enjoys this nonsingularity property, and it provides interpolating functions s that are employed routinely in many applications [16]. A disadvantage of the radial basis function method when n is large, however, is that hardly any sparsity occurs in the matrix of the system (1.2) for the usual choices of ϕ. Therefore some iterative procedures for calculating the coefficients λ_j, $j=1,2,\ldots,n$, in fewer than $\mathcal{O}(n^3)$ operations have been developed. The ideas that provide these procedures will be mentioned in the review of the radial basis function method that is presented in Section 4.

2 Piecewise polynomial interpolation

Let $\mathcal{A}\subset\mathcal{R}^d$ be the convex hull of the interpolation points $\underline{x}_i$, $i=1,2,\ldots,n$. In this section some algorithms will be described that divide $\mathcal{A}$ into regions and that let the interpolant s be a low order polynomial on each region, where s is continuous across the joins. Because the interpolation points are in general position, it is usual for each region to be a simplex, which means a triangle or a tetrahedron when $d=2$ or 3, respectively. More generally, a simplex in $\mathcal{R}^d$ is the convex hull of any $d+1$ points such that the volume of the convex hull is nonzero, so every simplex is closed. Each algorithm of this section begins by choosing simplices whose vertices are interpolation points. Of course, the union of the simplices should be $\mathcal{A}$ and they should have disjoint interiors. Further, the choice must be such that the only interpolation points in any simplex are the vertices. Then a highly useful way of constructing the interpolant $s(\underline{x})$, $\underline{x}\in\mathcal{A}$, is to define s on each simplex by linear polynomial interpolation to the data at the vertices. This method defines s unambiguously on the joins. Indeed, if $\mathcal{A}_1$ and $\mathcal{A}_2$ are adjacent simplices, then the intersection $\mathcal{A}_1\cap\mathcal{A}_2$ is the convex hull of the interpolation points in the intersection, and $s(\underline{x})$, $\underline{x}\in\mathcal{A}_1\cap\mathcal{A}_2$, is the linear or constant polynomial that interpolates the data at these points. Thus the piecewise linear function $s(\underline{x})$, $\underline{x}\in\mathcal{A}$, is continuous.

When the given right hand sides have the property $f_i=f(\underline{x}_i)$, $i=1,2,\ldots,n$, where $f(\underline{x})$, $\underline{x}\in\mathcal{A}$, is a function with bounded second derivatives, then it is elementary that the method of the previous paragraph provides the accuracy

$$|f(\underline{x})-s(\underline{x})| \leq c\,\|f''\|_\infty\, h^2, \quad \underline{x}\in\mathcal{A}, \tag{2.1}$$

where c is a constant, $\|f''\|_\infty$ is the largest modulus of a second derivative that occurs, and h is the greatest distance between two vertices of a simplex. Even when the accuracy (2.1) is adequate, another feature of the piecewise linear interpolant may be unacceptable. It is that the greatest and least values of

$s(\underline{x})$, $\underline{x} \in \mathcal{A}$, are just the greatest and least values of the data f_i, $i=1,2,\ldots,n$. Further, it is sometimes important for the joins of the pieces of s to satisfy the conditions for first derivative continuity. Therefore this section will also address some methods that allow higher order piecewise polynomials. First, however, we consider the initial partition of $\mathcal{A}$ into simplices.

The extension of the Delauney triangulation from $\mathcal{R}^2$ to $\mathcal{R}^d$ is recommended. Some properties and implementations of this method when $d=2$ are given by Lawson [17] and by Green and Sibson [14]. We describe it in terms of the Dirichlet tessellation. Specifically, for $j=1,2,\ldots,n$, let $\mathcal{T}_j$ be the set of points in $\mathcal{R}^d$ such that $\underline{x} \in \mathcal{T}_j$ if and only if $\|\underline{x}-\underline{x}_j\|_2$ is the least of the distances $\{\|\underline{x}-\underline{x}_i\|_2 : i=1,2,\ldots,n\}$, so $\mathcal{T}_j$ is a "tile" that surrounds the interpolation point $\underline{x}_j$. Further, each $\mathcal{T}_j$ is a convex polygon and there are points of $\mathcal{R}^d$ where $d+1$ tiles meet, which are the "vertices" of the Dirichlet tessellation. For simplicity, we make the nondegeneracy assumption that there are no points where $d+2$ tiles meet, which can be achieved by making notional infinitesimal changes to the positions of the interpolation points if necessary. Thus, if $\widehat{\underline{x}}$ is any vertex of the tessellation, then the least of the distances $\{\|\widehat{\underline{x}}-\underline{x}_i\|_2 : i=1,2,\ldots,n\}$ occurs for exactly $d+1$ values of i, and we let $\mathcal{I}(\widehat{\underline{x}})$ be the set of these integers. For each $\widehat{\underline{x}}$, we pick the simplex that is the convex hull of the points $\underline{x}_i$, $i \in \mathcal{I}(\widehat{\underline{x}})$. A simplex is chosen in this way for every vertex of the Dirichlet tessellation. It is stated by Green and Sibson [14] that for $d=2$ "these triangles can easily be shown to fit together into a triangulation of the convex hull of the generating points" and that "all of the above extends to higher dimensional cases". Nevertheless, we prove below that the simplices divide $\mathcal{A}$ into regions that have the properties that are required in the first paragraph of this section.

An example of this construction when $d=2$ is shown in Figures 1 and 2, where the points $\underline{x}_i \in \mathcal{R}^2$, $i=1,2,\ldots,n$, are denoted by "•". The dotted lines of Figure 1 are the boundaries of the Dirichlet tiles, so each line is a perpendicular bisector of two of the interpolation points. We see that the tessellation has 10 vertices. Therefore the method divides $\mathcal{A}$ into the 10 triangles whose edges are the solid lines of Figure 2.

We begin the proof of the efficacy of the method for general d by considering any one of the simplices. When $d=2$, there is a unique circle through its three vertices, and for general d there is a unique hypersphere through its $d+1$ vertices of the form $\{\underline{x} : \|\underline{x}-\widehat{\underline{x}}\|_2 = r\}$, where $\widehat{\underline{x}}$ and r are determined by the positions of the vertices. Further, the construction of the simplex implies that $\widehat{\underline{x}}$ is a vertex of the Dirichlet tessellation. Therefore different vertices of the tessellation provide different simplices. For each $\widehat{\underline{x}}$, we let the above hypersphere and its interior be the set

$$C(\widehat{\underline{x}}) = \{\underline{x} : \|\underline{x}-\widehat{\underline{x}}\|_2 \le \|\underline{x}_i-\widehat{\underline{x}}\|_2,\ i \in \mathcal{I}(\widehat{\underline{x}})\} \subset \mathcal{R}^d. \tag{2.2}$$

By the definition of $\mathcal{I}(\widehat{\underline{x}})$, the points $\underline{x}_i$, $i \in \mathcal{I}(\widehat{\underline{x}})$, are the only interpolation points in $C(\widehat{\underline{x}})$. Further, because $C(\widehat{\underline{x}})$ is convex, the simplex is a subset of $C(\widehat{\underline{x}})$. It follows that a simplex cannot include an interpolation point that is not one of its vertices.

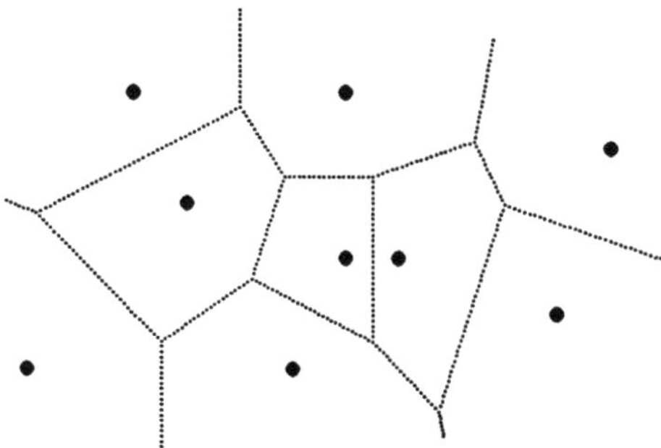

Figure 1: The Dirichlet tessellation of some data points

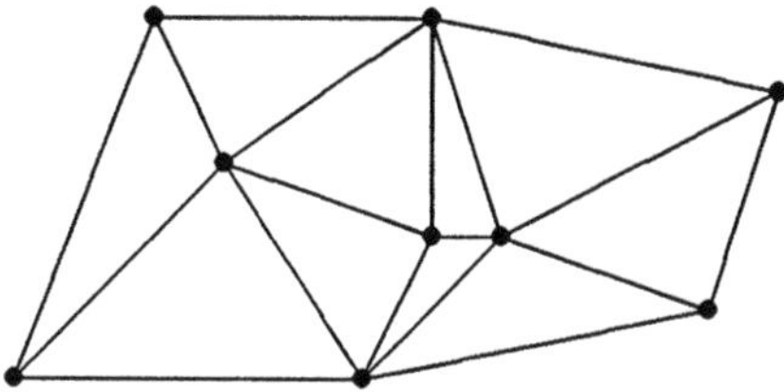

Figure 2: The Delauney triangulation of the data points

We also employ the properties of $C(\underline{\widehat{x}})$ to deduce that the chosen simplices have disjoint interiors. Indeed, if $\underline{\widehat{x}}$ and $\underline{\check{x}}$ are different vertices of the tessellation such that $\underline{x}$ is an interior point of the convex hulls of both $\underline{x}_i$, $i\in\mathcal{I}(\underline{\widehat{x}})$, and $\underline{x}_i$, $i\in\mathcal{I}(\underline{\check{x}})$, then the intersection of $C(\underline{\widehat{x}})$ and $C(\underline{\check{x}})$ is nonempty. Further, the points $\underline{x}_i$, $i\in\mathcal{I}(\underline{\check{x}})\backslash\mathcal{I}(\underline{\widehat{x}})$, are in $C(\underline{\check{x}})$ but not in $C(\underline{\widehat{x}})$, while the points $\underline{x}_i$, $i\in\mathcal{I}(\underline{\widehat{x}})\backslash\mathcal{I}(\underline{\check{x}})$ are in $C(\underline{\widehat{x}})$ but not in $C(\underline{\check{x}})$. It follows that the boundaries of $C(\underline{\check{x}})$ and $C(\underline{\widehat{x}})$ meet, so we can let $\mathcal{B}$ be the d–1 dimensional linear manifold in $\mathcal{R}^d$ that includes this common boundary, where we are making use of the fact that both $C(\underline{\check{x}})$ and $C(\underline{\widehat{x}})$ are spherical. Moreover, because $\underline{x}$ is an interior point, we can assume without loss of generality that it is not in $\mathcal{B}$. We concentrate on the half of $\mathcal{R}^d$ that contains $\underline{x}$ and is bounded by $\mathcal{B}$. Here $C(\underline{\check{x}})$ is entirely within $C(\underline{\widehat{x}})$ or *vice versa*, and we address only the former alternative. Now, as $\underline{x}$ is in the convex hull of $\underline{x}_i$, $i\in\mathcal{I}(\underline{\check{x}})$, at least one of the vertices of this convex hull must be on the side of $\mathcal{B}$ that is under consideration. Therefore this interpolation point is in $C(\underline{\widehat{x}})$, but it is not in the set $\underline{x}_i$, $i\in\mathcal{I}(\underline{\widehat{x}})$, because $\mathcal{B}$ includes all of the points $\underline{x}_i$, $i\in\mathcal{I}(\underline{\widehat{x}})\cap\mathcal{I}(\underline{\check{x}})$. As this conclusion contradicts one of the statements of the previous paragraph, the deduction is complete.

It remains to show that the chosen simplices cover $\mathcal{A}$ without leaving any gaps. If there is a gap, then the gap has a boundary, and part of the boundary is a face, $\mathcal{F}$ say, of one of the chosen simplices, where $\mathcal{F}$ includes some interior points of $\mathcal{A}$. Let the vertices of the simplex that has the face $\mathcal{F}$ be $\underline{x}_i$, $i \in \mathcal{I}(\widehat{\underline{x}})$, and let $\underline{x}_j$ be the vertex of the simplex that is opposite the face. Then we define $\mathcal{J}(\mathcal{F}) = \mathcal{I}(\widehat{\underline{x}}) \backslash \{j\}$, because $\mathcal{F}$ is the convex hull of the set $\underline{x}_i$, $i \in \mathcal{J}(\mathcal{F})$. Furthermore, the set of points in $\mathcal{R}^d$ that are equidistant from $\underline{x}_i$, $i \in \mathcal{J}(\mathcal{F})$, is a straight line that provides an edge of the Dirichlet tessellation, one of the end points of the edge being $\widehat{\underline{x}}$. It can be shown that the edge has another finite end point, $\check{\underline{x}}$ say, because $\mathcal{F}$ is not part of the boundary of $\mathcal{A}$. Thus $\check{\underline{x}}$ is a vertex of the tessellation, and $\mathcal{J}(\mathcal{F})$ is a subset of $\mathcal{I}(\check{\underline{x}})$. Therefore the simplex that has the vertices $\underline{x}_i$, $i \in \mathcal{I}(\check{\underline{x}})$ is included in our coverage of $\mathcal{A}$, and it also has the face $\mathcal{F}$. Now it is proved in the previous paragraph that the interiors of different simplices are disjoint. It follows that, if there is a gap, then the new simplex must help to fill it, which is a contradiction. Therefore the Dirichlet method for partitioning $\mathcal{A}$ into simplices does achieve the required conditions.

Another valuable technique for choosing simplices is proposed by Dyn, Levin and Rippa [10] in the case $d = 2$. They point out that it is sometimes highly advantageous to let the division of $\mathcal{A}$ depend on the right hand sides f_i, $i = 1, 2, \ldots, n$, although these data are ignored by the Dirichlet method that has been described already. Their technique begins with any division of $\mathcal{A}$ into triangles that is suitable for piecewise linear interpolation. Then one considers every pair of triangles that have a common edge, subject to the condition that the union of the triangles is a convex quadrilateral that does not degenerate to a single triangle. The current pair of triangles is separated by a diagonal of the quadrilateral, and it can be replaced by the pair of triangles that is separated by the other diagonal. Dyn et al. [10] propose several criteria that depend on the right hand sides for making this replacement, where each replacement has to provide a strict reduction in an objective function in order to prevent cycling. The objective function has the form $F(s) \in \mathcal{R}$, where s is now the piecewise linear function that is defined by piecewise linear interpolation to the data at the vertices of the current triangulation. Typically, $F(s)$ is the sum of moduli of the first derivative discontinuities of s. For example, if $\underline{\nabla} s$ takes the values $\underline{g}_1$ and $\underline{g}_2$ on two triangles that have a common edge, then it may be suitable if the edge makes a contribution of $\|\underline{g}_1 - \underline{g}_2\|_2$ to $F(s)$, in order that $F(s)$ is the sum of these contributions over all the internal edges of the current triangulation. The method makes exchanges of diagonals of convex quadrilaterals that reduce $F(s)$ until no further reductions can be achieved. It is particularly successful when there are regions of $\mathcal{R}^2$ where the data are values of a smooth underlying function $f(\underline{x})$, $\underline{x} \in \mathcal{R}^2$, and $\underline{\nabla} f$ changes rapidly in one direction but relatively slowly in the orthogonal direction.

We now turn to higher order piecewise polynomials. Let P_k^d be the linear space of polynomials from $\mathcal{R}^d$ to $\mathcal{R}$ of degree at most k, where d and k are any positive integers. The following relation between P_k^d and a grid on any simplex in $\mathcal{R}^d$ is very useful. Let the vertices of the simplex be $\underline{x}_j$, $j=1,2,\ldots,d+1$, and let the notation x_j, $j=1,2,\ldots,d$, denote the components of $\underline{x}$. Further, the numbers m_j, $j=1,2,\ldots,d$, are any nonnegative integers that satisfy $\sum_{j=1}^d m_j \le k$, and we define m_{d+1} to be the nonnegative integer $k-\sum_{j=1}^d m_j$. Thus the expressions

$$p(\underline{x}) = \prod_{j=1}^{d} x_j^{m_j}, \quad \underline{x}\in\mathcal{R}^d, \qquad \text{and} \qquad k^{-1}\sum_{j=1}^{d+1} m_j\,\underline{x}_j \tag{2.3}$$

are an element of P_k^d and a point in the simplex, respectively. We see that all possible choices of m_j, $j=1,2,\ldots,d$, provide not only linearly independent polynomials that span P_k^d but also a grid of distinct points in the simplex, which is the required relation. Therefore, if values of a function are given on this grid, and if we try to interpolate them by an element of P_k^d, then the number of coefficients is the same as the number of data. For example, a quadratic polynomial function of 2 variables has 6 coefficients, and the corresponding grid points in a triangle are the vertices and the mid-points of the sides, these grid points being shown in Figure 3 for the triangulation of Figure 2. Next we prove the well-known and fundamental result that interpolation on the grid from the linear space P_k^d has a unique solution.

Let i and j be any integers in the intervals $[0,k-1]$ and $[1,d+1]$. Then all grid points of the form (2.3) with the property $m_j=i$ lie in a hyperplane of $\mathcal{R}^d$. In particular, when $i=0$ this hyperplane is the face of the simplex that is opposite to the vertex $\underline{x}_j$, and the hyperplanes that occur for larger values of i are parallel to this face. We let q_{ij} be the linear polynomial from $\mathcal{R}^d$ to $\mathcal{R}$ such that the hyperplane is the set $\{\underline{x} : q_{ij}(\underline{x})=0\}$. Then, for the general point of the grid $\widehat{\underline{x}}=k^{-1}\sum_{j=1}^{d+1}\widehat{m}_j\,\underline{x}_j$, we consider the function

$$\ell(\underline{x}) = \prod_{i=0}^{\widehat{m}_1-1} q_{i1}(\underline{x}) \prod_{i=0}^{\widehat{m}_2-1} q_{i2}(\underline{x}) \cdots \prod_{i=0}^{\widehat{m}_{d+1}-1} q_{i\,d+1}(\underline{x}), \quad \underline{x}\in\mathcal{R}^d, \tag{2.4}$$

where any empty products on the right hand side are given the value one. We see that this function is in P_k^d and that it is nonzero at the grid point $\widehat{\underline{x}}$. If $\underline{x}=k^{-1}\sum_{j=1}^{d+1} m_j\,\underline{x}_j$ is any other grid point, however, then the product (2.4) is zero, because $m_j<\widehat{m}_j$ must occur for at least one value of j. It follows that a constant multiple of the product is the Lagrange function of interpolation for the grid point $\widehat{\underline{x}}$. The existence of these Lagrange functions implies that the interpolation problem has a unique solution.

In order to apply this method on each simplex of $\mathcal{A}$ when $k\ge 2$, it is necessary to augment the interpolation equations (1.1). We divide $\mathcal{A}$ into simplices as before, and then on each simplex we introduce the grid that has been mentioned for the current k. Therefore the new function values are required at the grid

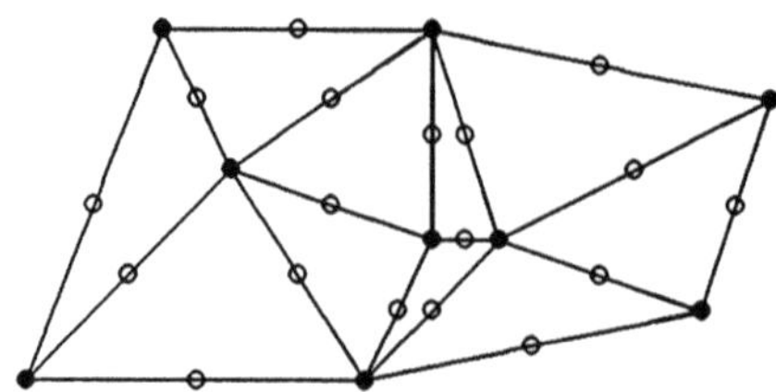

Figure 3: Grid points for piecewise quadratics

points that are not vertices, and usually they are generated by a preliminary calculation, which estimates them by local fits to the data (1.1). It is appropriate to choose fitting methods that are exact for all polynomials of degree k. Some of the algorithms that are studied in Section 3 are suitable for this preliminary work.

It should be clear from Figure 3 that the piecewise quadratic function that interpolates at the 6 points that are shown in each triangle is continuous across the triangle boundaries, because the data on each edge define the quadratic interpolant on the edge uniquely. This property is also enjoyed by the piecewise polynomial interpolation method that we are considering for general d and k. Indeed, the set of grid points on any face of each simplex in $\mathcal{R}^d$ is equivalent to the set of grid points of a complete simplex in $\mathcal{R}^{d-1}$, so s is defined uniquely on each face by the data of the face. It is straightforward to extend this remark to edges between faces and so on. Thus $s(\underline{x})$, $\underline{x} \in \mathcal{A}$, is a continuous function.

In some applications, however, it is important for the interpolant $s(\underline{x})$, $\underline{x} \in \mathcal{A}$, to have a continuous first derivative. Then, as mentioned in [17], local fitting methods may be used to pick suitable values of the gradients $\underline{\nabla} s(\underline{x}_i)$, $i = 1, 2, \ldots, n$, at the given interpolation points. We suppose that this has been done, and we address the task of determining s on each simplex of $\mathcal{A}$ by interpolation to function and first derivative values at the vertices of the simplex. Further, within each face of the simplex, we require s and $\underline{\nabla} s$ to be independent of the data at the opposite vertex. Thus the required continuity conditions can be achieved. Piecewise polynomial methods of this kind are proposed by Clough and Tocher [7] and by Powell and Sabin [26] for the $d = 2$ case. We consider them briefly. Therefore, in the remainder of this section, the simplices are triangles, the faces are edges of the triangles, and interpolation on the edges is a one-dimensional problem, the values and first derivatives of the interpolant being available at the two ends of the edge.

It is suitable to define s on each edge by cubic polynomial Hermite interpolation, which is the Clough–Tocher choice. Further, for convenience we force the normal derivative on the edge to vary linearly, which is the condition

$$\underline{n}^T \underline{\nabla} s(\underline{x}_i + \theta \{\underline{x}_j - \underline{x}_i\}) = (1-\theta)\, \underline{n}^T \underline{\nabla} s(\underline{x}_i) + \theta\, \underline{n}^T \underline{\nabla} s(\underline{x}_j), \quad 0 \le \theta \le 1, \tag{2.5}$$

where $\underline{x}_i$ and $\underline{x}_j$ are the end-points of the edge, and where $\underline{n}$ is any nonzero vector that is orthogonal to the edge. We see that the derivatives on the right hand side of expression (2.5) are known. If s were a cubic polynomial on the triangle, then it would have to satisfy 3 conditions at each vertex and one normal derivative condition on each edge, but in general this cannot be done because a cubic polynomial from $\mathcal{R}^2$ to $\mathcal{R}$ has only 10 coefficients. Therefore Clough and Tocher [7] let s be a *piecewise* cubic polynomial on each triangle. Specifically, they divide each triangle into 3 subtriangles by drawing straight lines between the centroid and the vertices, as shown on the left hand side of Figure 4. There is now a unique function s that is in P_3^2 on each subtriangle, that interpolates the data at the vertices of the original triangle, that satisfies condition (2.5) on the edges of the original triangle, and that has first derivative continuity across the internal boundaries of the subdivision. Of course the method is applied throughout $\mathcal{A}$. Thus one constructs a piecewise cubic polynomial function that satisfies the interpolation equations (1.1) and that has a continuous first derivative.

I was unable to find a simple proof in the literature of the existence and uniqueness claim of the previous paragraph, although Strang and Fix [30] explain that the Clough–Tocher subdivision of Figure 4 does provide 12 degrees of freedom in s, in order to satisfy the 12 conditions that have been mentioned. Therefore I proved for myself that the resultant 12×12 system of linear equations has a nonsingular matrix. I made use of the remark that, because uniqueness is independent of linear transformations of the variables, there is no loss of generality in assuming that the original triangle is equilateral.

Powell and Sabin [26] address the possibility of letting $s(\underline{x})$, $\underline{x} \in \mathcal{A}$, be a piecewise quadratic polynomial that has continuous first derivatives, so again each original triangle has to be subdivided. One of their schemes is shown on the right hand side of Figure 4. It splits each original triangle into 6 subtriangles that have a common vertex inside the original triangle. The subdivisions must have the property that, if we extend any internal boundary from the common vertex to an edge by a straight line, and if the edge is not part of the boundary of $\mathcal{A}$, then the extension is an internal boundary of the adjacent triangle subdivision, as suggested by the dotted lines of Figure 4. These subdivisions, and the use of piecewise quadratic functions that have continuous first derivatives across the internal boundaries of the subdivision, provide 9 degrees of freedom that allow the function and first derivative values at the vertices of the original triangle to be interpolated uniquely. The interpolant on each edge is a quadratic spline with an interior knot whose position is prescribed, so this part of the interpolant is defined by the data on the edge, which makes $s(\underline{x})$, $\underline{x} \in \mathcal{A}$, continuous. Further, if there is an internal edge of the original triangulation between $\underline{x}_i$ and $\underline{x}_j$, and if we now let $\underline{n}$ have the direction of the internal boundary that meets the edge at

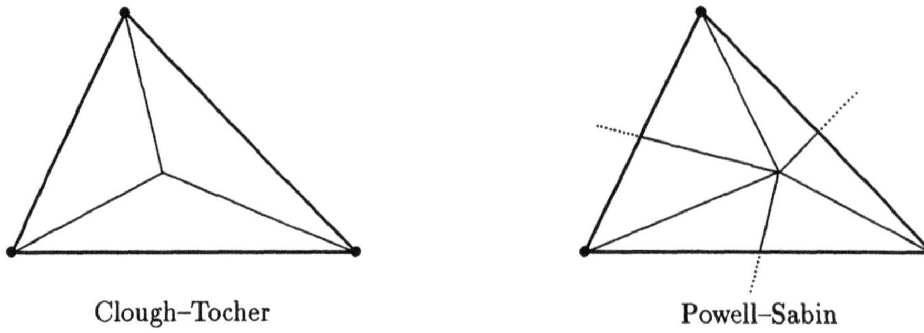

Figure 4: Subdivisions of a triangle

its knot, then equation (2.5) is satisfied, which provides the required continuity of $\underline{\nabla} s(\underline{x})$, $\underline{x} \in \mathcal{A}$.

3 Local interpolation schemes

So far we have taken the view that the interpolation conditions (1.1) have to provide a function s from $\mathcal{R}^d$ or $\mathcal{A}$ to $\mathcal{R}$, where $\mathcal{A}$ is still the convex hull of the interpolation points. On the other hand, if $s(\underline{x})$ is required for only a few particular values of $\underline{x}$, then it may be better to estimate each $s(\underline{x})$ individually. Further, when addressing this task, it may be appropriate to ignore data that are far from $\underline{x}$ and to give much weight to nearby data. We call methods that are developed in this way "local interpolation schemes". Of course each method should be able to define an interpolating function s, but now, instead of calculating $s(\underline{x})$ from a formula for s, we expect to derive each value of the interpolating function from the original data.

The prototype of such methods is usually attributed to Shepard [27]. He suggested the scheme

$$s(\underline{x}) = \sum_{i=1}^{n} w_i(\underline{x}) \, f_i \Big/ \sum_{i=1}^{n} w_i(\underline{x}), \quad \underline{x} \in \mathcal{R}^d, \tag{3.1}$$

where, for each $\underline{x}$, the weight functions $w_i(\underline{x})$, $i=1,2,\ldots,n$, are nonnegative and not all zero. Further, each w_i is continuous, except that $w_i(\underline{x})$ becomes infinite for $\underline{x} \to \underline{x}_i$, in order that the interpolation conditions (1.1) are satisfied. There is no need for $s(\underline{x})$, $\underline{x} \in \mathcal{R}^d$, to have a singularity at $\underline{x}_i$, however, because expression (3.1) can be written in the form

$$s(\underline{x}) = \Big\{ f_i + \sum_{\substack{j=1 \\ j \neq i}}^{n} [w_j(\underline{x})/w_i(\underline{x})] \, f_j \Big\} \Big/ \Big\{ 1 + \sum_{\substack{j=1 \\ j \neq i}}^{n} [w_j(\underline{x})/w_i(\underline{x})] \Big\}, \quad \underline{x} \in \mathcal{R}^d. \tag{3.2}$$

Shepard [27] gives particular attention to the choice

$$w_i(\underline{x}) = \|\underline{x}-\underline{x}_i\|_2^{-u}, \quad \underline{x}\in\mathcal{R}^d, \quad i=1,2,\ldots,n, \tag{3.3}$$

where u is a positive constant. Thus s is infinitely differentiable if u is an even integer.

As s is forced to interpolate the data, and as each $w_i(\underline{x})$, $i=1,2,\ldots,n$, is independent of the right hand sides f_j, $j=1,2,\ldots,n$, the definition (3.1) is an example of the Lagrange formula (1.3). Further, besides writing down the Lagrange functions

$$\ell_j(\underline{x}) = w_j(\underline{x}) \Big/ \sum_{i=1}^{n} w_i(\underline{x}), \quad \underline{x}\in\mathcal{R}^d, \quad j=1,2,\ldots,n, \tag{3.4}$$

explicitly, we know that formula (3.1) gives ℓ_j in the case $f_i=\delta_{ij}$, $i=1,2,\ldots,n$. Now the conditions $w_i(\underline{x})\geq 0$, $i=1,2,\ldots,n$, imply the bounds $0\leq\ell_j(\underline{x})\leq 1$, $\underline{x}\in\mathcal{R}^d$, so ℓ_j is at its upper or lower bound at every interpolation point. Therefore, when the interpolant (3.1) is differentiable for all right hand sides, the Lagrange functions must have the derivatives $\underline{\nabla}\ell_j(\underline{x}_i)=0$, $1\leq i,j\leq n$. Hence formula (1.3) implies the property $\underline{\nabla}s(\underline{x}_i)=0$, $i=1,2,\ldots,n$, for all right hand sides, which is unwelcome in an interpolation scheme.

This disadvantage is noted by several authors including Shepard. A useful remedy forces the interpolating function to satisfy $\underline{\nabla}s(\underline{x}_i) = \underline{g}_i$, $i=1,2,\ldots,n$, where the gradients $\underline{g}_i$, $i=1,2,\ldots,n$, are estimated by a preliminary calculation. Specifically, the weight functions are modified if necessary so that their singularities give the limits

$$\lim_{\underline{x}\to\underline{x}_i} \{\|\underline{x}-\underline{x}_i\|_2\, w_i(\underline{x})\}^{-1} = 0, \quad i=1,2,\ldots,n. \tag{3.5}$$

Then s is defined by the formula

$$s(\underline{x}) = \sum_{i=1}^{n} w_i(\underline{x}) \left\{ f_i + (\underline{x}-\underline{x}_i)^T \underline{g}_i \right\} \Big/ \sum_{i=1}^{n} w_i(\underline{x}), \quad \underline{x}\in\mathcal{R}^d. \tag{3.6}$$

Thus s interpolates both the given right hand sides and the prescribed gradients, which can be deduced from the analogue of expression (3.2).

A generalization of this technique can be found in [13]. It is derived from the remark that, if the data are values of a smooth function f, then the term in the braces of formula (3.6) can be regarded as an approximation to $f(\underline{x})$ that is suitable when $\|\underline{x}-\underline{x}_i\|_2$ is small. Let $a_i(\underline{x})$, $\underline{x}\in\mathcal{R}^d$, be a better estimate of $f(\underline{x})$ for small $\|\underline{x}-\underline{x}_i\|_2$. Then the generalized form of expression (3.6) is the function

$$s(\underline{x}) = \sum_{i=1}^{n} w_i(\underline{x})\, a_i(\underline{x}) \Big/ \sum_{i=1}^{n} w_i(\underline{x}), \quad \underline{x}\in\mathcal{R}^d, \tag{3.7}$$

where the condition (3.5) on the weight functions may be strengthened, in order that s inherits the accuracy of the approximation $a_i \approx f$ near $\underline{x}_i$, $i=1,2,\ldots,n$. The accuracy of this method is studied by Farwig [11] for the weight functions (3.3) when each a_i is a truncated Taylor series expansion of f about $\underline{x}_i$.

The schemes (3.1), (3.6) and (3.7) are "local" only if the weight functions have the property that each $w_i(\underline{x})$ is zero whenever $\|\underline{x}-\underline{x}_i\|_2$ is sufficiently large. Therefore Shepard [27] also recommends weights of the form

$$w_i(\underline{x}) = \psi(\|\underline{x}-\underline{x}_i\|_2), \quad \underline{x}\in\mathcal{R}^d, \quad i=1,2,\ldots,n, \tag{3.8}$$

where $\psi(r)$, $r \geq 0$, is a nonnegative function that satisfies $\psi(r) = 0$, $r \geq \widehat{r}$, in addition to $\psi(0)=+\infty$, the value of $\widehat{r}$ being given by the user. Specifically, he suggests the function

$$\psi(r) = \begin{cases} 1/r, & 0 \leq r \leq \widehat{r}/3, \\ (27/4)\,(r-\widehat{r})^2/\widehat{r}^3, & \widehat{r}/3 \leq r \leq \widehat{r}, \\ 0, & r \geq \widehat{r}, \end{cases} \tag{3.9}$$

which has a continuous first derivative. Moreover, it is pointed out by Levin [18] that the choice

$$\psi(r) = \begin{cases} \exp\left\{-\widehat{r}^2/(\widehat{r}-r)^2\right\} \Big/ \left\{\exp(r^2/h^2)-1\right\}, & 0 < r < \widehat{r}, \\ 0, & r \geq \widehat{r}, \end{cases} \tag{3.10}$$

provides infinite differentiability, where h is another positive parameter.

We now turn to local interpolation schemes of the form (3.1) that employ nonnegative functions w_i, $i=1,2,\ldots,n$, that are bounded. Therefore we satisfy the interpolation equations (1.1) by letting the weights have the properties

$$w_i(\underline{x}_i) > 0 \quad \text{and} \quad w_i(\underline{x}_j) = 0, \quad j\neq i, \quad 1\leq i,j\leq n. \tag{3.11}$$

This approach includes the piecewise linear scheme of the first paragraph of Section 2, provided that $\underline{x}$ is confined to the convex hull $\mathcal{A}$. Indeed, we let the weights w_i be the Lagrange functions of that method, which provides the identities

$$\sum_{i=1}^{n} w_i(\underline{x}) = \sum_{i=1}^{n} \ell_i(\underline{x}) = 1, \quad \underline{x}\in\mathcal{A}, \tag{3.12}$$

because the second sum is a piecewise linear function that interpolates the constant right hand sides $f_i = 1$, $i = 1,2,\ldots,n$. It follows that expression (3.1) is the Lagrange form of the interpolant on $\mathcal{A}$ for general right hand sides.

A disadvantage of piecewise linear interpolation is that there are discontinuities in the dependence of the weight functions w_i, $i=1,2,\ldots,n$, on the positions of the interpolation points $\underline{x}_i$, $i=1,2,\ldots,n$. Indeed, if the data points are moved gradually, then jumps may occur in the division of $\mathcal{A}$ into simplices. On the other hand, all the boundaries of the tiles of the Dirichlet tessellation vary continuously on any finite region of $\mathcal{R}^d$ that includes $\mathcal{A}$. Therefore another highly useful interpolation method, namely "natural neighbour interpolation" [29], is based on the tessellation.

In order to describe this scheme, we suppose that the Dirichlet tessellation of the data points $\underline{x}_i$, $i=1,2,\ldots,n$, has been constructed and that $\underline{x}$ is a general point of $\mathcal{R}^d$. We recall the notation $\mathcal{T}_i$ for the Dirichlet tile that surrounds $\underline{x}_i$. Let $\underline{x}$ be regarded as another interpolation point, and let $\mathcal{T}_{\underline{x}}$ be the tile that surrounds $\underline{x}$ if the tessellation is extended to include the new point. This tile is bounded if and only if $\underline{x}$ is strictly inside $\mathcal{A}$, and then the natural neighbour interpolation scheme defines $s(\underline{x})$ by the formula (3.1), where the weights have the values

$$w_i(\underline{x}) = \text{Volume}\,\{\mathcal{T}_{\underline{x}} \cap \mathcal{T}_i\}, \quad i=1,2,\ldots,n. \tag{3.13}$$

Clearly the right hand side of equation (3.1) is unchanged if the weights are scaled by a constant. Therefore, if $\mathcal{T}_{\underline{x}}$ is unbounded, we multiply the right hand sides of the definition (3.13) by a scaling factor, so that the weights become finite numbers that are not all zero. Then $s(\underline{x})$ is given the value (3.1) again, which completes the description of the method.

We consider some of the features of natural neighbour interpolation. Let $\underline{x}$ be any point of $\mathcal{R}^d$ and let $\|\underline{x}-\underline{x}_j\|_2$ be the least of the distances $\|\underline{x}-\underline{x}_i\|_2$, $i=1,2,\ldots,n$, so $\underline{x}$ is in the tile $\mathcal{T}_j$. If $\underline{y}$ is any point of $\mathcal{R}^d$ that is in the set

$$\mathcal{Y} = \left\{ \underline{y} : \|\underline{y}-\underline{x}_j\|_2 > \|\underline{x}-\underline{x}_j\|_2 + \min_{1\le i\le n} \|\underline{y}-\underline{x}_i\|_2 \right\}, \tag{3.14}$$

then the triangle inequality provides the bound

$$\|\underline{y}-\underline{x}\|_2 \ge \|\underline{y}-\underline{x}_j\|_2 - \|\underline{x}-\underline{x}_j\|_2 > \min_{1\le i\le n} \|\underline{y}-\underline{x}_i\|_2, \tag{3.15}$$

which shows that $\underline{y}$ is not in $\mathcal{T}_{\underline{x}}$. In other words, $\underline{y}$ is in $\mathcal{T}_{\underline{x}}$ only if the distance from $\underline{y}$ to $\underline{x}_j$ is within $\|\underline{x}-\underline{x}_j\|_2$ of the least distance from $\underline{y}$ to an interpolation point. Therefore it is usual for only a few of the weights (3.13) to be nonzero.

Moreover, let ε be any positive constant, let $\underline{x}$ satisfy $\|\underline{x}-\underline{x}_j\|_2 \le \varepsilon$, and let $\mathcal{T}_j(\varepsilon)$ be the set

$$\mathcal{T}_j(\varepsilon) = \left\{ \underline{y} : \|\underline{y}-\underline{x}_j\|_2 \le \varepsilon + \min_{1\le i\le n} \|\underline{y}-\underline{x}_i\|_2 \right\}. \tag{3.16}$$

Then, because $\underline{y} \notin \mathcal{T}_j(\varepsilon)$ implies $\underline{y} \in \mathcal{Y}$ which implies $\underline{y} \notin \mathcal{T}_{\underline{x}}$, we can write the definition (3.13) in the form

$$w_i(\underline{x}) = \text{Volume}\,\{\mathcal{T}_{\underline{x}} \cap \mathcal{T}_i \cap \mathcal{T}_j(\varepsilon)\}, \quad i=1,2,\ldots,n, \tag{3.17}$$

for sufficiently small $\|\underline{x}-\underline{x}_j\|_2$. Now $\mathcal{T}_j(\varepsilon)$ can be regarded as a slight extension of the tile $\mathcal{T}_j$, and the interiors of the tiles of the Dirichlet tessellation are disjoint. Thus expression (3.17) gives the limits

$$\lim_{\underline{x}\to\underline{x}_j} \{w_i(\underline{x})\,/\,w_j(\underline{x})\} = \delta_{ij}, \quad 1\le i\le n, \quad 1\le j\le n. \tag{3.18}$$

It follows from equation (3.1) that the natural neighbour scheme does satisfy the interpolation conditions (1.1), provided that if necessary we employ the freedom to scale all the weights by a constant.

If the right hand sides have the values $f_i = f(\underline{x}_i)$, $i=1,2,\ldots,n$, where f is any constant function, then it is clear that formula (3.1) provides the identity $s \equiv f$. Sibson [28] establishes a brilliant extension of this result, namely that natural neighbour interpolation also enjoys this property when f is any linear polynomial from $\mathcal{R}^d$ to $\mathcal{R}$. He also proves that any first derivative discontinuities of $s(\underline{x})$, $\underline{x} \in \mathcal{R}^d$, occur only at the interpolation points. Such behaviour is well-known when $d = 1$, because in this case the natural neighbour method is equivalent to piecewise linear interpolation. On the other hand, for $d \geq 2$, the natural neighbour scheme has two strong advantages over the piecewise linear method, namely better smoothness and the continuous dependence of $s(\underline{x})$, $\underline{x} \in \mathcal{R}^d$, on the positions of the data points $\underline{x}_i$, $i = 1,2,\ldots,n$. Some software for natural neighbour interpolation, written in C, can be found in [31].

The remainder of this section considers interpolation schemes that are exact when $f_i = f(\underline{x}_i)$, $i = 1,2,\ldots,n$, where f is any function in a finite dimensional linear space $\mathcal{F}$ that is chosen by the user. We recall the nonnegative weight functions $w_i(\underline{x})$, $\underline{x} \in \mathcal{R}^d$, $i=1,2,\ldots,n$, that are continuous except for the singularities $w_i(\underline{x}_i) = +\infty$, $i=1,2,\ldots,n$, the form (3.8) being usual. For each $\underline{x} \in \mathcal{R}^d$, let $\mathcal{I}(\underline{x})$ be the subset $\{i : w_i(\underline{x}) > 0\} \subset \{1,2,\ldots,n\}$. The schemes that will be considered provide a unique value of $s(\underline{x})$ if $\mathcal{F}$ has the property

$$f \in \mathcal{F} \quad \text{and} \quad f(\underline{x}_i) = 0, \quad i \in \mathcal{I}(\underline{x}), \quad \text{imply } f \equiv 0. \tag{3.19}$$

Therefore we assume that this condition holds for every $\underline{x}$ where the value of $s(\underline{x})$ may be required. Further, $\mathcal{F}$ has to include a function that is nonzero at every interpolation point.

The "moving least squares" method was proposed by McLain [20]. For any given $\underline{x}$ in $\mathcal{R}^d$, it begins the calculation of $s(\underline{x})$ by seeking the function in $\mathcal{F}$ that minimizes the weighted sum of squares

$$\sum_{i=1}^{n} w_i(\underline{x}) \{f(\underline{x}_i) - f_i\}^2, \quad f \in \mathcal{F}, \tag{3.20}$$

which is unique because of condition (3.19). Let this function be $f_{\underline{x}} \in \mathcal{F}$. Then the value

$$s(\underline{x}) = f_{\underline{x}}(\underline{x}) \tag{3.21}$$

is chosen. In order to show that interpolation is achieved, we pick $\underline{x} = \underline{x}_j$, where j is any integer from $[1,n]$. It follows from the singularity $w_j(\underline{x}) = +\infty$, and from the last statement of the previous paragraph, that $f_{\underline{x}}$ provides a finite value of expression (3.20) if and only if the equation $f_{\underline{x}}(\underline{x}_j) = f_j$ is obtained. Therefore the formula (3.21) and $\underline{x} = \underline{x}_j$ imply $s(\underline{x}_j) = f_j$ as required. In most applications, the dimension of $\mathcal{F}$ is much less than n and, for each $\underline{x}$, most of the weights

$w_i(\underline{x})$, $i=1,2,\ldots,n$, are zero. Thus the value of $s(\underline{x})$ is usually determined by the data at interpolation points that are relatively close to $\underline{x}$.

Another interesting scheme depends on a reformulation of condition (3.19). Therefore we choose a basis of $\mathcal{F}$, and, for every $\underline{y}\in\mathcal{R}^d$, we let $\underline{f}(\underline{y})\in\mathcal{R}^\alpha$ be the vector whose components are the basis functions calculated at $\underline{y}$, where α is the dimension of $\mathcal{F}$. Then condition (3.19) is equivalent to the statement that $\mathcal{R}^\alpha$ is spanned by the vectors $\underline{f}(\underline{x}_i)$, $i\in\mathcal{I}(\underline{x})$. In other words, if $\underline{v}$ is any element of $\mathcal{R}^\alpha$, then there exist coefficients $\theta_i\in\mathcal{R}$, $i\in\mathcal{I}(\underline{x})$, that provide the relation

$$\underline{v}=\sum_{i\in\mathcal{I}(\underline{x})}\theta_i\,\underline{f}(\underline{x}_i),\quad \underline{v}\in\mathcal{R}^\alpha. \tag{3.22}$$

This remark is important to the following algorithm, which is derived from the work of Backus and Gilbert [1]. If $s(\underline{x})$ is required for some $\underline{x}$ in $\mathcal{R}^d$, then the coefficients θ_i, $i\in\mathcal{I}(\underline{x})$, that minimize the expression

$$\sum_{i\in\mathcal{I}(\underline{x})}[w_i(\underline{x})]^{-1}\,\theta_i^2, \tag{3.23}$$

subject to the constraints

$$f(\underline{x})=\sum_{i\in\mathcal{I}(\underline{x})}\theta_i\,f(\underline{x}_i),\quad f\in\mathcal{F}, \tag{3.24}$$

are calculated. Thus the coefficients depend on $\underline{x}$ but not on f_i, $i=1,2,\ldots,n$. Then $s(\underline{x})$ is given the value

$$s(\underline{x})=\sum_{i\in\mathcal{I}(\underline{x})}\theta_i\,f_i. \tag{3.25}$$

We see that the constraints (3.24) are equivalent to equation (3.22) when $\underline{v}=\underline{f}(\underline{x})$, so they are consistent. Further, any freedom in the coefficients is removed by the minimization calculation, so each $s(\underline{x})$ is well-defined.

The reason for this procedure is shown by a comparison of equations (3.24) and (3.25). Indeed, equation (3.25) states that as usual we let $s(\underline{x})$ depend linearly on the right hand sides. Then equation (3.24) requires the coefficients to have the property $s(\underline{x})=f(\underline{x})$ if there is a function $f\in\mathcal{F}$ that satisfies $f(\underline{x}_i)=f_i$, $i\in\mathcal{I}(\underline{x})$. Further, the minimization of expression (3.23) helps to keep the moduli of the coefficients small in a way that takes account of the weights $w_i(\underline{x})$, $i\in\mathcal{I}(\underline{x})$. Actually, the range of i in the sum (3.23) can be replaced by $i\in\{1,2,\ldots,n\}$, because, if $w_i(\underline{x})=0$, then the minimization provides $\theta_i=0$ automatically. It is straightforward to deduce that s interpolates the data. Specifically, if $\underline{x}=\underline{x}_j$, then the coefficients $\theta_i=\delta_{ij}$, $i\in\mathcal{I}(\underline{x}_j)$, are optimal, because they satisfy the constraints (3.24), and the condition $w_j(\underline{x}_j)=+\infty$ reduces the objective function (3.23) to zero. It follows from the choice (3.25) that the interpolation equations (1.1) are achieved.

A surprising property of this algorithm is that it provides exactly the same interpolant as the "moving least squares" method, which has been described already. This result is mentioned by Levin [18], and a proof that is written for experts in functional analysis is given by Bos and Salkauskas [4]. An alternative proof is presented below.

Let m be the number of indices in $\mathcal{I}(\underline{x})$, let $\underline{\theta} \in \mathcal{R}^m$ be the vector whose components minimize expression (3.23) subject to the constraints (3.24), let Ω be the $m\times m$ diagonal matrix whose diagonal elements are $w_i(\underline{x})$, $i\in\mathcal{I}(\underline{x})$, which are positive and finite because we can restrict attention to the case when $\underline{x}$ is not an interpolation point, and let B be the $\alpha\times m$ matrix whose columns are the vectors $\underline{f}(\underline{x}_i)$, $i\in\mathcal{I}(\underline{x})$, of condition (3.22). Then $\underline{\theta}$ is the solution of the problem

$$\text{minimize } \underline{\theta}^T\Omega^{-1}\underline{\theta}, \quad \underline{\theta}\in\mathcal{R}^m, \quad \text{subject to } B\,\underline{\theta} = \underline{f}(\underline{x}). \tag{3.26}$$

Therefore, if $\underline{\theta}$ is optimal, there exists a vector $\underline{\lambda}\in\mathcal{R}^\alpha$ of Lagrange multipliers such that $\underline{\theta}$ and $\underline{\lambda}$ are defined by the linear equations

$$\Omega^{-1}\underline{\theta} = B^T\underline{\lambda} \qquad \text{and} \qquad B\,\underline{\theta} = \underline{f}(\underline{x}). \tag{3.27}$$

By eliminating $\underline{\theta}$, we find that $\underline{\lambda}$ is the vector $(B\,\Omega\,B^T)^{-1}\underline{f}(\underline{x})$. Therefore, remembering the choice (3.25), we deduce the identities

$$\begin{aligned} \underline{\theta} \;=\; \Omega\,B^T\underline{\lambda} \;&=\; \Omega\,B^T(B\,\Omega\,B^T)^{-1}\underline{f}(\underline{x}) \\ &\Longrightarrow s(\underline{x}) = \underline{f}^T\Omega\,B^T(B\,\Omega\,B^T)^{-1}\underline{f}(\underline{x}), \end{aligned} \tag{3.28}$$

where $\underline{f}\in\mathcal{R}^m$ has the components f_i, $i\in\mathcal{I}(\underline{x})$.

Turning to the moving least squares method, we recall that the components of the vector valued function $\underline{f}(\underline{y})$, $\underline{y}\in\mathcal{R}^d$, are a basis of $\mathcal{F}$. Therefore we can write $f_{\underline{x}}(\underline{y})=\underline{\mu}^T\underline{f}(\underline{y})$, $\underline{y}\in\mathcal{R}^d$, for each $\underline{x}$, where $\underline{\mu}$ is a vector in $\mathcal{R}^\alpha$ that depends on $\underline{x}$. Further, corresponding to expression (3.20), $\underline{\mu}$ provides the least value of the quadratic form

$$\sum_{i\in\mathcal{I}(\underline{x})} w_i(\underline{x})\,\{\underline{\mu}^T\underline{f}(\underline{x}_i) - f_i\}^2 = (B^T\underline{\mu}-\underline{f})^T\,\Omega\,(B^T\underline{\mu}-\underline{f}), \quad \underline{\mu}\in\mathcal{R}^\alpha, \tag{3.29}$$

which yields the normal equations

$$(B\,\Omega\,B^T)\,\underline{\mu} = B\,\Omega\,\underline{f}. \tag{3.30}$$

Hence formula (3.21) provides the function value

$$s(\underline{x}) = \underline{\mu}^T\underline{f}(\underline{x}) = \{(B\,\Omega\,B^T)^{-1}B\,\Omega\,\underline{f}\}^T\underline{f}(\underline{x}), \tag{3.31}$$

which is identical to the value (3.28), because both Ω and $B\,\Omega\,B^T$ are symmetric matrices. The proof is complete.

It is encouraging when two algorithms that were developed in different ways happen to be the same. Further, Levin [18] achieves excellent numerical results by combining the moving least squares method with the infinitely differentiable weight functions of equations (3.8) and (3.10).

4 The radial basis function method

We retain the assumption that the data points $\underline{x}_i$, $i=1,2,\ldots,n$, are all different, but we recall from Section 1 that, if $\mathcal{S}$ is any n-dimensional linear space of continuous functions from $\mathcal{R}^d$ to $\mathcal{R}$ that is independent of these points, then there may be no s in $\mathcal{S}$ that satisfies the given interpolation equations. Therefore the main assertion of the next paragraph is highly useful. It concerns the "radial basis function method", in which s has the form

$$s(\underline{x}) = \sum_{j=1}^{n} \lambda_j\, \phi(\|\underline{x}-\underline{x}_j\|_2) + \widehat{p}(\underline{x}), \quad \underline{x}\in\mathcal{R}^d, \tag{4.1}$$

where the notation is introduced in Section 1, except that $\widehat{p}$ is in the space P_k^d of polynomials of degree at most k from $\mathcal{R}^d$ to $\mathcal{R}$, the integer k being chosen by the user. The value $k=-1$ is reserved for the case when there is no polynomial term, so one can take the view in this case either that P_{-1}^d is empty or that $\widehat{p}$ is identically zero. The equations (1.1) provide the conditions

$$\sum_{j=1}^{n} \lambda_j\, \phi(\|\underline{x}_i-\underline{x}_j\|_2) + \widehat{p}(\underline{x}_i) = f_i, \quad i=1,2,\ldots,n, \tag{4.2}$$

on the coefficients of s, the number of coefficients being the sum of n and the dimension of P_k^d. Further, it is usual to impose the constraints

$$\sum_{j=1}^{n} \lambda_j\, p(\underline{x}_j) = 0, \quad p\in P_k^d, \tag{4.3}$$

on λ_j, $j=1,2,\ldots,n$. Thus, if $\widehat{p}$ is expressed as a linear combination of basis functions of P_k^d, and if p is set to each basis function in turn in the constraints (4.3), then the calculation of the coefficients of s becomes the solution of a linear system of equations, whose matrix is not only square but also symmetric.

This matrix is singular if a nonzero function of the form (4.1) vanishes at all the points $\underline{x}_i$, $i=1,2,\ldots,n$, and has coefficients that satisfy equation (4.3). Therefore, in order to achieve nonsingularity, we require the interpolation points and the integer k to have the property

$$p\in P_k^d \quad \text{and} \quad p(\underline{x}_i)=0, \quad i=1,2,\ldots,n, \quad \text{imply} \quad p\equiv 0. \tag{4.4}$$

We also assume $n\geq 2$. No further conditions are required on the positions of the data points. Thus, when $k=1$ for example, it is sufficient if the convex hull of the set $\{\underline{x}_i : i=1,2,\ldots,n\}$ has a nonzero volume in $\mathcal{R}^d$, as mentioned in the second paragraph of Section 1. We also have to pick a suitable ϕ, but $\phi(r)=r^2$, $r\geq 0$, for instance, would be disastrous when n is large, because then the function (4.1) would be confined to the linear space $P_{\max[k,2]}^d$. Fortunately,

Micchelli [21] establishes the following nonsingularity result. Let $\phi(r)$, $r \geq 0$, be continuous and let it be chosen so that the function

$$\eta(r) = \phi(r^{1/2}), \quad r \geq 0, \tag{4.5}$$

has the following properties:

1. $\eta(r)$, $r>0$, is infinitely differentiable,
2. the derivative $\eta^{(m)}(r)$, $r>0$, has no zeros for every integer m that exceeds $\max[k,0]=\kappa$, say, and
3. the signs of these derivatives alternate as m runs through the infinite sequence $\{\kappa+1, \kappa+2, \kappa+3, \ldots\}$.

Then, assuming that the points $\underline{x}_i$, $i=1,2,\ldots,n$, satisfy the conditions that have just been stated, the equations (4.2) and (4.3) have a unique solution for any right hand sides f_i, $i=1,2,\ldots,n$. A proof of this important result can also be found in [22].

The "multiquadric" choice $\phi(r)=(r^2+c^2)^{1/2}$, $r \geq 0$, is commended in Section 1. Then it is straightforward to verify that $\eta(r)=(r+c^2)^{1/2}$, $r \geq 0$, has the required properties without any restrictions on k. Similarly, there is no need for any polynomial terms in the cases $\phi(r)=\exp(-r^2)$, $\phi(r)=r$, and $\phi(r)=(r^2+c^2)^{-1/2}$, $r \geq 0$, which are called the "Gaussian", "linear" and "inverse multiquadric" radial functions, respectively. The choice $\phi(r)=r^{\gamma}$, $r \geq 0$, is also interesting, where γ is any positive constant. Indeed, we deduce from $\eta(r)=r^{\gamma/2}$, $r \geq 0$, that condition 1 holds because differentiability is not required at $r=0$. Further, condition 2 implies only that γ shall not be an even integer, and condition 3 provides the inequality $\frac{1}{2}\gamma-\kappa-1<0$, so a polynomial term is assumed if $\gamma>2$ but is optional otherwise. For example, it is usual for $\widehat{p}$ to be a linear polynomial in the "cubic" case $\phi(r)=r^3$, $r \geq 0$.

Radial basis functions arise naturally from some variational calculations. I am particularly interested in the work of Duchon [8] for $d=2$, which identifies the function $s(\underline{x})$, $\underline{x} \in \mathcal{R}^2$, that minimizes the integral

$$I(s) = \int\!\!\int_{\mathcal{R}^2} \left(\frac{\partial^2 s(\underline{x})}{\partial x^2}\right)^2 + 2\left(\frac{\partial^2 s(\underline{x})}{\partial x\, \partial y}\right)^2 + \left(\frac{\partial^2 s(\underline{x})}{\partial y^2}\right)^2 d\underline{x}, \tag{4.6}$$

subject to the interpolation conditions (1.1), where x and y are the components of $\underline{x}$. The resultant interpolant has the form (4.1), where ϕ is the function

$$\phi(0)=0 \quad \text{and} \quad \phi(r)=r^2 \log r, \quad r>0, \tag{4.7}$$

and where $\widehat{p}$ is in P_1^2. This choice of ϕ is called the "thin plate spline". Moreover, it follows from expression (4.1) that the conditions (4.3) when $d=2$ and $k=1$ are equivalent to the boundedness of the integral (4.6). Now this variational calculation has a unique solution whenever the points $\underline{x}_i$, $i=1,2,\ldots,n$, are all different and do not lie on a straight line (see [22], for instance). Therefore the choice (4.7) and $k=1$ should provide a function η that satisfies the conditions 1–3 for nonsingularity that are given above, which can be verified easily.

Many remarkable properties of the radial basis function method have been discovered by considering interpolation on the infinite grid $\mathcal{Z}^d$, which is the set of all vectors in $\mathcal{R}^d$ whose components are integers. These advances were made possible by the following results of Buhmann [5]. Let ϕ be any radial function that provides the conditions 1–3 on η that have been mentioned, and let $\widehat{\phi}(\|\underline{t}\|_2)$, $\underline{t}\in\mathcal{R}^d$, be the generalized Fourier transform of $\phi(\|\underline{x}\|_2)$, $\underline{x}\in\mathcal{R}^d$, so $\widehat{\phi}(r)$, $r\geq 0$, depends not only on ϕ but also on d. Further, let $\widehat{\ell}$ be the function

$$\widehat{\ell}(\underline{t}) = \widehat{\phi}(\|\underline{t}\|_2) \Big/ \sum_{\underline{j}\in\mathcal{Z}^d} \widehat{\phi}(\|\underline{t}-2\pi\underline{j}\|_2), \quad \underline{t}\in\mathcal{R}^d. \tag{4.8}$$

Then $\widehat{\ell}$ is absolutely integrable, so it has a well-defined inverse Fourier transform

$$\ell(\underline{x}) = (2\pi)^{-d} \int_{\underline{t}\in\mathcal{R}^d} \exp(i\underline{x}^T\underline{t})\,\widehat{\ell}(\underline{t})\,d\underline{t}, \quad \underline{x}\in\mathcal{R}^d. \tag{4.9}$$

Expression (4.8) implies not only that ℓ has the form

$$\ell(\underline{x}) = \sum_{\underline{j}\in\mathcal{Z}^d} \lambda_{\underline{j}}\,\phi(\|\underline{x}-\underline{j}\|_2), \quad \underline{x}\in\mathcal{R}^d, \tag{4.10}$$

but also that it satisfies the Lagrange conditions

$$\ell(0)=1 \qquad \text{and} \qquad \ell(\underline{i})=0, \quad \underline{i}\in\mathcal{Z}^d\backslash\{0\}. \tag{4.11}$$

Therefore the radial basis function solution to the interpolation equations $s(\underline{i})=f_{\underline{i}}$, $\underline{i}\in\mathcal{Z}^d$, is given by the formula

$$s(\underline{x}) = \sum_{\underline{i}\in\mathcal{Z}^d} \ell(\underline{x}-\underline{i})\,f_{\underline{i}}, \quad \underline{x}\in\mathcal{R}^d. \tag{4.12}$$

Generalized Fourier transforms occur in the definition (4.8), because of the suitability of radial functions $\phi(r)$, $r\geq 0$, that become unbounded as $r\to\infty$. For example, when $d=2$ and ϕ is the thin plate spline function (4.7), then equation (4.8) becomes the expression

$$\widehat{\ell}(\underline{t}) = \frac{\|\underline{t}\|_2^{-4}}{\sum_{\underline{j}\in\mathcal{Z}^2} \|\underline{t}-2\pi\underline{j}\|_2^{-4}} = \frac{1}{1+\sum_{\underline{j}\in\mathcal{Z}^2\backslash\{0\}} (\|\underline{t}\|_2/\|\underline{t}-2\pi\underline{j}\|_2)^4}, \quad \underline{t}\in\mathcal{R}^2, \tag{4.13}$$

the purpose of the last part being to show that $\widehat{\ell}$ is smooth in a neighbourhood of the origin. Similarly, smoothness is obtained at all the lattice points

$\underline{t}=2\pi\underline{j}$, $\underline{j}\in\mathcal{Z}^2$, of Fourier transform space. Thus, in this example, the function (4.9) enjoys an exponential rate of decay to zero as $\|\underline{x}\|_2\rightarrow\infty$. Hence the sum (4.12) is absolutely convergent for any $\underline{x}$ if the right hand sides have the values $f_{\underline{i}}=p(\underline{i})$, $\underline{i}\in\mathcal{Z}^d$, where p is any algebraic polynomial from $\mathcal{R}^2$ to $\mathcal{R}$. Therefore one can seek the greatest integer m such that the identity

$$p(\underline{x}) = s(\underline{x}) = \sum_{\underline{i}\in\mathcal{Z}^d} \ell(\underline{x}-\underline{i})\, p(\underline{i}), \quad \underline{x}\in\mathcal{R}^d, \tag{4.14}$$

is satisfied for every p in the linear space P_m^d.

At first sight one expects the identity (4.14) to fail when $p\equiv 1$, because, if we substitute the definition (4.10) into equation (4.14), in order to express the right hand side of the equation in the form $\sum_{\underline{j}\in\mathcal{Z}^d}\mu_{\underline{j}}\,\phi(\|\underline{x}-\underline{j}\|_2)$, then all the coefficients $\mu_{\underline{j}}$, $\underline{j}\in\mathcal{Z}^d$, must vanish if $s\equiv 1$ is achieved. On the other hand, if $d=1$ and if ℓ is the hat function of piecewise linear interpolation at the integers, then equation (4.14) holds for every p in P_1^1, although all the first derivative discontinuities of s must also be multiplied by zero. Thus first sight can be misleading. Actually, the "Strang and Fix conditions" show that the identity (4.14) is true for thin plate splines if $d=2$ and if p is any polynomial of degree at most 3. Indeed, one has only to verify that the function (4.13) has the Lagrange properties

$$\widehat{\ell}(0) = 1 \qquad \text{and} \qquad \widehat{\ell}(2\pi\underline{j}) = 0, \quad \underline{j}\in\mathcal{Z}^d\backslash\{0\}, \tag{4.15}$$

and that all derivatives of $\widehat{\ell}$ of degree at most 3 are zero at the lattice points $\underline{t}=2\pi\underline{j}$, $\underline{j}\in\mathcal{Z}^2$. Furthermore, for general d, the powers -4 and 4 of formula (4.13) are replaced by $-d-2$ and $d+2$, respectively, so equation (4.14) allows p to be any polynomial of degree $d+1$ when ϕ is the thin plate spline radial function. It is usual for the polynomial reproduction properties to become stronger if the divergence of $\lim_{r\rightarrow 0}\widehat{\phi}(r)$ to infinity becomes faster. Therefore it is advantageous if the integral of the function $\phi(\|\underline{x}\|_2)$, $\underline{x}\in\mathcal{R}^d$, is hugely unbounded. Indeed, the Gaussian radial function $\phi(r)=\exp(-r^2)$, $r\geq 0$, fails to reproduce any polynomials. The analysis that yields these conclusions and some more details are given by Buhmann [5] and by Powell [22]. Moreover, Buhmann, Dyn and Levin [6] show that these polynomial reproduction properties are preserved under perturbations of the interpolation points from the grid $\mathcal{Z}^d$, because of the vanishing of coefficients that is noted at the beginning of this paragraph.

Polynomial reproduction properties are important to orders of accuracy of approximations. In particular, if the data $f_{\underline{i}}$, $\underline{i}\in\mathcal{Z}^d$, in formula (4.12) are values of a smooth function $f(\underline{x})$, $\underline{x}\in\mathcal{R}^d$, on the infinite grid $\mathcal{Z}^d$, and if this formula reproduces all polynomials of degree m, then the error of the approximation $s\approx f$ remains the same if we subtract from f its m-th order Taylor series expansion about any point. Thus, if the variables are scaled so that interpolation on the new grid $\mathcal{Z}^d$ corresponds to interpolation to f on an infinite square grid of mesh size h, say, then we expect the accuracy $\|f-s\|_\infty=\mathcal{O}(h^{m+1})$. Some results of this kind can be found in [5] and in [22], for instance.

We now return to the situation when the number of interpolation points, namely n, is finite, and we ask whether the properties of radial basis function interpolation on an infinite grid have any relevance. Therefore we let the points $\underline{x}_i$, $i=1,2,\ldots,n$, form a regular square grid of mesh size h on the unit cube $[0,1]^d \subset \mathcal{R}^d$, so h^{-1} is an integer and n has the value $(1+h^{-1})^d$. Further, we assume $f_i = f(\underline{x}_i)$, $i=1,2,\ldots,n$, for some function $f(\underline{x})$, $\underline{x} \in \mathcal{R}^d$, that is smooth and bounded, where f may be an extension of a function that is given on $[0,1]^d$. Let s be a radial basis function interpolant to the data and let $\widehat{s}$ be defined by interpolation to f on the infinite grid $h\,\underline{i}$, $\underline{i} \in \mathcal{Z}^d$. Further, assuming $\|f-\widehat{s}\|_\infty = \mathcal{O}(h^{m+1})$, as suggested above, and that the degree of $\widehat{p}$ in expression (4.1) is at most m, we consider the error function

$$\begin{aligned} f(\underline{x})-s(\underline{x}) &= \widehat{s}(\underline{x})-s(\underline{x}) + f(\underline{x})-\widehat{s}(\underline{x}) \\ &= \widehat{s}(\underline{x})-s(\underline{x}) + \mathcal{O}(h^{m+1}), \quad \underline{x}\in[0,1]^d. \end{aligned} \tag{4.16}$$

Now the uniqueness properties of the interpolation method imply that $\widehat{s}-s$ is the interpolant to itself on the grid $h\,\underline{i}$, $\underline{i} \in \mathcal{Z}^d$. Thus the version of formula (4.12) for the grid of mesh size h provides the identity

$$\widehat{s}(\underline{x})-s(\underline{x}) = \sum_{\underline{i}\in\mathcal{Z}^d} \ell(h^{-1}\underline{x}-\underline{i})\,\{\widehat{s}(h\underline{i})-s(h\underline{i})\}, \quad \underline{x}\in\mathcal{R}^d, \tag{4.17}$$

where ℓ is still defined by equations (4.8) and (4.9). It follows from condition (4.16) and from the interpolation properties of $\widehat{s}$ and s that the error can be expressed in the form

$$f(\underline{x})-s(\underline{x}) = \sum_{\substack{\underline{i}\in\mathcal{Z}^d \\ h\underline{i}\notin[0,1]^d}} \ell(h^{-1}\underline{x}-\underline{i})\,\{f(h\underline{i})-s(h\underline{i})\} + \mathcal{O}(h^{m+1}), \quad \underline{x}\in[0,1]^d. \tag{4.18}$$

We see that none of the relevant values of $h\underline{i}$ are in $[0,1]^d$. Further, although $|f(h\underline{i})-s(h\underline{i})|$ can grow as $h\underline{i}$ moves away from the unit cube, it happens usually that $|\ell(h^{-1}\underline{x}-\underline{i})|$ decays very rapidly as $\|h^{-1}\underline{x}-\underline{i}\|_2$ increases. Therefore the $\mathcal{O}(h^{m+1})$ error bound of interpolation on the infinite grid is inherited sometimes by $|f(\underline{x})-s(\underline{x})|$, provided that $\underline{x}$ is far enough inside the unit cube. On the other hand, it is usual for some powers of h to be lost from this accuracy when $\underline{x}$ is near the grid boundary.

This deterioration in accuracy when $d=1$ and ϕ is a scaled version of the multiquadric $\phi(r) = (r^2+c^2)^{1/2}$, $r \geq 0$, is studied by Beatson and Powell [3]. They pick $n \geq 2$ and the equally spaced points $x_i = (i-1)\,h$, $i=1,2,\ldots,n$, where $h=1/(n-1)$. Indeed, they interpolate the values $f_i = f(x_i)$, $i=1,2,\ldots,n$, of a smooth function $f(x)$, $0 \leq x \leq 1$, by the form

$$s(x) = \sum_{i=1}^n \lambda_i\,\phi(\|h^{-1}x-h^{-1}x_i\|_2) = \sum_{i=1}^n \lambda_i\{h^{-2}(x-x_i)^2+c^2\}^{1/2}, \quad 0\leq x\leq 1, \tag{4.19}$$

where c is independent of h. The accuracy $\|f-s\|_\infty = \mathcal{O}(h^2)$ would occur on an infinite grid and it is also achieved at interior points of $[0,1]$ in the finite case. The edge effects of this case, however, cause the norm $\max\{|f(x)-s(x)| : 0 \le x \le 1\}$ to be only of magnitude h, except that $\mathcal{O}(h^2)$ occurs uniformly on $[0,1]$ if and only if f has the properties

$$f'(1) = f(0)+f(1) \qquad \text{and} \qquad f'(0)+f'(1) = 0. \tag{4.20}$$

Alternatively, one can force the better accuracy and preserve the interpolation conditions by giving s the form (4.19) plus a polynomial term $\widehat{p} \in P_1^1$. Thus expression (4.1) becomes relevant again, but it is important to note that we no longer remove the freedom in s by imposing the constraints (4.3), because in general this technique would fail to give the required accuracy. Instead, it is suitable to take up the freedom in any way that causes the first derivative errors $|f'(0)-s'(0)|$ and $|f'(1)-s'(1)|$ to be $\mathcal{O}(h)$. This work is mentioned because suitably chosen low order polynomials may assist the accuracy of other radial basis function interpolants on finite grids.

There are also some empirical reasons for studying the radial basis function method for interpolation. In particular, Franke [12] presents a heuristic comparison of 32 interpolation algorithms on 22 different data sets, the dimension $d=2$ being used throughout. The radial basis function methods in this comparison are the "thin plate spline", the "multiquadric", the "inverse multiquadric" and the "cubic" techniques that are mentioned soon after equation (4.5). Franke [12] awards marks for a range of attributes, the highest mark being an A. All of the radial methods achieve an A for "accuracy" and an A for "visual pleasantness" but the other algorithms are less successful. Therefore it is useful to many applications to be able to calculate the coefficients of the function (4.1) efficiently. Some suitable techniques are reviewed briefly in the remainder of this section.

Many researchers believe that such calculations are difficult, because often the relevant matrices are ill-conditioned. For example, condition numbers of 10^5 are mentioned by Dyn, Levin and Rippa [9]. Fortunately, however, it follows from the analysis of Micchelli [21] that one can avoid damage from ill-conditioning in the main part of the calculation, by reducing most of the work to the solution of a symmetric system of linear equations whose matrix is positive definite. Of course, no reformulation is necessary if the $n \times n$ matrix Φ with the elements

$$\Phi_{ij} = \phi(\|\underline{x}_i - \underline{x}_j\|_2), \quad 1 \le i,j \le n, \tag{4.21}$$

is already positive definite and if the polynomial $\widehat{p}$ is excluded from equation (4.1). Specifically, the positive definiteness of Φ is guaranteed for the Gaussian and inverse multiquadric radial functions, but negative eigenvalues occur in all the other cases that have been mentioned. In order to describe the reformulation, we recall the integer κ that is introduced after equation (4.5).

Let the polynomials p_j, $j=1,2,\ldots,\beta$, be a basis of P_κ^d, where $\beta = \dim P_\kappa^d$, and let P be the $n \times \beta$ matrix that has the elements

$$P_{ij} = p_j(\underline{x}_i), \quad 1 \le i \le n, \quad 1 \le j \le \beta. \tag{4.22}$$

Condition (4.4) implies that the columns of P are linearly independent if $\kappa = k$, and this property is trivial in the alternative case $\kappa > k$, because then the definition of κ gives $k=-1$, $\kappa=0$ and $\beta=1$. Further, we let Z be a nonsingular $n\times n$ matrix whose first $n-\beta$ columns are orthogonal to the columns of P, we define $\underline{\mu} \in \mathcal{R}^n$ by the equation $\underline{\lambda} = Z\,\underline{\mu}$, where the components of $\underline{\lambda}$ are the required coefficients λ_j, $j=1,2,\ldots,n$, and, when $k\geq 0$, we let $\underline{\nu}$ be the vector in $\mathcal{R}^\beta$ such that the polynomial part of the interpolant (4.1) is the function $\widehat{p}(\underline{x})=\sum_{j=1}^{\beta} \nu_j\, p_j(\underline{x})$, $\underline{x}\in\mathcal{R}^d$. Thus equation (4.2) takes the matrix form

$$\Phi\,\underline{\lambda} + P\,\underline{\nu} = \Phi\, Z\,\underline{\mu} + P\,\underline{\nu} = \underline{f}, \tag{4.23}$$

where $\underline{f}$ has the components f_i, $i=1,2,\ldots,n$, which provides the system

$$(Z^T \Phi\, Z)\,\underline{\mu} = Z^T \underline{f} - Z^T P\,\underline{\nu}. \tag{4.24}$$

The most important property of this system is that the leading $(n-\beta)\times(n-\beta)$ submatrix of $Z^T\Phi\, Z$ is a positive definite symmetric matrix, which can be deduced from the conditions 1–3 on η that are given after equation (4.5), except that the overall sign of $Z^T\Phi\, Z$ is inherited from ϕ [21].

The positive definiteness is highly useful to the calculation of $\underline{\mu}$ in all cases. Indeed, if $k\geq 0$, then, by writing condition (4.3) as $P^T\underline{\lambda}=0$, we deduce that $\underline{\lambda}$ is in the space that is spanned by the first $n-\beta$ columns of Z, so the last β components of $\underline{\mu}\in\mathcal{R}^d$ are zero. In other words, letting Q be the $n\times(n-\beta)$ matrix whose columns are the first $n-\beta$ columns of Z, and letting the components of $\widehat{\underline{\mu}}\in\mathcal{R}^{n-\beta}$ be the first $n-\beta$ components of $\underline{\mu}$, we have $\underline{\lambda}=Z\,\underline{\mu}=Q\,\widehat{\underline{\mu}}$. Further, the conditions on Z imply $Q^TP=0$. Therefore the first $n-\beta$ of the equations (4.24) provide the square system

$$(Q^T\Phi\, Q)\,\widehat{\underline{\mu}} = Q^T\underline{f}, \tag{4.25}$$

the matrix $Q^T\Phi\, Q$ being positive definite. Thus $\widehat{\underline{\mu}}$ and $\underline{\lambda}=Q\,\widehat{\underline{\mu}}$ are calculated before $\underline{\nu}\in\mathcal{R}^\beta$ is available. Then, after $\underline{\lambda}$ has been found, $\widehat{p}$ is extracted from the conditions (4.2), which requires little work for practical values of β. Alternatively, when $k=-1$, then the $Z^TP\,\underline{\nu}$ term of equation (4.24) is void, and the system can be solved for $\underline{\mu}$ without any pivoting, because $Z^T\Phi\, Z$ is nonsingular and its leading $(n-1)\times(n-1)$ submatrix is positive definite. Finally, we form $\underline{\lambda}=Z\,\underline{\mu}$.

The efficiency of this technique depends on the choice of Z. Therefore it is recommended that Z be a product of Givens rotations that puts zero elements throughout the first $n-\beta$ rows of Z^TP, and also in the first i positions of the $(n-i)$-th row of Z^TP for $i=1,2,\ldots,\beta-1$. Thus at most $\beta\,(n-\frac{1}{2}\beta-\frac{1}{2})$ rotations are needed, so the work of calculating all the elements of $Z^T\Phi\, Z$ is only $\mathcal{O}(\beta n^2)$. A bonus of orthogonal rotations is that they hardly ever lose accuracy unnecessarily. In some applications, however, the value of n is so large that it would be prohibitively expensive to generate $Z^T\Phi\, Z$ or $Q^T\Phi\, Q$ explicitly.

Therefore several papers have been published on iterative methods for calculating the coefficients of the function (4.1), particularly in the case of thin plate spline interpolation when $d=2$. This research is reviewed by Powell [24]. A strong advantage of such methods is shown by considering the solution of the system (4.25) by the conjugate gradient algorithm. Let Q be such that $Q\,\widehat{\underline{\mu}}$ can be evaluated in $\mathcal{O}(\beta n)$ operations for any $\widehat{\underline{\mu}}$, as in the previous paragraph. Further, careful attention to Q may provide $Q^T \Phi\, Q$ with properties that cause the conjugate gradient algorithm to yield a sufficiently accurate $\widehat{\underline{\mu}}$ in far fewer than $n-\beta$ iterations, the popular name for this endeavour being "pre-conditioning". We suppose that Q is given, however, and we address the amount of work of each iteration. A substantial gain is obtained over direct methods for sufficiently large n if this amount is much less than $\mathcal{O}(n^2)$.

The only task of a conjugate gradient iteration that may require about n^2 operations is the calculation of the product $(Q^T \Phi\, Q)\,\widehat{\underline{v}}$ for some $\widehat{\underline{v}} \in \mathcal{R}^{n-\beta}$. Usually none of the elements of $(Q^T \Phi\, Q)$ is zero, so we are seeking a special technique that depends on properties of Φ. Therefore, writing $\underline{v} = Q\,\widehat{\underline{v}} \in \mathcal{R}^n$ and letting v_j, $j=1,2,\ldots,n$, be the components of $\underline{v}$, we note that the components of $\Phi\,\underline{v}$ are the values of the function

$$t(\underline{x}) = \sum_{j=1}^{n} v_j\, \phi(\|\underline{x}-\underline{x}_j\|_2), \quad \underline{x} \in \mathcal{R}^d, \tag{4.26}$$

at the data points $\underline{x}_i$, $i=1,2,\ldots,n$. Now the fast multipole method of Greengard and Rokhlin [15] can provide an approximation $\widehat{t} \approx t$ such that the values $\widehat{t}(\underline{x}_i)$, $i=1,2,\ldots,n$, are estimates to prescribed accuracy of the required components of $\Phi\,\underline{v}$. Each value of $\widehat{t}(\underline{x})$ is constructed by replacing nearly all of the terms of the sum (4.26) by a few Laurent or Taylor series expansions. A typical expansion has a centre $\widehat{\underline{x}} \in \mathcal{R}^d$, say, and the term $v_j\, \phi(\|\underline{x}-\underline{x}_j\|_2)$ is included in the expansion only if $\underline{x}_j$ is sufficiently close to $\widehat{\underline{x}}$. Further, the order of each expansion is chosen to be suitable for the required accuracy. Details for the thin plate spline case when $d=2$ are given by Beatson and Newsam [2] and by Powell [23]. It is found that the amount of computation to form a suitable estimate of $\Phi\,\underline{v}$ in this way is close to $\mathcal{O}(n)$, and that savings over the direct calculation of $\Phi\,\underline{v}$ are achieved for $n>400$.

R.K. Beatson, G. Goodsell and I have developed some iterative methods for calculating the coefficients of s that make use of the Lagrange functions of interpolation, this approach being an alternative to conjugate gradients. The efficiency of one of these methods for thin plate spline interpolation when $d=2$ is beyond the dreams of avarice, even when the positions of the interpolation points are chosen randomly [25]. Indeed, the number of iterations to satisfy the interpolation equations (1.1) to ten decimal places seems to be fewer than 10 for all values of n. The main features of this algorithm are as follows.

We assume $n > q > 3$, where q is a prescribed integer whose purpose will become clear, the value $q=30$ being recommended. The interpolation points are ordered so that the area of the triangle with the vertices $\underline{x}_{n-2}$, $\underline{x}_{n-1}$ and $\underline{x}_n$ is relatively large, and then, for $k=1,2,\ldots,n-q$, the remaining freedom in the ordering is used to make the least of the distances $\{\|\underline{x}_i-\underline{x}_k\|_2 : i=k+1,k+2,\ldots,n\}$ as small as possible. Further, for each of these values of k, we let $\mathcal{L}_k$ be a set of exactly q of the indices $\{k,k+1,\ldots,n\}$, where k, $n-2$, $n-1$ and n are always in $\mathcal{L}_k$ and where the other elements of the set are chosen to minimize the distances $\|\underline{x}_i-\underline{x}_k\|_2$, $i\in\mathcal{L}_k$. Further, we let the function

$$\widehat{\ell}_k(\underline{x}) = \sum_{j\in\mathcal{L}_k} \lambda_{kj}\,\phi(\|\underline{x}-\underline{x}_j\|_2) + \widehat{p}_k(\underline{x}), \quad \underline{x}\in\mathcal{R}^2, \tag{4.27}$$

where $\widehat{p}_k\in P_1^2$, be the thin plate spline solution to the Lagrange conditions

$$\widehat{\ell}_k(\underline{x}_i) = \delta_{ik}, \quad i\in\mathcal{L}_k. \tag{4.28}$$

The coefficents $\{\lambda_{kj} : j\in\mathcal{L}_k\}$, $k=1,2,\ldots,n-q$, are calculated, which completes the preliminary work of the algorithm, apart from letting $s\equiv 0$ be the initial approximation to the required s.

Each iteration begins with an s that is an estimate of the required interpolant, and it makes $n-q+1$ changes to s. For $k=1,2,\ldots,n-q$, the number

$$\varepsilon_k = \sum_{j\in\mathcal{L}_k} \lambda_{kj}\,\{f_j - s(\underline{x}_j)\} \tag{4.29}$$

is found, and then $(\varepsilon_k/\lambda_{kk})\,\widehat{\ell}_k(\underline{x})$, $\underline{x}\in\mathcal{R}^2$, is added to s, before proceeding to the next value of k. Finally, the function σ, which is the thin plate spline solution of the q equations

$$\sigma(\underline{x}_i) = f_i - s(\underline{x}_i), \quad i=n-q+1,n-q+2,\ldots,n, \tag{4.30}$$

is calculated, and s is replaced by $s+\sigma$. It follows that the last q interpolation equations are satisfied by s at the end of each iteration.

The method depends on the remark that ε_k is a reasonable estimate of the error in the coefficient of $\phi(\|\underline{x}-\underline{x}_k\|_2)$ in the current s, so the addition of $(\varepsilon_k/\lambda_{kk})\,\widehat{\ell}_k$ to s makes the correction that is suggested by the estimate. Fortunately, the coefficient λ_{kk} is always positive, as it has the value $(8\pi)^{-1}I(\widehat{\ell}_k)$ where I is the integral (4.6). It has not been proved analytically that the algorithm converges, but it is known that the rate of convergence is linear, because each iteration multiplies the vector of residuals $f_i - s(\underline{x}_i)$, $i=1,2,\ldots,n$, by a matrix that is independent of the right hand sides. Applying the technique for other choices of ϕ should be straightforward, but more iterations are expected, because the decay of the Lagrange function (4.10) to zero as $\|\underline{x}\|\to\infty$ is particularly favourable for thin plate splines when $d=2$.

This review suggests that no single method for multivariable interpolation always has strong advantages over its competitors. Indeed, piecewise polynomials are very suitable for problems that arise from finite element calculations, the natural neighbour technique causes $s(\underline{x})$ for each $\underline{x}$ to depend only on nearby data in a way that does not demand decisions from the user, the enthusiasm in the recent report of Levin [18] is persuasive, and it is becoming possible to enjoy the "visual pleasantness" of thin plate splines when n is very large. Furthermore, the subject provides a wide range of opportunities for future research.

References

1. Backus, G. and Gilbert, F. (1968). The resolving power of gross earth data, *Geophys. J. Roy. Astr. Soc.*, *16*, 169–205.
2. Beatson, R. K. and Newsam, G. N. (1992). Fast evaluation of radial basis functions: I, *Comput. Math. Applic.*, *24*, 7–19.
3. Beatson, R. K. and Powell, M. J. D. (1992). Univariate interpolation on a regular finite grid by a multiquadric plus a linear polynomial, *IMA J. Numer. Anal.*, *12*, 107–133.
4. Bos, L. and Salkauskas, K. (1989). Moving least-squares are Backus–Gilbert optimal, *J. Approx. Theory*, *59*, 267–275.
5. Buhmann, M. D. (1990). Multivariate cardinal interpolation with radial-basis functions, *Constr. Approx.*, *6*, 225–255.
6. Buhmann, M. D., Dyn, N. and Levin, D. (1995). On quasi-interpolation by radial basis functions with scattered centres, *Constr. Approx.*, *11*, 239–254.
7. Clough, R. W. and Tocher, J. L. (1965). Finite element stiffness matrices for analysis of plates in bending. *Proc. Conf. Matrix Methods in Struct. Mech.*, Wright–Patterson Air Force Base, Ohio, 515–546.
8. Duchon, J. (1977). Splines minimizing rotation-invariant seminorms in Sobolev spaces. *Constructive Theory of Functions of Several Variables, Lecture Notes in Mathematics* *571*, eds. Schempp. W. and Zeller, K., Springer-Verlag (Berlin), 85–100.
9. Dyn, N., Levin, D. and Rippa, S. (1986). Numerical procedures for surface fitting of scattered data by radial functions, *SIAM J. Sci. Statist. Comput.*, *7*, 639–659.
10. Dyn, N., Levin, D. and Rippa, S. (1990). Data dependent triangulations for piecewise linear interpolation, *IMA J. Numer. Anal.*, *10*, 137–154.
11. Farwig, R. (1986). Rate of convergence of Shepard's global interpolation formula, *Math. Comp.*, *46*, 577–590.
12. Franke, R. (1982). Scattered data interpolation: tests of some methods, *Math. Comp.*, *38*, 181–200.
13. Franke, R. and Nielson, G. (1980). Smooth interpolation of large sets of scattered data, *Int. J. Numer. Meth. Eng.*, *15*, 1691–1704.
14. Green, P. J. and Sibson, R. (1978). Computing Dirichlet tessellations in the plane, *Computer Journal*, *21*, 168–173.
15. Greengard, L. and Rokhlin, V. (1987). A fast algorithm for particle simulations, *J. Comput. Phys.*, *73*, 325–348.
16. Hardy, R. L. (1990). Theory and applications of the multiquadric-biharmonic method, *Comput. Math. Applic.*, *19*, 163–208.
17. Lawson, C. L. (1977). Software for C^1 surface interpolation. *Mathematical Software III*, ed. Rice, J. R., Academic Press (New York), 161–194.
18. Levin, D. (1995). Near-best scattered-data approximations in $\mathcal{R}^d$, Technical Report, School of Math. Sciences, Tel-Aviv University.
19. Lorentz, G. G. (1966). *Approximation of Functions*, Holt, Rinehart & Winston (New York).

20. McLain, D. H. (1974). Drawing contours from arbitrary data points, *Computer Journal*, *17*, 318–324.

21. Micchelli, C. A. (1986). Interpolation of scattered data: distance matrices and conditionally positive definite functions, *Constr. Approx.*, *2*, 11–22.

22. Powell, M. J. D. (1992). The theory of radial basis function approximation in 1990. *Advances in Numerical Analysis Volume II: Wavelets, Subdivision Algorithms, and Radial Basis Functions*, ed. Light, W., Clarendon Press (Oxford), 105–210.

23. Powell, M. J. D. (1993). Truncated Laurent expansions for the fast evaluation of thin plate splines, *Numerical Algorithms*, *5*, 99–120.

24. Powell, M. J. D. (1996). A review of algorithms for thin plate spline interpolation in two dimensions, *Advanced Topics in Multivariate Approximations*, eds. Fontanella, F., Jetter, K. and Laurent, P. J., World Scientific (Singapore), 303–322.

25. Powell, M. J. D. (1997). A new iterative algorithm for thin plate spline interpolation in two dimensions, *Annals of Numerical Mathematics*, *4*, 519–527.

26. Powell, M. J. D. and Sabin, M. A. (1977). Piecewise quadratic approximations on triangles, *ACM Trans. Math. Software*, *3*, 316–325.

27. Shepard, D. (1968). A two-dimensional interpolation function for irregularly-spaced data. *Proc. 23rd Nat. Conf. ACM*, Brandon Systems Press Inc. (Princeton), 517–524.

28. Sibson, R. (1980). A vector identity for the Dirichlet tessellation, *Math. Proc. Cam. Phil. Soc.*, *87*, 151–155.

29. Sibson, R. (1981). A brief description of natural neighbour interpolation. *Interpreting Multivariate Data*, ed. Barnett, V., John Wiley & Sons (New York), 21–36.

30. Strang, G. and Fix, G. J. (1973). *An Analysis of the Finite Element Method*, Prentice–Hall (Englewood Cliffs).

31. Watson, D. (1994). *INNGRIDR: An Implementation of Natural Neighbor Interpolation*, PO Box 374, Claremont, WA 6010, Australia.

Large Scale Unconstrained Optimization

Jorge Nocedal

Department of Electrical and Computer Engineering, Northwestern University, Evanston, USA

Abstract

This paper reviews advances in Newton, quasi-Newton and conjugate gradient methods for large scale optimization. It also describes several packages developed during the last ten years, and illustrates their performance on some practical problems. Much attention is given to the concept of partial separability which is gaining importance with the arrival of automatic differentiation tools and of optimization software that fully exploits its properties.

1 Introduction.

This survey focuses on recent developments in large scale unconstrained optimization. I will not discuss advances in methods for small and medium scale problems because fairly comprehensive reviews of this work are given in [55] and [25]. I should stress, however, that small scale optimization remains an active area of research, and that advances in this field often translate into new algorithms for large problems.

The problem under consideration is

$$\min f(x), \tag{1.1}$$

where f is a smooth function of n variables. We assume that n is large, say $n > 500$, and we denote the gradient of f by g.

An important recent development has been the appearance of effective tools for automatically computing derivatives and for detecting partially separable structures. These programs are already having a significant impact in the practice of optimization and in the design of algorithms, and their influence is certain to grow with time. Automatic differentiation provides an excellent alternative for finite differences, which can be unreliable, and for hand-coded derivatives, which are error prone and labor intensive. The automatic detection of partially separable structure has only recently become available and its full impact is yet to be seen. It has the potential of making automatic differentiation practical for many large problems, and it may also popularize partially separable quasi-Newton methods.

This paper is organized around the three main classes of algorithms for unconstrained optimization: Newton, quasi-Newton and conjugate gradients. To set the stage for the description and analysis of algorithms, we begin with a discussion of some problem structures that arise in many important areas of application. We devote much attention to this topic because understanding the characteristics of the objective function is crucial in large scale optimization.

2 Problem structure

A well-known type of structure is that of sparsity in the Hessian matrix $\nabla^2 f$. Functions with sparse Hessians occur often in practice, and algorithms that exploit it have been developed since the 1970s [14,16]. Sparse finite difference techniques [15,20] have played a crucial role in making these methods economical and practical.

But many problems do not possess a sparse Hessian, and it is useful to divide these into two broad categories: problems that have some kind of structure, such as partial separability, and problems that do not possess any useful structure. In this section we will give a close look at partially separable problems because they are not as well understood as they should be, and because significant advances in solving them have been made in the last ten years. We will see that partial separability leads to economical problem representation, efficient automatic differentiation and effective quasi-Newton updating. Another attempt to identify useful problem structure is given in [46], and an interesting question is whether there are other types of structures that are as amenable to optimization as partial separability.

2.1 Partial separability

In many large problems the objective function can be written as a sum of much simpler functions; such an objective function is called partially separable. All functions with sparse Hessians can be shown to have this property, but so do many functions whose Hessian is not sparse. Some other problems cannot be written as the sum of simple functions, but still possess a structure similar to partial separability called group partial separability [17]. The goal of this section is to explain why this kind of structure is important in optimization.

The simplest form of partial separability arises when the objective function can be written as

$$f(x) = \sum_{i=1}^{ne} f_i(x), \tag{2.1}$$

where each of the *element functions* f_i depends only on a few variables.

An example is the minimum surface problem which is described in many calculus textbooks. We wish to find the surface of minimum area that interpolates a given continuous function on the boundary of the unit square. By discretizing this problem we obtain an objective function of the form Equation 2.1, where the element functions are defined as

$$f_i(x) = \frac{1}{m^2}\left[1 + \frac{m^2}{2}[(x_j - x_{j+m+1})^2 + (x_{j+1} - x_{j+m})^2]\right]^{\frac{1}{2}}. \qquad (2.2)$$

Thus each f_i, which is formally a function of all the variables $x_1, ..., x_n$, depends only on 4 variables, which are called the *element variables*. In Equation 2.2 m is a constant that determines the fineness of the discretization; the number of element functions ne and the total number of variables n are both equal to m^2. (The precise relationship between the indices i and j in Equation 2.2 is not important in the following discussion; it suffices to say that j is determined by i and m.) The gradient of f_i with respect to all the variables in the problem is

$$\nabla f_i(x) = \frac{1}{2m^2} f_i^{-1}(x) \begin{bmatrix} 0 \\ \vdots \\ x_j - x_{j+m+1} \\ x_{j+1} - x_{j+m} \\ \vdots \\ -x_{j+1} + x_{j+m} \\ -x_j + x_{j+m+1} \\ \vdots \\ 0 \end{bmatrix}. \qquad (2.3)$$

We note that, of the four non-zero components, two are negatives of each other. The Hessian of the element function f_i has the sparsity pattern

$$\begin{bmatrix} & & & & & \\ & * & * & & * & * \\ & * & * & & * & * \\ & & & & & \\ & * & * & & * & * \\ & * & * & & * & * \\ & & & & & \end{bmatrix}. \qquad (2.4)$$

A close examination shows that some of the nonzero entries differ only in sign, and that only three different magnitudes are represented. Moreover the 4×4 matrix formed by the non-zero elements happens to be singular.

The fact that repeated information is contained in the gradient and Hessian suggests that there is a compact representation that avoids these redundancies. The key observation is that f_i is *invariant* in the subspace

$$N_i = \{w \in R^n : w_j = w_{j+m+1} \quad \text{and} \quad w_{j+1} = w_{j+m}\}, \qquad (2.5)$$

which means that for any x and for any $w \in N_i$ we have that $f_i(x+w) = f_i(x)$. In other words, any move along a direction in N_i does not change the value of the function, and it is not useful to try to gather curvature information about f_i along N_i. Since dimension$(N_i) = n-2$, we seek a compact representation of f_i involving only two variables.

To accomplish this we define the *internal variables* u_j and u_{j+1} by

$$u_j = x_j - x_{j+m+1} \quad u_{j+1} = x_{j+1} - x_{j+m}, \tag{2.6}$$

and the *internal function* ϕ_i,

$$\phi_i(u_j, u_{j+1}) = \frac{1}{m^2}\left[1 + \frac{m^2}{2}(u_j^2 + u_{j+1}^2)\right]^{\frac{1}{2}}. \tag{2.7}$$

We have achieved our goal of finding a minimal representation of f_i since the gradient of ϕ has only two components and $\nabla^2 f_i$ has only 3 distinct elements.

It is useful to introduce the $2 \times n$ matrix U_i, which has zero elements except for

$$U_{1,j} = 1, \quad U_{1,j+m+1} = -1, \quad U_{2,j+1} = 1, \quad U_{2,j+m} = -1.$$

We can now write the objective function as

$$f(x) = \sum_{i=1}^{ne} \phi_i(U_i x). \tag{2.8}$$

The gradient and Hessian of f are given by

$$g(x) = \sum_{i=1}^{ne} U_i^T \nabla\phi_i(U_i x) \quad \text{and} \quad \nabla^2 f(x) = \sum_{i=1}^{ne} U_i^T \nabla^2\phi_i(U_i x) U_i, \tag{2.9}$$

which clearly exhibits the structure of the objective function. In Section 4.1 we describe a quasi-Newton method that updates approximations to each of the Hessians $\nabla^2\phi_i$. Since these are 2×2 matrices a good approximation can often be obtained after only a few iterations. We will also see that this representation of the gradient suggests efficient automatic differentiation techniques.

We now generalize the minimum surface problem and give the following definition.

> A function f is said to be *partially separable* if it is the sum of element functions, $f(x) = \sum_{i=1}^{ne} f_i(x)$, each of which has a nontrivial invariant subspace. This means that f can also be written in the form Equation 2.8, where the matrices U_i have dimension $n_i \times n$, with $n_i < n$.

In many practical problems the n_i are very small (say 2 or 4), and are independent of the number of variables n which can be in the thousands or more. To take advantage of partial separability we define the internal variables (and thus U_i) so that a change in any one of these variables causes a change in the element function f_i. It would be wasteful to define an internal variable that lies in the subspace N_i, since we know in advance that the derivatives of f_i with respect to this variable will be zero; moreover this would be harmful to the quasi-Newton method that exploits partial separability described in Section 4.1. Thus the number of internal variables equals the dimension of $N_i^{\perp}$, or

$$n - \text{dimension}(N_i).$$

Sometimes the invariant subspace N_i is easy to find, as in the minimum surface example, but this is not always the case. Software tools for automatically detecting partially separable structures are currently under development [30].

2.1.1 Sparsity vs partial separability

The concept of partial separability is more general than the notion of sparsity. It is shown in [35] that every twice continuously differentiable function $f : R^n \rightarrow R$ with a sparse Hessian is partially separable. But the converse is not true.

Consider the element function [17]

$$f_i(x) = (x_1 + \ldots + x_n)^2, \tag{2.10}$$

whose gradient and Hessian are dense. Since the invariant subspace of Equation 2.10 is the set

$$N_i = \{w \in R^n \,|\, e^T w = 0\}, \tag{2.11}$$

where $e = (1, 1, \ldots, 1)^T$, and since the dimension of N_i is $n - 1$, we see that f_i can be considered as a function of only one variable. Thus we can write it in the form Equation 2.8, where

$$U = [1, \ldots, 1], \qquad \text{and} \qquad \phi(u) = u^2.$$

This is an example of a function with a dense Hessian, but a very large invariant subspace and very simple structure.

A similar, but more realistic example, is a protein folding problem that has received much attention from the optimization community [39,52]. The objective is to minimize the energy of a configuration of atoms. If the positions of the atoms are expressed in Cartesian coordinates, then the element functions depend only on the differences of the coordinates of pairs of atoms, $p_i - p_j$, and every row of U_i contains only two nonzeros. Typical values for n_i are 3, 6 or 9. This protein folding problem is thus partially separable but one can show that its Hessian is completely dense. The components of the Hessian that correspond to atoms that end up being widely separated will be very small (and could be set to zero), but the locations of these small entries is not known beforehand.

2.1.2 Group partial separability

Partial separability is an important concept, but it is not quite as general as we would like. Consider a nonlinear least squares problem of the form

$$f(x) = \sum_{k=1}^{l} (f_k(x) + f_{k+1}(x) + c)^2, \tag{2.12}$$

where the functions f_j are partially separable and where c is a constant. The definition of partial separability given above forces us to regard the whole k-th term in the summation as the k-th element function. However there is clearly a lot of structure inside that term — it contains two element functions, grouped together and squared. We can extend the definition of partial separability slightly to make better use of this type of structure.

We say that a function $f : R^n \to R$ is *group partially separable* if it can be written in the form

$$f(x) = \sum_{k=1}^{l} \gamma_k(h_k(x)) \tag{2.13}$$

where γ_k is twice continuously differentiable function on the range of $h_k(\cdot)$, and where each h_k is a partially separable function from R^n to R.

We can use the form of f to find compact representations of the derivatives. By using the chain rule we have

$$\nabla[\gamma_k(h_k(x))] = \gamma_k'(h_k(x))\nabla h_k(x), \tag{2.14}$$

$$\nabla^2[\gamma_k(h_k(x))] = \gamma_k''(h_k(x))\nabla h_k(x)\nabla h_k(x)^T + \gamma_k'(h_k(x))\nabla^2 h_k(x). \tag{2.15}$$

We already know how to represent the derivatives of the partially separable function h_k (that is, ∇h_k and $\nabla^2 h_k$) in compact form. To get a compact representation of the derivatives of $\gamma_k(h_k(x))$, we simply have to include the two scalar quantities γ_k' and γ_k'', both evaluated at $h_k(x)$.

Group partial separability is a very general concept since it applies directly to nonlinear least squares problems and to penalty and merit functions arising in constrained optimization. The LANCELOT package [17] is designed to fully exploit its structure.

2.1.3 Automatic differentiation of partially separable functions

Automatic differentiation is based on the observation that any function f that can be evaluated by a computer program is executed as a sequence of elementary operations (such as additions, multiplications and trigonometric functions). By systematically applying the chain rule to the composition of these elementary functions, the gradient ∇f can be computed to machine accuracy [36]. There are two basic strategies for applying the chain rule: the *forward* and *reverse modes* of differentiation.

In the forward mode one proceeds in the direction determined by the evaluation of the function; as we evaluate f, we compute the derivatives of all the intermediate quantities with respect to the variables x of the problem. In the reverse mode one first evaluates the function; when this is completed we move backwards, computing the derivates of f with respect to the intermediate quantities arising in the evaluation of f, until we obtain the derivatives of f with respect to the problem variables x.

Various automatic differentiation codes have been developed (see [5] for references), and have proved to be effective in practice. I will now discuss two recent proposals on how to take advantage of partial separability to make automatic differentiation more efficient.

The approach described in [5] concerns the computation of the gradient g. It begins with the simple representation Equation 2.1, and associates with it the vector function

$$F(x) = \begin{pmatrix} f_1(x) \\ \vdots \\ f_{ne}(x) \end{pmatrix}.$$

Since the Jacobian $F'(x)$ is given by $F'(x)^T = (\nabla f_1(x), ..., \nabla f_{ne}(x))$ we have from Equation 2.1 that

$$g(x) = F'(x)^T e, \tag{2.16}$$

where $e = (1, 1, ..., 1)$. Therefore the gradient of a partially separable function could be computed via Equation 2.16 if we had the associated Jacobian $F'(x)$. It would seem that the latter is expensive to compute, but since each f_i depends only on a few variables, $F'(x)$ is sparse. By analyzing the sparsity structure of $F'(x)$ it is possible to find a set of structurally orthogonal columns [15,20], i.e. columns that do not have a nonzero entry in the same row. This information is provided to the automatic differentiation code, which is then able to compute $F'(x)$ in a compressed form that economizes storage and computation [5]. The gradient $g(x)$ is then computed by means of Equation 2.16.

Another recent proposal [30] we wish to discuss concerns the computation of the Hessian matrix. It is based on the fact that Hessian-vector products can easily be computed by automatic differentiation, without the need to calculate Hessians [36]. To compute $\nabla^2 f(x_k)d$ automatically is conceptually straightforward. We can simply apply backward automatic differentiation to compute the gradient of $d^T \nabla f(x)$, considering d constant.

The complete Hessian $\nabla^2 f(x)$ can therefore be obtained column by column by applying automatic differentiation to compute the n products $\nabla^2 f(x)e_i$. But it is preferable to consider the representation Equation 2.8 and compute the small internal Hessians $\nabla^2 \phi_i$ explicitly. This is done, for each element function ϕ_i, by means of a few products $\nabla^2 \phi_i e_j$. The total Hessian is obtained via

Equation 2.9, which is a sum of outer products [30]. A new FORTRAN 90 version of LANCELOT can compute the second derivatives of partially separable functions by automatic differentiation.

An automatic procedure for detecting the partially separable decomposition Equation 2.8 is also given in [30]. It consists of analyzing the composition of elementary functions forming f, and finding the transformations U_i and the internal functions ϕ_i. This procedure finds one particular partially separable representation of f (the "finest"), and it is too early to tell if it will be of wide applicability. If successful, it could popularize partially separable optimization methods.

We conclude this section on partial separability by noting that sparsity plays a major role in many existing codes for large scale optimization. Partial separability is now being proposed as a structure of wider applicability that could supersede sparsity, but as we have seen, the approach Equation 2.16 based on the compressed Jacobian matrices makes use of partial separability *and* sparsity. It is difficult to predict what the relative importance of these concepts will be ten years from now.

3 Newton's method

This is the most powerful algorithm for solving large nonlinear optimization problems. It normally requires the fewest number of function evaluations, is very good at handling ill-conditioning, and is capable of giving the most accurate answers. It may not always require the least computing time – this depends on the characteristics of the problem and the implementation of the Newton iteration – but it represents the most reliable method for solving large problems.

As is well known, to be able to apply Newton's method, the gradient g and Hessian $\nabla^2 f$ must be available. The gradient could be computed analytically or by automatic differentiation, and there are several options for providing the Hessian matrix: it could be supplied in analytic form, could be approximated by sparse finite differences [15,20], or could be computed by automatic differentiation techniques. Future problem-solving environments will allow the user to choose from these alternatives.

The newest implementations of Newton's method are simple and elegant, and deal well with the case when the Hessian matrix is indefinite. They construct a quadratic model of the objective function and define a trust region over which the model is considered to be reliable. The model is approximately minimized over the trust region by means of an ingenious adaptation of the conjugate gradient (CG) algorithm. Careful attention is given to the issue of preconditioning which is crucial for solving ill-conditioned problems. The routine SBMIN of the LANCELOT package, and the code TRCG that will be part of the Minpack-2 package follow this approach, which we now describe in some detail.

Let us denote the Hessian matrix by B, i.e. $\nabla^2 f(x_k) = B_k$. At the iterate x_k we formulate the trust region subproblem

$$\min\ \phi(p) = g_k^T p + \frac{1}{2} p^T B_k p \tag{3.1}$$

$$\text{subject to } \|p\|_2 \leq \Delta_k, \tag{3.2}$$

where the trust region radius Δ_k is updated at every iteration. Obtaining the exact solution of this subproblem is expensive when n is large – even when B_k is positive definite. It is more efficient to compute an approximate solution p that is relatively inexpensive (at least in the early stages of the algorithm) and gives a fast rate of convergence. This is done by first ignoring the trust region constraint and attempting to minimize the model Equation 3.1 by means of the CG method [34]. Thus we apply CG to the symmetric linear system

$$B_k p = -g_k, \tag{3.3}$$

starting with the initial guess $p_0 = 0$. The key points are how to take into account the trust region constraint and how to avoid failure of the CG iteration when B_k is not positive definite. For this purpose three different stopping steps are used, which we discuss later on. Once the approximate solution p has been obtained, we test whether $f(x_k + p_k)$ is sufficiently less than $f(x_k)$. If so, we define the new iterate as $x_k + p_k$, and update the trust region Δ_k according to how well the model ϕ estimates the nonlinear objective function f. On the other hand, if sufficient reduction in f is not obtained, we reduce the trust region radius Δ_k and find a new approximate solution to Equations 3.1 and 3.2.

We now focus on the case when the Hessian matrix B_k is not positive definite. The first stopping test stipulates that the CG iteration will be terminated as soon as negative curvature is detected. When this occurs, the direction of negative curvature is followed to the boundary of the trust region, and the resulting step is returned as the approximate solution. The second test monitors the length of the approximate solutions $\{p_j\}$ generated by the CG iteration and terminates when one of them exceeds the trust region radius. A third test is included to end the CG iteration if the linear system Equation 3.3 has been solved to the required precision.

This strategy for computing the search direction of the trust region Newton method is summarized in Algorithm 1. The iterates generated by the CG method are denoted by $\{p_j\}$, whereas the conjugate directions computed by the CG iteration are $\{d_j\}$. The differences between this algorithm and standard conjugate gradient are the two extra stopping conditions formulated as the first two IF statements within the LOOP. When negative curvature is encountered or when p_{j+1} violates the trust region constraint, a final estimate p is found by intersecting the current search direction with the the trust region boundary.

The third IF statement ensures that the CG iteration is terminated when the residual is sufficiently small compared to the initial residual – which equals the current gradient g_k of the objective function. Therefore as the Newton algorithm approaches the solution and g_k converges to zero, the termination test becomes ever more stringent, ensuring that the rate of convergence is fast [21]. Indeed, as the iterates approach the solution, the Hessian B_k will become positive definite, and one can show [26] that the standard strategies for updating the trust region radius [22] ensure that the trust region constraint becomes inactive. Thus, asymptotically, this method reduces to a pure truncated Newton method with unit steplengths.

Algorithm 1 Approximate computation of the Newton step by CG.

Constant $\epsilon > 0$ is given
Start with $p_0 = 0$, $r_0 = g_k$, and $d_0 = -r_0$
LOOP, starting with $j = 0$
  IF $d_j^T B_k d_j \le 0$
    THEN find τ so that $p = p_j + \tau d_j$ minimizes $\phi(p)$
      and satisfies $\|p\|_2 = \Delta$, and RETURN p
  $\alpha_j = r_j^T r_j \,/\, d_j^T B_k d_j$
  $p_{j+1} = p_j + \alpha_j d_j$
  IF $\|p_{j+1}\|_2 \ge \Delta$
    THEN find $\tau \ge 0$ so that $\|p_j + \tau d_j\|_2 = \Delta$,
      and RETURN $p = p_j + \tau d_j$
  $r_{j+1} = r_j + \alpha_j B_k d_j$
  IF $\|r_{j+1}\|_2 \,/\, \|r_0\|_2 < \epsilon$ THEN RETURN $p = p_{j+1}$
  $\beta_{j+1} = r_{j+1}^T r_{j+1} \,/\, r_j^T r_j$
  $d_{j+1} = r_{j+1} + \beta_{j+1} d_j$
CONTINUE, after incrementing j

Note that first estimate p_1 generated by the inner CG iteration is given by

$$p_1 = \alpha_0 d_0 = -\alpha_0 g_k,$$

where α_0 is the steplength that minimizes the quadratic model ϕ along the steepest descent direction $-g_k$ at the current iterate x_k. This is also called the Cauchy step, and guarantees that the trust region algorithm is globally convergent [59]. Subsequent inner CG iterations reduce ϕ and improve the quality of the search direction. But no attempt is made to try to ensure that the iterates converge only to solution points where the Hessian is positive definite.

It is important that the first estimate in the inner CG iteration be $p_0 = 0$, for in this case one can show [63] that each estimate is longer than the previous one. To be more precise, we have

$$0 = \|p_0\|_2 < \cdots < \|p_j\|_2 < \|p_{j+1}\|_2 < \cdots < \|p\|_2 \le \Delta.$$

This property shows that it is acceptable to stop iterating as soon as the trust region boundary is reached because all subsequent estimates will lie outside the trust region.

When the Hessian matrix B is positive definite, this approach is similar to the dogleg method [58] because the estimates generated by the inner CG iteration sweep out points that move on some interpolating path from the Cauchy step p_1 to the Newton step $p_N = -B_k^{-1} g_k$.

3.1 Preconditioning

An important question is how to precondition the CG iteration, i.e. how to find a nonsingular matrix D such that the eigenvalues of $D^{-T} B_k D^{-1}$ are clustered. The LANCELOT package [17] has the excellent feature of providing a suite of built-in preconditioners with which the user can experiment. The default is an 11-band modified Cholesky factorization, but many other choices can easily be activated. The choice of preconditioner has a marked effect on performance, but it is difficult to know in advance what the best choice is for a particular problem. This is a complex issue; for example there is a trade-off between the quality of the preconditioner and the computational work involved.

In LANCELOT the preconditioned conjugate gradient method is applied to Equation 3.3, the length of the estimates $||p_j||$ is monitored, and the CG iteration is terminated if $||p_j|| \geq \Delta_k$. However, this is not totally consistent with the ideas embodied in Algorithm 1 because when preconditioning is used the quantity that grows monotonically is $||Dp_j||$ and not $||p_j||$. Therefore it is possible for LANCELOT to terminate the inner CG iteration even though a subsequent iterate would fall inside the trust region and give a lower value of the model ϕ.

The routine TRCG from Minpack-2 takes a different approach, in that preconditioning is seen as a scaling of the trust region. Let D be any nonsingular matrix and consider the subproblem

$$\min_{p \in \mathbf{R}^n} \; \phi(p) = g_k^T p + \frac{1}{2} p^T B_k p \tag{3.4}$$

$$\text{subject to } ||Dp||_2 \leq \Delta_k. \tag{3.5}$$

By making the change of variables $\hat{p} = Dp$ and defining

$$\hat{g}_k = D^{-T} g_k, \quad \hat{B}_k = D^{-T} B_k D^{-1}$$

we obtain a problem of the form Equations 3.1 and 3.2 to which Algorithm 1 can be applied directly; the answer is then transformed back into the original variables p. The length of $||\hat{p}|| = ||Dp||$ is monitored, and this is the quantity that grows monotonically. Therefore, unlike the implementation in LANCELOT, once the CG iteration generates an estimate that leaves the trust region, it will never return. It is not clear how this difference in the handling of the trust region affects performance, and this question deserves some investigation.

Table 1. Performance of a preconditioned Newton method

Problem	n	iter	feval	time
GL2	2500	6	12	13.0
GL2	10,000	6	10	63.4
GL2	40,000	8	12	537.7
MSA	2500	6	7	2.8
MSA	10,000	6	7	2.8
MSA	40,000	10	14	91.1

Minpack-2 uses an incomplete (and possibly modified) Cholesky factorization as a preconditioner because computational experience in linear algebra has shown that this type of preconditioner can be effective for a large class of matrices. The incomplete Cholesky factorization of a positive definite matrix B_k finds a lower triangular factor L such that $B_k = LL^T - R$, where L reflects the sparsity of B_k. The scaling factor used in Equation 3.5 is set to $D = L$. Since in Newton's method B_k may not be positive definite, it may be necessary to modify it so that the Cholesky factorization can be performed. A typical strategy for doing this is described below. It begins by scaling the matrix B, since numerical experience indicates that this can be beneficial.

Incomplete and modified Cholesky factorization

The symmetric matrix B is given.

1. (Scale B.) Let $T = \text{diag}(\|Be_i\|)$, where e_i is the i-th unit vector. Define $\bar{B} = T^{-\frac{1}{2}}BT^{-\frac{1}{2}}$, and $\beta = \|\bar{B}\|$.
2. (Compute a shift to attempt to ensure positive definiteness.) If $\min(B_{i,i}) > 0$, set $\alpha_0 = 0$; otherwise set $\alpha_0 = \beta/2$.
3. Loop starting with $k = 0$.
   - Attempt to compute the incomplete Cholesky factorization [42] of $\bar{B} + \alpha_k I$. If the factorization is completed successfully exit and return L; otherwise set $\alpha_{k+1} = \max(2\alpha_k, \frac{1}{2}\beta)$, and continue loop.

We should note that scaling alters the incomplete Cholesky factorization, and that several other choices for the scaling matrix T can be used.

Table 1 presents results obtained by Bouaricha and Moré [7] on two problems from the Minpack-2 collection [4]. GL2 is a problem arising in the modeling of superconducting materials, and MSA is a minimum surface problem.

These are very good results in that the number of iterations stays nearly constant as the dimension of the problem increases, and computing time grows only a little faster than linearly. The results also compare very favorably with those obtained with the limited memory BFGS code described in Section 4.2, which has considerable difficulty solving the problems as the dimension increases. I do not know, however, if this kind of performance is typical of the Minpack-2 code, and more computational experience is necessary.

3.2 Current research and open questions

The Newton method in LANCELOT takes advantage of partial separability by computing the Hessians of the internal functions ϕ_i defined in Equation 2.8. This can give significant savings in storage in some problems, but the computational overhead required by this structured representation of the function can sometimes be onerous. Similarly, in the Minpack-2 code the preconditioned CG iteration can sometimes be quite expensive. Therefore, even though these Newton codes are highly successful, it is interesting to ask if the information generated in the course of the step computation can be used more thoroughly.

In the *Iterated Subspace Minimization Method* [18] a search direction is first computed using Algorithm 1. However, instead of accepting this step, one selects a few of the estimates $\{p_j\}$ generated during the inner CG iteration, in order to define a subspace S_k over which the nonlinear objective function f is minimized further. Since the subspace S_k is chosen to be of small dimension, this is a low-dimensional nonlinear optimization problem that is solved by means of the BFGS method (using second derivatives would require projections onto the subspace S_k which can be unnecessarily costly). The Iterated Subspace Minimization method therefore requires several evaluations of the objective function to produce a new iterate, and one can expect it to be most effective when the function is not too costly to evaluate. This approach is based on the assumption that CG determines important directions over which it is worth exploring the nonlinear objective function, and that the benefit of this exploration outweights the cost of additional function evaluations.

Numerical tests with the Iterated Subspace Minimization method show some promise, but a clear improvement in performance over the standard Newton method has not been observed. There is room, however, for further research. For example, there are many possible choices for the subspace S_k, and this selection could be crucial to the efficiency of the method. One could also try to develop an automatic mechanism that determines when the exploration of the subspace S_k is to be performed.

There are several open questions concerning the implementation of Algorithm 1. The first is the use of the residual in the stopping test

$$||r_{j+1}|| \leq \epsilon ||r_0||, \tag{3.6}$$

that determines that the linear system Equation 3.3 has been approximately solved. On ill-conditioned problems, the residual can oscillate greatly during the course of the CG iteration, and only drop sharply at the end. Some researchers (see for example [50]) propose stopping tests based on the reduction of the model ϕ, and others use an angle test [43], but no systematic comparison of these alternatives has been undertaken in the setting of nonlinear optimization. For a good recent discussion on CG stopping rules see [23].

Assuming that a residual-based stopping test is used, Equation 3.6 may not be its best implementation. It is known [40] that the test Equation 3.6 may be too stringent when ϵ is a small and when the approximate solution p of Equation 3.3 is large compared with r_0. It may be preferable to use $\|r_k\| \leq \epsilon(\|B_k\|\|p\| + \|r_0\|)$. The effects of rounding errors can also be important. Since r_k is not computed directly, but recurred, $\|r_k\|$ can differ from its true value by several orders of magnitude. A point in favor of a residual-based stopping test is that it allows us to easily control the rate of convergence of the optimization algorithm [21].

Another open question concerning the implementation of Algorithm 1 is the use of the conjugate gradient method. Since B_k can be indefinite, the CG iteration can become unstable [57], and it is not clear that the direction given by Algorithm 1 is always of good quality. An alternative [48] is to use the Lanczos iteration and continue past the point where negative curvature is first detected. It is also possible to alter the CG iteration, after encountering negative curvature, so that it can continue exploring other subspaces [3]. I do not know if significant improvements in performance can be realized with these proposals, but this question is important and deserves careful investigation.

Perhaps more important than any of these issues is the choice of preconditioner. The Hessian matrix B_k can change drastically during the course of the optimization iteration, and to use the same preconditioning strategy throughout the run is questionable. The idea of having a dynamic preconditioner is appealing but, to the best of my knowledge, has not been studied in the context of large scale nonlinear optimization.

3.3 Hessian free Newton method

This is a modification of Newton's method that allows us to solve problems where the Hessian $B_k \equiv \nabla^2 f(x_k)$ is not available. It is based on the observation that the conjugate gradient iteration in Algorithm 1 only needs the product of $\nabla^2 f(x_k)$ with certain displacement vectors d_j, and does not require the Hessian matrix itself. These products can be approximated by finite differences,

$$\nabla^2 f(x_k) d \approx \frac{g(x_k + \epsilon d) - g(x_k)}{\epsilon}, \tag{3.7}$$

where ϵ is a small differencing parameter. Since each iteration of the conjugate gradient method performs one product $\nabla^2 f(x_k) d_j$, this approach requires a new evaluation of the gradient g of the objective function at every CG inner iteration. A method that uses Equation 3.7 is called a discrete Newton method and has received considerable attention [49,50,56,61].

The finite-difference Equation 3.7 is unreliable and may deteriorate the performance of the Hessian free Newton method. An attractive alternative is to compute the product $\nabla^2 f(x_k)d$ by automatic differentiation [36], as discussed in Section 2.1.3. This has the important advantage of being accurate (in general, as accurate as the computation of the function).

In general, however, computing the Hessian-vector product by automatic differentiation will be as expensive (or more) as finite differences in terms of computing time. Therefore the Hessian-free Newton method can be made more reliable by automatic differentiation, but its overall cost remains high. Its efficiency will be highly dependent on the termination test used in the inner CG iteration. If only one CG iteration is performed, the step computation is inexpensive but the method reduces to steepest descent, whereas high accuracy in the CG iteration results in an approximation to Newton's method but a high computational cost [51]. Thus the proper termination of the inner CG iteration is even more delicate in this context than in the case when the Hessian is available.

An important open question is how to develop general purpose preconditioners for Hessian free Newton methods. In some applications it is useful to compute part of the Hessian (or a modification of it) and use it as a preconditioner. But in other applications this is not practical or effective, and the design of simple preconditioners that require no user intervention would greatly enhance the value of Hessian free Newton methods.

4 Quasi-Newton methods

There have been various attempts to extend quasi-Newton updating to the large-scale case, and two of them have proved to be very successful in different contexts. The first idea consists of exploiting the structure of partially separable functions by updating approximations to the Hessians of the internal functions Equation 2.8. This gives rise to a powerful algorithm of wide applicability, and whose only drawback is the need to fully specify a partially separable representation of the function. The second approach is that of limited memory updating in which only a few vectors are kept to represent the quasi-Newton approximation to the Hessian. Limited memory methods require minimal input from the user and are best suited for problems that do not possess a structure that can be exploited economically. They are not as robust and as rapidly convergent as partially separable quasi-Newton methods, but are probably much more widely used.

An approach that has not yet proved to be successful is that of designing sparse quasi-Newton updating formulae. But some new ideas that deserve attention have recently been proposed.

4.1 Partially separable quasi-Newton methods

Suppose that we know how to break a function f down into partially separable form Equation 2.8, i.e. that we have identified ne, the transformations U_i and ϕ_i. Rather than computing the Hessians of the internal functions ϕ_i, we can store and update quasi-Newton approximations B_i to each individual $\nabla^2\phi_i$. We then estimate the entire Hessian $\nabla^2 f(x)$ (see Equation 2.9) by

$$B = \sum_{i=1}^{ne} U_i^T B_i U_i. \tag{4.1}$$

The $n \times n$ matrix B can then be used in Algorithm 1, resulting in a partially separable quasi-Newton method using trust regions.

As for any quasi-Newton method, the approximations B_i are updated by requiring them to satisfy the following secant equation for each element:

$$B_i s_{[i]} = y_{[i]}. \tag{4.2}$$

Here

$$s_{[i]} = u^+_{[i]} - u_{[i]} \tag{4.3}$$

is the change in the internal variables corresponding to the i-th element function, and

$$y_{[i]} = \nabla\phi_i(u^+_{[i]}) - \nabla\phi(u_{[i]}) \tag{4.4}$$

is the corresponding change in gradients, where u^+ indicates the most recent iterate and u the previous one.

The success of this element-by-element updating technique can be understood by returning to the minimum surface problem Equation 2.2. In this case, the functions ϕ_i depend only on two internal variables, so that each Hessian approximation is 2×2. After just a few iterations, we will have sampled enough directions $s_{[i]}$ to make B_i an accurate approximation to $\nabla^2\phi_i$. Hence Equation 4.1 will be a very good approximation to $\nabla^2 f(x)$.

It is interesting to contrast this with a quasi-Newton method that ignores the partially separable structure of the objective function. This method will attempt to estimate the total average curvature (i.e. the sum of the individual curvatures) by constructing an $n \times n$ matrix. When the number of variables is large, after k iterations with $k << n$, this quasi-Newton matrix will not resemble the true Hessian well, and will not make very rapid progress towards the solution.

The partially separable quasi-Newton updating cannot always be performed by means of the BFGS formula because there is no guarantee that the curvature condition $s_{[i]}^T y_{[i]} > 0$ will be satisfied; this condition is needed to ensure that the BFGS approximation is well defined [22]. Even if a line search is used to guarantee that the total step $s_k = x_{k+1} - x_k$ satisfies the curvature condition, this would not imply that the changes of the individual components $u_{[i]}$ would satisfy it. In fact one of the element Hessians $\nabla^2\phi_i$ could be concave in a region around x_k.

One way to overcome this obstacle is to use SR1 [19] to update each of the element Hessians, with simple precautions to ensure that it is always well-defined. Or one could start with the BFGS update formula and, if at some stage the curvature condition for a particular element is not satisfied, then switch to the SR1 formula and use it until termination [65]. In LANCELOT, which is designed to take full advantage of partial separability, the user must choose the quasi-Newton update to be used, and SR1 turns out to be much more effective than BFGS in this context in which the curvature condition is often lacking Computational experience on the CUTE [6] test problem collection suggests that the partially separable SR1 method implemented in LANCELOT is nearly as effective as Newton's method.

4.2 Limited memory methods

Various limited memory methods have been proposed; some combine conjugate gradient and quasi-Newton steps [9], [10], and others are very closely related to quasi-Newton methods. The simplest implementation, and perhaps the most efficient, is the limited memory BFGS method (L-BFGS) [32,45,54]. It is a line search method in which the search direction has the form

$$d_k = -H_k g_k. \tag{4.5}$$

The inverse Hessian approximation H_k, which is not formed explicitly, is defined by a small number of BFGS updates. In the standard BFGS method, H_k is updated at every iteration by means of the formula

$$H_{k+1} = V_k^T H_k V_k + \rho_k s_k s_k^T, \tag{4.6}$$

where

$$\rho_k = 1/y_k^T s_k, \quad V_k = I - \rho_k y_k s_k^T, \tag{4.7}$$

and

$$s_k = x_{k+1} - x_k, \quad y_k = g_{k+1} - g_k.$$

The $n \times n$ matrices H_k are generally dense, so that storing and manipulating them is impractical when the number of variables is large. To circumvent this problem, the limited memory BFGS method does not form these matrices but only stores a certain number, say m, of pairs $\{s_k, y_k\}$ that define them implicitly through the BFGS update formula Equations 4.6 and 4.7. Two important features of the method, which we now describe, are a rescaling (or resizing) strategy, and the continuous refreshing of the curvature information.

Suppose that the current iterate is x_k and that we have stored the m pairs $\{s_i, y_i\}$, $i = k-m, ..., k-1$. We first define the basic matrix $H_k^{(0)} = \gamma_{k-1} I$ where

$$\gamma_{k-1} = \frac{s_{k-1}^T y_{k-1}}{y_{k-1}^T y_{k-1}}. \tag{4.8}$$

We then (formally) update $H_k^{(0)} m$ times using the BFGS formula Equations 4.6 and 4.7 and the m pairs $\{s_i, y_i\}$, $i = k-m, ..., k-1$. The product $H_k g_k$ is obtained by performing a sequence of inner products involving g_k and these m pairs $\{s_k, y_k\}$. After computing the new iterate, we save the most recent correction pair $\{s_k, y_k\}$ – unless the storage is full, in which case we first delete the oldest pair $\{s_{k-m}, y_{k-m}\}$ to make room for the newest one, $\{s_k, y_k\}$.

This approach is suitable for large problems because it has been observed in practice that small values of m, say $m \in [3, 20]$ often give satisfactory results [32,45]. The numerical performance of the limited memory method L-BFGS is illustrated in Table 2, where we compare it [67] with the Newton method provided by the LANCELOT package on a set of test problems from the CUTE [6] collection.
The total number of function and gradient evaluations is denoted by nfg, and time denotes the total execution time on a Sparcstation-2. The memory parameter for L-BFGS was set to $m = 5$, and LANCELOT was run with all its default settings. The four problems typify situations we have observed in practice. In CRAGGLVY Newton's method required many fewer function evaluations, but the execution times of the two methods were similar; this is a common occurrence. FMINSURF is highly atypical in that L-BFGS required fewer function and gradient evaluations. In the ill-conditioned problem DIXMAANI the advantage of the Newton method is striking in both measures, and illustrates a case in which L-BFGS performs quite poorly. Finally EIGENCLS represents a situation that is not uncommon: even though LANCELOT requires much fewer function evaluations it takes longer to solve the problem.

We can draw three conclusions from our computational experience with Newton and limited memory methods. First, it is clear that the quality of the limited memory matrix is rather poor compared with the true Hessian, as is shown by the wide gap in the number of iterations and function evaluations required for convergence. The second observation is that the relative cost of the L-BFGS iteration is so low that one cannot discount the possibility that it will require less computing time than the Newton method. Finally, L-BFGS is not as reliable as a Newton or partially separable quasi-Newton method. On problems with an unfavorable eigenvalue distribution L-BFGS may require a huge number of

Table 2. Unconstrained problems

Problem	n	L-BFGS		LANC/Newt	
		nfg	time	nfg	time
CRAGGLVY	1000	95	13	15	10
FMINSURF	1024	186	11	316	106
DIXMAANI	1500	1237	166	8	9
EIGENCLS	462	2900	563	543	2300

iterations [8], or may not achieve good accuracy in the answer. These difficulties are sometimes overcome by increasing m, say to 20 or 30, but this is not always the case.

An attractive feature of L-BFGS is that it can easily be generalized to solve bound constrained problems. But in order to obtain an efficient implementation in that case it is necessary to find new representations of limited memory matrices, which we now discuss.

4.2.1 Compact representations of limited memory matrices

The limited memory techniques described so far only store the difference vectors s_i and y_i, and avoid storing any matrices. We now show that limited memory updating can also be described using outer products of matrices. We begin by describing a result on BFGS updating that is interesting in its own right.

Let us define the $n \times k$ matrices S_k and Y_k by

$$S_k = [s_0, \ldots, s_{k-1}], \quad Y_k = [y_0, \ldots, y_{k-1}]. \tag{4.9}$$

It can be shown [11] that if H_0 is symmetric and positive definite, and if H_k is obtained by updating $H_0 k$ times using the BFGS formula Equation 4.6 and the pairs $\{s_i, y_i\}_{i=0}^{k-1}$, then

$$H_k = H_0 + \begin{bmatrix} S_k & H_0 Y_k \end{bmatrix} \begin{bmatrix} R_k^{-T}(D_k + Y_k^T H_0 Y_k) R_k^{-1} & -R_k^{-T} \\ -R_k^{-1} & 0 \end{bmatrix} \begin{bmatrix} S_k^T \\ Y_k^T H_0 \end{bmatrix}, \tag{4.10}$$

where R_k and D_k are $k \times k$ matrices given by

$$(R_k)_{i,j} = \begin{cases} s_{i-1}^T y_{j-1} & \text{if } i \leq j \\ 0 & \text{otherwise} \end{cases}, \tag{4.11}$$

and

$$D_k = \operatorname{diag}\left[s_0^T y_0, \ldots, s_{k-1}^T y_{k-1}\right]. \tag{4.12}$$

It is easy to describe a limited memory implementation based on this representation. We keep the m most recent difference pairs $\{s_i, y_i\}$ in the matrices S_k and Y_k, and H_0 stands for the basic matrix $H_k^{(0)}$ defined through Equation 4.8. The difference pairs are refreshed at every iteration by removing the oldest pair and adding a new one to S_k and Y_k. After this is done, the matrices R_k and D_k are updated to account for these changes.

Note that the inner matrix in Equation 4.10 is of size $2m \times 2m$, i.e. it is very small, so that the total storage of this representation is essentially the same as storing only the difference pairs $\{s_i, y_i\}$. One can show that updating the limited memory matrix and computing the search direction $H_k g_k$ using the compact representation Equation 4.10 costs roughly the same as in the approach described earlier, so that there is no clear benefit from using the compact form in the unconstrained case.

There are, however, many advantages to this approach if we wish to use updating formulae other than BFGS, or if we need to solve problems with bounds on the variables. For example products of the form $H_k A$, where A is a sparse matrix, occur often in constrained optimization and can be performed efficiently using Equation 4.10. In particular, when using the range-space or dual approach [33] to solve linearly constrained subproblems, we need to compute $A^T H_k A$, whose symmetry can be exploited to give further savings in computation.

In constrained optimization, however, it is more common to work with an approximation B_k to the Hessian matrix, rather than with an inverse approximation H_k. One can derive compact representations for the limited memory Hessian approximation B_k that are similar to Equation 4.10. These give rise to considerable savings compared with the simple-minded approach of storing only the correction vectors arising in BFGS updating. There are, in addition, compact representations for the symmetric rank-one (SR1) updating formula, which is particularly appealing in the constrained setting because it is not restricted by the positive definiteness requirement.

The recently developed code L-BFGS-B [12,67] uses a gradient projection approach together with compact limited memory BFGS matrices to solve the bound constrained optimization problem

$$\min f(x)$$

$$\text{subject to} \quad l \leq x \leq u.$$

Table 3 illustrates the performance of L-BFGS-B on bound constrained problems from the CUTE collection. Once more we use the Newton code of LANCELOT as a benchmark [67].

Here nbds denotes the number of active bounds at the solution. We find again that the Newton code of LANCELOT requires much fewer function evaluations, but in terms of computing time L-BFGS-B performs quite well. This may be due to the fact that the compact representations allow us to implement the projected gradient method with minimal computational cost. We should note, however, that L-BFGS-B fails to solve a few of the bound constrained problems in the CUTE collection to reasonable accuracy, and that the Newton method is more reliable in this respect.

Table 3. Bound constrained problems

			L-BFGS-B		LANC/Newt	
Problem	n	nbds	nfg	time	nfg	time
JNLBRNGA	15,625	5,657	332	740	22	1503
LINVERSE	999	338	291	57	28	150
OBSTCLAE	5,625	2,724	258	207	6	1423
TORSION6	14,884	12,316	362	707	9	130

4.2.2 Current research and open questions

We have devoted much attention so far to the L-BFGS method, but this may not be the most economical limited memory method. New algorithms designed to reduce the amount of storage without compromising performance are proposed in [24,28,44,62]; see also [1]. It is easy to see that if the BFGS method is started with an initial matrix that is a multiple of the identity, then $s_k \in span\{g_0, ..., g_k\}$. This suggests that there is some redundancy in storing both s_i and y_i in a limited memory method, and in most of the recent proposals storage is in fact cut in half.

A variety of new limited memory methods have been proposed to realize these savings; some of them are based on ingenious formulae for updating the information containing the quasi-Newton update information. Even though the algorithm proposed in [62] appears to give good performance compared with L-BFGS more analysis and testing is necessary.

Another recently proposed idea [13] is to combine the properties of Newton and limited memory in the *Discrete Newton Method with Memory.* This method attempts to reduce the computational cost of the Hessian free Newton method by saving information from the inner CG iteration and keeping it in the form of a limited memory matrix. Once this information has been gathered, a sequence of limited memory steps is performed until it is judged that a new Hessian free Newton step is needed.

Thus the algorithm interleaves limited memory and Hessian free Newton steps, but it does not simply alternate them. The key is to view the inner CG iteration in Algorithm 1 from the perspective of quasi-Newton methods, and to realize that it may probe the function f along directions of small curvature that would normally be ignored by a limited memory method; this information could improve the quality of the limited memory matrix. The Hessian free Newton step therefore serves the dual purpose of giving good progress towards the solution and of gathering important information for the subsequent limited memory steps.

Good results have been obtained with the Discrete Newton Method with Memory, when solving problems in a controlled setting [13]. In these experiments the eigenvalue distribution of the Hessian was known, and the inner CG iteration was designed to take advantage of it. Extensive numerical tests have not yet been performed, and it remains to be seen if these ideas – or variations of them – will prove to be valuable in practice.

4.3 Sparse quasi-Newton updates

An interesting idea that had been explored [64] and abandoned in the late 1970's has recently been resurrected [27,29]. It consists of developing quasi-Newton updates that mimic the sparsity pattern of the Hessian matrix $\nabla^2 f$.

In the approach described in [29], the goal is to construct a symmetric matrix B_{k+1} with the same sparsity pattern as $\nabla^2 f$, and which attempts to satisfy the secant conditions $B_{k+1}s_j = y_j$, $j = k - m + 1, ..., k$, as well as possible along

m past directions. The sparse quasi-Newton matrix is constructed using the following variational approach. Let S_k and Y_k denote the matrices containing the m most recent difference pairs, as in Equation 4.9, and let Ω specify the sparsity pattern of the Hessian matrix. The matrix B_{k+1} will be defined as the solution to

$$\min_B \|BS_k - Y_k\|_F^2$$

$$\text{subject to } B = B^T, \qquad \text{and} \qquad B_{i,j} = 0 \text{ for all } i, j \in \Omega,$$

where $\|\cdot\|_F$ denotes the Frobenius norm. Thus the secant equation $B_{k+1}s_j = y_j$, may not be satisfied even along the latest search direction s_k.

This convex optimization problem always has a solution, but to compute one is not easy. It is shown in [29] that the solution is unique if S_k satisfies a certain linear independence assumption. In this case B_{k+1} can be computed by solving a positive definite system – but B_{k+1} itself is not guaranteed to be positive definite. The analysis, which is quite novel, also reveals the minimum number m of difference pairs required to estimate a Hessian with a given sparsity pattern. Finding out this minimum number can be difficult because it requires consideration of all possible orderings of certain sparse matrices. Nevertheless some interesting cases are simple to analyze. For example, it is shown that two difference pairs are sufficient to estimate an arrowhead matrix.

A trust region method implementing these ideas is given in [29]. Since the number of elements of the Hessian approximation B_{k+1} that can be estimated is limited by the number m of difference pairs in S_k and Y_k, the algorithm begins by approximating only the diagonal; after the second iteration the diagonal plus one off-diagonal element per column is estimated, and so on. Once sufficient difference pairs have been saved to approximate all the nonzero elements in B_{k+1}, the oldest pair is replaced by a new one, as in limited memory updating.

One of the drawbacks of this approach is that the system that needs to be solved to obtain the new sparse quasi-Newton matrix B_{k+1} can be very large: its dimension equals the number of nonzeros in the lower triangular part of the Hessian. Preliminary numerical tests appear to indicate that this sparse quasi-Newton method requires fewer iterations than the L-BFGS method, but the difference seems to be too small to overcome the much larger expense of the iteration. It is not known if this approach can be made into a competitive algorithm. If this were to be the case, its main use could be in constrained optimization, where the sparse quasi-Newton matrix would be part of a KKT system.

5 Nonlinear conjugate gradient methods

Conjugate gradient methods remain very popular due to their simplicity and low storage requirements. They fall in the same category as limited memory and Hessian free Newton methods, in that they require only gradient information (and no information about the structure of the objective function), and use no matrix storage. Even though limited memory and Hessian free methods tend to be more predictable, robust and efficient, nonlinear CG methods require only a fraction of the storage. Most of the recent work in nonlinear CG methods has focused on global convergence properties and on the design of new line search strategies.

The search direction in all nonlinear conjugate gradient methods is given by

$$d_k = -g_k + \beta_k d_{k-1}. \tag{5.1}$$

There are two well-known choices [26] for β_k: the Fletcher-Reeves formula

$$\beta_k^{\mathrm{FR}} = \frac{\|g_k\|^2}{\|g_{k-1}\|^2},$$

and the more successful Polak-Ribière formula

$$\beta_k^{\mathrm{PR}} = \frac{g_k^T(g_k - g_{k-1})}{\|g_{k-1}\|^2}. \tag{5.2}$$

The new iterate is given by $x_{k+1} = x_k + \alpha_k d_k$, where the steplength α_k (usually) satisfies the *strong Wolfe conditions* [47]

$$f(x_k + \alpha_k d_k) \leq f(x_k) + \sigma_1 \alpha_k g_k^T d_k \tag{5.3}$$

$$|g(x_k + \alpha_k d_k)^T d_k| \leq -\sigma_2 g_k^T d_k, \tag{5.4}$$

where $0 < \sigma_1 < \sigma_2 < \frac{1}{2}$.

A major drawback of nonlinear CG methods is that the search directions tend to be poorly scaled, and the line search typically requires several function evaluations to obtain an acceptable steplength α_k. This is in sharp contrast with quasi-Newton and limited memory methods which accept the unit steplength most of the time. Nonlinear CG methods would therefore be greatly improved if we could find a means of properly scaling d_k. Many studies have suggested search directions of the form

$$d_k = -H_k g_k + \beta_k d_{k-1} \tag{5.5}$$

where H_k is a simple symmetric and positive definite matrix, often satisfying a secant equation [41]. However, if H_k requires several vectors of storage, the economy of the nonlinear CG iteration disappears, and its performance compared with limited memory methods is unlikely to be as good. This is because the second term in Equation 5.5 may prevent d_k from being a descent direction unless the line search is relatively accurate. In addition, the last term in Equation 5.5 can introduce bad scaling in the search direction. So far all attempts to derive an efficient method of the form Equation 5.5 have been unsuccessful.

An interesting framework for studying nonlinear CG, as well as quasi-Newton and Newton methods, is that of *Successive Affine Reduction* [53]. The idea is to make curvature estimates of the Hessian matrix in a low dimensional subspace formed by some of the most recent gradients and search directions. Quadratic termination properties of these methods have been studied in some detail, but practical implementations have not yet been fully developed, and their effectiveness in the context of large scale optimization remains to be demonstrated.

In an effort to improve nonlinear CG methods, some researchers have turned to global convergence studies to gain new insights into their behavior. This work is based on two important results concerning the Polak-Ribière [60] and Fletcher-Reeves methods [2]. In both cases it is assumed that the starting point x_0 is such that the level set $\mathcal{L} := \{x : f(x) \leq f(x_0)\}$ is bounded, and that in some neighborhood $\mathcal{N}$ of $\mathcal{L}$ the objective function f is continuously differentiable, and its gradient is Lipschitz continuous. The nonlinear CG method is also assumed to include no regular restarts.

It is shown in [60] that the Polak-Ribière method may fail to approach a solution point, in the sense that the sequence $\{||g_k||\}$ is bounded away from zero. In this analysis the line search always finds the first stationary point of the univariate function $\Psi(\alpha) = f(x_k + \alpha_k d_k)$. Recently, however, it has been shown [37] that these difficulties can be overcome by using a new line search strategy. More specifically, the Polak-Ribière iteration using this line search satisfies

$$\liminf_{k\to\infty} ||g_k|| = 0. \tag{5.6}$$

However numerical results appear to indicate that only a marginal improvement over the standard implementation of the Polak-Ribière method is obtained.

A different approach is motivated by the observation [60] that some undesirable behavior of the Polak-Ribière method occurs if the parameter β_k^{PR} becomes negative at regular intervals. It is shown in [31] that if β_k is defined as

$$\beta_k = \max\{\beta_k^{\text{PR}}, 0\}, \tag{5.7}$$

and if the line search satisfies a slight modification of the strong Wolfe conditions Equations 5.3 and 5.4, then the global convergence result Equation 5.6 can be established. This analysis has been generalized in [38] by allowing a more flexible line search. Numerical experiments again fail to show a significant improvement in performance over the standard Polak-Ribière method.

The analysis for the Fletcher-Reeves method is simpler. It is shown in [2] that if the line search satisfies the strong Wolfe conditions then the Fletcher-Reeves method is globally convergent in the sense that Equation 5.6 is satisfied. The same result is proved in [66] for all methods of the form Equation 5.1 with a line search satisfying the strong Wolfe conditions, and with any β_k such that $0 \leq \beta_k \leq \beta_k^{\text{FR}}$. The analysis is taken one step further in [31], where it is shown that global convergence is obtained for any method with $|\beta_k| \leq \beta_k^{\text{FR}}$. Moreover this result is tight in the following sense: there exists a smooth function f, a starting point, and values of β_k satisfying

$$|\beta_k| \leq c\beta_k^{\mathrm{FR}},$$

for some $c > 1$, such that the sequence of gradient norms $\{||g_k||\}$ is bounded away from zero.

Even though most of these theoretical results are interesting, and some of the proof techniques are innovative, these studies have not lead to significant practical advances in nonlinear CG methods. Their main contribution has been a better understanding of the crucial role played by line searches.

Acknowledgments

I would like to thank Marcelo Marazzi and Guanghui Liu for carefully reading the manuscript and suggesting many improvements. This work was supported by National Science Foundation Grant CCR-9400881, and by Department of Energy Grant DE-FG02-87ER25047.

References

1. Adams, L. and Nazareth, J.L. (Editors). (1996). *Linear and Nonlinear Conjugate Gradient-Related Methods*, SIAM Publications.
2. Al-Baali, M. (1985). Descent property and global convergence of the Fletcher-Reeves method with inexact line search. *IMA J. of Numerical Analysis*, **5**, 121–124.
3. Arioli, M., Chan, T.F., Duff, I.S., Gould, N.I.M. and Reid, J.K. (1993). Computing a search direction for large-scale linearly constrained nonlinear optimization calculations. *Technical Report TR/PA/93/94*, CERFACS Toulouse, France.
4. Averick, B.M., Carter, R.G., Moré, J.J. and Xue, G. (1992). The Minpack-2 test problem collection. *Preprint MCS-P153-0692*, Mathematics and Computer Science Division, Argonne National Laboratory.
5. Bischof, C.H., Bouaricha, A., Khademi, P.M. and Moré, J.J. (1995). Computing gradients in large scale optimization using automatic differentiation. *Report ANL/MCS-P488-0195*, Argonne National Laboratory.
6. Bongartz, I., Conn, A.R., Gould, N.I.M. and Toint, Ph.L. (1993). CUTE: Constrained and unconstrained testing environment. *Research Report*, IBM T.J. Watson Research Center, Yorktown Heights, New York.
7. Bouaricha, A. and Moré, J.J. (1996). A preconditioned Newton method for large-scale optimization. *Workshop on Linear Algebra in Optimization.*
8. Breitfeld, M.G. and Shanno, D.F. (1994). Computational experience with modified log-barrier functions for large-scale nonlinear programming. *Large Scale Optimization: State of the Art*, Editors: W.W. Hager, D.W. Hearn and P.M. Pardalos, Kluwer Academic Publishers.
9. Buckley, A. and LeNir, A. (1985). BBVSCG—A variable storage algorithm for function minimization. *ACM Transactions on Mathematical Software*, **11**, 103–119.
10. Buckley, A. (1989). Remark on algorithm 630. *ACM Transactions on Mathematical Software*, **15**, 262–274.

11. Byrd, R.H., Nocedal, J. and Schnabel, R.B. (1994). Representation of quasi-Newton matrices and their use in limited memory methods. *Mathematical Programming*, **63**, 129–156.

12. Byrd, R.H., Lu, P., Nocedal, J. and Zhu, C. (1995). A limited memory algorithm for bound constrained optimization. *SIAM J. on Scientific Computing*, **16**, 1190–1208.

13. Byrd, R.H., Nocedal, J. and Zhu, C. (1995). Towards a discrete Newton method with memory for large scale optimization. *Nonlinear Optimization and Applications*, Editors: G. Di Pillo and F. Giannessi, Plenum. (To Appear.)

14. Coleman, T.F. (1991). Large-scale numerical optimization: Introduction and overview. *Technical Report*, Cornell Theory Center, Advanced Computing Research Institute, **85**, 1–28.

15. Coleman, T.F. and Moré, J. (1984). Estimation of sparse Hessian matrices and graph coloring problems. *Mathematical Programming*, **28**, 243–270.

16. Coleman, T.F. (1984). Large sparse numerical optimization. *Lecture Notes in Computer Sciences*, **165**, Springer Verlag.

17. Conn, A.R., Gould, N.I.M. and Toint, Ph.L (1992). LANCELOT: A FORTRAN package for large-scale nonlinear optimization (Release A). *Computational Mathematics*, **17**, Springer-Verlag, New York.

18. Conn, A.R., Gould, N.I.M., Sartenaer, A. and Toint, Ph.L. (1996). On iterated-subspace minimization methods for nonlinear optimization. *Linear and Nonlinear Conjugate Gradient-Related Method*, Editors: L. Adams and L. Nazareth, SIAM Publications.

19. Conn, A.R., Gould, N.I.M. and Toint, Ph.L. (1991). Convergence of quasi-Newton matrices generated by the symmetric rank one update. *Math. Prog.*, **2**, 177–195.

20. Curtis, A., Powell, M.J.D. and Reid, J.K. (1974). On the estimations of sparse Jacobian matrices. *J.I.M.A.*, **13**, 117–120.

21. Dembo, R.S., Eisenstat, S.C. and Steihaug, T. (1982). Inexact Newton methods. *SIAM J. Numer. Anal.*, **19**, 400–408.

22. Dennis Jr., J.E. and Schnabel, R.B. (1983). *Numerical Methods for Unconstrained Optimization and Nonlinear Equations*, Prentice-Hall.

23. Eisenstat, S.C. and Walker, H.F. (1996). Choosing the forcing terms in an inexact Newton method. *SIAM J. on Scientific Computing*, **17**, 33–46.

24. Fenelon, M.C. (1981). Preconditioned conjugate-gradient-type methods for large-scale unconstrained optimization. *Ph.D. Dissertation*, Stanford University.

25. Fletcher, R. (1993). An overview of unconstrained optimization. *Technical Report NA/149*, Department of Mathematics and Computer Science, University of Dundee.

26. Fletcher, R. (1987). *Practical Methods of Optimization (2nd edition)*, John Wiley, New York.

27. Fletcher, R. (1995). An optimal positive definite update for sparse Hessian matrices. *SIAM J. on Optimization*, **5**, 192–218.

28. Fletcher, R. (1990). Low storage methods for unconstrained optimization. *Computational Solutions of Nonlinear Systems of Equations*, Editors: E.L. Allgower and K. Georg, **26**, AMS Publications, Providence, RI.

29. Fletcher, R., Grothey, A. and Leyffer, S. (1996). Computing sparse Hessian and Jacobian approximations with optimal hereditary properties. *Technical Report*, Department of Mathematics and Computer Science, University of Dundee.

30. Gay, D.M. (1996). More AD of nonlinear AMPL models: Computing Hessian information and exploiting partial separability. *Proc. of the Second Int. Workshop on Computational Differentiation.* (To Appear.)

31. Gilbert, J.C. and Nocedal, J. (1990). Global convergence properties of conjugate gradient methods for optimization. *SIAM J. on Optimization*, **2**, 21–42.

32. Gilbert, J.C. and Lemaréchal, C. (1989). Some numerical experiments with variable storage quasi-Newton algorithms. *Mathematical Programming*, **45**, 407–436.

33. Gill, P.E., Murray, W. and Wright, M.H. (1981). *Practical Optimization*, Academic Press, London.

34. Golub, G.H. and Van Loan, C.F. (1989). *Matrix Computations (2nd edition)*, The Johns Hopkins University Press, Baltimore and London.

35. Griewank, A. and Toint, Ph.L. (1982). On the unconstrained optimization of partially separable objective functions. *Nonlinear Optimization 1981*, Editor: M.J.D. Powell, Academic Press, London, 301–312.

36. Griewank, A. and Corliss, G.F. (Editors). (1991). *Automatic Differentiation of Algorithms: Theory, Implementation and Application*, SIAM Publications.

37. Grippo, L. and Lucidi, S. (1995). A globally convergent version of the Polak-Ribière gradient method. Dipartimento di Informatica e Sistematica, Universita degli Studi di Roma "La Sapienza", R. 08-95.

38. Han, J., Liu, G., Sun, D. and Yin, H. (1996). Relaxing sufficient descent condition for nonlinear conjugate gradient methods. *Technical Report*, Institute of Applied Mathematics, Academia Sinica, China.

39. Head-Gordon, T., Stillinger, F.H., Gay, D.M. and Wright, M.H. (1992). Poly(L-alanine) as a universal reference material for understanding protein energies and structures. *Proc. of the National Academy of Sciences*, **89**, 11513–11517.

40. Higham, N.J. (1996). *Accuracy and Stability of Numerical Algorithms*, SIAM Publications.

41. Hu, Y.F. and Storey, C. (1990). On unconstrained conjugate gradient optimization methods and their interrelationships. *Mathematics Report Number A129*, Loughborough University of Technology.

42. Jones, M.T. and Plassman, P. (1995). An improved incomplete Cholesky factorization. *ACM Transactions on Mathematical Software*, **21**, 5–17.

43. Karmakar, N.K. and Ramakrishnan, K.G. (1991). Computational results of an interior point algorithm for large scale linear programming. *Mathematical Programming*, **52**, 555–586.

44. Leonard, M.W. (1995). Reduced Hessian quasi-Newton methods for optimization. *Ph.D. Dissertation*, University of California, San Diego.

45. Liu, D.C. and Nocedal, J. (1989). On the limited memory BFGS method for large scale optimization. *Mathematical Programming*, **45**, 503–528.

46. McCormick, G.P. and Sofer, A. (1991). Optimization with unary functions. *Mathematical Programming*, **52**, 167–178.

47. Moré, J.J. and Thuente, D.J. (1994). Line search algorithms with guaranteed sufficient decrease. *ACM Transactions on Mathematical Software*, **20**, 286–307.

48. Nash, S.G. (1984). Newton-type minimization via the Lanczos method. *SIAM. J. Numerical Analysis*, **21**, 770–788.

49. Nash, S.G. (1984). User's guide for TN/TNBC: FORTRAN routines for nonlinear optimization. *Report 397*, Mathematical Sciences Department, The Johns Hopkins University.

50. Nash, S.G. (1985). Preconditioning of truncated-Newton methods. *SIAM J. on Scientific and Statistical Computing*, **6**, 599–616.

51. Nash, S.G. and Nocedal, J. (1991). A numerical study of the limited memory BFGS method and the truncated-Newton method for large scale optimization. *SIAM J. on Optimization*, **1**, 358–372.

52. Neumaier, A. (1996). Molecular modeling of proteins: A feasibility study of mathematical prediction of protein structure. *Manuscript*.

53. Nazareth, J.L. (1986). The method of successive affine reduction for nonlinear minimization. *Mathematical Programming*, **35**, 373–387.

54. Nocedal, J. (1980). Updating quasi-Newton matrices with limited storage. *Mathematics of Computation*, **35**, 773–782.

55. Nocedal, J. (1992). Theory of algorithms for unconstrained optimization. *Acta Numerica*, **1**, 199–242.

56. O'Leary, D.P. (1982). A discrete Newton algorithm for minimizing a function of many variables. *Mathematical Programming*, **23**, 20–33.

57. Paige, C.C. and Saunders, M.A. (1975). Solution of sparse indefinite systems of linear equations. *SIAM. J. Numerical Analysis*, **12**, 617–629.

58. Powell, M.J.D. (1970). A hybrid method for nonlinear equations. *Numerical Methods for Nonlinear Algebraic Equations*, Editor: P. Rabinowitz, Gordon and Breach, London, 87–114.

59. Powell, M.J.D. (1984). On the global convergence of trust region algorithm for unconstrained optimization. *Mathematical Programming*, **29**, 297–303.

60. Powell, M.J.D. (1984). Nonconvex minimization calculations and the conjugate gradient method. *Lecture Notes in Mathematics*, **1066**, Springer Verlag, Berlin, Germany, 122–141.

61. Schlick, T. and Fogelson, A. (1992). TNPACK—A truncated Newton package for large-scale problems: I. Algorithms and usage. *ACM Transactions on Mathematical Software*, **18**, 46–70.

62. Siegel, D. (1992). Implementing and modifying Broyden class updates for large scale optimization. *Report DAMTP 1992/NA12*, University of Cambridge.

63. Steihaug, T. (1983). The conjugate gradient method and trust regions in large scale optimization. *SIAM J. Numerical Analysis*, **20**, 626–637.

64. Toint, Ph.L. (1977). On sparse and symmetric matrix updating subject to a linear equation. *Mathematics of Computation*, **31**, 954–961.

65. Toint, Ph.L. (1983). *VE08AD—A Routine for Partially Separable Optimization with Bounded Variables*, Harwell Subroutine Library, A.E.R.E., United Kingdom.

66. Touati-Ahmed, D. and Storey, C. (1990). Efficient hybrid conjugate gradient techniques. *J. of Optimization Theory and Applications*, **64**, 379–397.

67. Zhu, C., Byrd, R.H., Lu, P. and Nocedal, J. (1995). L-BFGS-B: FORTRAN subroutines for large-scale bound constrained optimization. *Technical Report*, Department of Electrical Engineering and Computer Science, Northwestern University.

Interior Point Methods for Linear and Nonlinear Programming

David F. Shanno and Evangelia M. Simantiraki

Rutgers Center of Operations Research, Rutgers University, New Jersey, USA

Abstract

The paper first discusses recent advances in interior point methods for linear programming, including infeasibility detecting codes, target following, and better handling of dense columns. Computational results on the Netlib test suite using CPLEX 4.0 will be given. The paper then discusses two alternative methods for solving nonlinear programming problems using interior point methods, both designed to overcome the well-known difficulties with the classical barrier method. The first is a penalty–barrier method based on Polyak's modified barrier method, the second a method closely allied with sequential quadratic programming which handles nonlinear constraints via the classical barrier method applied to slack variables. Comparative numerical results will be given.

1 Introduction

Interior point methods for mathematical programming problems were created to solve the problem

$$\min f(x) \tag{1.1a}$$

$$\text{subject to} \quad c_i(x) \geq 0, \; i = 1, \cdots, m \tag{1.1b}$$

where $x = (x_1, \cdots, x_n)^T$. The interior point approach is to transform the problem (1.1) to the unconstrained optimization problem

$$\min B(x, \mu) = f(x) - \mu \sum_{i=1}^{m} \log(c_i(x)). \tag{1.2}$$

The method used to solve the problem is iterative, starting with $\mu_0 > 0$ and x^0 and letting x^k solve

$$x^k = \arg\min B(x, \mu_k).$$

μ_{k+1} is then chosen to satisfy

$$0 < \mu_{k+1} < \mu_k$$

and the process is repeated using x^k as the initial estimate for x^{k+1}. The methods were fully developed by Fiacco and McCormick [9], who showed that under sufficient conditions on $f(x)$ and $c_i(x)$, as $\mu_k \to 0$ the sequence $x^k \to \widehat{x}$, the solution to (1.1).

Despite the nice theoretical properties of interior point methods, in early implementations the methods proved unsatisfactory in practice, and dropped out of general use. The methods regained importance with the inception of Karmarkar's [13] interior point method for linear programming. It was quickly shown that Karmarkar's method was a special case of the logarithmic barrier method (1.2), and new algorithms for linear programming which proved significantly faster than the simplex method for large problems soon evolved. For a survey of these methods, see Lustig et.al. [14]. The success of interior point methods for linear programming has led to renewed interest in their application to nonlinear programming problems, as well as further development of linear programming algorithms. The paper will first discuss recent developments in linear programming, with numerical results from a state of the art code. Subsequent sections will deal with the extension to quadratic programming and two recent variations to solve general nonlinear programming problems. Preliminary results for the nonlinear algorithms will be given, as well as indications of problems for future research.

2 Barrier methods for linear programming

2.1 The primal–dual predictor–corrector method

The linear programming problem considered in this section is

$$\begin{aligned} &\min c^T x \\ &\text{subject to } Ax = b, \\ &x \geq 0, \end{aligned} \tag{2.1}$$

with the associated dual problem

$$\begin{aligned} &\max b^T y \\ &\text{subject to } A^T y + z = c \\ &z \geq 0. \end{aligned}$$

Interior point algorithms are derived by applying the logarithmic barrier method to the problem (2.1) and incorporating the equality constraints using Lagrange multipliers. The Lagrangian function of the transformed problem is

$$L(x, y, \mu) = c^T x - \mu \sum_{i=1}^{n} \log x_i - y^T (Ax - b). \tag{2.2}$$

The first order conditions for (2.2) are

$$\begin{aligned} c - \mu X^{-1}e - A^T y &= 0, \\ Ax - b &= 0, \end{aligned}$$

where X is the diagonal matrix with diagonal elements x_i and $e = (1, \cdots, 1)^T$. Letting $z = \mu X^{-1}e$ the first order conditions can be rewritten as

$$XZe = \mu e, \tag{2.3a}$$
$$Ax - b = 0, \tag{2.3b}$$
$$A^T y + z - c = 0, \tag{2.3c}$$

where again Z is the diagonal matrix with diagonal elements z_i. This is a system of nonlinear equations which can be solved iteratively using Newton's method. The Newton equations to determine search directions $\Delta x, \Delta y$, and Δz are

$$Z\Delta x + X\Delta z = \mu e - XZe, \tag{2.4a}$$
$$A\Delta x = b - Ax, \tag{2.4b}$$
$$A^T \Delta y + \Delta z = c - A^T y - z. \tag{2.4c}$$

The interior point algorithm is to choose an initial $x^0 > 0, y^0$, and $z^0 > 0$, and $\mu = \gamma \dfrac{x^T z}{n}$, $0 < \gamma < 1$ to solve the system (2.4) to obtain search directions, and to calculate a new approximation to the solution as

$$\widehat{x} = x + \alpha_P \Delta x,$$
$$\widehat{y} = y + \alpha_D \Delta y,$$
$$\widehat{z} = z + \alpha_D \Delta z,$$

where α_P and α_D are step lengths chosen to assure that $\widehat{x} > 0, \widehat{z} > 0$. The algorithm then continues iteratively until the complementarity gap $x^T z < \epsilon$ for an appropriate $\epsilon > 0$. We note in passing that the method is an infeasible method in that neither the primal nor the dual constraints are necessarily satisfied until optimality.

Rather than discuss Newton's method further at this time, we will look at a variant which leads directly to the method used in all current production codes, Mehrotra's [16] primal–dual predictor–corrector method. If we denote the primal and dual infeasibilities as

$$r_P = b - Ax$$

and

$$r_D = c - A^T y - z,$$

then we may view the problem as attempting to find $\Delta x, \Delta y, \Delta z$ which satisfy

$$\begin{aligned} (X + \Delta X)(Z + \Delta Z)e &= \mu e, \\ A\Delta x &= r_P, \\ A^T \Delta y + \Delta z &= r_D. \end{aligned}$$

Rearranging the first of these equations, we obtain the implicit system of equations

$$\begin{aligned} X\Delta z + Z\Delta x &= \mu e - XZe - \Delta X \Delta Z e, & \text{(2.5a)}\\ A\Delta x &= r_P, & \text{(2.5b)}\\ A^T\Delta y + \Delta z &= r_D, & \text{(2.5c)} \end{aligned}$$

where again ΔX and ΔZ are diagonal matrices. Mehrotra's method to solve (2.5) is to first solve the system

$$\begin{aligned} X\widetilde{\Delta z} + Z\widetilde{\Delta x} &= -XZe,\\ A\widetilde{\Delta x} &= r_P,\\ A^T\widetilde{\Delta y} + \widetilde{\Delta z} &= r_D, \end{aligned}$$

choose $\mu = \gamma\frac{x^Tz}{n}$ depending on how large a step can be taken along $\widetilde{\Delta x}, \widetilde{\Delta y}, \widetilde{\Delta z}$, (the actual choice here being γ), and then solve the system (2.5) with this choice of μ and $\widetilde{\Delta X}, \widetilde{\Delta Z}$ replacing $\Delta X, \Delta Z$ in the right hand side of the first equation. Note that both of the systems of equations of Mehrotra's method have the same coefficient matrix, which is identical to the coefficient matrix of Newton's method, so only a single factorization is required, but two backsolves are required for each iteration.

Simple algebraic manipulation of the system of equations shows that the solution is given by

$$\begin{aligned} \Delta y &= (A\Theta A^T)^{-1}\left[r_P + A\Theta(r_D + \mu e - XZe - \widetilde{\Delta X}\widetilde{\Delta Z}e)\right], & \text{(2.6a)}\\ \Delta x &= \Theta\left[A^T\Delta y - \mu e + XZe + \widetilde{\Delta X}\widetilde{\Delta Z}e - r_D\right], & \text{(2.6b)}\\ \Delta z &= \mu X^{-1}e - Ze - \Theta^{-1}\widetilde{\Delta X}\widetilde{\Delta Z}e - \Theta^{-1}\Delta x, & \text{(2.6c)} \end{aligned}$$

where $\Theta = XZ^{-1}$ is a diagonal matrix. Thus the major work of most iterations of an interior point method for linear programming lies in factoring the matrix $A\Theta A^T$.

2.2 The self–dual homogeneous model

As noted in the previous section, the primal–dual predictor–corrector method is an infeasible point method, as feasibility and optimality are approached simultaneously. This leads to difficulties with infeasible problems, which for the basic method are generally assumed to be the case if either the primal or the dual objective function appears to become unbounded. This is in general an unsatisfactory way of detecting infeasibility, and can take considerable time for some infeasible problems. An elegant solution to the problem of detecting infeasibility was proposed by Ye et. al. [23]. They considered the homogeneous and self–dual model

$$\begin{aligned} Ax - b\tau &= 0, && (2.7a)\\ -A^T y - z + c\tau &= 0, && (2.7b)\\ b^T y - c^T x - \kappa &= 0, && (2.7c) \end{aligned}$$

$$x \geq 0, z \geq 0, \tau \geq 0, \kappa \geq 0.$$

Here τ and κ are scalar variables. This system always has a solution, as setting all variables to 0 gives a solution. Ye et. al. prove that the original problem has a feasible solution if and only if any solution which satisfies

$$\begin{pmatrix} x+z \\ \tau+\kappa \end{pmatrix} > 0$$

has $\tau > 0$. The solution to the original problem is x/τ.

A predictor–corrector algorithm can be devised for the system (2.7) precisely as in the previous section. Here if we denote

$$r_G = c^T x - b^T y + \kappa$$

the system of equations that we wish to solve becomes

$$\begin{aligned} A\Delta x - b\Delta\tau &= \eta r_P, && (2.8a)\\ -A^T\Delta y - \Delta z + c\Delta\tau &= -\eta r_D, && (2.8b)\\ b^T\Delta y - c^T\Delta x - \Delta\kappa &= \eta r_G, && (2.8c)\\ X\Delta z + Z\Delta x &= \gamma\mu e - XZe - \Delta X\Delta Ze, && (2.8d)\\ \tau\Delta\kappa + \kappa\Delta\tau &= \gamma\mu e - \tau\kappa - \Delta\tau\Delta\kappa, && (2.8e) \end{aligned}$$

where

$$\mu = \gamma(x^T z + \tau\kappa)/(n+1)$$

and $0 < \gamma < 1$ and $\eta > 0$ are scalars. In [22] the authors let $\eta = 1-\gamma$. This system of equations has two more rows and two more columns than the system (2.5), and simple algebraic manipulation shows that to compute a normal equation solution akin to (2.6) leads to factoring a matrix of the form

$$\begin{bmatrix} A\Theta A^T & s \\ s^T & w \end{bmatrix}, \qquad (2.9)$$

where $A\Theta A^T$ is as in (2.6a). For any matrix of the form

$$W = \begin{bmatrix} M & s \\ s^T & w \end{bmatrix},$$

$$W^{-1} = \begin{bmatrix} M^{-1} - \theta M^{-1} s s^T M^{-1} & -\theta M^{-1} s \\ -\theta s^T M^{-1} & \theta \end{bmatrix},$$

where $\theta = (w - s^T M^{-1} s)^{-1}$. Hence if we have a factorization

$$LDL^T = A\Theta A^T,$$

the homogeneous system (2.8) can be solved using one additional backsolve to compute $M^{-1}s$. In practice, the homogeneous algorithm has proved very efficient at detecting infeasibility, and numerically very stable for ill–conditioned problems, but the extra backsolve does make the algorithm significantly slower than the basic algorithm (2.6). This will be demonstrated in the section on numerical results for linear programming.

2.3 Higher order methods

The predictor–corrector methods of the previous two sections compute only a single corrector direction. A higher order method can repeatedly substitute the search directions $\Delta x, \Delta y, \Delta z$ into the right hand side of (2.5), reusing the same factorization repeatedly at the cost of incorporating an extra backsolve for each higher order correction. This was studied in Carpenter et. al. [7], where it was shown that higher order methods do reduce the number of major iterations (factorizations), but that in general the cost of the extra backsolves offset the advantage of fewer iterations, and consequently in general, the optimal number of corrections was one.

Much of the analysis of logarithmic barrier methods for linear programming has been concerned with following the central path, that path along which $XZe = \mu e$. Another variant of a higher order method is for a given factorization and a fixed μ to iterate until $\|XZe - \mu e\|$ is sufficiently small. Gondzio [11] has suggested a variant of this strategy, which is to first compute the predictor–corrector direction, and then determine for the new point those terms where either $\hat{x}_i\hat{z}_i \geq \beta_{\max}\mu$ or $\hat{x}_i\hat{z}_i \leq \beta_{\min}\mu$. Conceptually, Gondzio's algorithm is to then compute higher order corrections by solving

$$\begin{aligned} Z\Delta x + X\Delta z &= r_C, \\ A\Delta x &= 0, \\ A^T\Delta y + \Delta z &= 0, \end{aligned}$$

where

$$r_{C_i} = \left\{ \begin{array}{ll} 0, & \beta_{\min}\mu \leq \hat{x}_i\hat{z}_i \leq \beta_{\max}\mu \\ \beta_{\max}\mu - \hat{x}_i\hat{z}_i, & \hat{x}_i\hat{z}_i > \beta_{\max}\mu \\ \beta_{\min}\mu - \hat{x}_i\hat{z}_i, & \hat{x}_i\hat{z}_i < \beta_{\min}\mu \end{array} \right\}.$$

These corrections are added to the predictor–corrector point to bring it closer to the central path. A critical feature of Gondzio's algorithm is that these corrections not be done if a backsolve is relatively expensive compared to the factorization. He provides a heuristic to determine the number of additional corrections which can be used based on the relative estimated costs of a backsolve and a factorization. This heuristic must be adjusted whenever the factorization algorithm is changed, but computational evidence suggests that the algorithm can be quite effective for large problems with expensive factorizations.

2.4 Dense columns

In using the normal equation approach to solve the linear system (2.5), it became immediately apparent that if any column of A is relatively dense, the resulting outer product $\theta_i a_i a_i^T$ in the matrix $A\Theta A^T$ will assure that this matrix is quite dense, thus making the factorization expensive. Schur complements have been used to handle dense columns for many problems (see, for example, [8]). A difficulty with the Schur complement approach has been that removal of a dense column can sometimes result in a row of all zeroes in the reduced matrix to be factored. This is a problem that occurs quite frequently in multistage stochastic programs, for example. A nice solution to this problem has been suggested by Andersen [1]. His solution is to partition A as

$$A = \left[\widehat{A} \mid E\right],$$

where E contains the dense columns. An additional matrix F is incorporated into the system, where F contains columns with only a single nonzero element which is chosen to assure that the matrix $\widehat{A}\widehat{\Theta}\widehat{A}^T + FF^T$ has strictly positive diagonal elements of sufficient size for numerical stability. The solution to any system of the form

$$A\Theta A^T x = b$$

is then computed by solving the system

$$\begin{bmatrix} \widehat{A}\widehat{\Theta}\widehat{A}^T + FF^T & E & F \\ E^T & -I & 0 \\ F^T & 0 & I \end{bmatrix} \begin{bmatrix} x \\ r \\ s \end{bmatrix} = \begin{bmatrix} b \\ 0 \\ 0 \end{bmatrix}. \tag{2.10}$$

The Schur complement solution to (2.10) is given by

$$LL^T = \widehat{A}\widehat{\Theta}\widehat{A}^T + FF^T, \tag{2.11a}$$

$$LV = E, \tag{2.11b}$$

$$LW = F, \tag{2.11c}$$

$$Lp = b, \tag{2.11d}$$

$$C = \begin{bmatrix} V^T \\ W^T \end{bmatrix} \begin{bmatrix} V & W \end{bmatrix} + \begin{bmatrix} I & 0 \\ 0 & -I \end{bmatrix}, \tag{2.11e}$$

$$C \begin{bmatrix} r \\ s \end{bmatrix} = \begin{bmatrix} V^T p \\ W^T p \end{bmatrix}, \tag{2.11f}$$

$$L^T x = p - Vr - Ws. \tag{2.11g}$$

This method has proved very successful in dealing with problems with dense columns when $\hat{A}$ is rank deficient. The effectiveness of the overall algorithm then depends on a good heuristic for identifying dense columns, which will again be dependent on the factorization algorithm. A satisfactory algorithm is given by Andersen [1].

2.5 Numerical results

All of the algorithms discussed in this section have been implemented in the barrier code contained in CPLEX 4.0 The homogeneous algorithm, the Gondzio algorithm, and the automatic handling of dense columns using Andersen's algorithm are all new features which were not present in CPLEX 3.0. In addition, the linear algebra in the new release has been made faster by better matrix ordering and improvements have been made in the preprocessing algorithm. This section documents the computational experience on the Netlib test suite (Gay [10]) of the default versions of CPLEX 3.0 and CPLEX 4.0 as well as the homogeneous algorithm with both the starting point $x^0 = (1, \cdots, 1)^T$, (Hom. Const.) and a more complex starting point (Hom. Est.) akin to what CPLEX uses for the default algorithm [14]. The tests were run on an SGI Indigo R4000 workstation under the Irix 5.3 operating system. Summary results are contained in Tables 1 and 2, and the detailed results are presented in Table 5. The times reported in Table 5 for *fit1p* and *fit2p* for CPLEX 3.0 were done with the dense column algorithm of that version, which was manually invoked as it was not a default algorithm.

Table 1. Total number of iterations on the Netlib test set

Algorithm	Iterations
CPLEX 3.0	1742
CPLEX 4.0	1671
Hom. Est.	1826
Hom. Const.	1926

Table 2. Total Run–Time on the Netlib test set

Algorithm	Run--Time
CPLEX 3.0	1694.01
CPLEX 4.0	1190.65
Hom. Est.	1766.53
Hom. Const.	1811.13

Table 3. Problems with difficulties on some run

	CPLEX 3.0		CPLEX 4.0		Hom. Est.		Hom. Const.	
Problem	time	iter.	time	iter.	time	iter.	time	iter.
dfl001	4231.07	46	2960.25	40	3771.78	52	718.57	10
modszk1	2.63	22	2.24	22	2.30	21	2.45	23

dfl001– Homogeneous algorithm with constant starting point converged to a point with small tau/kappa ratio.

modszk1– CPLEX 3.0 and 4.0 default algorithm restores a previous iterate due to lots of iterative refinement. With 4.0 problem is solved when Gondzio is turned on.

Table 4. Run–Times with and without Gondzio's Algorithm

	With Gondzio		Without Gondzio	
Problem	time	iter.	time	iter.
bnl2	17.16	21	17.83	28
d2q06c	56.18	27	49.05	29
d6cube	22.21	17	20.83	19
degen3	31.09	15	32.33	18
dfl001	2960.25	40	2930.90	41
israel	2.29	20	2.19	23
marosr7	115.71	9	140.25	13
pilot87	478.05	31	568.01	41
pilotja	18.18	30	20.16	41
pilotnov	10.99	18	9.89	20
pilots	111.33	26	128.05	35
stair	2.71	13	2.51	14

Table 5. Run–Times on the Netlib test set

	CPLEX 3.0		CPLEX 4.0		Hom. Est.		Hom. Const.	
Problem	time	iter.	time	iter.	time	iter.	time	iter.
25fv47	8.14	23	6.24	23	7.75	24	8.15	27
80bau3b	22.83	29	21.45	31	23.68	27	29.03	34
adlittle	0.09	12	0.09	12	0.11	11	0.13	14
afiro	0.02	9	0.03	9	0.03	7	0.03	9
agg	0.64	21	0.52	21	0.64	21	0.60	21
agg2	1.33	16	1.22	16	1.38	17	1.37	18
agg3	1.47	16	1.15	16	1.41	18	1.27	17
bandm	0.66	15	0.60	15	0.69	15	0.75	17
beaconfd	0.16	7	0.15	7	0.16	7	0.18	9
blend	0.12	12	0.11	11	0.12	10	0.10	8
bnl1	3.55	27	3.08	27	6.13	46	5.20	39
bnl2	27.06	33	17.16	21	19.60	28	21.36	32
boeing1	1.65	25	1.73	30	1.53	21	1.63	23
boeing2	0.36	15	0.33	15	0.36	14	0.35	14
bore3d	0.19	16	0.17	16	0.19	16	0.17	14
brandy	0.57	17	0.53	17	0.52	15	0.58	17
capri	0.72	17	0.65	17	0.72	16	0.80	18
cycle	13.52	27	7.90	28	9.39	28	10.48	33
czprob	4.08	33	3.31	34	3.72	30	4.79	40
d2q06c	67.80	29	56.18	27	70.96	38	89.02	49
d6cube	24.85	20	22.21	17	19.17	15	18.12	15
degen2	2.66	13	2.01	13	2.30	13	2.00	11
degen3	46.56	18	31.09	15	28.77	15	26.00	13
e226	0.69	16	0.57	16	0.69	16	0.85	19
etamacro	2.57	25	2.12	25	2.63	26	2.47	25
fffff800	3.58	30	2.89	30	2.71	26	2.58	25
finnis	1.03	20	0.93	19	1.45	24	1.08	18
fit1d	2.08	16	1.76	16	2.01	15	2.04	16
fit1p	2.82	15	2.65	15	5.28	27	3.39	17
fit2d	26.11	17	27.43	18	33.62	19	33.95	20
fit2p	25.12	19	24.43	19	33.06	22	28.01	18
forplan	0.96	21	0.86	20	1.41	29	1.30	27
ganges	1.95	14	1.72	14	3.07	25	2.00	15
gfrdpnc	0.83	17	0.84	17	0.96	14	1.69	27
greenbea	25.19	45	17.15	39	27.21	56	23.65	47
greenbeb	17.93	33	14.27	34	18.29	38	17.74	37
grow15	1.46	12	1.22	12	1.37	12	1.39	13
grow22	2.16	12	1.78	12	2.09	12	2.46	15
grow7	0.54	10	0.45	10	0.54	10	0.60	12
israel	2.66	23	2.29	20	2.20	21	2.56	26

Table 5. continued

Problem	CPLEX 3.0 time	CPLEX 3.0 iter.	CPLEX 4.0 time	CPLEX 4.0 iter.	Hom. Est. time	Hom. Est. iter.	Hom. Const. time	Hom. Const. iter.
kb2	0.10	17	0.09	18	0.10	15	0.09	16
lotfi	0.30	14	0.25	14	0.32	15	0.29	14
maros	3.83	27	3.15	27	3.25	23	3.39	25
marosr7	199.57	13	115.71	9	135.00	12	146.60	14
nesm	11.68	37	9.93	36	14.69	44	11.18	33
perold	7.08	33	5.95	33	7.26	33	6.75	32
pilot4	5.90	33	5.09	33	7.70	44	5.87	33
pilot87	698.31	41	478.05	31	868.30	62	825.80	60
pilotja	25.45	41	18.18	30	23.48	44	20.38	37
pilotnov	12.01	20	10.99	18	11.03	20	12.08	22
pilots	167.74	35	111.33	26	175.30	46	191.20	51
pilotwe	7.87	36	7.20	36	9.05	37	11.76	49
recipe	0.10	9	0.10	9	0.11	9	0.11	9
sc105	0.10	10	0.10	10	0.10	9	0.10	9
sc205	0.26	11	0.24	11	0.26	11	0.25	11
sc50a	0.04	9	0.04	9	0.04	8	0.04	8
sc50b	0.03	8	0.04	8	0.04	7	0.03	7
scagr25	0.55	14	0.48	14	0.61	14	0.74	18
scagr7	0.13	13	0.13	13	0.16	13	0.17	15
scfxm1	0.93	19	0.80	19	0.88	18	0.92	19
scfxm2	2.18	21	1.84	21	2.10	22	2.12	22
scfxm3	3.40	21	2.77	21	3.22	22	3.34	23
scorpion	0.22	10	0.20	9	0.23	9	0.27	12
scrs8	1.01	19	0.93	19	1.07	19	1.13	20
scsd1	0.33	10	0.31	10	0.35	9	0.35	9
scsd6	0.68	11	0.62	11	0.81	12	0.74	11
scsd8	1.42	9	1.17	9	1.37	9	1.33	9
sctap1	0.49	13	0.45	13	0.60	15	0.66	17
sctap2	2.64	15	2.37	15	2.85	16	2.65	15
sctap3	3.67	15	3.14	15	4.92	21	3.66	15
seba	0.13	8	0.12	8	0.13	8	0.12	6
share1b	0.43	22	0.42	22	0.45	23	0.61	32
share2b	0.20	14	0.17	14	0.18	13	0.16	11
shell	1.22	19	1.20	19	1.43	17	1.48	18
ship04l	1.48	13	1.11	13	1.41	14	1.41	14
ship04s	1.04	15	0.89	17	1.03	16	0.91	14
ship08l	2.55	14	1.85	14	2.40	16	2.72	19
ship08s	1.28	14	0.98	14	1.25	15	1.38	17
ship12l	4.38	21	2.96	19	3.54	18	4.69	26
ship12s	1.83	16	1.35	17	1.41	14	1.86	20

Table 5. continued

	CPLEX 3.0		CPLEX 4.0		Hom. Est.		Hom. Const.	
Problem	time	iter.	time	iter.	time	iter.	time	iter.
sierra	3.34	15	2.91	15	3.63	15	4.18	17
stair	3.06	14	2.71	13	2.64	13	2.52	14
standata	0.71	13	0.50	10	0.69	11	0.88	15
standmps	1.28	20	1.07	17	1.10	13	1.36	17
stocfor1	0.12	11	0.10	11	0.11	10	0.14	13
stocfor2	6.91	17	5.17	17	5.54	17	8.50	28
stocfor3	122.60	33	76.78	32	96.01	35	140.66	53
truss	14.23	17	11.63	17	13.15	17	12.90	17
tuff	1.31	22	1.21	18	1.22	17	1.20	18
vtpbase	0.10	10	0.09	10	0.10	8	0.11	10
wood1p	8.92	13	8.30	13	8.30	12	11.44	19
woodw	12.16	22	6.21	20	7.69	22	8.03	24

Table 2 clearly shows that on a problem set where all problems have a feasible solution, the homogeneous algorithm is significantly slower. Examination of the detailed results of Table 5 shows that for problems for which the difference is substantial, the homogeneous algorithm generally takes more iterations. This is almost certainly due to starting point. Interestingly, however, while there may be significant differences on a specific problem, the two different starting points for the homogeneous algorithm perform almost identically overall. It has been our general experience that this algorithm is much less sensitive to starting point than the standard algorithm. Finally, the results on *80bau3b* in Table 5 clearly show the cost of the extra backsolve required by the homogeneous algorithm, for the standard algorithm performs 31 iterations in less total time than the homogeneous algorithm takes for 27.

Table 3 contains the results on two problems where numerical difficulties arose with one or more of the variants.

The interesting result in Table 3 is that for small problems such as *modszk1*, Gondzio's algorithm is not invoked as a default, as the factorization is too inexpensive for the heuristic to allow for higher order corrections. When Gondzio's algorithm is invoked manually, the numerical difficulties disappear, which demonstrates that remaining close to the central path enhances numerical stability. Also, the homogeneous algorithm solves this problem with both variants. It appears generally to be somewhat more numerically stable than the standard variant.

Finally, Table 4 demonstrates the effect of Gondzio's algorithm for those Netlib problems where it is invoked by the heuristic. The problems were rerun with the algorithm disabled. The results in this table are not startling, as the problems in the Netlib set are generally too small for the effect to be highly noticeable. For larger problems, the effect becomes more pronounced. Even so, the effect is noticeable on *marosr7*, *pilot87*, and particularly on *dfl001*, where only five digits of accuracy could be obtained without Gondzio's algorithm. Clearly, the algorithm does reduce iteration count, and can improve accuracy.

2.6 Other topics

Two other topics have been the subject of recent interest concerning interior point methods for linear programming. The first is parallelization. In [15], Lustig and Rothberg discuss a parallel version of the CPLEX barrier code for the SGI R8000. They give full details of the parallelization, and show significant speedups on large test problems. The speedups are not fully scalable, but do show that for significantly large problems, up to 16 processors can be utilized successfully. In [15], they report a speedup factor of approximately nine using sixteen processors for the problem *gismondi*, with an effective rate of computation of two gigaflops. Improved performance also arises from the use of nested dissection ordering (Rothberg and Hendrickson [18]). The salient point here is that to date, similar speedups have not been achievable with simplex codes, so that as parallel machines become more available, the advantage of interior point methods for large linear programming problems seems likely to grow.

The other topic of recent interest is another approach to solving the linear system (2.5). The solution (2.6a) is, as previously noted, the normal equation solution. A mathematically equivalent, but numerically different approach, is to solve directly the reduced KKT equations

$$\begin{aligned} X^{-1}Z\Delta x - A^T\Delta y &= -r_D + X^{-1}(\mu e - XZe), && (2.12a)\\ -A\Delta x &= -r_P. && (2.12b) \end{aligned}$$

In this approach, the matrix which is ordered and factored is

$$\begin{bmatrix} X^{-1}Z & -A^T \\ -A & 0 \end{bmatrix}. \tag{2.13}$$

The advantages of this approach are automatic handling of dense columns during the ordering algorithm and easier handling of free variables. The major disadvantage has been numerical instability caused by the indefiniteness of the matrix (2.13). Vanderbei [21] in particular has studied this algorithm extensively, and recently Saunders [19] has addressed the problem of stable and efficient factorizations of (2.13). While much remains to be done, this approach certainly is of sufficient interest to merit extensive further research. Another motivation for studying this approach will be documented in the following section.

3 Quadratic programming

The extension of interior point methods to convex quadratic programming is straightforward. Here the problem to be studied is

$$\min \frac{1}{2}x^T Qx - x^T c$$

$$\text{subject to } Ax = b,$$

$$x \geq 0.$$

Applying the logarithmic barrier function to the nonnegativity constraints as in linear programming, the first order conditions for an optimum are

$$XZe = \mu e, \tag{3.1a}$$

$$Qx - c - A^T y - z = 0, \tag{3.1b}$$

$$Ax - b = 0. \tag{3.1c}$$

A primal–dual predictor–corrector method to solve these equations can be devised exactly as in linear programming, and in preliminary testing seems very efficient for convex quadratic programs [7]. The difficulty arises with the method of solution of the linear equations. These equations can be solved via the normal equation approach, but in this case the matrix factorization is

$$LL^T = A(Q + X^{-1}Z)^{-1}A^T. \tag{3.2}$$

If Q is diagonal, the work is identical to the work for linear programming. When Q is not diagonal, however, the normal equation approach can be disastrous. For example, if Q is tridiagonal, Q^{-1} is fully dense. Thus another approach is mandated.

Vanderbei [21] notes that if the reduced KKT system is solved, the matrix to be ordered and factored in this instance is

$$\begin{bmatrix} Q + X^{-1}Z & -A^T \\ -A & 0 \end{bmatrix},$$

and his experience with this approach is very promising.

As CPLEX to date uses the normal equation approach to solving the system (3.1), another approach was required. Here Q is first factored as

$$WW^T = Q,$$

where W is lower triangular. The problem is then restated as

$$\max \frac{1}{2}\, y^T y - c^T x$$

$$\text{subject to } \begin{bmatrix} Ax \\ Wx - y \end{bmatrix} = \begin{bmatrix} b \\ 0 \end{bmatrix},$$

$$x \geq 0.$$

This formulation clearly has a diagonal Hessian matrix, and the only rows which have to be appended to the constraint matrix correspond to rows of Q which have off-diagonal elements. To date, this approach has proved satisfactory in practice, but careful comparative testing awaits the implementation of an efficient and stable reduced KKT solver and the development of a good quadratic programming test set, which to date has been sadly lacking.

4 Nonlinear programming

4.1 The classic logarithmic barrier method

As noted in the introduction, logarithmic barrier methods were originally developed for nonlinear programming. The classic logarithmic barrier function $B(x, \mu)$ defined by (1.2) proved in practice to have very serious shortcomings. First, a feasible initial guess x^0 to the optimum was required, as otherwise for at least one i, $c_i(x^0) < 0$, and $\log(c_i(x^0))$ is not defined. As it is often as difficult to find a feasible solution as an optimal solution, this proved a serious drawback.

A second major drawback arose from ill-conditioning of the Hessian as the solution was approached. The first order condition for an optimal point for (1.2) is

$$\nabla_x B = \nabla f - \sum_{i=1}^{m} \frac{\mu}{c_i(x)} \nabla c_i(x) = 0 \tag{4.1}$$

and the KKT conditions imply that

$$\lim_{\mu \to 0} \frac{\mu}{c_i(x)} = \lambda_i^*,$$

where (x^*, λ^*) is an optimal primal-dual pair for the problem (1.1). Differentiating (4.1) again yields

$$\nabla_x^2 B(x^*, 0) = \nabla^2 f(x^*) - \sum_{i=1}^{m} \lambda_i^* \nabla^2 c_i(x^*) + \sum_{i=1}^{m} \frac{\lambda_i^*}{c_i(x^*)} \nabla c_i(x^*) \nabla c_i(x^*)^T,$$

so for any i with $c_i(x^*) = 0$, the Hessian of the barrier function has an eigenvalue of infinity. Thus the problem becomes extremely ill-conditioned as $\mu \to 0$ and the barrier is approached.

Other drawbacks associated with the classical barrier method include difficult line searches associated with the poles of the barrier function and a demonstrated sensitivity to the choice of μ. While many different approaches have been taken to attack these problems, we will not consider these here, but will rather discuss two recent approaches to overcome the stated problems and successfully apply logarithmic barrier methods to nonlinear programming.

As a final note on this section, the problem (1.1) has only inequality constraints. The general nonlinear programming problem is

$$\min f(x) \tag{4.2a}$$
$$\text{subject to} \quad c_i(x) \geq 0, i = 1, \cdots, m, \tag{4.2b}$$
$$g_i(x) = 0, i = 1. \cdots, p. \tag{4.2c}$$

Fiacco and McCormick incorporated the equality constraints by adding a penalty term to the barrier formulation, leading to the transformed problem

$$\min \ F(x,\mu) = f(x) - \mu \sum_{i=1}^{m} \log(c_i(x)) + \frac{1}{\mu} \sum_{i=1}^{p} (g_i(x))^2.$$

The addition of the penalty term assures that the equality constraints are driven to zero as $\mu \to 0$.

4.2 The modified penalty–barrier method

Polyak [17] proposed a modified barrier method for the problem (1.1) to attempt to overcome some of the difficulties associated with the classical logarithmic barrier method. His proposed barrier function is

$$B(x,\mu,\lambda) = f(x) - \mu \sum_{i=1}^{m} \lambda_i \log\left(s_i + \frac{c_i(x)}{\mu}\right), \tag{4.3}$$

where the λ_i are estimates to the Lagrange multipliers and the s_i are scalars used in scaling the problem. Breitfeld and Shanno [2], [3], [4] have studied this modified barrier method extensively, and have proposed replacing (4.3) with a modified penalty–barrier function

$$P(x,\mu,\lambda,\beta) = f(x) - \mu \sum_{i=1}^{m} \lambda_i \Phi(c_i(x)), \tag{4.4}$$

where

$$\Phi(c_i(x)) = \log\left(s_i + \frac{c_i(x)}{\mu}\right), \quad c_i(x) \geq -\beta\mu s_i,$$

$$\Phi(c_i(x)) = \frac{1}{2} q_i^a c_i(x)^2 + q_i^b c_i(x) + q_i^c, \quad c_i(x) < -\beta\mu s_i,$$

and

$$q_i^a = \frac{-1}{(s_i\mu(1-\beta))^2},$$
$$q_i^b = \frac{1-2\beta}{s_i\mu(1-\beta)^2},$$
$$q_i^c = \frac{\beta(2-3\beta)}{2(1-\beta)^2} + \log(s_i(1-\beta)).$$

The advantages of this modified penalty–barrier method are that it is defined for all values of x, feasible or infeasible, as long as $c_i(x)$ is defined, it has no poles and thus allows simple line searches, remains better conditioned for small values of μ, and is relatively insensitive to the initial choice of μ. The only inequality constraints not incorporated in this combined quadratic penalty–modified barrier function are simple bounds on the variables. these are incorporated in a classic log barrier function in order to assure that they are always satisfied, which is often necessary to assure that all $c_i(x)$ are defined. For a fuller explanation, see [2], [3], [4], or for a summary explanation see [20].

Equality constraints are incorporated into the problem via an augmented Lagrangian function. Here the function to be minimized at each iteration is

$$\Psi(x,\mu,\lambda,\beta) = P(x,\mu,\lambda,\beta) + \sum_{i=1}^{p} \lambda_{i+m} g_i(x) + \frac{1}{2\mu}\sum_{i=1}^{p} g_i(x)^2. \tag{4.5}$$

The algorithm then becomes

1. Choose x^0, μ_0, λ^0, and $0 < \beta < 1$. Set $k = 0$.
2. Find $x^{k+1} = \arg\min \Psi(x, \mu_k, \lambda^k, \beta)$
3. Set $\mu_{k+1} = \rho\mu_k$, $0 < \rho < \dfrac{1}{2}$.
4. Set $\lambda_i^{k+1} = \dfrac{\mu_k \lambda_i{}^k}{\mu_k s_i + c_i(x^k)}$, $i = 1, \cdots, m$, $\lambda_{i+m}^{k+1} = \lambda_{i+m}^k - \dfrac{c_i(x^k)}{\mu^k}$, $i = 1, \cdots, p$.
5. If a KKT point has been found, quit. Else, set $k = k+1$ and go to 2.

Two difficulties arise with the algorithm. First, the Lagrange multiplier updates are only first order updates, which can slow convergence near the optimum. More importantly, (4.5) is minimized for fixed values of λ and μ, and then these are updated. This can expend a great deal of work minimizing Ψ for poor initial estimates to λ. The next section explores an alternative method, more akin to the linear and quadratic programming algorithms described earlier in the paper.

4.3 A slack variable alternative

The linear and quadratic programming algorithms developed in the early sections of this paper only include nonnegativity constraints in the logarithmic barrier function. To develop a nonlinear programming extension of this method, we rewrite (4.2) as

$$\begin{aligned}
&\min\ f(x) && (4.6a)\\
\text{subject to}\quad & c_i(x) - z_i = 0, && (4.6b)\\
& g_i(x) = 0, && (4.6c)\\
& z_i \geq 0. && (4.6d)
\end{aligned}$$

The logarithmic barrier transformation for this problem is

$$\begin{aligned} &\min \; f(x) - \mu \sum_{i=1}^{m} \log(z_i) && \text{(4.7a)} \\ \text{subject to} \quad & c_i(x) - z_i = 0, && \text{(4.7b)} \\ & g_i(x) = 0. && \text{(4.7c)} \end{aligned}$$

The first order conditions for this problem are

$$\begin{aligned} \nabla f(x) - \sum_{i=1}^{m} \lambda_i \nabla c_i(x) - \sum_{i=1}^{p} y_i \nabla g_i(x) &= 0, && \text{(4.8a)} \\ \mu Z^{-1} e - \lambda &= 0, && \text{(4.8b)} \\ c_i(x) - z_i &= 0, && \text{(4.8c)} \\ g_i(x) &= 0. && \text{(4.8d)} \end{aligned}$$

Rewriting the second set of conditions in (4.8) as

$$Z \Lambda e = \mu e,$$

where $\Lambda = diag(\lambda_i)$, we can derive the nonlinear equivalent to the primal–dual method for linear programming by applying Newton's method to (4.8). El Bakry et. al. [5] developed this algorithm, giving conditions on μ sufficient to guarantee the reduction of the merit function, which is defined as follows. Let

$$F(x, z, \lambda, y) = \begin{pmatrix} \nabla f(x) - \sum_{i=1}^{m} \lambda_i \nabla c_i(x) - \sum_{i=1}^{p} y_i \nabla g_i(x) \\ c(x) - z \\ g(x) \\ Z \Lambda e \end{pmatrix}.$$

Then the merit function is defined as

$$\Phi(x, z, \lambda, y) = ||F(x, z, \lambda, y)||.$$

They also give restrictions on the step length to assure convergence to a KKT point under fairly general conditions. A difficulty with this algorithm is that convergence can be to any stationary point, not just a minimum. Further, applying an unmodified Newton's method to (4.8) can lead to all of the numerical difficulties which Newton's method is known to be prone to. El Bakry et. al. give limited computational experience with the algorithm. We have used numerical testing to fine tune some of the parameters of the El Bakry et. al. algorithm, and have sufficient numerical experience to both show the promising features of the algorithm and areas requiring further development. These results are contained in the next section.

4.4 Numerical results

We have run both of the nonlinear algorithms described in previous sections on a subset of the Hock and Schittkowski [12] test suite, which we accessed via CUTE [6]. The modified penalty–barrier algorithm, PENBAR, has in fact been extensively tested on the entire suite, as well as other problems available through CUTE. Several versions of this code are available, The one we used uses Newton's method to minimize the penalty–barrier function, so the work per iteration of the two algorithms is roughly comparable. As the results on the slack variable algorithm, NLCP (Nonlinear Complementarity), are preliminary, and do indicate problems as well as exceptional promise, we have only included those problems which we could access and analyze in our very limited testing environment. The results are contained in Table 6 at the end of the paper. For some problems the algorithm failed to converge because the matrix involved in the calculation of the search direction became nearly singular. This results in a search vector with a large norm which in turn results in very small step lengths and the algorithm halts. This behavior is indicated by the word "singular" in the table. All problems were run on Sun Sparcstations.

The most noticeable result in Table 6 is that when both algorithms correctly solved the problem, the slack variable algorithm generally took far fewer iterations. This is consistent with previous comparisons of sequential quadratic programming algorithms with augmented Lagrangian algorithms, although it should be pointed out that the increased number of function evaluations is typically compensated for by the much cheaper algebraic calculations required by augmented Lagrangian algorithms. Note, that the slack valiable algorithm resembles sequential quadratic programming algorithms in that it, too, involves solving a KKT system at every step. A second noticeable result is that the slack variable algorithm does on occasion find a local maximum (marked with "max" in the table) or other stationary point (marked with a *). Thus research is clearly necessary to assure that a minimizer is sought, not an arbitrary stationary point. Further, Problem 1 of the test suite uses Rosenbrock's function as the objective function. Here the performance of Newton's method is relatively poor. This can be improved by incorporating some form of trust region into the algorithm, but the exact mechanism for doing this remains for further research. All of these problems have been dealt with for the penalty–barrier algorithm, which is reasonably robust. It successfully solved all of the problems tested. Finally, for very large problems, obtaining and factoring a Hessian matrix may well be prohibitive. PENBAR has a variant which requires only first partial derivatives. Similar variants will be needed to allow the slack variable method to have general applicability. Overall, however, it appears that the barrier methods are becoming significantly improved and certainly merit further study in all areas of math programming. It is still early to conjecture if the remarkable successes on linear and quadratic problems can be extended to the general nonlinear case, but results to date certainly indicate that further research is worthwhile.

Table 6. The problem sizes and the results for NLCP and PENBAR

Problem	Size n	$\lvert E \cup I \rvert$	PENBAR $\tau = 10^{-8}$ o-it	f-ev	NLCP ITER
1	2	1	4	44	72
2	2	1	9	50	19
3	2	1	9	50	2
4	2	2	10	63	4
5	2	4	4	17	7
6	2	1	2	47	12
7	2	1	3	27	8
8	2	2	2	13	4
9	2	1	3	45	3 (max)
10	2	1	4	37	25
11	2	1	4	37	21
12	2	1	4	47	28
14	2	2	4	33	18
15	2	3	10	104	17
16	2	5	9	56	17
17	2	5	7	62	12
18	2	6	4	46	11
19	2	6	6	112	17
20	2	5	10	69	11
21	2	5	6	39	13
22	2	2	4	31	5
23	2	9	5	35	17
24	2	5	4	35	10 (max)
25	3	6	4	36	6 *
26	3	1	2	97	18
27	3	1	3	28	11
28	3	1	2	4	1
29	3	1	3	61	22
30	3	7	9	47	10
31	3	7	4	28	9
32	3	5	10	180	14
33	3	6	10	86	8
34	3	8	6	46	9
35	3	4	4	28	9
36	3	7	8	77	20 *
37	3	8	3	34	> 60
38	4	8	3	56	13 *
39	4	2	5	28	singular

Table 6. continued

Problem	Size n	Size $\|E \cup I\|$	PENBAR $\tau = 10^{-8}$ o-it	PENBAR $\tau = 10^{-8}$ f-ev	NLCP ITER
40	4	3	4	19	4
41	4	9	8	55	8
42	4	2	4	13	4
43	4	3	5	202	9
44	4	10	9	85	9 *
45	5	10	10	82	14 *
46	5	2	3	65	20
47	5	3	5	88	18
48	5	2	2	4	1
49	5	2	2	23	17
50	5	3	4	20	9
51	5	3	4	8	2
52	5	3	5	10	2
53	5	13	5	13	7
54	6	13	4	8	20
55	6	14	9	74	singular
56	7	4	4	23	6
57	2	3	4	19	9 *
59	2	7	3	16	> 60
60	3	7	4	25	6
61	3	2	4	83	singular
62	3	7	9	223	12
63	3	5	4	23	8
64	3	4	6	48	25 *
65	3	7	5	40	14
66	3	8	4	33	7
68	4	10	8	94	46
69	4	10	6	90	9
70	4	9	4	75	23 *
71	4	10	8	83	12
72	4	10	6	61	14
73	4	7	7	62	12
74	4	13	5	21	17
75	4	13	6	45	17
76	4	7	9	53	6
77	5	2	4	28	17
78	5	3	4	26	6
79	5	8	4	23	5

Table 6. continued

Problem	Size n	Size $\lvert E \cup I \rvert$	PENBAR $\tau = 10^{-8}$ o-it	PENBAR $\tau = 10^{-8}$ f-ev	NLCP ITER
80	5	13	4	88	7
81	5	13	4	56	7
86	5	15	5	48	10
87	6	16	5	60	singular
88	2	1	9	94	60
89	3	1	11	245	singular
90	4	1	9	257	singular
91	5	1	10	287	singular
92	6	1	9	87	57 *
93	6	8	5	47	7 max
95	6	16	11	106	25
96	6	16	11	108	26
97	6	16	11	287	39 *
98	6	16	11	276	20
99	7	16	8	123	12
100	7	4	4	60	21
105	8	17	9	112	26 *
106	8	22	5	7132	18
107	9	14	9	249	18
108	9	14	4	60	11 *
109	9	26	7	185	24
110	10	20	2	12	5
111	10	23	4	53	24
112	10	13	6	89	8 *
113	10	8	4	83	15
114	10	31	8	264	13
117	15	20	10	171	21
119	16	40	9	123	14

Acknowledgement

The authors wish to thank Irvin Lustig for providing the CPLEX run data. This research was sponsored by the Air Force Office of Scientific Research, Air Force System Command under Grant F49620-95-1-0110. The United States Government is authorized to reproduce and distribute reprints for governmental purposes notwithstanding any copyright notations thereon.

References

1. K. D. Andersen (1994). A modified Schur complement method for handling dense columns in interior point methods for linear programming. Technical Report, Department of Math. and Computer Sci., Odense University. Submitted to ACM Transaction on Mathematical Software.

2. M. G. Breitfeld and D. F. Shanno. Preliminary computational experience with modified log-barrier functions for large-scale nonlinear programming. In W. W. Hager and D. W. Hearn and P. M. Pardalos, eds., *Large Scale Optimization: State of the Art*, Kluwer Academic Publishers B.V., 1994.

3. ———(1995). Computational experience with penalty–barrier methods for nonlinear programming. RUTCOR Research Report RRR 17-93 (revised March 1994), Rutgers University, New Brunswick, New Jersey. To appear in *Annals of Operations Research.*

4. ———(1995). A globally convergent penalty–barrier algorithm for nonlinear programming and its computational performance. RUTCOR Research Report RRR 12-94 (revised September 1995), Rutgers University, New Brunswick, New Jersey, 1995.

5. A. S. El-Bakry, R. A. Tapia, T. Tsuchiya, and Y. Zhang (1992). On the formulation and theory of the primal–dual Newton interior–point method for nonlinear programming. Technical Report TR92-40, Department of Computational and Applied Mathematics, Rice University.

6. I. Bongartz, A. R. Conn, N. I. M. Gould, and Ph. L. Toint (1993). CUTE: Constrained and unconstrained testing environment.

7. T. J. Carpenter, I. J. Lustig, J. M. Mulvey, and D. F. Shanno (1993). Separable quadratic programming via a primal–dual interior–point method and its use in a sequential procedure, *ORSA Journal on Computing, 5*, 182–191.

8. I. C. Choi, C. L. Monma, and D. F. Shanno (1990). Further development of a primal–dual interior point method, *ORSA Journal on Computing, 2(4)*, 304–311.

9. A. V. Fiacco and G. P. McCormick (1968). *Nonlinear Programming: Sequential Unconstrained Minimization Techniques*, John Wiley & Sons, New York. Reprint: Volume 4 of *SIAM Classics in Applied Mathematics*, SIAM Publications, Philadelphia, Pennsylvania, 1990.

10. D. M. Gay (1985). Electronic mail distribution of linear programming test problems, *COAL Newsletter 13*, 10–12.

11. J. Gondzio (1994). Multiple centrality corrections in a primal–dual method for linear programming. Technical Report 1994.20, Logilab, HEC Geneva, Section of Management Studies, University of Geneva. Revised May 1995, to appear in Computational Optimization and Applications.

12. W. Hock and K. Schittkowski (1981). *Test Examples for Nonlinear Programming Codes*, vol. 187 of Lecture Notes in Economics and Mathematical Systems, Springer Verlag, Berlin.

13. N. K. Karmarkar (1984). A new polynomial–time algorithm for linear programming, *Combinatorica 4*, 373–395.

14. I. J. Lustig, R. E. Marsten, and D. F. Shanno (1994). Interior–point methods for linear programming: Computational state of the art, *ORSA Journal on Computing 6(1)*, 1–14.
15. I. J. Lustig and E.Rothberg (1995). Gigaflops in linear programming. Technical Report, Silicon Graphics, Inc.
16. S. Mehrotra (1992). On the implementation of a primal–dual interior point method, *SIAM Journal on Optimization 2(4)*, 575–601.
17. R. Polyak (1992). Modified barrier functions (theory and methods), *Mathematical Programming 54*, 177–222.
18. E. Rothberg, and B. Hendrickson (1996). Sparse matrix ordering methods for interior point linear programming. Technical Report, Silicon Graphics, Inc.
19. M. A. Saunders (1995). Cholesky–based methods for sparse least squares: The benefits of regularization. Technical Report, Systems Optimization Laboratory, Stanford University.
20. D. F. Shanno, M. Breitfeld, and E. Simantiraki (1995). Implementing Barrier Methods for Nonlinear Programming, RUTCOR Research Report RRR 39-95, to be included in the book *Interior Point Methods in Mathematical Programming*, Kluwer Academic Publisher.
21. R. J. Vanderbei (1994). An Interior–point code for quadratic programming. Report SOR–94–15, Dept. of Statistics and Operations Research, Princeton University, Princeton, NJ.
22. X. Xu, P. -F. Hung, and Y. Ye (1993). A simplified homogeneous and self–dual linear programming algorithm and its implementation. Technical Report, Department of Management Sciences, The University of Iowa.
23. Y. Ye, M. J. Todd, and S. Mizuno (1994). An $O(\sqrt{n}L)$–iteration homogeneous and self–dual linear programming algorithm, *Math. Oper. Res. 19*, 53–67.

Methods for Nonlinear Constraints in Optimization Calculations

Andrew R. Conn*, Nicholas I.M. Gould and Philippe L. Toint†**

**IBM T.J. Watson Research Center, New York, USA, **Department for Computation and Information, Rutherford Appleton Laboratory, Oxford, and †Department of Mathematics, Facultés Universitaires ND de la Paix, Namur, Belgium*

Abstract

Ten years ago, the broad consensus among researchers in constrained optimization was that sequential quadratic programming (SQP) methods were the methods of choice. While, in the long term, this position may be justified, the past ten years have exposed a number of difficulties with the SQP approach. Moreover, alternative methods have shown themselves capable of solving large-scale problems. In this paper, we shall outline the defects with SQP methods, and discuss the alternatives. In particular, we shall indicate how our understanding of the subproblems which inevitably arise in constrained optimization calculations has improved. We shall also consider the impact of interior-point methods for inequality constrained problems, described elsewhere in this volume, and argue that these methods likely provide a more useful Newton model for such problems than do traditional SQP methods. Finally, we shall consider trust-region methods for constrained problems, and the impact of automatic differentiation on algorithm design.

1 Introduction

In the previous assessment of the state of the art of constrained optimization, Powell (1987) presented powerful evidence that the future lay with sequential quadratic programming (SQP) methods. Powell's article focused on methods for problems with equality constraints. Perhaps, and with hindsight, it is possible to foresee the difficulties which arise as soon as inequality constraints are admitted. While we may be optimistic that SQP methods will still be the future methods of choice, the past decade has been a slightly sobering experience for those researchers working in constrained optimization, particularly for those interested in implementing algorithms. The overwhelming research thrust in optimization circles over the past ten years has been on interior-point methods for linear and, more recently, nonlinear programs. These methods offer an exciting alternative to the active-set methods which preceded them, but more importantly allow us to examine SQP methods in a new light.

In this paper, we shall try to outline the main advances that have taken place over the past decade. Following a brief review, we shall start where Powell (1987) left off, with methods for equality constraints. We then embark on a description of non-interior methods for handling inequality constraints. We caution the reader that the distinction between interior and non-interior methods is somewhat hazy, and we will sometimes delve into interior territory. Interior point methods are described elsewhere in this volume. For simplicity, we shall deliberately consider equality and inequality constraints separately, but remark that algorithms for problems with a mixture of constraints are normally a hybrid of those for the separate problems.

2 SQP methods

A thorough treatment of the history, theory and practice of SQP methods is given by Boggs and Tolle (1995).

2.1 Methods for equality constraints

We are concerned with finding the smallest value of the function $f(\boldsymbol{x})$ of the n real variables $\boldsymbol{x}$ in the case where $\boldsymbol{x}$ is required to satisfy a set of m equality constraints $\boldsymbol{c}(\boldsymbol{x}) = \boldsymbol{0}$. The first-order optimality or, as they are often known, Karush-Kuhn-Tucker (KKT) conditions for this problem are that

$$\nabla_x \ell(\boldsymbol{x}, \boldsymbol{y}) = \boldsymbol{0}, \quad \text{and} \quad \boldsymbol{c}(\boldsymbol{x}) = \boldsymbol{0}, \tag{2.1}$$

where the Lagrangian function $\ell(\boldsymbol{x}, \boldsymbol{y}) \stackrel{\text{def}}{=} f(\boldsymbol{x}) - \boldsymbol{c}(\boldsymbol{x})^T \boldsymbol{y}$ and where the components of the m-vector $\boldsymbol{y}$ are Lagrange multipliers.

A *sequential*, or *recursive*, quadratic programming (SQP) method is a method which seeks to improve an estimate $(\boldsymbol{x}, \boldsymbol{y})$ of the solution to (2.1) by finding corrections $(\boldsymbol{\Delta x}, \boldsymbol{\Delta y})$ by solving one (or more) quadratic programming problems. The next estimate of the required solution will be

$$\begin{pmatrix} \boldsymbol{x}^+ \\ \boldsymbol{y}^+ \end{pmatrix} = \begin{pmatrix} \boldsymbol{x} + \alpha_x \boldsymbol{\Delta x} \\ \boldsymbol{y} + \alpha_y \boldsymbol{\Delta y} \end{pmatrix}, \tag{2.2}$$

where the nonnegative stepsizes α_x and α_x may or may not be equal. The prototypical SQP method (see Pschenichny, 1970) finds $\boldsymbol{\Delta x}$ as the solution of the quadratic program

$$\underset{\Delta x \in \Re^n}{\text{minimize}} \quad \tfrac{1}{2} \boldsymbol{\Delta x}^T \boldsymbol{H} \boldsymbol{\Delta x} + \boldsymbol{\Delta x}^T \nabla_x \ell(\boldsymbol{x}, \boldsymbol{y}) \tag{2.3a}$$

$$\text{subject to} \quad \boldsymbol{A}(\boldsymbol{x}) \boldsymbol{\Delta x} + \boldsymbol{c}(\boldsymbol{x}) = \boldsymbol{0}, \tag{2.3b}$$

where $\boldsymbol{A}(\boldsymbol{x})$ is the Jacobian $\nabla_x \boldsymbol{c}(\boldsymbol{x})$, and $\boldsymbol{H}$ is a symmetric *approximation* to the Hessian of the Lagrangian function; $\boldsymbol{\Delta y}$ are taken as the Lagrange multipliers for (2.3). The step size α_x is found by requiring that $\psi(\boldsymbol{x}+\alpha_x\boldsymbol{\Delta x})$ is sufficiently smaller than $\psi(\boldsymbol{x})$ for some suitable *merit* function, and this is achieved by performing a backtracking (Armijo) *linesearch* with a unit initial stepsize. Finally α_y is either set to one or to α_x. The missing ingredients here are the choices of $\boldsymbol{H}$ and ψ, and it is mostly in these that the many proposed methods differ.

2.1.1 Hessian approximations

The first-order optimality conditions for (2.3) are that

$$\begin{pmatrix} \boldsymbol{H} & \boldsymbol{A}(\boldsymbol{x})^T \\ \boldsymbol{A}(\boldsymbol{x}) & 0 \end{pmatrix} \begin{pmatrix} \boldsymbol{\Delta x} \\ -\boldsymbol{\Delta y} \end{pmatrix} = - \begin{pmatrix} \nabla_x \ell(\boldsymbol{x},\boldsymbol{y}) \\ \boldsymbol{c}(\boldsymbol{x}) \end{pmatrix}. \tag{2.4}$$

Assuming that $\boldsymbol{A}(\boldsymbol{x})$ is of full rank, and letting $\boldsymbol{Y}(\boldsymbol{x})$ and $\boldsymbol{Z}(\boldsymbol{x})$ be matrices whose columns span, respectively, the range and null spaces of $\boldsymbol{A}(\boldsymbol{x})$, we may write $\boldsymbol{\Delta x} = \boldsymbol{Y}(\boldsymbol{x})\boldsymbol{\Delta x}_y+\boldsymbol{Z}(\boldsymbol{x})\boldsymbol{\Delta x}_z$. On substituting into (2.4) $\boldsymbol{\Delta x}_y$ is completely determined by the constraints, as the solution to the non-singular system

$$\boldsymbol{A}(\boldsymbol{x})\boldsymbol{Y}(\boldsymbol{x})\boldsymbol{\Delta x}_y = -\boldsymbol{c}(\boldsymbol{x}), \tag{2.5}$$

while $\boldsymbol{\Delta x}_z$ then satisfies

$$\boldsymbol{H}_{zz}\boldsymbol{\Delta x}_z = -\boldsymbol{Z}^T\nabla_x\ell(\boldsymbol{x},\boldsymbol{y}) - \boldsymbol{H}_{zy}\boldsymbol{\Delta x}_y, \tag{2.6}$$

where $\boldsymbol{H}_{zz} = \boldsymbol{Z}(\boldsymbol{x})^T\boldsymbol{H}\boldsymbol{Z}(\boldsymbol{x})$ and $\boldsymbol{H}_{zy} = \boldsymbol{Z}(\boldsymbol{x})^T\boldsymbol{H}\boldsymbol{Y}(\boldsymbol{x})$. Returning to (2.3) and performing the same substitution for $\boldsymbol{\Delta x}$ also yields (2.6), but the further requirement in (2.3) that a minimizer be sought suggests that the reduced Hessian $\boldsymbol{H}_{zz}$ should be positive semi-definite; to make this solution unique, the requirement is normally strengthened to insist that the reduced Hessian be positive definite.

Early SQP methods assumed that $\boldsymbol{H}$ was itself positive definite. This is clearly stronger than requiring that $\boldsymbol{H}_{zz}$ be definite. Most importantly, second-order optimality conditions for the original problem suggest that the reduced Hessian of the Lagrangian should be at least positive semi-definite, but that there is no reason for the Hessian of the Lagrangian itself to be definite. Advocates of this assumption cite simplicity, and were clearly keen to define $\boldsymbol{H}$ via one of the positive definite secant updating formulae which had proven so successful in unconstrained optimization. However, in our opinion, the contortions that were necessary to bend the secant updates into a suitable form (see, for example, Powell (1978)) underline the difficulties with the approach. In mitigation, when inequality constraints are introduced, the dimension of $\boldsymbol{Z}(\boldsymbol{x})$ may change dramatically from one iteration to the next, and it is then certainly convenient that $\boldsymbol{H}$ is positive definite. Remarkably, the very first SQP method (Wilson, 1963) used the exact Hessian of the Lagrangian, but until recently very few authors considered this choice (see Boggs, Tolle and Kearsley, 1994 and also Bonnans and Launay (1995) who sometimes modify the exact Hessian).

More recent methods have aimed at ensuring that $\boldsymbol{H}_{zz}$ is positive definite using positive-definite secant updates. However, this leads one to wonder how to handle the other matrices $\boldsymbol{H}_{zy}$ and $\boldsymbol{H}_{yy} \stackrel{\text{def}}{=} \boldsymbol{Y}(\boldsymbol{x})^T \boldsymbol{H} \boldsymbol{Y}(\boldsymbol{x})$, and it is here that most of the current proposals vary. Murray and Wright (1978) suggested that $\boldsymbol{H}_{zy}$ and $\boldsymbol{H}_{yy}$ should be set to zero. This gives what is known as a *reduced* Hessian method. With an appropriate secant update formula, such a scheme is two-step superlinearly convergent method so long as $\alpha_x = 1$ (Nocedal and Overton, 1985). A related reduced Hessian method, due to Coleman and Conn (1982a), replaces (2.5) by

$$\boldsymbol{A}(\boldsymbol{x})\boldsymbol{Y}(\boldsymbol{x})\boldsymbol{\Delta x}_y = -\boldsymbol{c}(\boldsymbol{x} + \boldsymbol{Z}\boldsymbol{\Delta x}_z), \tag{2.7}$$

in the vicinity of a stationary point or $\boldsymbol{\Delta x}_y = \boldsymbol{0}$ elsewhere. This method is also two-step superlinearly convergent. Perhaps more surprisingly, Byrd (1990) shows that the iterates $\boldsymbol{x} + \boldsymbol{Z}\boldsymbol{\Delta x}_z$ have in fact a one-step superlinear rate, and that such a rate is common for many SQP methods which involve the correction (2.7). Byrd and Nocedal (1991) show that these local results are not affected by global convergence concerns (see Section 2.1.2). Another possibility with the same theoretical convergence properties, proposed by Gilbert (1991), is to maintain a secant approximation to the inverse of $\boldsymbol{H}_{zz}$. A similar convergence rate is also achieved by methods which use Broyden-type secant methods to approximate the rectangular matrix $\boldsymbol{Z}(\boldsymbol{x})^T \boldsymbol{H} \left(\boldsymbol{Y}(\boldsymbol{x}) \quad \boldsymbol{Z}(\boldsymbol{x})\right)$ (see, Nocedal and Overton, 1985, or Fontecilla, Steihaug and Tapia, 1987). Gurwitz (1994) prefers updates which treat the portions $\boldsymbol{H}_{zz}$ and $\boldsymbol{H}_{zy}$ separately while maintaining a positive definite approximation to the former. Coleman and Fenyes (1992) propose a similar method, and also a second method which additionally maintains an approximation to $\boldsymbol{H}_{yy}$. These methods appear to perform slightly better than those which merely maintain a nonzero $\boldsymbol{H}_{zz}$. Finally, an interesting new proposal by Biegler, Nocedal and Schmid (1995) notes that (2.6) does not actually require $\boldsymbol{H}_{zy}$ but rather the vector $\boldsymbol{H}_{zy}\boldsymbol{\Delta x}_y$. They thus propose to approximate this term directly by either finite differences or by a Broyden update.

And what of the Lagrange multiplier estimates? The values $\boldsymbol{\Delta y}$ from (2.4) satisfy

$$\boldsymbol{Y}(\boldsymbol{x})^T \boldsymbol{A}(\boldsymbol{x})^T \boldsymbol{\Delta y} = \boldsymbol{Y}^T \nabla_x \ell(\boldsymbol{x}, \boldsymbol{y}) + \boldsymbol{H}_{zy}^T \boldsymbol{\Delta x}_z + \boldsymbol{H}_{yy} \boldsymbol{\Delta x}_y. \tag{2.8}$$

Clearly, setting $\boldsymbol{H}_{zy}$ and $\boldsymbol{H}_{yy}$ to zero imply that $\boldsymbol{y} + \boldsymbol{\Delta y}$ are least-squares multiplier estimates evaluated at $\boldsymbol{x}$. If included, these neglected terms would result in $\boldsymbol{y} + \boldsymbol{\Delta y}$ being approximations to least-squares multiplier estimates at $\boldsymbol{x} + \boldsymbol{\Delta x}$. Thus, rather than use these approximations, many authors prefer to use the current least-squares estimates

$$\boldsymbol{Y}(\boldsymbol{x}^+)^T \boldsymbol{A}(\boldsymbol{x}^+)^T \boldsymbol{y}^+ = \boldsymbol{Y}(\boldsymbol{x}^+)^T \nabla_x f(\boldsymbol{x}^+) \tag{2.9}$$

directly.

Finally, although we have argued that maintaining a positive definite approximation to the Hessian of the Lagrangian function is unnecessarily restrictive, it is more reasonable when approximating the Hessian of the augmented Lagrangian. Indeed, if ρ is a scalar, we can add the term $\rho\|\boldsymbol{A}(\boldsymbol{x})\boldsymbol{\Delta x} + \boldsymbol{c}(\boldsymbol{x})\|_2^2$ to the objective function of (2.3) without changing its solution. But, the Hessian of this modified problem is $\boldsymbol{H} + \rho\boldsymbol{A}(\boldsymbol{x})^T\boldsymbol{A}(\boldsymbol{x})$ which can be expected to be positive definite for sufficiently large ρ. Such a method was first proposed by Tapia (1977) and suitable secant update formulae for $\boldsymbol{H} + \rho\boldsymbol{A}(\boldsymbol{x})^T\boldsymbol{A}(\boldsymbol{x})$ are discussed by Byrd, Tapia and Zhang (1992).

2.1.2 Merit functions

The role of the merit function is to ensure convergence of the basic iteration (2.2) from arbitrary starting points. In unconstrained optimization, there is a natural merit function, the objective function. When there are constraints present, the conflicting goals of feasibility and optimality often preclude a natural choice, and most merit functions are attempts to balance these goals.

Early globally convergent SQP methods were based upon the l_1 exact penalty function

$$\psi_1(\boldsymbol{x}) = f(\boldsymbol{x}) + \rho\|c(\boldsymbol{x})\|_1 \tag{2.10}$$

(see Pschenichny, 1970, Han, 1977, and Powell, 1978). So long as the penalty parameter ρ is sufficiently large, the iteration (2.2) converges globally with many of the Hessian approximations discussed in Section 2.1.1. However, despite its simplicity, the iteration has one serious drawback, namely that the merit function (2.10) (or indeed any function of the form $f(\boldsymbol{x}) + w(c(\boldsymbol{x}))$ where $w \geq 0$ and $w(0) = 0$) may prohibit the step $\alpha_x = 1$ arbitrarily close to a KKT point. Thus the promise of a fast asymptotic rate may be denied by the merit function. This defect was first observed by Maratos (1978). A number of remedies have been proposed, falling broadly into two camps: modify the search direction or change the merit function.

The Maratos "effect" arises when the curvature of the constraints is not adequately represented by the linearized model (2.3b). Recognizing this, Mayne and Polak (1982) propose adding a second-order correction to the standard SQP direction (2.4); Coleman and Conn (1982a) prefer to use the step (2.7) directly. Both techniques allow asymptotic unit steps and hence encourage superlinear convergence. A variation on this theme has also been suggested by Fukushima (1986).

A variety of alternatives to (2.10) have been considered. Perhaps the simplest suggestion is that by Chamberlain, Lemaréchal, Pedersen and Powell (1982) in which the requirement that $\psi_1(\boldsymbol{x}^+)$ be smaller than $\psi_1(\boldsymbol{x})$ every iteration is replaced by the requirement that this should happen at least once every $t > 1$ iterations. Remarkably, this is sufficient to ensure that, so long as a unit step is always attempted and provided the iterate is reset to the last "satisfactory" value if more than t iterations pass without a "satisfactory" reduction, a unit step will eventually be "satisfactory" at least every other iteration. A related recent proposal by Panier and Tits (1991) is to replace the linesearch requirement that $\psi_1(\boldsymbol{x}^+)$ be sufficiently smaller than $\psi_1(\boldsymbol{x})$ by the weaker requirement that the new value be smaller than $\max\{\psi_1(\boldsymbol{x}), \psi_1(\boldsymbol{x}^-), \psi_1(\boldsymbol{x}^=)\}$, where $\boldsymbol{x}^-$ and $\boldsymbol{x}^=$ are the previous two iterates. They show that such a strategy does not asymptotically prevent unit steps; the use of a non-monotonic linesearch is reminiscent of the overlooked procedure for unconstrained minimization by Grippo, Lampariello and Lucidi (1986).

More recently, Fletcher's (1970) differentiable exact penalty function

$$\psi_d(\boldsymbol{x}) = f(\boldsymbol{x}) - \boldsymbol{c}(\boldsymbol{x})^T \boldsymbol{y}(\boldsymbol{x}) + \rho\|\boldsymbol{c}(\boldsymbol{x})\|_2^2, \tag{2.11}$$

where

$$\boldsymbol{A}(\boldsymbol{x})\boldsymbol{A}(\boldsymbol{x})^T \boldsymbol{y}(\boldsymbol{x}) = \boldsymbol{A}(\boldsymbol{x})\nabla_x f(\boldsymbol{x}), \tag{2.12}$$

has been considered as a merit function by Powell and Yuan (1986). The main theoretical drawbacks with this function are the expense of computing its derivatives as well as the danger that the multiplier function $\boldsymbol{y}(\boldsymbol{x})$ is not uniquely defined whenever $\boldsymbol{A}(\boldsymbol{x})$ is less than full rank. To circumvent the former problem, Powell and Yuan (1986) show that it is possible to replace $\boldsymbol{y}(\boldsymbol{x} + \alpha_x \boldsymbol{\Delta x})$ in the linesearch by the interpolant $\boldsymbol{y}(\boldsymbol{x}) + \alpha_x(\boldsymbol{y}(\boldsymbol{x} + \boldsymbol{\Delta x}) - \boldsymbol{y}(\boldsymbol{x}))$ while ensuring global convergence at a superlinear rate. Fletcher (1973) prefers the variant

$$\psi_f(\boldsymbol{x}) = f(\boldsymbol{x}) - \boldsymbol{c}(\boldsymbol{x})^T \boldsymbol{y}(\boldsymbol{x}) + \rho \boldsymbol{c}(\boldsymbol{x})^T \left(\boldsymbol{A}(\boldsymbol{x})\boldsymbol{A}(\boldsymbol{x})^T\right)^{-1} \boldsymbol{c}(\boldsymbol{x}) \tag{2.13}$$

of (2.11), which he shows can be rewritten as

$$\psi_f(\boldsymbol{x}) = f(\boldsymbol{x}) - \boldsymbol{c}(\boldsymbol{x})^T \boldsymbol{\lambda}(\boldsymbol{x}) \tag{2.14}$$

where

$$\boldsymbol{\lambda}(\boldsymbol{x}) = \arg\min_{\lambda} \tfrac{1}{2}\|\boldsymbol{A}^T(\boldsymbol{x})\boldsymbol{\lambda} - \nabla_x f(\boldsymbol{x})\|_2^2 + \rho \boldsymbol{\lambda}^T \boldsymbol{c}(\boldsymbol{x}). \tag{2.15}$$

Because of the expense of inverting $\boldsymbol{A}(\boldsymbol{x} + \alpha_x \boldsymbol{\Delta x})\boldsymbol{A}(\boldsymbol{x} + \alpha_x \boldsymbol{\Delta x})^T$ at trial points, Boggs and Tolle (1989) propose using a variant of (2.13) in which this term is approximated in the linesearch by $\boldsymbol{A}(\boldsymbol{x})\boldsymbol{A}(\boldsymbol{x})^T$. Further generalizations of (2.11) are possible, and most are covered by the function

$$\psi_g(\boldsymbol{x}) = f(\boldsymbol{x}) - \boldsymbol{c}(\boldsymbol{x})^T \boldsymbol{u}(\boldsymbol{x}) + \rho\|\boldsymbol{c}(\boldsymbol{x})\|_2^2 / a(\boldsymbol{x}), \tag{2.16}$$

where

$$\left(\boldsymbol{A}(\boldsymbol{x})\boldsymbol{A}(\boldsymbol{x})^T + \gamma \boldsymbol{I}\right) \boldsymbol{u}(\boldsymbol{x}) = \boldsymbol{A}(\boldsymbol{x})\nabla_x f(\boldsymbol{x}) \tag{2.17}$$

and $0 < a(\boldsymbol{x}) \le \alpha$, for some $\alpha > 0$ and $\gamma \ge 0$. Facchinei and Lucidi (1994) show that this merit function does not impede the quadratic convergence of SQP methods for which the exact derivatives $\boldsymbol{H} = \nabla_{xx}\ell(\boldsymbol{x}, \boldsymbol{u}(\boldsymbol{x}))$ are used.

One issue we have not considered so far is that the linearized constraints (2.3b) may be inconsistent. Fletcher (1981, Section 14.4) provides a useful alternative in which the model problem (2.3) is replaced by the problem of minimizing

$$m_1(\boldsymbol{\Delta x}) \stackrel{\text{def}}{=} \tfrac{1}{2}\boldsymbol{\Delta x}^T \boldsymbol{H} \boldsymbol{\Delta x} + \boldsymbol{\Delta x}^T \nabla_x f(\boldsymbol{x}) + \rho\|\boldsymbol{c}(\boldsymbol{x}) + \boldsymbol{A}(\boldsymbol{x})\boldsymbol{\Delta x}\|_1 \tag{2.18}$$

of the merit function (2.10). The problem of minimizing (2.18) may be reformulated as a quadratic programming problem, and has the desirable property that the subproblem is always consistent (this is also implicit in the algorithm of Coleman and Conn, 1982b. Nonetheless, Fletcher (1982) and Yuan (1985a) observe that the Maratos "effect" may still occur if the search direction is computed by minimizing (2.18) but can be prevented if a second-order correction of the form (2.7) is made. Fast local convergence properties of such a method are examined by Womersley (1985) and Yuan (1985b), while Wright (1987, 1989b) shows that it is not necessary to minimize (2.18) to full accuracy to achieve fast convergence.

2.2 Trust region methods

Linesearch methods aim *a posteriori* to control a bad choice of step $\boldsymbol{\Delta x}$ by comparing $\psi(\boldsymbol{x} + \alpha\boldsymbol{\Delta x})$ with $\psi(\boldsymbol{x})$. Trust-region methods, on the other hand, aim *a priori* to ensure that the step is adequate by imposing extra restrictions on the model from which the step is derived. The simplest example would be to consider the model problem (2.3) but to require additionally that

$$||\boldsymbol{\Delta x}|| \leq \Delta \tag{2.19}$$

for some scalar $\Delta > 0$. The extra constraint (2.19) is known as the *Trust-region* constraint and the scalar Δ is the trust-region *radius*. The size of the radius is controlled by comparing the actual reduction in the merit function when the step is taken with the value predicted by a model of this function for which $\boldsymbol{\Delta x}$ is a good step. Normally Δ will be increased if there is good agreement and the trust-region constraint is active, and decreased when the agreement is poor. The introduction of a trust region allows considerable extra freedom when specifying $\boldsymbol{H}$ as (2.19) stops inappropriate choices of $\boldsymbol{H}$ leading to unbounded steps.

Fletcher (1982) includes a trust-region constraint when minimizing the model function (2.18) of (2.10). It is particularly convenient in this case to choose the infinity norm for (2.19) as the resulting model problem may then still be posed as a quadratic program. There are a number of problems however if we try to impose a trust region on (2.3).

Firstly, the linear constraints (2.3b) and the trust-region (2.19) may have no common feasible point. A number of attempts have been made to overcome this defect. Vardi (1985) and Byrd, Schnabel and Schultz (1987) suggests replacing (2.3b) by constraints of the form

$$\boldsymbol{A}(\boldsymbol{x})\boldsymbol{\Delta x} + \theta \boldsymbol{c}(\boldsymbol{x}) = \boldsymbol{0}, \tag{2.20}$$

where $\theta \in (0, 1]$ is chosen so that the new constraints and the trust region have a common feasible point. (A similar device was proposed by Powell, 1978, to handle inconsistent constraints in the basic subproblem (2.3).) Another possibility is to replace (2.3b) by

$$||\boldsymbol{A}(\boldsymbol{x})\boldsymbol{\Delta x} + \boldsymbol{c}(\boldsymbol{x})|| \leq \theta, \tag{2.21}$$

where θ is chosen so that the intersection of (2.19) and (2.21) has a solution. (Once again, a similar device was proposed by Burke and Han (1989) to handle inconsistent constraints in the basic subproblem (2.3).) Celis, Dennis and Tapia (1985) choose

$$\min_{\|d\|\leq\Delta} \|A(x)d + c(x)\| \leq \theta \leq \|c(x)\|, \tag{2.22}$$

while Powell and Yuan (1990) prefer

$$\min_{\|d\|\leq\beta_1\Delta} \|A(x)d + c(x)\| \leq \theta \leq \min_{\|d\|\leq\beta_2\Delta} \|A(x)d + c(x)\|, \tag{2.23}$$

where $0 < \beta_2 \leq \beta_1 < 1$. Both sets of authors suggest particular choices of θ which satisfy their restrictions when the two-norm is used. It is important to note that it often suffices to obtain an approximate solution to the given model problem, and a general theory which covers this possibility is given by Dennis, El-Alem and Maciel (1997). Burke (1992) studies methods of this sort in a very general setting, and allows for the possibility that the original problem may be infeasible by showing convergence to a "nearest" infeasible KKT point.

Secondly it is not clear what model of the merit function should be used. The most common approach is to model the merit function by taking first or second-order approximations of constituent terms. Both El-Alem (1995), for the subproblem based on (2.20), and El-Alem (1991) and Powell and Yuan (1990), for that based on (2.21), use the merit function (2.11) and model this by

$$\begin{aligned}\psi_d(x) + \Delta x^T \nabla_x \ell(x, y(x)) - (y(x+\Delta x) - y(x))^T \left(c(x) + \tfrac{1}{2}A(x)\Delta x\right) \\ + \tfrac{1}{2}\Delta x^T H \Delta x_n + \rho\left(\|c(x) + A(x)\Delta x\|_2^2 - \|c(x)\|_2^2\right),\end{aligned} \tag{2.24}$$

where Δx_n is the orthogonal projection of Δx into the null-space of $A(x)$. The authors provide schemes for automatically adjusting the penalty parameter ρ and establish global and locally superlinear convergence.

Thirdly, even when the constraints are consistent, the resulting subproblem may not be easy to solve. When the infinity norm is used, the resulting subproblem is inevitably a quadratic program. When the two-norm is used, Yuan (1990) gives an algorithm for solving the subproblem involving (2.19) and (2.21), that is of minimizing a quadratic function in a region defined by the intersection of two balls. Zhang (1992) simplifies this scheme in the case where H_{zz} is positive semi-definite, while Heinkenschloss (1994) does the same in the general case. Finally, recent work by Moré (1993) and Stern and Wolkowicz (1995) has generalized this to the case where (2.19) is replaced by the condition $\Delta_1 \leq \Delta x^T C \Delta x \leq \Delta_2$ and where C may be indefinite.

When a reduced Hessian method is used, the obvious trust-region generalization is to choose the step Δx_z by solving a subproblem of the form

$$\underset{\|\Delta x_z\|\leq\Delta}{\text{minimize}} \ \tfrac{1}{2}\Delta x_z^T H_{zz} \Delta x_z + \Delta x_z^T Z^T \nabla_x \ell(x, y). \tag{2.25}$$

Such a scheme is proposed by Zhang and Zhu (1990), while methods which allow approximate solutions of (2.25) are considered by Zhang, Zhu and Fan (1993).

2.3 Methods for inequality constraints

Nonlinear programming problems rarely exclusively involve equality constraints, but typically involve a mixture of equations and inequalities. We thus turn to the inequality constrained problem. The Karush-Kuhn-Tucker (first-order optimality) conditions for the inequality problem

$$\underset{\boldsymbol{x}\in\Re^n}{\text{minimize}} \ f(\boldsymbol{x}) \ \text{ subject to } \ \boldsymbol{c}(\boldsymbol{x}) \geq \boldsymbol{0} \tag{2.26}$$

are that

$$\nabla_x \ell(\boldsymbol{x}, \boldsymbol{\lambda}) = \boldsymbol{0}, \ \boldsymbol{c}(\boldsymbol{x}) \geq \boldsymbol{0}, \ \boldsymbol{\lambda} \geq \boldsymbol{0}, \ \text{ and } \ \boldsymbol{c}(\boldsymbol{x})^T\boldsymbol{\lambda} = 0, \tag{2.27}$$

where the Lagrangian function $\ell(\boldsymbol{x}, \boldsymbol{\lambda}) = f(\boldsymbol{x}) - \boldsymbol{c}(\boldsymbol{x})^T\boldsymbol{\lambda}$.

At the time of the last conference, algorithms for (2.26) were primarily of the active-set variety. An *active set* method is a method which aims to solve (2.26) by predicting which of the inequalities will be active (ie, which of the $c_i(\boldsymbol{x}) = 0$) and which are inactive (ie, $c_i(\boldsymbol{x}) > 0$) at the solution. Once these sets are known, the problem can be solved as if it involves only equality constraints, namely those deemed to be active at the solution. The main justification, therefore for much of the work described in Section 2.1 is as a tool for analyzing active-set methods.

The principal differences between active set methods is in the way that the active set is assigned. In inequality-(constrained) quadratic programming (IQP) methods, no *a priori* choice of the active set is made when choosing the correction $\boldsymbol{\Delta x}$, rather $\boldsymbol{\Delta x}$ is obtained by solving the quadratic programming problem of minimizing (2.3a) subject to a linear approximation $\boldsymbol{A}(\boldsymbol{x})\boldsymbol{\Delta x} + \boldsymbol{c}(\boldsymbol{x}) \geq \boldsymbol{0}$ of all of the constraints. The active set for this problem is taken as a prediction of that for (2.26). Robinson (1974) provides theoretical justification for such an approach. In equality-(constrained) quadratic programming (EQP) methods, the active set is assigned prior to the selection of $\boldsymbol{\Delta x}$ (primarily on the basis of inequalities which are close to being active and whose Lagrange multiplier estimates are positive) and $\boldsymbol{\Delta x}$ is found directly by solving (2.3a) subject to the linear equality constraints $\boldsymbol{A}(\boldsymbol{x})_{\mathcal{A}}\boldsymbol{\Delta x} + \boldsymbol{c}_{\mathcal{A}}(\boldsymbol{x}) = \boldsymbol{0}$, where the subscript $\mathcal{A}$ denotes those constraints which are considered to be active. Which strategy is preferable is a matter for some debate (see, for example, Murray and Wright, 1982).

The merit function must also account for inequality constraints. Normally, this is merely a matter of replacing the term(s) which handle equality constraints with a similar term for the inequalities. For instance, the analog of (2.10) for the problem (2.26) is the function

$$f(\boldsymbol{x}) + \rho\|\boldsymbol{c}(\boldsymbol{x})_-\|_1, \tag{2.28}$$

where c_- gives, componentwise, the smaller of c_i and zero. Pantoja and Mayne (1991) and Heinz and Spellucci (1994) prefer the infinity to the one-norm in (2.28), and propose a line-search method based on the model problem

$$\underset{\Delta x \in \Re^n}{\text{minimize}} \quad \tfrac{1}{2}\Delta x^T H \Delta x + \Delta x^T \nabla_x f(x) + \rho \|(c(x) + A(x)\Delta x)_-\|_\infty. \tag{2.29}$$

Yuan (1995) analyses a similar method in which a trust region (2.19) is imposed on the model (2.29).

Generalizations of the functions (2.11) or (2.13) are more interesting. Fletcher (1973) gives a simple generalization of the function (2.13) in the inequality case by replacing (2.15) with

$$\lambda(x) = \underset{\lambda \geq 0}{\arg\min} \; \tfrac{1}{2}\|A^T(x)\lambda - \nabla_x f(x)\|_2^2 + \rho \lambda^T c(x). \tag{2.30}$$

Boggs, Tolle and Kearsley (1991) prefer to introduce extra slack variables s in order to replace the inequality constraints with the equations $c_i(x) - \frac{1}{4}s_i^2 = 0$. Introducing these variables into (2.13) and setting $z_i = \frac{1}{4}s_i^2 \geq 0$, they propose the merit function

$$f(x) - (c(x) - z)^T \pi(x) + \rho (c(x) - z)^T \left(A(x)A(x)^T + Z\right)^{-1} (c(x) - z), \tag{2.31}$$

where Z is the diagonal matrix with entries z_i and

$$(A(x)A(x)^T + Z)\pi(x) = A(x)\nabla_x f(x). \tag{2.32}$$

It is straightforward to design updates for z which ensure that $z \geq 0$. A version of (2.16) appropriate for inequality constraints is given by Di Pillo, Facchinei and Grippo (1992).

2.4 Difficulties

Having surveyed the main developments in SQP methods, we now consider the difficulties with the approach. Many of the difficulties are directly attributable to a lack of coherence between the step calculation and the merit function. In unconstrained optimization, there is a direct relationship between the merit function — in this case, invariably the objective function — and the calculation of the step. The prototypical method, Newton's method, may be viewed both as a method which attempts to satisfy the first-order optimality conditions *and* as a method which aims to reduce the merit function through a Taylor's series approximation. In the constrained case, while the SQP direction (2.3) may be viewed as an attempt to satisfy the KKT conditions, it is not directly related to any of the merit functions that have been proposed, although it does often provide a descent direction for them. Perhaps the only satisfactory methods from the point of consistency are those which directly attempt to link the merit function and the step. Fletcher's (1982) and Coleman and Conn's (1982b) methods based on (2.10) and (2.18) and their generalizations are in this class, but, as we have noted, even these have disadvantages.

A second drawback is that so few of the suggestions we have considered are appropriate if the number of variables is large. In particular, unless function values are expensive, the dominant cost of the methods tends to be in solving linear systems; for inequality problems this may be particularly acute as each subproblem may require the solution of a sequence of such systems. If n is large, there is little hope unless either the required systems are small or sparse. There are two important cases where this is so. Firstly, if the number of equality or active constraints is close to n, reduced Hessian methods, such as those proposed by Coleman and Conn (1982a), Gilbert (1991), and Biegler et al. (1995), which maintain the matrix $\boldsymbol{H}_{zz}$ but ignore $\boldsymbol{H}_{zy}$ and $\boldsymbol{H}_{yy}$, may be successful. The only systems which need to be solved involve $\boldsymbol{H}_{zz}$ (small) and $\boldsymbol{A}(\boldsymbol{x})\boldsymbol{Y}(\boldsymbol{x})$ and its transpose (sparse, we hope). Secondly, if the matrix $\boldsymbol{H}$ is sparse, sparse methods for linear systems (when an EQP method is used) or quadratic programming (for IQP methods) may be employed. This will often be the case if $\boldsymbol{H}$ is chosen as the Hessian of the Lagrangian function, or from a structured or sparse secant updating formula (see, for example, Toint, 1977, Conn, Gould and Toint, 1990, and Fletcher, 1995).

Finally, one of the main advances in methods for the unconstrained minimization of large problems was the recognition that a Newton-like direction need not computed very accurately when far from a stationary point (see Dembo, Eisenstat and Steihaug, 1982). Clearly, when equality constraints are present a similar result would be valuable, but there has been remarkably little work on this topic. For equality constrained problems involving relatively few constraints, Fontecilla (1990) notes that (2.5) and (2.9) are small systems, and the only large system is (2.6). He thus proposes a method which initially solves (2.6) to low accuracy. Alas, for fast asymptotic convergence (2.6) must eventually be solved to high accuracy which limits the effectiveness of this proposal. When the constraints are inequalities and an active-set IQP method is used, Murray and Prieto (1995) show that it is possible to stop the solution of the QP subproblem at the first stationary point encountered rather than solving the problem to completion.

3 Optimality-condition based methods

Since the mid 1980s there has been a revolution in the way in which the optimality conditions (2.27) have been viewed. In essence, active set methods ultimately aim to satisfy the dual feasibility requirement $\nabla_x \ell(\boldsymbol{x}, \boldsymbol{\lambda}) = \boldsymbol{0}$, while ensuring that the remaining feasibility requirements, $\boldsymbol{c}(\boldsymbol{x}) \geq \boldsymbol{0}$ and $\boldsymbol{\lambda} \geq \boldsymbol{0}$, and complementarity condition, $\boldsymbol{c}(\boldsymbol{x})^T \boldsymbol{\lambda} = 0$, are always (effectively) satisfied. This is achieved by the simple combinatorial expedient of ensuring that, for each constraint, either $c_i(\boldsymbol{x}) = 0$ or $\lambda_i = 0$, but suggests that in the worst case, all combinations may be examined. By contrast, the newer interior-point methods try for optimality by ensuring that the feasibility requirements are always (effectively) satisfied while aiming ultimately to satisfy the complementarity condition. If we consider, for a moment, the case where (2.26) is a linear or quadratic program, the feasibility requirements are linear while the complementarity condition is nonlinear (quadratic). Thus the active set methods may be viewed as trying to hide this nonlinearity within a combinatorial problem, while the interior-point methods confront the nonlinearity directly. Significantly, the interior-point approach has been shown to have strong complexity advantages on many classes of convex problems (see Nesterov and Nemirovskii, 1994), and there is some evidence that these advantages transfer to improved practical performance (see, for instance, Jarre and Saunders, 1995). Gill, Murray, Saunders, Tomlin and Wright (1986) were quick to make the connection between the new-looking interior-point methods and the long-discarded barrier-function methods, and thus lead the community to re-examine barrier methods in this new light. Clearly, IQP based methods are already able to take advantage of interior-point technology as the quadratic subproblems are often convex. It remains to be seen whether the same is true of EQP based methods.

While barrier function and other interior-point methods are discussed elsewhere in this volume, we briefly mention a couple of other possibilities. Friedlander, Martínez and Santos (1994) propose a method for solving linearly-constrained problems by minimizing

$$\|\nabla_x \ell(\boldsymbol{x}, \boldsymbol{\lambda})\|_2^2 + (\boldsymbol{c}(\boldsymbol{x})^T \boldsymbol{\lambda})^2$$

over the feasible set $\boldsymbol{c}(\boldsymbol{x}) \geq \mathbf{0}$. They show that this formulation does not introduce unnecessary local minimizers when f is convex.

Kanzow and Kleinmichel (1995) observe that the optimality conditions (2.27) may be replaced by the nonlinear system

$$\nabla_x \ell(\boldsymbol{x}, \boldsymbol{\lambda}) = 0, \quad \text{and} \quad \phi(c_i(\boldsymbol{x}), \lambda_i) = 0, \tag{3.1}$$

where ϕ is any function for which

$$\phi(u, v) = 0 \quad \text{if and only if} \quad u \geq 0, \;\; v \geq 0 \;\; \text{and} \;\; uv = 0. \tag{3.2}$$

A simple example is the function $\phi(u, v) = \sqrt{u^2 + v^2} - u - v$. Their proposal is now to apply Newton's method to (3.1), and they provide a local analysis of such a method. However, the Jacobian of such a system will be singular whenever the solution is degenerate. Pang (1991, 1994) prefers the choice $\phi(u, v) = \min\{u, v\}$, and shows that this choice may be made the basis of a globally convergent method using the merit function

$$f(\boldsymbol{x}) + \mu \left(\|\nabla_x \ell(\boldsymbol{x}, \boldsymbol{\lambda})\|_2^2 + \sum_i |\phi(c_i(\boldsymbol{x}), \lambda_i)|^2 \right).$$

4 Other methods for nonlinear constraints

Now that barrier functions are back in fashion, it is worth evaluating the status of other methods which SQP algorithms were supposed to have succeeded. One of the earliest methods for solving equality constrained problems was to minimize the quadratic penalty function

$$\psi_2(\boldsymbol{x}) = f(\boldsymbol{x}) + \rho\|\boldsymbol{c}(\boldsymbol{x})\|_2^2, \tag{4.1}$$

for a sequence of scalars ρ approaching infinity. Although this method was dismissed in the 1970s, it has been seen in a more favourable light since then. Firstly, perceived difficulties with ill-conditioning were shown to be benign provided sufficient care is taken (Broyden and Attia, 1984, Gould, 1986, Coleman and Hempel, 1990). Secondly, the requirement that (4.1) be minimized is easily relaxed. Moreover, Gould (1989) shows that asymptotically at most two Newton-like steps are required for each value of ρ and that this results in a globally and (two-step) superlinearly convergent method. This result is generalized by Dussault (1995) to problems involving both equations and inequalities.

These methods and the succeeding augmented Lagrangian methods have three significant advantages over most SQP methods. Firstly, the choice of the correction $\boldsymbol{\Delta x}$ is normally intimately connected to the merit function — it is usually obtained by minimizing a second-order model of the function. Secondly, the second derivative matrices of these functions can normally be expected to be positive (semi-)definite in some neighbourhood of the solution, and thus it is reasonable to approximate these derivatives via positive-definite secant formulae. Thirdly, the Newton-like systems which arise are usually easier to handle when n is large and the problem sparse than for SQP methods — this is a consequence of the matrices being (relatively) sparse and definite. It is this third point which may explain why many of the currently most successful codes for large-scale nonlinear programming are based on these methods (see Section 7). However, these methods do have some drawbacks, namely that a sequence of problems has to be solved, the iterates will often initially move away from the solution, and, significantly, no advantage is taken of linear or other simple constraints.

Augmented Lagrangian methods solve the equality constrained problem by minimizing a sequence of problems of the form

$$\psi_a(\boldsymbol{x}) = f(\boldsymbol{x}) - \boldsymbol{c}(\boldsymbol{x})^T\boldsymbol{y} + \rho\|\boldsymbol{c}(\boldsymbol{x})\|_2^2, \tag{4.2}$$

where $\boldsymbol{y}$ are estimates of the Lagrange multipliers and ρ a positive penalty parameter (see, Hestenes, 1969, and Powell, 1970). Convergence is assured by adjusting $\boldsymbol{y}$ and ρ, and it is not necessary for ρ to approach infinity. When first-order multiplier updates are used, the minimizers of (4.2) converge linearly, the rate being proportional to $1/\rho$. Faster rates are possible using higher-order multiplier updates, but first-order updates are convenient and effective for large-scale problems. Bartholomew-Biggs (1987) notes that the Newton equations for (4.2) may be reformulated as a quadratic programming problem, and this problem is a perturbation of (2.3). The method is easily generalized for inequality constraints (see, Rockafellar, 1974).

Realistic nonlinear programming problems often involve a mixture of linear and nonlinear constraints. Conn, Gould and Toint (1991) consider the case where there are simple bounds $\boldsymbol{l} \leq \boldsymbol{x} \leq \boldsymbol{u}$ on the variables in addition to equality constraints. Their algorithm finds a sequence of minimizers of (4.2) where the simple bounds are explicitly enforced and only approximate minimizations are performed. The convergence results obtained match that for the case without simple bounds. Inequality constraints are handled by introducing slack variables, but Conn, Gould and Toint (1994) show that these slack variables need not affect the linear algebra costs. More significantly, Conn, Gould and Toint (1992b) show that a single Newton-like step eventually suffices in the approximate minimization, and thus the iterates are globally convergent at a (fast) linear rate. This strategy is generalized by Conn, Gould, Sartenaer and Toint (1996) for the case where the constraint set is a mixture of equality and linear inequality constraints; a sequence of approximate minimizers of (4.2) subject to the linear constraints are sought. This theory also allows for independent penalty parameters for each of the penalized constraints.

Another class of important methods, the direct ancestors of SQP methods, are the sequential *linear* programming (SLP) methods (see Griffith and Stewart, 1961) in which linear approximations are taken of both objective and constraint functions. The resulting linear program may then be solved using either simplex or interior point methods. Modern versions are based on linear approximations of the merit function (2.10) (see, for example, Zhang, Kim and Lasdon, 1985), while Fletcher and Sainz de la Maza (1989) propose a hybrid method which tries a Newton-like (Coleman and Conn, 1982a) step for (2.10) but falls back on the SLP correction whenever the Newton step is unsuccessful. Clearly, unless some form of second-order acceleration is used the convergence of these methods will typically be rather slow, but the mature state of linear programming algorithms means that the subproblems can be solved efficiently.

At the other extreme, Maany (1987) proposes a method in which quadratic approximations are taken of both objective and constraints. Such an approach is appropriate for highly-curved constraints, but has the disadvantage that the subproblems are hard to solve.

Finally, an interesting class of feasible-point SQP methods have been developed by Panier and Tits (1987, 1993) and Bonnans, Panier, Tits and Zhou (1992) for inequality constrained problems. As the iterates are feasible, the objective function may be used as a merit function. The methods require the solution of two linear or quadratic programming problems at each iteration to generate corrections; a backtracking linesearch is performed along a quadratic arc defined by these directions, and the first step which sufficiently reduces f and satisfies the constraints is accepted. The methods are shown to be globally and two-step superlinearly convergent.

5 Linear and convex constraints

We have seen that most methods for nonlinear constraints solve a sequence of subproblems involving simpler constraints. For instance, SLP and SQP methods solve linear and quadratic programs, while other methods differentiate between "difficult" constraints (normally the nonlinear ones) and "easy" ones (normally the linear or convex ones), treating the easy constraints directly in the subproblem. In this section, we consider methods specifically designed for problems involving linear or convex constraints; we exclude a discussion of interior-point methods as they are covered elsewhere in this volume.

5.1 Simple bounds, gradient projection and projected gradients

Two classes of problems have attracted attention here, those for which the objective function is a quadratic and those with a general objective. Standard active-set methods for linearly constrained optimization problems typically refine the active set slowly, perhaps one constraint leaving or entering the active set at each iteration. When the constraint set only involves simple bounds, however, it is far easier to add or delete many constraints at each iteration, and the best mechanism for achieving this is the gradient projection algorithm.

The *gradient projection* algorithm (Levitin and Polyak, 1966) simply chooses iterates according to

$$\boldsymbol{x}^{+} = P_{\Omega}[\boldsymbol{x} - \alpha_x \nabla_x f(\boldsymbol{x})], \tag{5.1}$$

where Ω is the set of feasible points, $P_{\Omega}[\boldsymbol{v}]$ is the projection of $\boldsymbol{v}$ into Ω and α_x is a suitable stepsize (see, for instance, Bertsekas, 1976, or Dunn, 1981). When the constraints are simple bounds, $\Omega = \{\boldsymbol{x} : \boldsymbol{l} \leq \boldsymbol{x} \leq \boldsymbol{u}\}$, and the projection is easily computed as $P_{\Omega}[\boldsymbol{v}] = \text{mid}(\boldsymbol{l}, \boldsymbol{v}, \boldsymbol{u})$, where mid denotes the vector whose components are the medians of l_i, v_i and u_i. However, as the gradient projection algorithm is just a constrained variant of the method of steepest descent, it is clear that some form of acceleration is needed if the method is to be practical.

Moré and Toraldo (1991) propose that the active sets for consecutive iterates be compared, and if the sets are identical, or if little progress is being made on the "face" defined by the current set, the current face should be explored using a higher-order method. To this end, the authors propose that the conjugate-gradient method should be used to find an approximate stationary point $\boldsymbol{x}^c$ on the face, and then a linesearch performed on the piecewise-linear arc $P_{\Omega}[\boldsymbol{x} + \alpha_x(\boldsymbol{x}^c - \boldsymbol{x})]$.

The justification for such a scheme is that the gradient projection algorithm is guaranteed to determine the optimal active set for nondegenerate problems in a finite number of iterations (see Berksekas, 1976). Calamai and Moré (1987) improve this considerably by noticing that the result is true for *any* algorithm for which the projected gradient converges to zero; the *projected gradient* is

$$\nabla_{\Omega} f(\boldsymbol{x}) = P_{T(x)}[-\nabla_x f(\boldsymbol{x})], \tag{5.2}$$

where the tangent cone $T(\boldsymbol{x})$ is the closure of the cone of all feasible directions at $\boldsymbol{x}$. While the projected gradient may be difficult to calculate in general, for simple bounds it is given (componentwise) by

$$\nabla_{\Omega} f(\boldsymbol{x})_i = \begin{cases} \min\{\nabla_x f(\boldsymbol{x})_i, 0\} & \text{if} \quad x_i = l_i \\ \nabla_x f(\boldsymbol{x})_i & \text{if} \quad l_i < x_i < u_i \\ \max\{\nabla_x f(\boldsymbol{x})_i, 0\} & \text{if} \quad x_i = u_i \end{cases} \tag{5.3}$$

This result has subsequently been generalized. Dunn (1987) and Moré (1988) relax the non-degeneracy assumption to one of requiring that there is a set of strictly complementary Lagrange multipliers while Burke and Moré (1994) obtain similar results in a more general geometric setting. De Angelis and Toraldo (1993) show that the simplified gradient projection scheme of Dem'yanov and Rubinov (1970), in which (5.1) is replaced by an iteration of the form $\boldsymbol{x}^+ = \boldsymbol{x} + \alpha_x(P_{\Omega}[\boldsymbol{x} - \eta\nabla_x f(\boldsymbol{x})] - \boldsymbol{x})$ for some η, inherits the active-constraint identification property of its predecessor. Burke (1990) generalizes the previous analysis for nonconvex problems, and shows that projected gradient of a suitable linearization of the problem correctly identifies the optimal active set. This result, and a similar analysis by Wright (1989a), suggests why EQP strategies for SQP and linearizations of (2.28) are successful.

Moré and Toraldo (1991) show that their algorithm for quadratic objectives is necessarily finite under a nondegeneracy assumption. Friedlander and Martínez (1994) provide a mechanism for leaving an unpromising face and ensuring that their algorithm will not return to the face unless a substantial improvement in the value of the objective is possible.

An interesting observation drives the method of Coleman and Hulbert (1993). They note that the optimality conditions may be expressed as

$$\nabla_x f(\boldsymbol{x}) \cdot (\boldsymbol{x} - \tfrac{1}{2}(\boldsymbol{l} + \boldsymbol{u}) + \tfrac{1}{2}(\boldsymbol{u} - \boldsymbol{l}) \cdot \text{ sign } (\nabla_x f(\boldsymbol{x}))) = \mathbf{0}, \tag{5.4}$$

where $\text{sign}(\boldsymbol{v})_i$ is -1 if $v_i \leq 0$ and 1 otherwise, and $\boldsymbol{v} \cdot \boldsymbol{w}$ is the vector whose components are $v_i w_i$. They then derive a Newton-like correction for the nonlinear system (5.4), while recognizing that the function is non-differentiable whenever a component of $\nabla_x f(\boldsymbol{x}))$ is zero. A related merit function is provided, and they are able to show global and superlinear convergence under a suitable nondegeneracy assumption.

For general objective functions, Conn, Gould and Toint (1988) provide a class of trust-region methods in which a quadratic model is minimized within a region defined by the intersection of the simple bounds and a trust region. An approximate solution of the model problem is found, of which only a local minimizer of the model along the arc

$$\boldsymbol{x}(\alpha) = P_{\Omega \cap \{y : \|y - x\| \leq \Delta\}}[\boldsymbol{x} - \alpha\nabla_x f(\boldsymbol{x})] \tag{5.5}$$

is required for global convergence. Convergence is accelerated by continuing the model minimization using conjugate gradients in the face determined by the solution to (5.5). As before, so long as the problem is non-degenerate, the active set at the solution is identified by that of (5.5) after a finite number of iterations, and thus the speed of convergence is determined by the accuracy required in the conjugate-gradient step. Lescrenier (1991) shows that the non-degeneracy assumption is unnecessary. Byrd, Lu, Nocedal and Zhu (1995) give a line-search variant of Conn et al.'s (1988) algorithm, using an efficient limited-memory Hessian approximation. Finally, Toint (1988) shows that the frameworks of Calamai and Moré (1987) and Conn et al. (1988) may be extended to cover infinite-dimensional problems in a general Hilbert-space setting.

5.2 General linear and convex constraints

One of the earliest active set methods for linearly constrained minimization was Rosen's (1960) gradient projection algorithm (not to be confused with the method (5.1) of the same name). At each stage of this prototypical feasible-point active-set method, a step is taken in the direction of the gradient projected orthogonally into the null-space of those constraints which are currently considered active. When this projected gradient is small compared to the largest of the least-squares multiplier estimate (2.9), the step may leave the current face. Although the method was succeeded long ago by SQP and other methods which incorporate curvature, remarkably the method has only recently been shown to be globally convergent (Du and Zhang, 1986, 1989).

Many active-set methods have been proposed for convex quadratic programming, most of which differ in their linear algebra requirements rather than in the iterates they generate (Best, 1984). There have been relatively fewer methods for non-convex problems. Fletcher (1987) assessed the state of the art ten years ago. The only real advances since then, aside from those with interior-point methods, are for large-scale problems. Gill, Murray, Saunders and Wright (1990, 1991) and Gould (1991) consider versions of Fletcher's (1971) method for non-convex problems which are able to exploit sparsity. Boggs, Domich and Rogers (1995) suggest investigating a low-dimensional subspace of "interesting" directions at each iteration. This reduction in dimension means that the sub-problems are small and can thus be tackled with any of the good algorithms for small problems.

Concerns over the past decade for general linearly constrained problems have focused on how to cope with problems which involve considerably more inequality constraints than variables. Powell (1989), wishing to avoid the increase in dimensionality which would occur if slack variables were added to the inequalities, prefers to treat the inequalities directly. He proposes that constraints which are "close" to active are not allowed to approach the constraint boundaries until absolutely necessary, and this allows larger steps to be taken in early iterations than would be possible in a conventional active set method. The state of the art for large-scale linearly constrained minimization has changed very little, the method proposed by Murtagh and Saunders (1978) still being pre-eminent.

One of the main difficulties which arises in active-set methods for linearly constrained optimization may occur when the constraint gradients in the active set at $\boldsymbol{x}$ are linearly dependent. The correction $\boldsymbol{\Delta x}$ will normally be chosen orthogonal to a linearly independent subset of the the active constraints, but in the degenerate case a nonzero step along $\boldsymbol{\Delta x}$ may not be possible because of the remaining active constraints. In the worst case this can cause an algorithm to cycle infinitely through subsets of the active constraints. Thus finding a $\boldsymbol{\Delta x}$ which is capable of moving away from $\boldsymbol{x}$ is crucial, especially as degenerate active sets are extremely common in practice. Of late, attention has focused on methods which are capable of dealing with this situation when floating-point computations are performed, and the methods of Fletcher (1988), Ryan and Osborne (1988), Dax (1989) and Gill, Murray, Saunders and Wright (1989) have all proved effective in practice.

Burke, Moré and Toraldo (1990) and Conn, Gould, Sartenaer and Toint (1993) give a general theory of trust-region methods for problems involving convex constraints. As solving the trust-region problem may now be an expensive calculation, Conn et al. (1993) show that an approximate solution suffices, and provide an algorithm which delivers such an approximation. Sartenaer (1995) indicates that this approach is effective in the case of network constraints. Martinez and Santos (1995) extend this class of methods to handle general, nonconvex domains.

6 Other topics

In this short survey, we naturally have to be selective on what to include and what to leave out. One trend that has been noticeable over the past ten years has been the increasing cross-fertilization between nonlinear optimization and other branches of numerical analysis and applied mathematics. In this section, we briefly mention a few topics we feel deserve more attention from the numerical analysis community.

The promise of automatic differentiation, that is the automatic accumulation of derivatives directly from codes which provide function values (see Griewank, 1989), has been a long time coming. Dixon (1991) argues that automatic differentiation will revitalize second derivative methods, as there is then little reason to rely on secant approximations. This has profound implications as, for example, many SQP methods depend upon properties which do not hold for exact derivatives (see Section 2.1.1). The main drawback in the past was the lack of efficient software for automatic differentiation, but that has dramatically changed over the past five years (see the WWW page http://www.mcs.anl.gov/Projects/autodiff /AD_Tools for details of currently available packages). A number of optimization packages now make direct use of automatic derivatives.

A fundamental issue when using optimization algorithms is how important it is that function and derivative information is accurate. Kupferschmid and Ecker (1987) compare the Ellipsoid and SQP algorithms when function values and gradients are inaccurate, and observe that the former is much less susceptible to inaccuracies. Toint (1988) and Carter (1991) examine the convergence of trust-region methods when the gradient is inaccurate, while Conn et al. (1993) do the same for inaccurate function and gradient values.

An important class of problems not covered above are those for which derivatives may not exist but for which it is possible to calculate subgradients. These non-differentiable problems are frequently solved by methods which build local piecewise affine models of the functions concerned. For instance, each convex function $f(\boldsymbol{x})$ may be modelled as $\max f(\boldsymbol{x}_k) + \boldsymbol{s}_k^T(\boldsymbol{x} - \boldsymbol{x}_k)$, where the $\boldsymbol{x}_i$ are (not necessarily distinct) previous iterates and $\boldsymbol{s}_k$ is a subgradient evaluated at $\boldsymbol{x}_k$. Modelling each function in this way, we may then obtain a new trial iterate by minimizing the modelled objective subject to the modelled constraints and a stabilizing trust region (or sometimes a penalty on the objective). The set of previous values $(\boldsymbol{x}_k, \boldsymbol{s}_k)$ which are included is known as a *bundle* and the determination of consecutive bundles is fundamental to the success of the method. An introduction to such methods is provided by Kiwiel (1989), Hiriart-Urruty and Lemaréchal (1993) and Lemaréchal and Zowe (1994), while numerical evidence that they are effective is given by Schramm and Zowe (1992).

Finally, we have said very little about infinite-dimensional problems. We particularly regret having no space to mention the significant advances in algorithms for network optimization, optimal control, and parallel optimization, but merely point to Bertsekas (1991), Hager (1990) and Schnabel (1994) as "tasters" for progress in these fields.

7 Software

We should not forget that the main reason for designing and analyzing algorithms is to enable others to solve "real" optimization problems, and one of the best ways of doing this is for researchers to provide quality software which implements their ideas. Frankly, we were surprised when researching this paper quite how few papers contained numerical results which justified their author's optimistic analytic assessments, or indeed any numerical results at all! Fortunately, there is a fair amount of good software available particularly for small problems. Major sources are the Harwell, Hatfield, NAG, and Visual Numerics (formerly IMSL) subroutine libraries. The book by Moré and Wright (1993) provides a thorough assessment of the state and scope of optimization software.

Possibly the biggest change over the past ten years has been the size of nonlinear problems that can be, and are now being, solved. It was rare to find results for problems involving, say, more than 50 unknowns at the time of the last conference, but now highly nonlinear problems involving, say, 20,000 unknowns and similar numbers of constraints can be solved in reasonable times on current desktop computers. Of current codes capable of handling such problems CONOPT (Drud, 1985) and LSGRG2 (Smith and Lasdon, 1992) are generalized reduced gradient methods, MINOS (Murtagh and Saunders, 1982) and

LANCELOT (Conn, Gould and Toint, 1992a) are based on augmented Lagrangian functions, ETR (Lalee, Nocedal and Plantega, 1993) is an SQP method for equality constraints, while SNOPT (Gill, Murray and Saunders, 1996) and that by Boggs et al. (1994) is a general SQP method.

8 Conclusion

The past ten years in nonlinear optimization have been a time of consolidation rather than inspiration. The energy that has been devoted to interior-point methods — particularly for linear programs — have left the community slightly exhausted when the challenges of nonlinearity arise. However, we certainly have a better understanding of when and why the methods we considered ten years ago work. Furthermore, we are now capable of solving far larger problems than before, primarily because of our better exploitation of problem structure.

It is not difficult to see how the field will develop in the short term. Interior-point methods will be extended to handle nonlinear and nonconvex problems, and many of the subproblems currently solved using active set methods will be tackled with interior methods. Moreover, the wider availability of second (and higher order) derivatives must result in a reappraisal of our current "favourite" approaches.

Some areas remain vastly understudied. The global effect of strong nonlinearity on algorithms has not been considered in any depth, most algorithms retreating to tiny steps and/or variants of (constrained) steepest descent under these circumstances. Little is really known about how modern algorithms compare, especially on large or highly nonlinear problems. We have many preconceptions but, as the resurrection of barrier methods shows, folklore should not necessarily be trusted. Another area which deserves more attention is the effects of noise on minimization algorithms, particularly as so many industrial problems involve noisy functions. Yet further topics which have only recently received attention are nonlinear mixed integer and global optimization problems. And finally, we tend to rely heavily on matrix factorization as a tool, but there are many classes of large problems for which this impossible. We must concern ourselves more in the future on methods for which approximate solutions to model problems are sought.

In our experience a considerable number of users want to solve large problems, while perfectly adequate methods are now available for small problems so long as derivatives are available. So we end with a plea to the optimization research community: if the "new" method you are considering is not applicable to large problems, consider seriously whether it really is worth investigating. Perhaps, in this way, in ten years time, we shall be able to report a narrowing of the gap between the needs of the user community and the provisions of researchers.

References

1. Bartholomew-Biggs, M. C. 1987 Recursive quadratic-programming methods based on the augmented Lagrangian. *Mathematical Programming Studies* **31**, 21–41.
2. Bertsekas, D. P. 1976 On the Goldstein-Levitin-Polyak gradient projection method. *IEEE Transactions on Automatic Control* **21**, 174–184.
3. Bertsekas, D. P. 1991 *Linear Network Optimization: Algorithms and Codes*. Cambridge, Massachusetts, USA: MIT Press.
4. Best, M. J. 1984 Equivalence of some quadratic-programming algorithms. *Mathematical Programming* **30**, 71–87.
5. Biegler, L., Nocedal, J., & Schmid, C. 1995 A reduced Hessian method for large-scale constrained optimization. *SIAM Journal on Optimization* **5**, 314–347.
6. Boggs, P. T., & Tolle, J. W. 1989 A strategy for global convergence in a sequential quadratic programming algorithm. *SIAM Journal on Numerical Analysis* **26**, 600–623.
7. Boggs, P. T., & Tolle, J. W. 1995 Sequential quadratic programming. *Acta Numerica* **4**, 1–51.
8. Boggs, P. T., Domich, P. D., & Rogers, J. E. 1995 An interior point method for general large-scale quadratic programming problems. Internal report NISTIR 5406 Applied and Computational Mathematics Division, National Institute of Standards and Technology, Gaithersburg, Maryland, USA.
9. Boggs, P. T., Tolle, J. W., & Kearsley, A. J. 1991 A merit function for inequality constrained nonlinear programming problems. Internal report NISTIR 4702 Applied and Computational Mathematics Division, National Institute of Standards and Technology, Gaithersburg, Maryland, USA.
10. Boggs, P. T., Tolle, J. W., & Kearsley, A. J. 1994 A practical algorithm for general large scale nonlinear optimization problems. Internal report NISTIR 5407 Applied and Computational Mathematics Division, National Institute of Standards and Technology, Gaithersburg, Maryland, USA.
11. Bonnans, J. F., & Launay, G. 1995 Sequential quadratic programming with penalization of the displacement. *SIAM Journal on Optimization* **5**, 792–812.
12. Bonnans, J. F., Panier, E. R., Tits, A. L., & Zhou, J. L. 1992 Avoiding the Maratos effect by means of a nonmonotone line search II. inequality constrained problems – feasible iterates. *SIAM Journal on Numerical Analysis* **29**, 1187–1208.
13. Broyden, C. G., & Attia, N. F. 1984 A smooth sequential penalty function method for nonlinear programming. In: *11th IFIP Conference on System Modelling and Optimization* (A. V. Balarkishnan and M. Thomas, eds). Lecture Notes in Control and Information Sciences 59. Heidelberg, Berlin, New York: Springer Verlag 237–245.
14. Burke, J. V. 1990 On the identification of active constraints II: The nonconvex case. *SIAM Journal on Numerical Analysis* **27**, 1081–1102.
15. Burke, J. V. 1992 A robust trust region method for constrained nonlinear programming problems. *SIAM Journal on Optimization* **2**, 324–347.
16. Burke, J. V., & Han, S. P. 1989 A robust sequential quadratic-programming method. *Mathematical Programming* **43**, 277–303.
17. Burke, J. V., & Moré, J. J. 1988 On the identification of active constraints. *SIAM Journal on Numerical Analysis* **25**, 1197–1211.
18. Burke, J. V., & Moré, J. J. 1994 Exposing constraints. *SIAM Journal on Optimization* **4**, 573–595.
19. Burke, J. V., Moré, J. J., & Toraldo, G. 1990 Convergence properties of trust region methods for linear and convex constraints. *Mathematical Programming* **47**, 305–336.

20. Byrd, R. H. 1990 On the convergence of constrained optimization methods with accurate Hessian information on a subspace. *SIAM Journal on Numerical Analysis* **27**, 141–153.

21. Byrd, R. H., & Nocedal, J. 1991 An analysis of reduced Hessian methods for constrained optimization. *Mathematical Programming* **49**, 285–323.

22. Byrd, R. H., Lu, P., Nocedal, J., & Zhu, C. 1995 A limited memory algorithm for bound constrained optimization. *SIAM Journal on Scientific Computing* **16**, 1190–1208.

23. Byrd, R. H., Schnabel, R. B., & Schultz, G. A. 1987 A trust region algorithm for nonlinearly constrained optimization. *SIAM Journal on Numerical Analysis* **24**, 1152–1170.

24. Byrd, R. H., Tapia, R. A., & Zhang, Y. 1992 An SQP augmented Lagrangian BFGS algorithm for constrained optimization. *SIAM Journal on Optimization* **2**, 210–241.

25. Calamai, P. H., & Moré, J. J. 1987 Projected gradient methods for linearly constrained problems. *Mathematical Programming* **39**, 93–116.

26. Carter, R. G. 1991 On the global convergence of trust region methods using inexact gradient information. *SIAM Journal on Numerical Analysis* **28**, 251–265.

27. Celis, M. R., Dennis, J. E., & Tapia, R. A. 1985 A trust region strategy for nonlinear equality constrained optimization. In: *Numerical Optimization 1984* (P. T. Boggs, R. H. Byrd and R. B. Schnabel, eds). Philadelphia: SIAM 71–82.

28. Chamberlain, R. M., Powell, M. J. D., Lemaréchal, C., & Pedersen, H. C., 1982 The watchdog technique for forcing convergence in algorithms for constrained optimization. *Mathematical Programming Studies* **16**, 1–17.

29. Coleman, T. F., & Conn, A. R. 1982*a* Nonlinear programming via an exact penalty function method: Asymptotic analysis. *Mathematical Programming* **24**, 123–136.

30. Coleman, T. F., & Conn, A. R. 1982*b* Non-linear programming via an exact penalty-function: Global analysis. *Mathematical Programming* **24**, 137–161.

31. Coleman, T. F., & Fenyes, P. A. 1992 Partitioned quasi-Newton methods for nonlinear equality constrained optimization. *Mathematical Programming* **53**, 17–44.

32. Coleman, T. F., & Hempel, C. 1990 Computing a trust region step for a penalty function. *SIAM Journal on Scientific and Statistical Computing* **11**, 180–201.

33. Coleman, T. F., & Hulbert, L. A. 1993 A globally and superlinearly convergent algorithm for convex quadratic programs with simple bounds. *SIAM Journal on Optimization* **3**, 298–321.

34. Conn, A. R., Gould, N. I. M., Sartenaer, A., & Toint, Ph. L. 1993 Global convergence of a class of trust region algorithms for optimization using inexact projections of convex constraints. *SIAM Journal on Optimization* **3**, 164–221.

35. Conn, A. R., Gould, N. I. M., Sartenaer, A., & Toint, Ph. L. 1996 Convergence properties of an augmented Lagrangian algorithms for optimization with a combination of general equality and linear constraints. *SIAM Journal on Optimization* **6**, 674–703.

36. Conn, A. R., Gould, N. I. M., & Toint, Ph. L. 1988 Global convergence of a class of trust region algorithms for optimization with simple bounds. *SIAM Journal on Numerical Analysis* **25**, 433–460. See also same journal **26**, 764-767, 1989.

37. Conn, A. R., Gould, N. I. M., & Toint, Ph. L. 1990 An introduction to the structure of large scale nonlinear optimization problems and the LANCELOT project. In: *Computing Methods in Applied Sciences and Engineering* (R. Glowinski and A. Lichnewsky, eds). Philadelphia: SIAM 42–54.

38. Conn, A. R., Gould, N. I. M., & Toint, Ph. L. 1991 A globally convergent augmented Lagrangian algorithm for optimization with general constraints and simple bounds. *SIAM Journal on Numerical Analysis* **28**, 545–572.

39. Conn, A. R., Gould, N. I. M., & Toint, Ph. L. 1992*a* LANCELOT*: a Fortran package for large-scale nonlinear optimization (Release A)*. Springer Series in Computational Mathematics 17. Heidelberg, Berlin, New York: Springer Verlag.

40. Conn, A. R., Gould, N. I. M., & Toint, Ph. L. 1992*b* On the number of inner iterations per outer iteration of a globally convergent algorithm for optimization with general nonlinear equality constraints and simple bounds. In: *Proceedings of the 14th Biennial Numerical Analysis Conference Dundee 1991* (D. F. Griffiths and G. A. Watson, eds). Longmans 49–68.

41. Conn, A. R., Gould, N. I. M., & Toint, Ph. L. 1994 A note on exploiting structure when using slack variables. *Mathematical Programming* **67**, 89–97.

42. Dax, A. 1989 The minimax solution of linear equations subject to linear constraints. *IMA Journal of Numerical Analysis* **9**, 95–109.

43. P. L., & Toraldo, G. 1993 On the identification property of a projected gradient method. *SIAM Journal on Numerical Analysis* **30**, 1483–1497.

44. Dembo, R. S., Eisenstat, S. C., & Steihaug, T. 1982 Inexact-Newton methods. *SIAM Journal on Numerical Analysis* **19**, 400–408.

45. Dem'yanov, V. F., & Rubinov, A. M. 1970 *Approximation Methods in Optimization Problems*. Amsterdam: Elsevier.

46. Dennis, J. E, El-Alem, M., & Maciel, M. C. 1997 A global convergence theory for general trust-region based algorithms for equality constrained optimization. *SIAM Journal on Optimization* **7**, 177–207.

47. G., Facchinei, F., & Grippo, L. 1992 An RQP algorithm using a differentiable exact penalty function for inequality constrained problems. *Mathematical Programming* **55**, 49–68.

48. Dixon, L. C. W. 1991 On the impact of automatic differentiation on the relative performance of parallel truncated Newton and variable metric algorithms. *SIAM Journal on Optimization* **1**, 475–486.

49. Drud, A. 1985 CONOPT - a GRG code for large sparse dynamic nonlinear optimization problems. *Mathematical Programming* **31**, 153–191.

50. Du, D. Z., & Zhang, X. S. 1986 A convergence theorem of Rosen's gradient projection method. *Mathematical Programming* **36**, 135–144.

51. Du, D. Z., & Zhang, X. S. 1989 Global convergence of Rosen's gradient projection method. *Mathematical Programming* **44**, 357–366.

52. Dunn, J. C. 1981 Global and asymptotic convergence rate estimates for a class of projected gradient processes. *SIAM Journal on Control and Optimization* **19**, 368–400.

53. Dunn, J. C. 1987 On the convergence of projected gradient processes to singular critical points. *Journal of Optimization Theory and Applications* **55**, 203–216.

54. Dussault, J.-P. 1995 Numerical stability and efficiency of penalty algorithms. *SIAM Journal on Numerical Analysis* **32**, 296–317.

55. El-Alem, M. 1991 A global convergence theory for the Celis-Dennis-Tapia trust region algorithm for constrained optimization. *SIAM Journal on Numerical Analysis* **28**, 266–290.

56. El-Alem, M. 1995 A robust trust-region algorithm with a nonmonotonic penalty parameter scheme for constrained optimization. *SIAM Journal on Optimization* **5**, 348–378.

57. Facchinei, F., & Lucidi, S. 1994 Local properties of a Newton-like direction for equality constrained minimization problems. *Optimization Methods and Software* **3**, 13–26.

58. Fletcher, R. 1970 A class of methods for nonlinear programming with termination and convergence properties. In: *Integer and nonlinear programming* (J. Abadie, ed). Amsterdam: North Holland.

59. Fletcher, R. 1971 A general quadratic programming algorithm. *Journal of the Institute of Mathematics and its Applications* **7**, 76–91.

60. Fletcher, R. 1973 An exact penalty function for nonlinear programming with inequalities. *Mathematical Programming* **5**, 129–150.

61. Fletcher, R. 1981 *Practical Methods of Optimization: Constrained Optimization*. New-York: J. Wiley and Sons.

62. Fletcher, R. 1982 Second order corrections for non-differentiable optimization. In: *Numerical Analysis, Proceedings Dundee 1981* (D. F. Griffiths, ed). Lecture Notes in Mathematics 912. Heidelberg, Berlin, New York: Springer Verlag 85–114.

63. Fletcher, R. 1987 Recent developments in linear and quadratic programming. *in* Iserles and Powell (1987) 213-243.

64. Fletcher, R. 1988 Degeneracy in the presence of roundoff. *Linear Algebra and its Applications* **106**, 149–183.

65. Fletcher, R. 1995 An optimal positive definite update for sparse Hessian matrices. *SIAM Journal on Optimization* **5**, 192–217.

66. Fletcher, R., & Sainz de la Maza, E. 1989 Nonlinear programming and nonsmooth optimization by successive linear programming. *Mathematical Programming* **43**, 235–256.

67. Fontecilla, R. 1990 Inexact secant methods for nonlinear constrained optimization. *SIAM Journal on Numerical Analysis* **27**, 154–165.

68. Fontecilla, R., Steihaug, T., & Tapia, R. A. 1987 A convergence theory for a class of quasi-Newton methods for constrained optimization. *SIAM Journal on Numerical Analysis* **24**, 1133–1151.

69. Friedlander, A., & Martínez, J. M. 1994 On the maximization of a concave quadratic function with box constraints. *SIAM Journal on Optimization* **4**, 177–192.

70. Friedlander, A., Martínez, J. M., & Santos, S. A. 1994 On the resolution of linearly constrained convex minimization problems. *SIAM Journal on Optimization* **4**, 331–339.

71. Fukushima, M. 1986 A successive quadratic-programming algorithm with global and superlinear convergence properties. *Mathematical Programming* **35**, 253–264.

72. Gilbert, J. C. 1991 Maintaining the positive definiteness of the matrices in reduced secant methods for equality constrained optimization. *Mathematical Programming* **50**, 1–28.

73. Gill, P.E., Murray, W. & Saunders, M.A. 1996 SNOPT: an SQP algorithm for large-scale constrained optimization. Technical Report SOL96-0, Department of Operations Research, Stanford University, Stanford, California 94305, USA.

74. Gill, P. E., Murray, W., Saunders, M. A., & Wright, M. H. 1989 A practical anticycling procedure for linearly constrained optimization. *Mathematical Programming, Series B* **45**, 437–474.

75. Gill, P. E., Murray, W., Saunders, M. A., & Wright, M. H. 1990 A Schur-complement method for sparse quadratic programming. In: *Reliable Scientific Computation* (M. G. Cox and S. J. Hammarling, eds). Oxford: Oxford University Press 113–138.

76. Gill, P. E., Murray, W., Saunders, M. A., & Wright, M. H. 1991 Inertia-controlling methods for general quadratic programming. *SIAM Review* **33**, 1–36.

77. Gill, P. E., Murray, W., Saunders, M. A., Tomlin, J. A., & Wright, M. H. 1986 On projected Newton barrier methods for linear programming and an equivalence to Karmarkar's projective method. *Mathematical Programming* **36**, 183–209.

78. Gould, N. I. M. 1986 On the accurate determination of search directions for simple differentiable penalty functions. *IMA Journal of Numerical Analysis* **6**, 357–372.

79. Gould, N. I. M. 1989 On the convergence of a sequential penalty function method for constrained minimization. *SIAM Journal on Numerical Analysis* **26**, 107–128.

80. Gould, N. I. M. 1991 An algorithm for large-scale quadratic programming. *IMA Journal of Numerical Analysis* **11**, 299–324.

81. Griewank, A. 1989 On automatic differentiation. *in* Iri and Tanabe (1989) 83–108.

82. Griffith, R. E., & Stewart, R. A. 1961 A nonlinear programming technique for the optimization of continuous processing systems. *Management Science* **7**, 379–392.

83. Grippo, L., Lampariello, F., & Lucidi, S. 1986 A nonmonotone line search technique for Newton's method. *SIAM Journal on Numerical Analysis* **23**, 707–716.

84. Gurwitz, C. 1994 Local convergence of a two-piece update of a projected Hessian matrix. *SIAM Journal on Optimization* **4**, 461–485.

85. Hager, W. W. 1990 Multiplier methods for nonlinear optimal control. *SIAM Journal on Numerical Analysis* **27**, 1061–1080.

86. Han, S. P. 1977 A globally convergent method for nonlinear programming. *Journal of Optimization Theory and Applications* **15**, 319–342.

87. Heinkenschloss, M. 1994 On the solution of a two ball trust region subproblem. *Mathematical Programming* **64**, 249–276.

88. Heinz, J., & Spellucci, P. 1994 A successful implementation of the Pantoja-Mayne SQP method. *Optimization Methods and Software* **4**, 1–28.

89. Hestenes, M. R. 1969 Multiplier and gradient methods. *Journal of Optimization Theory and Applications* **4**, 303–320.

90. Hiriart-Urruty, J.-B., & Lemaréchal, C. 1993 *Convex Analysis and Minimization Algorithms. Part 2: Advanced Theory and Bundle Methods.* Heidelberg, Berlin, New York: Springer Verlag.

91. Iri, M., & Tanabe, K., eds 1989 *Mathematical Programming: Recent Developments and Applications.* Dordrecht: Kluwer Academic Publishers.

92. Iserles, A., & Powell, M. J. D., eds 1987 *The State of the Art in Numerical Analysis.* Oxford: Oxford University Press.

93. Jarre, F., & Saunders, M. A. 1995 Practical interior-point method for convex programming. *SIAM Journal on Optimization* **5**, 149–171.

94. Kanzow, C., & Kleinmichel, H. 1995 A class of Newton-type methods for equality and inequality constrained optimization. *Optimization Methods and Software* **5**, 173–198.

95. Kiwiel, K. C. 1989 A survey of bundle methods for non-differentiable optimization. *in* Iri and Tanabe (1989) 263–282.

96. Kupferschmid, M., & Ecker, J. G. 1987 A note on solution of nonlinear-programming problems with imprecise function and gradient values. *Mathematical Programming Studies* **31**, 129–138.

97. Lalee, M., Nocedal, J., & Plantega, T. 1993 On the implementation of an algorithm for large-scale equality optimization. Technical Report NAM 09-93 Department of Electrical Engineering and Computer Science, Northwestern University, Evanston, Illinois.

98. Lemaréchal, C., & Zowe, J. 1994 A condensed introduction to bundle methods in nonsmooth optimizations. *in* Spedicato (1994) 357–382.

99. Lescrenier, M. 1991 Convergence of trust region algorithms for optimization with bounds when strict complementarity does not hold. *SIAM Journal on Numerical Analysis* **28**, 476–495.

100. Levitin, E. S., & Polyak, B. T. 1966 Constrained minimization problems. *USSR Computational Mathematics and Mathematical Physics* **6**, 1–50.

101. Maany, Z. A. 1987 A new algorithm for highly curved constrained optimization. *Mathematical Programming Studies* **31**, 139–154.

102. Maratos, N. 1978 Exact penalty function algorithms for finite-dimensional and control optimization problems. PhD thesis University of London England.

103. Martinez, J. M., & Santos, S. A. 1995 A trust-region strategy for minimization on arbitrary domains. *Mathematical Programming* **68**, 267–301.

104. Mayne, D. Q., & Polak, E. 1982 A superlinearly convergent algorithm for constrained optimization problems. *Mathematical Programming Studies* **16**, 45–61.

105. Moré, J. J. 1993 Generalizations of the trust region problem. *Optimization Methods and Software* **2**, 189–209.

106. Moré, J. J., & Toraldo, G. 1991 On the solution of large quadratic programming problems with bound constraints. *SIAM Journal on Optimization* **1**, 93–113.

107. Moré, J. J., & Wright, S. J. 1993 *Optimization Software Guide*. Frontiers in Applied Mathematics 14. Philadelphia: SIAM.

108. Murray, W., & Prieto, F. J. 1995 A sequential quadratic programming algorithm using an incomplete solution of the subproblem. *SIAM Journal on Optimization* **5**, 590–640.

109. Murray, W., & Wright, M. H. 1978 Projected Lagrangian methods based on the trajectories of penalty and barrier functions. Technical Report SOL78-23 Department of Operations Research, Stanford University, Stanford, California 94305, USA.

110. Murray, W., & Wright, M. H. 1982 Computation of a search direction in constrained optimization algorithms. *Mathematical Programming Studies* **16**, 62–83.

111. Murtagh, B. A., & Saunders, M. A. 1978 Large-scale linearly constrained optimization. *Mathematical Programming* **14**, 41–72.

112. Murtagh, B. A., & Saunders, M. A. 1982 A projected Lagrangian algorithm and its implementation for sparse non-linear constraints. *Mathematical Programming Studies* **16**, 84–117.

113. Nesterov, Y., & Nemirovskii, A. 1994 *Interior-point polynomial algorithms in convex programming*. Philadelphia: SIAM.

114. Nocedal, J., & Overton, M. L. 1985 Projected Hessian updating algorithms for nonlinearly constrained optimization. *SIAM Journal on Numerical Analysis* **22**, 821–850.

115. Pang, J. S. 1991 A B-differentiable equation-based, globally and locally quadratically convergent algorithm for nonlinear programs, complementarity and variational inequality problems. *Mathematical Programming* **51**, 101–131.

116. Pang, J.-S. 1994 Serial and parallel computation of Karush-Kuhn-Tucker points via nonsmooth equations. *SIAM Journal on Optimization* **4**, 872–893.

117. Panier, E. R., & Tits, A. L. 1987 A superlinearly convergence feasible method for the solution of inequality constrained optimization problems. *SIAM Journal on Control and Optimization* **25**, 934–950.

118. Panier, E. R., & Tits, A. L. 1991 Avoiding the Maratos effect by means of a nonmontone line search I. general constrained problems. *SIAM Journal on Numerical Analysis* **28**, 1183–1195.

119. Panier, E. R., & Tits, A. L. 1993 On combining feasibility, descent and superlinear convergence in inequality constrained optimization. *Mathematical Programming* **59**, 261–276.

120. Pantoja, J. F. A., & Mayne, D. Q. 1991 Exact penalty functions with simple updating of the penalty parameter. *Journal of Optimization Theory and Applications* **69**, 441–467.

121. Powell, M. J. D. 1970 A new algorithm for unconstrained optimization. In: *Nonlinear Programming* (J. B. Rose, O. L. Mangasarian and K. Ritter, eds). London and New York: Academic Press.

122. Powell, M. J. D. 1978 A fast algorithm for nonlinearly constrained optimization calculations. In: *Numerical Analysis, Dundee 1977* (G. A. Watson, ed). Lecture Notes in Mathematics 630. Heidelberg, Berlin, New York: Springer Verlag 144–157.

123. Powell, M. J. D. 1987 Methods for nonlinear constraints in optimization calculations. *in* Iserles and Powell (1987) 325–358.

124. Powell, M. J. D. 1989 A tolerant algorithm for linearly constrained optimization calculations. *Mathematical Programming, Series B* **45**, 547–566.

125. Powell, M. J. D., & Yuan, Y. 1986 A recursive quadratic-programming algorithm that uses differentiable exact penalty-functions. *Mathematical Programming* **35**, 265–278.

126. Powell, M. J. D., & Yuan, Y. 1990 A trust region algorithm for equality constrained optimization. *Mathematical Programming* **49**, 189–211.

127. Pschenichny, B. N. 1970 Algorithms for general problems of mathematical programming. *Kibernetica* **6**, 120–125.

128. Robinson, S. M. 1974 Perturbed Kuhn-Tucker points and rates of convergence for a class of nonlinear programming algorithms. *Mathematical Programming* **7**, 1–16.

129. Rockafellar, R. T. 1974 Augmented Lagrangian multiplier functions and duality in nonconvex programming. *SIAM Journal on Control and Optimization* **12**, 268–285.

130. Rosen, J. B. 1960 The gradient projection method for nonlinear programming, Part I - linear constraints. *SIAM Journal on Applied Mathematics* **8**, 181–217.

131. Ryan, D. M., & Osborne, M. R. 1988 On the solution of highly degenerate linear programs. *Mathematical Programming* **41**, 385–392.

132. Sartenaer, A. 1995 A class of trust region methods for nonlinear network optimization problems. *SIAM Journal on Optimization* **5**, 379–407.

133. Schnabel, R. B. 1994 Parallel nonlinear optimization: limitations, challenges and opportunities. *in* Spedicato (1994) 531–559.

134. Schramm, H., & Zowe, J. 1992 A version of the bundle idea for minimizing a nonsmooth function: Conceptual idea, convergence analysis, numerical results. *SIAM Journal on Optimization* **2**, 121–152.

135. Smith, S., & Lasdon, L. 1992 Solving large-sparse nonlinear programs using GRG. *ORSA Journal of Computing* **4**, 1–15.

136. Spedicato, E., ed 1994 *Algorithms for continuous optimization. The state of the art.* Dordrecht: Kluwer Academic Publishers.

137. Stern, R. J., & Wolkowicz, H. 1995 Indefinite trust region subproblems and nonsymmetric eigenvalue perturbations. *SIAM Journal on Optimization* **5**, 286–313.

138. Tapia, R. A. 1977 Diagonalized multiplier methods and quasi-newton methods for constrained optimization. *Journal of Optimization Theory and Applications* **22**, 135–194.

139. Toint, Ph. L. 1977 On sparse and symmetric matrix updating subject to a linear equation. *Mathematics of Computation* **31**, 954–961.

140. Toint, Ph. L. 1988 Global convergence of a class of trust region methods for nonconvex minimization in Hilbert space. *IMA Journal of Numerical Analysis* **8**, 231–252.

141. Vardi, A. 1985 A trust region algorithm for equality constrained minimization: convergence properties and implementation. *SIAM Journal on Numerical Analysis* **22**, 575–591.

142. Wilson, R. B. 1963 A simplicial algorithm for concave programming. PhD thesis Harvard University Massachusetts, USA.

143. Womersley, R. S. 1985 Local properties of algorithms for solving nonsmooth composite functions. *Mathematical Programming* **32**, 69–89.

144. Wright, S. J. 1987 Local properties of inexact methods for minimizing nonsmooth composite functions. *Mathematical Programming* **37**, 232–252.

145. Wright, S. J 1989*a* Convergence of SQP-like methods for constrained optimization. *SIAM Journal on Control and Optimization* **27**, 13–26.

146. Wright, S. J. 1989*b* An inexact algorithm for composite nondifferentiable optimization. *Mathematical Programming* **44**, 221–234.

147. Yuan, Y. 1985*a* Conditions for convergence of trust region algorithms for nonsmooth optimization. *Mathematical Programming* **31**, 220–228.

148. Yuan, Y. 1985*b* On the superlinear convergence of a trust region algorithm for nonsmooth optimization. *Mathematical Programming* **31**, 269–285.

149. Yuan, Y. 1990 On a subproblem of trust region algorithms for constrained optimization. *Mathematical Programming* **47**, 53–63.

150. Yuan, Y. 1995 On the convergence of a new trust region algorithm. *Numerische Mathematik* **70**, 515–539.

151. Zhang, J., Kim, N.-H., & Lasdon, L. 1985 An improved successive quadratic programming algorithm. *Management Science* **31**, 1312–1331.

152. Zhang, J., & Zhu, D. 1990 Projected quasi-Newton algorithm with trust-region for constrained optimization. *Journal of Optimization Theory and Applications* **67**, 369–393.

153. Zhang, J., Zhu, D., & Fan, Y. 1993 A practical trust region method for equality constrained optimization problems. *Optimization Methods and Software* **2**, 45–68.

154. Zhang, Y. 1992 Computing a Celis-Dennis-Tapia trust-region step for equality constrained optimization. *Mathematical Programming* **55**, 109–124.

Stabilization Techniques and Subgrid Scales Capturing

F. Brezzi*, L.P. Franca, T.J.R. Hughes† and A. Russo ††**

**Dipartimento di Matematica, Università di Pavia, Italy, **Department of Mathematics, University of Colorado at Denver, USA, †Division of Applied Mechanics, Stanford University, USA and ††Istituto di Analisi Numerica del CNR, Italy*

Abstract

We present an overview of stabilized finite element methods and of the standard Galerkin method enriched with *residual-free* bubble functions. The inadequacy of the standard Galerkin method using piecewise polynomials is discussed for different applications; the treatment using stabilized methods in their different versions is reviewed; and the connection to the standard Galerkin method with richer subspaces follows using the subgrid method or the residual-free-bubbles viewpoint. We close with a discussion on how to approximate the exact problem suggested by residual-free bubbles.

1 The standard Galerkin method and some of its failures

The standard Galerkin method can be roughly described as being an approximation of the variational formulation of a PDE (or system of PDE's) in a space of functions that is spanned by piecewise polynomials. This simple idea presents several advantages: first, the discrete system of equations that arise from such an approximation is going to be "banded" since the piecewise polynomials can be constructed to have a "small" support, and therefore the matrices involved are sparse. Second, taking derivatives and integrating polynomials is a very attractive task for any first year calculus student, and the simplicity of the implementation of the method for the most cumbersome PDE or system of PDE's seems straightforward. Third, the mathematical analysis seems to be possible without a lot of sophistication (at least if we have an elliptic problem, and we disregard technicalities referring to domain shape, etc.).

No wonder there was a boost of this methodology in the mid 60's and early 70's, and a general feeling that this was the way to approximate PDE's in general, and a confirmation of the expectations were available for a variety of structural problems which are elliptic.

However in the midst of this success the experts were aware that there were problems in applying this recipe to all problems under the sun. We will describe a few of those examples as the failures of the standard Galerkin method.

First let us describe the Galerkin method for an abstract boundary value problem given by

$$Au = f \qquad u \in V, \tag{1.1}$$

where for concreteness we take $V = H_0^1(\Omega)$. We then have

$$< Au, v >= a(u, v) =< u, A^* v > \qquad \forall u, v \in V, \tag{1.2}$$

and the variational formulation corresponding to (1.1) is given by:

$$\begin{cases} \text{find } u \in V \text{ such that} \\ a(u, v) = (f, v) \text{ for all } v \in V. \end{cases} \tag{1.3}$$

The classical Galerkin method then consists of taking a finite-dimensional subspace of V, say $V_h \subset V$, which is spanned by continuous piecewise polynomials (often the choice is piecewise linears) and using the same variational formulation, given by (1.3), in V_h, namely:

$$\begin{cases} \text{find } u_h \in V_h \text{ such that} \\ a(u_h, v_h) = (f, v_h) \text{ for all } v_h \in V_h. \end{cases} \tag{1.4}$$

Error estimates are readily available if we can find two constants $\alpha > 0$ and $0 < M < \infty$ such that

$$\alpha ||v||_V^2 \leq a(v, v) \qquad \forall v \in V, \tag{1.5}$$

$$a(v, w) \leq M ||v||_V ||w||_V \qquad \forall v, w \in V. \tag{1.6}$$

Indeed from (1.5) and (1.6) we have the optimal error bound:

$$||u - u_h||_V \leq \frac{M}{\alpha} \inf_{v_h \in V_h} ||u - v_h||_V, \tag{1.7}$$

and the recipe is successful if $M/\alpha \approx 1$. An example of a problem where this theory is immediately applicable is the Poisson equation that governs several problems of physical interest. However the list does not go on very far without complications, and we now briefly illustrate some of the failures of the standard Galerkin method.

A) Failure # 1: Advection-dominated problems

Here the problem consists in finding a scalar valued function $u(x)$ (temperature, for example) in a domain Ω, such that

$$-\varepsilon\Delta u + \boldsymbol{a}\cdot\nabla u = f \quad \text{in}\Omega, \tag{1.8}$$

subject (for the sake of simplicity) to the boundary condition

$$u = 0 \qquad \text{on}\,\partial\Omega, \tag{1.9}$$

where the flow velocity field $\boldsymbol{a}$ and the source function f are given. The variational formulation of this problem is:

$$\begin{cases} \text{find } u \in H_0^1(\Omega) \text{ such that} \\ \varepsilon\displaystyle\int_\Omega \nabla u\cdot\nabla v\, d\Omega + \int_\Omega (\boldsymbol{a}\cdot\nabla u)\; v\, d\Omega = \int_\Omega f v\, d\Omega \quad \forall v \in H_0^1(\Omega). \end{cases} \tag{1.10}$$

When we take the subspace of piecewise linears we discover that the method produces spurious oscillations throughout the domain if $h|\boldsymbol{a}| >> \varepsilon$. One indication why we should expect trouble comes from the realization that the optimal error bound (1.7) is practically useless in this case since by definition our bilinear form (in the simplest case when $\boldsymbol{a}$ is constant) yields:

$$a(v,v) = \varepsilon||\nabla u||^2_{L^2}, \tag{1.11}$$

and therefore, from (1.5), $\alpha = \varepsilon$ times the Poincaré's constant (which is of order one), while M in (1.6) is of the same size as $|\boldsymbol{a}|$. Thus, for small values of ε the right-hand-side of (1.7) is very large, implying that the error can be very large! (And it is, with few special exceptions!!!)

B) Failure # 2: The Stokes' problem

Here we wish to compute the velocity $\mathbf{u}$ and the pressure p such that

$$\begin{cases} -\Delta\mathbf{u} + \nabla p = \mathbf{f} & \text{in}\,\Omega \\ \nabla\cdot\mathbf{u} = 0 & \text{in}\,\Omega \\ \mathbf{u} = 0 & \text{on}\,\partial\Omega, \end{cases} \tag{1.12}$$

where the source function $\mathbf{f}$ is given and the viscosity is taken to be one, for simplicity. The variational formulation of this problem is:

$$\begin{cases} \text{find } u \in (H_0^1(\Omega))^n \text{ and } p \in L^2(\Omega) \text{ such that} \\ \displaystyle\int_\Omega \nabla\mathbf{u}\cdot\nabla\mathbf{v}\, d\Omega - \int_\Omega p\nabla\cdot\mathbf{v}\, d\Omega = \int_\Omega \mathbf{f}\cdot\mathbf{v}\, d\Omega \qquad \forall\mathbf{v} \in (H_0^1(\Omega))^n \\ \displaystyle\int_\Omega q\,\nabla\cdot\mathbf{u}\, d\Omega = 0 \qquad \forall q \in L^2(\Omega), \end{cases} \tag{1.13}$$

where n is the number of space dimensions. If, say, $\mathbf{u}$ and p are approximated by continuous piecewise linears then the Galerkin method *fails for every mesh-size h!* As in the previous example we may be tempted to look at the optimal error bound (1.7) and in this case the bilinear form given by the sum of the left-hand-sides of (1.13) gives

$$a(\,(\mathbf{u},p)\,,\,(\mathbf{u},p)\,) = ||\nabla \mathbf{u}||^2_{L^2(\Omega)}, \tag{1.14}$$

which does not give any information about the pressure variable p. We know from numerical experiments that this choice of approximations for velocity-pressure yields a poor method. The mathematical theory to analyze constrained problems is known as mixed method theory, and it provides a better framework for proving stability and convergence of suitable pairs of velocity-pressure (this theory is not covered by the simple-minded estimate (1.14)). The mixed method theory will also confirm that the present choice of equal-order linears leads to an unstable method (for example, see [6,10,29,31,41,48] and references therein).

C) Failure # 3: Reissner-Mindlin Plates

Here we are interested in computing the rotations $\boldsymbol{\theta}$, vertical displacement w and shear strain $\boldsymbol{\gamma}$ of a plate of thickness t governed by:

$$\begin{cases} \nabla \cdot \mathbf{C}\boldsymbol{\varepsilon}(\boldsymbol{\theta}) - \boldsymbol{\gamma} = \mathbf{0} & \text{in}\,\Omega \\ \nabla \cdot \boldsymbol{\gamma} = f & \text{in}\,\Omega \\ \nabla w - \boldsymbol{\theta} - t^2\boldsymbol{\gamma} = \mathbf{0} & \text{in}\,\Omega, \end{cases} \tag{1.15}$$

where f is the distributed load on the surface of the plate, $\mathbf{C}$ is the fourth-order tensor of elastic moduli and $\boldsymbol{\varepsilon}(\boldsymbol{\theta})$ is the symmetric part of the gradient of the rotation vector, given by

$$\boldsymbol{\varepsilon}(\boldsymbol{\theta}) = \frac{\nabla\boldsymbol{\theta} + (\nabla\boldsymbol{\theta})^T}{2}. \tag{1.16}$$

We append to (1.15) the boundary conditions:

$$\boldsymbol{\theta} = \mathbf{0} \qquad \text{on}\,\partial\Omega \tag{1.17}$$

$$w = 0 \qquad \text{on}\,\partial\Omega, \tag{1.18}$$

and we set, for the sake of simplicity, a certain number of physical constants to one. The variational formulation of this problem is:

$$\begin{cases} \text{find } \boldsymbol{\theta} \in (H^1_0(\Omega))^2,\ w \in H^1_0(\Omega) \text{ and } \boldsymbol{\gamma} \in (L^2(\Omega))^2 \text{ such that} \\ \displaystyle\int_\Omega \mathbf{C}\boldsymbol{\varepsilon}(\boldsymbol{\theta}) \cdot \boldsymbol{\varepsilon}(\boldsymbol{\eta})\,d\Omega + \int_\Omega \boldsymbol{\gamma} \cdot (\nabla v - \boldsymbol{\eta})\,d\Omega = \int_\Omega f v\,d\Omega \\ \qquad\qquad\qquad \text{for all } \boldsymbol{\eta} \in (H^1_0(\Omega))^2,\ v \in H^1_0(\Omega) \\ \displaystyle -\int_\Omega \boldsymbol{\delta} \cdot (\nabla w - \boldsymbol{\theta})\,d\Omega + t^2 \int_\Omega \boldsymbol{\delta} \cdot \boldsymbol{\gamma}\,d\Omega = 0 \qquad \text{for all } \boldsymbol{\delta} \in (L^2(\Omega))^2. \end{cases} \tag{1.19}$$

If $\boldsymbol{\theta}$ and w are approximated by continuous piecewise linears and $\boldsymbol{\gamma}$ by either continuous piecewise linears or piecewise constants then the Galerkin method *fails* for $h >> t$, a situation common to thin plates. As in the previous example the bilinear form given by the sum of the left-hand-sides of (1.19) yields

$$a(\,(\boldsymbol{\theta},w,\boldsymbol{\gamma})\,,\,(\boldsymbol{\theta},w,\boldsymbol{\gamma})\,) = (\mathbf{C}\boldsymbol{\varepsilon}(\boldsymbol{\theta}),\boldsymbol{\varepsilon}(\boldsymbol{\theta})) + t^2\|\boldsymbol{\gamma}\|^2_{L^2(\Omega)}, \tag{1.20}$$

which does not give any information on the displacement variable w and has a coefficient t^2 (that may be small for thin plates) multiplying the L^2-norm of the strains $\boldsymbol{\gamma}$. We may be led to believe that this method does not have stability for displacements and has poor stability for strains. Again referring to the more sophisticated mixed method analysis this particular choice of approximation is confirmed to be inadequate (see [6,10,31,48] and references therein). For specially designed elements that will work in this case see for example [1,7,11].

D) Failure # 4: Dirichlet problem with Lagrange multipliers

Here we wish to find the scalar variable u solution of the Dirichlet problem:

$$\begin{cases} -\Delta u = f & \text{in}\,\Omega \\ \quad\; u = g & \text{on}\,\Gamma = \partial\Omega, \end{cases} \tag{1.21}$$

and the Lagrange multiplier

$$\lambda = \frac{\partial u}{\partial n} \qquad \text{on}\,\Gamma, \tag{1.22}$$

where f and g are given functions. The variational formulation for this problem is (see [2]):

$$\begin{cases} \text{find } u \in H^1(\Omega) \text{ and } \lambda \in H^{-1/2}(\Gamma) \text{ such that} \\ \displaystyle\int_\Omega \nabla u \cdot \nabla v \, d\Omega - \int_\Gamma \lambda v \, ds = \int_\Omega f v \, d\Omega \qquad \forall v \in H^1(\Omega), \\ \displaystyle\int_\Gamma u\mu \, ds = \int_\Gamma g\mu \, ds \qquad \forall \mu \in H^{-1/2}(\Gamma). \end{cases} \tag{1.23}$$

The Galerkin method will fail if the "grid for λ" is too fine compared with the "grid for u" (see always [2]). Also as in the previous example the bilinear form given by the sum of the left-hand-sides of (1.23) yields

$$a(\,(u,\lambda)\,,\,(u,\lambda)\,) = \|\nabla u\|^2_{L^2(\Omega)}, \tag{1.24}$$

which does not give any information on the Lagrange multiplier variable λ.

In the next two sections we will shortly discuss two strategies to deal with the shortcomings pointed out above, that have been developed in the past decade (although the first appearance of SUPG goes back to 1979). The first one consists of modifying the bilinear form a associated with the problem so that enhanced numerical stability is achieved without compromising consistency. This approach is known under several names, starting with SUPG for example 1 above, then Galerkin-least-squares and more recently as stabilized methods (see [4,5,9,14,16,17,18,20,21,23,24,28,30,33,34,35,36,37,38,45,46,47] and references therein). The second approach consists of using the standard Galerkin method enriched with special functions (and in particular we shall focus our attention on the so-called residual-free bubbles). The idea is to enlarge the space of functions to deal with the particular problem at hand so that our mesh, say coarse mesh, is able to deal with the effects of the unresolvable scales. This is also known as a subgrid model approach (see [3,8,12,13,15,19,25,26,27,32,39,40,42,43,44] and references therein).

2 Stabilized methods

To roughly describe the general idea of the method and its variations let us reconsider our abstract variational problem

$$Au = f \qquad u \in V, \tag{2.1}$$

and its variational formulation

$$a(u,v) = (f,v) \qquad \forall v \in V. \tag{2.2}$$

The idea is to modify the bilinear form $a(u,v)$ so that consistency is preserved and stability is increased. We consider two basic variants:

$$a(u_h, v_h) + \sum_K \tau_K (Au_h - f, Av_h)_K = (f, v_h) \qquad \forall v_h \in V_h\,, \tag{2.3}$$

or,

$$a(u_h, v_h) - \sum_K \tau_K (Au_h - f, A^* v_h)_K = (f, v_h) \qquad \forall v_h \in V_h\,, \tag{2.4}$$

where τ_K in either case is a stability parameter that depends on the element size and on the application.

If we apply these strategies to each problem discussed in the previous section we obtain:

A) Advection-dominated problems:

$$\begin{cases} -\varepsilon \Delta u + \boldsymbol{a} \cdot \nabla u = f & \text{in } \Omega \\ u = 0 & \text{on } \partial\Omega. \end{cases} \tag{2.5}$$

The stabilized variational formulation is:

$$\begin{cases} \text{find } u_h \in V_h \text{ such that} \\ \varepsilon \int_\Omega \nabla u_h \cdot \nabla v_h \, d\Omega + \int_\Omega \boldsymbol{a} \cdot \nabla u_h \, v_h \, d\Omega + S = \int_\Omega f v_h \, d\Omega \qquad \forall v_h \in V_h, \end{cases} \tag{2.6}$$

where S is the additional stabilizing term that, for this example, should add stability in the streamline direction, i.e. a term of the form

$$\tau \int_\Omega (\boldsymbol{a} \cdot \nabla u_h)(\boldsymbol{a} \cdot \nabla v_h)\, d\Omega . \tag{2.7}$$

A term of the form (2.7) is needed for stability, but as it stands it would destroy consistency. Thus we consider either

$$S = \tau \sum_K \int_K (-\varepsilon \Delta u_h + \boldsymbol{a} \cdot \nabla u_h - f)(\boldsymbol{a} \cdot \nabla v_h)\, d\Omega , \tag{2.8}$$

or

$$S = \tau \sum_K \int_K (-\varepsilon \Delta u_h + \boldsymbol{a} \cdot \nabla u_h - f)(-\varepsilon \Delta v_h + \boldsymbol{a} \cdot \nabla v_h)\, d\Omega , \tag{2.9}$$

or

$$S = -\tau \sum_K \int_K (-\varepsilon \Delta u_h + \boldsymbol{a} \cdot \nabla u_h - f)(-\varepsilon \Delta v_h - \boldsymbol{a} \cdot \nabla v_h)\, d\Omega . \tag{2.10}$$

Either of these options will work, as surveyed in [21]. Notice that, here and in the following examples, for the sake of simplicity, we took the simpler form $\tau \sum_K (\ldots)$ rather than the (better) $\sum_K \tau_K(\ldots)$ as in (2.3) and (2.4).

B) The Stokes' problem:

$$\begin{cases} -\Delta \mathbf{u} + \nabla p = \mathbf{f} & \text{in } \Omega \\ \nabla \cdot \mathbf{u} = 0 & \text{in } \Omega \\ \mathbf{u} = \mathbf{0} & \text{on } \partial\Omega . \end{cases} \tag{2.11}$$

The stabilized variational formulation for this problem is:

$$\int_\Omega \nabla \mathbf{u}_h \cdot \nabla \mathbf{v}_h \, d\Omega - \int_\Omega p_h \nabla \cdot \mathbf{v}_h \, d\Omega + S_1 = \int_\Omega \mathbf{f} \cdot \mathbf{v}_h \, d\Omega \qquad \forall \mathbf{v}_h , \tag{2.12}$$

$$\int_\Omega q_h \nabla \cdot \mathbf{u}_h \, d\Omega + S_2 = 0 \qquad \forall q_h , \tag{2.13}$$

where

$$S_1 = \tau_1 \sum_K \int_K (-\Delta \mathbf{u}_h + \nabla p_h - \mathbf{f})(-\Delta \mathbf{v}_h)\, d\Omega , \tag{2.14}$$

$$S_2 = \tau_2 \sum_K \int_K (-\Delta \mathbf{u}_h + \nabla p_h - \mathbf{f}) \cdot \nabla q_h \, d\Omega , \tag{2.15}$$

or some variants (see [23] for a survey of stabilized methods for the Stokes problem). The term that is crucial for stability is contained in S_2 and it is essentially

$$\tau_2 \sum_K \int_K \nabla p_h \cdot \nabla q_h \, d\Omega . \tag{2.16}$$

C) Reissner-Mindlin Plates:

$$\begin{cases} \nabla \cdot \mathbf{C}\boldsymbol{\varepsilon}(\boldsymbol{\theta}) - \boldsymbol{\gamma} = \mathbf{0} & \text{in } \Omega \\ \nabla \cdot \boldsymbol{\gamma} = f & \text{in } \Omega \\ \nabla w - \boldsymbol{\theta} - t^2 \boldsymbol{\gamma} = \mathbf{0} & \text{in } \Omega, \end{cases} \tag{2.17}$$

where

$$\boldsymbol{\varepsilon}(\boldsymbol{\theta}) = \frac{\nabla \boldsymbol{\theta} + (\nabla \boldsymbol{\theta})^T}{2}, \tag{2.18}$$

and with the boundary conditions

$$\boldsymbol{\theta} = \mathbf{0} \qquad \text{on } \partial\Omega, \tag{2.19}$$

$$w = 0 \qquad \text{on } \partial\Omega. \tag{2.20}$$

The stabilized variational formulation for this problem is:

$$\int_\Omega \mathbf{C}\boldsymbol{\varepsilon}(\boldsymbol{\theta}_h) \cdot \boldsymbol{\varepsilon}(\boldsymbol{\eta}_h)\, d\Omega + \int_\Omega \boldsymbol{\gamma}_h \cdot (\nabla v_h - \boldsymbol{\eta}_h)\, d\Omega + S_1 = \int_\Omega f v_h \, d\Omega \quad \forall \boldsymbol{\eta}_h, \forall v_h, \tag{2.21}$$

$$-\int_\Omega \boldsymbol{\delta}_h \cdot (\nabla w_h - \boldsymbol{\theta}_h)\, d\Omega + t^2 \int_\Omega \boldsymbol{\delta}_h \cdot \boldsymbol{\gamma}_h \, d\Omega + S_2 = 0 \qquad \forall \boldsymbol{\delta}_h, \tag{2.22}$$

where

$$S_1 = \tau_1 \sum_K \int_K (\nabla w_h - \boldsymbol{\theta}_h - t^2 \boldsymbol{\gamma}_h) \cdot (\nabla v_h \ldots)\, d\Omega, \tag{2.23}$$

$$S_2 = \tau_2 \sum_K \int_K (\Delta \boldsymbol{\theta}_h - \boldsymbol{\gamma}_h) \cdot \boldsymbol{\delta}_h \, d\Omega, \tag{2.24}$$

or variants (see [34] for a family of methods and [11,45] for some convergent low-order stabilized elements).

D) Dirichlet problem with Lagrange multipliers:

$$\begin{cases} -\Delta u = f & \text{in } \Omega \\ u = g & \text{on } \Gamma = \partial\Omega \\ \lambda = \dfrac{\partial u}{\partial n} & \text{on } \Gamma. \end{cases} \tag{2.25}$$

The stabilized variational formulation for this problem is:

$$\int_\Omega \nabla u_h \cdot \nabla v_h \, d\Omega - \int_\Gamma \lambda_h v_h \, ds + S_1 = \int_\Omega f v_h \, d\Omega \qquad \forall v_h, \tag{2.26}$$

$$\int_\Gamma u_h \mu_h \, ds + S_2 = \int_\Gamma g \mu_h \, ds \qquad \forall \mu_h. \tag{2.27}$$

The key term for stability is in S_2 and it is given by $\tau_1 \int_\Gamma \lambda_h \mu_h \, ds$ (which is not consistent). Then we may consider

$$(\Delta u_h - f, \Delta v_h)_K, \tag{2.28}$$

$$(\frac{\partial u_h}{\partial n} - \lambda_h, \frac{\partial v_h}{\partial n} - \mu_h)_{\Gamma \cap \partial K}, \tag{2.29}$$

$$(u - g, v)_{\Gamma \cap \partial K}. \tag{2.30}$$

All of these additional terms are suggested in [5] and a method using this approach in the framework of domain decomposition methods is presented in [4] with an abstract theory for stabilized methods.

In our discussion above we avoided the crucial question: How do we select τ?

The general strategy to find it may be described as follows: first you guess the form of it, then you perform an error analysis to confirm that the order of the parameter yields optimal or quasi-optimal estimates. If this is not the case, guess again, until you get the best possible estimate. Then perform numerical experiments for your choices and try to cover a wide range of problems of interest. If all works then you have a method!

While this may seem unsatisfactory, this is how most of these methods have been designed. The interesting result is that the methods obtained by this procedure sometimes may be robust enough so that changing the choice of the parameters does not affect the accuracy of the results. For some versions of the method this will not be true, but again for sufficiently fine meshes (or meshes that are adaptively refined) all versions will lead to accurate methods with very minor differences in the numerical results.

A wide variety of problems have been treated by stabilized methods and various authors advocate one parameter or another. We refer to [21,22,23,24,33,34,35,36,37,38] for some choices of τ available.

3 Bubble stabilization

The other main strategy developed in the past decade is based in the same variational formulation as the continuous problem, but now approximated by a richer subspace than the original one.

The idea can be roughly described as follows: given a grid $\mathcal{C}_h$, and a finite element space V_h, try to increase V_h (suitably) in order to increase stability.

Typically this is achieved by adding *bubbles* to each element, which are eliminated afterwards by *static condensation*. To make ideas precise let us first define the term *bubble*:

Definition 1. *A bubble (in V, on $\mathcal{C}_h$) is a function $\phi \in V$ such that supp(ϕ) is contained in a single element.*

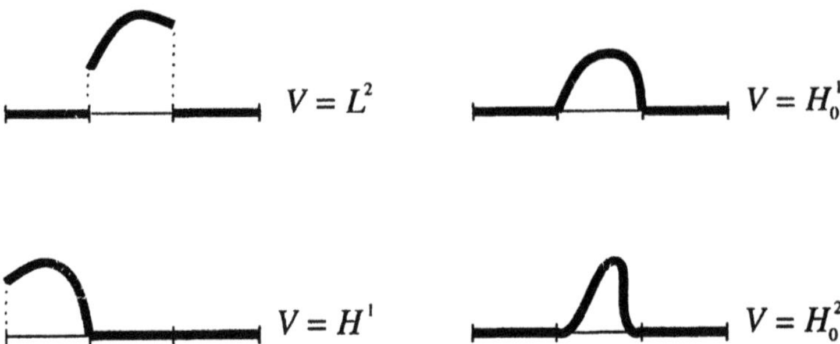

Figure 1. Some bubble functions

The definition stresses the dependence on the underlying space V that describes the variable to be approximated. In Figure 1 we illustrate some possibilities depending on the space V.
Now we wish to consider the static condensation procedure to shed some light on the effect of eliminating the bubbles and simply writing the final method in terms of the reduced space of polynomials that we started with. If we denote the new augmented space of functions by V_h, then

$$V_h = V_L \oplus V_B\,, \tag{3.1}$$

where V_L is spanned by continuous piecewise linears and V_B, by bubble functions. Then members of V_h can be decomposed as follows:

$$u_h = u_L + u_B, \qquad\qquad v_h = v_L + v_B\,. \tag{3.2}$$

If we substitute into the variational formulation

$$a(u_h, v_h) = (f, v_h) \qquad \forall v_h \in V_h\,, \tag{3.3}$$

then we have

$$a(u_L + u_B, v_L + v_B) = (f, v_L + v_B) \qquad \forall v_L \in V_L\,, \forall v_B \in V_B\,. \tag{3.4}$$

If we take first $v_L = 0$ then the variational formulation reduces to

$$\begin{aligned} a(u_B, v_B) &= (f, v_B) - a(u_L, v_B) \\ &= < f - Au_L, v_B > \qquad \forall v_B \in V_B, \end{aligned} \tag{3.5}$$

which can be "solved", for any $u_L \in V_L$, with solution given by

$$u_B = M_B(f - Au_L)\,, \tag{3.6}$$

where M_B is a bounded linear operator from V' to V_B. If we now take $v_B = 0$ in (3.4) and use (3.6) we get a method in terms of the reduced space V_L only:

$$a(u_L, v_L) + < M_B(f - Au_L), A^* v_L > = (f, v_L)\,. \tag{3.7}$$

If for a minute we disregard the second term on the left-hand side, the method reduces to the standard Galerkin method using piecewise linears. The second term represents the effect of *adding V_B and then eliminating it by static condensation.*

The question now is: what does M_B look like?

For the advective-diffusive equation, $-\varepsilon\Delta u + \boldsymbol{a}\cdot\nabla u = f$, if f and $\boldsymbol{a}$ are piecewise constants and V_B is spanned by a single function b_K in each element K, then

$$< M_B(f - Au_L), A^* v_L > = \sum_K \frac{(\int_K b_K\, d\Omega)^2}{\varepsilon(\int_K |\nabla b_K|^2\, d\Omega)} \frac{1}{|K|} \int_K (f - Au_L)(A^* v_L)\, d\Omega\,. \tag{3.8}$$

The term multiplying the last integral may be identified with the stability parameter τ_K of SUPG as shown in [8]. More generally we have that (Baiocchi-Brezzi-Franca [3]) with a suitable choice of the bubble space V_B, it is possible to make $< M_B u, v >$ equal to any bilinear form ρ, provided

$$0 \le \rho \le c_0 I\,, \tag{3.9}$$

with c_0 depending on the problem and on V_L. Also, if we consider $M_B = \tau I$ where I is the identity operator then we see that bubble stabilization suggests perturbation terms of the form

$$-\sum_K \int_K \tau_K (Au - f) A^* v\, d\Omega\,, \tag{3.10}$$

where τ_K depends on V_B.

As the presentation of the stabilized method left open the definition of the stability parameter τ, here we now have the basic question: *How to select the bubbles?*

A recent (and promising) point of view can be justified by the following argument:

1. The major reason of failure is the inadequate treatment of fine scales.

2. If rich enough, the bubbles (or, more generally, V_B) should be able to deal adequately with fine scales.

3. The stabilizing term represents the effect of *fine* (unresolvable) scales onto *coarse* (resolvable) ones.

4. Conceptually, we should take V_B as large as possible.

This last item suggests, ideally, that $V_L \oplus V_B = V$. In this case u_L would be the interpolant of the exact solution. This is, in particular, possible for all linear one-dimensional problems, but not, in general, in two or three dimensions. On the other hand, the basic idea of having V_B made of bubbles is very convenient, since it allows an *element-by-element* computation of the additional stabilizing term (the second term in the left-hand-side of (3.7)), although some experiments with spaces V_B made with functions having support in two or more elements indicate that this might also be an interesting line of future developments.

If, however, we stick to the idea of one-element bubbles (as in our definition), the largest choice we can take for V_B is $V_B = \oplus_K H_0^1(K)$. We can easily see that this is equivalent to selecting V_B as the space spanned by *residual-free-bubbles*, which are defined to satisfy the governing equations in strong form, i.e.,

$$Au_B = -(Au_L - f) \qquad \text{in } K, \tag{3.11}$$

subject to zero Dirichlet boundary condition on the element boundary

$$u_{B|K} = 0 \qquad \text{on } \partial K\,. \tag{3.12}$$

The problem given by (3.11) and (3.12) is addressed by solving instead [25,26,27]:

$$\begin{cases} A\varphi_{i,K} = -A\psi_{i,K} & \text{in } K, \\ \varphi_{i,K} = 0 & \text{on } \partial K, \end{cases} \tag{3.13}$$

where the $\psi_{i,K}$'s are the n_{en} local basis functions for u_L and

$$\begin{cases} A\varphi_{f,K} = f & \text{in } K, \\ \varphi_{f,K} = 0 & \text{on } \partial K. \end{cases} \tag{3.14}$$

Thus, if $u_{L|K} = \sum_{i=1}^{n_{en}} c_{i,K}\psi_{i,K}$ then

$$u_{B|K} = \sum_{i=1}^{n_{en}} c_{i,K}\varphi_{i,K} + \varphi_{f,K}\,, \tag{3.15}$$

with the same coefficients $c_{i,K}$'s.

Thus given a problem we should solve (3.13) and (3.14) to find the bubble basis functions φ's to determine the space of residual-free-bubbles V_B which, in turn, will produce the "optimal form" for the stabilizing term. This presents a systematic procedure to derive discretizations without the aforementioned open questions.

Notice that, in several cases, the functions $\{\varphi_{1,K}, \ldots, \varphi_{i,n_{en}}, \varphi_{f,K}\}$ might not be linearly independent, so that in practice we have to deal with less than $n_{en}+1$ bubbles. In particular, for the model problem (2.5), with piecewise constant $\boldsymbol{a}$ and f, we only need one function (that we call b_K^r) for every K in (3.13) and (3.14). The function b_K^r satisfies

$$a(b_K^r, v) = (1, v) \qquad \forall v \in H_0^1(K), \tag{3.16}$$

so that (3.8) now simply becomes

$$< M_B(f - Au_L), A^* v_L >= \sum_K \widehat{\tau}_K \int_K (f - Au_L)(A^* v_L)\, d\Omega, \tag{3.17}$$

where

$$\widehat{\tau}_K = \frac{1}{|K|} \int_K b_K^r \, d\Omega. \tag{3.18}$$

The parameter $\widehat{\tau}_K$ multiplying the last integral in (3.17) may be identified with the stability parameter τ_K of SUPG in the advective limit as shown in [15]. For the limit case $\varepsilon \to 0$, a reasonable value for $\widehat{\tau}_K$ can be guessed (as in [15]) by taking, instead of b_K^r in (3.18) the (limit) solution of $\boldsymbol{a} \cdot \nabla b_K = -1$ in K with $b_K = 0$ on the inflow part of ∂K. This design of $\widehat{\tau}_K$ is effective.

However, for more general ε, even this simple case presents relevant difficulties for the actual computation of (3.18).

A possible alternative can be represented by the use of the pseudo residual-free-bubbles introduced in [13]. Very briefly, the basic ideas are the following:

1. For every P internal to K, we partition K into three triangles K_i having a common vertex in P, and we set b_P to be the bubble which is linear in each K_i.

2. We solve (approximately) (3.16) in the one-dimensional space spanned by b_P. The solution $B_P = \alpha(P) b_P$ is given by

$$\alpha(P) = \frac{(1, b_P)}{a(b_P, b_P)}. \tag{3.19}$$

3. For choosing P, we now minimize

$$J(P) = \int_K |-\varepsilon \Delta B_P + \boldsymbol{a} \cdot \nabla B_P - 1| \, d\Omega\,, \tag{3.20}$$

   where the integral is in the sense of measures.

4. The actual minimization is done on a single oriented segment, depending on $\boldsymbol{a}$ and K. Among the minimizers, we take "the first one."

5. If P^* is the chosen minimizer of (3.20), we then define the "pseudo residual-free-bubble" as B_{P^*} and we have

$$\widehat{\tau}_K^{\psi} = \frac{1}{|K|} \int B_P^* \, d\Omega\,, \tag{3.21}$$

   similarly to (3.18).

For further details, we refer to [13]. We just explicitly remark that the practical computation of P^* is very cheap, and that, in the limit for $\varepsilon \to 0$ (for fixed K and $\boldsymbol{a}$) (3.18) and (3.21) give the same result.

References

1. D. N. Arnold and R. S. Falk, *A uniformly accurate finite element method for the Reissner-Mindlin plate*, SIAM J. Numer. Anal., 26 (1989), pp. 1276–1290.
2. I. Babuška, *The finite element method with Lagrangian multipliers*, Numer. Math., 20 (1973), pp. 179–192.
3. C. Baiocchi, F. Brezzi, and L. Franca, *Virtual bubbles and the Galerkin-least-squares method*, Comput. Methods Appl. Mech. Engrg., 105 (1993), pp. 125–141.
4. C. Baiocchi, F. Brezzi, and D. Marini, *Stabilization of Galerkin methods and applications to domain decomposition*, in Future Tendencies in Computer Science, Control and Applied Mathematics, A. Bensoussan and J.-P. Verjus, eds., vol. 653 of Lecture Notes in Computer Science, Springer-Verlag, 1992, pp. 345–355. Proceedings of the International Conference on the Occasion of the 25th Anniversary of INRIA, Paris, France, December 1992.
5. H. Barbosa and T. J. R. Hughes, *Boundary Lagrange multipliers in finite element methods: error analysis in natural norms*, Numer. Math., 62 (1992), pp. 1–16.
6. K.-J. Bathe, *Finite Element Procedures*, Prentice-Hall, Englewood Cliffs, New Jersey, 1996.
7. F. Brezzi, K.-J. Bathe, and M. Fortin, *Mixed interpolated elements for Reissner-Mindlin plates*, Int. J. Numer. Methods Eng., 28 (1989), pp. 1787–1801.
8. F. Brezzi, M. Bristeau, L. Franca, M. Mallet, and G. Rogé, *A relationship between stabilized finite element methods and the Galerkin method with bubble functions*, Comput. Methods Appl. Mech. Engrg., 96 (1992), pp. 117–129.
9. F. Brezzi and J. Douglas, *Stabilized mixed methods for the Stokes problem*, Numer. Math., 53 (1988), pp. 225–236.
10. F. Brezzi and M. Fortin, *Mixed and Hybrid Finite Element Methods*, vol. 15 of Springer Series in Computational Mathematics, Springer-Verlag, Berlin, New-York, 1991.
11. F. Brezzi, M. Fortin, and R. Stenberg, *Quasi-optimal error bounds for approximation of shear-stresses in Mindlin-Reissner plate models*, Math. Models Meth. Appl. Sci., 1 (1991).
12. F. Brezzi, L. P. Franca, T. J. R. Hughes, and A. Russo, $b = \int g$, Comput. Methods Appl. Mech. Engrg., (1996). To appear.
13. F. Brezzi, D. Marini, and A. Russo, *Pseudo residual-free bubbles and stabilized methods*. To appear in the Proceedings of the Third ECCOMAS Computational Fluid Dynamics Conference, September 9–13, 1996, Paris, France.
14. F. Brezzi and J. Pitkäranta, *On the stabilization of finite element approximations of the Stokes problem*, in Efficient Solutions of Elliptic Systems, Notes on Numerical Fluid Mechanics, W. Hackbusch, ed., vol. 10, Viewig, 1984, pp. 11–19.
15. F. Brezzi and A. Russo, *Choosing bubbles for advection-diffusion problems*, Math. Models Meth. Appl. Sci., 4 (1994), pp. 571–587.
16. A. N. Brooks and T. J. R. Hughes, *Streamline upwind/Petrov-Galerkin formulations for convection dominated flows with particular emphasis on the incompressible Navier-Stokes equations*, Comput. Methods Appl. Mech. Engrg., 32 (1982), pp. 199–259.
17. J. Douglas and J. Wang, *An absolutely stabilized finite element method for the Stokes problem*, Math. Comp., 52 (1989), pp. 495–508.
18. L. P. Franca and E. G. D. do Carmo, *The Galerkin gradient least-squares method*, Comput. Methods Appl. Mech. Engrg., 74 (1989), pp. 41–54.
19. L. P. Franca and C. Farhat, *Bubble functions prompt unusual stabilized finite element methods*, Comput. Methods Appl. Mech. Engrg., 123 (1995), pp. 299–308.
20. L. P. Franca and S. L. Frey, *Stabilized finite element methods: II. The incompressible Navier-Stokes equations*, Comput. Methods Appl. Mech. Engrg., 99 (1992), pp. 209–233.

21. L. P. Franca, S. L. Frey, and T. J. R. Hughes, *Stabilized finite element methods: I. Application to the advective-diffusive model*, Comput. Methods Appl. Mech. Engrg., 95 (1992), pp. 253–276.

22. L. P. Franca and T. J. R. Hughes, *Two classes of mixed finite element methods*, Comput. Methods Appl. Mech. Engrg., 69 (1988), pp. 89–129.

23. L. P. Franca, T. J. R. Hughes, and R. Stenberg, *Stabilized finite element methods for the Stokes problem*, in Incompressible Computational Fluid Dynamics-Trends and Advances, M. D. Gunzburger and R. Nicolaides, eds., Cambridge University Press, 1993, pp. 87–107.

24. L. P. Franca and A. L. Madureira, *Element diameter free stability parameters for stabilized methods applied to fluids*, Comput. Methods Appl. Mech. Engrg., 105 (1993), pp. 395–403.

25. L. P. Franca and A. Russo, *Deriving upwinding,mass lumping and selective reduced integration by residual-free bubbles*, Math. Models Meth. Appl. Sci., (1996). To appear.

26. ———, *Mass lumping emanating from residual-free bubbles*, Comput. Methods Appl. Mech. Engrg., (1996). To appear.

27. ———, *Unlocking with residual-free bubbles*, Comput. Methods Appl. Mech. Engrg., (1996). To appear.

28. L. P. Franca and R. Stenberg, *Error analysis of some Galerkin least squares methods for the elasticity equations*, SIAM J. Numer. Anal., 28 (1991), pp. 1680–1697.

29. V. Girault and P. A. Raviart, *Finite Element Methods for Navier-Stokes Equations, Theory and Algorithms*, vol. 5 of Springer Series in Computational Mathematics, Berlin, New-York, 1986.

30. P. Hansbo and A. Szepessy, *A velocity-pressure streamline diffusion finite element method for the incompressible Navier-Stokes equation*, Comput. Methods Appl. Mech. Engrg., 84 (1990), pp. 175–192.

31. T. J. R. Hughes, *The Finite Element Method: Linear Static and Dynamic Finite Element Analysis*, Prentice-Hall, Englewood Cliffs, New Jersey, 1987.

32. ———, *Multiscale phenomena: Green's functions, the Dirichlet-to-Neumann formulation, subgrid scale models, bubbles and the origin of stabilized methods*, Comput. Methods Appl. Mech. Engrg., 127 (1995), pp. 387–401.

33. T. J. R. Hughes and L. P. Franca, *A new finite element formulation for computational fluid dynamics: VII. The Stokes problem with various well-posed boundary conditions: symmetric formulations that converge for all velocity/pressure spaces*, Comput. Methods Appl. Mech. Engrg., 65 (1987), pp. 85–96.

34. ———, *A mixed finite element formulation for Reissner-Mindlin plate theory: uniform convergence of all higher-order spaces*, Comput. Methods Appl. Mech. Engrg., 67 (1988), pp. 223–240.

35. T. J. R. Hughes, L. P. Franca, and M. Balestra, *A new finite element formulation for computational fluid dynamics: V. Circumventing the babuška-brezzi condition: A stable Petrov-Galerkin formulation of the Stokes problem accommodating equal-order interpolations*, Comput. Methods Appl. Mech. Engrg., 59 (1986), pp. 85–99.

36. T. J. R. Hughes, L. P. Franca, and G. M. Hulbert, *A new finite element formulation for computational fluid dynamics: VIII. The Galerkin-least-squares method for advective-diffusive equations*, Comput. Methods Appl. Mech. Engrg., 73 (1989), pp. 173–189.

37. C. Johnson, *Numerical solution of partial differential equations by the finite element method*, Cambridge University Press, Cambridge, 1987.

38. C. Johnson, U. Nävert, and J. Pitkäranta, *Finite element methods for linear hyperbolic problem*, Comput. Methods Appl. Mech. Engrg., 45 (1984), pp. 285–312.

39. M. Lesoinne, C. Farhat, and L. P. Franca, *Unusual stabilized finite element methods for second order linear differential equations*, in Proceedings of the Ninth International Conference on Finite Elements in Fluids - New Trends and Applications, M. M. Cecchi, K. Morgan, J. Periaux, B. A. Schrefler, and O. C. Zienkiewicz, eds., Venice, Italy, October 1995, pp. 377–386.
40. R. Pierre, *Simple C^0 approximations for the computation of incompressible flows*, Comput. Methods Appl. Mech. Engrg., 68 (1988), pp. 205–227.
41. O. Pironneau, *Finite Element Methods for Fluids*, John Wiley,New York, 1989.
42. A. Russo, *Residual free bubbles and stabilized methods*, in Proceedings of the Ninth International Conference on Finite Elements in Fluids - New Trends and Applications, M. M. Cecchi, K. Morgan, J. Periaux, B. A. Schrefler, and O. C. Zienkiewicz, eds., Venice, Italy, October 1995, pp. 1607–1615.
43. ———, *Bubble stabilization of finite element methods for the linearized incompressible Navier-Stokes equations*, Comput. Methods Appl. Mech. Engrg., (1996). To appear.
44. ———, *A posteriori error indicators via bubble functions*, Math. Models Meth. Appl. Sci., 6 (1996), pp. 33–41.
45. R. Stenberg, *A new finite element formulation for the plate bending problem*, in Asymptotic Methods for Elastic Structures, P. Ciarlet, L. Trabucho, and J. M. Viano, eds., Walter de Gruyter & Co., 1995, pp. 209–221.
46. T. E. Tezduyar, J. Liou, and M. Behr, *A new strategy for finite element computations involving moving boundaries and interfaces – The DSD/ST procedure: I. The concept and the preliminary numerical tests*, Comput. Methods Appl. Mech. Engrg., 94 (1992), pp. 339–352.
47. T. E. Tezduyar, J. Liou, M. Behr, and S. Mittal, *A new strategy for finite element computations involving moving boundaries and interfaces – The DSD/ST procedure: II. Computation of free-surface flows, two-liquid flows, and flows with drifting cylinders*, Comput. Methods Appl. Mech. Engrg., 94 (1992), pp. 353–372.
48. O. C. Zienkiewicz and R. L. Taylor, *The Finite Element Method*, McGraw-Hill, London, 4th ed., 1989.

Approximation of Curvature Dependent Interface Motion

C.M.Elliott

Centre for Mathematical Analysis and Its Applications, School of Mathematical Sciences, University of Sussex

Abstract

Let Γ_t be an evolving hypersurface in $\mathbb{R}^N$ of co-dimension 1. This paper is concerned with the numerical approximation of the evolution law $V = -\kappa + g$ (and various generalisations), where V is the normal velocity of Γ_t in the direction $\mathbf{n}$, κ is the mean curvature of Γ_t and g is a prescribed driving force. Here Γ_t is the boundary of a bounded domain $\Omega^I(t)$, $\mathbf{n}$ is pointing out of $\Omega^I(t)$ and κ is positive when $\Omega^I(t)$ is a ball. Such equations arise naturally in differential geometry and physical applications such as the motion of phase boundaries. Three mathematical approaches and their realisations as numerical methods are surveyed together with applications.

1 Introduction

The propagation of interfaces, surfaces and free boundaries arises naturally in many applications as well as being a subject of compelling mathematical interest and beauty. In this paper we survey numerical methods for the evolution of surfaces where normal velocity is strongly dependent on the mean curvature of the surface. A widely studied problem is 'motion by mean curvature'. Here Γ_t is an evolving closed hypersurface in $\mathbb{R}^N$ of co-dimension 1, whose interior is a bounded open set $\Omega_I(t)$, which evolves so that the normal velocity V of Γ_t in the normal direction $\mathbf{n}$ pointing away from $\Omega_I(t)$ is given by

$$V = -\kappa, \tag{1.1}$$

where κ is the sum of the $(N-1)$ principal curvatures of Γ_t. We call κ the mean curvature rather than the arithmetic mean of the principal curvatures. Our sign convention for the curvature is that κ is positive when $\Omega_I(t)$ is a ball. This is a purely geometrical problem. It is obvious that a convex domain $\Omega_I(t)$ shrinks.

We will concentrate on three different methods:

(1) Direct Approach: Partial differential equations can be formulated for a parameterisation $X(s,t)$ of Γ_t, s belongs to the unit sphere of $\mathbb{R}^N$, and these are discretised directly. This can be viewed as a systematic front tracking method.

(2) Level Set Approach: A degenerated nonlinear parabolic equation for a scalar $\omega : \mathbb{R}^N \times (0,T) \to \mathbb{R}$ can be written down, which implies that the level

surfaces of ω each evolve according to a mean curvature evolution law. This equation is discretised and Γ_t is approximated by a discrete level surface of the numerical solution.

(3) Phase Field Approach: Singularly perturbed reaction diffusion equations for a bi-stable potential with a small parameter ε can be formulated such that the zero level surface of u_ε, the solution of the equation, approximates motion by mean curvature. This equation is then solved numerically.

Section §2 introduces some problems, concepts and notations for the evolution of simple closed curves in the plane and smooth surfaces in $I\!R^N$. The three approaches are described successively in the following three sections. A survey of applications is given in section §6. We conclude with some remarks in section §7.

The style of the paper is expository with the intention of covering as much material and as many ideas as possible in the available space. Few proofs are given. However, an extensive list of references is provided.

Acknowledgement: This research has been supported by the EPSRC grants GR/J283884 and GR/H61445, the EU HCM project 'Phase transition and surface tension' and NATO. I gratefully acknowledge the assistance of CMAIA colleagues in preparing this paper in particular: K.Deckelnick, A.R.Gardiner, T.Kuhn and V.Styles.

2 Preliminaries

2.1 Curve shrinking

An evolving planar closed curve Γ_t is a map $\mathbf{x} : I\!R \setminus 2\pi \times (0,T) \to I\!R^2$ such that $\mathbf{x}(s,t)$ is a point on the curve Γ_t and $\mathbf{x}(s+2\pi,t) = \mathbf{x}(s,t)$ for all s. Note that s is not the arclength. Γ_t is said to be simple (or embedded) if it has no intersections. In this case Γ_t can be a material interface separating its interior and exterior. We choose a clockwise parameterization so that the unit tangent to the curve is $\boldsymbol{\tau} = \mathbf{x}_s/|\mathbf{x}_s|$ and the unit normal $\mathbf{n} = \mathbf{x}_s^\perp/|\mathbf{x}_s|$ where $(\alpha_1,\alpha_2)^\perp = (-\alpha_2,\alpha_1)$ is pointing away from the curve. It follows by the Frenet formulae that the curvature κ is defined by the curvature vector $-\kappa\mathbf{n}$ satisfying

$$a) \; -\kappa\mathbf{n} = \frac{1}{|\mathbf{x}_s|}\boldsymbol{\tau}_s, \quad b) \; \kappa\boldsymbol{\tau} = \frac{1}{|\mathbf{x}_s|}\mathbf{n}_s \tag{2.1}$$

Here we have chosen the sign convention for the curvature that κ is positive for the curve being a circle. The velocity of the curve is given by

$$\mathbf{x}_t := V\mathbf{n} + V_{\boldsymbol{\tau}}\boldsymbol{\tau}, \tag{2.2}$$

where V is the normal velocity and $V_{\boldsymbol{\tau}}$ is the tangential velocity.

We consider motion by the mean curvature vector so that

$$\mathbf{x}_t = -\kappa\mathbf{n}. \tag{2.3}$$

This yields the nonlinear parabolic system

$$\mathbf{x}_t = \frac{1}{|\mathbf{x}_s|}\left(\frac{\mathbf{x}_s}{|\mathbf{x}_s|}\right)_s \tag{2.4}$$

and the interface motion law

$$< \mathbf{x}_t, \mathbf{n} >= -\kappa \quad \text{or} \quad V = -\kappa, \tag{2.5}$$

which is called mean curvature flow for the curve.

Note that a curve may have many parameterisations which will have the same normal, tangent and curvature. But the evolution of the curve depends only on the normal velocity and different parameterisations will have different tangential velocities. For example a curve evolving with the velocity given by

$$\mathbf{x}_t = \frac{\mathbf{x}_{ss}}{|\mathbf{x}_s|^2} \tag{2.6}$$

satisfies

$$\mathbf{x}_t = \frac{1}{|\mathbf{x}_s|}\left(\frac{\mathbf{x}_s}{|\mathbf{x}_s|}\right)_s + \frac{\langle \mathbf{x}_{ss}, \mathbf{x}_s \rangle \mathbf{x}_s}{|\mathbf{x}_s|^4} = -\kappa \mathbf{n} + \frac{\langle \mathbf{x}_{ss}, \mathbf{x}_s \rangle \boldsymbol{\tau}}{|\mathbf{x}_s|^3}$$

and so has normal velocity $V =< \mathbf{x}_t, \mathbf{n} >= -\kappa$. Evolving the initial data $\mathbf{x}_0(s)$ by the system (2.4) and (2.6) yields the same evolving curve but with different parameterisations. Also one has many choices for the parameterisations of the initial curve to define the initial data for either of the systems (2.4) and (2.6).

The element of arc length is $|\mathbf{x}_s|\mathrm{d}s$ and the length of the curve is

$$L(t) = \int_0^{2\pi} |\mathbf{x}_s|\mathrm{d}s. \tag{2.7}$$

It follows that, using the periodicity of $\mathbf{x}$,

$$\begin{aligned} \frac{\mathrm{d}L}{\mathrm{d}t} &= \int_0^{2\pi} \frac{\langle \mathbf{x}_{st}, \mathbf{x}_s \rangle}{|\mathbf{x}_s|}\mathrm{d}s = -\int_0^{2\pi} \left\langle \mathbf{x}_t, \left(\frac{\mathbf{x}_s}{|\mathbf{x}_s|}\right)_s \right\rangle \mathrm{d}s = \int_0^{2\pi} \langle \mathbf{x}_t, \kappa \mathbf{n} \rangle |\mathbf{x}_s|\mathrm{d}s \\ &= -\int_0^{2\pi} \kappa^2 |\mathbf{x}_s|\mathrm{d}s \; < \; 0 \end{aligned} \tag{2.8}$$

for a closed curve. Alternatively we have for (2.4)

$$\frac{\mathrm{d}L}{\mathrm{d}t} = -\int_0^{2\pi} |\mathbf{x}_t|^2 |\mathbf{x}_s|\mathrm{d}s$$

and for (2.6) that

$$\frac{\mathrm{d}L}{\mathrm{d}t} = -\int_0^{2\pi} \left[|\mathbf{x}_t|^2 - \frac{\langle \mathbf{x}_t, \mathbf{x}_s \rangle^2}{|\mathbf{x}_s|^2} \right] |\mathbf{x}_s|\mathrm{d}s$$

Thus the length of the curve strictly decreases and this motivates the label of '*curve shrinking*' for curves evolving with $< \mathbf{x}_t, \mathbf{n} >= -\kappa$.

Actually for (2.4) a local version of (2.8) holds. Observing the identity

$$|\mathbf{x}_s|\frac{d|\mathbf{x}_s|}{dt} = \frac{1}{2}\frac{d}{dt} < \mathbf{x}_s, \mathbf{x}_s > = < \mathbf{x}_s, \mathbf{x}_{st} >$$

we have using the Frenet formulae that

$$\begin{aligned} \frac{d}{dt}|\mathbf{x}_s| &= < \tau, \tfrac{\partial}{\partial s}(-\kappa \mathbf{n}) > \\ &= - < \tau, \kappa_s \mathbf{n} + \kappa \mathbf{n}_s > \\ &= -\kappa^2 |\mathbf{x}_s| = -|\mathbf{x}_t|^2 |\mathbf{x}_s|. \end{aligned} \tag{2.9}$$

The simplest example is that of a shrinking circle of radius $R(t)$ so that $\mathrm{d}R/\mathrm{d}t$ is the normal velocity and R^{-1} is the curvature which yields the ordinary differential equation

$$\frac{\mathrm{d}R}{\mathrm{d}t} = -\frac{1}{R}, \qquad t > 0 \quad R(0) = R_0.$$

This has the solution

$$R(t) = (R_0^2 - 2t)^{1/2} \quad \text{for } \ t \in [0, t_f],$$

where $t_f = \frac{1}{2}R_0^2$ is the finite time of extinction.

In fact in a series of celebrated papers it has been shown that convex closed curves remain convex and that any simple closed curve shrinks to a point in finite time and that it becomes circular as it approaches the extinction time [67, 70, 76].

The equation of motion (2.4) can also be applied to curves $\mathbf{x} : \mathbb{R}^1 \backslash 2\pi \to \mathbb{R}^N$ of co-dimension $N - 1 > 0$ and again the curve shrinking property holds. Furthermore it can be applied to non-embedded curves with self-intersection. In these cases we do not have the notion of an 'material' interface and in the case of self-intersection the mathematical model describing the relevant phenomenon usually breaks down at such an occurrence.

2.2 Forced and advected mean curvature flow

Let $g : \mathbb{R}^2 \times (0, T) \to \mathbb{R}$ and $\mathbf{q} : \mathbb{R}^2 \times (0, T) \to \mathbb{R}^2$ be given sufficiently smooth functions of space and time. Consider the evolution of a curve in the plane by the law of motion

$$V = -\kappa + \langle \mathbf{q}, \mathbf{n} \rangle + g. \tag{2.10}$$

Thus g corresponds to a forcing in the normal direction and $\mathbf{q} \cdot \mathbf{n}$ is a normal forcing due to advection. When $\mathbf{q} = 0$, (2.10) is an eikonal-curvature equation which is a curvature regularisation of the first order eikonal equation. (Properly the eikonal equation is

$$|\nabla T| g = 1$$

where $T(x)$ is the time at which the front passes through x.) Two simple examples are:

(1) A circle of radius $R(t)$ evolving according to (2.10) with constant forcing $g = 1$ and zero advection $\mathbf{q} = \mathbf{0}$ is defined by

$$\frac{\mathrm{d}R(t)}{\mathrm{d}t} = -\frac{1}{R} + 1, \qquad R(0) = R_0.$$

Clearly, if $R_0 > 1$ then the circle continually expands with the long time normal velocity tending to 1, if $R_0 < 1$ then the circle shrinks to zero in finite time and $R_0 = 1$ is an unstable equilibrium.

(2) Let $\mathbf{q}$ be a constant vector. If $\mathbf{x} = \mathbf{x}(s,t)$, $s \in I\!R \setminus 2\pi$, is a solution of $V = -\kappa$ then $\mathbf{x}_q := \mathbf{x} + t\mathbf{q}$ has the same normal as $\mathbf{x}$ and solves

$$V = -\kappa + \langle \mathbf{q}, \mathbf{n} \rangle.$$

This corresponds to the curve $\mathbf{x}$ being advected by the constant velocity field $\mathbf{q}$ to yield $\mathbf{x}_q$.

Parametric description of curves evolving according to (2.10) can be defined as in §2.1 . With the same notation, if $\mathbf{x} = \mathbf{x}(s,t)$, $s \in I\!R \setminus 2\pi$, is an evolving curve which satisfies either

$$\mathbf{x}_t = \frac{1}{|\mathbf{x}_s|}\left(\frac{\mathbf{x}_s}{|\mathbf{x}_s|}\right)_s + \mathbf{q} + g\frac{\mathbf{x}_s^{\perp}}{|\mathbf{x}_s|} \tag{2.11}$$

or

$$\mathbf{x}_t = \frac{\mathbf{x}_{ss}}{|\mathbf{x}_s|^2} + \mathbf{q} + g\frac{\mathbf{x}_s^{\perp}}{|\mathbf{x}_s|} \tag{2.12}$$

then

$$V = \langle \mathbf{x}_t, \mathbf{n} \rangle = \langle \mathbf{x}_t, \mathbf{x}_s^{\perp} \rangle / |\mathbf{x}_s| = -\kappa + \langle \mathbf{q}, \mathbf{n} \rangle + g.$$

2.3 Anisotropy

Let $\gamma \in C^2(I\!R^2 \setminus \{0\})$ be positively homogeneous of degree one, i.e. $\gamma(\lambda\mathbf{p}) = \lambda\gamma(\mathbf{p})$, $\lambda > 0$. It follows that the first derivative $\gamma'(\mathbf{p}) = \{\partial_i \gamma(\mathbf{p})\} \in I\!R^2$, γ' is homogeneous of degree zero, $\gamma'(\lambda\mathbf{p}) = \gamma'(\mathbf{p})$, and $\langle \gamma'(\mathbf{p}), \mathbf{p} \rangle = \gamma(\mathbf{p})$. Also the second derivative $D^2\gamma(\mathbf{p}) = \{\partial_{ij}^2 \gamma(\mathbf{p})\} \in I\!R^{2\times 2}$, $D^2\gamma$ is homogeneous of degree minus one, $D^2\gamma(\lambda\mathbf{p}) = \frac{1}{\lambda}D^2\gamma(\mathbf{p})$ and $D^2\gamma(\mathbf{p})\mathbf{p} = 0$. Observe that $D^2\gamma(\mathbf{p})\mathbf{q} = \langle D^2\gamma(\mathbf{p})\mathbf{q}, \mathbf{p}^{\perp} \rangle \mathbf{p}^{\perp}$ $\forall \mathbf{q}$ and $|\mathbf{p}| = 1$. We assume that γ is strictly convex in the sense that

$$\langle \mathbf{p}^{\perp}, D^2\gamma(\mathbf{p})\mathbf{p}^{\perp} \rangle = a(\mathbf{p}) \geq a_0, \quad |\mathbf{p}| = 1.$$

Then anisotropic surface energy may be associated with a simple closed curve Γ by

$$\mathcal{E}(\Gamma) := \int_{\Gamma} \gamma(\mathbf{n}), \tag{2.13}$$

where $\mathbf{n}$ is the unit normal to the curve and γ is the surface energy density which depends on the orientation of the curve. If $\gamma(\mathbf{p}) = |\mathbf{p}|$ then we are in the isotropic case and $\mathcal{E}(\Gamma)$ is just the length of the curve. Furthermore we know from §2.1 that the curve shortening flow is the gradient flow for this energy functional.

For $\Gamma_t = \{\mathbf{x}(s,t) : s \in I\!R \setminus 2\pi\}$ we have

$$\begin{aligned}
\frac{\mathrm{d}}{\mathrm{d}t}\mathcal{E}(\Gamma_t) &= \frac{\mathrm{d}}{\mathrm{d}t}\int_0^{2\pi} \gamma\left(\frac{\mathbf{x}_s^\perp}{|\mathbf{x}_s|}\right)|\mathbf{x}_s|\mathrm{d}s = \frac{\mathrm{d}}{\mathrm{d}t}\int_0^{2\pi}\gamma(\mathbf{x}_s^\perp)\mathrm{d}s = \int_0^{2\pi}\langle\gamma'(\mathbf{x}_s^\perp), \mathbf{x}_{st}^\perp\rangle \mathrm{d}s \\
&= -\int_0^{2\pi}\frac{1}{|\mathbf{x}_s|}\left\langle\frac{\partial}{\partial s}\gamma'(\mathbf{x}_s^\perp), \mathbf{x}_t^\perp\right\rangle|\mathbf{x}_s|\mathrm{d}s.
\end{aligned}$$

Thus we can identify a gradient flow for (2.13) as being

$$\mathbf{x}_t = \frac{1}{|\mathbf{x}_s|}\frac{\partial}{\partial s}\left[\gamma'(\mathbf{x}_s^\perp)\right]^\perp. \tag{2.14}$$

This can be rewritten as

$$\begin{aligned}
\mathbf{x}_t^\perp &= \frac{1}{|\mathbf{x}_s|}D^2\gamma\left(\mathbf{x}_s^\perp\right)\mathbf{x}_{ss}^\perp = D^2\gamma\left(\frac{\mathbf{x}_s^\perp}{|\mathbf{x}_s|}\right)\frac{\mathbf{x}_{ss}^\perp}{|\mathbf{x}_s|^2} = D^2\gamma(\mathbf{n})\left[-\kappa\mathbf{n} + \frac{\langle\mathbf{x}_{ss}, \mathbf{x}_s\rangle\boldsymbol{\tau}}{|\mathbf{x}_s|^3}\right]^\perp \\
&= -\kappa D^2\gamma(\mathbf{n})\boldsymbol{\tau}.
\end{aligned}$$

It follows that $\mathbf{x}_t$ is in the normal direction and

$$V = \langle\mathbf{x}_t, \mathbf{n}\rangle = -\langle\boldsymbol{\tau}, D^2\gamma(\mathbf{n})\boldsymbol{\tau}\rangle\kappa = -a(\mathbf{n})\kappa. \tag{2.15}$$

Equation (2.15) is known as anisotropic (or weighted) mean curvature flow for a curve. A more familiar form is obtained by setting $\hat{\gamma}(\theta) = \gamma(\cos\theta, \sin\theta)$ and $\mathbf{n} = \mathbf{N}(\theta) = (\cos\theta, \sin\theta)$, where θ is the angle the normal makes with the x_1-axis. It follows that

$$a(\mathbf{n}) = \langle\boldsymbol{\tau}, D^2\gamma(\mathbf{n})\boldsymbol{\tau}\rangle = \hat{\gamma}''(\theta) + \hat{\gamma}(\theta).$$

An anisotropic kinetic mobility $\beta(\mathbf{n}) \geq b_0 > 0$ may be introduced in front of the velocity to yield, [6, 28, 81, 122, 125],

$$\beta(\mathbf{n})V = -\kappa_\gamma.$$

where $\kappa_\gamma := (\hat{\gamma}'' + \hat{\gamma})\kappa$ is the anisotropic mean curvature.

The Wulff-problem, [81], is:

$$\inf\int_{\partial\Omega}\gamma(\mathbf{n}) \quad \text{subject to} \quad |\Omega| = A_0,$$

where Ω is a bounded domain with boundary $\partial\Omega$ and A_0 is a given constant. It is known that for γ being strictly convex that there is a unique solution W which is a dilation of the Wulff-region

$$\mathcal{W} := \{\mathbf{x} \in I\!R^n : \langle\mathbf{x}, \mathbf{N}(\theta)\rangle \leq \hat{\gamma}(\theta), \quad \theta \in I\!R\}.$$

Furthermore W is convex and the boundary ∂W is parameterised by

$$\mathbf{x}_W(\theta) = \hat{\gamma}(\theta)\mathbf{N}(\theta) - \hat{\gamma}'(\theta)\mathbf{T}(\theta), \quad \theta \in \mathbb{R},$$

with $\mathbf{T}(\theta) = \mathbf{N}^{\perp}(\theta)$. Its curvature is given by

$$\kappa_W(\theta) = [\hat{\gamma}(\theta) + \hat{\gamma}''(\theta)]^{-1}$$

so that the Wulff curve ∂W has constant anisotropic (weighted) mean curvature. It is easy to see that Wulff-curves shrink self-similarly for the evolution law

$$\tfrac{1}{\hat{\gamma}(\theta)} V = -(\hat{\gamma} + \hat{\gamma}'')(\theta)\kappa - g,$$

where g is a constant and the kinetic mobility factor is $\hat{\gamma}^{-1}(\theta)$. Then $R(t)\mathbf{x}_W(\theta)$ is a solution provided

$$\frac{\mathrm{d}R}{\mathrm{d}t} = -\frac{1}{R} - g, \quad R(0) = R_0.$$

As in the isotropic case § 2.2 we have finite time extinction if $g > 0$ or $R_0 < -g$ when $g < 0$ and monotone expansion if $R_0 > -g$ for $g < 0$. Furthermore it is known that in the case $g = 0$ any convex curve shrinks in finite time to a point and it tends to be a Wulff curve as it shrinks, [69].

2.4 Crystalline curvature

It is natural to crystal growth problems to consider the evolution of polygonal curves in the plane (polyhedral surfaces in $\mathbb{R}^3$). Suppose that the Wulff-curve W of section 2 is no longer smooth but is a convex polygon. The evolution of simple polygonal curves in the plane according to crystalline curvature can be formulated. Let Γ_t be a polygonal curve and have sides $\{S_i\}$ with normals ν_i and lengths $l_i(t)$, such that adjacent side facets are not collinear. The crystalline curvature of a facet is defined to be

$$\kappa_W(S_i) = \sigma_i \Lambda(\nu_i)/l_i(t)$$

where $\Lambda(\nu_i)$ is the length of the facet of the Wulff-shape W with normal ν_i. If no facet of W has normal ν_i then $\Lambda(\nu_i) = 0$. The constant σ_i is taken to be $+1$, -1 or 0 depending on whether the interior of Γ_t with respect to the three adjacent sides S_{i-1}, S_i, S_{i+1} is convex, concave or neither. Let $d_i(t)$ be the distance of S_i from the origin. Then crystalline curvature motion is defined to be

$$\dot{d}_i(t) = -\sigma_i \Lambda(\nu_i)/l_i(t).$$

This formulation is due, independently, to Taylor [122, 123] and Angenent and Gurtin [6].

2.5 The distance function and mean curvature flow in $\mathbb{R}^N$

Let Γ be a C^2 surface of co-dimension 1 in $\mathbb{R}^N$. The signed (oriented) distance function $d(\mathbf{x})$ is defined by

$$\mathbf{y} \in \Gamma, \ |\mathbf{x}-\mathbf{y}| = \text{dist}(\mathbf{x}, \Gamma), \ \mathbf{x} = \mathbf{y} + d(\mathbf{x})\mathbf{n}$$

and $\mathbf{n}$ is the unit normal to Γ at $\mathbf{y}$. It is easily shown that d is Lipschitz continuous. Without loss of generality, we can write Γ as a graph

$$x_N = F(x_1, \cdots, x_{N-1}) := F(\mathbf{x}')$$

in a neighbourhood of $\mathbf{y} = (\mathbf{y}', y_N)$ with $DF(\mathbf{y}') = 0$ and the x_N-axis being in the normal direction. The eigenvalues of D^2F at $\mathbf{y}$ are labelled $-\kappa_1, -\kappa_2, \cdots$, $-\kappa_{N-1}$ and the κ_i are called the principal curvatures of Γ at $\mathbf{y}$. There exists a δ-neighbourhood of Γ in which $d \in C^2$ and $|\nabla d| = 1$ and for $\mathbf{x}$ in that neighbourhood and lying on the normal to $\mathbf{y} \in \Gamma$, [72]

$$\Delta d(\mathbf{x}) = \sum_{i=1}^{N-1} \frac{\kappa_i}{1 + \kappa_i d}.$$

It follows that

$$\nabla d|_\Gamma = \mathbf{n}, \ \Delta d|_\Gamma = \sum_{i=1}^{N-1} \kappa_i := \kappa,$$

and we call κ (rather than $1/(N-1)\sum_{i=1}^{N-1} \kappa_i$) the mean curvature of Γ.

When Γ is the boundary of a bounded domain in $\mathbb{R}^N$ we use the convention that $\mathbf{n}$ is the outward pointing normal. Thus if Γ is the boundary of a ball then F is strictly concave and D^2F has negative eigenvalues yielding positive principal curvature and mean curvature.

Let Γ_t be an evolving surface with signed distance function $d(\mathbf{x}, t)$. For points $\mathbf{x}(t)$ on Γ_t,

$$0 = d(\mathbf{x}(t), t)$$

which yields

$$0 = d_t + \nabla d \cdot \mathbf{x}_t(t)$$

and the normal velocity is $V = -d_t$. Hence for Γ_t evolving by mean curvature flow

$$V = -\kappa$$

it follows that

$$(d_t - \Delta d)|_{\Gamma_t} = 0.$$

For existence results for hypersurfaces in $\mathbb{R}^N$ we refer to [84, 85] for example. As in the plane, convex surfaces shrink and become more like spheres, [84]. However non-convex surfaces for $N \geq 3$ can develop 'pinching off' singularities and exhibit topological change. A famous example is that of a dumbbell, [77]. The handle of the dumbbell has two principal curvatures of opposite sign and the inner radius of the neck maybe sufficiently small so that the mean curvature is always positive, leading to pinching off. This is in contrast to the two dimensional dumbbell where the handle has just one principal curvature which is negative. A possible continuation is curvature flow for two separated components. See [47, 115] for relevant computations.

In general when evolving surfaces develop singularities such as self intersection, pinching off and finite time extinction the classical formulation breaks down. Possible ways of defining a notion of solution which 'integrates' through these events yielding global in time existence results are:- the seminal work of Brakke on varifolds [20], the level set formulation described in §4, [35, 62, 104] and the variational approach of Almgren et. al [4].

3 Direct approach

3.1 Curve shortening (1)

We consider the evolution of a closed curve $\mathbf{x}(\cdot,t) : I\!R \setminus 2\pi \to I\!R^2$ whose velocity is given by the curvature vector so that

$$\mathbf{x}_t = \frac{1}{|\mathbf{x}_s|}\left(\frac{\mathbf{x}_s}{|\mathbf{x}_s|}\right)_s \qquad s \in I = [0, 2\pi],\ t \in (0,T), \tag{3.1}$$

$$\mathbf{x}(s,t) = \mathbf{x}(s+2\pi, t), \quad \mathbf{x}(\cdot, 0) = \mathbf{x}_0. \tag{3.2}$$

Assume that the initial data is such that the solution of (3.1) and (3.2) is sufficiently smooth for our numerical analysis and that $|\mathbf{x}_s| \geq c_0 > 0$ for $t \in [0,T]$. A weak formulation for (3.1) is

$$\int_I \mathbf{x}_t |\mathbf{x}_s| \eta + \int_I \frac{\mathbf{x}_s}{|\mathbf{x}_s|}\eta_s = 0, \tag{3.3}$$

for test functions $\eta \in \mathrm{H}^1(I\!R \setminus 2\pi)$. This is the basis for the numerical scheme introduced by Dziuk [48].

Let S^h be the finite element space of periodic continuous piece-wise linear functions on a uniform partition of I with mesh size $\Delta s = 2\pi/M$ and basis functions $\{\phi_j\}$ for $j = 0, 1, \cdots, M$. A semi-discrete approximation $\mathbf{X}(t) = \sum_{j=0}^{M} \mathbf{X}_j(t)\phi_j(s)$ with $\mathbf{X}_0 = \mathbf{X}_M$ is defined by

$$\int_I \mathbf{X}_t |\mathbf{X}_s| \phi_j + \int_I \frac{\mathbf{X}_s}{|\mathbf{X}_s|}\phi_{j,s} = 0 \quad j = 1, \cdots, M \tag{3.4}$$

$$\mathbf{X}(\cdot, t) = \mathbf{x}_0^{\Delta}, \tag{3.5}$$

where $\mathbf{x}_0^{\Delta}$ interpolates the initial data of $\mathbf{x}_0$. A consequence of the geometric form of (3.4) is that it may be written as the finite difference scheme

$$\frac{1}{6}\dot{\mathbf{X}}_{j-1}h_j + \frac{1}{3}\dot{\mathbf{X}}_j(h_j + h_{j+1}) + \frac{1}{6}\dot{\mathbf{X}}_{j+1}h_{j+1} = \frac{\mathbf{X}_{j+1} - \mathbf{X}_j}{h_{j+1}} - \frac{\mathbf{X}_j - \mathbf{X}_{j-1}}{h_j}, \tag{3.6}$$

where $\dot{\mathbf{X}}_j \equiv \frac{d}{dt}\mathbf{X}_j$ and $h_j = |\mathbf{X}_j - \mathbf{X}_{j-1}|$. Thus Δs does not explicitly appear in (3.6) and h_j is the distance between successive vertices of the polygonal curve defined by $\mathbf{X}(t)$. If appropriate numerical integration is used on the first term of (3.4) then the following lumped scheme is obtained

$$\frac{1}{2}(h_j + h_{j+1})\dot{\mathbf{X}}_j = \frac{\mathbf{X}_{j+1} - \mathbf{X}_j}{h_{j+1}} - \frac{\mathbf{X}_j - \mathbf{X}_{j-1}}{h_j}. \tag{3.7}$$

An interesting property of (3.7) is that for $h_j > 0$

$$\dot{h}_j = -\tfrac{1}{4}(h_{j-1} + h_j)\,|\dot{\mathbf{x}}_{j-1}|^2 - \tfrac{1}{4}(h_j + h_{j+1})\,|\dot{\mathbf{x}}_j|^2 \leq 0, \tag{3.8}$$

which is the discrete analogue of (2.9). Thus the sides of the polygonal curve $\mathbf{X}(t)$ decrease in length during the time evolution.

Inequality (3.8) is used in the proof of the error bounds of Dziuk. He showed that for h sufficiently small there is a unique solution to (3.7) which satisfies

$$\max_{t\in[0,T]} \|x - \mathbf{X}\| + \left(\int_0^T \|\mathbf{x}_s - \mathbf{X}_s\|^2 dt\right)^{1/2} \leq C\Delta s,$$

$$\max_{t\in[0,T]} \|x_t - \mathbf{X}_t\| + \left(\int_0^T \|\mathbf{x}_{ts} - \mathbf{X}_{ts}\|^2 dt\right)^{1/2} \leq C\Delta s,$$

where $\|\cdot\|$ denotes the $L^2(I)$-norm. The analysis uses the homotopy

$$\mathbf{x}_t^\sigma = \frac{1}{|\mathbf{x}_s^\sigma|_\sigma}\left(\frac{\mathbf{x}_s^\sigma}{|\mathbf{x}_s^\sigma|_\sigma}\right)_s, \quad |\mathbf{x}_s^\sigma|_\sigma = \left(\sigma^2 + (1-\sigma^2)|\mathbf{x}_s^\sigma|^2\right)^{1/2}$$

and relies on estimating $\mathbf{x} - \mathbf{X}$ and $\mathbf{x}_t - \mathbf{X}_t$ simultaneously.

A natural time discretisation of (3.7) is the semi-implicit scheme

$$\frac{1}{2}(h_j^{k-1} + h_{j+1}^{k-1})\left(\frac{\mathbf{X}_j^k - \mathbf{X}_j^{k-1}}{\Delta t}\right) = \frac{\mathbf{X}_{j+1}^k - \mathbf{X}_j^k}{h_{j+1}^{k-1}} - \frac{\mathbf{X}_j^k - \mathbf{X}_{j-1}^k}{h_j^{k-1}}. \tag{3.9}$$

Evolving the curve over a time step is achieved by solving a positive definite tridiagonal system. This results in a very fast algorithm.

The generalisation of this approach to (2.11) is

$$\int_I \mathbf{X}_t|\mathbf{X}_s|\phi_j + \int_I \frac{\mathbf{X}_s}{|\mathbf{X}_s|}\phi_{j,s} = \int_I \mathrm{q}(\mathbf{X})|\mathbf{X}_s|\phi_j + \int_I g(\mathbf{X})\mathbf{X}_s^\perp\phi_j.$$

In the case $g = 0$, error bounds similar to these of Dziuk have been obtained [87].

3.2 Curve shortening (2)

As an alternative to the approach of §3.1 we may seek to approximate the curvature flow (2.5) by discretising (2.6). A weak formulation is

$$\int_I |\mathbf{x}_s|^2 \mathbf{x}_t \eta + \int_I \mathbf{x}_s \eta_s = 0$$

for test functions $\eta \in \mathrm{H}^1(I\!R \setminus 2\pi)$. Using the same finite element space as in §3.1 we obtain a semi-discrete approximation $\mathbf{X}(t) = \sum_{j=0}^{M} \mathbf{X}_j(t)\phi_j(s)$ with $\mathbf{X}_0 = \mathbf{X}_M$ defined by

$$\int_I \mathbf{X}_t |\mathbf{X}_s|^2 \phi_j + \int_I \mathbf{X}_s \phi_s = 0 \qquad \forall j.$$

The equivalent finite difference scheme is

$$\frac{1}{6} h_j^2 \dot{\mathbf{X}}_{j-1} + \frac{1}{3}(h_j^2 + h_{j+1}^2)\dot{\mathbf{X}}_j + \frac{1}{6}\dot{\mathbf{X}}_{j+1} h_{j+1}^2 = \mathbf{X}_{j-1} - 2\mathbf{X}_j + \mathbf{X}_{j+1} \quad \forall j.$$

Here we have used the same notation as in the previous subsection. Again there is no dependence on Δs. The lumped mass scheme is

$$\tfrac{1}{2}\left(h_j^2 + h_{j+1}^2\right)\dot{\mathbf{X}}_j = \mathbf{X}_{j-1} - 2\mathbf{X}_j + \mathbf{X}_{j+1}. \tag{3.10}$$

As in §3.1 a fast algorithm is obtained by discretising implicit in time and evaluating h_j at the old time level. This yields a linear system to solve and is a fast method.

This scheme was analysed by Deckelnick and Dziuk [41] who proved error bounds similar to those of §3.1. Improved bounds have been obtained for a more general situation, see the next subsection.

3.3 Curves attached to boundaries

Let Γ_t be an evolving simple curve in the plane whose end points are attached normally to two fixed C^2 boundaries Γ_L and Γ_R. Suppose that Γ_L and Γ_R can be described by two level sets so that $\Gamma_L = \{x : F(x) = 0\}$ and $\Gamma_R = \{x : G(x) = 0\}$. In fact, one can also consider the case Γ_L and Γ_R being the same level set of F. The evolution of Γ_t by curvature is then described by the system.

$$\mathbf{x}_t = \frac{\mathbf{x}_{ss}}{|\mathbf{x}_s|^2}, \quad s \in I = (0,1),\ t \in (0,T),$$

$$F(\mathbf{x}) = 0,\ \langle \mathbf{x}_s, \nabla F(\mathbf{x})^{\perp}\rangle = 0 \text{ at } s = 0,\quad G(\mathbf{x}) = 0,\ \langle \mathbf{x}_s, \nabla G(\mathbf{x})^{\perp}\rangle = 0 \text{ at } s = 1.$$

A semi-discrete finite element approximation using the same setting as in §3.1, §3.2 is:

$$\int_I \langle \mathbf{X}_t, \eta\rangle |\mathbf{X}_s|^2 + \int_I \langle \mathbf{X}_s, \eta_s\rangle = 0$$

where $\mathbf{X}(t) = \sum_{j=0}^{M} \mathbf{X}_j(t)\phi_j(s)$ (satisfying $\mathbf{X}_0 \in \Gamma_L$, $\mathbf{X}_M \in \Gamma_R$) and $\boldsymbol{\eta}$ is any finite element function $\boldsymbol{\eta} = \sum_{j=0}^{M} \boldsymbol{\eta}_j \phi_j(s)$ satisfying $\langle \boldsymbol{\eta}_0, \nabla F(\mathbf{X}_0)\rangle = \langle \boldsymbol{\eta}_M, \nabla G(\mathbf{X}_m)\rangle = 0$. Lumped mass integration yields (3.10) for interior nodes $j = 1, \cdots, M-1$ and at the boundaries

$$\tfrac{1}{2}h_1^2\langle \dot{\mathbf{X}}_0, \nabla F(\mathbf{X}_0)^{\perp}\rangle = \langle \mathbf{X}_1 - \mathbf{X}_0, \nabla F(\mathbf{X}_0)^{\perp}\rangle, \quad \langle \dot{\mathbf{X}}_0, \nabla F(\mathbf{X}_0)\rangle = 0,$$

$$\tfrac{1}{2}h_M^2\langle \dot{\mathbf{X}}_M, \nabla G(\mathbf{X}_M)^{\perp}\rangle = \langle \mathbf{X}_{M-1} - \mathbf{X}_M, \nabla G(\mathbf{X}_M)^{\perp}\rangle, \quad \langle \dot{\mathbf{X}}_M, \nabla G(\mathbf{X}_M)\rangle = 0.$$

This scheme yields the optimal order error bound, [44]

$$||\mathbf{x} - \mathbf{X}|| + \Delta s||\mathbf{x}_s - \mathbf{X}_s|| \leq C\Delta s^2.$$

3.4 Motion of a graph

Let Γ_t be an evolving hypersurface in $I\!R^N$ which can be written as the graph $\{(x, x_N) : x \in \Omega \subset I\!R^{N-1}, x_N = u(x,t)\}$. It follows that the normal to Γ_t is $\mathbf{n} = (-\nabla u, 1)/\sqrt{1+|\nabla u|^2}$, the normal velocity is $V = u_t/\sqrt{1+|\nabla u|^2}$ and the mean curvature is $\nabla.(\nabla u/\sqrt{1+|\nabla u|^2})$. Thus mean curvature flow for Γ_t is defined by the partial differential equation

$$u_t = \sqrt{1+|\nabla u|^2}\;\nabla.\left(\frac{\nabla u}{\sqrt{1+|\nabla u|^2}}\right) \quad \text{in } \Omega \times (0,T). \tag{3.11}$$

This is supplemented by the initial condition $u(\cdot, 0) = u_0$ and an appropriate boundary condition. For example we take $u = g$ on $\partial\Omega \times (0,T)$. Thus the surface is fixed on the boundary of the domain Ω. Another possibility is to take $\frac{\partial u}{\partial \nu} = 0$ where ν is the outward pointing normal to $\partial\Omega$ so that the surface is orthogonal at the boundary to the walls of the cylinder $\Omega \times \{x_N\}$. Global existence of classical solutions in $C^{2,\alpha}(\Omega \times (0,\infty))$ for u_0, $g \in C^{2,\alpha}(\overline{\Omega})$ and $\partial\Omega \in C^{2,\alpha}$ was proved by Huisken [85].

A natural semi-discrete finite element approximation is given by the weak form: $U(t) \in S_0^h$

$$\int_\Omega \frac{\dot{U}\varphi + \nabla U \nabla \varphi}{\sqrt{1+|\nabla U|^2}} = 0 \qquad \forall \varphi \in S_0^h,$$

$$U(0) = u_0^h \in S_0^h.$$

Here we assume for convenience that $g = 0$ and that the finite element space $S_0^h \subset C^0(\overline{\Omega})$ is the space of continuous piecewise linear functions on a quasi-uniform triangulation of the domain $\Omega \subset I\!R^2$ with curved edges for triangles adjacent to the boundary.

This scheme was analysed by Deckelnick and Dziuk [42] where u_0^h was taken to be the finite element numerical surface projection defined by

$$\int_\Omega \frac{\nabla u_0^h \nabla \varphi}{\sqrt{1+|\nabla u_0^h|^2}} = \int_\Omega \frac{\nabla u_0 \nabla \varphi}{\sqrt{1+|\nabla u|^2}} \qquad \forall \varphi \in S_0^h.$$

They proved that for $t \in [0,T]$,

$$\|u-U\| \le Ch^2|\log h|^2, \quad \|u_t - U_t\| \le Ch|\log h|, \quad \left(\int_0^t \|\nabla(u-U)\|^2\right)^{1/2} \le Ch.$$

3.5 Miscellaneous

- Anisotropy and crystalline curvature

  The approach of §3.1 is easily extended to the anisotropic equation (2.14). Indeed error bounds have been obtained and the method applied to a non-smooth surface energy density yielding crystalline curvature flow, [49]. Computation and analysis of crystalline curvature flow can be found in the paper of Taylor [123].

  Girão and Kohn [73, 74, 75] have developed a crystalline algorithm for computing mean curvature and anisotropic curvature flow for planar curves. Smooth Wulff shapes are approximated by polygons and they derive ordinary differential equations for the motion of the vertices. It's essentially a kind of nonlinear Galerkin scheme. Error bounds have been obtained for graphs and convex curves as the polygonal Wulff shape converges to the smooth one. Fukui and Giga [66] developed a theory for the evolution of an arbitrary planar graph by crystalline curvature. Further analysis and computations with a convergent discretisation may be found in [54].

- Surfaces in $\mathbb{R}^3$

  Let Γ_t be a compact two dimensional surface in $\mathbb{R}^3$ without boundary which can be represented as a map $\mathbf{x} : \Omega \times (0,T) \to \mathbb{R}^3$ where Ω is a bounded domain in $\mathbb{R}^2$. Let Δ_{Γ_t} be the Laplace-Beltrami operator on Γ_t, [72]. It follows that

  $$-\Delta_{\Gamma_t}\mathbf{x} = \kappa\mathbf{n}$$

  where $\mathbf{n}$ is the outward pointing normal to Γ_t and

  $$\mathbf{x}_t = \Delta_{\Gamma_t(t)}\mathbf{x} \quad \text{on } \Omega \times (0,T) \tag{3.12}$$

  is the natural generalisation of the curve shortening equation (2.4) which yields

  $$V = <\mathbf{x}_t, \mathbf{n}> = -\kappa.$$

  Dziuk [47] has developed a numerical discretisation of (3.12) based on a finite element approximation of the Laplace-Beltrami operator.

- Triple points

  When there are three or more phases present in a two dimensional physical system triple points naturally occur at the junction of three phase boundaries. The simplest problem to consider is the motion of three curves $\{\mathbf{x}_i\}$ $i = 1, 2, 3$, attached normally to a fixed boundary $\Gamma = \{x : F(x) = 0\}$ and meeting at an evolving triple point $\mathbf{x}_p$ with prescribed angles $\{\theta_{ij}\}$ between the curves. The problem is:-

$$\mathbf{x}^i_t = \mathbf{x}^i_{ss}/|\mathbf{x}^i_s|^2 \quad s \in (0,1)\ i = 1,2,3$$

$$F(\mathbf{x}^i) = 0, \ < \mathbf{x}^i_s, \nabla F^{\perp}(\mathbf{x}^i) >= 0, \ s = 0\ i = 1,2,3$$

$$\mathbf{x}^1 = \mathbf{x}^2 = \mathbf{x}^3 = \mathbf{x}_p, \ < \mathbf{x}^i_s, \mathbf{x}^j_s >= \cos\theta_{ij}|\mathbf{x}^i_s||\mathbf{x}^j_s|, \ i \neq j, \ s = 1$$

$$\theta_{12} + \theta_{23} + \theta_{31} = 2\pi.$$

Extensive computations using the approach of §3.2 have been performed in [23] for networks of curves using ad hoc surgery methods for the singularities which occur when triple points or phase boundaries vanish.

4 Level set approach

4.1 Formulation

Let $\omega : I\!R^N \times (0,T) \to I\!R$ be sufficiently smooth. Consider the level set $\Gamma_t = \{\mathbf{x} \in I\!R^N : \omega(\mathbf{x},t) = 0\}$ and suppose that it is a smooth closed hypersurface of co-dimension 1 with interior $\Omega_I(t) = \{\mathbf{x} \in I\!R^N : \omega(\mathbf{x},t) < 0\}$. For $\mathbf{x} = \mathbf{x}(t) \in \Gamma_t$ we have $\omega(\mathbf{x}(t),t) = 0$ which implies that

$$\omega_t + \mathbf{x}_t(t) \cdot \nabla\omega = 0 \quad \text{on} \quad \Gamma_t.$$

Since $\mathbf{n} = \nabla\omega/|\nabla\omega|$ is the unit normal to Γ_t pointing out of $\Omega_I(t)$ we have $V := \mathbf{x}_t(t) \cdot \mathbf{n} = -\omega_t/|\nabla\omega|$. Furthermore since $\kappa = \mathrm{div}(\mathbf{n})$ is the sum of the principal curvatures of Γ_t (positive for convex $\Omega_I(t)$) it follows that Γ_t will evolve according to the mean curvature flow $V = -\kappa$ provided ω solves

$$\frac{\omega_t}{|\nabla\omega|} = \mathrm{div}\left(\frac{\nabla\omega}{|\nabla\omega|}\right). \tag{4.1}$$

Indeed any smooth level set of a function ω solving (4.1) evolves according to motion by mean curvature. This approach was introduced by Osher and Sethian [104] for quite general geometric evolution laws.

To propagate a surface Γ_0 by the level set approach we choose initial data $\omega_0(\mathbf{x})$ such that $\Gamma_0 = \partial\Omega_I(0) = \{\mathbf{x} : \omega_0(\mathbf{x}) = 0\}$ and $\Omega_I(0) = \{\mathbf{x} : \omega_0(\mathbf{x}) < 0\}$ and then solve (4.1) with $\omega_0(\mathbf{x})$ as initial data. A choice for the initial data is the signed distance function to Γ_0.

The motion of any level set depends only on its geometry and not on any other level set. This results in the observation that $F(\omega)$ also solves the same partial differential equation as ω for sufficiently smooth F. In a sense the equation is uniform parabolic along each level set but is degenerated in the normal direction to level sets. The equation is difficult to consider in a weak sense because it is not in divergence form and integration by parts does not remove the division by $|\nabla\omega|$. Thus the analysis will always be difficult because $|\nabla\omega| = 0$ occurs. For example if ω is smooth and negative inside a surface Γ_t then at least at one point $\nabla\omega = 0$. Furthermore we are particularly interested in using this formulation for surfaces which self-intersect and this naturally gives rise to the vanishing of the gradient of ω.

A good notion of solutions is provided by the concept of '*weak sub- and super- solutions*' which is also known as '*viscosity solutions*'. However, this is not the same notion of viscosity as used in hyperbolic conservation laws. A function $\omega \in \mathrm{C}\left(I\!R^N \times [0,\infty)\right) \cap \mathrm{L}^\infty\left(I\!R^N \times [0,\infty)\right)$ is a '*weak sub-solution*' of (4.1) provided that:

if $\omega - \phi$ has a local maximum at a point $(\mathbf{x}_0, t_0) \in I\!R^N \times (0,\infty)$ for each $\phi \in \mathrm{C}^\infty\left(I\!R^N \times [0,\infty)\right)$ then

$$\phi_t \leq |\nabla\phi| \mathrm{div}\left(\tfrac{\nabla\phi}{|\nabla\phi|}\right) \quad \text{at } (\mathbf{x}_0, t_0), \quad \text{if } \nabla\phi(\mathbf{x}_0, t_0) \neq 0,$$

and

$$\phi_t \leq (\delta_{ij} - \eta_i\eta_j)\, \phi_{x_i x_j} \quad \text{at } (\mathbf{x}_0, t_0) \text{ for some } \eta \in I\!R^N, \; |\eta| \leq 1, \quad \text{if } \nabla\phi(\mathbf{x}_0, t_0) = 0.$$

A '*weak supersolution*' is defined similarly by reversing the inequalities. Then ω is said to be a weak (viscosity) solution if it is both a weak sub- and super-solution.

The global existence and uniqueness of such solutions has been proved by Evans and Spruck [62] for motion by mean curvature and more generally for forced anisotropic mean curvature flow by Chen, Giga, Goto [35]. Comparison principles are important in the analysis of these equations. Existence is proved in [35] using Perron's method whereas in [62] they considered the regularised equation

$$\omega_t^\varepsilon = \left(|\nabla\omega^\varepsilon|^2 + \varepsilon^2\right)^{1/2} \nabla . \left(\frac{\nabla\omega^\varepsilon}{(|\nabla\omega^\varepsilon|^2 + \varepsilon^2)^{1/2}}\right), \tag{4.2}$$

which is rather close to the equation of motion of a graph by mean curvature.

It is straightforward to construct level set formulations of more general velocity laws using the basic formulae for the normal, curvature and normal velocity of a level set of ω in terms of its derivatives. For example, let $\mathbf{q} : I\!R^N \times (0,T) \to I\!R^N$ and $g : I\!R^N \times (0,T) \to I\!R$ be prescribed sufficiently smooth fields. Let $\beta \in \mathrm{C}^0\left(I\!R^N \setminus \{0\}\right)$ be homogeneous of degree zero and $\gamma \in \mathrm{C}^2\left(I\!R^N \setminus \{0\}\right)$ be homogeneous of degree 1. Consider the degenerate quasi-linear parabolic equation

$$\tfrac{\beta(\nabla\omega)}{\gamma(\nabla\omega)}\omega_t = \mathrm{tr}\left(D^2\gamma(\nabla\omega)D^2\omega\right) + \langle \mathbf{q}, \tfrac{\nabla\omega}{|\nabla\omega|}\rangle + g. \tag{4.3}$$

It follows that the level set of ω evolves according to the velocity law

$$\tfrac{\beta(\mathbf{n})}{\gamma(\mathbf{n})}V = -\kappa_\gamma - \langle \mathbf{q}, \mathbf{n}\rangle - g, \tag{4.4}$$

where κ_γ is the anisotropic mean curvature of the level set which is given by

$$\kappa_\gamma = \mathrm{tr}\left(D^2\gamma(\nabla d)D^2 d\right),$$

where d is the signed distance function to the level set. Here we use $\mathrm{tr}(X)$ to denote the trace of a matrix X. Global in time existence and uniqueness results are proved in [35].

Equation (4.3) gives a notion of solution which allows self-intersection and other singularities. Another phenomenon is that of 'fattening'. A level set of ω which initially has no interior can thicken or fatten. This can happen for mean curvature flow in $\mathbb{R}^2$ when the initial surface has a figure of eight shape [62]. A thorough analysis of 'fattening' is still under development. See [8, 9] for example.

4.2 Discretisation

The level set approach has been extensively developed by Sethian and co-workers [1, 2, 37, 36, 104, 115, 116, 119] as a numerical method for a wide variety of applications involving geometric front evolution. The body of this work is surveyed in the article [117] and more thoroughly in the book [118]. There are many open problems in the rigorous numerical analysis of discretisation.

For the eikonal equation, a first order (Hamilton-Jacobi) equation,

$$\omega_t + F|\nabla\omega| = 0$$

where F is a given function of position, it is appropriate to employ monotone up-winding finite difference schemes developed for first order conservation laws.

A first order scheme in two space dimensions on a uniform grid with mesh size h is

$$\omega_{ij}^{n+1} = \omega_{ij}^{n} - \Delta t([F_{ij}]_+\nabla^+(\omega_{ij}^n) + [F_{ij}]_-\nabla^-(\omega_{ij}^n))$$

where

$$\nabla^+ = ([D_{ij}^{-x}]_+^2 + [D_{ij}^{+x}]_-^2 + [D_{ij}^{-y}]_+^2 + [D_{ij}^{+y}]_-^2)^{1/2}$$

$$\nabla^- = ([D_{ij}^{+x}]_+^2 + [D_{ij}^{-x}]_-^2 + [D_{ij}^{+y}]_+^2 + [D_{ij}^{-y}]_-^2)^{1/2}$$

$$D_{ij}^{+x}\omega_{ij} = \frac{\omega_{i+1j} - \omega_{ij}}{h}, \quad D_{ij}^{-x}\omega_{ij} = \frac{\omega_{ij} - \omega_{i-1j}}{h}, \text{ etc.}$$

Here we use the notation $[v]_+ = \max(v,0)$ and $[v]_- = \min(v,0)$.

For the eikonal-curvature equation

$$\omega_t + F|\nabla\omega| = |\nabla\omega|\nabla(\frac{\nabla\omega}{|\nabla\omega|}) \tag{4.5}$$

Osher and Sethian [104] propose the use of second order central finite differences approximation on the right hand side and up-winding schemes on the left.

There will be problems in the discrete curvature term when the discrete gradient $\nabla\omega_{ij}$ vanishes. If this happens at an isolated point Osher and Sethian [104] recommend considering the limit $|\nabla\omega|\kappa$ for concentric spheres. Computations of Sethian and co-workers show that the level set approach can be used to successfully integrate through geometric singularities such as pinching off for a dumbell in three dimensions into two disconnected components which then shrink and become sphere-like.

As described above the level set method increases the space dimension by one. Hence it is inherently more time consuming than the direct approach of §3 but with the crucial advantage of being able to naturally handle geometric singularities. In some instances one might need all level sets of ω. If one is interested in just one level surface then there is a natural localization. Choose an annular domain in $I\!R^N$ containing the initial surface. Equation (4.5) can be solved with homogenous Neumann data and with the signed distance function being the initial data. The zero level set evolves with the correct law of motion and when it approaches the boundary of the narrow annular band, the band can be updated. This is the basis of the fast level set method proposed in [1]. This approach now combines the relatively low computational demands of front tracking with the ability to integrate through topological change.

A natural approach to overcome the difficulties associated with $\nabla\omega = 0$ is to discretise the regularised equation (4.2). A finite element approximation in the case of homogenous Neumann boundary data is:- find $\omega^h(\cdot,t) \in S^h$ such that

$$\int_\Omega \frac{\omega_t^h \chi + \nabla\omega^h \nabla\chi}{(\varepsilon^2 + |\nabla\omega^h|^2)^{1/2}} dx = 0 \quad \forall \chi \in S^h$$

where S^h is a suitable finite element space. This has been implemented in [65] where numerical results indicating $\mathcal{O}(h^2)$ convergence in the case of piecewise linear elements and $\varepsilon = h^2$ are presented for smoothly evolving flows.

5 Phase field approach

5.1 The Allen-Cahn equation

Let $W(r) = \frac{1}{4}(r^2-1)^2$ be an equal double well potential, ε be a small parameter and $u : I\!R^N \times (0,T) \to I\!R$ be a solution of

$$\varepsilon u_t = \varepsilon \Delta u - \tfrac{1}{\varepsilon} W'(u). \tag{5.1}$$

The flow of the ordinary differential equation $u_t = -\frac{1}{\varepsilon^2}W'(u)$ drives positive values to $+1$ and negative values to -1. Thus one might expect the solution of the equation (5.1) after a short transient time to have developed narrow transition layers connecting regions where u is close to ± 1. This expectation can be mathematically justified and furthermore it can be shown that the motion of the transition layers approximates motion by mean curvature. To motivate this consider a closed surface Γ_t with interior $\Omega_I(t)$ evolving by mean curvature flow and let $d(\mathbf{x},t)$ be the signed distance function to Γ_t. Thus $\Omega_I(t) = \{\mathbf{x} : d(\mathbf{x},t) < 0\}$, $I\!R^N \setminus \overline{\Omega_I(t)} = \{\mathbf{x} : d(\mathbf{x},t) > 0\}$ and $\Gamma_t = \partial\Omega_I(t)$. It follows that $\mathbf{n} = \nabla d$, $|\nabla d(\mathbf{x},t)| = 1$, $-d_t$ is the normal velocity of Γ_t in the direction $\mathbf{n}$ and $\kappa = \Delta d|_{\Gamma_t}$. Setting

$$\mathcal{P}(u) := \varepsilon u_t - \varepsilon \Delta u + \tfrac{1}{\varepsilon} W'(u)$$

we find that for $\psi : I\!R \to I\!R$,

$$\mathcal{P}(\psi(\tfrac{d}{\varepsilon})) = \psi'(d_t - \Delta d) - \tfrac{1}{\varepsilon}(\psi'' - W'(\psi)).$$

Let $\psi = \psi(z)$ be the unique solution to

$$-\psi'' + W'(\psi) = 0 \quad z \in I\!R$$

$$\psi(0) = 0, \ \psi'(z) > 0, \ \psi(z) \to \pm 1 \ z \to \pm\infty,$$

which yields for $W'(r) = r^3 - r$ that

$$\psi(z) = \tanh\frac{z}{\sqrt{2}}.$$

Since $(d_t - \Delta d)|_{\Gamma_t} = 0$ and $\psi'(\frac{d}{\varepsilon})$ is only large in an ε-neighbourhood of Γ_t and tends to zero away from Γ_t, it follows that $\psi(\frac{d}{\varepsilon})$ is close to being a solution of (5.1). Using comparison theory for parabolic equations it is possible to construct upper and lower solutions, based on perturbations of the form $\psi(\frac{d}{\varepsilon})$, to (5.1) with initial data

$$u(x,0) = \psi(\frac{d(x,0)}{\varepsilon}).$$

This yields convergence and error bounds for smoothly evolving curvature flow. An $\varepsilon|\log\varepsilon|$ error was originally proved by Chen [32]. This was refined to $\varepsilon^2|\log\varepsilon|^2$ by Belletini and Paolini [10] even for the Allen-Cahn equation with right hand-side

$$\varepsilon u_t = \varepsilon\Delta u - \frac{1}{\varepsilon}W'(u) + \varepsilon c_W g,$$

where $c_W = \frac{1}{2}\int_{-1}^{1}\sqrt{2W(r)}dr$, which converges to the forced mean curvature flow

$$V = -\kappa + g.$$

Formal asymptotics were first used to derive the relation between the Allen-Cahn equation and mean curvature flow in [3, 110]. The first proof of convergence was based on a rigorous analysis of an asymptotic development of u in powers of ε by de Mottoni and Schatzman [46]. See also [22]. The most general results are the convergence to viscosity solutions of the level set formulation in the absence of fattening proved by Evans, Soner and Souganidis [61] using upper and lower solutions and the convergence to Brakke's notion of solution proved by Ilmanen [86], see also [8].

5.2 The double obstacle phase field model

The solution of the Allen-Cahn equation u_ε with the smooth double well potential $W(u) = \frac{1}{4}(u^2-1)^2$ has a transition layer of width $\mathcal{O}(\varepsilon|\log\varepsilon|)$ defined by the level sets $\pm 1 \mp \varepsilon^k$ for any integer $k \geq 1$ and is not constant away from the layer. This is inconvenient for numerical computations because (5.1) has to be solved in the large domain Ω in order to approximate mean curvature flow even though the action takes place in a small region of measure $|\Gamma_t||\varepsilon|\log\varepsilon|$. This may be overcome by using the double obstacle potential

$$W(u) = \begin{cases} \frac{1}{2}(1-u^2) & \text{if } u \in [-1,1] \\ +\infty & \text{if } |u| > 1, \end{cases}$$

i.e. $W(u) = \frac{1}{2}(1-u^2) + \mathrm{I}_{[-1,1]}(u)$, where $\mathrm{I}_{[-1,1]}(u) = 0$ for $|u| \le 1$, $\mathrm{I}_{[-1,1]}(u) = +\infty$ for $|u| > 1$. Set $W'(u) := -u + \beta(u)$, $\beta(u) = \partial \mathrm{I}_{[-1,1]}$ is the sub-differential of $\mathrm{I}_{[-1,1]}$ and

$$\beta(u) = \begin{cases} (-\infty, 0] & \text{if } u = -1 \\ 0 & \text{if } |u| < 1 \\ [0, \infty) & \text{if } u = 1. \end{cases}$$

This double obstacle was introduced in the gradient theory of phase transitions in [12, 13, 14, 15, 16], see also [128]. The double obstacle Allen-Cahn equation is, informally,

$$\varepsilon u_t = \varepsilon \Delta u - \tfrac{1}{\varepsilon} W'(u) + c_W g.$$

More precisely it can be written as the inclusion

$$\varepsilon u_t - \varepsilon \Delta u - \tfrac{1}{\varepsilon} u - c_W g \in -\beta(u)/\varepsilon,$$

or the parabolic variational inequality: $u(\cdot, t) \in \mathcal{K}$

$$\varepsilon(u_t, \eta - u) + \varepsilon(\nabla u, \nabla \eta - \nabla u) \ge \tfrac{1}{\varepsilon}(u, \eta - u) + (c_W g, \eta - u) \quad \forall \eta \in \mathcal{K}, \tag{5.2}$$

where $\mathcal{K} = \{\eta \in \mathrm{H}^1(\Omega) : |\eta| \le 1\}$. Again when the interface Γ is bounded away from $\partial\Omega$ we can replace $\mathcal{K}$ by $\mathcal{K} = \{\eta \in \mathrm{H}^1(\Omega) : |\eta| \le 1, \ \eta|_{\partial\Omega} = 1\}$. The interface profile is defined by the unique solution of:

$$-\psi'' + W'(\psi) = 0, \quad \psi(z) \to \pm 1 \ \text{ for } \ z \to \pm\infty, \quad \psi(0) = 0, \quad \psi'(z) \ge 0,$$

given by

$$\psi(z) = \begin{cases} +1 & z \ge \frac{\pi}{2} \\ \sin z & |z| \le \frac{\pi}{2} \\ -1 & z \le -\frac{\pi}{2}, \end{cases} \tag{5.3}$$

and $c_W = \frac{1}{2}\int_{-1}^{1} \sqrt{2W(u)} = \frac{\pi}{4}$.

Thus we expect solutions to have narrow transition layers of width $\mathcal{O}(\varepsilon)$ separating large regions where u_ε is ± 1 and the transition layer approximates mean curvature flow. This was first proved rigorously by Chen and Elliott [33] in the case $g = 0$. The proof is quite short and we repeat it here.

Theorem *Let Γ_t be a mean curvature flow for $t \in [0, T]$ with interior $\Omega^-(t)$ and exterior $\Omega^+(t)$ such that $dist(\Gamma_t, \partial\Omega) \ge \delta > 0$ and sufficiently smooth so that*

$$|d_t - \Delta d| \le D_0 |d| \quad \textit{for } |d| \le \delta, \ t \in [0, T].$$

Here $d(\mathbf{x}, t)$ is the signed distance function to Γ_t, positive in $\Omega^+(t)$ and negative in $\Omega^-(t)$. Let ε be sufficiently small such that

$$\tfrac{1}{2}\pi\varepsilon \le \delta(1 + 2e^{2D_0 T})^{-1}.$$

Let $u^\varepsilon(\mathbf{x}, t)$ be the unique solution of (5.2) ($g = 0$) with initial data

$$u^\varepsilon(\mathbf{x}, 0) = \psi\left(\frac{d(\mathbf{x}, 0)}{\varepsilon}\right).$$

Then for all $t \in [0,T]$,

$$d(\mathbf{x},t) \geq \tfrac{1}{2}\pi\varepsilon\left(1+2e^{2D_0 t}\right) \implies u^\varepsilon(\mathbf{x},t)=1 \tag{5.4}$$

$$d(\mathbf{x},t) \leq -\tfrac{1}{2}\pi\varepsilon\left(1+2e^{2D_0 t}\right) \implies u^\varepsilon(\mathbf{x},t)=-1. \tag{5.5}$$

Proof. Set $z^\varepsilon(\mathbf{x},t) := d(\mathbf{x},t) - \pi\varepsilon e^{2D_0 t}$ and $\Gamma_t^\varepsilon := \{\mathbf{x}\in\Omega : |z^\varepsilon(\mathbf{x},t)| < \frac{1}{2}\varepsilon\pi\}$. It follows for $\mathbf{x}\in\Gamma_t^\varepsilon$ that $|d(\mathbf{x},t)| \leq \delta$. Consequently, for all $\mathbf{x}\in\Gamma_t^\varepsilon$ and $t\in[0,T]$,

$$\begin{aligned} z_t^\varepsilon - \Delta z^\varepsilon &= d_t - \Delta d - 2D_0\pi\varepsilon e^{2D_0 t} \leq D_0\left(|d| - 2\pi\varepsilon e^{2D_0 t}\right) \\ &\leq D_0\left(|z| - \pi\varepsilon e^{2D_0 t}\right) \leq D_0\left(\tfrac{1}{2}\varepsilon\pi - \pi\varepsilon e^{2D_0 t}\right) \leq 0. \end{aligned}$$

Set $v^\varepsilon(\mathbf{x},t) = \psi\left(\frac{z^\varepsilon(\mathbf{x},t)}{\varepsilon}\right)$. Clearly, $v^\varepsilon(\mathbf{x},0) > -1$ implies that $d(\mathbf{x},0) > \frac{1}{2}\varepsilon\pi$. Hence, $-1 \leq v^\varepsilon(\mathbf{x},0) \leq u^\varepsilon(\mathbf{x},0)$.

Furthermore, for any $\varphi \in L^2(0,T;H^1(\Omega))$

$$\int_0^T\int_{\Gamma_t^\varepsilon} v_t^\varepsilon\varphi + \nabla v^\varepsilon\nabla\varphi - \tfrac{1}{\varepsilon^2}v^\varepsilon\varphi\, dx dt = \int_0^T\int_{\Gamma_t^\varepsilon}(v_t^\varepsilon - \Delta v^\varepsilon - \tfrac{1}{\varepsilon^2}v^\varepsilon)\varphi\, dx dt$$

since on $\partial\Gamma_t^\varepsilon$, $|z^\varepsilon| = \frac{1}{2}\varepsilon\pi$ and $\nabla v^\varepsilon = \psi'(\frac{z^\varepsilon}{\varepsilon})\frac{\nabla z^\varepsilon}{\varepsilon} = 0$. Direct calculations yield for $\mathbf{x}\in\Gamma_t^\varepsilon$, $t\in[0,T]$ that

$$v_t^\varepsilon - \Delta v^\varepsilon - \tfrac{1}{\varepsilon^2}v^\varepsilon = \psi'\tfrac{z_t^\varepsilon}{\varepsilon} - \psi'\tfrac{\Delta z^\varepsilon}{\varepsilon} - \psi''\tfrac{|\nabla z^\varepsilon|^2}{\varepsilon^2} - \tfrac{1}{\varepsilon^2}\psi,$$

and since $\psi'' = -\psi$, $\psi' > 0$ and $|\nabla z^\varepsilon| = |\nabla d| = 1$ we have that

$$v_t^\varepsilon - \Delta v^\varepsilon - \tfrac{1}{\varepsilon^2}v^\varepsilon = \tfrac{1}{\varepsilon}\psi'(z_t^\varepsilon - \Delta z^\varepsilon) \leq 0.$$

We wish to show that $(v^\varepsilon - u^\varepsilon)_+ = 0$. From the above we know that

$$\int_0^T\int_\Omega v_t^\varepsilon(v^\varepsilon - u^\varepsilon)_+ + \nabla v^\varepsilon\nabla(v^\varepsilon-u^\varepsilon)_+ - \tfrac{1}{\varepsilon^2}v^\varepsilon(v^\varepsilon-u^\varepsilon)_+\, dx dt \leq 0,$$

since for $\mathbf{x}\notin\Gamma^\varepsilon(t)$, $-v^\varepsilon(v^\varepsilon-u^\varepsilon)_+ \leq 0$ and $v^\varepsilon(\mathbf{x},t)$ is either identical $=1$ or -1. Taking $\eta = u^\varepsilon + (v^\varepsilon - u^\varepsilon)_+ \in \mathcal{K}$ in (5.2) we have

$$\int_0^T\int_\Omega u_t^\varepsilon(v^\varepsilon - u^\varepsilon)_+ + \nabla u^\varepsilon\nabla(v^\varepsilon-u^\varepsilon)_+ - \tfrac{1}{\varepsilon^2}u^\varepsilon(v^\varepsilon-u^\varepsilon)_+\, dx dt \geq 0,$$

and adding these inequalities yields

$$\int_0^T \tfrac{1}{2}\tfrac{d}{dt}\|(v^\varepsilon-u^\varepsilon)_+\|^2 + \|\nabla(v^\varepsilon-u^\varepsilon)_+\|^2 \leq \int_0^T \tfrac{1}{\varepsilon^2}\|(v^\varepsilon-u^\varepsilon)_+\|^2.$$

Integrating this inequality and using the fact that $(v^\varepsilon - u^\varepsilon)_+(\mathbf{x},0) = 0$, yields the result. Hence $u^\varepsilon(\mathbf{x},t) \geq v^\varepsilon(\mathbf{x},t)$ for all $\mathbf{x}\in\Omega$, $t\in[0,T]$ and since $d(\mathbf{x},t) \geq \frac{1}{2}\pi\varepsilon + \pi\varepsilon e^{2D_0 t}$ implies that $z \geq \frac{1}{2}\pi\varepsilon \Rightarrow v^\varepsilon(\mathbf{x},t) = 1$ we have (5.4). A similar argument with $z^\varepsilon(\mathbf{x},t) = d(\mathbf{x},t) + \varepsilon\pi e^{2D_0 t}$ yields (5.5). □

A consequence of this theorem is that the transition layer is of finite width bounded by $c(t)\varepsilon = \frac{1}{2}\pi(1+e^{2D_0 t})\varepsilon$ and that the zero level set of $u^\varepsilon(\mathbf{x},t)$ is within $c(t)\varepsilon$ of Γ_t.

In fact a more refined error analysis of Nochetto, Paolini and Verdi [96, 98, 99] yields an $\mathcal{O}(\varepsilon^2)$ interface estimate for smoothly evolving flows even for $g \neq 0$. Furthermore convergence to viscosity solutions of the level set formulation in the absence of fattening has been established, [101]. The inclusion of advection in (5.2) is considered in [89].

5.3 Discretisation of the Allen-Cahn equation

Let Ω be a polyhedral domain in $\mathbb{R}^N$ ($N = 2,3$) for ease of exposition. Let $\mathcal{J}^h$ be a quasi-uniform partitioning of Ω into disjoint open simplices where diameters are comparable with h. We denote by S^h the finite element space of continuous piecewise linear functions with canonical basis $\{\phi_i\}_{i=1}^I$ associated with the points $\{\mathbf{x}_i\}_{i=1}^I$. We assume that $\mathcal{J}^h$ is acute so that for any nodes i,j belonging to the same element

$$(\nabla\phi_i, \nabla\phi_j) \leq 0 \quad i \neq j. \tag{5.6}$$

For computational convenience we use a discrete inner product $(\cdot,\cdot)_h$ on $C(\overline{\Omega})$ defined by

$$(\chi,\phi)_h := \int_\Omega \Pi^h\left(\chi(x)\phi(x)\right) dx = \sum_{i=1}^I M_i \chi(\mathbf{x}_i)\phi(\mathbf{x}_i),$$

where $\Pi^h : C(\overline{\Omega}) \to S^h$ is the usual interpolation operator for S^h.

Natural explicit ($\theta = 1$) and implicit ($\theta = 0$) time stepping finite element discretisation of the Allen-Cahn equation are: For $n \geq 1$ seek $U^n \in S^h$

$$\varepsilon(\partial U^n, \chi)_h + \varepsilon(\nabla U^{n-\theta}, \nabla\chi) = (\tfrac{1}{\varepsilon}W'(U^{n-\theta}) + g^{n-\theta}, \chi)_h \quad \forall \chi \in S^h,$$

where $\partial U^n = \frac{U^n - U^{n-1}}{\Delta t}$, $g^n := g(\cdot, t_n)$, $t_n = n\Delta t$ and $\Delta t = T/M$. One can also use a Dirichlet boundary condition and take $U^n \in S^h(1)$ and $\chi \in S^h(0)$ where $S^h(\alpha) = \{\chi \in S^h : \chi|_{\partial\Omega} = \alpha\}$. A suitable initial datum U_0 is required which for approximating interface motion might be $U^0 = \pi^h\psi(\frac{d(\cdot,0)}{\varepsilon})$.

Both implicit and explicit schemes require time step restrictions :-

$$\Delta t \leq Ch^2 \quad \text{explicit} \tag{5.7}$$

$$\Delta t \leq \alpha\varepsilon^2 \quad \text{implicit.} \tag{5.8}$$

Restriction (5.7) is the usual stability restriction for the explicit method for semi-linear parabolic equations. The constant C depends on the mesh and on the maximum norm of the initial data through the magnitude of $|W''|$, see [34, 60]. Restriction (5.8) yields a unique solution for the implicit scheme, where $\alpha \geq -W''(r)\ \forall r$. A consequence of (5.6), (5.7) and (5.8) is that the comparison principle holding for the Allen-Cahn equation also holds for the discrete approximation. This is important in proving convergence results to curvature flow as $\varepsilon, h \to 0$ by the method of lower and upper solutions.

A convergence analysis for zero forcing has recently been given, [34], in the special case of a uniform grid which yields a finite difference scheme. Specifically let

$$\Sigma_h := \{h(i_1, \cdots, i_N) \in \Omega;\; i_1, \cdots, i_N \in \mathbb{Z}\} \quad \text{and} \quad \Sigma_{\Delta t} := \{n\Delta t;\; n \in \mathbb{Z}\}$$

$\mathbf{e}_i$ be the unit vector in the $\vec{i}$-direction,

$$\partial U(\mathbf{x}, t) := \frac{U(\mathbf{x}, t+\Delta t) - U(\mathbf{x}, t)}{\Delta t} \quad \text{and} \quad \Delta_h U(\mathbf{x}) := \sum_{i=1}^{N} \frac{U(\mathbf{x}+h\mathbf{e}_i) - 2U(\mathbf{x}) + U(\mathbf{x}-h\mathbf{e}_i)}{h^2}.$$

Consider the explicit finite difference scheme

$$\begin{aligned} &\varepsilon \partial U - \varepsilon \Delta_h U + \tfrac{1}{\varepsilon} W'(U) = 0 && (\mathbf{x}, t) \in \Sigma_h \times \Sigma_\Delta \\ &U = -1 && (\mathbf{x}, t) \in \partial\Sigma_h \times \Sigma_{\Delta t} \\ &U = U_0^\varepsilon && \mathbf{x} \in \Sigma_h,\; t = 0 \end{aligned} \tag{5.9}$$

where

$$U_0^\varepsilon = \begin{cases} 1 & \mathbf{x} \in \Omega_0^+ \\ -1 & \mathbf{x} \in \Omega_0^-. \end{cases}$$

They proved the following result: Let $p > 1$ be a fixed constant. Assume that h, Δt and ε satisfy

$$h \leq \varepsilon^p, \quad \Delta = \frac{h}{\varepsilon}, \quad \frac{\Delta t}{h^2} \leq \frac{1}{2N + \Delta^2 K},$$

where K depends only on W''. Let $U_{h,\Delta t}^\varepsilon$ be the solution of (5.9). Let Γ_t be a smooth evolving interface with interior Ω_t^- and exterior Ω_t^+ and normal velocity $V = -\kappa$. Assume Γ_t is bounded away from $\partial\Omega$. Then for any $(\mathbf{x}, t) \in \Sigma_h \times \Sigma_{\Delta t}$,

$$\lim_{\varepsilon \to 0} U_{h,\Delta t}^\varepsilon(\mathbf{x}, t) = \begin{cases} 1 & \mathbf{x} \in \Omega_t^+ \\ -1 & \mathbf{x} \in \Omega_t^- \end{cases}$$

and the zero level set of $U_{h,\Delta t}^\varepsilon$ at time t is within a distance $\mathcal{O}(\Delta + \varepsilon + \Delta t/\varepsilon)$ from Γ_t. Actually the Allen-Cahn equation with convection is also considered. Numerical computations simulating curvature flow by this method can be found in [26, 34].

A steady state numerical analysis of a finite element approximation showing convergence to a prescribed curvature problem with an $\mathcal{O}(\varepsilon^2 |\ln \varepsilon|^2)$ interface error bound has been given in [106]. See also [12, 17] .

5.4 Discretisation of the double obstacle phase field model

Consider the finite element setting of §5.3. The double obstacle version of (5.2) is

$$\varepsilon(\partial U^n, \chi - U^n)_h + \varepsilon(\nabla U^{n-\theta}, \nabla\chi - \nabla U^n) \geq \tfrac{1}{\varepsilon}(U^{n-\theta}, \chi - U^n)_h + (c_W g^n, \chi - U^n)$$

for all $\chi \in \mathcal{K}^h$, where $\mathcal{K}^h = \{\chi \in S^h : |\chi| \leq 1\}$ or $\mathcal{K}^h = \{\chi \in S^h : |\chi| \leq 1, \chi|_{\partial\Omega} = 1\}$. The initial data is chosen to be

$$U^0 := \Pi^h \psi\left(\frac{d(\mathbf{x}, 0)}{\varepsilon}\right),$$

where ψ is the double obstacle profile (5.3) and d is the signed distance function to the evolving interface Γ_t. Again we have the time step constraints (5.7) and (5.8).

Explicit time stepping is easily carried out by the procedure

$$\underline{U}^{n+1/2} = \left((1 + \tfrac{\Delta t}{\varepsilon^2})I - \Delta t A\right)\underline{U}^n + c_W \tfrac{\Delta t}{\varepsilon}\underline{g}^n,$$
$$\underline{U}^{n+1} = P\underline{U}^{n+1/2},$$

where $P : \mathbb{R}^I \to \mathbb{R}^I$ is the component-wise projection on $[-1,1]$ defined by $(P\underline{U})_j = \max(-1, \min(1, U_j))$. The implicit time stepping scheme leads to the algebraic problem:

$$\langle(1 - \tfrac{\Delta t}{\varepsilon^2}\underline{U}^n + \Delta t A\underline{U}^n - \underline{U}^{n-1} - c_W \tfrac{\Delta t}{\varepsilon}\underline{g}^n, \chi - \underline{U}^n\rangle \geq 0, \quad \forall\chi \text{ with } |\chi_i| \leq 1.$$

This is easily solved by projected SOR, see [56] for example.

A remarkable feature of the explicit scheme is that if U^n is equal to 1 (or respectively -1) at a node $\mathbf{x}_j$ and at all the nearest neighbours on the mesh then

$$U_j^{n+1/2} = 1 + \tfrac{\Delta t}{\varepsilon^2}(1 + c_W \varepsilon g^n)$$

and $U_j^{n+1} = 1$ (or respectively -1) provided $|g| \leq \frac{4}{\varepsilon\pi}$. Thus the discrete transition layer can not move faster than one element per time step. Furthermore, it is only necessary to compute U^{n+1} on the closure of the transition layer. This has been exploited in the two dimensional dynamic mesh algorithm (DMA) of Nochetto, Paolini and Verdi [100]. Essentially they carry a mesh only in the transition layer and add and remove triangles where necessary. An advancing front algorithm is used to generate a new mesh when triangles are added or removed. The stiffness matrix and varying mass matrix requires updating when the mesh changes.

An alternative approach of Elliott and Gardiner [52, 53] is to fix the mesh and only compute in the transition layer. It is possible to store nodal values only in the transition layer. Storage can be further reduced by using a uniform grid so that the stiffness and mass matrices require minimal storage (just stencil values). This so called 'mask' method has been successfully used for geometric curvature flow and the curvature flow coupled with the heat equation for dendritic growth of supercooled liquid [52].

An error analysis for these schemes has been carried out by Nochetto and Verdi [102, 103]. In [102] they considered the case of the implicit scheme without numerical integration and using a strongly acute triangulation so that a discrete maximum principle still holds. Taking $\Delta t, h^2 \approx \mathcal{O}(\varepsilon^3)$ and assuming that there is no fattening they showed convergence to the viscosity solution. A linear rate of convergence $\mathcal{O}(\varepsilon)$ was obtained for the Hausdorff distance between the smoothly evolving interface Γ_t and the zero level set of U provided $\Delta t, h^2 \approx \mathcal{O}(\varepsilon^4)$. The explicit scheme was considered in [103]. Convergence to the viscosity solution in the absence of fattening was proved under $\Delta t, h^2 \approx \mathcal{O}(\varepsilon^4)$. If exact integration is used in the potential term $\frac{1}{\varepsilon^2}(U, \chi)$ they showed an interface estimate of $\mathcal{O}(\varepsilon^2)$ for smooth flows when $\Delta t, h^2 \approx \mathcal{O}(\varepsilon^5)$. The analysis of [103] has been extended to the advected double obstacle Allen-Cahn model

$$\varepsilon(\partial_t u, \eta - u) + \varepsilon(\nabla u, \nabla\eta - \nabla u) - \varepsilon(\mathbf{q} \cdot \nabla u, \eta - u)$$
$$\geq \frac{1}{\varepsilon}(u, \eta - u) + \frac{\pi}{4}(g, \eta - u)$$

by Kuhn [89].

It is possible to replace ε by a space-time dependent relaxation parameter $\varepsilon a(x, t)$ where a is bounded and uniformly positive. This gives the possibility of having a narrower interface in the neighbourhood of singularities yielding better resolution, [97, 107]. Double obstacle phase field computations of smooth evolution and flows with pinching off singularities may be found in [17, 53, 94, 97, 100, 107].

5.5 Anisotropy

Consider the phase field model

$$\varepsilon\beta(\nabla u)u_t - \varepsilon\nabla(A'(\nabla u)) + \frac{1}{\varepsilon}W'(u) = c_W g$$

where $\beta \in C^0(I\!R^N \backslash \{o\})$ and $A \in C^{1,1}(I\!R^N)$ are respectively homogenous of degree zero and two. The potential W is taken to be either of the two forms of §5.1 and §5.2. For small ε the zero level set of u approximates a surface Γ_t which evolves according to the anisotropic curvature flow

$$\frac{\beta(\mathbf{n})}{\gamma(\mathbf{n})}V = -\mathrm{tr}(D^2\gamma(\mathbf{n})\mathrm{div}\mathbf{n}) - g$$

where $\gamma := \sqrt{2A}$. In the case $N = 2$ we can take $A(\mathbf{p}) = \frac{1}{2}\hat{\gamma}(\theta)^2|\mathbf{p}|^2$ where $\hat{\gamma}$ is defined in §2.3 and $\theta = \arctan(p_2/p_1)$.

Formal asymptotics for smooth W, [90, 129] and for double obstacle W with $\beta = 1$, [11] show the link between anisotropic phase field models and the geometric interface motion. Elliott and Schatzle [59] proved convergence in the double obstacle case to the viscosity solution in the absence of fattening. $\mathcal{O}(\varepsilon^2)$ interface estimates for smooth evolving flow have been established in [57, 58]. Recently a generalised double obstacle phase field model was considered in [53] with a potential which 'interpolates' between the classical Ginzburg-Landau smooth potential and the double obstacle potential introduced by Blowey and Elliott and which has a right hand side $c_W g \rho(u)$ for a suitable ρ. Other results for smooth potentials and variable relaxation parameter may be found in [89]. Finite element computations based on this approach are described in [53, 89, 105]

6 Applications

- Grain boundary motion and crystal growth

  Grain boundaries in alloys, perhaps separating crystals of different phases or crystals of the same phase with differing orientations, may be modelled by evolution laws which depend on local structure and which decrease the surface energy of the grain boundaries [3, 92, 125]. For isotropic surface energy density the simplest model is motion by mean curvature,

$$V = -\kappa.$$

  For anisotropic surface energy γ and a kinetic mobility β which also depends on the orientation of the surface the simplest model is

$$\beta(\mathbf{n})V = -\kappa_\gamma + g,$$

  where g is a local driving force arising from the change in volume of the interior of the surface, [125]. The growth of crystals into a nutrient phase can be modelled by similar laws where g might depend on temperature or concentration, [81, 114, 131].

- Stefan problem

  Let Γ_t be the evolving interface between an interior solid region $\Omega_S(t)$ and an exterior liquid region $\Omega_L(t)$. The forced mean curvature flow

$$\delta_v \beta(\mathbf{n})V = -\delta_c \kappa_\gamma - \theta \tag{6.1}$$

  arises as a Gibbs-Thompson relation with kinetic undercooling for the temperature θ at the interface. Here δ_v and δ_c are small positive constants and κ_γ is the anisotropic mean curvature. Away from the interface the temperature satisfies the heat equation. The system is completed with fixed boundary conditions and the energy balance

$$LV = -[\mathbf{q}] \cdot \mathbf{n} \tag{6.2}$$

  at the interface, where L is the latent heat and $[\mathbf{q}]$ is the jump in heat flux across Γ_t. This system models the solidification of liquid initially undercooled to below its melting temperature. This is an unstable situation and gives rise to complex morphology of the solid/liquid interface. Of particular interest is dendritic growth which is computationally challenging. We refer to [5, 7, 50, 51, 52, 88, 119, 113, 121, 129] for numerical simulations.

- Hele-Shaw flow

A model similar to the Stefan problem is Hele-Shaw flow for the motion of an interface separating two immiscible viscous fluids lying in a narrow gap between two parallel plates, where the field variable is the pressure p which is harmonic on either side of the interface. However p is discontinuous at the interface and θ in (6.1) is replaced by the jump in pressure $[p]$. On the other hand the normal component of the fluid velocity is continuous at the interface and (6.2) is replaced by

$$V = -k_1 \frac{\partial p}{\partial n} = -k_2 \frac{\partial p}{\partial n}$$

where the k_i are constants depending on the fluids. See [43, 112] and also [83] for a review of numerical methods for this and related problems in fluid mechanics.

- Waves in excitable media

Travelling waves of chemical, physical or biological activity in excitable media have been shown to be modelled by the two dimensional eikonal-curvature law

$$V = g - \kappa.$$

This law of motion is derived from formal asymptotics of systems of reaction diffusion equations. Fascinating patterns of so-called target and spiral waves arise depending on initial condition, the geometry of the medium and possible spatial dependence of g. We refer to [21, 63, 78, 126, 127] for further analysis, computations and applications.

- Image processing

Let $\omega_0(x)$, $x \in I\!R^2$, denote the density of greyness of an image. The theory of filtering to yield smoother images $\omega(\cdot, t)$ 'evolving' from $\omega_0(\cdot)$ leads to the level set equation for motion by mean curvature and other nonlinear partial differential equations with geometric interpretation. This is an exciting area of application developed in the last decade and is reviewed in another article in this volume [80].

- Superconductivity

The motion of a line vortex in a three dimensional superconducting medium can be modelled by the geometric law

$$V = \kappa \mathbf{n} + \mathbf{j} \wedge \boldsymbol{\tau}$$

where $\mathbf{j}$ is a prescribed current density and the other terms have their usual meanings, [31]. The motion of a planar vortex when $\mathbf{j}$ is normal to the plane is then described by forced mean curvature flow. Some numerical calculations and mathematical analysis of the long time asymptotics for a graph may be found in [45].

- Surface Diffusion

  The surface Γ_t is said to evolve by surface diffusion if

$$V = \Delta_{\Gamma_t}\kappa$$

  where Δ_{Γ_t} is the surface Laplacian, κ is its mean curvature and v is its normal velocity, [29, 40, 93]. When Γ_t is the boundary of a bounded domain the evolution is volume preserving. It is known in the plane that circles are stable [55]. Formal asymptotics, [27], show that the equation arises from a degenerate Cahn-Hilliard equation. Some numerical computation can be found in [38, 132]. A crystalline version of surface diffusion has been defined in [30] where numerical computations can also be found. Another volume preserving flow is, [68],

$$V = -\kappa + \bar{\kappa}$$

  where $\bar{\kappa}$ is the mean value of κ on the evolving surface. It arises as an asymptotic limit of a non local Allen-Cahn equation [111] and double obstacle phase field computations can be found in [17].

- Geochemical Transport

  The advected curvature flow

$$V = -\kappa + < \mathbf{q}, \mathbf{n} >$$

  arises as a model for the evolution of sharp reaction fronts in geochemical systems, [79]. The velocity field $\mathbf{q}$ may be determined by solving the equations

$$\mathrm{div}\ \mathbf{q} = 0, \quad \mathbf{q} = -k\nabla\mathbf{p}$$

  for D'Arcy flow and k might be different on either side of the front.

- Diffuse interface models

  The Allen-Cahn equation frequently arises as one equation of a system of phase field equations modelling phase transformations with interfaces between phases having a finite non-zero thickness, [3, 7, 15, 16, 17, 18, 19, 26, 24, 25, 27, 52, 64, 88, 90, 108, 120, 129, 130]. These equations are models in their own right but can be thought of as approximations to sharp interface models which is the point of view taken here in §5. An isotropic phase field system is

$$\alpha u_t = \gamma\Delta u - W'(u) + \delta\theta$$

$$c\theta_t + \frac{l}{2}u_t = \kappa\Delta\theta.$$

By taking α, c and δ to be zero or non-zero one has the Allen-Cahn, Viscous Cahn-Hilliard or the Cahn-Hilliard equation, [7]. By scaling various constants with a small parameter ε one can obtain in the limit $\varepsilon \to 0$ a variety of sharp interface models including motion by mean curvature and the Stefan problem with the Gibbs-Thomson and kinetic undercooling law, [24, 25, 64, 90, 120].

7 Concluding remarks

Consider a closed simple curve or a curve attached to boundaries in the plane evolving according to (2.8). The method described in section 3 will be the most efficient (actually very fast) for computing the evolution. The level set and phase field method can be made very efficient and close to front tracking schemes by only computing in narrow bands around the curve but they still involve computing in a two dimensional domain. However, consider the case of several growing curves which will eventually intersect. The methods of section 3 can not handle this automatically. An efficient method of determining the points of self-intersection needs to be found and a definition of evolution beyond the singularities is required. The level set and phase field methods handle this situation automatically. Furthermore, their definition of solution agrees provided fattening does not occur. In many cases the phase field equations give a 'physical' meaning to the evolution of interfaces beyond singularities.

There are many further challenges including the analysis of discretisation and the application of these methods to complicated geometries, inhomogeneous media and coupling with other field equations. It is likely that the method of choice will depend both on taste and the application.

References

1. Adalsteinsson, D., Sethian, J.A. (1995) A fast level set method for propagating interfaces. J. Comp. Physics. **118**, 269-277.

2. Adalsteinsson, D., Sethian, J.A. (1995) A unified level set approach to etching, deposition and lithography I: Algorithms and two dimensional simulations. J. Comp. Physics. **120**, 128-144.

3. Allen, S., Cahn, J. (1979) A microscopic theory for antiphase boundary motion and its application to antiphase domain coarsing. Acta. Metall. **27**,1084-1095.

4. Almgren, F., Taylor, J.E., Wang, L. (1993) Curvature-driven flows: a variational approach. SIAM J. Control Optim. **31**, 387-437.

5. Almgren, R. (1993) Variational algorithms and pattern formation in dendritic solidification. J. Comput. Phys. **106**, 337-354.

6. Angenent, S., Gurtin, M. (1989) Multiphase Thermomechanics with Interfacial Structure 2. Evolution of an Isothermal Interface. Archive for Rational Mechanics and Analysis **108**, 323-391.

7. Bai, F., Elliott, C.M., Gardiner, A.R., Spence, A., Stuart, A. (1995) The viscous Cahn-Hilliard equation Part I: Computations. Nonlinearity **8**, 131-160.

8. Barles, G., Soner, H.M., Souganidis, P.E. (1993) Front propagation and phase field theory. SIAM J Control Optim. **31**, 439-469.

9. Bellettini, G., Paolini, M. (1994) Two examples of fattening for the curvature flow with a driving force. Atti Accad. Naz. Lincei Cl. Sci. Fis. Mat. Natar. Ren. (9) Mat.Appl. **5**, 229-236.

10. Bellettini, G., Paolini, M. (1995) Quasi-optimal error estimates for the mean-curvature flow with forcing term. Differential Integral Equations **8**, 735-752.

11. Bellettini, G., Paolini, M. (to appear) Anisotropic motion by mean curvature in the context of Finsler geometry. Hokkaido Math. J.

12. Bellettini, G., Paolini, M., Verdi, C. (1990) Γ-convergence of discrete approximations to interfaces with prescribed mean curvature. Rend. Atti. Naz. Lincei. **1**, 317-328.

13. Bellettini, G., Paolini, M., Verdi, C. (1991) Numerical minimization of geometrical type problems related to calculus of variations. Calcolo **27**, 251-278.

14. Blowey, J.F. (1990) The Cahn-Hilliard gradient theory for phase separation with non-smooth free energy. D.Phil thesis, University of Sussex.

15. Blowey, J.F., Elliott, C.M. (1991) The Cahn-Hilliard gradient theory for phase separation with non-smooth free energy Part 1: Mathematical Analysis. European J. Applied Mathematics **2**, 233-280.

16. Blowey, J.F., Elliott, C.M. (1992) The Cahn-Hilliard gradient theory for phase separation with non-smooth free energy Part II: Numerical Analysis. European J. Applied Mathematics **3**, 147-179.

17. Blowey, J.F., Elliott, C.M. (1993) Curvature dependent phase boundary motion and parabolic obstacle problems. Proc. IMA Workshop on Degenerate Diffusion (1991) (ed. W. Ni, L. Peletier, J.L. Vasquez) Springer Verlag IMA **47**, 19-60.

18. Blowey, J.F., Elliott, C.M. (1994) A phase field model with a double obstacle potential. 'Motion by mean curvature and related topics', ed. G. Buttazzo and A. Visintin, de Gruyter, New York, 1-22.

19. Blowey, J.F., Copetti, M., Elliott C.M. (1996) Numerical analysis of multi-component phase separation. I.M.A. Journal of Numerical Analysis **16**, 111-139.

20. Brakke, K.A. (1978) The motion of a surface by its mean curvature. Princeton University Press.

21. Brazhnik, P.K. (1996) Exact solutions for the kinematic model of autowaves in two dimensional excitable media. Physica D **94**, 205-220.

22. Bronsard, L., Kohn, R.V. (1991) Motion by mean curvature as the singular limit of a Ginzburg-Landau model. J. Diff. Eqns. **90**, 211-237.

23. Bronsard, L. Wetton, B. (1995) A numerical method for tracking curve networks moving with curvature motion. J. Comp. Phys. **120**, 66-87.

24. Caginalp, G., (1986) An analysis of a phase-field model of a free boundary. Archive for Rational Mechanics and Analysis **92**, 205-245.

25. Caginalp, G. (1989) Stefan and Hele-Shaw type models as asymptotic limits of the phase-field equations. Physical Review A **39**, No. 11, 5887-5896.

26. Caginalp, G. Socolovsky, E. (1994) Phase field computations of single-needle crystals, crystal growth, and motion by mean curvature. SIAM J. Sci. Comput. **15**, 106-126.

27. Cahn, J.W., Elliott, C.M., Novick-Cohen A. (1996) The Cahn-Hilliard equation with a concentration dependent mobility: motion by minus the Laplacian of the mean curvature. Euro. J. Appl. Math. **7**, 287-301.

28. Cahn, J.W., Handwerker, C.A., Taylor, J.E. (1992) Geometric models of crystal growth. Acta. Metall. **40**, 1443-1474.

29. Cahn, J.W., Taylor, J.E. (1994) Surface motion by surface diffusion. Acta metall. mater. **42**, No. 4, 1045-1063.

30. Carter, W.C., Roosen, A.R., Cahn, J.W., Taylor, J.E. (1995) Shape evolution by surface diffusion and surface attachment limited kinetics on completely faceted surfaces. Acta Metal. Mater. **43**, 4309-4323.

31. Chapman, S.J., Richardson, G. (1995). Motion of vortices in type-II superconductors. SIAM J. Appl. Math. **55**, 1275-1296.

32. Chen, X. (1992) Generation and propagation of interface in reaction-diffusion equations. J.Diff.Eqns. **96**, 116-141.

33. Chen, X., Elliott, C.M. (1994) Asymptotics for a parabolic double obstacle problem. Proc. Roy. Soc. London Ser. A. **444**, 429-445.

34. Chen, X., Elliott, C.M., Gardiner, A.R., Zhao, J.J. (1996) Convergence of numerical solutions to the Allen-Cahn equation. University of Sussex CMAIA Research report 96-05.

35. Chen, Y., Giga Y., Goto, S. (1991) Uniqueness and Existence of Viscosity Solutions of generalized Mean Curvature Flow Equations. Journal of Differential Geometry **33**, 749-786.

36. Chopp, D.L. (1993) Computing minimal surfaces via level set curvature flow. J. Comp. Physics **106**, 77-91.

37. Chopp, D.L. (1994) Numerical computation of self-similar solutions for mean curvature flow. J. Exper. Math. **3**, 1-15.

38. Coleman, B.D., Falk, R.S., Moakher, M (1994) Space-Time Finite Element Methods for Surface Diffusion with Applications to the Theory of the Stability of Cylinders, submitted to SIAM J. Scientific Computing.

39. Crandall, M.G., Ishi, H., Lions, P.-L. (1992) User's Guide to Viscosity Solutions of second Order Partial Differential Equations. Bulletin of the American Mathematical Society **27**, 1-67.

40. Davi, F., Gurtin, M.E. (1990) On the motion of a phase interface by surface diffusion. J. Appl. Math. Phys. **41**, 782-811.

41. Deckelnick, K., Dziuk, G. (1994) On the approximation of the curve shortening flow. Bandle, C., Bemelmans, J., Chipot, M., Saint Jean Paulin, J., Shafrir, I., (eds). Calculus of Variations, Applications and Computations: Pout-a-Mousson. Pitman Research Notes in Mathematics Series, 100-108.

42. Deckelnick, K., Dziuk, G. (1995) Convergence of a finite element method for non-parametric mean curvature flow. Num. Math. 72, 197-222.

43. Deckelnick, K., Elliott, C.M. (1996) Local and global existence results for anisotropic Hele-Shaw flows. University of Sussex CMAIA Research report 96-06.

44. Deckelnick, K., Elliott, C.M. (1996) Finite element error bounds for curve shrinking with prescribed normal contact to a fixed boundary. University of Sussex CMAIA Research report 96-20.

45. Deckelnick, K., Elliott, C.M., Richardson, G (1997) Long time asymptotics for forced curvature flow with application to the motion of a superconducting vortex. University of Sussex CMAIA Research report 96-18. Nonlinearity (to appear).

46. de Mottoni, P., Schatzmann, M. (1989) Evolution géométrique d'interfaces. CRAS **309**, 453-458.

47. Dziuk, G. (1991) An algorithm for evolutionary surfaces. Numer. Math. **58**, 603-611.

48. Dziuk, G. (1994) Convergence of a semi-discrete scheme for the curve shortening flow. Math. Models Meth. Appl Sci. **4**, 589-606.

49. Dziuk, G. (in preparation).

50. Elliott, C.M., Gardiner, A.R. (1994) One dimensional phase field computations. Numerical Analysis 1993, Proceedings of Dundee Conference, ed. D.F. Griffiths and G.A. Watson, Longman Scientific and Technical, 56-74.

51. Elliott, C.M. Gardiner, A.R. (1994) Numerical analysis of the phase field equations and phase boundary motion. Computational Techniques and Applications: CTAC93 ed. D. Stewart, H. Gardener and D. Singleton, World Scientific, 12-25.

52. Elliott, C.M., Gardiner, A.R. (1996) Double obstacle phase field computations of dendritic growth. University of Sussex CMAIA Research report 96-19.

53. Elliott, C.M. Gardiner, A.R., Kuhn, T. (1996) Generalised double obstacle phase field approximation of the anisotropic mean curvature flow. University of Sussex CMAIA Research report 96-17.

54. Elliott, C.M., Gardiner, A.R., Schatzle, R. (1998) Crystalline curvature flow in a variational setting. Advances in Mathematical Sciences and Applications **8** (to appear).

55. Elliott, C.M., Garke, H. (1997) Existence results for diffusive surface motion laws. Adv. Math. Sci. Appl. **7**, 465–488.

56. Elliott, C.M., Ockendon, J.R. (1982) Weak and variational methods for moving boundary problems. Pitman, London.

57. Elliott, C.M., Paolini, M., Schatzle, R. (1996) Interface estimates for the fully anisotropic Allen-Cahn equation and anisotropic mean curvature flow. Math. Models Methods Appl. Sci., **6**, 1103–1118.

58. Elliott, C.M., Schätzle, R. (1996) The limit of the anisotropic double-obstacle Allen-Cahn equation. Proceedings of the Royal Society of Edinburgh **126**, 1217–1234.

59. Elliott, C.M., Schätzle, R. (1997) The limit of the fully anisotropic double-obstacle Allen-Cahn equation in the non-smooth case. SIAM Journal on Mathematical Analysis, (to appear).

60. Elliott, C.M., Stuart, A. (1993) The global dynamics of discrete semilinear parabolic equations. SIAM J. Numer. Anal. **30**, 1622-1663.

61. Evans, L.C., Soner, H.M., Souganidis, P.E. (1992) Phase transitions and generalized motion by mean curvature, Communications on Pure and Applied Mathematics **XLV**, 1097-1123.

62. Evans, L.C., Spruck, J. (1991) Motion of level sets by mean curvature, I, J. Differential Geom. **33**, 635-681.

63. Fife, P.C. (1988) Dynamics of internal layers and diffusive interfaces, SIAM **53**, CB MS-NSF Regional Conf. Ser. Appl. Math.

64. Fife, P.C., Penrose, O. (1995) Interfacial dynamics for thermodynamically consistent phase field models with non-conserved order parameter. Electronic J. Diff. Eqns. **1995**, No. 16, 1-49.

65. Fried, J.M. (1993) Berechnung des KrümmungsFlusses von Niveauflächen, Diplomarbeit, Universität Freiburg.

66. Fukui, T., Giga, Y. (1995) Motion of a graph by nonsmooth weighted curvature. Proceedings of the First World Congress of Nonlinear Analysts (ed V. Lakshmikantham) deGruyter, Berlin, 47-56.

67. Gage, M. (1984) Curvature shortening makes convex curves circular. Inventiones Mathematica **76**, 357-364.

68. Gage, M. (1986) On an area preserving evolution equation for plane curves. Contemporary Math. **51**, 51-62.

69. Gage, M. (1993). Evolving plane curves by curvature in relative geometries. Duke Math. J. **72**, 441-466.

70. Gage, M. and Hamilton, R.S. (1986). The heat equation shrinking convex plane curves. J. Differential Geom. **23**, 69-96.

71. Giga, Y., Gurtin, M.E., Matais, J. (1996) On the dynamics of crystalline motion. Hokkaido Unversity preprint series no. 336.

72. Gilbarg, D., Trudinger, N.S. (1977) Elliptic partial differential equations of second order, Springer Verlag.

73. Girão, P.M., Kohn, R.V. (1994) Convergence of a crystalline algorithm for the heat equation in one dimension and for the motion of a graph by weighted curvature, Numerische Mathematik, **67**, pp. 41-70.

74. Girão, P.M., Kohn, R.V. (to appear) The crystalline algorithum for computing motion by curvature in Variational Methods for Discontinuous Structures ed R.Serapioni and F.Tomarelli, Birkhäuser.

75. Girão, P.M. (1995) Convergence of a crystalline algorithm for the motion of a simple closed convex curve by weighted curvature. SIAM J. Numer. Anal. **32**, 886-889.

76. Grayson, M.A. (1987) The heat equation shrinks embedded plane curves to round points. J. Differential Geom. **26**, 285-314.

77. Grayson, M. (1989) A short note on the evolution of surfaces via mean curvature. Duke Math. Journal **58**, 555-558.

78. Grindrod, P. (1991) Patterns and waves: the theory and applications of reaction diffusion equations, Clarendon Press, Oxford.

79. Grindrod, P. (1995) Geometry and motion of sharp fronts within geochemical transport problems. Proc. Roy. Soc. London Ser(A) **449**, 123-138.

80. Guichard, F., Morel, J. M. (1996) Partial differential equations and image iterative filtering. State of the Art in Numerical Analysis 1996, IMA Conference Proceedings.

81. Gurtin, M.E. (1993) Thermomechanics of evolving phase boundaries in the plane. Oxford Mathematical Monographs.

82. Gurtin, M., Soner, H. M., Souganidis, P. E. (1995) Anisotropic motion of an interface relaxed by the formation of infinitesimal wrinkles. J. Diff. Eq. **119**, 54-108.

83. Hou, T.Y. (1995) Numerical solutions to free boundary problems. Acta Numerica, 335-415.

84. Huisken, G. (1984) Flow by mean curvature of convex surfaces into spheres. Journal of Differential Geometry **20**, 237-266.

85. Huisken, G. (1989) Non-parametric mean curvature evolution with boundary conditions. J. Diff. Eqns. **77**, 369-378.

86. Ilmanen, T. (1993) Convergence of the Allen-Cahn equation to Brakke's motion by mean curvature. J. Diff. Geom. **38**, 417-461.

87. Juden, J. (1996) CMAIA (in preparation).

88. Kobayashi, R. (1993) Modelling and numerical simulations of dendritic crystal growth. Physica D **63**, 410-423.

89. Kuhn, T. (1996) DPhil thesis, University of Sussex.

90. McFadden, G.B., Wheeler, A.A., Braun, R.J., Coriell, S.R., Sekerka, R.F. (1993) Phase-field models for anisotropic interfaces. Physical Review E **48**, No. 3, 2016-2024.

91. Merriman, B., Bence, J., Osher, S. (1994) Motion of multiple junctions: A level set approach. J. Comp. Phys. **112**, 334-363.

92. Mullins, W. W. (1956) Two dimensional motion of idealised grain boundaries. J. Appl. Phys. **27**, 900-904.

93. Mullins, W. W. (1957) Theory of thermal grooving. J. Appl. Phys. **28**, 333-339.

94. Nochetto, R.H., Paolini, M., Rovida, S., Verdi, C. (1994) Variational aproximation of the geometric motion of fronts. Motion by Mean Curvature and Related Topics (G. Buttazzo and A. Visintin, eds), de Gruyter, Berlin, 124-149.

95. Nochetto, R.H., Paolini, M., Verdi, C. Numerical analysis of geometric motion of fronts; manuscript.

96. Nochetto, R.H., Paolini, M., Verdi, C. (1993) Sharp error analysis for curvature dependent evolving fronts. Math. Models Methods Appl. Sci. **3**, 711-723.

97. Nochetto, R.H., Paolini, M., Verdi, C. (1995) Double obstacle formulation with variable relaxation parameter for the smooth geometric front evolutions: asymptotic interface error estimate. Asymptotic Anal. **10**, 173-198.

98. Nochetto, R.H., Paolini, M., Verdi, C. (1994) Quadratic rate of convergence for curvature dependent smooth interfaces: a simple proof. Appl. Math. Letters **7**, 59-63.

99. Nochetto, R.H., Paolini, M., Verdi, C. (1994) Optimal interface error estimates for the mean curvature flow. Ann. Scuola Norm. Sup. Pisa Cl. **21**, 4, 193-212.

100. Nochetto, R.H., Paolini, M. Verdi, C. (1996) A dynamic mesh method for curvature dependent evolving interfaces. J. Comput. Phys. **123**, 296-310.

101. Nochetto, R.H., Verdi, C. (1996) Convergence of double obstacle problems to the geometric motions of fronts. SIAM J. Math. Anal. **26**, 1514-1526.

102. Nochetto, R.H., Verdi, C. (to appear) Convergence past singularities for a fully discrete approximation of curvature driven interfaces. SIAM J. Numer.Anal.

103. Nochetto, R.H., Verdi, C. (1996) Combined effect of the explicit time-stepping and quadrature for curvature driven flows. Numer. Math. **74**, 105-136.

104. Osher, S., Sethian, J.A. (1988) Fronts propagating with curvature dependent speed: algorithms based on Hamilton-Jacobi formulations. J. Comp. Phys. **79**, 12-49.

105. Paolini, M. (1994) An efficient algorithm for computing anisotropic evolution by mean curvature. Proc. of Motion by Mean Curvature and Appns.

106. Paolini, M. (1996) A quasi-optimal error estimate for a discrete singularly perturbed approximation to the prescribed curvature problem. Math Comp. (to appear).

107. Paolini, M., Verdi, C. (1992) Asymptotic and numerical analysis of the mean curvature flow with a space dependent relaxation parameter. Asympt. Anal. **5**, 553-574.

108. Penrose, O., Fife, P.C. (1990) Thermodynamically consistent models of phase-field type for kinetics of phase transitions. Physica D **43**, 44-60.

109. Roosen, A, Taylor, J. (1994) Modelling crystal growth in a diffusion field using fully faceted interfaces. J. Comp. Phys. **114**, 113-128.

110. Rubinstein, J., Sternberg, P., Keller, J.B. (1989) Fast reaction, slow diffusion and curve shorting. SIAM J. Appl. Math. **49**, 116-133.

111. Rubinstein, J., Sternberg, P. (1992) Nonlocal reaction-diffusion equations and nucleation. IMA J. Appl. Math. **48**, 249-268.

112. Saffman, P.G. (1986) Viscous fingering in Hele-Shaw cells. J. Fluid Mechanics **173**, 73-94.

113. Schmidt, A. (1996) Computation of three dimensional dendrites with finite elements. J. Comp. Phys. **125**, 293-312.

114. Sekerka, R.F. (1993) Role of instabilities in determination of the shapes of growing crystals. Journal of Crystal Growth **128**, 1-12.

115. Sethian, J.A. (1990) Numerical algorithms for propagating interfaces: Hamilton-Jacobi equations and conservation laws. J. Differential Geom. **31**, 131-161.

116. Sethian, J.A. (1985) Curvature and the evolution of fronts. Comm. Math. Phys. **101**, 487-499.

117. Sethian, J.A. (1996) Theory, algorithm and applications of level set methods for propagating interfaces. Acta Numerica, 309-395.

118. Sethian, J.A. (1996) Level Set Methods: Evolving interfaces in Geometry, Fluid Mechanics, Computer Vision and Material Science. Cambridge Monographs on Applied and Computational Mathematics, Cambridge University Press.

119. Sethian, J.A., Strain, J. (1992) Crystal growth and dendrite solidification. J. Comp. Phys. **98**, 231.

120. Soner, H.M. (1995) Convergence of the Phase Field Equations to the Mullins-Sekerka Problem With Kinetic Undercooling. Arc. Rat. Mech. Anal.,**131**, 139-197.

121. Strain, J. (1989) A boundary integral approach to unstable solidification. J. Comp. Phys. **85**, 342.

122. Taylor, J.E. (1992) Mean curvature and weighted mean curvature. Acta metall. mater. **40**, 1475-1485.

123. Taylor, J.E. (1993) Motion of curves by crystalline curvature, including triple junctions and boundary points. Proc. of Symposia in Pure Math. **54**, 417-438.

124. Taylor, J.E., Cahn, J.W. (1993) Linking Anisotropic Sharp and Diffuse Surface Motion Laws via Gradient Flows. Journal of Statistical Physics **77**, 183-197.

125. Taylor, J.E., Cahn, J.W., Handwerker, C.A. (1992) Geometric Models of Crystal Growth. Acta metall. mater. **40**, 1443-1474.

126. Tyson, J.J., Keener, J.P. (1988) Singular perturbation theory of travelling waves in excitable media (a review). Physica (D) **32**, 327-361.

127. Swinney, H.L., Krinsky, V.I. (1991) Waves and patterns in Chemical and Biological media. Physica (D) **49** Nos 1 and 2. (Collection of articles), 1-256.

128. Visintin, A. (1988) Surface tension effects in phase transition. Material Instabilities in Continuums Mechanics (J.Ball ed.). Clarendon Press, Oxford, 505-537.

129. Wheeler, A.A., McFadden, G.B. (1994) A ξ-Vector Formulation of Anisotropic Phase-Field Models: 3-D Asymptotics. Euro.J.Appl.Math. (1996) **7**, 367-381.

130. Wheeler, A.A., Murray, B.T., Schaefer, R.J. (1993) Computations of dendrites using a phase field model. Physica D **66**, 243-262.

131. Yokoyama, E., Sekerka, R.F. (1992) A numerical study of the combined effect of anisotropic surface tension and interface kinetics on pattern formation during the growth of two-dimensional crystals. Journal of Crystal Growth **125**, 389-403.

132. Zhang, W. (1996) Using MOL to solve a high order nonlinear PDE boundary in the simulation of a sintering process. Applied Numerical Mathematics **20**, 235-244.

Finite Element Methods for Hyperbolic Problems: *a Posteriori* Error Analysis and Adaptivity

Endre Süli and Paul Houston

Computing Laboratory, University of Oxford

1 Introduction

Partial differential equations of hyperbolic type are of fundamental importance in many areas of applied mathematics, particularly fluid dynamics and electromagnetics. Solutions to these equations exhibit localised phenomena, such as propagating discontinuities and sharp transition layers, and their reliable numerical approximation presents a challenging computational task; indeed, in order to resolve such localised features in an accurate and efficient manner it is essential to use adaptively refined computational meshes whose construction is governed by sharp *a posteriori* error bounds. Over the last decade the *a posteriori* error analysis of finite element methods for partial differential equations of hyperbolic and nearly-hyperbolic character has been a subject of intensive research. For an overview of current activity in this area we refer to the article of Johnson [11]; see also [10–13], and references therein. The approach proposed in those papers, and reviewed in the next section, rests on performing an elliptic regularisation of the hyperbolic problem and exploiting the smoothing properties of the resulting adjoint problem in conjunction with Galerkin orthogonality.

Subsequent sections of the paper are devoted to discussing an alternative, more direct, approach to the *a posteriori* error analysis of finite element approximations of hyperbolic problems which avoids the need for regularisation. Because of the inherent lack of strong smoothing properties in hyperbolic problems, particular care has then to be taken about existence of traces and integration-by-parts formulae which the *a posteriori* error analysis relies on; the necessary functional-analytic prerequisites that concern these issues are reviewed in the third section. By exploiting this theory, we derive *a posteriori* bounds on the global error in the H^{-1} norm for a general class of finite element methods (including certain finite volume schemes, interpreted as Petrov-Galerkin methods) for symmetric hyperbolic systems. For related work, see [15–17,21,25].

Controlling the computational error in a given *norm*, however, is not always the ultimate goal in practice. Indeed, in many engineering problems the accurate approximation of particular linear functionals of the solution (such as the lift, the drag, and the flux across the boundary of the domain) is at least as important as the reliable approximation of the solution in a given norm over the entire computational domain. We show, by considering the example of estimating the normal flux through the boundary, how problems of this kind may be approximated in an efficient way by exploiting a hyperbolic duality argument in conjunction with theoretical results concerning the propagation of singularities.

In the final part of the paper we discuss unsteady hyperbolic problems. In particular, we consider the *a posteriori* error analysis of a class of evolution-Galerkin methods for multi-dimensional scalar hyperbolic equations on a spatial domain Ω and finite time interval $[0,T]$. Here, we derive an *a posteriori* bound on the global error in the norm of the space $L_\infty(0,T;H^{-1}(\Omega))$. We conclude by indicating the practical relevance of this work and by illustrating the implementation of the *a posteriori* analysis into an adaptive evolution-Galerkin algorithm.

2 Johnson's paradigm

In this brief section we review, in the context of first-order hyperbolic systems, the general theoretical framework of *a posteriori* error estimation pursued by Johnson and his co-workers; for a comprehensive account, see [11,13].

Suppose that $\mathcal{M}$ is a Hilbert space with inner product $(\cdot,\cdot)$ and let $A : \mathcal{M} \to \mathcal{M}$ be a linear operator on $\mathcal{M}$ with domain $D(A) \subset \mathcal{M}$ and range $R(A) = \mathcal{M}$ (in our case a system of first-order hyperbolic differential operators). Given that $f \in \mathcal{M}$, consider the problem of finding $u \in D(A)$ such that

$$Au = f.$$

In order to construct a Galerkin approximation to this problem, we consider a sequence of finite-dimensional spaces $\{\mathcal{U}^h\}$, parameterised by the positive discretisation parameter h; for the sake of simplicity we shall suppose that $\mathcal{U}^h \subset D(A)$ for each h. Simultaneously, consider a sequence of finite-dimensional spaces $\{\mathcal{M}^h\}$, with $\mathcal{M}^h$ contained in $\mathcal{M}$ for each h. For the purposes of this paper, $\mathcal{U}^h$ and $\mathcal{M}^h$ can be thought of as standard finite element spaces consisting of piecewise polynomial functions on a partition, of granularity h, of the computational domain. Let Π_h denote the orthogonal projector in $\mathcal{M}$ onto $\mathcal{M}^h$. The Galerkin approximation u_h of u is then sought in $\mathcal{U}^h$ as the solution of the finite-dimensional problem

$$\Pi_h A u_h = \Pi_h f.$$

In order to obtain a computable bound on the *global error* $e_h = u - u_h$ in terms of the *residual* r_h, defined by

$$r_h = f - Au_h,$$

we begin by observing the *Galerkin orthogonality* property

$$(r_h, v_h) = 0 \qquad \forall v_h \in \mathcal{M}^h.$$

This will be a key ingredient in the analysis. In addition, denoting by A^* the adjoint of A, we consider the following auxiliary problem, referred to as the dual problem:

$$A^*\phi = u - u_h.$$

The *a posteriori* error analysis is based on a duality argument, and it proceeds as follows:

$$\begin{aligned}\|u - u_h\|^2 &= (u - u_h, u - u_h) = (u - u_h, A^*\phi) \\ &= (A(u - u_h), \phi) = (Au - Au_h, \phi) \\ &= (f - Au_h, \phi) = (r_h, \phi).\end{aligned}$$

By virtue of the Galerkin orthogonality property $(r_h, \phi_h) = 0$ for any $\phi_h \in \mathcal{M}^h$, and therefore

$$\|u - u_h\|^2 = (r_h, \phi - \phi_h) = (h^s r_h, h^{-s}(\phi - \phi_h)),$$

where s is a non-negative real number, to be chosen below. By virtue of the Cauchy-Schwarz inequality,

$$\|u - u_h\|^2 \leq \|h^s r_h\| \, \|h^{-s}(\phi - \phi_h)\|.$$

While the first term on the right-hand side is of the desired form, involving the (computable) residual r_h multiplied by an appropriate power of the discretisation parameter, the second term incorporates ϕ, the solution to the dual problem. Since the dual-problem has the global error as data, ϕ is unknown and has to be eliminated from the analysis by relating it to $u - u_h$; furthermore a suitable choice of ϕ_h has to be made. This is achieved as follows: suppose that $\{\mathcal{W}_\sigma\}_{\sigma \geq 0}$ is a scale of Hilbert spaces, with associated norms $|||\cdot|||_\sigma$, such that $\mathcal{W}_0 = \mathcal{M}$ and $\mathcal{W}_{\sigma_2}$ is continuously embedded into $\mathcal{W}_{\sigma_1}$ whenever $\sigma_2 \geq \sigma_1$. Furthermore, we assume that there exist $\phi_h \in \mathcal{M}^h$ and a positive constant $C_\ddagger$ such that, for $\phi \in \mathcal{W}_s$,

$$\|h^{-s}(\phi - \phi_h)\| \leq C_\ddagger |||\phi|||_s.$$

For finite element methods, this hypothesis is easily fulfilled by choosing $\mathcal{W}_s$ as a hilbertian Sobolev space of index s and referring to standard approximation properties of piecewise polynomial functions in Sobolev spaces, with ϕ_h taken as the projection, the interpolant or the quasi-interpolant of ϕ from $\mathcal{M}^h$. Thereby we arrive at the bound

$$\|u - u_h\|^2 \leq C_\ddagger \|h^s r_h\| \, |||\phi|||_s.$$

This brings us to the final, and most important, step in the *a posteriori* error analysis. The norm $|||\phi|||_s$ appearing on the right-hand side of the last inequality has to be eliminated in terms of $u - u_h$ by recalling the relationship between ϕ and $u - u_h$, namely that $A^*\phi = u - u_h$. Thus, to proceed, we assume that A^* is invertible and that $(A^*)^{-1}$ is a bounded linear operator from $\mathcal{M}$ to $\mathcal{W}_s$; hence

$$|||\phi|||_s = |||(A^*)^{-1}(u - u_h)|||_s \leq C_\star \|u - u_h\|,$$

where $C_\star$ is a positive constant, greater than or equal to the norm of $(A^*)^{-1}$. Finally, combining the last two bounds we arrive at the desired *a posteriori* bound on the global error $e_h = u - u_h$ in terms of the computable residual r_h:

$$\|u - u_h\| \leq C_\ddagger C_\star \|h^s r_h\|.$$

In order to turn this into an error bound that can be used in practice, the constants $C_\ddagger$ and $C_\star$ have to be determined. Estimating $C_\ddagger$ is a relatively simple matter using readily available results from approximation theory (see, for example, Exercise 3.1.2 in Ciarlet's monograph [5] for an explicit formula for $C_\ddagger$ in the case of standard finite element spaces consisting of continuous piecewise polynomials on simplices, or the work of Handscomb [7] for very much sharper estimates of $C_\ddagger$ for piecewise linear finite elements on triangles). On the other hand, supplying $C_\star$ is much harder, involving an analytical study of the well-posedness of the dual problem. Since any value of $C_\star$ that is arrived at through such general analytical arguments is necessarily a considerable overestimate of the ratio $|||\phi|||_s/\|u - u_h\|$, in practice the stability constant $C_\star$ is determined computationally for the problem at hand, as part of the process of *a posteriori* error estimation.

Finally, we have to fix s, the exponent of h in the error bound. As one would like the *a posteriori* bound to reflect the approximation property of the space $\mathcal{M}^h$ to its full extent, one would wish to choose s as large as possible; unfortunately, for linear first-order hyperbolic systems consisting of m equations on a domain $\Omega \subset \mathbb{R}^d$, $(A^*)^{-1}$ is a bounded operator from $\mathcal{M} = [L_2(\Omega)]^m$ to $\mathcal{M} = \mathcal{W}_0 = [L_2(\Omega)]^m$ only, so s cannot exceed 0; thus we end up with the bound

$$\|u - u_h\| \leq C_\ddagger C_\star \|r_h\|.$$

In fact, we note that when $s = 0$ there is no benefit in exploiting Galerkin orthogonality, and we may simply take $\phi_h = 0$ in our argument to improve this bound to

$$\|u - u_h\| \leq C_\star \|r_h\|.$$

One way or the other, in the case of a first-order hyperbolic system, the *a posteriori* error bound that we arrive at on the basis of the reasoning outlined above is unsatisfactory in that it fails to display the approximation properties of the test space $\mathcal{M}^h$. Worse still, when the data are discontinuous, linear hyperbolic systems may possess solutions that are discontinuous across characteristic hypersurfaces and, under mesh refinement, $\|r_h\|$ will then converge to 0 very slowly, if at all; consequently, in the absence of the compensating factor h^s, any adaptive algorithm driven by this error bound is likely to be inefficient.

The problem may be rectified by perturbing the first-order hyperbolic operator A^* (or, indeed, both A and A^*) through the addition of a second-order elliptic term with small coefficient; this then provides additional regularity which, in favourable circumstances, allow one to take $s = 2$. This approach, however, is associated with undesirable complications related to the fact that artificial boundary conditions have to be supplied for the resulting second-order operator in such a way that the features of the solution to the hyperbolic system are retained in the vicinity of the boundary; a further difficulty with elliptic regularisation of non-dissipative hyperbolic systems, such as the Maxwell system of electro-magnetism, is that it may introduce a physically unacceptable level of damping into the model.

Thus, in the rest of the paper, we consider an alternative approach.

3 *A posteriori* error analysis for steady hyperbolic PDEs

In this section we develop the *a posteriori* error analysis of a general class of finite element methods for steady linear multi-dimensional hyperbolic systems. We begin with a review of basic results from the theory of symmetric hyperbolic systems in the sense of Friedrichs [6].

3.1 Symmetric hyperbolic systems

Throughout this section Ω will denote a bounded open set in $\mathbb{R}^d$ with Lipschitz continuous boundary $\partial\Omega$. Suppose that A_i, $i = 1, \dots, d$, and C are (matrix-valued) mappings from $\overline{\Omega}$ into $\mathbb{R}^{m\times m}$, $m \geq 1$; we shall assume that, for each i, the entries of A_i are continuously differentiable on $\overline{\Omega}$ and the components of C are continuous on $\overline{\Omega}$. We consider the hyperbolic system of linear first-order partial differential equations

$$Lu \equiv \sum_{i=1}^{d} \frac{\partial}{\partial x_i}(A_i \mathbf{u}) + C\mathbf{u} = \mathbf{f} \text{ in } \Omega. \tag{3.1}$$

The system (3.1) is said to be *symmetric positive* if the following conditions hold:

(a) the matrices A_i, for $i = 1, \dots, d$, are symmetric, i.e. $A_i = A_i^*$;

(b) there exists $\alpha \geq 0$ and a unit vector $\xi \in \mathbb{R}^d$, such that the symmetric part of the matrix

$$K_\xi = C + \frac{1}{2}\sum_{i=1}^{d} \frac{\partial A_i}{\partial x_i} + \alpha \sum_{i=1}^{d} \xi_i A_i$$

is positive definite, uniformly on $\overline{\Omega}$, i.e. there exists a positive constant $c_0 = c_0(\Omega)$ such that

$$\frac{1}{2}(K_\xi(x) + K_\xi^*(x)) \geq c_0 I \tag{3.2}$$

for all x in $\overline{\Omega}$.

Since $\partial\Omega$ is Lipschitz continuous, the outer unit normal vector field $\hat{\nu} = (\hat{\nu}_1, ..., \hat{\nu}_d)$ is defined almost everywhere on $\partial\Omega$ with respect to the $(d-1)$-dimensional measure on $\partial\Omega$. Consider the symmetric matrix

$$B = \hat{\nu}_1 A_1 + ... + \hat{\nu}_d A_d.$$

We shall suppose that B is non-singular almost everywhere on $\partial\Omega$, that is, the boundary of Ω is non-characteristic for L. Since B is symmetric and of full rank it can be decomposed as $B = B^+ + B^-$, where B^+ is positive semi-definite and B^- is negative semi-definite. Given that $\mathbf{g}$ is a sufficiently smooth function defined on $\partial\Omega$, an admissible boundary condition for equation (3.1) is given by

$$B^-\mathbf{u}\big|_{\partial\Omega} = B^-\mathbf{g} \text{ on } \partial\Omega; \tag{3.3}$$

we shall also consider the homogeneous counterpart of this boundary condition:

$$B^-\mathbf{u}\big|_{\partial\Omega} = 0 \text{ on } \partial\Omega. \tag{3.4}$$

At this stage these boundary conditions are to be understood formally; below we shall state a trace theorem which assigns a precise meaning to $B^-\mathbf{u}|_{\partial\Omega}$. First, however, we define the *graph space* of the operator L as the linear space

$$H(L,\Omega) = \{\mathbf{v} \in [L_2(\Omega)]^m \; : \; L\mathbf{u} \in [L_2(\Omega)]^m\}.$$

When equipped with the norm

$$|||\mathbf{v}|||_{\Omega,\xi} = \left(\|e^{-\alpha(\xi\cdot x)}\mathbf{v}\|^2 + \|e^{-\alpha(\xi\cdot x)}L\mathbf{v}\|^2\right)^{\frac{1}{2}}$$

the graph space $H(L,\Omega)$ is a Hilbert space. Here and throughout the paper $\|\cdot\|$ will stand for the L_2 norm over Ω induced by the $L_2(\Omega)$ inner product $(\cdot,\cdot)$. Consider L^*, the formal adjoint of L, defined by

$$L^*\mathbf{v} = -\sum_{i=1}^{d} A_i \frac{\partial \mathbf{v}}{\partial x_i} + C^*\mathbf{v};$$

the associated graph space $H(L^*,\Omega)$ and graph norm $|||\cdot|||_{*,\Omega,\xi}$ are defined analogously as for L, but with the weight-function $e^{-\alpha(\xi\cdot x)}$ replaced by $e^{\alpha(\xi\cdot x)}$.

Let $\Gamma_{0,\partial\Omega} : [H^1(\Omega)]^m \to [H^{1/2}(\partial\Omega)]^m$ signify the usual trace operator (which to each element of $[H^1(\Omega)]^m$ assigns its restriction to $\partial\Omega$). We denote by $[H^{-1/2}(\partial\Omega)]^m$ the dual space of $[H^{1/2}(\partial\Omega)]^m$. The following trace theorem will play a key rôle in our error analysis (see [17]).

Theorem 1. *Assuming that $\partial\Omega$ is non-characteristic for the operator L, the mapping $\Gamma_{B,\partial\Omega} : \mathbf{v} \mapsto B(\Gamma_{0,\partial\Omega}\mathbf{v})$ defined on $[H^1(\Omega)]^m$ can be extended by continuity to a linear and continuous mapping, still denoted $\Gamma_{B,\partial\Omega}$, from $H(L,\Omega)$ into $[H^{-1/2}(\partial\Omega)]^m$. Moreover, for any $\mathbf{u} \in H(L,\Omega)$ and $\mathbf{v} \in [H^1(\Omega)]^m$ the following Green's formula holds*

$$(L\mathbf{u},\mathbf{v}) - (\mathbf{u}, L^*\mathbf{v}) = \langle \Gamma_{B,\partial\Omega}\mathbf{u}, \Gamma_{0,\partial\Omega}\mathbf{v}\rangle.$$

An analogous result is valid for $H(L^*,\Omega)$.

Here, $\langle\cdot,\cdot\rangle$ denotes the duality pairing between the function spaces $[H^{-1/2}(\partial\Omega)]^m$ and $[H^{1/2}(\partial\Omega)]^m$.

The splitting $B = B^+ + B^-$ induces a natural decomposition of the trace operator $\Gamma_{B,\partial\Omega}$ which leads to the definition of the partial trace operators

$$\Gamma_{B^\pm,\partial\Omega} : H(L,\Omega) \to [H^{-1/2}(\partial\Omega)]^m$$

with

$$\Gamma_{B,\partial\Omega} = \Gamma_{B^+,\partial\Omega} + \Gamma_{B^-,\partial\Omega}.$$

Indeed, letting

$$\Gamma_{B^\pm,\Omega}\mathbf{u} = B^\pm(\Gamma_{0,\partial\Omega}\mathbf{u}) \qquad \forall \mathbf{u} \in [H^1(\Omega)]^m,$$

on the dense subspace $[H^1(\Omega)]^m$ of $H(L,\Omega)$ this decomposition can be constructed by splitting the trace of $\mathbf{u} \in [H^1(\Omega)]^m$; the linear operators $\Gamma_{B^\pm,\partial\Omega} : H(L,\Omega) \to [H^{-1/2}(\partial\Omega)]^m$ are then defined by continuous extension from the dense subspace $[H^1(\Omega)]^m$ to all of $H(L,\Omega)$. One can proceed in the same way for $H(L^*,\Omega)$. Thus, in the case of the homogeneous boundary value problem (3.1), (3.4), we can define precisely the domains of the operators L and L^*:

$$D(L,\Omega) = \{\mathbf{u} \in H(L,\Omega) \quad : \quad \Gamma_{B^-,\partial\Omega}\mathbf{u} = \mathbf{0} \text{ on } \partial\Omega\},$$

$$D(L^*,\Omega) = \{\mathbf{u} \in H(L^*,\Omega) \quad : \quad \Gamma_{B^+,\partial\Omega}\mathbf{u} = \mathbf{0} \text{ on } \partial\Omega\};$$

when equipped with the associated graph-norms $|||\cdot|||_{\xi,\Omega}$, $|||\cdot|||_{*,\xi,\Omega}$, $D(L,\Omega)$ and $D(L^*,\Omega)$ are Hilbert subspaces of $H(L,\Omega)$ and $H(L^*,\Omega)$, respectively.

Given that $\mathbf{f} \in [L_2(\Omega)]^m$, a function $\mathbf{u} \in [L_2(\Omega)]^m$ satisfying

$$(\mathbf{u}, L^*\phi) = (\mathbf{f},\phi) \qquad \forall\phi \in D(L^*,\Omega) \cap [H^1(\Omega)]^m$$

is called a *weak solution* of the homogeneous boundary value problem (3.1), (3.4). A weak solution belonging to $H(L,\Omega)$ is called a *strong solution.* We note that the requirement that $\mathbf{u}$ be a strong solution does not preclude the possibility of $\mathbf{u}$ being discontinuous; indeed, since only $L\mathbf{u} \in [L_2(\Omega)]^m$, discontinuities in $\mathbf{u}$ may occur across characteristic hypersurfaces.

Theorem 2. *Assuming that $\partial\Omega$ is a non-characteristic hypersurface for L, and given that $\mathbf{f} \in [L_2(\Omega)]^m$, the homogeneous boundary value problem (3.1), (3.4) has a unique strong solution $\mathbf{u} \in D(L,\Omega)$. In addition the linear operator L is a continuous bijection from $D(L,\Omega)$ onto $[L_2(\Omega)]^m$ with a continuous inverse $L^{-1} : [L_2(\Omega)]^m \to D(L,\Omega)$. An analogous result holds for L^*.*

Proof. The proof is based on Banach's Closed Range Theorem. Given that $\mathbf{v} \in [H^1(\Omega)]^m$, upon taking the inner product of $L\mathbf{v}$ with $\mathrm{e}^{-2\alpha(\xi\cdot x)}\mathbf{v}$, integrating by parts using Theorem 1 and splitting the trace operator $\Gamma_{B,\partial\Omega}$, we have that

$$\langle e^{-\alpha(\xi\cdot x)}\Gamma_{B+,\partial\Omega}\mathbf{v}, e^{-\alpha(\xi\cdot x)}\Gamma_{0,\partial\Omega}\mathbf{v}\rangle + \left(\frac{1}{2}(K_\xi + K_\xi^*)e^{-\alpha(\xi\cdot x)}\mathbf{v}, e^{-\alpha(\xi\cdot x)}\mathbf{v}\right)$$
$$= -\langle e^{-\alpha(\xi\cdot x)}\Gamma_{B-,\partial\Omega}\mathbf{v}, e^{-\alpha(\xi\cdot x)}\Gamma_{0,\partial\Omega}\mathbf{v}\rangle + (e^{-\alpha(\xi\cdot x)}L\mathbf{v}, e^{-\alpha(\xi\cdot x)}\mathbf{v}).$$

Thence, recalling (3.2) of hypothesis (b), we deduce the following Gårding inequality:

$$c_0\| e^{-\alpha(\xi\cdot x)}\mathbf{v}\| \le \| e^{-\alpha(\xi\cdot x)}L\mathbf{v}\| \qquad \forall \mathbf{v} \in [H^1(\Omega)]^m \cap D(L,\Omega). \tag{3.5}$$

Noting that $[H^1(\Omega)]^m$ is dense in $H(L,\Omega)$, it follows that $[H^1(\Omega)]^m \cap D(L,\Omega)$ is dense in $D(L,\Omega)$, so that

$$\|e^{-\alpha(\xi\cdot x)}L\mathbf{v}\| \ge c_1 |||\mathbf{v}|||_{\xi,\Omega} \qquad \forall \mathbf{v} \in D(L,\Omega), \tag{3.6}$$

where $c_1 = (1 + c_0^{-2})^{-1/2}$. Now (3.6) implies that L is an injective operator from $D(L,\Omega)$ onto its range space $\mathcal{R}(L)$, and that the inverse of L is continuous. Hence L is an isomorphism from $D(L,\Omega)$ onto $\mathcal{R}(L)$; this implies that $\mathcal{R}(L)$ is a closed linear subspace of $[L_2(\Omega)]^m$. By virtue of hypothesis (b), the transpose ${}^tL$ of L has trivial kernel, i.e. $\mathrm{Ker}({}^tL) = \{0\}$. The Closed Range Theorem then implies that $\mathcal{R}(L) = \mathrm{Ker}({}^tL)^\perp = [L_2(\Omega)]^m$. Hence L is an isomorphism from $H(L,\Omega)$ onto $[L_2(\Omega)]^m$.

We deduce from this theorem that (weak and strong) solutions of the homogeneous boundary value problem (3.1), (3.4) satisfy the stability estimate

$$\|\mathbf{u}\| \le \frac{1}{c_0} e^{2\alpha L}\|\mathbf{f}\|,$$

where $L = \mathrm{diam}(\Omega)$.

More generally, consider the non-homogeneous boundary value problem (3.1), (3.3), where $\mathbf{f} \in [L_2(\Omega)]^m$ and $\mathbf{g} \in [L_2(\partial\Omega)]^m$. A function $\mathbf{u} \in [L_2(\Omega)]^m$ obeying

$$(\mathbf{u}, L^*\phi) + \langle B^-\mathbf{g}, \phi\rangle = (\mathbf{f}, \phi) \qquad \forall \phi \in D(L^*,\Omega) \cap [H^1(\Omega)]^m$$

is called a *weak solution* of the boundary value problem. A weak solution $\mathbf{u}$ of the non-homogeneous problem (3.1), (3.3) which belongs to $H(L,\Omega)$ is called a *strong solution*. Lax and Phillips [14] proved, under the assumption that $\partial\Omega$ is non-characteristic for L, that every weak solution is a strong solution. Furthermore, assuming that $\mathbf{f} \in [H^1(\Omega)]^m$, $\mathbf{g} \in [H^1(\partial\Omega)]^m$ and the entries of C are continuously differentiable on $\overline{\Omega}$, the following regularity result holds:

$$\|\mathbf{u}\|_{H^1(\Omega)} \le c_1 \left(\|\mathbf{f}\|_{H^1(\Omega)} + \|\mathbf{g}\|_{H^1(\partial\Omega)}\right)$$

(see Lax and Phillips [14]). An analogous bound holds for the problem

$$L^*\phi \quad = \quad \mu \text{ in } \Omega,$$

$$B^+\phi\mid_{\partial\Omega} \quad = \quad B^+\zeta \text{ on } \partial\Omega,$$

with corresponding stability constant c_1'; namely,

$$\|\phi\|_{H^1(\Omega)} \leq c_1'(\|\mu\|_{H^1(\Omega)} + \|\zeta\|_{H^1(\partial\Omega)}). \tag{3.7}$$

3.2 *A posteriori* bound on the global error

Suppose that $\mathcal{U}^h \subset H(L,\Omega)$ is a finite element trial space consisting of piecewise polynomial functions on a partition $(\kappa_i)_{i=1,\ldots,N_h}$ of $\overline{\Omega}$; here h is the mesh parameter. Further, let $\mathcal{M}^h \subset [L_2(\Omega)]^m$ be a finite element test space on this partition, and let Π_h denote the orthogonal projector in $[L_2(\Omega)]^m$ onto $\mathcal{M}^h$. We consider the following general class of Galerkin approximations of the boundary value problem (3.1), (3.3): find $\mathbf{u} \in \mathcal{U}^h$ such that

$$\Pi_h L\mathbf{u}_h = \Pi_h \mathbf{f} \text{ on } \Omega, \qquad B^-\mathbf{u}_h\big|_{\partial\Omega} = B^-\mathbf{g}, \tag{3.8}$$

where, for the sake of simplicity, we have assumed that the function $\mathbf{g}$ belongs to the restriction of the finite element trial space $\mathcal{U}^h$ to the boundary. We shall suppose that (3.8) has a unique solution. Particular examples of such discretisations, including finite element and finite volume methods, will be considered below.

We denote by $\mathbf{e}_h = \mathbf{u} - \mathbf{u}_h$ the *global error* and define the *residual* $\mathbf{r}_h = \mathbf{f} - L\mathbf{u}_h$. Since L is a linear operator, it follows that $\mathbf{e}_h$ is the unique solution of the boundary value problem

$$L\mathbf{e}_h = \mathbf{r}_h \text{ on } \Omega, \qquad B^-\mathbf{e}_h\big|_{\partial\Omega} = \mathbf{0}.$$

This is a key relationship between the (computable) residual $\mathbf{r}_h$ and the global error $\mathbf{e}_h$ which allows us to obtain computable bounds on $\mathbf{e}_h$ in terms of $\mathbf{r}_h$ by bounding the inverse of the operator L.

We shall further suppose that the finite element test space $\mathcal{M}^h$ possesses the following standard approximation property:

(c) There exists a positive constant c_2, independent of h, such that for each $\phi \in [H^1(\Omega)]^m$ there is $\phi_h \in \mathcal{M}^h$ with

$$\left(\sum_\kappa \|h^{-1}(\phi - \phi_h)\|^2_{L_2(\kappa)}\right)^{\frac{1}{2}} \leq c_2|\phi|_{H^1(\Omega)}.$$

Theorem 3. *Suppose that hypotheses (a), (b) and (c) hold, and that the entries of C are continuously differentiable on $\overline{\Omega}$. Then,*

$$\|\mathbf{u} - \mathbf{u}_h\|_{H^{-1}(\Omega)} \leq c_1'c_2\left(\sum_\kappa \|h\mathbf{r}_h\|^2_{L_2(\kappa)}\right)^{\frac{1}{2}}.$$

Proof. Given $\psi \in [C_0^\infty(\Omega)]^m$, consider the dual problem

$$L^*\phi = \psi \text{ on } \Omega, \qquad B^+\phi\big|_{\partial\Omega} = \mathbf{0}.$$

Then,

$$(\mathbf{u} - \mathbf{u}_h, \psi) = (\mathbf{u} - \mathbf{u}_h, L^*\phi) = (L(\mathbf{u} - \mathbf{u}_h), \phi) = (r_h, \phi).$$

Recalling (3.8), it follows that

$$(r_h, \phi_h) = 0;$$

so, for each $\phi_h \in \mathcal{M}^h$, we have that

$$\begin{aligned}(\mathbf{u} - \mathbf{u}_h, \psi) &= (\mathbf{r}_h, \phi - \phi_h) \\ &\leq \left(\sum_\kappa \|h\mathbf{r}_h\|^2_{L_2(\kappa)}\right)^{\frac{1}{2}} \left(\sum_\kappa \|h^{-1}(\phi - \phi_h)\|^2_{L_2(\kappa)}\right)^{\frac{1}{2}} \\ &\leq c_2 \left(\sum_\kappa \|h\mathbf{r}_h\|^2_{L_2(\kappa)}\right)^{\frac{1}{2}} |\phi|_{H^1(\Omega)}. \qquad (3.9)\end{aligned}$$

According to the regularity result (3.7), with $\mu = \psi$,

$$|\phi|_{H^1(\Omega)} \leq \|\phi\|_{H^1(\Omega)} \leq c_1' \|\psi\|_{H^1(\Omega)}.$$

Substituting this into (3.9), dividing both sides by $\|\psi\|_{H^1(\Omega)}$, and taking the supremum over all $\psi \in [C_0^\infty(\Omega)]^m$, we obtain the desired error bound.

3.3 Application to the cell vertex finite volume method

In this section we present an *a posteriori* error bound for a cell vertex finite volume approximation of (3.1), (3.3); this is a generalisation of a similar result derived, in the case of homogeneous boundary conditions, in [25].

For the sake of simplicity we suppose that $\Omega = (0,1)^2$, and consider the Friedrichs system in conservation form:

$$\begin{aligned} L\mathbf{u} \equiv \textstyle\sum_{i=1}^2 \frac{\partial}{\partial x_i}(A_i\mathbf{u}) + C\mathbf{u} &= \mathbf{f} \text{ in } \Omega, \\ B^-\mathbf{u}|_{\partial\Omega} &= B^-\mathbf{g} \text{ on } \partial\Omega. \end{aligned}$$

The cell vertex finite volume approximation of this boundary value problem is obtained by subdividing Ω using a structured mesh that consists of convex quadrilateral elements, and integrating the differential equation over each element; using Gauss' theorem, integrals over cells are converted into contour integrals which are then approximated by the trapezium rule. This construction gives rise to a conservative four-point finite difference scheme, called the cell vertex scheme. Equivalently, the cell vertex scheme can be formulated as a Petrov-Galerkin finite element method based on continuous piecewise bilinear trial functions and piecewise constant test functions (see [2,22–24]). It is this latter interpretation that is adopted here.

Let $\mathcal{F} = \{\mathcal{T}_h\}$, $h > 0$, be a family of structured partitions $\mathcal{T}_h$ of $\Omega = (0,1)^2$ into convex quadrilateral elements. We shall assume that the partition is *quasi-parallel* in the sense that, for each element $\kappa_{ij} \in \mathcal{T}_h$, the distance between the midpoints of the diagonals is bounded by $c_\star \mathrm{meas}(\kappa_{ij})$, where $c_\star$ is a fixed positive constant. In order to introduce the relevant finite element spaces, we define the reference square $\hat{\kappa} = (-1,1)^2$, and denote by $F_{\kappa_{ij}}$ the bilinear function that maps $\hat{\kappa}$ onto the "finite volume" κ_{ij}. Let $\mathcal{Q}_1(\hat{\kappa})$ be the set of bilinear functions on $\hat{\kappa}$, and $\mathcal{Q}_0(\hat{\kappa})$ the set of constant functions on $\hat{\kappa}$. We define

$$\begin{aligned} \mathcal{M}^h &= \left\{ \mathbf{q} \in [L_2(\Omega)]^m : \mathbf{q} = \hat{\mathbf{q}} \circ F_{\kappa_{ij}}^{-1},\ \hat{\mathbf{q}} \in [\mathcal{Q}_0(\hat{\kappa})]^m,\ \kappa_{ij} \in \mathcal{T}_h \right\}, \\ \mathcal{U}^h &= \left\{ \mathbf{v} \in [H^1(\Omega)]^m : \mathbf{v} = \hat{\mathbf{v}} \circ F_{\kappa_{ij}}^{-1},\ \hat{\mathbf{v}} \in [\mathcal{Q}_1(\hat{\kappa})]^m,\ \kappa_{ij} \in \mathcal{T}_h \right\}, \end{aligned}$$

as well as $\mathcal{U}^h_-$, the set of all functions $\mathbf{v}$ in $\mathcal{U}^h$ such that $B^-\mathbf{v}|_{\partial\Omega} = B^-\mathbf{g}$ on $\partial\Omega$; here we have assumed, for the sake of simplicity, that $\mathbf{g}$ belongs to the restriction of the finite element trial space $\mathcal{U}^h$ to the boundary of the domain. Let us denote by $\Pi_h : [L_2(\Omega)]^m \to \mathcal{M}^h$ the L_2-orthogonal projector onto $\mathcal{M}^h$, and by $\mathcal{I}_h : (H(L,\Omega) \cap [C^0(\overline{\Omega})]^m)^2 \to \mathcal{U}^h \times \mathcal{U}^h$ the interpolation projector onto $\mathcal{U}^h \times \mathcal{U}^h$.

The cell vertex finite volume approximation is defined as follows: find $\mathbf{u}_h \in \mathcal{U}^h_-$ satisfying

$$(\operatorname{div} \mathcal{I}_h(\mathcal{A}\mathbf{u}_h), \mathbf{q}_h) + (C\mathbf{u}_h, \mathbf{q}_h) = (\mathbf{f}, \mathbf{q}_h) \qquad \forall \mathbf{q}_h \in \mathcal{M}^h, \tag{3.10}$$

where $\mathcal{A} = (A_1, A_2)$. We define the *cell vertex operator* $L_h : \mathcal{U}^h \longrightarrow \mathcal{M}^h$ by

$$L_h : \mathbf{v}_h \longmapsto \Pi_h (\operatorname{div} \mathcal{I}_h(\mathcal{A}\mathbf{v}_h)) + \Pi_h(C\mathbf{v}_h).$$

With this definition, we can reformulate (3.10) as follows: find $\mathbf{u}_h \in \mathcal{U}^h_-$ satisfying

$$L_h \mathbf{u}_h = \Pi_h \mathbf{f}.$$

We shall adopt the following hypothesis:

(b') The matrices A_i, $i = 1, 2$, are positive-definite, uniformly on $\overline{\Omega}$.

Clearly (b') is sufficient for (b) when $d = 2$. Under this assumption $\mathcal{U}^h_-$ and $\mathcal{M}^h$ have the same dimension and L_h is then easily seen to be bijective. For theoretical results concerning the stability and convergence of the cell vertex scheme we refer to [2,18,19,22–24]. Here we derive an *a posteriori* error bound for the method.

Theorem 4. *Suppose that hypothesis (b') holds, that $\mathcal{F} = \{\mathcal{T}_h\}$ is a family of structured quasi-parallel partitions of Ω, and that, given s, $1 < s \leq 3$, the* *coefficients of the matrices A_i, $i = 1, 2$, belong to $H^s(\Omega) \cap C^1(\overline{\Omega})$ and those of C to $C^1(\overline{\Omega})$. Then, the cell vertex scheme obeys the following a posteriori error bound:*

$$\|\mathbf{u} - \mathbf{u}_h\|_{H^{-1}(\Omega)} \leq c_3 \left(\sum_{\kappa} \|h\mathbf{r}_h\|^2_{L_2(\kappa)} + \sum_{\kappa} |h^{s-1} \mathcal{A}\mathbf{u}_h|^2_{H^s(\kappa)} \right)^{\frac{1}{2}},$$

where c_3 is a computable constant.

Proof. Suppose that $\psi \in [C_0^\infty(\Omega)]^m$ and consider the dual problem

$$L^*\phi = \psi \text{ on } \Omega, \qquad B^+\phi\big|_{\partial\Omega} = \mathbf{0}.$$

Then,

$$\begin{aligned}
(\mathbf{u} - \mathbf{u}_h, \psi) &= (\mathbf{u} - \mathbf{u}_h, L^*\phi) = (L(\mathbf{u} - \mathbf{u}_h), \phi) \\
&= (\mathbf{f} - L\mathbf{u}_h, \phi - \phi_h) + (\mathbf{f} - L\mathbf{u}_h, \phi_h) \\
&= (\mathbf{r}_h, \phi - \phi_h) + (\Pi_h(\operatorname{div} \mathcal{I}_h(\mathcal{A}\mathbf{u}_h) - \operatorname{div}(\mathcal{A}\mathbf{u}_h)), \phi_h).
\end{aligned}$$

Thus, choosing $\phi_h = \Pi_h\phi$ and exploiting the approximation property (c), we have that

$$\begin{aligned}
|(\mathbf{u} - \mathbf{u}_h, \psi)| \leq & \left(\sum_{\kappa} \|\Pi_h(\operatorname{div} \mathcal{I}_h(\mathcal{A}\mathbf{u}_h) - \operatorname{div}(\mathcal{A}\mathbf{u}_h))\|^2_{L_2(\kappa)} \right)^{\frac{1}{2}} \|\phi\| \\
& + c_2 \left(\sum_{\kappa} \|h\mathbf{r}_h\|^2_{L_2(\kappa)} \right)^{\frac{1}{2}} |\phi|_{H^1(\Omega)}. \qquad (3.11)
\end{aligned}$$

By virtue of a superconvergence result stated in [2] (for the special case of a uniform finite difference mesh; the extension to quasi-parallel meshes is straightforward),

$$\|\Pi_h(\operatorname{div} \mathcal{I}_h(\mathcal{A}\mathbf{u}_h) - \operatorname{div}(\mathcal{A}\mathbf{u}_h))\|_{L_2(\kappa)} \leq c_4 |h^{s-1} \mathcal{A}\mathbf{u}_h|_{H^s(\kappa)}, \qquad 1 < s \leq 3.$$

Inserting this into (3.11), recalling the hyperbolic regularity estimate (3.7),

$$|\phi|_{H^1(\Omega)} \leq \|\phi\|_{H^1(\Omega)} \leq c_1' \|\psi\|_{H^1(\Omega)},$$

dividing both sides of the inequality by $\|\psi\|_{H^1(\Omega)}$ and taking the supremum over all ψ in $[C_0^\infty(\Omega)]^m$, we obtain the desired result, with $c_3 = c_1' \max(c_2, c_4)$.

4 *A posteriori* error estimation for linear functionals

In engineering applications it is frequently the case that linear functionals of the solution (such as the lift, the drag, or the flux of the solution across the boundary) are of prime concern. In such instances it is unlikely that *a posteriori* error bounds of the kind stated in the last two theorems will be of use in the design of adaptive algorithms. Our aim in this section is to propose an approach to deriving *a posteriori* error bounds on linear functionals, without attempting to obtain an upper bound on the error in a *norm* in which the functional is bounded. In order to illustrate the key ideas we discuss a specific example: the *a posteriori* estimation of the error in the outflow flux across the boundary (or part of the boundary) for a symmetric hyperbolic system.

Let us consider the non-homogeneous boundary value problem (3.1), (3.3), where $\mathbf{f} \in [L_2(\Omega)]^m$ and $\mathbf{g} \in [L_2(\partial\Omega)]^m$. For the sake of simplicity, we assume that $\mathbf{g}$ belongs to the restriction of the trial space $\mathcal{U}^h$ to the boundary, and we approximate the non-homogeneous problem by the following method: find $\mathbf{u}_h$ in $\mathcal{U}^h$ such that

$$\begin{aligned} \Pi_h L\mathbf{u}_h &= \Pi_h \mathbf{f} \text{ in } \Omega, \\ B^- \mathbf{u}_h|_{\partial\Omega} &= B^- \mathbf{g} \text{ on } \partial\Omega. \end{aligned}$$

Given any $\psi \in [L_2(\Omega)]^m$, consider the linear functional

$$N_\psi(\mathbf{v}) = \int_{\partial\Omega} B^+ \mathbf{v}\,\psi ds, \qquad \mathbf{v} \in [H^1(\Omega)]^m;$$

here ψ plays the rôle of a weight function that can be chosen freely (for example ψ can be taken to have its support contained in a subset of $\partial\Omega$, etc.). We seek to approximate the (weighted) outflow flux $N_\psi(\mathbf{u})$ by $N_\psi(\mathbf{u}_h)$.

Theorem 5. *Suppose that hypotheses (a)–(c) hold, and that the entries of C are continuously differentiable on $\overline{\Omega}$. Then, for each $\psi \in [H^1(\partial\Omega)]^m$, assuming that $\mathbf{u}$ and $\mathbf{u}_h$ belong to $[H^1(\Omega)]^m$, we have that*

$$|N_\psi(\mathbf{u}) - N_\psi(\mathbf{u}_h)| \le c_1' c_2 \left(\sum_\kappa \|h\mathbf{r}_h\|^2_{L_2(\kappa)} \right)^{1/2} \|\psi\|_{H^1(\partial\Omega)}.$$

Proof. Consider the dual problem

$$\begin{aligned} L^*\phi &= 0 \text{ in } \Omega, \\ B^+\phi|_{\partial\Omega} &= B^+\psi \text{ on } \partial\Omega. \end{aligned}$$

Since $\mathbf{u} - \mathbf{u}_h$ is in $D(L,\Omega) \cap [H^1(\Omega)]^m$, Green's formula from Theorem 1 gives

$$0 = (\mathbf{u} - \mathbf{u}_h, L^*\phi) = (L(\mathbf{u} - \mathbf{u}_h), \phi) - N_\psi(\mathbf{u} - \mathbf{u}_h).$$

Consequently, for any $\phi_h \in \mathcal{M}_h$,

$$N_\psi(\mathbf{u}) - N_\psi(\mathbf{u}_h) = (\mathbf{r}_h, \phi - \phi_h).$$

Exploiting hypothesis (c), it follows that

$$|N_\psi(\mathbf{u}) - N_\psi(\mathbf{u}_h)| \le c_2 \left(\sum_\kappa \|h\mathbf{r}_h\|^2_{L_2(\kappa)} \right)^{1/2} |\phi|_{H^1(\Omega)},$$

and therefore, by the hyperbolic regularity theorem

$$|N_\psi(\mathbf{u}) - N_\psi(\mathbf{u}_h)| \le c_1' c_2 \left(\sum_\kappa \|h\mathbf{r}_h\|^2_{L_2(\kappa)} \right)^{1/2} \|\psi\|_{H^1(\partial\Omega)}.$$

A particularly relevant case concerns the estimation of the flux through an open subset $\gamma \subset \partial\Omega$. In this case it is tempting to take $\psi = \chi_\gamma$ where χ_γ is the characteristic function of γ; unfortunately such ψ does not belong to $[H^1(\partial\Omega)]^m$ so Theorem 5 does not apply.

We shall illustrate on a simple example of a scalar hyperbolic problem how a refined analysis may be pursued to obtain a bound on the flux across $\gamma \subset \Omega$ by recalling that singularities propagate along (bi-)characteristics.

Consider the scalar hyperbolic equation

$$Lu \equiv \nabla \cdot (\mathbf{a}u) + cu = f \text{ in } \Omega \tag{4.1}$$

for Ω the unit square $(0,1)^2$, where $\mathbf{a} = (a_1, a_2)$ is a vector function with positive, continuously differentiable entries defined on $\overline{\Omega}$, c is a continuous function of $\overline{\Omega}$, and $f \in L_2(\Omega)$.

Let $b = \hat{\nu}_1 a_1 + \hat{\nu}_2 a_2$, where $\hat{\nu} = (\hat{\nu}_1, \hat{\nu}_2)$ is the unit outward normal vector field to the boundary of Ω; at the corners of the square we define $\hat{\nu}$ to be the outward pointing unit vector collinear with the diagonal passing through that corner. Clearly b is non-zero almost everywhere on $\partial\Omega$, i.e. $\partial\Omega$ is non-characteristic for L. We consider the partial differential equation (4.1) subject to the homogeneous boundary condition

$$b^- u = 0 \text{ on } \partial\Omega; \tag{4.2}$$

here $b^- = \min(b, 0)$ denotes the negative part of b. The boundary condition (4.2) is equivalent to requiring that $u = 0$ along the *inflow* part of the boundary,

$$\partial_-\Omega = \{x = (x_1, x_2) \in \partial\Omega \,:\, \mathbf{a}(x) \cdot \hat{\nu}(x) < 0\}.$$

The complement of the inflow boundary,

$$\partial_+\Omega = \{x = (x_1, x_2) \in \partial\Omega \,:\, \mathbf{a}(x) \cdot \hat{\nu}(x) \ge 0\},$$

is called the *outflow* part of the boundary. Let us suppose that γ is a connected, relatively open measurable subset of the outflow boundary $\partial_+\Omega$, and that we wish to approximate the normal outflow flux along γ, defined as

$$N_\psi(u) = \int_{\partial_+\Omega} (\mathbf{a}\cdot\hat{\nu})u\,\psi\,\mathrm{d}s, \qquad u \in H^1(\Omega)\cap D(L,\Omega),$$

where $\psi = \chi_\gamma\zeta$, and ζ is a fixed element in $H^1(\gamma)$ which plays the rôle of a weight function (for example $\zeta \equiv 1$ on γ). Now Sobolev's embedding theorem implies that ζ is continuous on $\overline{\gamma}$, and therefore ψ has possible discontinuities only at the points α and β, say, that comprise the boundary of γ. Consider the characteristic curves C_α and C_β of L in $\overline{\Omega}$ that contain the points α and β, respectively; these two curves divide Ω into three disjoint regions, one of which, denoted Ω_0, has γ as its outflow boundary, and two other regions over each of which ϕ, the solution of the dual problem,

$$\begin{aligned} L^*\phi \equiv -\mathbf{a}\cdot\nabla\phi + c\phi &= 0 \text{ in } \Omega \\ \phi|_{\partial_+\Omega} &= \psi \text{ on } \partial_+\Omega, \end{aligned}$$

is identically zero. Let $\Omega_{\alpha,h}$ be a characteristic strip of cross-section $c_7^2 h/2$ (with c_7 a positive constant; and $\tilde{h}$ a small positive constant, commensurable with the granularity of the mesh) which contains the characteristic C_α in its interior, with $\Omega_{\beta,h}$ defined analogously, so that the elements in the partition of Ω that are in contact with C_α (respectively C_β) are contained in $\Omega_{\alpha,h}$ (respectively $\Omega_{\beta,h}$). We define $\gamma_{\alpha,h}$ (respectively $\gamma_{\beta,h}$) as the intersection of the closure of $\Omega_{\alpha,h}$ (respectively $\Omega_{\beta,h}$) with $\partial_+\Omega$.

Theorem 6. *Let u and u_h belong to $H^1(\Omega)\cap D(L,\Omega)$, and assume that the finite element test space possesses the following approximation property:*

(c') for each $\phi \in L_2(\Omega)$ that is piecewise H^1 on Ω, there exists $\phi_h \in \mathcal{M}^h$ and a positive constant c_5, independent of h, such that

$$\|h^{-k}(\phi - \phi_h)\|_{L_2(\kappa)} \le c_5|\phi|_{H^k(\kappa)},$$

where $k = 0$ if $\phi \in L_2(\kappa)$ and $k = 1$ if $\phi \in H^1(\kappa)$.

Then we have the following error bound:

$$\begin{aligned} |N_\psi(u) - N_\psi(u_h)| \le\; & c_5c_6c_7\left(\sum_{\kappa\subset\Omega_{\alpha,h}\cup\Omega_{\beta,h}} \|\tilde{h}^{1/2}r_h\|^2_{L_2(\kappa)}\right)^{\frac{1}{2}} \|\psi\|_{L_\infty(\gamma)} \\ & + c_5c_6\left(\sum_{\kappa\not\subset\Omega_{\alpha,h}\cup\Omega_{\beta,h}} \|hr_h\|^2_{L_2(\kappa)}\right)^{\frac{1}{2}} \|\psi\|_{H^1(\gamma)}, \end{aligned}$$

where c_5, c_6 and c_7 are positive constants, independent of h, and κ signifies any finite element in the partition of Ω whose closure has nonempty intersection with $\overline{\Omega}_0$, the domain of influence for $\overline{\gamma}$.

Proof. Since $\phi \equiv 0$ in the complement of $\overline{\Omega}_0$, it will only be necessary to consider those elements κ in the partition of the domain Ω whose closure has nonempty intersection with $\overline{\Omega}_0$. In order to simplify the notation we shall implicitly assume throughout the proof that, for each element κ that we refer to, $\overline{\kappa} \cap \overline{\Omega}_0$ is nonempty. Arguing in the same manner as in Theorem 5, we deduce that

$$\begin{aligned}
|N_\psi(u) - N_\psi(u_h)| &= |(r_h, \phi - \phi_h)| \leq \sum_\kappa |(r_h, \phi - \phi_h)_\kappa| \\
&= \sum_{\kappa \subset \Omega_{\alpha,h} \cup \Omega_{\beta,h}} |(r_h, \phi - \phi_h)_\kappa| + \sum_{\kappa \not\subset \Omega_{\alpha,h} \cup \Omega_{\beta,h}} |(r_h, \phi - \phi_h)_\kappa| \\
&\leq \sum_{\kappa \subset \Omega_{\alpha,h} \cup \Omega_{\beta,h}} \|r_h\|_{L_2(\kappa)} \|\phi - \phi_h\|_{L_2(\kappa)} \\
&\quad + \sum_{\kappa \not\subset \Omega_{\alpha,h} \cup \Omega_{\beta,h}} \|h r_h\|_{L_2(\kappa)} \|h^{-1}(\phi - \phi_h)\|_{L_2(\kappa)} \\
&\leq c_5 \left(\sum_{\kappa \subset \Omega_{\alpha,h} \cup \Omega_{\beta,h}} \|r_h\|^2_{L_2(\kappa)} \right)^{\frac{1}{2}} \|\phi\|_{L_2(\Omega_{\alpha,h} \cup \Omega_{\beta,h})} \\
&\quad + c_5 \left(\sum_{\kappa \not\subset \Omega_{\alpha,h} \cup \Omega_{\beta,h}} \|h r_h\|^2_{L_2(\kappa)} \right)^{\frac{1}{2}} |\phi|_{H^1(\Omega \setminus (\Omega_{\alpha,h} \cup \Omega_{\beta,h}))}.
\end{aligned}$$

Thus, exploiting hyperbolic regularity on $\Omega_{\alpha,h} \cup \Omega_{\beta,h}$ and its complement, we deduce that

$$\begin{aligned}
|N_\psi(u) - N_\psi(u_h)| &\leq c_5 c_6 \left(\sum_{\kappa \subset \Omega_{\alpha,h} \cup \Omega_{\beta,h}} \|r_h\|^2_{L_2(\kappa)} \right)^{\frac{1}{2}} \|\psi\|_{L_2(\gamma_{\alpha,h} \cup \gamma_{\beta,h})} \\
&\quad + c_5 c_6 \left(\sum_{\kappa \not\subset \Omega_{\alpha,h} \cup \Omega_{\beta,h}} \|h r_h\|^2_{L_2(\kappa)} \right)^{\frac{1}{2}} \|\psi\|_{H^1(\gamma \setminus (\gamma_{\alpha,h} \cup \gamma_{\beta,h}))} \\
&\leq c_5 c_6 c_7 \left(\sum_{\kappa \subset \Omega_{\alpha,h} \cup \Omega_{\beta,h}} \|\tilde{h}^{1/2} r_h\|^2_{L_2(\kappa)} \right)^{\frac{1}{2}} \|\psi\|_{L_\infty(\gamma)} \\
&\quad + c_5 c_6 \left(\sum_{\kappa \not\subset \Omega_{\alpha,h} \cup \Omega_{\beta,h}} \|h r_h\|^2_{L_2(\kappa)} \right)^{\frac{1}{2}} \|\psi\|_{H^1(\gamma)},
\end{aligned}$$

and hence the desired result.

Now choosing $\psi = \chi_\gamma$, the characteristic function of γ, we deduce from Theorem 6 that

$$|\int_\gamma (\mathbf{a}\cdot\hat{\nu})u\,ds - \int_\gamma (\mathbf{a}\cdot\hat{\nu})u_h\,ds| \leq c_5c_6c_7\left(\sum_{\kappa\subset\Omega_{\alpha,h}\cup\Omega_{\beta,h}} \|\tilde{h}^{1/2}r_h\|^2_{L_2(\kappa)}\right)^{\frac{1}{2}}$$

$$+c_5c_6|\gamma|^{1/2}\left(\sum_{\kappa\not\subset\Omega_{\alpha,h}\cup\Omega_{\beta,h}} \|hr_h\|^2_{L_2(\kappa)}\right)^{\frac{1}{2}},$$

where $|\gamma|$ denotes the one-dimensional measure of the set γ. This error bound indicates that in order to reduce the size of the error in the normal flux across γ one has to refine only those elements κ that have non-empty intersection with the domain of influence for γ. Note also, that since, at least on uniformly regular partitions, the number of entries in the two sums on the right are, respectively, $O(h^{-1})$ and $O(h^{-2})$, the two terms on the right-hand side are commensurable.

Due to theoretical results about finite speed of propagation of singularities along bi-characteristics and the microlocal regularity theorem of Hörmander, one would expect that a bound akin to Theorem 6 holds more generally for a large class of hyperbolic systems.

5 Unsteady problems

In this section we develop the *a posteriori* error analysis of a class of evolution-Galerkin finite element methods for unsteady scalar hyperbolic problems. For a detailed overview of the (unconditional) stability and accuracy properties of evolution-Galerkin methods we refer to [20].

Given a final time $T > 0$, a function $f \in L_2(I; L_2(\Omega))$ with $I = (0, T]$, and $u_0 \in L_2(\Omega)$, we consider the problem

$$\frac{\partial u}{\partial t} + \mathbf{a}\cdot\nabla u = f, \qquad \mathbf{x}\in\Omega,\ t\in I, \tag{5.1}$$

$$u(\mathbf{x},0) = u_0(\mathbf{x}), \quad \mathbf{x}\in\Omega, \tag{5.2}$$

where $\Omega = \mathbb{R}^d$. For the sake of simplicity we assume that the velocity field $\mathbf{a}$ is in $C([0,T];[C_0^1(\Omega)]^d)$, that it is incompressible, i.e. $\nabla\cdot\mathbf{a} = 0$ on $\Omega\times[0,T]$, and that the supports of u_0 and f are compact subsets in Ω and $\Omega\times[0,T]$, respectively. This problem has a unique weak solution u in $L_\infty(I; L_2(\Omega))$; moreover, if $u_0 \in H^1(\Omega)$ and $f \in L_2(I; H^1(\Omega))$ then u belongs to $L_2(I; H^1(\Omega))$.

Let $0 = t^0 < t^1 < \ldots < t^M < t^{M+1} = T$ be a subdivision (not necessarily uniform) of I, with corresponding time intervals $I^n = (t^{n-1}, t^n]$ and time steps

$k_n = t^n - t^{n-1}$. For each n, let $T^n = \{\kappa\}$ be a partition of Ω into closed simplices κ, with corresponding mesh function h_n, satisfying

$$C_\dagger h_\kappa^2 \leq \text{meas}(\kappa) \qquad \forall \kappa \in T^n, \tag{5.3}$$

$$C_* h_\kappa \leq h_n(\mathbf{x}) \leq h_\kappa \qquad \forall \mathbf{x} \in \kappa \qquad \forall \kappa \in T^n, \tag{5.4}$$

where h_κ is the diameter of κ, and $C_\dagger$ and C_* are positive constants. Further, h will denote the global mesh function given by $h(\mathbf{x}, t) = h_n(\mathbf{x})$ for $(\mathbf{x}, t)$ in $\Omega \times I^n$, and we define the corresponding time step function $k = k(t)$ by $k(t) = k_n$, $t \in I^n$.

Let $\Lambda^n = \Omega \times I^n$; for $p, q \in \mathbb{N}_0$, let

$$\begin{aligned}
S^{h_n} &= \{v \in H^1(\Omega) : v|_\kappa \in \mathbf{P}_p(\kappa) \quad \forall \kappa \in T^n\}, \\
V^{h_n} &= \left\{v : v(\mathbf{x}, t)|_{\Lambda^n} = \textstyle\sum_{j=0}^q t^j v_j, v_j \in S^{h_n}\right\}, \\
V^h &= \{v : v(\mathbf{x}, t)|_{\Lambda^n} \in V^{h_n}, n = 1, \ldots, M+1\},
\end{aligned}$$

where $\mathbf{P}_p(\kappa)$ denotes the set of polynomials of degree at most p over κ.

In the following, we shall assume that $p = 1$ and $q = 0$. We note that if $v \in V^{h_n}$ for $n = 1, \ldots, M+1$, then v is continuous in space at any time, but may be discontinuous in time at the discrete time levels t^n. To account for this, we introduce the notation $v_\pm^n := \lim_{s \to 0^+} v(t^n \pm s)$, and $[v^n] := v_+^n - v_-^n$.

The evolution-Galerkin approximation of (5.1), (5.2) makes use of the particle trajectories (or characteristics) associated with equation (5.1): the path $\mathbf{X}(\mathbf{x}, s; \cdot)$ of a particle located at position $\mathbf{x} \in \Omega$ at time $s \in [0, T]$ is defined as the solution of the initial value problem

$$\frac{\mathrm{d}}{\mathrm{d}t}\mathbf{X}(\mathbf{x}, s; t) = \mathbf{a}(\mathbf{X}(\mathbf{x}, s; t), t), \tag{5.5}$$

$$\mathbf{X}(\mathbf{x}, s; s) = \mathbf{x}. \tag{5.6}$$

For u smooth enough, the *material derivative* $D_t u$ is then defined by

$$\begin{aligned}
D_t u(\mathbf{x}, s) &:= \frac{\mathrm{d}}{\mathrm{d}t} u(\mathbf{X}(\mathbf{x}, s; t), t)\, |_{t=s} \\
&= \frac{\partial}{\partial t} u(\mathbf{x}, s) + \mathbf{a}(\mathbf{x}, s) \cdot \nabla u(\mathbf{x}, s) \quad \forall \mathbf{x} \in \Omega,\ s \in I.
\end{aligned} \tag{5.7}$$

The evolution-Galerkin time-discretisation involves approximating the material derivative by a divided difference operator. The simplest appropriate discretisation is the backward Euler method, giving, for $n = 0, \ldots, M$,

$$D_t u(\cdot, t^{n+1}) \approx \frac{u(\cdot, t^{n+1}) - u(\mathbf{X}(\cdot, t^{n+1}; t^n), t^n)}{k_{n+1}}.$$

If we now let u_h^n denote the Galerkin finite element approximation to $u(\cdot, t^n)$ at time t^n, then applying the finite element method in space yields what is known as the evolution-Galerkin discretisation of (5.1), (5.2): find $u_h^{n+1} \in S^{h_{n+1}}$, for $n = 0, \ldots, M$, such that

$$\left(\frac{u_h^{n+1} - u_h^n(\mathbf{X}(\cdot, t^{n+1}; t^n))}{k_{n+1}}, v\right) = (\bar{f}, v) \qquad \forall v \in S^{h_{n+1}}, \tag{5.8}$$

$$(u_h^0, v) = (u_0, v) \qquad \forall v \in S^{h_0}, \tag{5.9}$$

where $\bar{f}|_{\Lambda^{n+1}} := f(\cdot, t^{n+1})$. Alternatively, by integrating equation (5.8) with respect to t over I^{n+1}, we obtain the following equivalent formulation: find u_h such that, for $n = 0, 1, \ldots, M$, $u_h|_{\Lambda^{n+1}} \in V^{h_{n+1}}$ and satisfies

$$(D_t^h u_h, v)_{n+1} = (\bar{f}, v)_{n+1} \qquad \forall v \in V^{h_{n+1}}, \tag{5.10}$$

$$(u_{h-}^0, v) = (u_0, v) \qquad \forall v \in V^{h_0}, \tag{5.11}$$

where

$$D_t^h u_h|_{\Lambda^{n+1}} = (u_{h-}(\mathbf{X}(\mathbf{x}, t^{n+1}; t^{n+1}), t^{n+1}) - u_{h-}(\mathbf{X}(\mathbf{x}, t^{n+1}; t^n), t^n))/k_{n+1};$$

here, for $v, w \in L_2(I^{n+1}; L_2(\Omega))$, we have used the notation

$$(v, w)_{n+1} = \int_{t^n}^{t^{n+1}} (v, w) \mathrm{d}t.$$

We remark that in (5.10), (5.11) the space discretisation may vary in both space and time, but the time steps are only variable in time and not in space. Hence, the corresponding mesh will not be fully optimal.

5.1 *A posteriori* error analysis

In this subsection we derive an *a posteriori* estimate for the error, $e = u - u_h$, in the $L_\infty(0, T; H^{-1}(\Omega))$ norm, where u and u_h are the solutions of (5.1), (5.2) and (5.10), (5.11), respectively. We shall first introduce the following notation: for v and w in $L_2(0, T; L_2(\Omega))$ and $Q := \Omega \times I$, we define

$$(v, w)_Q := \sum_{n=0}^{M} \int_{t^n}^{t^{n+1}} (v, w) \mathrm{d}t,$$

$$\|v\|_Q := ((v, v)_Q)^{1/2},$$

where $(\cdot, \cdot)$ is the inner product in $L_2(\Omega)$; throughout this section $\|\cdot\|$ denotes the $L_2(\Omega)$ norm.

5.1.1 Error representation

Given that $\psi \in C_0^\infty(\Omega)$, consider the dual problem: find ϕ such that

$$
\begin{aligned}
-\phi_t - \nabla\cdot(\mathbf{a}\phi) &= 0, \qquad \mathbf{x}\in\Omega,\ t\in[0,T), && (5.12)\\
\phi(\mathbf{x},T) &= \psi(\mathbf{x}), \quad \mathbf{x}\in\Omega. && (5.13)
\end{aligned}
$$

Multiplying equation (5.12) by e and integrating by parts in both space and time gives

$$
\begin{aligned}
(e(T),\psi) &= (e(T),\phi(T))\\
&= \int_0^T (e,\phi_t)\mathrm{d}t + \sum_{n=0}^{M}\int_{t^n}^{t^{n+1}} (e_t,\phi)\mathrm{d}t - \sum_{n=0}^{M}([u_h^n],\phi(t^n)) + (u_0 - u_{h-}^0,\phi(0))\\
&= \sum_{n=0}^{M}\int_{t^n}^{t^{n+1}} (f - \mathbf{a}\cdot\nabla u_h,\phi)\mathrm{d}t - \sum_{n=0}^{M}([u_h^n],\phi(t^n)) + (u_0 - u_{h-}^0,\phi(0)).
\end{aligned}
$$

If we now let $\Phi \in V^h$, then using equation (5.10) we have

$$
\begin{aligned}
(e(T),\psi) &= \sum_{n=0}^{M}\int_{t^n}^{t^{n+1}} ([u_h^n]/k_{n+1} + \mathbf{a}\cdot\nabla u_h - f, \Phi - \phi)\mathrm{d}t\\
&\quad + \sum_{n=0}^{M}\int_{t^n}^{t^{n+1}} (D_t^h u_h - ([u_h^n]/k_{n+1} + \mathbf{a}\cdot\nabla u_h), \Phi)\mathrm{d}t\\
&\quad + \sum_{n=0}^{M}\int_{t^n}^{t^{n+1}} ([u_h^n]/k_{n+1}, \phi - \phi(t^n))\mathrm{d}t + \sum_{n=0}^{M}\int_{t^n}^{t^{n+1}} (f - \bar{f}, \Phi)\mathrm{d}t\\
&\quad + (u_0 - u_{h-}^0, \phi(0))\\
&\equiv \mathrm{I} + \mathrm{II} + \mathrm{III} + \mathrm{IV} + \mathrm{V} \quad \forall\Phi\in V^h. \qquad (5.14)
\end{aligned}
$$

It remains to estimate each of the terms I to V.

5.1.2 Interpolation/projection estimates for the dual problem

We shall now choose $\Phi \in V^h$ in equation (5.14) as a suitable approximation to ϕ. To do so, consider the quasi-interpolation operator $P_n : L_2(\Omega) \to S^{h_n}$ in the spatial variables (see, for example, Bernardi [3]), and the L_2 orthogonal projector $\pi_n : L_2(I^n) \to \mathbf{P}_0(I^n)$ in time defined by

$$
\int_{t^{n-1}}^{t^n} (\pi_n\phi - \phi)v\mathrm{d}t = 0 \quad \forall v \in \mathbf{P}_0(I^n); \qquad (5.15)
$$

then, we define (locally) $\Phi|_{\Lambda^n} \in V^{h_n}$ by letting

$$
\Phi|_{\Lambda^n} = P_n\pi_n\phi = \pi_n P_n\phi \in V^{h_n},
$$

where $\phi = \phi|_{\Lambda^n}$. Further, we introduce P and π by

$$(P\phi)|_{\Lambda^n} = P_n(\phi|_{\Lambda^n}), \tag{5.16}$$

$$(\pi\phi)|_{\Lambda^n} = \pi_n(\phi|_{\Lambda^n}), \tag{5.17}$$

and define $\Phi \in V^h$ as

$$\Phi = P\pi\phi = \pi P\phi \in V^h. \tag{5.18}$$

It follows from the properties of the quasi-interpolant that

$$\|\Phi\|_Q \leq C_2^i\|\phi\|_Q, \tag{5.19}$$

with C_2^i a positive constant, and that the following theorem holds.

Theorem 7. *Let $n \in \mathbb{N}_0$ and suppose that T^n satisfies conditions (5.3) and (5.4). Let $w \in H^1(\Omega)$ have compact support; then, with $C_1^i > 0$ a constant,*

$$\left(\sum_\kappa \|h_n^{-1}(P_n w - w)\|^2_{L_2(\kappa)}\right)^{1/2} \leq C_1^i C_*^{-1}|w|_{H^1(\Omega)}, \tag{5.20}$$

$$\left(\sum_\kappa \|P_n w\|^2_{L_2(\kappa)}\right)^{1/2} \leq C_2^i\|w\|. \tag{5.21}$$

In the next two lemmas we provide bounds for the operators P and π, in order to estimate $\phi - \Phi = \phi - P\pi\phi$.

Lemma 8. *Suppose that $R \in L_2(I; L_2(\Omega))$; then*

$$|(R, P\phi - \phi)_Q| \leq C_1^i C_*^{-1}\|hR\|_Q\|\nabla\phi\|_Q. \tag{5.22}$$

Proof. Using the Cauchy-Schwarz inequality gives

$$|(R, P\phi - \phi)_Q| \leq \|hR\|_Q\|h^{-1}(P\phi - \phi)\|_Q.$$

Now

$$\|h^{-1}(P\phi - \phi)\|_Q = \left(\sum_{n=0}^{M}\int_{t^n}^{t^{n+1}} \|h_{n+1}^{-1}(P_{n+1}\phi - \phi)\|^2\right)^{1/2}, \tag{5.23}$$

and

$$\|h_{n+1}^{-1}(P_{n+1}\phi - \phi)\| \leq C_1^i C_*^{-1}\|\nabla\phi\|.$$

Therefore

$$\|h^{-1}(P\phi - \phi)\|_Q \leq C_1^i C_*^{-1}\|\nabla\phi\|_Q.$$

The proof of the next Lemma is given in [8].

Lemma 9. *Suppose that* $R \in L_2(I; L_2(\Omega))$*; then*

$$
\begin{aligned}
|(R, P(\pi\phi - \phi))_Q| &\leq C_2^i C_3^i \|kR\|_Q \|\phi_t\|_Q, &(5.24)\\
|\sum_{n=0}^{M} \int_{t^n}^{t^{n+1}} (R, \phi^n - \phi)\,dt| &\leq \|kR\|_Q \|\phi_t\|_Q, &(5.25)
\end{aligned}
$$

where $C_3^i = 1/\sqrt{2}$ *and* $\phi^n = \phi(\mathbf{x}, t^n)$.

5.1.3 Strong stability of the dual problem

This brief subsection summarises the stability estimates for the solution ϕ of the dual problem (5.12), (5.13) in terms of the corresponding final datum ψ.

Lemma 10. *There exists a constant* C_1^s *such that*

$$\|\phi\|_{L_\infty(I;L_2(\Omega))} \leq C_1^s \|\psi\|.$$

Lemma 11. *There exists a constant* $C_2^s = C_2^s(T, \mathbf{a})$ *such that*

$$\|\nabla\phi\|_Q \leq C_2^s \|\nabla\psi\|.$$

Lemma 12. *There exists a constant* $C_3^s = C_3^s(T, \mathbf{a})$ *such that*

$$\|\phi_t\|_Q \leq C_3^s \|\psi\|_{H^1(\Omega)}.$$

Recalling that $\nabla \cdot \mathbf{a} = 0$, a straightforward energy analysis shows that $C_1^s = 1$, $C_2^s = \max(1, \mathrm{e}^{-\alpha T})$ where α is any real number such that $\nabla\mathbf{a} \geq \alpha I$, and $C_3^s = C_2^s \|\mathbf{a}\|_{L_\infty(\Omega)}$.

5.1.4 Completion of the proof of the a *posteriori* error estimate

We shall now proceed to estimate the terms I–V on the right-hand side of equation (5.14). For the first term I, we have

$$
\begin{aligned}
\mathrm{I} &= \sum_{n=0}^{M} \int_{t^n}^{t^{n+1}} ([u_h^n]/k_{n+1} + \mathbf{a} \cdot \nabla u_h - f, \Phi - \phi)dt \\
&= (R_1, \Phi - \phi)_Q = (R_1, P\phi - \phi)_Q + (R_1, P(\pi\phi - \phi))_Q \\
&\equiv \mathrm{I}_1 + \mathrm{I}_2,
\end{aligned}
$$

where P and π are as defined by (5.16) and (5.17), and

$$R_1|_{\Lambda^{n+1}} = [u_h^n]/k_{n+1} + \mathbf{a} \cdot \nabla u_h - f.$$

By Lemma 8 we have

$$|\mathrm{I}_1| \leq C_1^i C_*^{-1} \|hR_1\|_Q \|\nabla\phi\|_Q \leq C_1^i C_2^s C_*^{-1} \|hR_1\|_Q \|\psi\|_{H^1(\Omega)}.$$

Similarly, using Lemma 9, we have

$$|\mathrm{I}_2| \leq C_2^i C_3^i \|kR_1\|_Q \|\phi_t\|_Q \leq C_2^i C_3^s C_3^i \|kR_1\|_Q \|\psi\|_{H^1(\Omega)}.$$

Hence,

$$|\mathrm{I}| \leq \left(C_1^i C_2^s C_*^{-1} \|hR_1\|_Q + C_2^i C_3^s C_3^i \|kR_1\|_Q\right) \|\psi\|_{H^1(\Omega)}.$$

Next we consider term II:

$$\begin{aligned} \mathrm{II} &\leq \|kR_2\|_Q \|\Phi\|_Q \leq C_2^i \|kR_2\|_Q \|\phi\|_Q \leq C_2^i \sqrt{T} \|kR_2\|_Q \|\phi\|_{L_\infty(I;L_2(\Omega))} \\ &\leq C_1^s C_2^i \sqrt{T} \|kR_2\|_Q \|\psi\|, \end{aligned} \tag{5.26}$$

where

$$R_2|_{\Lambda^{n+1}} = (D_t^h u_h - ([u_h^n]/k_{n+1} + \mathbf{a} \cdot \nabla u_h))/k_{n+1}.$$

For the term III, using Lemma 9 and Lemma 12, we have

$$|\mathrm{III}| \leq \|kR_3\|_Q \|\phi_t\|_Q \leq C_3^s \|kR_3\|_Q \|\psi\|_{H^1(\Omega)},$$

where

$$R_3|_{\Lambda^{n+1}} = [u_h^n]/k_{n+1} = (u_h^{n+1} - u_h^n)/k_{n+1}.$$

Now consider the term IV:

$$\begin{aligned} |\mathrm{IV}| &\leq \|f - \bar{f}\|_Q \|\Phi\|_Q \leq C_2^i \|f - \bar{f}\|_Q \|\phi\|_Q \leq C_2^i \sqrt{T} \|f - \bar{f}\|_Q \|\phi\|_{L_\infty(I;L_2(\Omega))} \\ &\leq C_1^s C_2^i \sqrt{T} \|f - \bar{f}\|_Q \|\psi\|. \end{aligned}$$

Finally, for term V, the Cauchy-Schwarz inequality and Lemma 10 give the following chain of inequalities:

$$|\mathrm{V}| \leq \|u_0 - u_{h-}^0\| \, \|\phi(0)\| \leq \|u_0 - u_{h-}^0\| \, \|\phi\|_{L_\infty(I;L_2(\Omega))} \leq C_1^s \|u_0 - u_{h-}^0\| \, \|\psi\|.$$

Substituting the bounds for the terms I to V into equation (5.14), dividing both sides by $\|\psi\|_{H^1(\Omega)}$ and taking the supremum over all ψ in $C_0^\infty(\Omega)$, we obtain the following *a posteriori* error bound.

Theorem 13. *Let u and u_h be solutions of (5.1), (5.2) and (5.10), (5.11), respectively. Then*

$$\|u - u_h\|_{L_\infty(0,T;H^{-1}(\Omega))} \leq \mathring{\mathcal{E}}(u_h, h, k, f), \tag{5.27}$$

where

$$\overset{\circ}{\mathcal{E}}(u_h, h, k, f) = \mathcal{E}(u_h, h, k, f) + \mathcal{E}_0(u_0, u_{h-}^0, h),$$

$$\mathcal{E}(u_h, h, k, f) = C_1\|hR_1\|_Q + C_2\|kR_1\|_Q$$

$$+C_3\|kR_2\|_Q + C_4\|kR_3\|_Q + C_5\|kR_4\|_Q, \quad (5.28)$$

$$\mathcal{E}_0(u_0, u_{h-}^0, h) = C_6\|u_0 - u_{h-}^0\|, \quad (5.29)$$

and

$$R_1|_{\Lambda^{n+1}} = [u_h^n]/k_{n+1} + \mathbf{a}\cdot\nabla u_h - f,$$

$$R_2|_{\Lambda^{n+1}} = (D_t^h u_h - ([u_h^n]/k_{n+1} + \mathbf{a}\cdot\nabla u_h))/k_{n+1},$$

$$R_3|_{\Lambda^{n+1}} = [u_h^n]/k_{n+1},$$

$$R_4 = (f - \bar{f})/k,$$

and C_i, $i = 1, ..., 6$, *are (computable) positive constants; namely,* $C_1 = C_1^i C_2^s C_*^{-1}$, $C_2 = C_2^i C_3^s C_3^i$, $C_3 = C_1^s C_2^i \sqrt{T}$, $C_4 = C_3^s$, $C_5 = C_1^s C_2^i \sqrt{T}$, *and* $C_6 = C_1^s$.

5.2 Computational implementation

In Theorem 13 we stated an *a posteriori* error estimate in the $L_\infty(H^{-1})$ norm, for the evolution-Galerkin approximation of the unsteady hyperbolic problem (5.1), (5.2), of the form:

$$\|u - u_h\|_{L_\infty(I;H^{-1}(\Omega))} \leq \overset{\circ}{\mathcal{E}}(u_h, h, k, f), \quad (5.30)$$

where $\overset{\circ}{\mathcal{E}}(u_h, h, k, f)$ is a computable global error estimator.

We shall now consider the problem of implementing the error bound (5.30) into an adaptive evolution-Galerkin finite element algorithm to ensure global error control (and thereby reliability) with respect to a prescribed tolerance, TOL, i.e.

$$\|u - u_h\|_{L_\infty(I;H^{-1}(\Omega))} \leq \text{TOL}, \quad (5.31)$$

with the minimum amount of computational effort.

We begin, in Section 5.2.1 by designing an adaptive algorithm to determine the mesh parameters h and k in such a way that (5.31) holds. Then, in Section 5.2.2, we describe how the space-time mesh may be locally adapted so that the actual mesh parameters h and k are "close" to those predicted by the adaptive algorithm.

5.2.1 Adaptive algorithm

For a given tolerance, TOL, we consider the problem of finding a discretisation in space and time $\mathcal{S}^h = \{(T^n, t^n)\}_{n\geq 0}$ such that:

1. $\|u - u_h\|_{L_\infty(I;H^{-1}(\Omega))} \leq \text{TOL}$;
2. $\mathcal{S}^h$ is optimal in the sense that the number of degrees of freedom is minimal.

In order to satisfy these criteria we shall use the *a posteriori* error estimate (5.30) to choose $\mathcal{S}^h$ such that:

1. $\overset{\circ}{\mathcal{E}}(u_h, h, k, f) \leq \text{TOL}$;
2. The number of degrees of freedom in $\mathcal{S}^h$ is minimal.

The term $\mathcal{E}_0(u_0, u^0_{h-}, h)$ is easily controlled at the start of a computation; so here we shall only consider the problem of constructing $\mathcal{S}_h$ in an efficient way to ensure that

$$\mathcal{E}(u_h, h, k, f) \leq \text{TOL}',$$

where $\text{TOL} = \text{TOL}' + \mathcal{E}_0(u_0, u^0_{h-}, h)$. To do so, we first write $\mathcal{E}$ symbolically in terms of two residual terms: one that controls the spatial mesh and one that controls the temporal mesh, i.e. we let

$$\mathcal{E}(u_h, h, k, f) \equiv C_1'\|hR_1'\|_Q + C_2'\|kR_2'\|_Q. \tag{5.32}$$

Simultaneously, we split the tolerance TOL' into a spatial part, TOL_h, and a temporal part, TOL_k. Thus, for reliability we require that the following conditions hold:

$$C_1'\|hR_1'\|_Q \leq \text{TOL}_h, \tag{5.33}$$

$$C_2'\|kR_2'\|_Q \leq \text{TOL}_k. \tag{5.34}$$

To design the space-time mesh $\mathcal{S}^h$, at each time level t^n we decompose the norm in (5.33) into norms over elements $\kappa \in T^n$, and the norm in (5.34) into norms over time slabs as follows:

$$\begin{aligned}
C_1'\|hR_1'\|_Q &\leq C_1'\sqrt{T} \max_{1\leq n\leq M+1} \|h_n R_1'(u_h^n)\| \\
&\leq C_1'\sqrt{NT} \max_{1\leq n\leq M+1} \left(\max_{\kappa\in T^n} \|h_n R_1'(u_h^n)\|_{L_2(\kappa)} \right), \\
C_2'\|kR_2'\|_Q &\leq C_2'\sqrt{T} \max_{1\leq n\leq M+1} \|k_n R_2'(u_h^n)\|,
\end{aligned}$$

where N is the number of elements in the spatial mesh at time t^n. Thus, if

$$C_1'\sqrt{NT}\|h_n R_1'(u_h^n)\|_{L_2(\kappa)} \leq \mathrm{TOL}_h \quad \forall\kappa\in T^n, \text{ for } n=1,\dots,M+1,$$

$$C_2'\sqrt{T}\|k_n R_2'(u_h^n)\| \leq \mathrm{TOL}_k, \quad \text{for } n=1,\dots,M+1,$$

then (5.33) and (5.34) will automatically hold.

For the practical implementation of this method, we consider the following adaptive algorithm for constructing $\mathcal{S}^h$, under the assumption that the final time T is fixed: for each $n=1,2,\dots,M+1$, with T_0^n a given initial mesh and $k_{n,0}$ an initial time step, determine meshes T_j^n with N_j elements of size $h_{n,j}(\mathbf{x})$ and time steps $k_{n,j}$ and the corresponding approximate solution $u_{h,j}$ defined on I_j^n such that, for $j=0,1,\dots,\hat{n}-1$,

$$C_1\|h_{n,j+1}R_1(u_{h,j}^n)\|_{L_2(\kappa)} = \frac{\mathrm{TOL}_h}{\sqrt{N_jT}} \quad \forall\kappa\in T_j^n, \tag{5.35}$$

$$\begin{aligned} C_2\|k_{n,j+1}R_1(u_{h,j}^n)\| + C_3\|k_{n,j+1}R_2(u_{h,j}^n)\| & \\ +C_4\|k_{n,j+1}R_3(u_{h,j}^n)\| + C_5\|k_{n,j+1}R_4(u_{h,j}^n)\| &= \frac{\mathrm{TOL}_k}{\sqrt{T}}, \end{aligned} \tag{5.36}$$

where $I_j^n=(t^{n-1},t^{n-1}+k_{n,j}]$ and $\mathrm{TOL}'=\mathrm{TOL}_h+\mathrm{TOL}_k$. We define $T^n=T_{\hat{n}}^n$, $k_n=k_{n,\hat{n}}$ and $h_n=h_{n,\hat{n}}$, where for each n, the number of trials $\hat{n}$ is the smallest integer such that for $j=\hat{n}$, the following stopping condition is satisfied:

$$C_1\|h_{n,\hat{n}}R_1(u_{h,\hat{n}}^n)\|_{L_2(\kappa)} \leq \frac{\mathrm{TOL}_h}{\sqrt{N_{\hat{n}}T}} \quad \forall\kappa\in T_{\hat{n}}^n, \tag{5.37}$$

$$\begin{aligned} C_2\|k_{n,\hat{n}}R_1(u_{h,\hat{n}}^n)\| + C_3\|k_{n,\hat{n}}R_2(u_{h,\hat{n}}^n)\| & \\ +C_4\|k_{n,\hat{n}}R_3(u_{h,\hat{n}}^n)\| + C_5\|k_{n,\hat{n}}R_4(u_{h,\hat{n}}^n)\| &\leq \frac{\mathrm{TOL}_k}{\sqrt{T}}. \end{aligned} \tag{5.38}$$

By construction, this stopping condition will guarantee reliability of the adaptive algorithm; for efficiency, we try to ensure that (5.37) and (5.38) are satisfied with near equality. Since the final time T is fixed, the time step given by (5.36) may need to be limited to ensure that $t^M+k_{M+1,\hat{n}}=T$. For the implementation of this adaptive algorithm, we shall assume that $T_0^n=T^{n-1}$ for $n=2,3,\dots$.

5.2.2 Mesh modification strategy

In Section 5.2.1 we described the construction of the space-time mesh $\mathcal{S}^h$ to achieve the required error control,

$$\|u-u_h\|_{L_\infty(I;H^{-1}(\Omega))} \leq \mathrm{TOL}.$$

However, to obtain the desired mesh we must employ a suitable mesh modification technique. Temporal adaptation is quite straightforward, since the time step can just be set equal to $k_{n,j+1}$ given by (5.36). For constructing the spatial mesh in two space dimensions we use the red-green isotropic refinement strategy of Bank [4]. Here, the user must first specify a (coarse) *background* mesh upon which any future refinement will be based. Red refinement corresponds to dividing a certain triangle (father) into four similar triangles (sons) by connecting the midpoints of the sides. Green refinement is only temporary and is used to remove any hanging nodes caused by a red refinement. We note that green refinement is only used on elements which have one hanging node. For elements with two or more hanging nodes a red refinement is performed. The advantage of this refinement strategy is that the degradation of the "quality" of the mesh is limited since red refinement is obviously harmless and green triangles can *never* be further refined.

Within this mesh modification strategy it is also possible to de-refine the mesh by removing redundant elements, provided that these do not belong to the original background mesh. Thus, to prevent an overly refined mesh in regions where the solution is smooth the background mesh should be chosen suitably coarse. For the practical implementation of this mesh modification strategy we have used the FEMLAB package developed by K. Eriksson.

5.3 Numerical experiments

In this section, we present some numerical experiments to illustrate the performance of the adaptive algorithm (5.35), (5.36) on the model hyperbolic test problem:

$$\frac{\partial u}{\partial t} + \mathbf{a} \cdot \nabla u = f, \qquad \mathbf{x} \in \Omega,\ t \in I, \tag{5.39}$$

$$u(\mathbf{x}, 0) = u_0(\mathbf{x}), \quad \mathbf{x} \in \Omega, \tag{5.40}$$

where $\Omega = (0,1)^2$, $f = 0$, $\mathbf{a} = (2,1)^T$, subject to the boundary condition $u(0,y) = 1$ for $0 \le y \le 1$, $u(x,0) = (\delta - x)^+/\delta$ for $0 \le x \le 1$. The function u_0 appearing in the initial condition is defined as follows: $u_0(\mathbf{x}) = 0$ for $\mathbf{x} \in \Omega_\delta = (\delta, 1) \times (0,1)$; and for $\mathbf{x} \in \Omega \backslash \Omega_\delta$, $u_0(\mathbf{x})$ is chosen to be the linear function that satisfies the boundary conditions at inflow. We note that initially, for δ small, the solution to this problem has a boundary layer along $x = 0$; this layer then propagates into the domain Ω, and eventually exits through $x = 1$. In the following, we shall let $\delta = 7.8125 \times 10^{-3}$ and $T = 0.55$.

First, we specify the background mesh to be the 5×5 triangular mesh shown in Figure 1(a). This mesh is initially refined in order to properly resolve the boundary layer along $x = 0$ at time $t = 0$, i.e. so that $\mathcal{E}_0(u_0, u^0_{h-}, h) = 0$ (see Figure 1(b)). Numerical results are presented in Figure 2 for $\mathrm{TOL}_h = 0.01$ and $\mathrm{TOL}_k = 0.25$. The estimated error is $\overset{\circ}{\mathcal{E}}(u_h, h, k, f) = 2.4742 \times 10^{-1}$.

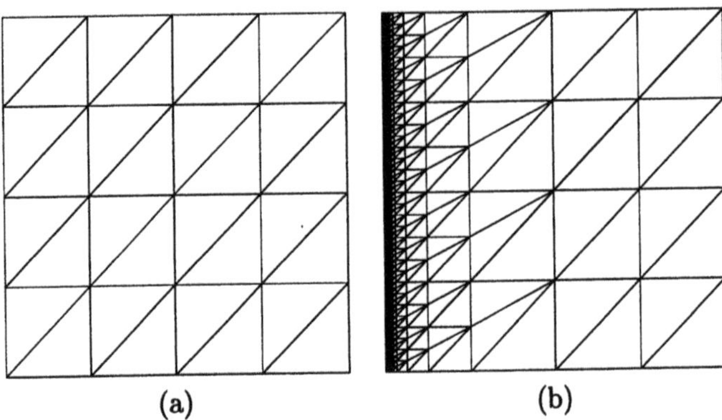

Figure 1. (a) Background mesh, with 25 nodes and 32 elements, and (b) Background mesh adapted to resolve the initial condition, with 526 nodes and 900 elements

In Figures 2(a), 2(b) and 2(c), 2(d) we see that the adaptive algorithm has refined the spatial mesh in parts of the domain where the solution has a steep layer, and has kept the mesh coarse elsewhere. Figures 2(e) and 2(f) show the history of the number of nodes in the spatial mesh against time, and the size of the time step against time, respectively.

We note that the artificial diffusion model introduced in [9] was used in the above experiment with $C_1^{\hat{\epsilon}} = C_2^{\hat{\epsilon}} = 0.2$ and $\hat{\epsilon}_{\max} = 5.0 \times 10^{-4}$.

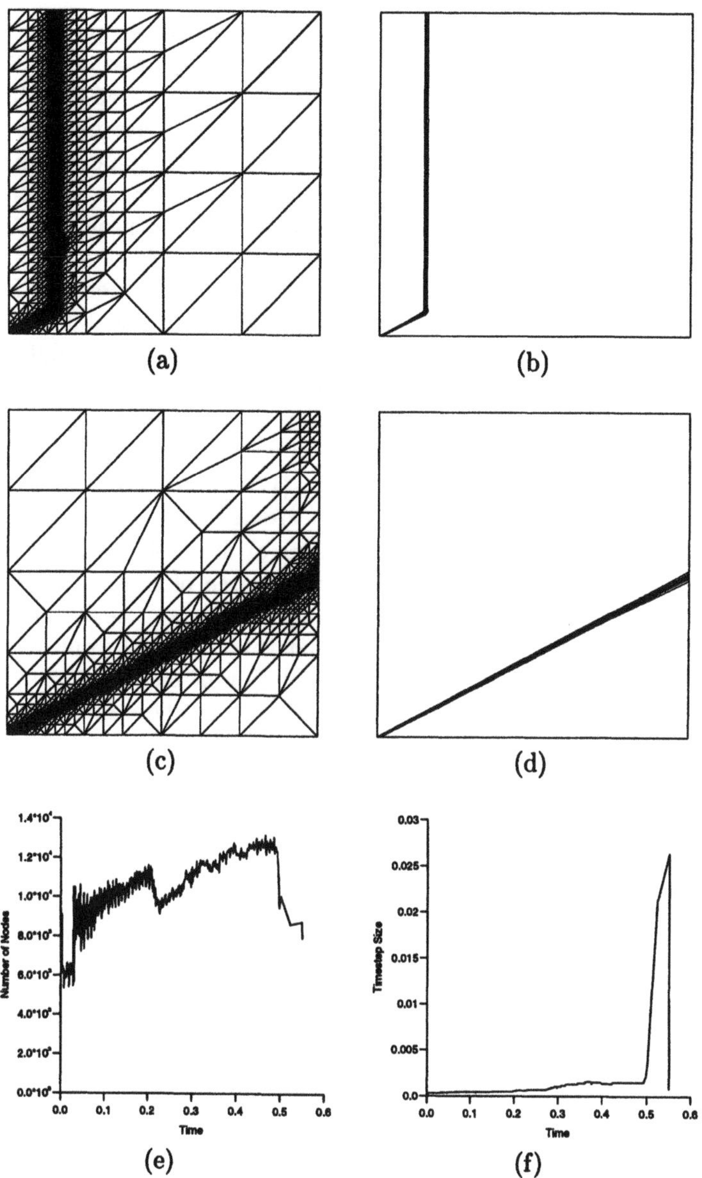

Figure 2. Layer problem for $\mathrm{TOL}_h = 0.01$ and $\mathrm{TOL}_k = 0.25$ with $T = 0.55$, (a) and (b) Mesh and solution (respectively) at time, $t = 0.0714$, with 8425 nodes and 16769 elements, (c) and (d) Mesh and solution (respectively) at final time, $t = 0.55$, with 7914 nodes and 15736 elements, (e) History of nodes against time, and (f) History of time step size against time

6 Conclusions

In this paper we have presented an overview of recent developments that concern the *a posteriori* error analysis of finite element approximations to first-order hyperbolic problems. While for elliptic equations there is a well-established theoretical framework of *a posteriori* error estimation that has been successfully implemented into working adaptive algorithms (see, for example, the monograph of Ainsworth and Oden [1] for an extensive list of literature in this area), very much less is known about hyperbolic and nearly-hyperbolic problems. The main problem both from the theoretical and the practical point of view is the estimation of the stability constant of the dual problem that enters the final error bound. Current work in this area focuses on calculating the stability constant during the course of the computation, rather than using general analytical results which tend to overestimate its size by several orders of magnitude.

Another important area where little progress has been made concerns the derivation of two-sided *a posteriori* error bounds for finite element approximations of hyperbolic problems (see, however, [17] where two-sided *a posteriori* bounds have been established for the locally created part of the global error).

Predictably, the *a posteriori* error analysis of finite element and finite volume methods will be an important and fruitful area of research over the next decade; theoretical work in this field is already making an impact on large-scale engineering computations, and it is clear that this trend will continue. Indeed, guaranteed error control for the numerical solution of partial differential equations will be the norm rather than the exception for, to quote Babuška, one must have the confidence "to sign the blueprint".

This article is merely a tiny scratch on the surface of a large body of theory that will, we hope, soon be unearthed.

References

1. Ainsworth, M. and Oden, J.T. (1996). *A Posteriori Error Estimation in Finite Element Analysis, Series in Computational and Applied Maths.*, Elsevier.

2. Balland, P. and Süli, E. (1994). Analysis of the cell vertex scheme for hyperbolic problems with variable coefficients. *Oxford University Computing Laboratory Technical Report NA94/01. SIAM J. of Numerical Analysis.* (To Appear.)

3. Bernardi, C. (1989). Optimal finite-element interpolation on curved domains. *SIAM J. of Numerical Analysis*, **26**, 1212–1240.

4. Bank, R. (1985). *PLTMG user's guide. Technical Report*, **4**, University of California, San Diego.

5. Ciarlet, P.G. (1978). *The Finite Element Method for Elliptic Problems*, North Holland, Amsterdam.

6. Friedrichs, K.O. (1958). Symmetric positive linear differential equations. *Communications in Pure and Applied Maths.*, **11**, 333–418.

7. Handscomb, D.C. (1995). Error of linear interpolation on a triangle. *Oxford University Computing Laboratory Technical Report NA95/09.*

8. Houston, P. and Süli, E. (1995). Adaptive Lagrange-Galerkin methods for unsteady convection–dominated diffusion problems. *Oxford University Computing Laboratory Technical Report NA95/24.*

9. Houston, P. and Süli, E. (1996). On the design of an artificial diffusion model for the Lagrange-Galerkin method on unstructured triangular grids. *Oxford University Computing Laboratory Technical Report NA96/07.*

10. Johnson, C. (1990). Adaptive finite element methods for diffusion and convection problems. *Computer Methods in Applied Mechanics and Engineering*, **82**, 301–322.

11. Johnson, C. (1994). A new paradigm for adaptive finite element methods. *The Maths. of Finite Elements and Applications. Highlights 1993*, Editor: J.R. Whiteman, John Wiley, 105–120.

12. Johnson, C. and Hansbo, P. (1992). Adaptive finite element methods in computational mechanics. *Computer Methods in Applied Mechanics and Engineering*, **101**, 143–181.

13. Estep, D., Eriksson, K., Hansbo, P. and Johnson, C. (1995). Introduction to adaptive methods for differential equations. *Acta Numerica*, Cambridge University Press, 105–158.

14. Lax, P.D. and Phillips, R.S. (1960). Local boundary conditions for dissipative symmetric linear differential operators. *Communications in Pure and Applied Maths.*, **13**, 427–455.

15. Mackenzie, J., Sonar, T. and Süli, E. (1994). Adaptive finite volume methods for hyperbolic problems. *The Mathematics of Finite Elements and Applications Highlights 1993*, Editor: J.R. Whiteman, John Wiley, 289–298.

16. Mackenzie, J., Süli, E. and Warnecke, G. (1994). A posteriori error estimates for the cell-vertex finite volume method. *Adaptive Methods: Algorithms, Theory and Applications*, Editors: W. Hackbusch and G. Wittum, Vieweg, Braunschweig, **44**, 221–235.

17. Mackenzie, J., Süli, E. and Warnecke, G. (1995). A posteriori error analysis of Petrov-Galerkin approximations of Friedrichs systems. *Oxford University Computing Laboratory Technical Report. NA95/01.* (Submitted for Publication.)

18. Morton, K.W. and Süli, E. (1991). Finite volume methods and their analysis. *IMA J. of Numerical Analysis*, **11**, 241–260.

19. Morton, K.W. and Süli, E. (1994). A posteriori and a priori error analysis of finite volume methods. *The Maths. of Finite Elements and Applications Highlights 1993*, Editor: J.R. Whiteman, John Wiley, 267–288.

20. Morton, K.W. and Süli, E. (1995). Evolution Galerkin methods and their supraconvergence. *Numerische Math.*, **71**, 331–355.

21. Sonar, T. and Süli, E. (1994). A dual graph-norm refinement indicator for finite volume approximations of the Euler equations. *Oxford University Computing Laboratory Technical Report NA94/09.* (Submitted for Publication.)

22. Süli, E. (1989). Finite volume methods on distorted meshes: stability, accuracy, adaptivity. *Oxford University Computing Laboratory Technical Report NA89/06.*

23. Süli, E. (1992). The accuracy of cell vertex finite volume methods on quadrilateral meshes. *Maths. of Computation*, **59**, 359–382.

24. Süli, E. (1991). The accuracy of finite volume methods on distorted partitions. *The Maths. of Finite Elements and Applications VII*, Editor: J.R. Whiteman, Academic Press, London, 253–260.

25. Süli, E. (1996). A posteriori error analysis and global error control for adaptive finite element approximations of hyperbolic problems. *Proc. 16th Int. Conf. on Numerical Analysis, Pitman Research Notes in Maths.*, Longman Scientific and Technical, Harlow.

Approximation of Multidimensional Hyperbolic Partial Differential Equations

K.W. Morton

Oxford University Computing Laboratory

1 Introduction

Ten years ago, in the corresponding State of the Art lecture (see [34]), attention was properly focussed on the important development of solution-adaptive algorithms for systems of first order conservation laws in one dimension, that is

$$\mathbf{w}_t + \mathbf{f}_x = \mathbf{0} \quad \text{or} \quad \mathbf{w}_t + A\mathbf{w}_x = \mathbf{0}, \tag{1.1}$$

where $\mathbf{f} \equiv \mathbf{f}(\mathbf{w})$ is the vector of fluxes and $A \equiv A(\mathbf{w})$ is the Jacobian matrix $\partial\mathbf{f}/\partial\mathbf{w}$. A key objective was to obtain second or third order accuracy where the solution is smooth, but to combine this with oscillation-free approximation of the discontinuities such as shocks which are characteristic of such nonlinear conservation laws. Extensions to practically important two- and three-dimensional problems were regarded as "conceptually easy" and implemented in a straightforward way, even on unstructured meshes. Indeed, these methods have by this means become standard tools in the field. This is particularly true in aeronautical engineering where, in combination with Runge–Kutta time-stepping, they have become one of the two commonest approaches to modelling the steady equations of inviscid, compressible gas flow.

However, the practical and theoretical limitations of this approach to multidimensional problems has been widely recognised in recent years. In this article we will describe steps that have been taken to overcome these limitations which at the same time address three related issues:

- the need to approximate evolution operators which require characteristic cones as well as characteristic paths for their definition — as for the second order wave equation in two dimensions;
- introduction of a more general framework for steady problem algorithms which includes the main alternative to the above methods, namely finite volume schemes;
- widening of the viewpoint to include algorithms designed for steady and those for unsteady problems.

2 Scalar problems in 1D and 2D

2.1 Linear advection in 1D

This special case of (1.1), which we write as $u_t + au_x = 0$ where a is constant, has been called by Ron Mitchell [26] the "ultimate embarrassment of numerical analysts" when approximated on a fixed, uniform mesh. Any of the standard finite difference or finite element methods, for which typical stencils are shown in Figure 1, suffer from one or more serious deficiencies such as amplitude damping, dispersion, spurious oscillation or limited stability. For specialised problems, the Moving Finite Element method (see [3]) provides the perfect answer; but the availability of a good fixed mesh approximation is necessary as a basis for the general class of hyperbolic problems in both one and more dimensions.

It is now generally accepted that the best approach is to make some use of the Lagrangian formulation, in which we write

$$\frac{Du}{Dt} = 0, \quad \text{where} \quad \frac{D}{Dt} \equiv \frac{\partial}{\partial t} + a\frac{\partial}{\partial x}; \tag{2.1}$$

that is, constancy of the solution along the characteristics $\mathrm{d}X/\mathrm{d}t = a$ is taken as the basis for approximating the evolution operator. In the finite difference framework, this leads to the semi-Lagrangian methods (where the "semi" refers to the use of a fixed mesh, as opposed to the moving mesh of the free-Lagrangian methods); and in a finite element framework we obtain the Lagrange–Galerkin and characteristic-Galerkin methods — see [28] for an introduction to these methods and further references.

At the present time the state of the art for this problem, for methods that have the potential of extension to general systems of equations in multi-dimensions, consists of three main elements:

(i) a basic piecewise constant or piecewise linear approximation to give $U^n(x)$ to approximate $u(x, t_n)$;

(ii) recovery (or reconstruction) by continuous piecewise parabolic functions, giving $\tilde{u}^n = RU^n$ and yielding third order accuracy for smooth solutions;

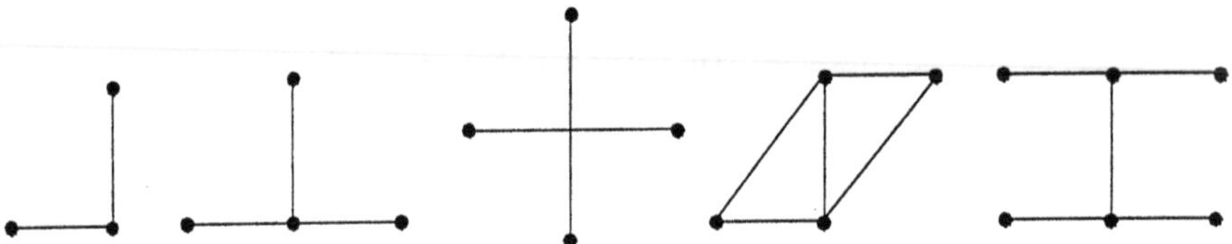

Figure 1. Meshpoint stencils for, from left to right, an upwind, Lax-Wendroff, leap-frog, angled-derivative and Crank-Nicolson difference scheme

(iii) solution-adaptivity, to deal with propagating discontinuities in the solution, either in the recovery process (for example by slope limiters) or by means of artificial dissipation terms.

The original scheme of this type is the Piecewise Parabolic Method (PPM) of Colella and Woodward [10], but there are now several schemes which are very similar in form. There are also two main formulations for such methods. The first is based on a direct approximation of the evolution operator

$$E(\Delta t) : u(\cdot, t) \mapsto u(\cdot, t + \Delta t); \tag{2.2a}$$

and we denote by E_Δ the transport of a value along a straight characteristic $\mathrm{d}X/\mathrm{d}t = a$,

$$(E_\Delta v)(\cdot) = v(\cdot - a\Delta t), \tag{2.2b}$$

which is exact in the present case but is only an approximation if the characteristics have curved trajectories. Then if P represents a projection onto the basic approximation space, the whole algorithm can be written as

$$U^{n+1} = PE_\Delta RU^n, \tag{2.3}$$

and is hence called a PERU scheme. It essentially consists of two steps: the recovery process which may be solution-adaptive (for example by imposing monotonicity preservation, positivity or the TVD condition) and would therefore then be nonlinear; and the projection and evolution carried out as one step. In a finite difference scheme, such as a semi-Lagrangian method, the projection would typically consist of interpolation at the mesh points; and hence the characteristics would only need to be drawn back from these points to give

$$U_j^{n+1} = (E_\Delta \tilde{u}^n)(x_j) = \tilde{u}^n(x_j - a\Delta t). \tag{2.4}$$

In a finite element setting, the projection would normally be the L_2 projection, although for a continuous approximation space a mixed norm projection might be used. For a survey of these methods see [28] or [27].

In the alternative formulation, we work from the conservation law form $u_t + f_x = 0$, with $f = au$, and by integrating over a $\Delta x \times \Delta t$ box in (x,t)-space obtain the formula

$$(U_i^{n+1} - U_i^n)\Delta x + (F_{i+\frac{1}{2}}^{n+\frac{1}{2}} - F_{i-\frac{1}{2}}^{n+\frac{1}{2}})\Delta t = 0. \tag{2.5a}$$

Here U_i^n represents the average of U between $x_{i-\frac{1}{2}}$ and $x_{i+\frac{1}{2}}$ at time level t_n and $F_{i-\frac{1}{2}}^{n+\frac{1}{2}}$ the average of $f(U)$ at $x_{i-\frac{1}{2}}$ between time levels t_n and t_{n+1}. Thus this is most easily regarded as a finite volume formulation using piecewise constant approximations in each cell; but the same schemes are often interpreted as finite difference schemes, as they were in [34]. In order to relate to the first formulation, we could also write

$$F_{i-\frac{1}{2}}^{n+\frac{1}{2}} = \frac{1}{\Delta t}\int_0^{\Delta t} f(E(\tau)\tilde{u}_{i-\frac{1}{2}}^n)\mathrm{d}\tau, \tag{2.5b}$$

where $E(\tau)\tilde{u}_{i-\frac{1}{2}}^n = (E(\tau)\tilde{u}^n)(x_{i-\frac{1}{2}})$, and in general $E(\tau)$ would need to be approximated as in (2.2b).

A useful and demanding test problem has been suggested in [19] and [20], consisting of the convection of various waveforms across a domain of width 2.25 with periodic boundary conditions. In Figure 2 we show the result of computations over 100 time steps with the CFL number 0.45 for the first order upwind scheme, the second order Lax–Wendroff scheme and for a solution-adaptive mix of the two — namely, a PERU scheme obtained from recovery by discontinuous piecewise linear functions using the Davis slope limiter. The last is typical of the schemes which were current ten years ago and clearly represents a major improvement over the first two schemes, combining the best features of both. However, after 1070 time steps we see from the top picture of Figure 3 that there is a significant loss of accuracy with this scheme at even quite smooth peaks. General arguments put forward by Leonard (see [19] and [20]) and others suggest that schemes with an odd order of accuracy on a uniform mesh are to be preferred for advection problems; principally, this is because the leading error term then has a damping rather than a dispersive effect, so that waves that can be properly represented on the mesh travel at close to the correct speed and higher frequency waves which might be aliassed are damped. This is confirmed by the family of Lagrange–Galerkin schemes based on splines of order s, which on a uniform mesh have order of accuracy $2s - 1$ (see Childs and Morton [8]); in particular, use of continuous piecewise linear basis functions yields the well-known third order accurate scheme, the results for which are given as the second picture of Figure 3. These are remarkably accurate for a non-adaptive scheme; moreover, exactly the same results can be obtained by using a basic piecewise constant approximation followed by recovery with quadratic splines — see [8]. This is a strong motivation for the PPM scheme; for by using continuous piecewise parabolic functions in the recovery we can maintain an accuracy comparable with that obtained from quadratic splines but leave free one parameter on each mesh interval to allow some solution-adaptivity. The result of such a scheme is shown as the bottom picture of Figure 3. It clearly again combines the best features of the top two schemes, and can be taken as representing the present state of the art. It has been obtained by using a fourth order difference, as in a fourth order artificial dissipation, to distinguish a smooth peak from a spurious oscillation and switching to the RCFVM scheme when the latter is detected — see [28] and [13] for details.

Before we end this section, we should note that the third order accuracy attained by the Lagrange–Galerkin method with piecewise linear basis functions is an example of superconvergence; and it is only attained on a uniform mesh with

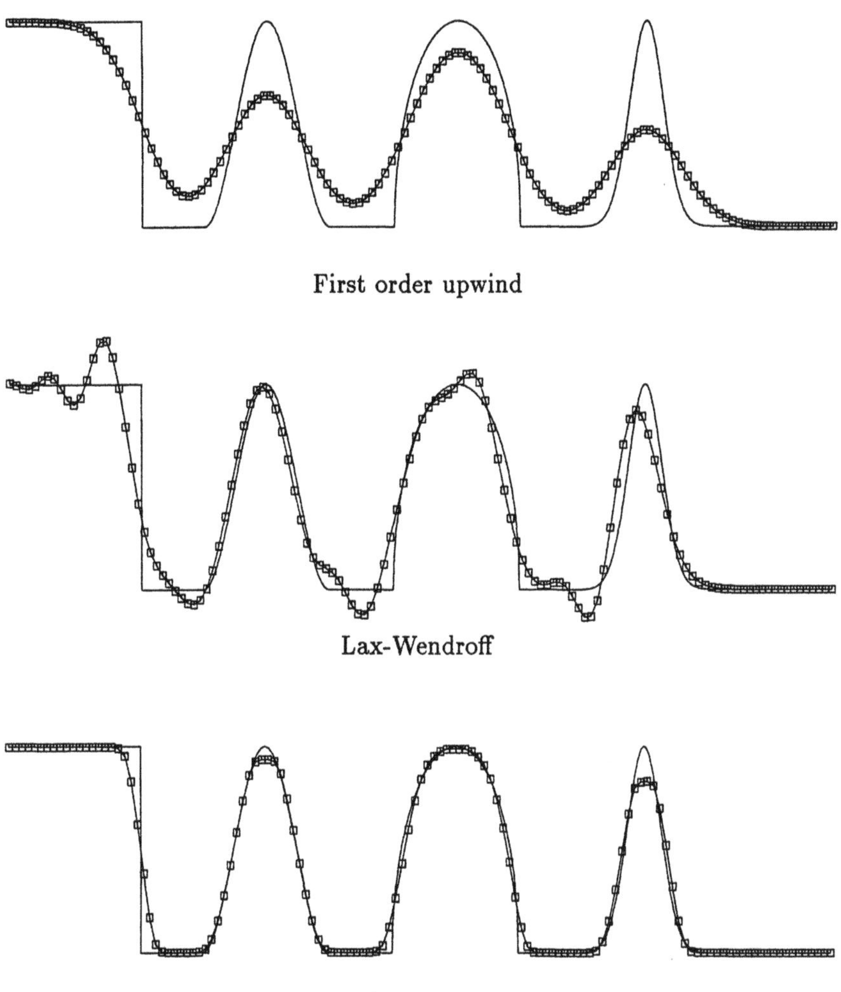

Figure 2. Results after 100 time steps at $\nu = 0.45$ for the Leonard–Niknafs test problem and three standard schemes; RCFVM $\equiv$ recovered characteristic finite volume method

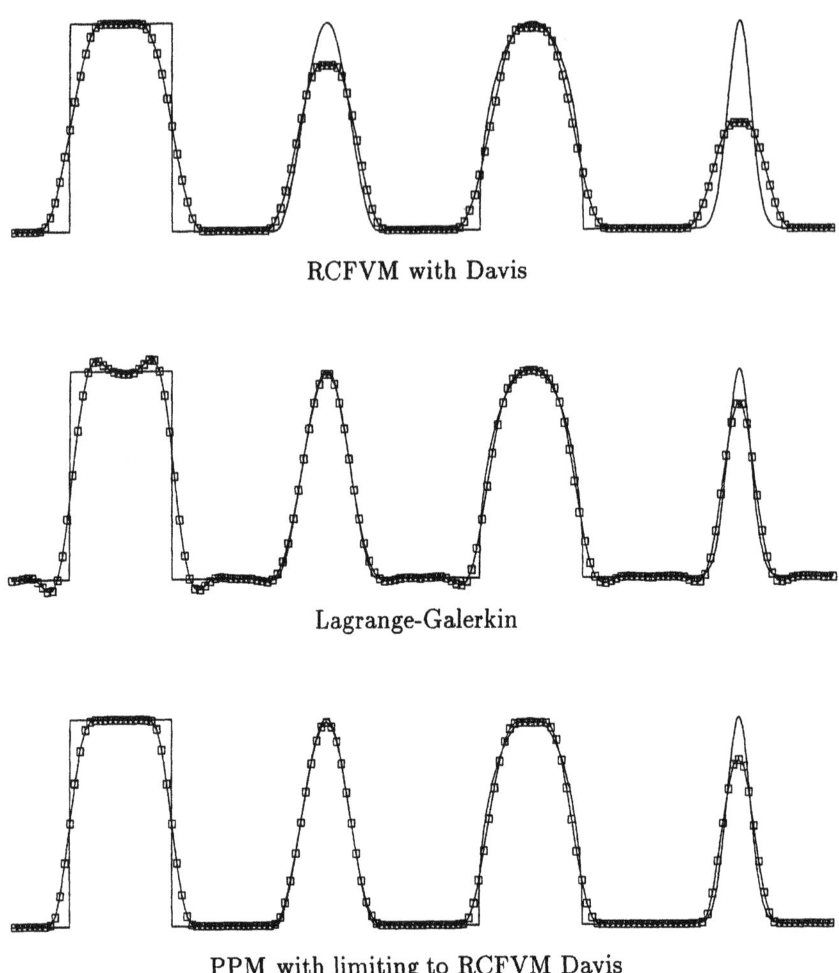

Figure 3. As for Figure 2, but after 1070 time steps and with the RCFVM scheme, the Lagrange–Galerkin method based on continuous piecewise linear basis functions, and a Piecewise Parabolic Method (PPM) limited by a monitor function

constant coefficient linear advection. On a general mesh and with a smoothly varying advection speed, only the second order accuracy to be anticipated with this approximation space can be established. In [31] a number of results are proved to show that in a PERU scheme and for smooth solutions the order of approximation attained is that provided by the recovery approximation space.

2.2 Convection in two dimensions

In the previous section, which concentrated on constant coefficient advection, the two formulations described before the test problem gave essentially the same algorithms; but here there are many more options for us to consider. We shall also assume a general characteristic velocity field $\mathbf{v}$ and suppose the convective terms are only part of the full equation system; thus we write

$$\frac{\partial u}{\partial t} + \mathbf{v}\cdot\nabla u = s, \quad \text{or} \quad \frac{Du}{Dt} = s, \tag{2.6}$$

where s may represent a source term or other differential terms whose approximation is of less concern to us.

We first outline two examples of the direct use of the evolution operator. One is the semi-Lagrangian method, used for example for weather forecasting and pollutant transport in the atmosphere on a structured quadrilateral mesh. As indicated on the left of Figure 4, the core of the algorithm is to draw back a straight characteristic from a mesh point (x_i, y_i) and to interpolate for the value at its foot (x_i', y_i'). Interpolation is typically cubic, either cubic spline, Hermite cubic, cubic Lagrange or a monotonicity-preserving variant; and the velocity

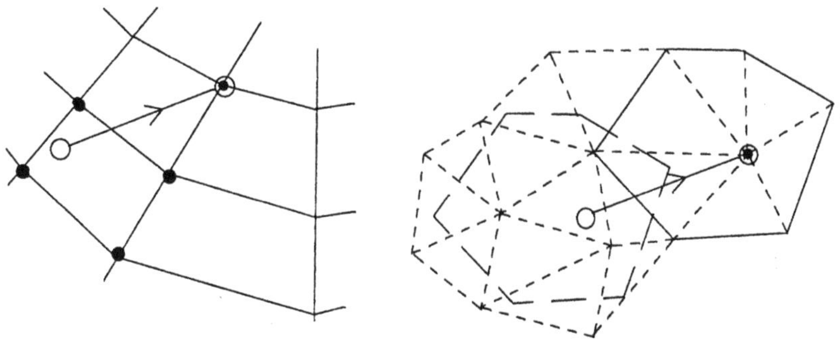

Figure 4. Mapping back straight line characteristics: on the left, for the semi-Lagrangian scheme on a structured quadrilateral mesh; on the right, for a Lagrange-Galerkin scheme on an unstructured triangular mesh

for the characteristic path may be based on an explicit Euler method, i.e. using $\mathbf{v}(x_i, y_i, t_n)$, or on a midpoint rule which would involve an iteration in a nonlinear problem to find the midpoint between (x_i, y_i) and (x_i', y_i').
The result will be an explicit update of the form

$$U_i^{n+1} = \sum_{(j)} c_{ij} U_j^n + \sum_{(j)} d_j S_j^n. \tag{2.7}$$

Note that an important advantage of this scheme is that there is no fundamental stability limit on the time step; in practice, time steps are limited by accuracy considerations although in principle a stability limitation could arise from the characteristic path computation. Note that this property of the characteristic-based schemes was not exploited in the previous section because of the wish to compare on equal terms with other schemes.

The other example of directly using the evolution operator is the Lagrange–Galerkin method as applied to the incompressible Navier–Stokes equations, for which it has long been a popular scheme — see, for example, [6] and [38]. As illustrated on the right of Figure 4, it is based on piecewise linear approximations over a general triangular mesh, with basis functions $\phi_i(\mathbf{x})$; the unknown to which it is applied is the flow velocity itself, given by the equation

$$\frac{\partial \mathbf{v}}{\partial t} + \mathbf{v}\cdot\nabla\mathbf{v} + \nabla p = \nu\nabla^2\mathbf{v}, \tag{2.8}$$

where p is the pressure and ν the viscosity. This is then approximated by

$$\begin{aligned}\frac{1}{\Delta t}(\mathbf{V}^{n+1}, \phi_i \mathbf{e}) \quad + \quad (\nu\nabla\mathbf{V}^{n+1} - P^{n+1}, \nabla\phi_i\mathbf{e}) \\ = \frac{1}{\Delta t}\int \mathbf{e}\cdot\mathbf{V}^n(\mathbf{x}')\phi_i(\mathbf{x})\mathrm{d}\mathbf{x},\end{aligned} \tag{2.9}$$

where P^{n+1} should be regarded as a scalar-valued matrix and $\mathbf{e} = (1,0)^T$ or $(0,1)^T$ to give the two components of $\mathbf{V}^n$. The foot of the characteristic $\mathbf{x}'$ drawn back from $\mathbf{x}$ is given by

$$\mathbf{x}' = \mathbf{x} - \Delta t\mathbf{V}^n(\mathbf{x}), \tag{2.10}$$

and the core of the method lies in the integral on the right-hand side of (2.9). That on the left, for $\mathbf{V}^{n+1}$, can be carried out exactly to give a mass matrix; but that on the right, though of similar form, involves integration of the basis function $\phi_i(\mathbf{x})$ against shifted basis functions $\phi_j(\mathbf{x}')$ and has to be approximated by a carefully chosen quadrature scheme. The remaining terms on the left, arising from the viscous terms and the pressure, together with a discretisation of the incompressibility condition $\nabla\cdot\mathbf{v} = 0$, yield a linear Stokes problem for $\mathbf{V}^{n+1}$ and P^{n+1} which has to be solved at each time step. Thus use of the Lagrange–Galerkin formulation has removed the nonlinearity in the original equation (2.9) and yielded an accurate, stable approximation — see [46] for an analysis of the scheme.

Let us now consider a finite volume formulation for the scalar conservation law

$$u_t + f_x + g_y = 0, \tag{2.11}$$

where $\partial f/\partial u = a$, $\partial g/\partial u = b$. On a general triangular mesh we also construct a dual mesh, the most common being the median dual shown in Figure 5 because of its rôle in the interpretation of the Galerkin finite element method. Suppose that V_i is the measure of the dual mesh cell Ω_i around the vertex or node (x_i, y_i) and Γ_i its boundary. Then integration of (2.11) over this cell and application of Gauss' theorem gives the equation

$$\frac{\partial \bar{u}_i}{\partial t} + \int_{\Gamma_i} [f\mathrm{d}y - g\mathrm{d}x] = 0, \tag{2.12}$$

where $\bar{u}_i = (1/V_i)\int_{\Omega_i} u\mathrm{d}x\mathrm{d}y$. This forms the basis for generalising the scheme (2.5) to two dimensions, with further generalisation to a tetrahedral mesh in three dimensions being straightforward. The greatest practical use of such schemes has been in aerodynamics, for the Euler equation system of conservation laws which model inviscid compressible flow. We shall give the decomposition of this system into scalar laws in the next section, and the main basis of the methods has been the one-dimensional TVD schemes reviewed in [34]. So here we will summarise only their key features. The time-stepping of (2.12) is usually carried out with a Runge–Kutta scheme and special TVD-preserving schemes

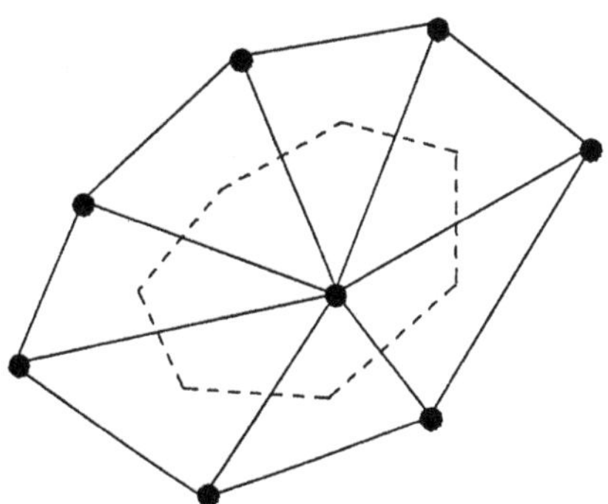

Figure 5. A triangular mesh and its median dual as used in a vertex-centred finite volume scheme

have been developed for this purpose in [9], [42] and [43]. With the resulting relatively short intermediate time steps, little use is made of tracing back the characteristics in order to approximate the evolution operator in a formula of the type given in (2.5b), although the extended stencils in an ENO scheme reflect this same approach. Instead, various approximate Riemann solvers are used as in the one-dimensional schemes of [34]; for this purpose, the boundary integral in (2.12) is broken up into straight line segments of Γ_i to give contributions of the form

$$h(u)\Delta l \equiv [n_x f(u) + n_y g(u)]\Delta l, \tag{2.13}$$

where (n_x, n_y) is the unit outward normal to the segment and Δl its length. The Riemann solver is then applied to the one-dimensional normal flux $h(u)$. That is, no account is taken in this approximation to the evolution operator of the tangential propagation of waves from the interface; this effect appears only through the intermediate time steps of the Runge–Kutta scheme. This contrasts with the full characteristic–Galerkin method based on piecewise constants, and using the Brenier transport collapse operator to approximate the evolution operator in the case of a nonlinear flux function — see [8] and [23]: thus in one dimension and for small time steps, this characteristic–Galerkin method is identical with the present scheme; but in two dimensions on a rectangular mesh, mixed derivative terms arise from tangentially propagating waves and the effects of the cell corners — see [24] and [28]. This first example of the shortcomings arising from the straightforward application of one-dimensional difference schemes has also been pointed out and remedied by LeVeque [21], [22].

However, to continue the description of these finite volume methods which we shall want to refer to in later sections, we need to consider the recovery stage, for without recovery the piecewise constant scheme can give only first order accuracy. Many alternative approximations have been proposed and used: Abgrall in [1] and [2] has extended the ENO reconstruction procedures to triangular meshes and Sonar [44] has made use of radial basis functions, as well as other approximations; for the present discussion, the approach of Barth in [5] and [4] is most convenient. The averages in (2.12) define an averaging operator A by

$$\bar{u}_i \equiv (Au)_i := \frac{1}{V_i}\int_{\Omega_i} u \mathrm{d}\Omega, \tag{2.14}$$

where Ω_i is the dual mesh cell around the node (x_i, y_i). Let $R^{(k)}$ be the recovery operator which defines a local k^{th} order polynomial over each cell Ω_i, to be obtained from the set of cell averages $\bar{u} \equiv \{\bar{u}_i\}$. That is, we have

$$u^{(k)} \equiv R^{(k)}\bar{u}, \quad \text{where } u^{(k)} = \sum_{m+n\le k} \alpha_{m,n}(x - x_c)^m (y - y_c)^n \tag{2.15}$$

and (x_c, y_c) is the centroid of the cell; the coefficients here will be different for each cell and will be calculated by making use of the cell averages over a patch of cells. A key requirement is that

$$\bar{u} = Au^{(k)}, \quad \text{or} \quad AR^{(k)} = I; \tag{2.16}$$

that is, on each cell Ω_i the average $\bar{u}_i$ is reproduced from $u^{(k)}$ so that $R^{(k)}$ is a right-inverse of A. Then the recovery operator is called k-exact if it recovers exactly a kth degree polynomial u substituted into (2.14) to obtain the cell averages; that is, $R^{(k)}$ is a left-inverse of A restricted to the domain $\mathbb{P}_k$ of kth degree polynomials,

$$R^{(k)}A|_{\mathbb{P}_k} = I. \tag{2.17}$$

In [5] more than the minimum number of neighbouring cells is used in the construction of $u^{(k)}$; if the set around Ω_i is denoted by $\mathcal{N}_i$, then a least squares procedure is used to obtain the coefficients $W_{m,n,j}$ in the formula

$$\alpha_{m,n} = \sum_{j \in \mathcal{N}_i} W_{m,n,j}\bar{u}_j. \tag{2.18}$$

This can be carried out before the flow computation starts, and needs to be done once only for a mesh if the coefficients can be stored.

The resulting recovered functions will usually be discontinuous across the cell boundary Γ_i, hence defining a one-dimensional Riemann problem at each point of the boundary. Any of the techniques described in [34] can then be applied to these problems, and for higher order recovery this choice is relatively unimportant; it is important, however, to use at least two-point Gaussian quadrature along each edge of Γ_i. For completeness and comparison with other schemes defined below, we note that one preferred scheme for the scalar problem is to decompose the $h(u)$ defined in (2.13) into its monotone increasing and decreasing parts, $h(u) \equiv h^+(u) + h^-(u)$; then if $\tilde{u}_{in}$ denotes the interior value at the boundary point given by the recovery process, and $\tilde{u}_{out}$ the exterior value in the neighbouring cell, the outgoing flux to be used in the update derived from (2.12) is

$$h^+(\tilde{u}_{in}) + h^-(\tilde{u}_{out}). \tag{2.19}$$

In the original presentation by Osher and Solomon [33] of an unrecovered scheme for a system of conservation laws in two dimensions, this decomposition was generalised by using the set of simple waves that are supported by the system.

2.3 Decomposition of the Euler equations into scalar problems

As we shall use this system of conservation laws to illustrate many of the techniques described in this article, we introduce here some of its key properties. In two dimensions, the system consists of four conservation laws for mass, two components of momentum, and energy: in terms of the density ρ, the two components of velocity u and v, and the specific total energy e, it is written

$$\mathbf{w}_t + \mathbf{f}_x + \mathbf{g}_y = \mathbf{0}, \tag{2.20}$$

where

$$\mathbf{w} = \begin{bmatrix} \rho \\ \rho u \\ \rho v \\ \rho e \end{bmatrix}, \quad \mathbf{f} = \begin{bmatrix} \rho u \\ \rho u^2 + p \\ \rho uv \\ \rho uH \end{bmatrix}, \quad \mathbf{g} = \begin{bmatrix} \rho v \\ \rho uv \\ \rho v^2 + p \\ \rho vH \end{bmatrix} \tag{2.21}$$

and, for an adiabatic gas with gas constant γ, the enthalpy H and pressure p are given by

$$H = e + \frac{p}{\rho} = \frac{\gamma}{\gamma - 1}\frac{p}{\rho} + \tfrac{1}{2}(u^2 + v^2); \tag{2.22}$$

for several purposes it is also useful to introduce the sound speed c for which we have $c^2 = \gamma p/\rho$. The Mach number M is defined as $(u^2 + v^2)^{\frac{1}{2}}/c$ and the flow is subsonic if $M < 1$ and supersonic if $M > 1$. The flux $\mathbf{h}(\mathbf{w}) \equiv n_x\mathbf{f}(\mathbf{w}) + n_y\mathbf{g}(\mathbf{w})$, defined as in (2.13), can be expressed in terms of the flow velocity in the same direction, $q \equiv n_x u + n_y v$, as

$$\mathbf{h}(\mathbf{w}) = [\rho q,\ \rho q u + n_x p,\ \rho q v + n_y p,\ \rho q H]^T. \tag{2.23}$$

In this way the decomposition associated with $\mathbf{h}$ can be deduced from that for the one-dimensional problem.

In the one-dimensional case, the key formulae that we shall need below are as follows. With now $\mathbf{w} = [\rho, \rho u, \rho e]^T$ and $\mathbf{f}$ similarly reduced to $[\rho u, \rho u^2 + p, \rho uH]^T$, the Jacobian $A \equiv \partial\mathbf{f}/\partial\mathbf{w}$ can be diagonalised, which we express in terms of the matrix of right eigenvectors as

$$AR = R\Lambda := \begin{bmatrix} 1 & 1 & 1 \\ u - c & u & u + c \\ H - uc & \frac{1}{2}u^2 & H + uc \end{bmatrix} \begin{bmatrix} u - c & 0 & 0 \\ 0 & u & 0 \\ 0 & 0 & u + c \end{bmatrix}. \tag{2.24}$$

Thus when the flow is subsonic, one characteristic speed is negative, one is positive and the other has the sign of u; when the flow is supersonic all have the same sign. The corresponding characteristic form of the equations is

$$\frac{\partial\mathbf{q}}{\partial t} + \Lambda\frac{\partial\mathbf{q}}{\partial x} = \mathbf{0}, \tag{2.25a}$$

where the characteristic variables are given by

$$\partial\mathbf{q} = \begin{bmatrix} \partial p/\rho c - \partial u \\ \partial p/c^2 - \partial\rho \\ \partial p/\rho c + \partial u \end{bmatrix}; \tag{2.25b}$$

note that when these differential relations can be integrated they yield the Riemann invariants associated with each set of characteristics.

The right eigenvectors $[\mathbf{r}^{(1)}, \mathbf{r}^{(2)}, \mathbf{r}^{(3)}]$ making up the columns of R in (2.24) can be used to decompose the vector of unknowns $\mathbf{w}$ as

$$\mathbf{w} = \sum_{s=1}^{3} \alpha_s \mathbf{r}^{(s)}$$
$$\equiv \frac{\rho}{2\gamma}\begin{bmatrix} 1 \\ u-c \\ H-uc \end{bmatrix} + \frac{\rho(\gamma-1)}{\gamma}\begin{bmatrix} 1 \\ u \\ \frac{1}{2}u^2 \end{bmatrix} + \frac{\rho}{2\gamma}\begin{bmatrix} 1 \\ u+c \\ H+uc \end{bmatrix}; \quad (2.26a)$$

and the corresponding decomposition of the flux vector is then

$$\mathbf{f} = \sum_{s=1}^{3} \alpha_s \lambda^{(s)} \mathbf{r}^{(s)} \tag{2.26b}$$

where $\lambda^{(1)}, \lambda^{(2)}, \lambda^{(3)}$ are the diagonal elements of Λ.

We shall also be concerned with the steady system obtained from (2.20) when $\mathbf{w}_t \equiv 0$; introducing the Jacobian $B = \partial\mathbf{g}/\partial\mathbf{w}$ for the flux $\mathbf{g}$ this can be written

$$A\mathbf{w}_x + B\mathbf{w}_y = 0. \tag{2.27}$$

Now let us suppose A is nonsingular, for example $0 < u < c$ or $u > c$. Then diagonalising $A^{-1}B$ we obtain

$$A^{-1}B \equiv L^{-1}\Lambda L \Rightarrow L\mathbf{w}_x + \Lambda L\mathbf{w}_y = 0. \tag{2.28}$$

This system always has two real streamline characteristics, $\lambda = v/u$, with characteristic variables given by $\partial\mathbf{q} := L^{-1}\partial\mathbf{w}$, that is

$$\partial p/c^2 - \partial\rho \quad \text{and} \quad \partial p + \rho(u\partial u + v\partial v), \tag{2.29}$$

which together carry the entropy and the enthalpy along the streamline. But the other pair depend on $\beta := (M^2 - 1)^{\frac{1}{2}}$ and are therefore complex when the flow is subsonic:

$$\lambda_\pm = \frac{\beta v \pm u}{\beta u \mp v}, \quad \partial q_\pm = \partial p \pm \beta^{-1}\rho(u\partial v - v\partial u), \tag{2.30}$$

where we should note that $u\partial v - v\partial u = (u^2 + v^2)\partial\theta$ for $\tan\theta = v/u$. Thus the steady Euler equations form a full hyperbolic system when the flow is supersonic and can be marched forward as in the unsteady one-dimensional case; but when the flow is subsonic the system has a mixed elliptic/hyperbolic character and cannot be solved by a marching algorithm.

We shall add to these formulae in later sections as we adopt a more truly multi-dimensional approach; but it is hoped that this outline description of the Euler equation system is sufficient for the reader to understand the straightforward extension of one-dimensional schemes that we have so far described, and to see that this has its limitations in the case of subsonic flow.

3 Evolutionary hyperbolic problems in several space dimensions

In the general linear system of equations in two dimensions

$$\mathbf{w}_t + A\mathbf{w}_x + B\mathbf{w}_y = \mathbf{0}, \tag{3.1}$$

the assumption of hyperbolicity means that every linear combination $n_x A + n_y B$ has real eigenvalues, and a complete system of linearly independent eigenvectors; however, in general A and B do not commute so that they cannot be diagonalised simultaneously and the system of equations cannot be reduced to a set of scalar problems, as was the case in one dimension — witness the example just described. It is the recognition of and response to this problem in the last few years that we regard as the central theme in this article.

3.1 The wave equation in 2D

The generic problem is the second order wave equation, particularly when posed in an even number of space dimensions, for example in two dimensions,

$$\phi_{tt} = c^2(\phi_{xx} + \phi_{yy}). \tag{3.2}$$

Then the solution at a point $P \equiv (x, y, t + \Delta t)$ depends on that in the whole of the base of the characteristic cone drawn back from P to time level t, that is all points (x', y', t) such that $(x' - x)^2 + (y' - y)^2 \leq c^2(\Delta t)^2$ — see Figure 6. This is in contrast to the situation in three space dimensions when the solution depends only on that on the characteristic sphere $(x' - x)^2 + (y' - y)^2 + (z' - z)^2 = c^2(\Delta t)^2$; this distinction is embodied in Huygens' Principle — see [16], [37] or similar texts. Indeed, Kirchoff's formula provides an explicit expression for the solution of the 3D problem at time level $t + \Delta t$ in terms of the data at level t; and the corresponding formula in 2D is obtained from this by Hadamard's method of descent, as is generally the case in an even number of dimensions.

However, these formulae seem not to have been used as the basis for generating numerical approximations; while similar formulae for the evolution operator of the equivalent first order system of equations have been exploited by a number of authors. Thus for ease of generalisation to the system (3.1), we write the wave equation as the first order system

$$\left.\begin{aligned} &\phi_t + c(u_x + v_y) = 0 \\ &u_t + c\phi_x = 0 \\ &v_t + c\phi_y = 0 \end{aligned}\right\} \tag{3.3}$$

for the unknowns (ϕ, u, v). Along a bicharacteristic, such as PQ in Figure 6, we have

$$\frac{\mathrm{d}}{\mathrm{d}\sigma} \equiv \frac{\partial}{\partial t} + c\cos\theta\frac{\partial}{\partial x} + c\sin\theta\frac{\partial}{\partial y}; \tag{3.4}$$

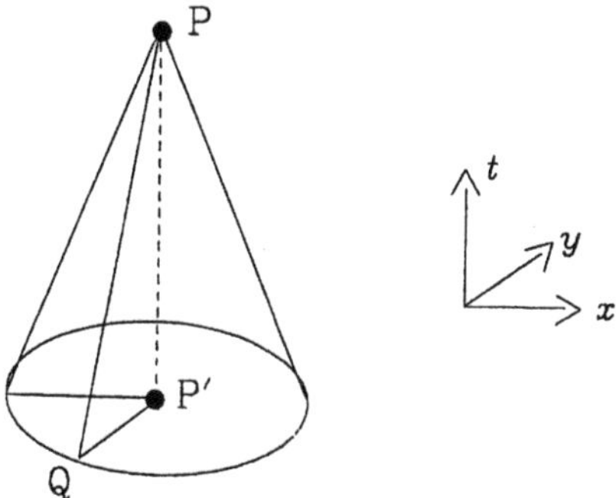

Figure 6. Characteristic cone for the 2D wave equation

and hence by combining the three equations of (3.3) with weights $(1, \cos\theta, \sin\theta)$ we obtain

$$\frac{\mathrm{d}}{\mathrm{d}\sigma}(\phi + u\cos\theta + v\sin\theta) = -S \tag{3.5a}$$

where

$$S = c[u_x \sin^2\theta - (u_y + v_x)\sin\theta\cos\theta + v_y\cos^2\theta]. \tag{3.5b}$$

That is, this linear combination is not constant along the bicharacteristic but has the awkward source function given by (3.5b), and hence gives

$$[\phi]_Q^P + [u]_Q^P\cos\theta + [v]_Q^P\sin\theta = -\int_Q^P S(t+\tau,\theta)\mathrm{d}\tau \tag{3.6}$$

in the notation of Figure 6. Integrating this around the cone then yields

$$\begin{aligned} \phi_P &= \frac{1}{2\pi}\int_0^{2\pi}[\phi_Q + u_Q\cos\theta + v_Q\sin\theta]\mathrm{d}\theta \\ &\quad - \frac{1}{2\pi}\int_0^{\Delta t}\int_0^{2\pi} S(t+\tau,\theta)\mathrm{d}\theta\mathrm{d}\tau. \end{aligned} \tag{3.7a}$$

In the same way, by increasing all the weights by a factor $\cos\theta$ or $\sin\theta$, one obtains the relations

$$\begin{aligned} u_P &= \frac{1}{\pi}\int_0^{2\pi}[\phi_Q\cos\theta + u_Q\cos^2\theta + v_Q\sin\theta\cos\theta]\mathrm{d}\theta \\ &\quad - \frac{1}{\pi}\int_0^{\Delta t}\int_0^{2\pi} S(t+\tau,\theta)\cos\theta\mathrm{d}\theta\mathrm{d}\tau \end{aligned} \tag{3.7b}$$

$$v_P = \frac{1}{\pi}\int_0^{2\pi}[\phi_Q \sin\theta + u_Q \sin\theta\cos\theta + v_Q \sin^2\theta]\mathrm{d}\theta$$
$$\qquad - \frac{1}{\pi}\int_0^{\Delta t}\int_0^{2\pi} S(t+\tau, \mathrm{d}\theta)\sin\theta\mathrm{d}\theta\mathrm{d}\tau. \tag{3.7c}$$

This form of evolution relation, with an integral over the characteristic cone rather than one over its base, seems to have first been exploited by Butler [7] and then later and more explicitly by Prasad and Ravindran [39]; more recently it has been derived, analysed and used as the basis of an evolution Galerkin method by Ostkamp [35].

Approximation of the source term surface integrals in (3.7) makes for considerable complication, but their complete neglect can be shown to lead to an inconsistent method. However, their awkward dependence on the gradients of u and v can be overcome: it is readily seen that these terms only contain derivatives tangential to the cone; and an integration by parts will yield, for instance, over the base circle of radius $c\Delta t$

$$\Delta t\int_0^{2\pi} S(t,\theta)\mathrm{d}\theta + \int_0^{2\pi}[u_Q\cos\theta + v_Q\sin\theta]\mathrm{d}\theta = 0. \tag{3.8}$$

This suggests approximating the time integral by the rectangle rule using the value at $\tau = 0$, to give

$$\phi_P = \frac{1}{2\pi}\int_0^{2\pi}[\phi_Q + 2(u_Q\cos\theta + v_Q\sin\theta)]\mathrm{d}\theta + O(\Delta t^2), \tag{3.9}$$

and corresponding formulae for u_P and v_P. These are very convenient formulae, eliminating both the dependence on the gradients of u and v and the need to integrate over the surface of the cone. One can also integrate along the axis of the cone to give

$$u_P = u_{P'} + \frac{1}{\pi}\int_0^{2\pi}\phi_Q\cos\theta\mathrm{d}\theta + O(\Delta t^2), \tag{3.10}$$

where ϕ_x at P' has been replaced by its average over the base of the cone and then changed into a line integral by application of Gauss' theorem. The combination of (3.9) and (3.10) has formed the basis of the numerical schemes developed in [35], and by an independent derivation has also been used by Fey [15].

However, the earlier schemes in [7] and [40] achieved second order accuracy by applying the trapezoidal rule in the time integration of the source terms, giving for example

$$\frac{1}{2\pi}\int\int S\mathrm{d}\theta\mathrm{d}\tau = \frac{\Delta t}{2}\Big[\frac{c}{2}(u_x + v_y)_P$$
$$\qquad - \frac{1}{2\pi\Delta t}\int_0^{2\pi}(u_Q\cos\theta + v_Q\sin\theta)\mathrm{d}\theta\Big] + O(\Delta t^3). \tag{3.11}$$

If the trapezoidal rule is also used to integrate the first equation of (3.3) along the cone axis, the value of $u_x + v_y$ at P can be eliminated; and if as in (3.10) the value of $u_x + v_y$ at P' is replaced by a line integral, we obtain the formula

$$\phi_P = \frac{1}{\pi}\int_0^{2\pi} [\phi_Q + u_Q \cos\theta + v_Q \sin\theta] \mathrm{d}\theta - \phi_{P'} + O(\Delta t^3), \tag{3.12}$$

to be compared with (3.9). The difference schemes of [7] and [40] used nine-point bivariate interpolation at time level t to apply these formulae. It would be interesting to see their full exploitation in an evolution–Galerkin scheme.

3.2 General hyperbolic systems

Suppose we have the system

$$\frac{\partial \mathbf{w}}{\partial t} + \sum_{k=1}^{d} A_k \frac{\partial \mathbf{w}}{\partial x_k} = 0 \tag{3.13}$$

in d dimensions. Although in general the d matrices A_k cannot be simultaneously diagonalised, by the assumption of hyperbolicity any linear combination has real eigenvalues and can be diagonalised. So we introduce and diagonalise $A \equiv A(\mathbf{n})$, corresponding to the direction vector $\mathbf{n} \equiv (n_1, n_2 \ldots, n_d)^T$, by

$$A := \sum_{k=1}^{d} n_k A_k, \quad \text{with} \quad AR = R\Lambda \tag{3.14}$$

giving its matrix of right-eigenvectors R and diagonal matrix of (real) eigenvalues Λ. Then we approximately diagonalise each matrix A_k with the similarity transformation using R, to obtain

$$M_k := R^{-1} A_k R = \Lambda_k + M_k', \tag{3.15}$$

where the Λ_k are diagonal matrices.

In terms of what would be the characteristic variables given by $\partial \mathbf{q} = R^{-1}\partial \mathbf{w}$ if all the M_k' were zero, we have

$$\frac{\partial \mathbf{q}}{\partial t} + \sum_{k=1}^{d} M_k \frac{\partial \mathbf{q}}{\partial x_k} = 0; \tag{3.16}$$

and this can be rewritten as

$$\frac{\partial \mathbf{q}}{\partial t} + \sum_{k=1}^{d} \Lambda_k \frac{\partial \mathbf{q}}{\partial x_k} = -\mathbf{S} \tag{3.17a}$$

where

$$\mathbf{S} = \sum_{k=1}^{d} M_k' \frac{\partial \mathbf{q}}{\partial x_k}. \tag{3.17b}$$

This is a direct generalisation of the formulae given in the previous section for the second order wave equation.

The bicharacteristics given by

$$\frac{\mathrm{d}\mathbf{X}_s}{\mathrm{d}t} = ((\Lambda_1)_{ss}, (\Lambda_2)_{ss}, \ldots, (\Lambda_d)_{ss})^T, \tag{3.18}$$

where $(\Lambda_k)_{ss}$ is the sth diagonal element of Λ_k, when drawn through the point $P \equiv (\mathbf{x}, t + \Delta t)$ to time level t give points $Q_s(\mathbf{n})$ at their feet. Hence we obtain

$$q_s(P) = q_s(Q_s(\mathbf{n})) - \int_0^{\Delta t} S_s \mathrm{d}t, \tag{3.19}$$

which should be integrated over $\mathbf{n}$, i.e. the unit sphere in $\mathbb{R}^d$. So far we have not specified whether the system (3.13) is linear, has constant or variable coefficients, or is quasilinear. In the linear, constant coefficient case, the derivation of a second or third order accurate approximation to the evolution operator from (3.17) is reasonably straightforward, although the convenience of the resulting formulae will depend very much on the eigenstructure of A — see [39] and [35] for generalisations of the wave equation system. Of greater practical and theoretical interest are a number of quasilinear systems, exemplified by the Euler equations which we consider in the next section.

3.3 The Euler equations in 2D

This system of equations was given in Section 2.3, and with $A := \partial\mathbf{f}/\partial\mathbf{w}$, $B := \partial\mathbf{g}/\partial\mathbf{w}$ it can be written in the form (3.1). As in (3.14) but with a minor change of notation, we have in the direction (n_x, n_y)

$$A_n := An_x + Bn_y. \tag{3.20}$$

The eigenvalues and right eigenvectors of this Jacobian matrix have a similar form to those given in (2.24) for the one-dimensional case; thus $A_n R_n = R_n \Lambda_n$, where with $u_n := un_x + vn_y$ and $q^2 = u^2 + v^2$, we have

$$R_n := \begin{bmatrix} 1 & 1 & 0 & 1 \\ u - cn_x & u & cn_y & u + cn_x \\ v - cn_y & v & -cn_x & v + cn_y \\ H - cu_n & \frac{1}{2}q^2 & cun_y - cvn_x & H + cu_n \end{bmatrix}, \tag{3.21a}$$

$$\Lambda_n := \begin{bmatrix} u_n - c & 0 & 0 & 0 \\ 0 & u_n & 0 & 0 \\ 0 & 0 & u_n & 0 \\ 0 & 0 & 0 & u_n + c \end{bmatrix}. \tag{3.21b}$$

Applying the transformation with R_n to each of the individual flux Jacobians then gives

$$R_n^{-1}AR_n = \begin{bmatrix} u - cn_x & 0 & cn_y & 0 \\ 0 & u & 0 & 0 \\ \frac{1}{2}cn_y & 0 & u & \frac{1}{2}cn_y \\ 0 & 0 & cn_y & u + cn_x \end{bmatrix}, \tag{3.22a}$$

$$R_n^{-1}BR_n = \begin{bmatrix} v - cn_y & 0 & -cn_x & 0 \\ 0 & v & 0 & 0 \\ -\frac{1}{2}cn_x & 0 & v & -\frac{1}{2}cn_x \\ 0 & 0 & -cn_x & v + cn_y \end{bmatrix}. \tag{3.22b}$$

It is clear that the first is diagonal when $n_y = 0$, its only off-diagonal elements being proportional to n_y; while the second has off-diagonal elements proportional to n_x. The middle two of the four bicharacteristics are in the direction of the fluid velocity (u, v) and the outer two form the characteristic cone with this direction as its axis, that is with directions $(u \pm cn_x, v \pm cn_y)$. This could be used as the basis of an evolution Galerkin method, in terms of the characteristic variables given by

$$\partial \mathbf{q}_n := R_n^{-1}\partial \mathbf{w},$$

as in (3.17) and (3.19). However, an alternative approach is possible in this case and has been used instead.

This alternative arises from the fact that the Euler equations are homogeneous of order one. Thus $A\mathbf{w} = \mathbf{f}(\mathbf{w})$ and $B\mathbf{w} = \mathbf{g}(\mathbf{w})$; and hence

$$\begin{aligned} \mathbf{f}n_x + \mathbf{g}n_y &= (An_x + Bn_y)\mathbf{w} = A_n\mathbf{w} \\ &= R_n(\Lambda_n R_n^{-1}\mathbf{w}) \\ &= \sum_{(s)} \lambda_n^{(s)}\alpha_s \mathbf{r}_n^{(s)}, \end{aligned} \tag{3.23}$$

where $\lambda_n^{(s)}, \mathbf{r}_n^{(s)}$ are the eigenvalues and right eigenvectors of A_n given in (3.21), while α_s are the coefficients in the decomposition of $\mathbf{w}$ as in (2.26),

$$\mathbf{w} = \sum_{s=1}^{4} \alpha_s \mathbf{r}_n^{(s)}, \tag{3.24a}$$

$$\boldsymbol{\alpha} = \left(\frac{\rho}{2\gamma}, \frac{\rho(\gamma-1)}{\gamma}, 0, \frac{\rho}{2\gamma}\right)^T. \tag{3.24b}$$

This flux vector splitting was used by Steger and Warming [45] as the basis of a finite difference method. Here the simultaneous decomposition of the unknowns and the fluxes, and the association of each component $\mathbf{r}_n^{(s)}$ with a set of bicharacteristics $\lambda_n^{(s)}$, has been used by Ostkamp [35] to develop an evolution–Galerkin method; that is, compared with (3.17) and (3.19), the characteristic variable transformation $\partial \mathbf{q} = R^{-1}\partial \mathbf{w}$ has been replaced by $\boldsymbol{\alpha} = R_n^{-1}\mathbf{w}$.

The resultant approximate evolution operator corresponds to that in (3.9) for the wave equation and can therefore be the basis of a first order approximation to the Euler equations. It has the form

$$\begin{aligned} \mathbf{w}_P &= \frac{1}{2\pi}\int_0^{2\pi}\sum_{s=1}^{4}\alpha_s \mathbf{r}_n^{(s)}(Q_s)\mathrm{d}\theta \\ &+ \frac{1}{2\pi}\int_0^{2\pi}[\alpha_4\mathbf{r}_n^{(4)}(Q_4) - \alpha_1\mathbf{r}_n^{(1)}(Q_4)]\mathrm{d}\theta, \end{aligned} \tag{3.25}$$

where $P \equiv (x, y, t + \Delta t)$ is the vertex of the characteristic cone and Q_4 is the foot of the bicharacteristic using $(n_x, n_y) = (\cos\theta, \sin\theta)$.

Moreover, Ostkamp [35] was able to show that when this was coupled to piecewise constant approximation on a rectangular mesh to give an evolution Galerkin method, the result was identical to the scheme developed by Fey [15]. Fey's method is much more directly based on an eigenvector decomposition, and it is instructive even in one dimension to see how his decomposition differs from the conventional decomposition, as in [45]. In the latter, with the decomposition explicitly given in (2.26a), the three eigenvectors are associated with the characteristic speeds $u - c, u$ and $u + c$ respectively, and a scalar advection scheme is applied to each component. In the Fey scheme, with the notation of (2.26a), the splitting is into the three components

$$\alpha_2\mathbf{r}^{(2)}, \quad \tfrac{1}{2}(\alpha_1\mathbf{r}^{(1)} + \alpha_3\mathbf{r}^{(3)}) \quad \text{and} \quad \pm\tfrac{1}{2}(\alpha_3\mathbf{r}^{(3)} - \alpha_1\mathbf{r}^{(1)}). \tag{3.26}$$

The first is evolved with a scalar advection scheme using the speed u; the second is treated similarly with both speeds $u + c$ and $u - c$; and for the third, the two signs apply to the two speeds $u \pm c$. Of course, for a simple linear problem treated by a linear method one obtains the same result from the two decompositions; but that is not true when adaptive methods are needed. Moreover, the Fey decomposition generalises immediately to two dimensions: the decomposition is into the four components given by (3.24); $\alpha_2\mathbf{r}^{(2)}$ is transported with velocity vector $\mathbf{u} = (u, v)$; and both $\frac{1}{2}(\alpha_1\mathbf{r}_n^{(1)} + \alpha_4\mathbf{r}_n^{(4)})$ and $\frac{1}{2}(\alpha_4\mathbf{r}_n^{(4)} - \alpha_1\mathbf{r}_n^{(1)})$ are distributed with density $\frac{1}{2\pi}$ and velocity vector $\mathbf{u} + c\mathbf{n}$. Comparing with (3.25) one can see how the two approaches can coincide.

4 Steady and nearly steady problems

In many fields of engineering, for instance aerospace engineering, by far the commonest computations are for steady problems, or those where the temporal variation is quite slow. Yet for such fluid flow problems the PDE's are often hyperbolic in character. Thus the method developed by Butler [7] and described above for the wave equation, was actually developed and used by him to solve the steady three-dimensional flow past a supersonic aircraft as well as to study plane unsteady flow, and that was true for many related methods that were developed at that time. It was pointed out in Section 2.3 that the steady Euler equations in two dimensions contain an elliptic system when the flow is subsonic, but that when it is supersonic the system has four real characteristics and can be decomposed into four characteristic equations. Similarly, for steady supersonic three-dimensional flow there is a cone of real bicharacterisitcs around the flow direction as its axis, and the situation is very similar to that for plane unsteady flow as described in the last section.

However, for the numerical analyst the link between unsteady and steady flow problems is even closer: the discrete equations in the latter subsonic case are very nonlinear and it is usual to solve them by an iteration technique closely based on time-stepping; that is, the situation is similar to that which held for fully elliptic equations before direct solution methods were introduced. Moreover, in this field there is the very great advantage that the solution technique is then the same whether the flow is wholly subsonic, wholly supersonic, or changes character over part of the domain — the commercially very important transonic case. Finally, for slowly varying flows it will be most economic to use an implicit discretisation, so that the steady flow solver is then needed to solve the implicit equations at each time level.

Thus in the following sections we shall describe some of the recent developments in discretising steady flow problems. There are two main techniques, fluctuation splitting methods and finite volume methods. We shall concentrate on the former because the key developments there have mirrored those for unsteady problems, in their recognition of the need to distinguish the hyperbolic/elliptic character of the operators.

4.1 Fluctuation splitting in 1D

This novel approach introduced by Roe [41] can be considered first for 1D conservation laws of the form (1.1). Suppose in the scalar case there is a linear variation between W_i^n and W_{i+1}^n at the time level n between mesh points x_i and x_{i+1}. The fluctuation over the cell $i+\frac{1}{2}$, of width $\Delta x_{i+\frac{1}{2}} = x_{i+1} - x_i$ and midpoint $x_{i+\frac{1}{2}}$, is defined as

$$\Phi_{i+\frac{1}{2}} := A_{i+\frac{1}{2}}(W_{i+1}^n - W_i^n), \tag{4.1}$$

and is used to update the nodal values at either end; with distribution factors that sum to unity, $\beta_{i+\frac{1}{2}}^i + \beta_{i+\frac{1}{2}}^{i+1} = 1$, we obtain

$$W_i^n \to W_i^n - \frac{\Delta t}{\Delta x_i}\beta_{i+\frac{1}{2}}^i \Phi_{i+\frac{1}{2}}, \quad W_{i+1}^n \to W_{i+1}^n - \frac{\Delta t}{\Delta x_i}\beta_{i+\frac{1}{2}}^{i+1} \Phi_{i+\frac{1}{2}}, \tag{4.2}$$

where $\Delta x_i = \frac{1}{2}(\Delta x_{i-\frac{1}{2}} + \Delta x_{i+\frac{1}{2}}) \equiv x_{i+\frac{1}{2}} - x_{i-\frac{1}{2}}$. This update is looped over all cells and the result labelled W_i^{n+1} at each node. To maintain conservation, that is the sum of $\Delta x_i W_i^n$, we impose the condition

$$A_{i+\frac{1}{2}}(W_{i+1}^n - W_i^n) = F_{i+1}^n - F_i^n \equiv f(W_{i+1}^n) - f(W_i^n). \tag{4.3}$$

In this scalar case this fully determines $A_{i+\frac{1}{2}}$, except when the fluctuation is zero, and the simplest update is fully upwinded:

$$\begin{aligned} \beta_{i+\frac{1}{2}}^{i} &= 0, \quad \beta_{i+\frac{1}{2}}^{i+1} = 1 \quad \text{if} \quad A_{i+\frac{1}{2}} \geq 0 \\ \beta_{i+\frac{1}{2}}^{i} &= 1, \quad \beta_{i+\frac{1}{2}}^{i+1} = 0 \quad \text{if} \quad A_{i+\frac{1}{2}} < 0. \end{aligned} \tag{4.4}$$

For a typical set of data, most of the nodes will receive an update from a single cell on one side, but some will receive updates from both sides (shock nodes) and some from neither (rarefaction nodes). When first introduced this formulation, which overlaps with several others at this simple level, was an important departure from attempting to write formulae for each new W_i^{n+1} directly; its value was first appreciated for one-dimensional systems of equations and more recently for multidimensional systems.

For a system of equations, we decompose the cell gradient

$$\frac{\mathbf{W}_{i+1}^n - \mathbf{W}_i^n}{x_{i+1} - x_i} = \sum_{(s)} \alpha_{i+\frac{1}{2}}^{(s)} \mathbf{r}_{i+\frac{1}{2}}^{(s)}, \tag{4.5a}$$

using the right eigenvectors of the cell Jacobian $A_{i+\frac{1}{2}}$,

$$A_{i+\frac{1}{2}} \mathbf{r}_{i+\frac{1}{2}}^{(s)} = \lambda_{i+\frac{1}{2}}^{(s)} \mathbf{r}_{i+\frac{1}{2}}^{(s)}. \tag{4.5b}$$

Then the fluctuation is given as

$$\boldsymbol{\Phi}_{i+\frac{1}{2}} = \sum_{(s)} \boldsymbol{\Phi}_{i+\frac{1}{2}}^{(s)} := \sum_{(s)} \Delta x_{i+\frac{1}{2}} \lambda_{i+\frac{1}{2}}^{(s)} \alpha_{i+\frac{1}{2}}^{(s)} \mathbf{r}_{i+\frac{1}{2}}^{(s)}, \tag{4.6}$$

and $\boldsymbol{\Phi}_{i+\frac{1}{2}}^{(s)}$ updates W_i^n or $W_{i+\frac{1}{2}}^n$ according to the sign of $\lambda_{i+\frac{1}{2}}^{(s)}$ as for the scalar case.

For the Euler equations, the vector form of the conservation condition is achieved by means of the Roe linearisation [41]. Intermediate variables $\mathbf{z}$ are used instead of $\mathbf{w}$, in terms of which both $\mathbf{w}$ and the fluxes are quadratic quantities so that (4.3) can be achieved by using average values in A and hence in $\mathbf{r}$ and λ. Thus we define $\mathbf{z} := \sqrt{\rho}(1, u, H)^T$ and set $\mathbf{Z} = \frac{1}{2}(\mathbf{Z}_i + \mathbf{Z}_{i+1})$ to obtain

$$\Delta \mathbf{F} = \left(\frac{\partial \mathbf{f}}{\partial \mathbf{z}}\right)_{av} \Delta \mathbf{Z} = \left(\frac{\partial f}{\partial \mathbf{z}}\right)_{av} \left(\frac{\partial \mathbf{w}}{\partial \mathbf{z}}\right)_{av}^{-1} \Delta \mathbf{W} \equiv A(\bar{\mathbf{Z}}) \Delta \mathbf{W}; \tag{4.7}$$

and as in (2.24) and (2.26), we have

$$\{\lambda^{(s)}\} = \tilde{u}, \tilde{u} \pm \tilde{c}; \quad \{\tilde{\mathbf{r}}^{(s)}\} = \begin{pmatrix} 1 \\ \tilde{u} \\ \frac{1}{2}\tilde{u}^2 \end{pmatrix}, \begin{pmatrix} 1 \\ \tilde{u} \pm \tilde{c} \\ \tilde{H} + \tilde{u}\tilde{c} \end{pmatrix} \tag{4.8}$$

where, for example, the notation $\tilde{u}$ means the value of u obtained from the average vector $\bar{\mathbf{Z}}$. The wave component intensities in (4.8) are similarly given by $\tilde{\rho}(\gamma-1)/(\gamma\Delta x)$ and $\tilde{\rho}/(2\gamma\Delta x)$.

4.2 Fluctuation splitting in 2D

We consider the conservation law system

$$\mathbf{w}_t + \mathbf{f}_x + \mathbf{g}_y = \mathbf{0}, \tag{4.9}$$

approximated by a piecewise linear $\mathbf{W}$ on a triangular mesh. We begin with the scalar case and on a triangle Ω_α define the fluctuation as

$$\begin{aligned}\Phi_\alpha \approx \int_{\Omega_\alpha} \nabla\cdot(f,g)\mathrm{d}x\mathrm{d}y &= \int_{\partial\Omega_\alpha} f\mathrm{d}y - g\mathrm{d}x \\ &= \int_{\partial\Omega_\alpha} (\mathcal{F}\cdot\mathbf{n})\mathrm{d}s,\end{aligned} \tag{4.10}$$

where $\mathcal{F} \equiv (f,g)$ and the trapezoidal rule is used to approximate the boundary integrals. Then the update to each of the vertices i_1, i_2 and i_3 is allocated by formulae of the form

$$W_{i_1}^n \to W_{i_1}^n - \frac{\Delta t}{V_{i_1}}\beta_\alpha^{i_1}\Phi_\alpha, \tag{4.11}$$

where V_{i_1} is the measure of the median dual cell about i_1 and

$$\beta_\alpha^{i_1} + \beta_\alpha^{i_2} + \beta_\alpha^{i_3} = 1 \tag{4.12}$$

to maintain conservation — see Figure 7. We introduce an average advection speed for the cell by

$$\boldsymbol{\lambda}_\alpha \approx \frac{1}{V_\alpha}\int_{\Omega_\alpha}\left(\frac{\partial f}{\partial w}, \frac{\partial g}{\partial w}\right)^T \mathrm{d}x\mathrm{d}y, \tag{4.13}$$

where V_α is the measure of cell Ω_α; then the fluctuation can be written

$$\Phi_\alpha = V_\alpha \boldsymbol{\lambda}_\alpha \cdot \nabla W^n = \tfrac{1}{2}\sum_{k=1}^{3}(\boldsymbol{\lambda}_\alpha \cdot \mathbf{N}_{i_k})\, W_{i_k}^n, \tag{4.14}$$

where $\mathbf{N}_{i_k}$ is in the inward normal direction to the cell edge opposite i_k and has magnitude equal to its length.

In the terminology of Deconinck *et al* [14], the allocation scheme (4.11) is *linear* if all the coefficients β_α^i are independent of $\{W_i^n\}$, it is *positive* if they are all positive, and it is *linearity-preserving* if $\Phi_\alpha = 0 \Rightarrow$ a zero update. The authors then show that no linear scheme is both positive and linearity preserving — a result similar to that for monotone schemes in one dimension. The following three schemes will illustrate the possible choices.

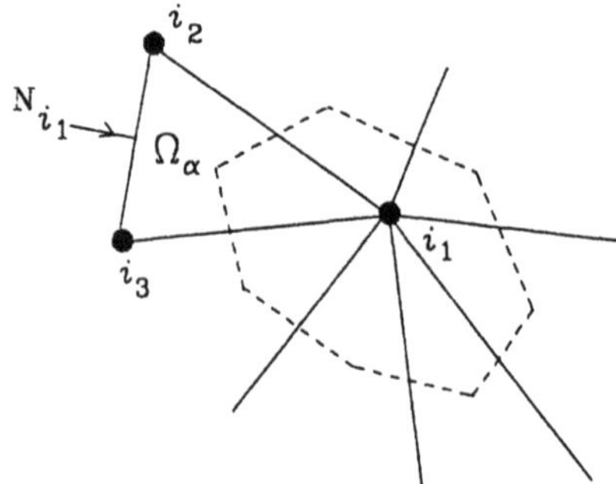

Figure 7. Notation for allocation of fluctuation for cell Ω_α to triangle vertices

In each scheme the allocation depends on whether the triangle has one inflow edge, $\boldsymbol{\lambda}_\alpha \cdot \mathbf{N}_i > 0$, or two; for all schemes in the former case, say $\boldsymbol{\lambda}_\alpha \cdot \mathbf{N}_{i_1} > 0$, we set $\beta_\alpha^{i_1} = 1$ and $\beta_\alpha^{i_2} = \beta_\alpha^{i_3} = 0$. In the latter case, say $\boldsymbol{\lambda}_\alpha \cdot \mathbf{N}_{i_1} > 0$ and $\boldsymbol{\lambda}_\alpha \cdot \mathbf{N}_{i_2} > 0$, and where of course $\mathbf{N}_{i_1} + \mathbf{N}_{i_2} + \mathbf{N}_{i_3} = \mathbf{0}$, in each of the three typical schemes we consider we set $\beta_\alpha^{i_3} = 0$. In the low diffusion A scheme (LDA), we set

$$^{\mathrm{LDA}}\beta_\alpha^{i_1} = -\frac{\boldsymbol{\lambda}_\alpha \cdot \mathbf{N}_{i_1}}{\boldsymbol{\lambda}_\alpha \cdot \mathbf{N}_{i_3}}, \quad ^{\mathrm{LDA}}\beta_\alpha^{i_2} = -\frac{\boldsymbol{\lambda}_\alpha \cdot \mathbf{N}_{i_2}}{\boldsymbol{\lambda}_\alpha \cdot \mathbf{N}_{i_3}}, \tag{4.15}$$

which is linearity-preserving but not positive. In the N-scheme we set

$$^{N}\beta_\alpha^{i_1}\Phi_\alpha = \tfrac{1}{2}(\boldsymbol{\lambda}_\alpha \cdot \mathbf{N}_{i_1})(W_{i_1}^n - W_{i_3}^n), \quad ^{N}\beta_\alpha^{i_2}\Phi_\alpha = \tfrac{1}{2}(\boldsymbol{\lambda}_\alpha \cdot \mathbf{N}_{i_2})(W_{i_2}^n - W_{i_3}^n), \tag{4.16}$$

which is positive but not linearity-preserving; that is, it fails to preserve the integrity of the fluctuation while still satisfying (4.12) in the sense that $\sum_{(k)} {}^{N}\beta_\alpha^{i_k}\Phi_\alpha = \Phi_\alpha$. Both of these schemes are linear; but a nonlinear scheme that is both positive and linearity-preserving is the positive streamwise invariant (PSI) scheme of [14], for which

$$^{\mathrm{PSI}}\beta_\alpha^{i_1}\Phi_\alpha = \Psi\left({}^{N}\beta_\alpha^{i_1}\right)\Phi_\alpha, \quad ^{\mathrm{PSI}}\beta_\alpha^{i_2}\Phi_\alpha = \Psi\left({}^{N}\beta_\alpha^{i_2}\right)\Phi_\alpha, \tag{4.17}$$

where $\Psi(\cdot)$ is the minmod limiter function.

A Roe linearisation can also be used to apply these ideas to the Euler equations. As in one dimension, intermediate variables defined by $\mathbf{z} := \sqrt{\rho}(1, u, v, H)^T$ are introduced and the average value on a cell is given by $\bar{\mathbf{Z}}_\alpha = \frac{1}{3}(\mathbf{Z}_{i_1} + \mathbf{Z}_{i_2} + \mathbf{Z}_{i_3})$. Then, considering $\mathbf{Z}$ as linear over the cell, we have

$$\boldsymbol{\Phi}_\alpha = V_\alpha \left[\left(\frac{\partial \mathbf{f}}{\partial \mathbf{z}}\right)_{\bar{\mathbf{Z}}_\alpha} \mathbf{Z}_x + \left(\frac{\partial \mathbf{g}}{\partial \mathbf{z}}\right)_{\bar{\mathbf{Z}}_\alpha} \mathbf{Z}_y \right]; \tag{4.18}$$

and also

$$\overline{\mathbf{W}}_x := \left(\frac{\partial \mathbf{w}}{\partial \mathbf{z}}\right)_{\bar{\mathbf{Z}}_\alpha} \mathbf{Z}_x, \quad \overline{\mathbf{W}}_y := \left(\frac{\partial \mathbf{w}}{\partial \mathbf{z}}\right)_{\bar{\mathbf{Z}}_\alpha} \mathbf{Z}_y, \tag{4.19}$$

$$\bar{A}_\alpha := \left(\frac{\partial \mathbf{f}}{\partial \mathbf{z}}\right)_{\bar{\mathbf{Z}}_\alpha} \left(\frac{\partial \mathbf{w}}{\partial \mathbf{z}}\right)^{-1}_{\bar{\mathbf{Z}}_\alpha}, \quad \bar{B}_\alpha := \left(\frac{\partial \mathbf{g}}{\partial \mathbf{z}}\right)_{\bar{\mathbf{Z}}_\alpha} \left(\frac{\partial \mathbf{w}}{\partial \mathbf{z}}\right)^{-1}_{\bar{\mathbf{Z}}_\alpha}. \tag{4.20}$$

Hence we obtain

$$\boldsymbol{\Phi}_\alpha = V_\alpha \left[\bar{A}_\alpha \overline{\mathbf{W}}_x + \bar{B}_\alpha \overline{\mathbf{W}}_y\right], \tag{4.21}$$

to which various approximate decompositions have been applied. These have generally been based on wave models of the form

$$\boldsymbol{\Phi}_\alpha = V_\alpha \sum_1^m (\boldsymbol{\lambda}^{(l)} \cdot \overline{\nabla \mathbf{Q}}^{(l)} + \bar{\mathbf{q}}^{(l)}) \bar{\mathbf{r}}^{(l)}, \tag{4.22}$$

where the $\boldsymbol{\lambda}^{(l)}$ are the ray speed vectors, $\mathbf{Q}$ is the characteristic variable, $\bar{\mathbf{q}}^{(l)}$ is a coupling term, and typically $m = 6$ in the absence of source terms. The close similarity with the approximate diagonalisation given in Sections 3.2 and 3.3 should be noted here; and the scheme used by Butler [7], which used four points round the base of the characteristic cone and integrated two equations along its axis, could readily be interpreted as an early six-wave model.

4.3 The cell vertex finite volume method

This method is primarily designed to solve steady problems, though it is also used with implicit time-stepping methods to solve unsteady problems with slow time variation, so-called low reduced frequency problems. It is also most distinctive when applied to a quadrilateral or hexahedral mesh — see for example [32], [18], [29], [12]; but it has recently been used on triangles and tetrahedra in [11]. Starting from the steady form of (4.9), the key first step is to define the cell residual $\mathbf{R}_\alpha(\mathbf{W})$ as in (4.10),

$$\begin{aligned} \frac{1}{V_\alpha}\int_{\Omega_\alpha} \nabla\cdot(\mathbf{f},\mathbf{g})\mathrm{d}\Omega &= \frac{1}{V_\alpha}\int_{\partial\Omega_\alpha} \mathbf{f}\mathrm{d}y - \mathbf{g}\mathrm{d}x \\ \approx \mathbf{R}_\alpha(W) &:= \frac{1}{2V_\alpha}[(\mathbf{F}_1 - \mathbf{F}_3)\delta y_{24} + (\mathbf{F}_2 - \mathbf{F}_4)\delta y_{31} \\ &\quad - (\mathbf{G}_1 - \mathbf{G}_3)\delta x_{24} - (\mathbf{G}_2 - \mathbf{G}_4)\delta x_{31}] \end{aligned} \tag{4.23}$$

using the notation of Figure 8, and with $\delta y_{24} := y_2 - y_4$ etc. For simple problems, and for most cells in complicated problems, setting $\mathbf{R}_\alpha(\mathbf{W}) = \mathbf{0}\ \forall\alpha$ determines $\mathbf{W}$ uniquely.

However, to cope with the general situation (and triangular meshes), it is necessary to define nodal residuals, as follows

$$\mathbf{N}_i(\mathbf{W}) := \sum_{\alpha,\beta,\gamma,\delta} V_\alpha(D_{\alpha,i}\mathbf{R}_\alpha + \mathbf{A}_{\alpha,i}) / \sum_{\alpha,\beta,\gamma,\delta} V_\alpha, \tag{4.24}$$

again using the notation of Figure 8; here the $D_{\alpha,i}$ are distribution matrices, similar in function to the allocation factors β^i_α in (4.11); and the $\mathbf{A}_{\alpha,i}$ are artificial dissipation vectors needed to eliminate spurious solution modes, and to overcome under-specification arising from boundary conditions and local critical points. These residuals are then used to define a pseudo-time stepping or relaxation update procedure,

$$\mathbf{W}_i^{n+1} = \mathbf{W}_i^n - \omega_i \mathbf{N}_i(\mathbf{W}^n), \tag{4.25}$$

which is usually applied within a multigrid framework.

The commonest distribution matrices in practical applications are those based on the Lax–Wendroff method, although more sophisticated schemes are also used. At convergence of the iteration (4.25) we have $\mathbf{N}_i(\mathbf{W}) = \mathbf{0} \ \ \forall i$, and the rôle of the distribution matrices and artificial dissipation is to ensure that

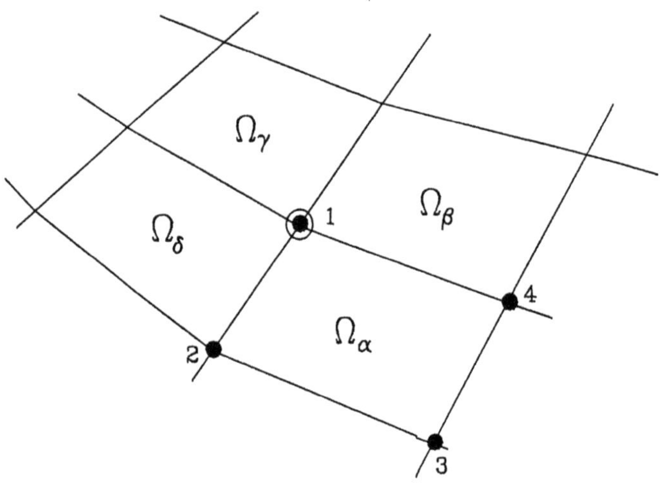

Figure 8. Notation for the cell vertex method

(i) $\mathbf{R}_\alpha(\mathbf{W}) = o(h^2)$ for "most" α;

(ii) $\mathbf{W}$ is free of spurious oscillation modes, particularly the chequerboard mode which lies in the null space of the cell residual;

(iii) the convergence is suitably fast.

Note the similarity of (i) to the linearity preservation sought after in the fluctuation splitting schemes. Here, (i) is usually achieved in all cells except those that straddle shocks — see [30].

4.4 Elliptic/hyperbolic decomposition

Motivated by the work of Ta'asan [47], Roe and Deconinck together with their collaborators ([25], [36]) have recently introduced an elliptic/hyperbolic decomposition to their fluctuation splitting schemes. Consider the system of m equations

$$A\mathbf{w}_x + B\mathbf{w}_y = \mathbf{0}, \tag{4.26}$$

and suppose that A is non-singular and $A^{-1}B$ is diagonalisable with k pairs of complex eigenvalues and $m - 2k$ real eigenvalues. Then we can write

$$A = R\Lambda_A Q, \quad B = R\Lambda_B Q, \tag{4.27a}$$

where Λ_A is diagonal, with real elements, and Λ_B has the form

$$\Lambda_B = \begin{bmatrix} \begin{pmatrix} \lambda_R^{(1)} & \lambda_I^{(1)} \\ -\lambda_I^{(1)} & \lambda_R^{(1)} \end{pmatrix} & & & & 0 \\ & \ddots & & & \\ & & \begin{pmatrix} \lambda_R^{(k)} & \lambda_I^{(k)} \\ -\lambda_I^{(k)} & \lambda_R^{(k)} \end{pmatrix} & & \\ & 0 & & \lambda^{(2k+1)} & \\ & & & & \ddots \\ & & & & & \lambda^{(m)} \end{bmatrix}, \tag{4.27b}$$

also with real elements. Note that the key construction, finding $Q \in \mathbb{R}^{m\times m}$ such that $A^{-1}B = Q^{-1}\Lambda_A^{-1}\Lambda_B Q$, is closely related to decompositions used in numerical algorithms for the generalised eigenvalue problem, such as the real Schur decomposition and the block diagonal decomposition — see [17]. The resulting differential system represents k 2×2 elliptic systems plus $m-2k$ scalar hyperbolic equations. An example is provided by the subsonic Euler equations; the resulting matrices can be developed from those in Sections 2.3 and 3.3, but the simpler forms given in [25] and [36] are based on using streamline coordinates.

To exploit this decomposition, either for the unsteady problem or for the steady problem using a time-stepping iteration, we reintroduce the time derivative and use the variables given by $\partial\mathbf{q} := Q\partial\mathbf{w}$ to obtain

$$(QR)^{-1}\mathbf{q}_t + \Lambda_A\mathbf{q}_x + \Lambda_B\mathbf{q}_y = \mathbf{0}. \tag{4.28}$$

As used in [25] and [36] for steady problems, the $(QR)^{-1}$ can be incorporated into the very effective preconditioning step developed by van Leer *et al.* [48]; then the hyperbolic subsystems are updated using the fluctuation splitting techniques of Section 4.2, while the elliptic subsystems use the cell vertex method with simple Lax–Wendroff distribution matrices. If this same decomposition were used for unsteady problems in an evolution–Galerkin scheme, the natural choice would be for the methods of Section 2.1 to be applied to the hyperbolic subsystems and those of Sections 3.1 and 3.2 to the elliptic subsystems.

5 Conclusions

The differences in objectives when approximating steady and unsteady PDE problems are too great for methods in the two cases not to have very distinctive features and differing emphases. However, we have seen that there is much in common in recent developments for hyperbolic or mixed elliptic/hyperbolic systems in multi-dimensions. Key points in the development include:

- recognition of the importance of identifying and approximating associated evolution operators;
- decomposing spatial multi-dimensional operators into their elliptic and hyperbolic subsystems;
- for conservation laws, a finite volume formulation using simple basis functions plus solution-adaptive recovery by higher order approximations yields a very powerful class of methods of wide-ranging utility.

References

1. R. Abgrall. An essentially non-oscillatory reconstruction procedure on finite-element type meshes: application to compressible flows. *Comput. Methods Appl. Mech. Engrg.*, 116:95–101, 1994.

2. R. Abgrall. On essentially non-oscillatory schemes on unstructured meshes: Analysis and implementation. *J. Comput. Phys.*, 114:45–58, 1994.

3. M.J. Baines. *Moving Finite Elements*. Monographs on Numerical Analysis. Clarendon Press, 1994.

4. T.J. Barth. Aspects of unstructured grids and finite-volume solvers for the Euler and Navier-Stokes equations. In *Special Course on Unstructured Grid Methods for Advection Dominated Flows*, Report 787, pages 6.1–6.61. AGARD, 1992.

5. T.J. Barth and P.O. Frederickson. Higher order solution of the Euler equations on unstructured grids using quadratic reconstruction. AIAA Paper-90-0013, 1990.

6. J.P. Benqué, G. Labadie, and J. Ronat. A new finite element method for the Navier-Stokes equations coupled with a temperature equation. In T. Kawai, editor, *Proceedings of the Fourth International Symposium on Finite Element Methods in Flow Problems*, pages 295–301. North-Holland, 1982.

7. D.S. Butler. The numerical solution of hyperbolic systems of partial differential equations in three independent variables. *Proceedings of the Royal Society of London*, 255A:233–252, 1960.

8. P.N. Childs and K.W. Morton. Characteristic Galerkin methods for scalar conservation laws in one dimension. *SIAM J. Numer. Anal.*, 27:553–594, 1990.

9. B. Cockburn and C.W. Shu. TVB Runge-Kutta local projection discontinuous Galerkin finite element method for scalar conservation laws II: General framework. *Math. Comp.*, 52:411–435, 1989.

10. P. Colella and P. Woodward. The piecewise parabolic method (PPM) for gas-dynamical simulations. *J. Comput. Phys.*, 54:174–201, 1984.

11. P. Crumpton and M.B. Giles. Aircraft computations using multigrid and an unstructured parallel library. AIAA Paper 95-0210, 1995.

12. P.I. Crumpton, J.A. Mackenzie, and K.W. Morton. Cell vertex algorithms for the compressible Navier-Stokes equations. *J. Comput. Phys.*, 109(1):1–15, 1993.

13. I.R. Dawkins. *Development of practical evolution Galerkin algorithms on unstructured meshes*. PhD thesis, Oxford University Computing Laboratory, 1996.

14. H. Deconinck, P.L. Roe, and R. Struijs. A multi-dimensional generalisation of Roe's flux difference splitter for the Euler equations. *Comput. & Fluids*, 22:215–222, 1993.

15. M. Fey. Ein echt mehrdimensionales ver fahren zur lösung der Euler gleichungen, 1993. Dissertation, ETH Zürich.

16. P.R. Garabedian. *Partial Differential Equations*. J. Wiley & Sons, 1964.

17. G.H. Golub and C.F. Van Loan. *Matrix Computations*. North Oxford Academic, Oxford, 2nd edition, 1986.

18. M.G. Hall. Cell vertex schemes for the solution of the Euler equations. In K.W. Morton and M.J. Baines, editors, *Proceedings of the Conference on Numerical Methods for Fluid Dynamics, University of Reading, U.K.*, pages 303–345. Oxford University Press, 1986.

19. B.P. Leonard. The ULTIMATE conservative difference scheme applied to unsteady one-dimensional advection. *Comput. Methods Appl. Mech. Engrg.*, 88:17–74, 1991.

20. B.P. Leonard and H.S. Niknafs. Sharp monotonic resolution of discontinuities without clipping of narrow extrema. *Comput. & Fluids*, 19(1):141–154, 1991.

21. R.J. LeVeque. High resolution finite volume methods on arbitrary grids via wave propagation. *J. Comput. Phys.*, 78:36–63, 1988.

22. R.J. LeVeque. High-resolution conservative algorithms for advection in incompressible flow. *SIAM J. Numer. Anal.*, 33(2):627–665, 1996.

23. P. Lin, K.W. Morton, and E. Süli. Euler characteristic Galerkin scheme with recovery. M^2AN, 27(7):863–894, 1993.

24. P. Lin, K.W. Morton, and E. Süli. Characteristic Galerkin schemes for conservation laws in two and three space dimensions. To appear in the SIAM J. Numer. Anal., 1996.

25. L.M. Mesaros and P.L. Roe. Multidimensional fluctuation-splitting schemes based on decomposition methods, 1995. AIAA Paper 95-1699 also in AIAA CP 956 San Diego.

26. A.R. Mitchell. Recent developments in the finite element method. In J. Noye and C.A.J. Fletcher, editors, *Computational Techniques and Applications: CTAC-83*, pages 2–14. Elsevier, 1984.

27. K.W. Morton. Lagrange–Galerkin and Characteristic–Galerkin methods and their applications. In B. Engquist and B. Gustafsson, editors, *Third International Conference on Hyperbolic Problems. Theory, Numerical Methods and Applications*, volume II, pages 742–755. Studentlitteratur, 1991.

28. K.W. Morton. *Numerical Solution of Convection-Diffusion Problems*, volume 12 of *Applied Mathematics and Mathematical Computation*. Chapman & Hall, 1996.

29. K.W. Morton and M.F. Paisley. A finite volume scheme with shock fitting for the steady Euler equations. *J. Comput. Phys.*, 80:168–203, 1989.

30. K.W. Morton, M.A. Rudgyard, and G.J. Shaw. Upwind iteration methods for the cell vertex scheme in one dimension. *J. Comput. Phys.*, 114(2):209–226, 1994.

31. K.W. Morton and E. Süli. Evolution-Galerkin methods and their supraconvergence. *Numer. Math.*, 71:331–355, 1995.

32. R.-H. Ni. A multiple grid scheme for solving the Euler equations. *AIAA Journal*, 20(11):1565–1571, Nov 1981.

33. S. Osher and F. Solomon. Upwind difference schemes for hyperbolic systems of conservation laws. *Math. Comp.*, 38:339–374, 1982.

34. S. Osher and P.K. Sweby. Recent developments in the numerical solution of nonlinear conservation laws. In A. Iserles and M.J.D. Powell, editors, *The State of the Art in Numerical Analysis*, pages 681–701. Clarendon Press, 1987.

35. S. Ostkamp. *Multidimensional characterisitic Galerkin schemes and evolution operators for hyperbolic systems.* PhD thesis, Universität Hannover, 1995.

36. H. Paillère, E. van der Weide, and H. Deconinck. Multidimensional upwind methods for inviscid and viscous flows. In H. Deconinck, editor, *Computational Fluid Dynamics*, number 1995-02 in Lecture Series, pages 5.1–5.56. von Karman Institute, 1995.

37. I.G. Petrovsky. *Lectures on Partial Differential Equations.* Interscience, 1954.

38. O. Pironneau. On the transport-diffusion algorithm and its application to the Navier-Stokes equations. *Numer. Math.*, 38:309–332, 1982.

39. P. Prasad and R. Ravindran. Canonical form of a quasilinear hyperbolic system of first order equations. *J. Math. Phys. Sci.*, 18(4):361–364, 1984.

40. A.S. Reddy, V.G. Tikekar, and P. Prasad. Numerical solution of hyperbolic equations by method of bicharacteristics. *Journal of Mathematical and Physical Sciences*, 16(6):575–603, 1982.

41. P.L. Roe. Approximate Riemann solvers, parameter vectors, and difference schemes. *J. Comput. Phys.*, 43:357–372, 1981.

42. C.-W. Shu. Total-variation-diminishing time discretisations. *SIAM J. Sci. Statist. Comput.*, 9:1073–1084, 1988.

43. C.-W. Shu and S. Osher. Efficient implementation of essentially non-oscillatory shock-capturing schemes. *J. Comput. Phys.*, 77:439–471, 1988.

44. T. Sonar. Multivariate rekonstruktionsverfahren zur numerischen berechnung hyperbolischer erhaltungsgleichengen. Technical Report 95-02, DLR Forschungsbericht, 1995.

45. J.L. Steger and R.F. Warming. Flux vector splitting of the inviscid gas dynamic equations with applications to finite-difference methods. *J. Comput. Phys.*, 40:263–293, 1981.

46. E. Süli. Convergence and nonlinear stability of the Lagrange-Galerkin method for the Navier-Stokes equations. *Numer. Math.*, 53:459–483, 1988.

47. S. Ta'asan. Canonical-variables multigrid method for Euler equations. In S.M. Deshpande, S.J. Desai, and R. Narasimha, editors, *Fourteenth International Conference on Numerical Methods in Fluid Dynamics*, Lecture Notes in Physics, pages 173–177. Springer-Verlag, 1995.

48. B. van Leer, W-T. Lee, and P. Roe. Characteristic time-stepping or local preconditioning of the Euler equations. AIAA-91-1552-CP, 1991.

Algorithms in Tomography

Frank Natterer

Institut für Numerische und instrumentelle Mathematik, Universität Münster, Germany

1 Introduction

The basic problem in computerized tomography (CT) is the reconstruction of a function from its line or plane integrals. Applications come from diagnostic radiology, astronomy, electron microscopy, seismology, radar, plasma physics, nuclear medicine and many other fields. More recent kinds of tomography replace the straight line model by an inverse problem for a partial differential equation.

The outline of this paper is as follows. In Section 2 we survey the mathematical models used in tomography. In Section 3 we give a fairly detailed survey of $2D$ reconstruction algorithms, which are still the work horse of tomography. In Section 4 we describe recent developments in $3D$ reconstruction. In Section 5 we make a few remarks on the beginning development of algorithms for non-straight-line tomography.

2 Mathematical models in tomography

In the description of the mathematical models we restrict ourselves to those features which are important for the mathematical scientist. For physical and medical aspects see [35], [6], [32].

(a) Transmission CT.

This is the original and simplest case of CT. In transmission tomography one probes an object with non-diffracting radiation, e.g. x-rays for the human body. If I_0 is the intensity of the source, $a(x)$ the linear attenuation coefficient of the object at point x, L the ray along which the radiation propagates, and I the intensity past the object, then

$$I = I_0 \, e^{-\int\limits_L a(x)dx} \tag{2.1}$$

In the simplest case the ray L may be thought of as a straight line. Modelling L as a strip or cone, possibility with a weight factor to account for detector inhomogeneities may be more appropriate. In (2.1) we neglect the dependence of a on the energy (beam hardening effect) and other non-linear phenomena (e.g. partial volume effect).

The mathematical problem in transmission tomography is to determine a from measurements of I for a large set of rays L. If L is simply the straight line connecting the source x_0 with the detector x_1, (2.1) gives rise to the integrals

$$\ln \frac{I}{I_0} = -\int_{x_0}^{x_1} a(x)dx \tag{2.2}$$

where dx is the restriction to L of the Lebesgue measure in R^n. We have to compute a in a domain $\Omega \subseteq R^n$ from the values of (2.2) where x_0, x_1 run through certain subsets of $\partial\Omega$.

For $n = 2$, (2.2) is simply a reparameterization of the Radon transform

$$(Ra)(\theta, s) = \int_{x \cdot \theta = s} a(x)dx \tag{2.3}$$

where $\theta \in S^{n-1}$, the $(n-1)$-dimensional unit sphere in the n dimensional space. Thus our problem is in principle solved by Radon's inversion formula

$$a = R^* K g \ , \qquad g = Ra \tag{2.4}$$

where

$$(R^* g)(x) = \int_{S^{n-1}} g(\theta, x \cdot \theta)d\theta \tag{2.5}$$

is the so-called backprojection and

$$K = \frac{1}{2}(2\pi)^{1-n} \begin{cases} (-1)^{(n-2)/2} \, H \frac{\partial^{n-1}}{\partial s^{n-1}} & , \quad n \text{ even} \\ (-1)^{(n-1)/2} \, \frac{\partial^{n-1}}{\partial s^{n-1}} & , \quad n \text{ odd} \end{cases} \tag{2.6}$$

with H the Hilbert transform [39]. In fact the numerical implementation of (2.4) leads to the filtered backprojection algorithm which is the standard algorithm in commercial CT scanners, see Section 3.

For $n = 3$, the relevant integral transform is the x-ray transform

$$(Pa)(\theta, x) = \int_{R^1} a(x + s\theta)ds$$

where $\theta \in S^{n-1}$ and $x \in \theta^\perp$. P admits a similar inversion formula as R, to wit

$$a = P^* K g \qquad g = Pf \tag{2.7}$$

with K very similar to (2.6) and

$$(P^* g)(x) = \int_{S^{n-1}} g(\theta, E_\theta x)d\theta$$

where E_θ is the orthogonal projection onto $\theta^\perp$. Unfortunately, (2.7) is not as useful as (2.4). The reason is that (2.7) requires g for all θ and $y \in \theta^\perp$, i.e. (2.2) has to be available for all x_0, $x_1 \in \partial\Omega$. This is not practical. Also, it is not necessary for unique reconstruction of a. In fact it can be shown that a can be recovered uniquely from (2.2) with sources x_0 on a circle surrounding supp(a) and $x_1 \in S^{n-1}$. Unfortunately the determination of a in such an arrangement is, though uniquely possible, highly unstable. The condition of stability is the following: each plane meeting supp(a) must contain at least one source [16]. This condition is obviously violated for sources on a circle. Cases in which the condition is satisfied include the helix and a pair of orthogonal circles. A variety of inversion formulae has been derived, see Section 4.

If scatter is to be included, a transport model is more appropriate. Let $u(x,\theta)$ be the density of the particles at x travelling (with speed 1) in direction θ. Then,

$$\begin{aligned} \theta \cdot \nabla u(x,\theta) + a(x)u(x,\theta) &= \int_{S^{n-1}} \eta(x,\theta,\theta')u(x,\theta')d\theta' \\ &\quad +\delta(x - x_0)\,. \end{aligned} \tag{2.8a}$$

Here, $\eta(x,\theta,\theta')$ is the probability that a particle at x travelling in direction θ is scattered in direction θ'. Again we neglect dependence on energy. δ is the Dirac δ-function modelling a source of unit strength. (2.2a) holds in a domain Ω of R^n ($n = 2$ or 3), and $x_0 \in \partial\Omega$. Since no radiation comes in from outside we have

$$u(x,\theta) = 0\,, \quad x \in \partial\Omega\,, \quad \nu_x \cdot \theta \leq 0 \tag{2.8b}$$

where ν_x is the exterior normal on $\partial\Omega$ at $x \in \partial\Omega$. (4.1) is now replaced by

$$I(x_1, x_0, \theta) = I_0 u(x_1,\theta)\,, \quad x_1 \in \partial\Omega\,, \quad \nu_{x_1} \cdot \theta \geq 0\,. \tag{2.8c}$$

The problem of recovering a from (2.8) is much harder. An explicit formula for a such as (2.4) has not become known and is unlikely to exist. Nevertheless numerical methods have been developed for special choices of η [3], [7]. The situation gets even more difficult if one takes into account that, strictly speaking, η is object dependent and hence not known in advance. (2.8) is a typical example of an inverse problem for a partial differential equation. In an inverse problem one has to determine the differential equation - in our case a, η - from information about the solution - in our case (2.8c).

(b) Emission CT.

In emission tomography one determines the distribution f of radiating sources in the interior of an object by measuring the radiation outside the object in a tomographic fashion. Let again $u(x,\theta)$ be the density of particles at x travelling in direction θ with speed 1, and let a be the attenuation distribution of the object. (This is the quantity which is sought for in transmission CT). Then,

$$\theta \cdot \nabla u(x,\theta) + a(x)u(x,\theta) = f(x)\,. \tag{2.9a}$$

This equation holds in the object region $\Omega \subseteq R^n$ for each $\theta \in S^{n-1}$. Again there exists no incoming radiation, i.e.

$$u(x,\theta) = 0 \quad , \quad x \in \partial\Omega \, , \quad \nu_x \cdot \theta \leq 0 \tag{2.9b}$$

while the outgoing radiation

$$u(x,\theta) = g(x,\theta) \, , \quad x \in \partial\Omega \, , \quad \nu_x \cdot \theta \geq 0 \tag{2.9c}$$

is measured and hence known. (2.9) again constitutes an inverse problem for a transport equation. For a known, (2.9a-b) is readily solved to yield

$$u(x,\theta) = \int\limits_{x-\infty\cdot\theta}^{x} e^{-\int\limits_{y}^{x} a ds} f(y) dy \, .$$

Thus (2.9c) leads to the integral equation

$$g(x,\theta) = \int\limits_{x-\infty\cdot\theta}^{x} e^{-\int_y^x a ds} f(y) dy \tag{2.10}$$

for f. Apart from the exponential factor, (2.10) is identical (up to notation) to the integral equation in transmission CT. Except for very special cases - e.g. a constant in a known domain [52], [42] no explicit inversion formulas are available. Numerical techniques have been developed but are considered to be slow. Again the situation becomes worse if scatter is taken into account. This can be done by simply adding the scattering integral in (2.8a) to the right hand side of (2.9a).

What we have described to far is called SPECT (= single particle emission CT). In PET (= positron emission tomography) the sources eject the particles pairwise in opposite directions. They are detected in coincidence mode, i.e. only events with two particles arriving at opposite detectors at the same time are counted. (1.10) has to be replaced by

$$\begin{aligned} g(x,\theta) &= \int\limits_{x-\infty\cdot\theta}^{x} e^{-\int\limits_{y}^{x} a ds - \int\limits_{x-\infty\cdot\theta}^{y} a ds} f(y) dy \\ &= e^{-\int\limits_{x-\infty\cdot\theta}^{x} a ds} \int\limits_{x-\infty\cdot\theta}^{x} f(y) dy \, . \end{aligned} \tag{2.11}$$

Thus PET is even closer to the case of transmission CT. If a is known, we simply have to invert the x-ray transform. Inversion formulae which can make use of the data collected in PET are available, see Section 3.

The main problems in emission CT are unknown attenuation, noise, and scatter. For the attenuation problem, the ideal mathematical solution would be a method for determining f and a in (2.9) simultaneously. Under strong assumptions on a (e.g. a constant in a known region [26], a affine distortion of a prototype [37], a close to a known distribution [8]) encouraging results have been obtained. Theoretical results based on the transport formulation have been obtained, even for models including scatter [43]. But a clinically useful way of determining a from the emission data has not yet been found.

Noise and scatter are stochastic phenomena. Thus, besides models using integral equations, stochastic models have been set up for emission tomography [48]. These models are completely discrete. We subdivide the reconstruction region into m pixels or *voxels*. The number of events in pixel/voxel j is a Poisson random variable φ_j whose mathematical expectation $f_j = E\varphi_j$ is a measure for the activity in pixel/voxel j. The vector $f = (f_1, \ldots, f_n)$ is the sought-for quantity. The vector $g = (g_1, \ldots, g_n)$ of measurements is considered as a realization of the random variable $\gamma = (\gamma_1, \ldots, \gamma_n)$ where γ_i is the number of events detected in detector i. The model is determined by the $(n \times m)$-matrix $A = (a_{ij})$ whose elements are

$$a_{ij} = P\text{ (event in pixel/voxel } j \text{ detected in detector } i),$$

where P denotes probability. We have $E(\gamma) = Af$. f is determined from g by the maximum likelihood method. A numerical method for doing this is the EM (= expectation maximisation) algorithm. In its basic form it reads

$$f^{k+1} = f^k A^* \frac{g}{Af^k}, \qquad k = 0, 1, \ldots,$$

where division and multiplication are to be understood component wise. The problem with the EM-algorithm is that it is only semi-convergent, i.e. noise is amplified at high iteration numbers. This is known as the checkerboard effect. Various suggestions have been made to get rid of this effect. The most exciting and interesting ones use "prior information" and attempt to maximize "posterior likelihood". Thus f is assumed to have a prior probability distribution, called a Gibbs-Markov random field $\pi(f)$, which gives preference to certain functions f [46], [21], [18]. Most prior π simply add a penalty term to the likelihood function to account for correction between neighboring pixels and do not use biological information. However if π is carefully chosen so that piecewise constant functions f with smooth boundaries forming the region of constancy are preferred then the noise amplification at high iteration numbers can be avoided. The question remains as to whether this conclusion will remain valid for function f which are assigned low probability by π - or, more to the point - whether "real" emission densities f will be well-resolved by this Bayesian method. An ROC study (e.g. double-blind trials where radiologists are to find lesions from images produced by two different algorithms) concluded that maximum likelihood methods were superior to the filtered backprojection algorithm (see Section 3) in certain clinical applications. The same type of study is needed to determine whether or not Gibbs priors will improve the maximum likelihood reconstruction (stopped short of convergence to avoid noise amplification) on real data.

(c) Ultrasound CT

X-rays travel along straight lines. For other sources of radiation, such as ultrasound and microwaves, this is no longer the case. The paths are no longer straight, and their exact shape depends on the internal structure of the object. We can no longer think in terms of simple projections and linear integral equations. More sophisticated non-linear models have to be used.

In the following we consider an object $\Omega \subseteq R^n$ with refractive index n_{index}. We assume $n_{\text{index}} = 1$ outside the object. The object is probed by a plane wave

$$e^{-ikt}u_\theta(x)\,, \quad u_\theta(x) = e^{ikx\cdot\theta}$$

with wave number $k = \frac{2\pi}{\lambda}$, λ the wave length, travelling in the direction θ. The resulting wave $e^{-ikt}u(x)$ satisfies the reduced wave equation

$$\Delta u + k^2(1+f)u = 0\,, \quad f = n_{\text{index}}^2 - 1\,, \tag{2.12a}$$

plus suitable boundary conditions at infinity. The inverse problem to be solved is now the following. Assume that

$$g(x,\theta) = u(x)\,, \quad \theta \in S^{n-1} \tag{2.12b}$$

is known outside Ω. Determine f inside Ω!

Uniqueness and stability of the inverse problem (2.12) have recently been settled [36]. However, stability is only logarithmic [2], i.e. a datum error of size δ results in a reconstruction error $1/\log(1/\delta)$. Numerical algorithms did not emerge from this work.

Numerical methods for (2.12) are mostly based on linearizations, such as the Born and Rytov approximation [13]. In order to derive the Born approximation, one rewrites (2.12a) as

$$u(x) = u_\theta(x) - k^2 \int_\Omega G(x-y)f(y)u(y)dy \tag{2.13}$$

where G is an appropriate Green's function. For $n = 3$, we have

$$G(x) = \frac{e^{ik|x|}}{4\pi|x|}\,. \tag{2.14}$$

The Born approximation (simply one step of Picard iteration applied to (2.13)) is now obtained by assuming $u \sim u_\theta$ in the integral in (2.13). With this approximation, (2.12b) reads

$$g(x,\theta) = u_\theta(x) - k^2 \int_\Omega G(x-y)u_\theta(y)f(y)dy\,, \quad x \notin \Omega\,. \tag{2.15}$$

This is a linear integral equation for f, valid for all x outside the object and for all measured directions θ.

Numerical methods based on (2.15) - and a similar equation for the Rytov approximation - have become known as diffraction tomography. Unfortunately, the assumptions underlying the Born and Rytov approximations are not satisfied in medical imaging. Thus, the reconstructions of f obtained from (2.15) are very poor. However, we may use (2.15) to get some encouraging information about stability. For $|x|$ large, (2.15) assumes the form

$$\begin{aligned} g(x,\theta) &= u_\theta(x) - \frac{k^2}{4\pi|x|} e^{ik|x|} \int e^{ik(\theta - \frac{x}{|x|})\cdot y} f(y)dy \\ &= u_\theta(x) - \frac{k^2}{4\pi|x|} e^{ik|x|} (2\pi)^{3/2} \hat{f}(k(\frac{x}{|x|} - \theta)) \end{aligned} \tag{2.16}$$

with $\hat{f}$ the Fourier transform of f. (2.16) determines $\hat{f}$ within a ball of radius $\sqrt{2}k$ from the data (2.12b) in a completely stable way. We conclude that the stability of the inverse problem (2.12) is much better than logarithmic. If the resolution is restricted to spatial frequencies below $\sqrt{2}k$ - which is perfectly reasonable from a physical point of view - then we can expect (2.12) to be perfectly stable.

So far we have considered plane wave irradiation at fixed frequency, and we worked in the frequency domain. Time domain methods are conceivable as well. We start out from the wave equation

$$\frac{\partial^2 u}{\partial t^2} = c^2 \Delta u \tag{2.17a}$$

with the propagation speed c assumed to be 1 outside the object. With x_0 a source outside the object we consider the initial conditions

$$u(x,0) = 0\,, \quad \frac{\partial u}{\partial t}(x,0) = \delta(x - x_0)\,. \tag{2.17b}$$

We want to determine c inside the object from knowing

$$g(x_0, x_1, t) = u(x_1, t)\,, \quad t > 0 \tag{2.17c}$$

for many sources x_0 and receivers x_1 outside the object. In the one dimensional case, the inverse problem (2.17) can be solved by the famous Gelfand-Levitan method in a stable way. It is not clear how Gelfand-Levitan can be extended to dimensions two and three. The standard methods use sources and receivers on all of the boundary of the object. This is not practical in medical imaging. However, for reduced data sets, comparable to those in 3D X-ray tomography, we do not know how to use Gelfand-Levitan, nor do we know anything about stability.

Of course one can always solve the nonlinear problem (2.17) by a Newton type method. Such methods have been developed [23], [30]. They suffer from excessive computing time and from their apparent inability to handle large wave numbers k.

(d) Optical tomography

Here one uses NIR (= near infra-red) lasers for the illumination of the body. The process is now described by the transport equation

$$\frac{\partial u}{\partial t}(x,\theta,t)+\theta\cdot\nabla u(x,\theta,t)+a(x)u(x,\theta,t)$$

$$=b(x)\int_{S^{n-1}}\eta(\theta\cdot\theta')u(x,\theta',t)d\theta'+f(x,\theta,t) \tag{2.18a}$$

for the density $u(x,\theta,t)$ of the particles at $x\in\Omega$ flying in direction $\theta\in S^{n-1}$ at time t. a and b are the sought-for tissue parameters. The scattering kernel η is assumed to be known. The source term f is under the control of the experimenter. Together with the initial and boundary conditions

$$\begin{aligned} u(x,\theta,0) &= 0 \text{ in } \Omega\times S^{n-1}\,, \\ u(x,\theta,t) &= 0 \text{ on } \partial\Omega\times S^{n-1}\times R^1\,, \quad \nu_x\cdot\theta\le 0 \end{aligned} \tag{2.18b}$$

(1.18a) has a unique solution under natural conditions on a, b, η and f. As in (1.1) we pose the inverse problem. Assume that we know the outward radiation

$$g(x,\theta,t)=u(x,\theta,t) \text{ on } \partial\Omega\times S^{n-1}\times R^1\,, \quad \nu_x\cdot\theta\ge 0\,, \tag{2.18c}$$

can we determine one or both the quantities a, b?

There are essentially three methods for illuminating the object, i.e. for choosing the source term f in (2.18a). In the stationary case one puts $f=\delta(x-x_0)$ where $x_0\in\partial\Omega$ is a source point. u is considered stationary, too. A second possibility is the light flash $f=\delta(x-x_0)\delta(t)$. Finally one can also use time harmonic illumination, in which case $f=\delta(x-x_0)\,e^{i\omega t}$. This case reduces to the stationary case with a replaced by $a+i\omega$. In all three cases, the data function g of (2.18c) is measured at $x\in\partial\Omega$, possibly averaged over one or both of the variables θ, t.

Light tomography is essentially a scattering phenomenon. This means that the scattering integral in (2.18a) is essential. It can no longer be treated merely as a perturbation as in x-ray CT. Thus the mathematical analysis and the numerical methods are expected to be quite different from what we have seen in other types of tomography.

The mathematical theory of the inverse problem (2.18) is in a deplorable state. There exist some Russian papers on uniqueness [4]. General methods have been developed, too, but apparently they have been applied to 1D problems only

[44]. Nothing seems to be known about stability. The numerical methods which have become known are of the Newton type, either applied directly to the transport equation or to the so-called diffusion approximation [5], [31]. The diffusion approximation is an approximation to the transport equation by a parabolic differential equation. Since inverse problems for parabolic equations are severely ill-posed, this approach is questionable. Higher order approximations [29], [20] are hyperbolic, making the inverse problem much more stable.

As an alternative to the transport equation one can also model light tomography by a discrete stochastical model [22]. In the 2D case, break up the object into a rectangular arrangement of pixels labelled by indices i, j with $a \leq i \leq b$ and $c \leq j \leq d$. Attach to each pixel the quantities f_{ij}, b_{ij}, r_{ij}, ℓ_{ij} meant to denote the probability of a forward, backward, rightward or leftward transition out of the pixel i, j with respect to the direction used to get into this pixel. For each pair of boundary pixels i, j and i', j' let $P_{ij,i'j'}$ be the probability that a particle that enters the object at pixel i, j will eventually leave the object at pixel i', j'. The problem is to determine the quantities f_{ij}, b_{ij}, r_{ij}, ℓ_{ij} from the values of $P_{ij,i'j'}$ for all boundary pixels. Preliminary numerical tests show that this is possible, at least in principle. However, the computations are very time consuming. More seriously, they reveal a very high degree of instability.

(e) Electrical Impedance Tomography

Here, the sought-for quantity is the electrical impedance σ of an object Ω. Voltages are applied via electrodes on $\partial\Omega$, and the resulting currents at these electrodes are measured. With u the potential in Ω, we have

$$\begin{aligned} \operatorname{div}(\sigma \cdot \nabla u) &= 0 \quad \text{in} \quad \Omega \\ u &= g \quad , \quad \sigma \frac{\partial u}{\partial \nu} = f \quad \text{on} \quad \partial\Omega \,. \end{aligned} \tag{2.19}$$

Knowing many voltage-current pairs g, f on $\partial\Omega$, we have to determine σ from (2.19).

Uniqueness for the inverse problem (2.19) has recently been settled [36]. Unfortunately, the stability properties are very bad [2]. Numerical methods based on Newton's method, linearization, simple backprojection, and layer stripping have been tried. All these methods suffer from the severe ill-posedness of the problem. There seems to be no way to improve stability by purely mathematical means.

(f) Magnetic Resonance Imaging (MRI)

The physical phenomena exploited here is the precession of the spin of a proton in a magnetic field of strength H about the direction of that field. The frequency of this precession is the Larmor frequency γH where γ is the gyromagnetic ratio. By making the magnetic field H space dependent in a controlled way the local magnetization $M_0(x)$ (together with the relaxation times $T_1(x)$, $T_2(x)$) can be imaged. In the following we derive the imaging equations [27].

The magnetization $M(x,t)$ caused by a magnetic field $H(x,t)$ satisfies the Bloch equation

$$\frac{\partial M}{\partial t} = \gamma M \times H - \frac{1}{T_2}(M_1 e_1 + M_2 e_2) - \frac{1}{T_1}(M_3 - M_0)e_3 \,. \tag{2.20}$$

Here, M_i is the i-th component of M, and e_i is the i-th unit vector $i = 1,2,3$. The significance of T_1, T_2, M_0 become apparent if we solve (2.20) with the static field $H = H_0 e_3$ and with initial values $M(x,0) = M^0(x)$. We obtain with $\omega_0 = \gamma H_0$

$$\begin{aligned} M_1(x,t) &= e^{-t/T_2} \quad (M_1^0 \cos(\omega_0 t) + M_2^0 \sin(\omega_0 t)) \\ M_2(x,t) &= e^{-t/T_2} \quad (-M_1^0 \sin(\omega_0 t) + M_2^0 \cos(\omega_0 t)) \\ M_3(x,t) &= e^{-t/T_1} \quad M_3^0 + (1 - e^{-t/T_1})M_0. \end{aligned} \tag{2.21}$$

Thus the magnetization rotates in the $x_1 - x_2$-plane with Larmor frequency ω_0 and returns to the equilibrium condition $(0,0,M_0)$ with speed controlled by T_2 in the $x_1 - x_2$-plane and by T_1 in the x_3-direction.

In an MRI scanner one generates a field

$$H(x,t) = (H_0 + G(t) \cdot x)e_3 + H_1(t)(\cos(\omega_0 t)e_1 + \sin(\omega_0 t)e_2)$$

where G and H_1 are under control. In the jargon of MRI, $H_0 e_3$ is the static field, G the gradient, and H_1 the radio frequency (RF) field. The input G, H_1 produces in the detecting system of the scanner the output signal

$$S(t) = -\frac{d}{dt} \int_{R^3} M(x,t) B(x) dx \tag{2.22}$$

where B characterizes the detecting system. Depending on the choice of H_1 various approximations to S can be derived.

(i) H_1 is constant in the small interval $[0,\tau]$ and $\gamma \int_0^\tau H_1 dt = \frac{\pi}{2}$ (Short $\frac{\pi}{2}$ pulse). In that case,

$$S(t) = \int_{R^3} M_0(x) e^{-i\gamma \int_0^t G(t')dt' \cdot x - t/T_2(x)} dx \,.$$

Choosing G constant for $\tau \le t \le \tau + T$ and zero otherwise we get for $T \ll T_2$

$$\begin{aligned} S(t) &= \int_{R^3} M_0(x) e^{-i\gamma(t-\tau)G\cdot x} dx \\ &= (2\pi)^{3/2} \hat{M}_0(\gamma(t-\tau)G) \end{aligned} \tag{2.23}$$

where $\hat{M}_0$ is the 3D Fourier transform of M_0.

From here we can proceed in several ways. We can either use (2.23) to determine the 3D Fourier transform $\hat{M}_0$ of M_0 and to compute M_0 by an inverse 3D Fourier transform. This requires $\hat{M}_0$ to be known on a Cartesian grid, which can be achieved by a proper choice of the gradients or by interpolation. We can also evoke the central slice theorem to obtain the 3D Radon transform RM_0 of M_0 by a series of 1D Fourier transforms. M_0 is recovered in turn by inverting the 3D Radon transform.

The main numerical problem in MRI is to reconstruct a function from imperfect measurements of its Fourier transform [34]. Not much has been achieved so far.

(ii) H_1 is the shaped pulse

$$H_1(t) = \phi(t\gamma G)e^{i\gamma G x_3 t}$$

where ϕ is a smooth positive function supported in $[0, \tau]$. Then, with x', G' the first two components of x, G, respectively, we have

$$S(t) = \int_{R^2} M_0'(x', x_3) e^{-i\gamma \int_0^t G'(t')dt' \cdot x' - t/T_2(x', x_3)} dx' \tag{2.24}$$

where

$$M_0'(x', x_3) = \int M_0(x', y_3) Q(x_3' - y_3) dy_3$$

with a function Q essentially supported in a small neighborhood of 0. (2.24) is the 2D analogue of (2.23). So we have again the choice between Fourier imaging (i.e. computing the 2D Fourier transform from (2.24) and doing an inverse 2D Fourier transform) and projection imaging (i.e. doing a series of 1D Fourier transforms in (2.24) and inverting the 2D Radon transform).

Some of the mathematical problems of MRI, e.g. interpolation in Fourier space, are common to other techniques in medical imaging. An interesting mathematical problem occurs if the magnets do not produce suffiently homogeneous fields [33]. It calls for the reconstruction of a function from integrals over slightly curved manifolds. Even though this is a problem of classical integral geometry it has not yet found a satisfactory solution.

(g) Vector Tomography

If the domain under consideration contains a moving fluid, then the Doppler shift can be used to measure the velocity $u(x)$ of motion. Assume the time harmonic signal $e^{i\omega_0 t}$ is transmitted along the oriented line L. This signal is reflected by particles travelling with speed ν in the direction of L as $e^{i(\omega_0 - k\nu)t}$, where $k = 2\omega_0/c$, c the speed of the probing signal, i.e. $k\nu$ is the Doppler shift. Let $S(L, \nu)$ be the Lebesgue measure of these particles on L which move with speed

$< \nu$, i.e. $u(x) \cdot e_L < \nu$, e_L the tangent vector on L. Then the total response is

$$g(L,t) = \int_{-\infty}^{+\infty} e^{i(\omega_0 - k\nu))t} S(L,\nu) d\nu \ .$$

Thus S can be recovered from g by a Fourier transform. The problem is to recover u from S.

Not much is known about uniqueness. However, the first moment of S,

$$\int_{-\infty}^{+\infty} \nu S(L,\nu) d\nu = \int_L u(x) \cdot e_L dx = (Ru)(L) \tag{2.25}$$

is similar to the Radon transform. One can show that curl u can be computed from Ru, and an inversion formula similar to the Radon inversion formula exists. Numerical simulations are given in [51].

(h) Tensor Tomography

As an immediate extension of transmission CT to non-isotropic media we consider a matrix valued attenuation $a(x) = (a_{ij}(x))$, i, $j = 1, \ldots, n$. We solve the vector differential equation

$$\frac{du(t)}{dt} = -a(x(t))u(t) \ , \quad x(t) = (1-t)x_0 + tx_1 \ , \quad 0 \le t \le 1 \tag{2.26}$$

for the vector valued function $u(t) = (u_i(t))_{i=1,\ldots,n}$. Let a be defined in a convex domain Ω, and let x_0, $x_1 \in \partial\Omega$. Then, $u(x_1) = U(x_0, x_1)u(x_0)$ with a nonlinear map $U(x_0, x_1)$ depending on a. The problem is to recover a in Ω from the knowledge of $U(x_0, x_1)$ for x_0, $x_1 \in \partial\Omega$. For $n = 1$ we regain (2.1). Applications of (2.25) for $n > 1$ have become known in photoelasticity [1], but applications to medicine are not totally out of the question.

In a further extension we let a depend on the direction $\xi = (x_1 - x_0)/|x_1 - x_0|$. Such problems occur in the polarization of harmonic electromagnetic and elastic waves in anisotropic media. In linearized form these problems give rise to the transverse x-ray transform

$$(Ja)(x, \theta, \omega) = \int \omega^T a(x + t\theta)\omega dt \ , \quad \omega \perp \theta \tag{2.27}$$

and to the longitudinal x-ray transform

$$(Ja)(x, \theta) = \int \theta^T a(x + t\theta)\theta dt \ . \tag{2.28}$$

(2.27) can easily be reduced to the $(n-1)$-dimensional x-ray transform in the plane $H_{\omega,s} = \{y : y \cdot \omega = s\}$. We only have to introduce the function $a_\omega(y) =$

$\omega^T a(y)\omega$. Then, (2.27) provides for $x \in H_{\omega,s}$ all the line integrals of a_ω on $H_{\omega,s}$. For (2.28), the situation is not so easy. We decompose a in its solenoidal and potential part, i.e.

$$a = a_1 + \nabla a_2 \quad \text{div } a_1 = 0 \quad a_2 = 0 \quad \text{on } \partial\Omega .$$

It can be shown that a_1 can be recovered uniquely from (2.28), but a_2 is completely undetermined [47]. This is reminiscent of vector tomography.

3 Basic algorithms in $2D$ tomography

The numerical implementation of Radon's inversion formula (2.6) is now well understood [39], [25], [28]. We consider only the simplest case of parallel scanning. This means that $g(\theta, s) = (Ra)(\theta, s)$ is sampled at

$$\begin{aligned} \theta = \theta_j &= \begin{pmatrix} \cos\varphi_j \\ \sin\varphi_j \end{pmatrix} , \qquad \varphi_j = \pi j/p , \quad j = 0, \dots, p-1 , \\ s = s_\ell &= \ell \frac{\rho}{q} , \qquad \ell = -q, \dots, q . \end{aligned} \tag{3.1}$$

The reconstruction region is $|x| < \rho$.

We need some tools from sampling theory. For a function f in R^n we define the Fourier transform by

$$\hat{f}(\xi) = (2\pi)^{-n/2} \int e^{-ix\cdot\xi} f(x) dx .$$

We call a function (essentially) Ω-band-limited if $\hat{f}(\xi)$ vanishes (approximately) for $|\xi| > \Omega$. Shannon's sampling theorem states that an Ω-band-limited function can be recovered exactly from the samples $f(hk)$, $k \in Z^n$, if $|h| \le \pi/\Omega$. If f_1, f_2 are both Ω-band-limited, then

$$\int_{R^n} f_1 f_2 dx = h^n \sum_{k\in Z^n} f_1(hk) f_2(hk) . \tag{3.2}$$

The convolution of two functions is denoted by $f_1 \star f_2$, independently of the dimension.

Rather than using (2.6) we start out from

$$V \star a = R^*(v \star g) \tag{3.3}$$

if $V = R^* v$. Here, R^* is the $2D$ backprojection from (2.5), and the convolution on the right hand side is with respect to the second argument only. In order to make (3.3) equivalent to (2.6) one has to choose v such that $V = \delta$, the Dirac δ-function. One can show that this is the case iff

$$\hat{v}(\sigma) = \frac{1}{2}(2\pi)^{-3/2}|\sigma| \,. \tag{3.4}$$

Thus v is a distribution rather than a function. For the numerical evaluation of (3.3) we have to regularize v, i.e. we replace it by

$$\hat{v}_\Omega(\sigma) = \frac{1}{2}(2\pi)^{-3/2}|\sigma|\hat{\phi}(\sigma/\Omega)$$

where ϕ is a low-pass filter, i.e. $\pi(\sigma) = 0$ for $|\sigma| \geq 1$. For the ideal low-pass $\phi(\sigma) = 1$, $|\sigma| \leq 1$, we have

$$v_\Omega(s) = \frac{\Omega^2}{4\pi^2}u(\Omega s)\,, \quad u(s) = sinc(s) - \frac{1}{2}(sinc(s/2))^2$$

with the *sinc*-function $sinc(x) = \sin(x)/x$. This filter produces reconstructions with maximal spatial resolution but with unpleasant artefacts. Hence one prefers a filter with a smaller jump at the transition point, such as the much used Shepp and Logan filter

$$\hat{\phi}(\sigma) = \begin{cases} sinc|\sigma|\pi/2) & , \quad 0 \leq |\sigma| \leq 1\,, \\ 0 & , \quad |\sigma| \geq 1\,. \end{cases}$$

The convolution on the right hand side of (3.3) is now approximated by the trapezoidal rule

$$(v_\Omega \star g)(\theta, s) = h \sum_{\ell=-q}^{q} v_\Omega(s - s_\ell)g(\theta, s_\ell)\,. \tag{3.5}$$

At this stage we assume a to be essentially Ω-band-limited. Since

$$(Ra)^\wedge(\theta, \sigma) = (2\pi)^{1/2}\hat{a}(\sigma\theta) \tag{3.6}$$

where $(Ra)^\wedge$ means the $1D$ Fourier transform with respect to the second argument, Ra is essentially Ω-band-limited, too, and (3.5) is correct with good accuracy for $\frac{\ell}{q} \leq \frac{\pi}{\Omega}$, as can be seen from (3.2). For the backprojection we use the trapeziodal rule again, obtaining as approximation to $a(x)$

$$a(x) = \frac{\pi h}{p}\sum_{j=0}^{p-1}\sum_{\ell=-q}^{q} v_\Omega(x \cdot \theta_j - s_\ell)g(\theta_j, s_\ell)\,. \tag{3.7}$$

It can be shown that $(v_\Omega \star g)(\theta, \theta \cdot x)$ as a function of φ, $\theta = (\cos\varphi, \sin\varphi)^T$, is essentially $\Omega\rho$-band-limited. Thus, (3.7) is satisfied with good accuracy provided that $\frac{\pi}{p} \leq \frac{\pi}{\Omega\rho}$. The sampling theorem requires a to be evaluated on a grid with stepsize $\frac{\pi}{\Omega}$. Thus the evaluation of (3.6) needs $O(\Omega^4)$ operations. This number can be reduced to $O(\Omega^3)$ by introducing the functions

$$b_j(s) = \sum_{\ell=-q}^{q} v_\Omega(s - s_\ell) g(\theta_j, s_\ell) , \tag{3.8}$$

evaluating these functions for $s = s_k$ and computing $b_j(x \cdot \theta_j)$, which is needed in (3.7), by interpolation. Practical experience shows that linear interpolation suffices.

(3.6) is called the filtered backprojection algorithm. The algorithm depends on the parameters p, q, Ω and the filter ϕ. We make some remarks concerning the role of these parameters.

1. Ω controls the spatial resolution of the algorithm. According to the sampling theorem, details of size $2\pi/\Omega$ (and no smaller ones) can be accurately reconstructed provided that $p \geq \Omega\rho$ and $q \geq \frac{1}{\pi}\Omega\rho$. Thus the number of data needed for reconstructing an essentially Ω-band-limited function in $|x| \leq \rho$ is essentially $2qp = \frac{2}{\pi}\Omega^2\rho^2$.

It can be shown that $\frac{1}{\pi}\Omega^2\rho^2$ pieces of data suffice. Without losing resolution we can drop $g(\theta_j, s_\ell)$ for $j + \ell = p \bmod 2$. In that case the interpolation step for (3.7) has to be done with care, e.g. with a stepsize much smaller than ρ/q [15].

2. The condition $\frac{\rho}{q} \leq \frac{\pi}{\Omega}$ has to be satisfied strictly. If it is violated, (3.5) is not valid even approximately, making the reconstruction completely unacceptable.

3. If $\frac{\pi}{p} \leq \frac{\pi}{\Omega\rho}$ is not satisfied, then the number p of directions permits artefact free reconstructions only within a circle of radius $\rho' < \rho$ where $\frac{\pi}{p} = \frac{\pi}{\Omega\rho'}$. This means that artefacts occur in a distance of $2\rho'$ from high density objects. These can be avoided if we sacrifice resolution by choosing $\Omega' < \Omega$ such that $\frac{\pi}{p} = \frac{\pi}{\Omega'\rho}$ in the filter ϕ.

4. Only filters with vanishing kernel sum, i.e.

$$\sum_{\ell} v_\Omega(s_\ell) = 0$$

should be used.

5. Choosing p, q in an optimal way leads to $p = \pi q$. In practice this relation is not strictly observed. Usually, p is chosen smaller. According to 3. this leads to artefacts. But these artefacts are usually outside the region of interest.

Besides the filtered backprojection algorithm there exists a plethora of other algorithms which, however, are much less used. The direct Fourier algorithm is based on (3.6). Theoretically the complexity of this algorithm is $O(\Omega^2 \log \Omega)$, but the accurate and efficient implementation is in no way easy. It seems that the gridding method [45] works satisfactorily, but the theory behind this method is not well understood. Direct algebraic algorithms [39] compute a minimal norm solution in L_2 for finitely many data by FFT techniques. Iterative methods work on discrete versions of linear reconstruction problems [10], [24]. ART (= algebraic reconstruction technique [24], [25]) is based on the Kaczmarz iteration for linear systems and can be viewed as an SOR method [14]. The EM iteration is described in Section 2(b).

The filtered backprojection algorithm can also be used for fan beam geometry, in which the X-ray tube sits on a circle surrounding the patient, with a detector array on the opposite side. The sampling requirements are discussed in [38].

4 Formulas for $3D$ reconstruction

In $3D$ CT one reconstructs a from the values of

$$g(\theta, v) = \int a(v + t\theta)dt$$

where v runs through the source curve V outside supp(a) and $\theta \in S^2$. Most reconstruction formulae make use of an intermediate function

$$G(\theta, v) = \int_{S^2} g(\omega, v)h(\omega \cdot \theta)d\omega \tag{4.1}$$

where h is homogeneous of degree -2, and a function F on $S^2 \times R^1$ derived from G by

$$F(\theta, v \cdot \theta) = G(v, \theta) . \tag{4.2}$$

The reconstruction formula then reads

$$a(x) = \frac{1}{16\pi^3} \int_{S^2} (F \star k)(\theta, \theta \cdot x)d\theta \tag{4.3}$$

where the convolution with the $1D$ function k in the second argument. It can be shown that (4.1-3) is in fact a reconstruction formula provided that

$$\hat{h}(\sigma)\hat{k}(\sigma) = \sigma^2 + \text{ odd function of } \sigma . \tag{4.4}$$

For the proof we start out from

$$G(\theta, v) = (h \star Ra)(\theta, v \cdot \theta)$$

with the $3D$ Radon transform [49], hence $F = h \star Ra$. Then, (4.3) is compared with (2.4) for $n = 3$, i.e.

$$a(x) = -\frac{1}{8\pi^2} \int_{S^2} (Rf)''(\theta, x \cdot \theta)d\theta .$$

This coincides with (4.3) if (4.4) holds.

The simplest choice for h, k is $\hat{h} = -(-2\pi)^{-1/2} i\sigma$, $\hat{k} = (2\pi)^{1/2} i\sigma$. In this case, both $h = -\delta'$, $k = 2\pi\delta'$ are local. We obtain Grangeat's inversion formula [19]

$$G(\theta, v) = \int_{S^2 \cap \theta^\perp} \frac{\partial}{\partial\theta} g(v, \omega) d\omega , \tag{4.5}$$

$$a(x) = \frac{1}{8\pi^2} \int_{S^2} F'(\omega, \omega \cdot x) d\omega \tag{4.6}$$

where $\frac{\partial}{\partial\theta}$ is the derivative in direction θ with respect to the second argument and F' is the partial derivative with respect to the second argument. In order to apply (4.5-6), $F(\omega, s)$ is needed for each plane $\omega \cdot x = s$ meeting supp(a). Since F is obtained from G by means of (4.2) we need for each such plane a source v in that plane. In view of (4.5) this means that g is available for a neighbourhood of the fan in that plane converging to the source v. This is Grangeat's completeness condition.

The inversion formulae of Tuy [53], B. Smith [50] and Gelfand and Goncharov [17] can be obtained by putting $\hat{h}(\sigma) = (2\pi)^{1/2}(|\sigma| - \sigma)$, $\hat{k}(\sigma) = (2\pi)^{-1/2}\sigma$ and $\hat{h}(\sigma) = (2\pi)^{1/2}|\sigma|$, $\hat{k}(\sigma) = (2\pi)^{-1/2}|\sigma|$, respectively [12]. These formulae are not as useful as Grangeat's formula since h is no longer local.

In practice $g(v, \cdot)$ is measured on a detector plane $D_v = p_v - v + (p_v - v)^\perp$ where p_v is the orthogonal projection of v onto D_v. Putting $g_v(y) = g(v, y - v)$, (4.1) assumes the form

$$G(v, \theta) = |p_v - v| \int_{D_v} g_v(y) h(\theta \cdot (y - v)) dy .$$

Introducing an orthogonal system $W_v = (w_1, w_2)$ in $(p_v - v)^\perp$ we obtain in the Grangeat case (4.5)

$$G(v, \theta) = \frac{|p_v - v|}{|\theta_v|^3} \, \theta_v \cdot (Rg_v') \left(\frac{\theta_v}{|\theta_v|}, \frac{(v - p_v) \cdot \theta}{|\theta_v|} \right) \tag{4.7}$$

where $\theta_v = w^T \theta$, R is the $2D$ Radon transform and g_v' the gradient of g_v. Thus Grangeat's formula can be implemented by computing line integrals in the detector plane, followed by a $3D$ backprojection (4.6). An implementation analogous to the filtered backprojection algorithm of $2D$ tomography can be found in [12].

In $3D$ emission CT, the requirements are quite different. In PET one puts the object into a vertical cylinder whose interior surface is covered by detectors. With such an arrangement one measures the X-ray transform for all lines joining two points on the mantle of the cylinder. In principle one could do the reconstruction layer by layer, using only horizontal lines in each layer. However, all the information contained in the oblique rays would be lost.

A formula which at least partially copes with this situation is

$$f(x) = -\frac{1}{4\pi^2}\Delta \int_G \int_{\theta^\perp} \frac{Pf(\theta, x-y)}{|y|L(\theta, y)} dy d\theta \tag{4.8}$$

where $G \subseteq S^2$ is a spherical zone around the equator and $L(\theta, y)$ is the length of the intersection of G and the plane spanned by θ, y [41]. With (4.8) one still has problems near the openings of the cylinder. More satisfactory reconstruction formulae based on the principle of the stationary phase have been given in [11].

5 Algorithm for more general inverse problems

The problem of CT is probably the most simple inverse problem. In general one has to solve the inverse problem for a partial differential equation. The numerical analysis of such inverse problems is still in its infancy. Of course one can always use a Newton method for solving the inverse problem simply as a nonlinear problem [31]. A method for solving the inverse problem (2.12) of ultrasound CT by an ART type method (compare Section 3) is as follows [40]. For each direction θ_j, $j = 1, \ldots, p$, define the nonlinear (Radon-type) operator $R_j = L_2(\Omega) \to L_2(\partial\Omega)$

$$R_j(f) = u|_{\partial\Omega_j}$$

where u is the solution to (2.12a) for $\theta = \theta_j$ with boundary values $\frac{\partial}{\partial\nu}u(x) = \frac{\partial}{\partial\nu}g(x, \theta_j)$. Then one has to find an approximate solution f to

$$R_j(f) = g_j \quad , \quad g_j(x) = g(x, \theta_j) \text{ on } \partial\Omega \, , \quad j = 1, \ldots, p \, .$$

This can be done by a Kaczmarz-type iteration. Starting out from an initial approximation f_0 we put

$$f_j = f_{j-1} - \omega R_j'(f_{j-1})^* C_j (R_j(f_{j-1}) - g_j) \quad , \quad j = 1, \ldots, p \tag{5.1}$$

with some positive definite operator C_j and a relaxation parameter ω. After p steps one repeats the whole process. It turns out that for practical purposes, one can take C_j to be the identity. In this form each step of (5.1) requires the solution of (2.12a) with boundary values as specified above (this yields $R_j(f_{j-1})$), and the solution of an adjoint boundary value problem of the same structure for the application of $R_j'(f_{j-1})^*$. In the engineering literature this is known as the adjoint field method [9]. Nothing is known about convergence, and the performance - even though it is much faster than Newton-type methods - leaves much to be desired. Presently, $2D$ problems with a resolution 128×128 can be solved in a few minutes on a workstation, but $3D$ problems in a clinical environment are totally out of the question. The same applies to equations such as (2.8) and to the equations of optical tomography (2.18).

References

1. Aben, H.K.: Integrated Photoelasticity, Valgus, Talin 1975 (Russian).
2. Allesandrini, G.: Stable Determination of Conductivity by Boundary Measurements, *Appl. Analysis*, **27**, 153-172, (1988).
3. Anikonov, D.S. - Prokhorov, I.V. - Kovtanyuk, E.E.: Investigation of Scattering and Absorbing Media by the Methods of X-ray Tomography, *J. Inv. Ill-Posed Problems*, **1**, 259-281; (1993).
4. Anikonov, D.S.: Uniqueness of Simultaneous Determination of Two Coefficients of the Transport Equation, *Soviet Math. Dokl.* **30**, 149-151, (1984).
5. Arridge, S.R. - Van der Zee, P. - Cope, M. - Delpy, D.T.: Reconstruction Methods for Infrared Absorption Imaging, *Poc. SPIE* **1431**, 204-215, (1991).
6. Barrett, H.H. - Swindell, S.: Radiological Imaging, Vol. I, II, Academic Press 1981.
7. Bondarenko, A.- Antyufeev, V.: X-Ray Tomography in Scattering Media, *Institute of Mathematics*, Novosibirsk, Russia (1990).
8. Bronnikov, A.V.: Degration Transform in Tomography, *Pattern Recognition Letters* **15**, 527-592, (1994).
9. Bui, H.D.: Inverse Problems in the Mechanics of Materials. CRC Press 1994.
10. Censor, Y.: Finite Series-Expansion Reconstruction Methods, *Proc. IEEE* **71**, 409-419, (1983).
11. Defrise, M. et al.: Performance Study of 3D Reconstruction Algorithms for Positron Emission Tomography, International Meeting on Fully Three-Dimensional Image Reconstruction in Radiology and Nuclear Medicine, June 23-25, 1993, Snowbird, Utah, USA.
12. Defrise, M. - Clack, R.: A Cone-Beam Reconstruction Algorithm Using Shift-Variant Filtering and Cone-Beam Backprojection, IEEE Transactions on Medical Imaging **13**, 186-195 (1995).
13. Devaney, A.J.: A Filtered Backpropagation Algorithm for Diffraction Tomography, *Ultrasonic Imaging*, **4**, 336-350, (1982).
14. Elving, T.: Block-iterative methods for consistent and inconsistent linear equations, Numer. Math. **35**, 1-12 (1980).
15. Faridani, A.: Reconstructing From Efficiently Sampled Data in Parallel-Beam Computed Tomography, in: Inverse Problems and Imaging, G.F. Roach, ed., *Pitman Res. Notes Math. Ser.* **245**, 68-102, (1991).
16. Finch, D.V.: Cone Beam Reconstruction with Sources on a Curve. *SIAM J. Appl. Math.*, **45**, 665-673,(1985).
17. Gelfand, I.M. - Goncharov, A.B.: Recovery of a Compactly Supported Function Starting from its Integrals over Lines Intersecting a Given Set of Points in Space, *Doklady* **290** (1986), English Translation in Soviet Math.. Doklady **34**, 373-376 (1987).
18. Geman, S. - McClure, D.: Statistical Methods for Tomographic Image Reconstruction, ISI Tokio session, *Bull. Int. Statist. Inst.*, **LII(4)**, 5-21, 1987.
19. Grangeat, P.: Mathematical Framework of Cone Beam 3D Reconstruction via the First Derivative of Radon Transform, in: Herman et al. (eds.): *Mathematical methods in tomography*, Springer 1991.
20. Gratton, E. et al.: A novel approach to laser tomography, *Bioimaging*, **1**, 40-46 (1993).
21. Green, P.J.: Bayesian Reconstructions from Emission Tomography Data Using a Modified EM Algorithm, *IEEE Transactions on Medical Imaging*, **9**(1), 84-93, März 1990.
22. Grünbaum, F.A. - Kohn, P.D. - Latham, G.A. - Singer, J.R. - Zubelli, J.P.: Diffuse Tomography, *Proc. SPIE*, **1431**, 232-238 (1991).

23. Gutman, S. - Klibanov, M.V.: Regularized Quasi-Newton Method for Inverse Scattering Problems, *Mathl. Comput. Modelling* **18**, No. 1, 5-31, Pergamon Press Ltd. (1993).

24. Herman, G.T. - Lent, A.: Iterative Reconstruction Algorithms, *Comput. Biol. Med.* **6**, 273-294, (1976).

25. Herman, G.: Image Reconstruction From Projections. The Fundamentals of Computerized Tomography. Academic Press 1980.

26. Hertle, A.: The Identification Problem for the Constantly Attenuation Radon Transform, *Math. Z.* **197**, 13-9, (1988).

27. Hinshaw, W.S. - Lent, A.H.: An Introduction to NMR Imaging: From the Bloch Equation to the Imaging Equation, *Proc. IEEE* **71**, 338-350 (1983).

28. Kak, A.C. - Slaney, M.: Principle of Computerized Tomography Imaging. IEEE Press 1987.

29. Kaltenbach, J.-M. - Kaschke, M.: Frequency- and time-domain modelling of light transport in random media, *Technical Report, Carl Zeiss, PF 1980*, Oberkochen, Germany, 1992.

30. Kleinman, R.E. - van den Berg, P.M.: A Modified Gradient Method for Two-Dimensional Problems in Tomography, *J. Comp. Appl. Math.*, **42**, 17-35, (1992).

31. Klibanov, M.V. - Gutman, S. - Barbour, R. - Chang, J. - Malinsky, J. - Alfano, R.R.: Consideration of Solutions to the Inverse Scattering Problem for Biomedical Applications, *Proc. SPIE* **1887**, (1993).

32. Krestel, E. (ed.): Imaging Systems for Medical Diagnostics, Siemens Aktiengesellschaft, 1990.

33. Lai, C-M.: Reconstructing NMR Images from Projections Under Inhomogeneous Magnetic Field and Non-Linear Field Gradient, *Phys. Med. Biol.* **8**, 925-938, (1983).

34. Liang, Z.-P. - Boada, F.E. - Constable, R.T. - Haacke, E.M. - Lauterbur, P.C. - Smith, M.R.: Constrained Reconstruction Methods in MR Imaging, Rev Magn. Reson. Med. **4**, 67-185, (1992).

35. Louis, A.K.: Medical Imaging: State of the Art and Future Development, *Inverse Problems* **8**, 709-738 (1992).

36. Nachman, A.I.: Global Uniqueness for a Two-Dimensional Inverse Boundary Value Problem, *Department of Mathematics, Preprint Series*, Number **19**, University of Rochester (1993).

37. Natterer, F.: Determination of Tissue Attenuation in Emission Tomography of Optically Dense Media, *Inverse Problems* **9**, 731-736 (1993).

38. Natterer, F.: Sampling in Fan Beam Tomography, *SIAM J. Appl. Mathematics* **53**, 358-380 (1993).

39. Natterer, F.: The Mathematics of Computerized Tomography. Wiley-Teubner 1986.

40. Natterer, F. - Wübbeling, F.: A propagation-backpropagation method for ultrasound tomography, *Inverse Problems* **11**, 1225-1232 (1995).

41. Orlov, S.S.: Theory of Three Dimensional Reconstruction. II. The Recovery Operator, Sov. Phys. Crystallogr. **20**, 429-433 (1976).

42. Palamodov, V.: An Inversion Method for Attenuated X-Ray Transform in Space, submitted to *SIAM J. Appl. Math.*

43. Romanov, V.G.: Conditional Stability Estimates for the Problem of Recovering of Absorption Coefficients and Right Hand Side of Transport Equations (Russian), to appear in *Siberia Math. J.*

44. Sanchez, R. - McCormick, N.J.: General Solutions to Inverse Transport Problems, *J. Math. Phys.* **22**, 847-855, (1981).

45. Schomberg, H. - Timmer, J.: The Gridding Method for Image Reconstruction by Fourier Transformation, *IEEE Transactions on Medical Imaging* **14**, 596-607 (1995).

46. Setzepfandt, B.: ESNM: Ein rauschunterdrückendes EM - Verfahren für die Emissionstomographie. Thesis, Fachbereich Mathematik der Universität Münster, Germany 1992.

47. Sharafutdinov, V.A.: Integral Geometry of Tensor Fields. *Nauka, Novosibirsk 1993* (Russian).

48. Shepp, L.A. - Vardi, Y.: Maximum Liklihood Reconstruction for Emission Tomography, *IEEE Trans. Med. Imag.* **1**, 113-121, (1982).

49. Smith, K.T. - Solmon, D.C. - Wagner, S.L.: Practical and Mathematical Aspects of the Problem of Reconstructing Objects From Radiographs, *Bull* AMS **83**, 1227-1270 (1977).

50. Smith, B.D.: Image Reconstruction from Cone-Beam Projections: Necessary and Sufficient Conditions and Reconstruction Methods, *IEEE Transactions on Medical Imaging* **4**, 14-25 (1985).

51. Sparr, G. - Stråklén, K. - Lindström, K. - Persson, W.: Doppler tomography for vector fields, *Inverse Problems*, **11**, 1051-1061 (1995). **1**, 113-121, (1982).

52. Tretiak, O.J. - Metz, C.: The Exponential Radon Transform, *SIAM J. Appl. Math.*, **39**, 341-354, (1980).

53. Tuy, H.K.: An Inversion Formula for Cone-Beam reconstruction, *SIAM J. Appl. Math.* **43**, 546-552 (1983).

Partial Differential Equations and Image Iterative Filtering

Frédéric Guichard and Jean-Michel Morel

Ceremade, Université Paris IX Dauphine, France

Abstract

It is well known that a conveniently rescaled iterated convolution of a linear positive kernel converges to a gaussian. As a consequence, all iterative linear smoothing methods of a signal or an image boil down to the application to the signal of the heat equation. This survey explains how a similar analysis can be performed for image iterative smoothing by a wide class of nonlinear operators, the morphological operators. These monotone operators have a property which makes them suitable for image analysis: they commute with contrast changes of the images. Among them, the median operator which computes a local "mean value" independent of constrast changes. We prove that all morphological (i.e. monotone, constrast invariant) operators, are asymptotically equivalent (when they become more and more local) to a motion of the image by its curvature. The iteration of these filters is equivalent to the application to the image of a nonlinear heat equation. We give in parallel a classification of all image multiscale smoothing methods (the so called "scale space" methods in Computer Vision). It is shown that both approaches (classification of iterative filtering methods or of "scale spaces") yield the same, recently discovered, partial differential equations. Experiments are presented with both classical and new, contrast invariant and monotone numerical schemes.

1 The bottom-up approach : from local filters to partial differential operators

Let $u_0(\mathbf{x})$ be a real function defined on the plane, which we interpret throughout this paper as a real image, $\mathbf{x}$ being a point of the image plane and $u_0(\mathbf{x})$ the "grey level" at $\mathbf{x}$. In signal and image processing, "filtering" means an operation, applied to the image, which tends to smooth out asperities, yields a more reliable value at each point and permits the computations of some derivatives of the image. Linear filtering methods are often defined by a modification of the Fourier spectrum of the image, which attenuates high frequencies. It is therefore a global operation, whose local result depends on global events such as the size and location of the observation window of the image (in practice, an image is

Figure 1. Zoom on noise...

defined in a rectangle of the plane). There has been a growing tendency in image processing to define filtering methods which are as independent as possible of the observation window, which in practice means that they are *local*. A local operator is, intuitively, an operator T which transforms the initial image u_0 into a new image $(Tu_0)(\mathbf{x})$ and such that the value $Tu_0(\mathbf{x})$ only depends on the values $u_0(\mathbf{y})$ where $\mathbf{y}$ lies in an *a priori* fixed neighborhood of $\mathbf{x}$. This neighborhood should be as small as possible.

1.1 The linear case

As for an obvious (but useful) example, let us mention the case where $Tu_0(\mathbf{x}) = (M_h u_0)(\mathbf{x})$ is obtained as the mean value of u_0 in a neighborhood of $\mathbf{x}$,

$$M_h u_0(\mathbf{x}) = \frac{1}{\pi h^2} \int_{D(\mathbf{x},h)} u_0(\mathbf{y})d\mathbf{y}, \tag{1.1}$$

In this case, the locality means that only values of u_0 inside a disk $D(\mathbf{x}, h)$ around $\mathbf{x}$ matter. When h tends to zero, we get a more and more local estimate of the value of u_0 at $\mathbf{x}$. The parameter h characterizes the "locality" of the considered operator, that is, the size of the neigborhood involved in the filtering operation. The question arises of whether such an operation can be made independent of h. A first answer is to analyse the asymptotic development of $M_h u_0(\mathbf{x})$ as h tends to zero. Clearly, if u_0 is continuous at $\mathbf{x}$, we have $M_h u_0(\mathbf{x}) \to u_0(\mathbf{x})$. If we assume that u_0 is twice continuously differentiable (C^2) at $\mathbf{x}$, then it is easily seen that in fact [1]

[1] We always denote in this text by $o(h)$ a function of h which tends to zero faster than h, by $O(h)$ a function such that, for some constant C, $|O(h)| \leq C|h|$, and by $\varepsilon(h)$ a function tending to zero as h tends to zero. Thus an $o(h)$ function can also be written : $h\varepsilon(h)$.

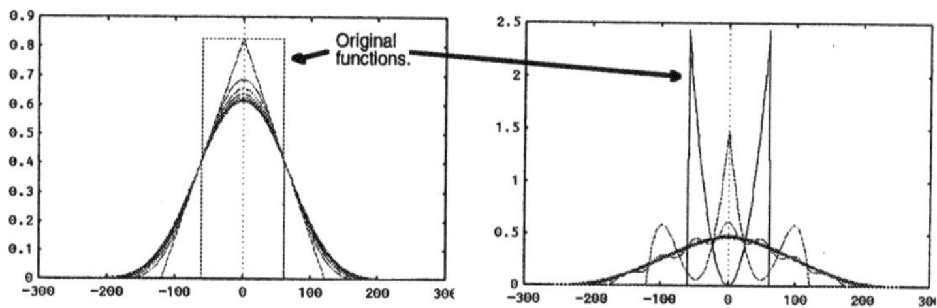

Figure 2. In this experiment on one dimensional functions, it can be appreciated how fast an iterated convolution of a positive kernel converges to a gaussian. On the left, we display nine iterations of the convolution of a characteristic function with itself, with the adequate rescaling. On the right, the same experiment is repeated with a by far more irregular kernel. Convergence to the gaussian is almost equally fast. (See Theorem 1)

$$\frac{M_h u_0(\mathbf{x}) - u_0(\mathbf{x})}{h^2} = \frac{1}{8}\Delta u_0(\mathbf{x}) + \varepsilon(h) \tag{1.2}$$

This formula suggests the following result : consider for any t a sequence of real numbers h and integers n such that $nh^2 \to t$. Set $u^n = T_h^n u_0$. Then it is to be expected that $u^n(\mathbf{x}) \to u(t, \mathbf{x})$ where $u(t, \mathbf{x})$ satisfies the heat equation associated with (1.2),

$$\frac{\partial u}{\partial t}(t, \mathbf{x}) = \frac{1}{8}\Delta u(t, \mathbf{x}), \quad u(0) = u_0. \tag{1.3}$$

This convergence result is easily obtained by a Fourier analysis argument in $\mathbb{R}^N$. It is important to notice that the result is fairly independent of the kind of linear smoothing effectuated. More precisely, setting $\mathbf{x} = (x_1, ..., x_N) \in \mathbb{R}^N$, and denoting by $(g(\mathbf{x}))^{n*}$ the convolution of a function $g(\mathbf{x})$ with itself n times, one can prove the following theorem.

Theorem 1. *Let $g(\mathbf{x}) = g(|\mathbf{x}|)$, $\mathbf{x} \in \mathbb{R}^N$, be a radial function, nonnegative, compactly supported and piecewise C^1, satisfying the normalization conditions $\int_{\mathbb{R}^N} g(\mathbf{x})d\mathbf{x} = 1$ and $\int_{\mathbb{R}^N} x_1^2 g(\mathbf{x})d\mathbf{x} = 2$. Then for every $t > 0$ and every sequence h of real numbers and of integers n such that $t = nh^2$ and $n \to \infty$, one has*

$$(g_h(\mathbf{x}))^{n*} \to \frac{1}{(4\pi t)^{\frac{N}{2}}} e^{-\frac{|\mathbf{x}|^2}{4t}}, \tag{1.4}$$

the convergence being verified pointwise, uniformly and in $L^2(\mathbb{R}^N)$. As a consequence, defining, for every initial bounded image $u_0(\mathbf{x})$, $(L_h u_0) = g_h * u_0$, we can assert that

$$(L_h)^n u_0 \to T_t u_0 \quad \textit{uniformly},$$

where $(T_t u_0)(\mathbf{x}) = u(t, \mathbf{x})$ *and* $u(t, \mathbf{x})$ *is solution of the heat equation*

$$\frac{\partial u}{\partial t} = \Delta u, \quad u(0, \mathbf{x}) = u_0(\mathbf{x}). \tag{1.5}$$

In other terms, the preceding theorem means that from the asymptotic, scale independent viewpoint, all local linear "low pass" iterated filters are equivalent to the heat equation (and to the convolution with gaussians with increasing width).

1.2 Morphological filters

Our aim here is to generalize the above linear results to a wide class of nonlinear filters, the so called morphological filters. A filter T is called morphological if it satisfies the following properties. (We denote by u, v, ... real functions on

Figure 3. On the first row, we display left : a grey level image seabird1, photograph of a seabird, whose width is 410 and height is 270, and right : its level lines for level multiples of 12. On the second row, the heat equation, or, equivalently, a convolution with a gaussian has been applied to the original image. The variance of the gaussian is 4, which means that its spatial range is comparable to a disk with radius 4. The image gets blurred by the convolution, which mixes grey level values and removes all sharp edges. This can be appreciated on the right, where we have displayed all level lines for levels multiples of 12. We can see how the level lines on the boundaries of the image have splitted into parallel level lines which go away from each other. The image has become smooth and has lost its structure

$\mathbb{R}^N$ representing N-dimensional images. T transforms a function u into a new function Tu.)

a- Monotonicity. If $u \geq v$, then $Tu \geq Tv$. This means that the filter acts as the convolution with a positive kernel.

b- Contrast invariance. If $g(s)$ is a nondecreasing continuous real function, which we call a "contrast change", then $T(g(u)) = g(T(u))$. This means that the action of T does not depend upon the contrast of the image ; an assumption which is of course false for linear filters.

c- Locality. Basically, this assumption means that $(Tu)(\mathbf{x})$ only depends upon the values of u in a fixed neighborhood of $\mathbf{x}$, for instance a disk $D(\mathbf{x}, h)$.

The contrast invariance assumption was proposed in Mathematical Morphology, an image analysis rigorous doctrine founded by Matheron [41]. Mathematical Morphology has benefited from several recent expositions [38], [53] and its methods are by now classical in Image Processing. From the morphological viewpoint, an image $u(\mathbf{x})$ defined on $\mathbb{R}^N$ with values in $\mathbb{R}$ is not a reliable information since it has been submitted to an unknown (and arbitrary) contrast change g, where g is an increasing continuous function. When we apply an arbitrary contrast change g to u, what is left of u ? Essentially the family of level sets,

$$\chi_\lambda(u) = \{\mathbf{x}, u(\mathbf{x}) \geq \lambda\}.$$

Indeed, it is obviously true that if g is an increasing contrast change, then

$$\chi_{g(\lambda)}(gou) = \{\mathbf{x}, g(u(\mathbf{x})) \geq g(\lambda)\} = \chi_\lambda(u). \tag{1.6}$$

Conversely, a function u can be recovered from its level sets $\chi_\lambda(u)$ by

$$u(\mathbf{x}) = \sup\{\lambda, \mathbf{x} \in \chi_\lambda(u)\} \tag{1.7}$$

and it is easily seen that if two functions u and v have the same level sets, then they can be deduced from each other by a contrast change. Matheron defined morphological operators T on binary images (or, equivalently, on subsets of $\mathbb{R}^N$, which we denote by X, Y,...) by the fundamental property

Set Monotonicity. If $X \subset Y$ then $TX \subset TY$

This property permits a formal extension to a grey level image u by simply applying T to each level set $\chi_\lambda(u)$ of u and then constructing Tu from the $T(\chi_\lambda u)$ by the formula deduced from (1.7),

$$Tu(\mathbf{x}) = \sup\{\lambda, \mathbf{x} \in T(\chi_\lambda(u))\}, \tag{1.8}$$

so that the level sets of Tu are the images by T of the level sets of u. (Without risk of ambiguity, we take the same notation for T acting on sets or on functions). Notice that this reconstruction is only possible because of the Monotonicity, which preserves the obvious inclusion property of level sets,

$$\text{If} \quad \lambda \geq \mu, \quad \text{then} \quad \chi_\lambda(u) \subset \chi_\mu(u). \tag{1.9}$$

Conversely, one can extend T to compact sets when it is defined on continuous functions in the following way. Let X be a compact subset of $\mathbb{R}^N$ and let $u(\mathbf{x})$ be a continuous function such that

$$X = \{\mathbf{x},\ u(\mathbf{x}) \leq 0\} \text{ and } u(\mathbf{x}) \geq C > 0 \text{ for } \mathbf{x} \text{ large enough.} \tag{1.10}$$

(X is the zero level set of $-u$). It is easily seen that the distance to X, that is $d(\mathbf{x}, X) = \inf\{|\mathbf{x} - \mathbf{y}|, \mathbf{y} \in X\}$ defines such a function. We can define $T(X)$, by using (1.10) in the following way.

Theorem and Definition 1.1. *(Evans-Spruck). Let T be a monotone and morphological operator (that is, commuting with continuous nondecreasing contrast changes) which is defined on a set $\mathcal{F}$ containing all continuous functions. Then we can extend T to all closed and bounded subsets of $\mathbb{R}^N$ by setting*

$$T(X) = \{\mathbf{x}, Tu(\mathbf{x}) \leq 0\}. \tag{1.11}$$

The set $T(X)$ so defined does not depend upon the particular choice of u satisfying (1.10).

Thanks to the preceding results, we can consider as abundantly proved that we can canonically associate with a morphological monotone operator defined on continuous functions a monotone operator defined on sets and conversely. Let us now see how to give an analytic form to monotone morphological operators.

There is a theorem, due to Matheron, giving a characterization of all morphological filters.

Theorem 2. *Let T be a translation invariant monotone operator acting on a set of subsets of $\mathbb{R}^N$. Then there exists a family of sets $\mathbb{B} \subset \mathcal{P}(\mathbb{R}^N)$, which can be defined as $\mathbb{B} = \{X, 0 \in T(X)\}$, such that*

$$T(X) = \bigcup_{B \in \mathbb{B}} \bigcap_{\mathbf{y} \in B} (X - \mathbf{y}) = \{\mathbf{x},\ \exists B \in \mathbb{B},\ \mathbf{x} + B \subset X\} \tag{1.12}$$

Let T be a translation invariant monotone and morphological operator defined on a space of continuous and bounded functions. Then there exists a family of closed sets $\mathbb{B} \subset \mathcal{P}(\mathbb{R}^N)$, which can be defined as $\mathbb{B} = \{X, 0 \in T(X)\}$, such that

$$Tu(\mathbf{x}) = \sup_{B \in \mathbb{B}} \inf_{\mathbf{y} \in \mathbf{x}+B} u(\mathbf{y}). \tag{1.13}$$

Conversely, if $\mathbb{B}$ is a set of subsets of $\mathbb{R}^N$, Relation (1.12) defines a monotone and translation invariant operator and (1.13) defines a morphological monotone translation invariant operator.

Figure 4. We display a grey level image seabird 2, photograph of a seabird, whose width is 256 and height 172, with different contrasts. All of three images have exactly the same level sets level lines, but associated with three different scales of grey levels which are displayed on the right : the graphs on the right are the graphs of the functions $u \to g(u)$ which have been applied to the initial grey levels. The first one is convex and enhances the darker parts of the image. The second one is the identity, leaving thus the image untouched. The third one is concave and enhances the brighter part of the image

In the terminology of the mathematical morphology school, the operator T is an "erosion of u by the family of structuring elements, $\mathbb{B}$". Every statement about T can be adapted to the "dual" operators of the kind

$$T'u = \inf_{B \in \mathbb{B}} \sup_{\mathbf{y} \in \mathbf{x}+B} u(\mathbf{y}) \tag{1.14}$$

which are called in the same way "dilation by $\mathbb{B}$". Indeed, $T'(u) = -T(-u)$, so that all discussions about dilations or erosions are equivalent with the obvious adaptations. When $\mathbb{B}$ is reduced to a single element, B, we call T "erosion, (resp. dilation) by B". A particular, important, case is when $B = D(0,t)$ is a disk centered at 0 with radius t. In that case, we define

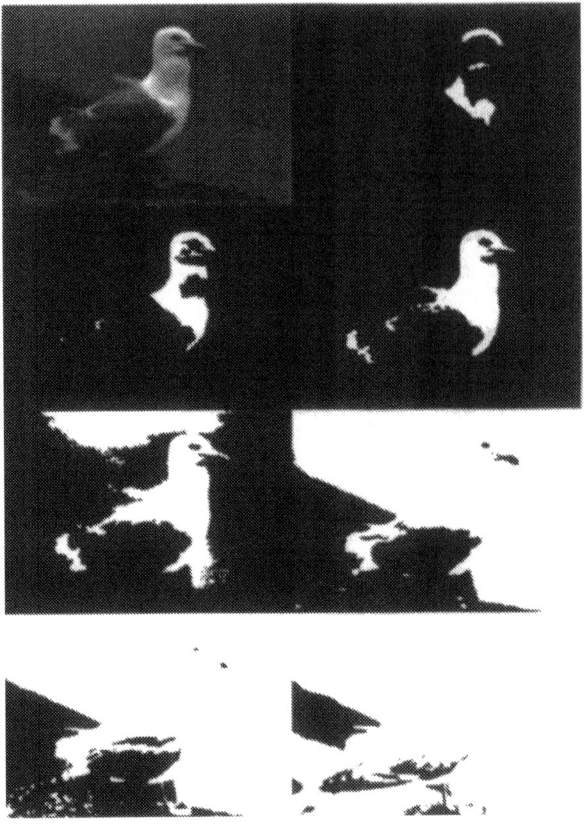

Figure 5. In this figure, we display the image seabird2, which has a range of grey levels from 0 to 255, and then all level sets whose level is a multiple of 32, in decreasing order from 224 to 32. Notice how essential features of the shapes are contained in the boundaries of level sets, the level lines. Each level set (in white) is contained in the next, (formula 1.9)

• Dilation with radius t : $TX = X^t$, the t-neighborhood of X, or, in an equivalent way,

$$T_t u(\mathbf{x}) = \sup_{\mathbf{y} \in D(\mathbf{x},t)} u(\mathbf{y});$$

• Erosion with radius t : $TX = X^{-t}$, the t-interior neighborhood of X, or, in an equivalent way,

$$T_t u(\mathbf{x}) = \inf_{\mathbf{y} \in D(\mathbf{x},t)} u(\mathbf{y}).$$

Another very important and useful class of morphological filters are the so called median filters.

Figure 6. On the first row, seabird1, and its level lines for all level multiple of 12. On the second row, an erosion with radius 4 has been applied. On the right, the resulting level lines where the circular shape of the structuring element (a disk with radius 4) appears arround each local minimum of the original image. Erosion removes local maxima (in particular, all small white spots) and expands minima, in particular, all dark spots like the eye. The third row displays the effect of the operator dual to the erosion, the dilation and the resulting level lines. We see how local minima are removed (see for example the eye of the bird) and how white spots on the tail expand. Here again, circular level lines appear around all local maxima of the original image

Median filters.

The median value of a function u on a bounded set B is the result of an attempt to define a "mean value" which does not depend upon contrast changes. We define for any function u defined (for simplicity) in $\mathbb{R}^2$ and every set B with bounded measure, the *median filter on* B by

$$\mathrm{med}_B u = \inf\left\{\lambda, \mathrm{meas}\{\mathbf{x}; u(\mathbf{x}) \leq \lambda\} \geq \frac{\mathrm{meas}(B)}{2}\right\} \tag{1.15}$$

In order to make intuitive the proposed definition, we remark that if $\text{meas}\{\mathbf{x}, u(\mathbf{x}) = \text{med}_B u\} = 0$, then $\text{meas}\{\mathbf{x}, u(\mathbf{x}) \geq \text{med}_B u\} = \text{meas}\{\mathbf{x}, u(\mathbf{x}) \leq \text{med}_B u\}$. With the above definition, we do not necessarily have

$$\text{med}_B u = -\text{med}_B(-u).$$

This relation is true, again, if and only if $\text{meas}\{u(\mathbf{x}) = \text{med}_B u\} = 0$. In order to understand the consequences of the definition in case of functions which have "flat" parts and jumps, it is useful to consider the following examples of functions defined on $B = [0, 2]$:

- $u(x) = x$ if $x \leq 1$, $u(x) = 1$ if $x > 1$: $\text{med}_B u = 1$.
- $u(x) = \frac{x}{4}$ if $x \leq 1$, $u(x) = 1$ if $x > 1$: $\text{med}_B u = \frac{1}{4}$.
- $u(x) = 0$ if $x \leq \frac{3}{4}$, $u(x) = 1$ if $\frac{3}{4} < x < \frac{4}{3}$, $u(x) = 2$ if $x \geq \frac{4}{3}$: $\text{med}_B u = 1$.

Our choice for the definition of the median filter is further justified by the following property, which ensures that the median filter can be expressed as an inf sup-operator. It is easily checked that

$$\text{med}_B u = \inf_{\text{meas}\, B' \geq \frac{\text{meas} B}{2}, B' \subset B} \ \sup_{\mathbf{x} \in B'} u(\mathbf{x}). \tag{1.16}$$

Of course, the fact that the median filter can be put in the "inf sup" form can be directly deduced from Theorem 2.

Weighted median filter.
An infinite number of variants for the median filter can be obtained by changing the measure on $\mathbb{R}^N$. Considering that we wish to define the median value of u at some point $\mathbf{x}$, it is natural to let the measure depend upon the distance to $\mathbf{x}$. We define for any function u defined (for simplicity) in $\mathbb{R}^2$ and every radial, nonnegative, continuous function k satisfying $\int_{\mathbb{R}^2} k(\mathbf{x}) d\mathbf{x} = 1$, the measure associated with k by

$$\underset{k}{\text{meas}}(E) = \int_E k(\mathbf{x}) d\mathbf{x}$$

and the *weighted median filter associated with* k by

$$\text{med}_k u = \inf \left\{ \lambda, \underset{k}{\text{meas}}\{u(\mathbf{x}) \leq \lambda\} \geq \frac{1}{2} \right\} \tag{1.17}$$

Is is again easily seen that we can rewrite the weighted median filter as an "infsup", morphological operator,

$$\text{med}_k u = \inf_{\text{meas}_k\, B' \geq \frac{1}{2}, B' \subset \mathbb{R}^2} \ \sup_{\mathbf{x} \in B'} u(\mathbf{x}). \tag{1.18}$$

1.3 Relevant differential operators in mathematical morphology

Before arriving to statements for the morphological operators which will play the same role for morphological filters as Theorem 1 plays for linear filters, we need to define some differential operators which will be canonically associated with those filters, in the same way as the Laplacian $\Delta u = u_{xx} + u_{yy}$ is associated with all linear local low pass filters. From now on, we assume that the dimension N is 2, because of the particular attention we pay to image processing operators. Let us assume that u is twice differentiable (C^2) in a neighborhood of $\mathbf{x}$. We denote by $Du(\mathbf{x}) = (u_x, u_y)(\mathbf{x})$ the gradient of u at $\mathbf{x}$, by $|Du(\mathbf{x})| = (u_x^2 + u_y^2)^{\frac{1}{2}}(\mathbf{x})$ its euclidian norm. We call "orientation of the gradient" the normalized vector

$$\vec{n}(\mathbf{x}) = -\frac{Du(\mathbf{x})}{|Du(\mathbf{x})|}. \tag{1.19}$$

We consider the "level" line of u passing by $\mathbf{x}$, defined as the set

$$\{\mathbf{y}, u(\mathbf{y}) = u(\mathbf{x})\}$$

If we assume that $Du(\mathbf{x}) \neq 0$, by the Implicit Function Theorem, we know that the "level line" indeed is a curve, which inherits the smoothness of u in a

Figure 7. In this display, we compute curvature of the original image seabird2 after it has been smoothed by mean curvature motion. The first image displays the smoothed version of seabird 1 at a small scale. The second image displays the absolute value of the curvature, with the convention the darkest points have the largest curvature. We have displayed the curvature only at points where the gradient of the image was larger than 5. (The image grey levels range from 0 to 255). On the second row, we display on separate images the positive part of curvature and the negative part

neighborhood of $\mathbf{x}$. We set $Du^{\perp} = (-u_y, u_x)$, a vector orthogonal to Du and denote by $curv(u)(\mathbf{x})$ the curvature of the level line of u passing by $\mathbf{x}$. Then

$$curv(u)(\mathbf{x}) = \frac{1}{|Du|^3} D^2u(Du^{\perp}, Du^{\perp}) = \frac{u_{xx}u_y^2 - 2u_{xy}u_xu_y + u_{yy}u_x^2}{(u_x^2 + u_y^2)^{\frac{3}{2}}}. \quad (1.20)$$

Easy calculations show that this last formula can be written in a compact way as

$$curv(u)(\mathbf{x}) = div(\frac{Du}{|Du|})(\mathbf{x}), \quad (1.21)$$

where we define, as usual, $div = \frac{\partial}{\partial x} + \frac{\partial}{\partial y}$. In the remainder of this text, we shall be essentially concerned with the above mentioned differential operators for a very simple reason : they are invariant with respect to contrast changes. Indeed, level lines do not depend upon the contrast for, if $u(\mathbf{x}) = u(\mathbf{y})$, then $g(u(\mathbf{x})) = g(u(\mathbf{y}))$. The tangent vector to the level line and the normal vector are therefore contrast-insensitive and so is the curvature of the level line. This can be of course directly checked by substituting $g(u)$ to u in the formula (1.20) defining the curvature. As will be clear from the next section, the curvature plays the same role in mathematical morphology as the Laplacian in the linear theory.

1.4 Asymptotic behaviour of morphological filters

The next theorem fixes what happens asymptotically with any morphological filter which is isotropic and associated with a uniformly bounded family of structuring elements. We know from the results mentionned above that any morphological operator can be given a "sup inf" or the "inf sup" form. Thus, in the following, we consider operators directly defined in this form.

Theorem 3. *Let $\mathbb{B}$ be a family of structuring elements in $\mathbb{R}^2$ which is bounded ($\forall B \in \mathbb{B}$, $B \subset D(0, M)$) and isotropic (if $B \in \mathbb{B}$, then $RB \in \mathbb{B}$ for every linear isometry R of $\mathbb{R}^2$).*

Let $Tu(\mathbf{x}) = \sup_{B \in \mathbb{B}} \inf_{\mathbf{y} \in \mathbf{x}+B} u(\mathbf{y})$ and define the rescaled operator

$$T_h u(\mathbf{x}) = SI_h u(\mathbf{x}) = \sup_{B \in h\mathbb{B}} \inf_{\mathbf{y} \in \mathbf{x}+B} u(\mathbf{y})$$

Then, considering the real function $H(h) = T(x + hy^2)(0)$ and assuming that u is a C^3 function on $\mathbb{R}^2$,

(i) if $H(0) \neq 0$, then

$$(T_h u)(\mathbf{x}) = SI_h u(\mathbf{x}) = u(\mathbf{x}) + hH(0)|Du|(\mathbf{x}) + O(h^2)$$

and we obtain an erosion or a dilation depending on the sign of $H(0)$;

(ii) if $H(0) = 0$, and $Du(\mathbf{x}) \neq 0$, then

$$T_h u(\mathbf{x}) = SI_h u(\mathbf{x}) = u(\mathbf{x}) + h|Du(\mathbf{x})|H(\frac{1}{2}hcurv(u)) + O(h^{2+\frac{1}{3}});$$

and we obtain a curvature motion;

(iii) if $H(0) = 0$ and $Du(\mathbf{x}) = 0$, then the above formulas are not valid, but

$$T_h u(\mathbf{x}) = SI_h u(\mathbf{x}) = u(\mathbf{x}) + h^2 \sup_{B \in \mathbb{B}} \inf_{\mathbf{y} \in B} (\lambda_1 x^2 + \lambda_2 y^2) + O(h^3),$$

where λ_1 and λ_2 are the eigenvalues of $D^2 u(\mathbf{x})$.

Proof. Since T is monotone, H is a nondecreasing function, which is, in addition, Lipschitz. Indeed, by the monotonicity of T and the boundedness of $\mathbb{B}$,

$$T(x + h_1 y^2) - |h_2 - h_1| M^2 \leq H(h_2) \leq T(x + h_1 y^2) + |h_2 - h_1| M^2$$

and therefore

$$|H(h_2) - H(h_1)| \leq M^2 |h_2 - h_1|, \tag{1.22}$$

where we have used the fact that all elements of $\mathbb{B}$ are contained in a disk $D(0, M)$.

We notice that $T(u - u(\mathbf{x})) = Tu - u(\mathbf{x})$ and that, T being translation and isometry invariant, we can choose the origin 0 at $\mathbf{x}$, and the orthogonal axes $(\vec{i}, \vec{j})$ so that $\vec{i} = \frac{Du}{|Du|}(0)$ and $\vec{j} = (Du^{\perp}/|Du|)(0)$, if $Du \neq 0$. Since u is C^3, we can therefore write

$$u(\mathbf{x}) = px + ax^2 + by^2 + cxy + O(|\mathbf{x}|^3), \tag{1.23}$$

where $p = |Du|(0) \geq 0$, and, if $p > 0$,

$$b = \frac{1}{2}\frac{\partial^2 u}{\partial y^2}(0) = \frac{1}{2} D^2 u(\vec{j}, \vec{j})$$

$$a = \frac{1}{2}\frac{\partial^2 u}{\partial x^2}(0) = \frac{1}{2} D^2 u(\vec{i}, \vec{i}) \tag{1.24}$$

$$c = \frac{\partial^2 u}{\partial x \partial y}(0) = D^2 u(\vec{j}, \vec{i})$$

so that, if $p = |Du| \neq 0$ we have by Formula 1.20

$$b = \frac{1}{2}|Du| curv(u)(0). \tag{1.25}$$

From (1.23) we deduce that for every $\mathbf{x}$ in $hD(0, M)$, we have

$$px + ax^2 + by^2 + cxy - O(h^3) \leq u(\mathbf{x}) \leq px + ax^2 + by^2 + cxy + O(h^3).$$

Therefore, since every hB is contained in $D(0, Mh)$,

$$T_h u(0) = T_h(px + ax^2 + by^2 + cxy)(0) + O(h^3) \tag{1.26}$$

Let us first consider the case where $pH(0) = T(px)(0) \neq 0$. We then have from (1.26) and the monotonicity of T_h :

$$-(|a|+|b|+|c|)M^2h^2 - O(h^3) \leq T_h u(0) - T_h(px)(0) \leq (|a|+|b|+|c|)M^2h^2 + O(h^3) \tag{1.27}$$

Thus $T_h u(0) = hT(px)(0) + O(h^2) = phH(0) + O(h^2)$, which proves (i), since $p = |Du(0)|$.

From now on, we assume that $T(px)(0) = 0$, so that (1.27) implies in particular that $|T_h u(0)| \leq O(h^2)$ as h tends to zero. The reader may think of the median filter as a canonical example for this case. Indeed, the median filter does not alter linear functions and we therefore have $\mathrm{med}(px)(0) = 0$. An obvious rescaling yields

$$T_h u(0) = hT(px + ahx^2 + bhy^2 + chxy)(0) + O(h^3)$$

In order to prove the theorem, we see that we have to show that c and a play no role in the asymptotic behaviour of $T_h u(0)$. We begin by getting rid of the cxy term. For every $\epsilon > 0$, we have, in virtue of the inequality $|st| \leq s^2 + t^2$,

$$-|c|\epsilon y^2 - \frac{|c|}{\epsilon}x^2 \leq cxy \leq |c|\epsilon y^2 + \frac{|c|}{\epsilon}x^2.$$

Thus choosing $\epsilon = h^\alpha$,

$$T_h u(0) \leq hT(px + (a + |c|h^{-\alpha})hx^2 + (b + h^\alpha|c|)hy^2)(0) + O(h^3) \tag{1.28}$$

$$T_h u(0) \geq hT(px + (a - |c|h^{-\alpha})hx^2 + (b - h^\alpha|c|)hy^2)(0) - O(h^3) \tag{1.29}$$

Setting $a(h) = (a +$ or $- |c|h^{-\alpha})h$ and $b(h) = (b +$ or $- |c|h^\alpha)h$, we are led to study expressions of the type

$$T(px + a(h)x^2 + b(h)y^2)(0),$$

where $a(h) = O(h^{1-\alpha})$ and $b(h) = O(h)$ as h tends to zero. In fact, we shall prove that such expressions are asymptotically independent of $a(h)$. In order to do that, we use the morphological character of T : $g \circ T = T \circ g$, where we fix $g(s) = s - \frac{a(h)}{p^2}s^2$. This function is increasing in a neighborhood of zero. Thus, since, as noticed above, $T(px + a(h)x^2 + b(h)y^2)(0)$ tends to zero as h tends to zero, using $goT = Tog$:

$$T(px + a(h)x^2 + b(h)y^2)(0) =$$

$$g^{-1}(T(\ px + b(h)y^2 - \frac{a(h)}{p^2}((a(h)x^2 + b(h)y^2)^2 + 2px(a(h)x^2 + b(h)y^2)))\ )(0))$$

Taking into account that $a(h) = O(h^{1-\alpha})$ and $b(h) = O(h)$, we obtain

$$T(px + a(h)x^2 + b(h)y^2)(0) = g^{-1}(T(px + b(h)y^2)(0) + O(h^{2-2\alpha})).$$

We know that $g^{-1}(s) = s + O(\frac{a(h)}{p^2}s^2)$. Thus, since $|T(px + b(h)y^2)(0)| \leq T(px)(0) + M^2|b(h)| = M^2|b(h)|$, we have

$$\begin{aligned} T(px + a(h)x^2 + b(h)y^2)(0) &= T(px + b(h)y^2)(0) + O(h^{2-2\alpha}) \\ &= pT(x + \frac{b(h)}{p}y^2)(0) + O(h^{2-2\alpha}), \end{aligned}$$

for any $\alpha > 0$. Returning to (1.28-1.29) and the definition of $a(h)$ and $b(h)$, we finally obtain

$$\begin{aligned} T_h u(0) \leq phT(x + \tfrac{bh+|c|h^{1+\alpha}}{p}y^2)(0) + O(h^{3-2\alpha}) \\ T_h u(0) \geq phT(x + \tfrac{bh-|c|h^{1+\alpha}}{p}y^2)(0) + O(h^{3-2\alpha}) \end{aligned} \tag{1.30}$$

for any $\alpha > 0$. Thus, using the definition of H,

$$phH(\frac{bh - |c|h^{1+\alpha}}{p}) - O(h^{3-2\alpha}) \leq T_h u(0) \leq phH(\frac{bh + |c|h^{1+\alpha}}{p}) + O(h^{3-2\alpha}) \tag{1.31}$$

Using the fact that H is Lipschitz (1.22), we have

$$H(\frac{bh \pm |c|h^{1+\alpha}}{p}) = H(\frac{bh}{p}) \pm O(h^{1+\alpha}).$$

Thus (1.31) yields

$$T_h u(0) = phH(\frac{bh}{p}) + O(h^{2+\alpha} + h^{3-2\alpha}).$$

Choosing $\alpha = 1/3$, we obtain

$$T_h u(0) = phH(\frac{bh}{p}) + O(h^{2+\frac{1}{3}}).$$

Since, by Formula (1.25), $b = \frac{1}{2}curv(u)|Du|(0)$ and $p = |Du|(0)$, we obtain (ii). The proof of (iii) is obvious by choosing for axes the eigenvalues of $D^2u(\mathbf{x})$ and returning to the definition of T. ∘

The application of the preceding theorem to the above mentioned filters (median, weighted median, etc.) is straightforward : one only needs to compute in each case the associated function H. This is, in the considered examples, an easy task. In order to perform the scaling associated with the last theorem, we "shrink" $k(\mathbf{x})$ into $k_h(\mathbf{x}) = h^{-2}k(h^{-1}\mathbf{x})$, which corresponds to scaling the structuring elements by h, as usual.

Theorem 4. *(Ishii) There exists a constant $c(k)$ such that if u is a C^3 function in $\mathbb{R}^2$, then*

$$\text{med}_{k_h} u = u(0) + c(k)curv(u)|Du|(0)h^2 + h^2O(h)$$

if $Du(0) \neq 0$ *and*

$$|\mathrm{med}_{k_h} - u(0)| \leq |\Delta u(0)|h^2 + h^2 O(h)$$

if $Du(0) = 0$.

Lemma 1. *The function* $H(h) = \mathrm{med}_k(x + hy^2)$ *associated with the weighted median filter satisfies*

$$H(h) = c(k)h + o(h),$$

where

$$c(k) = \frac{\int_{-\infty}^{+\infty} y^2 k(y)dy}{\int_{-\infty}^{+\infty} k(y)dy}.$$

Proof. We notice that the function $m(h) = med_{k_h}(x + hy^2)$ is defined by

$$\int_{-\infty}^{+\infty} \int_{0}^{m(h)-hy^2} k((x^2 + y^2)^{\frac{1}{2}})dxdy = 0. \tag{1.32}$$

This formula uniquely defines $m(h)$ because its first member can be viewed as a strictly increasing, continuous function of $m(h)$. In this proof, we set, in order to reduce the size of formulas,

$$\varphi(a,b) = \int_{-\infty}^{+\infty} y^2 k((y^2 + (a-by^2)^2)^{\frac{1}{2}})dy, \quad \psi(a,b) = \int_{-\infty}^{+\infty} k((y^2 + (a-by^2)^2)^{\frac{1}{2}})dy \tag{1.33}$$

These functions are well defined, continuous and positive because of the assumptions on k. Formally differentiating (1.32) with respect to h yields

$$m'(h)\psi(m(h),h) - \varphi(m(h),h) = 0,$$

Figure 8. The median filter does a smoothing linked to the curvature of the level set. Left : image of a simple shape. Right : difference of this image and the same smoothed by one iteration of the median filter. We see in black, the points which have changed. The width of the difference is proportional to the curvature, (Theorem 5)

so that, formally again,

$$m'(0) = \frac{\varphi(0,0)}{\psi(0,0)} = \frac{\int_{-\infty}^{+\infty} y^2 k(y)dy}{\int_{-\infty}^{+\infty} k(y)dy} = c(k). \tag{1.34}$$

This formal computation is easy to justify. We simply introduce the ordinary differential equation

$$\tilde{m}'(h) = \frac{\psi(\tilde{m}(h),h)}{\varphi(\tilde{m}(h),h)}, \quad \tilde{m}(0) = 0, \tag{1.35}$$

which has, by Peano Theorem, a solution on some interval $[0, h_0]$. Indeed, the second member of (1.35) is continuous with respect to h and $\tilde{m}(h)$. Multiplying both members of (1.35) by $\varphi(\tilde{m}(h),h)$ and integrating between 0 and h yields

$$\int_{-\infty}^{+\infty} dy \int_0^{\tilde{m}(h)-hy^2} k((x^2+y^2)^{\frac{1}{2}})dx = 0. \tag{1.36}$$

Thus $\tilde{m}(h)$ satisfies (1.32) and therefore $m(h) = \tilde{m}(h)$ on $[0, h_0]$. As a consequence, (1.34) is true and we obtain

$$m(h) = med_{k_h}(x + hy^2) = c(k)h + o(h).$$

$\circ$

Proof of Theorem 4. By Lemma 1.18, the operator

$$Tu(\mathbf{x}) = \mathrm{med}_{k_h(\mathbf{y}-\mathbf{x})} u(\mathbf{y})$$

satisfies the conditions of Theorem 3. By Lemma 1, the function H associated with T satisfies $H(0) = 0$, and we obtain the conclusions (ii) and (iii) of Theorem 3. Using Lemma 1 yields

$$H(h) = c(k)h + o(h^3)$$

and therefore, by Theorem 3(ii),

$$\mathrm{med}_{k_h} u = u(0) + h|Du(0)|(c(k)h\,curv(u)) + h^2 0(h)$$

if $|Du(0)| \neq 0$. If $Du(0) = 0$, we obtain from Theorem 3(iii) :

$$|\mathrm{med}_{k_h} u - u(0)| \leq |\Delta u(0)|h^2 + O(h^3).$$

$\circ$

Applying the preceding theorem to the case of the classical median filter, one easily obtains for any u which is a C^3 function in $\mathbb{R}^2$,

$$\mathrm{med}_{D(\mathbf{x}_0,h)} u = u(\mathbf{x}_0) + \frac{1}{6} curv(u)|Du|(\mathbf{x}_0)h^2 + O(h^{2+\frac{1}{3}})$$

if $Du(\mathbf{x}_0) \neq 0$.

As another direct application of Theorem 3, we consider the case where the structuring elements all are segments with the same length and centered at zero, which yields an operator introduced by F. Catté, F. Dibos and G. Koepfler [16]. Surprisingly enough, this operator has the same effect as the median filter. More precisely, we deduce the following result, also proved in [16].

Theorem 5. *Let $\mathbb{B}$ as the set of all segments of the plane with length 2 and centered at zero. Let u be a C^3 function in $\mathbb{R}^2$. And, set for every real function $u(\mathbf{x})$ defined on the plane,*

$$SI_h u(\mathbf{x}) = \sup_{B \in h\mathbb{B}} \inf_{\mathbf{y} \in \mathbf{x}+B} u(\mathbf{y})$$

$$IS_h u(\mathbf{x}) = \inf_{B \in h\mathbb{B}} \sup_{\mathbf{y} \in \mathbf{x}+B} u(\mathbf{y}).$$

Then

$$\frac{1}{2}(IS_h + SI_h)u(\mathbf{x}_0) = u(\mathbf{x}_0) + h^2 \frac{1}{4} curv(u)|Du|(\mathbf{x}_0) + O(h^{2+\frac{1}{3}})$$

if $Du(\mathbf{x}_0) \neq 0$ and

$$|\frac{1}{2}(IS_h + SI_h)u(\mathbf{x}_0) - u(\mathbf{x}_0)| \leq |\Delta u(\mathbf{x}_0)|h^2 + O(h^3)$$

if $Du(\mathbf{x}_0) = 0$.

Proof. The operator IS_h associated with $\mathbb{B}$ satisfies the conditions of Theorem 3. Let us compute the function H_1 associated with IS_h. We have

$$H_1(h) = \inf_{-\frac{\pi}{2} \leq \theta \leq \frac{\pi}{2}} \sup_{-1 \leq r \leq 1} (r \cos\theta + hr^2 \sin^2\theta).$$

Since the function $r \rightarrow r\cos\theta + hr^2 \sin^2\theta$ is convex when $h \geq 0$, we deduce

$$H_1(h) = \inf_{-\frac{\pi}{2} \leq \theta \leq \frac{\pi}{2}} (\cos\theta + h \sin^2\theta) = h$$

if $h \geq 0$ is small enough. Let us show that

$$H_1(h) = 0$$

when $h \leq 0$. Indeed, $r\cos\theta + hr^2\sin^2\theta \leq r\cos\theta$. By doing exactly the same computations for SI_h, we get in that second case for the associated function $H_2(h) = h^-$. We notice that in both cases, $H_i(0) = 0$, so that the conclusions (ii) and (iii) of Theorem 3 apply. Adding the relations thus obtained for IS_h and SI_h yields the announced results. ∘

Let us now consider the case of unbounded families of structuring elements, which was excluded in Theorem 3. One could think it spurious to consider

Figure 9. On the first row, seabird1 and its level lines for all levels multiple of 12. On the second row, a median filter on a disk with radius 2 has been iterated twice on seabird2. On the third row, an infsup and then a supinf filter based on segments have been applied. On the right : the corresponding level lines of the results, which, according to the theoretical results, move at a normal speed proportional to their curvature, (Theorems 5 and 6)

such families for defining local operations on the image. This is, however, not the case when families of structuring elements are "naturally" unbounded ; our main example in view here is the case of affine invariant families. The interest of such families is to define filtering operators on an image which are -in a restrictive sense- projective invariant : indeed, projection invariance reduces to affine invariance when photographed plane objects are far enough with respect to the focal distance.

Asymptotic behaviour of affine and morphologically invariant filters. Let $\mathbb{B}$ be a family of plane closed nonempty bounded sets which is invariant by the special affine group $SL(\mathbb{R}^2)$, that is, $\mathrm{A}B \in \mathbb{B}$ whenever $\det(\mathrm{A}) = 1$ and $B \in \mathbb{B}$. An obvious example of such a family is the family of ellipses with area 1, and, more generally, we can consider the set $\{\mathrm{A}B, \mathrm{A} \in SL(\mathbb{R}^2)\}$ where B is an arbitrary bounded set of the plane.

In what follows, we write $\mathbf{x}, \mathbf{y}, \mathbf{z}$ for elements of $\mathbb{R}^2$ and $\mathbf{x} = (x, y)$. We consider two scaling parameters s and h such that $s = h^{\frac{3}{2}}$, and we set $\mathbb{B}_s = \{s^{\frac{1}{2}}B, B \in \mathbb{B}\}$. The meaning of this scaling is that s behaves like the area of elements of $\mathbb{B}_s$, while the use of h makes statements on the asymptotic behaviour simpler. Without risk of ambiguity, we shall sometimes write $\mathbb{B}_h$ instead of $\mathbb{B}_s$, in which case the relation $s = h^{\frac{3}{2}}$ is assumed to be satisfied.

Set, for every real function $u(\mathbf{x})$ defined on the plane,

$$SI_h u(\mathbf{x}) = \sup_{B \in h\mathbb{B}} \inf_{\mathbf{y} \in \mathbf{x}+B} u(\mathbf{y})$$

$$IS_h u(\mathbf{x}) = \inf_{B \in h\mathbb{B}} \sup_{\mathbf{y} \in \mathbf{x}+B} u(\mathbf{y}).$$

$SI_h u$ is understood as an "affine erosion" of u and $IS_h u$ as an "affine dilation". Our main concern is the behaviour of $IS_h u(\mathbf{x})$ and $SI_h u(\mathbf{x})$ when $u(\mathbf{x})$ is a smooth (C^3) function and $h \to 0$. Under fairly general assumptions on $\mathbb{B}$, there is a theorem proving that all affine invariant morphological operators have one and the same behaviour. As an example, if $\mathbb{B} = \{AB, A \in SL(\mathbb{R}^2)\}$ and B is a bounded connected set whose convex hull contains 0, then there are constants $c^+_{\mathbb{B}} > 0$ and $c^-_{\mathbb{B}} > 0$ such that

$$\lim_{h \to 0} \frac{(IS_h u)(\mathbf{x}_0) - u(\mathbf{x}_0)}{h} = |Du| g(curv(u))(\mathbf{x}_0),$$

$$\text{where } g(r) = c^+_{\mathbb{B}}\left(\tfrac{r}{2}\right)^{\frac{1}{3}} \quad \text{if} \quad r > 0$$
$$= c^-_{\mathbb{B}}\left(-\tfrac{r}{2}\right)^{\frac{1}{3}} \quad \text{if} \quad r < 0.$$

Alternate schemes

Alternate schemes have been considered in Mathematical Morphology in order to restore the reverse contrast invariance $T(-u) = -T(u)$. Such an invariance is desirable for an image analysis operator because it is shocking to have a different analysis for, say, a white shape on black background and a black shape on white background. Now, erosions and dilations do not satisfy this invariance property and the main idea for restoring it is to make infinite products of alternate erosions and dilations. This is what we do now in the affine invariant case. We need more restrictive assumptions on the set of structuring elements $\mathbb{B}$ which will make it easy to obtain asymptotic theorems on what we shall call "alternate schemes", that is, infinite products of the operator $IS_h SI_h$. We can obtain for such alternate schemes a convergence theorem equivalent to the result stated at the end of the preceding section. If we consider an alternate operator like $T = IS_h SI_h$, it is not true that $T(-u) = -T(u)$. (In fact, $IS_h SI_h(-u) = -SI_h IS_h(u)$). We shall prove, however, that if we let $h \to 0$ and consider iterations $(IS_h SI_h)^n$ with $n \to \infty$, then the limit operators do satisfy this property. The restrictions we do on the set $\mathbb{B}$ are following : Every element $B \in \mathbb{B}$ is a convex set, with area 1 and symmetric with respect to 0, that is, $x \in B$ if and only if $-x \in B$. Then one can prove, with a proof close to the proof of Theorem3, that there exists a constant $c_{\boldsymbol{B}} > 0$ such that for every u which is C^3 in a neighbourhood of $\mathbf{x}_0$,

$$\lim_{h \to 0} \frac{(T_h u)(\mathbf{x}_0) - u(\mathbf{x}_0)}{h} = c_{\boldsymbol{B}} |Du| g(curv(u))(\mathbf{x}_0)$$

$$\begin{aligned} \text{where } g(r) &= (r^-)^{\frac{1}{3}} \quad \text{if} \quad T_s = SI_s \\ &= (r^+)^{\frac{1}{3}} \quad \text{if} \quad T_s = SI_s \\ &= (r)^{\frac{1}{3}} \quad \text{if} \quad T_s = SI_s\, IS_s \end{aligned}$$

This achieves what we had to say on the infinitesimal behaviour of linear and morphological smoothing filters. In order to go further, we shall consider the problem of iterating such filters and looking what happens when the number of iterations tends to infinity and simultaneously the scale parameter of the filters tends to zero. We shall see that limit flows are obtainable, and these limit flows correspond to the various theories of "scale space" which have been proposed in Vision theory. (From the mathematical viewpoint, all of this discussion is classical and we can describe the questions answered in this section and the next as "What is the infinitesimal generator of the semigroup", "What is the underlying partial differential equation of the semigroup" ?)

2 The top-down approach : scale-space and PDE's

2.1 What basic principles must obey a scale space?

In this section, we consider an abstract framework, the "scale space", which at the end boils down, from the algorithmic viewpoint, to iterated filtering. Now, this framework will make it easier to classify and modelize the possible asymptotic behaviours of iterated filtering. We define a "scale space" as an abstract family of image smoothing operators T_t, depending on a scale parameter t. Given an image $u_0(\mathbf{x})$, $(T_t u_0)(\mathbf{x}) = u(t, \mathbf{x})$ is the "image u_0 analysed (in fact : smoothed) at scale t". Formal, but natural and classical, assumptions on the smoothing process follow.

The pyramidal structure. We assume that the output at scale t can be computed from the output at a scale $t - h$ for very small h. This is natural, since a coarser analysis of the original picture is likely to be deduced from a finer one without any dependence upon the original picture. By that way the finest picture smoothing is the identity. T_t is obtained by composition of "transition filters", which we denote by $T_{t+h,t}$.

Definition 1. *We shall say that a scale space is* **pyramidal** *if*

$$T_{t+h} = T_{t+h,t} T_t, \quad T_0 = Id. \tag{2.1}$$

A strong version of the pyramidal structure assumption is the semigroup property

Figure 10. A multiscaled world ... This series of images is an experiment in the relative stability of perception of objects seen at different distances. Each photograph has been obtained in a park by stepping forward and taking a snapshot at a much closer distance than the former one. We display by a rectangle in each image the part of the object which has been photographed in the next. As one can appreciate, when getting closer, the visual aspect of objects changes and new structures arise. Thus, the computing of primitives in an image is always a scale-dependent task, quite dependent on the distance to objects. When we look at a certain distance at an object, we get rid of too much finer structure : for instance, leaves are not to be seen in the two first photographs, as well as in the two last ones when we got too close. Multiscale smoothing tries to emulate this physical and perceptual fact by defining a smoothing of a digital image at different scales which gets rid of the finer structures, but minimally modifies the image at scales above the considered scale

Definition 2. *We shall say that a scale space is* **recursive** *if*

$$T_0 = Id, \qquad T_s \circ T_t(f) = T_{s+t}(f) \text{ on } \mathbb{R}^N, \text{ for all } s,t \geq 0 \text{ and } f \text{ in } \mathcal{F}. \quad (2.2)$$

If the recursivity is satisfied, T_t can be deduced from the n-times iteration of $T_{t/n}$. Let us continue with an intuitive requirement which is called in image processing "causality". Since the visual pyramid is assumed to yield more and more global information about the image and its features, it is clear that when the scale increases, no new feature should be created by the scale space : the image at scale t'>t must be simpler than the image at scale t. Furthermore, the operators (or transition filters) $T_{t+h,t}$ are assumed to act "locally", that is, to look at a small part of the processed image. In other terms, $(T_{t+h,t}u_0)(\mathbf{x})$ must essentially depend upon the values of $u_0(\mathbf{y})$ when $\mathbf{y}$ lies in a small neighborhood of $\mathbf{x}$. We condense the locality and the simplification assumptions in a *local comparison principle* : if an image u is locally brighter than another v, then this order must be conserved some time by the analysis (prevalence of local order on global order).

Definition 3. *A scale space satisfies the* **Local Comparison Principle,** *if for all u and v such that $u(\mathbf{y}) > v(\mathbf{y})$ for $\mathbf{y}$ in a neighborhood of $\mathbf{x}$ and $\mathbf{y} \neq \mathbf{x}$, then for h small enough we have*

$$(T_{t+h,t}u)(\mathbf{x}) \geq (T_{t+h,t}v)(\mathbf{x})$$

In order to propose a classification of scale spaces, we finally need some assumption stating that a very smooth image must evolve in a smooth way with the scale space. As can be proven (see [3]), it is enough to make this assumption for quadratic functions. From the mathematical viewpoint, the next assumption implies the existence of an infinitesimal generator for the semigroup T_t.

Definition 4. *Let $u(\mathbf{y}) = \frac{1}{2} \prec A(\mathbf{y}-\mathbf{x}), \mathbf{y}-\mathbf{x} \succ + \prec p, \mathbf{y}-\mathbf{x} \succ + c$ be a quadratic form of $\mathbb{R}^N$. ($A = D^2u(\mathbf{x})$ is a $N * N$ matrix, $p = Du(\mathbf{x})$ a vector of $\mathbb{R}^N$, $c = u(\mathbf{x})$ a constant.)*

We shall say that a scale space is **regular** *if there exists a function $F(A, p, \mathbf{x}, c, t)$, continuous with respect to A, such that*

$$\frac{(T_{t+h,t}u - u)(\mathbf{x})}{h} \to F(A, p, \mathbf{x}, c, t) \quad \text{when} \quad h \to 0. \quad (2.3)$$

Since we shall in the following constantly assume that the considered scale spaces are pyramidal, regular, and satisfy the local comparison principle, it is opportune for this set of basic properties to be given a single name.

Definition 5. *We call* **causal** *any scale space which satisfies the* **Local Comparison Principle** *and is pyramidal and local.*

As we shall prove, the causality assumption is enough to imply that the scale space is governed by a PDE.

Theorem 6. *(Fundamental Scale Space Theorem) If a scale space T_t is causal, then it satisfies*

$$((T_{t+h,t}u - u)/h)(\mathbf{x}) \to F(D^2u(\mathbf{x}), Du(\mathbf{x}), u(\mathbf{x}), \mathbf{x}, t) \tag{2.4}$$

as h tends to 0^+ for all u and $\mathbf{x}$ where u is C^2, where F is the real function associated with the regularity assumption (2.3). In addition, F is nondecreasing with respect to its first argument :

$$\text{If } A \geq \tilde{A}, \text{ for the ordering of symmetric matrices }, \tag{2.5}$$

$$\text{then, } \quad F(A, p, c, \mathbf{x}, t) \geq F(\tilde{A}, p, c, \mathbf{x}, t)).$$

This easy to prove theorem reduces the classification of all iterated nonlinear low pass filtering (or, in other words, of all scale spaces) to the classification of all interesting functions F. In dimension 2, these real functions have nine arguments. This number, however, will be drastically reduced when we impose obvious and rather necessary and useful invariance properties to the associated scale space. We shall now list these properties and then give the resulting classification of "interesting" functions F. Let us begin with the linear case, where Theorem 6 yields a "top-down" characterization of the heat equation $\partial u/\partial t = \Delta u$ as the unique linear and isometrically invariant scale space. This model, classical in image processing comes from Marr-Hildreth [39] and has been formalized by Witkin [57], Koenderink [32] and J. Babaud, A. Witkin, A. Baudin and R. Duda [8]. The basic step of the scale space is the convolution of the original image with gaussians of increasing variance. Koenderink [32] noticed that the convolution of the signal with gaussians at each scale was equivalent to the solution of the heat equation with the signal as initial datum. Denoting by u_0 this datum, the "scale space" analysis associated with u_0 consists in solving the system

$$\frac{\partial u(t,\mathbf{x})}{\partial t} = (\Delta u)(t, \mathbf{x}), \tag{2.6}$$

$$u(0, \mathbf{x}) = u_0(\mathbf{x}).$$

Theorem 7. *Let T_t be a causal and translation invariant scale space. If the $T_{t,s}$ are* **linear** *and* **euclidean invariant**, *then (up to a rescaling $t' = h(t)$), $u(t, \mathbf{x}) = (T_t u_0)(\mathbf{x})$ is the solution of the heat equation $\partial u/\partial t - \Delta u = 0$ in $\mathbb{R}^N * [0, \infty[$, $u(0, .) = u_0(.)$ in $\mathbb{R}^N$.*

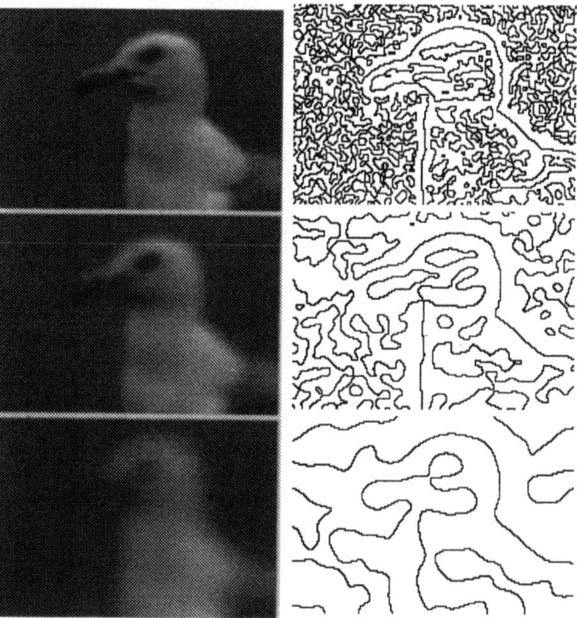

Figure 11. In this diplay, we illustrate the original scale space theory, as it is for instance developped in the founding book by David Marr, Vision. In order to extract more global structure from an original image, convolutions with gaussians with variances which are powers of 2 are effectuated. One computes the laplacian of the resulting smooth images and display the lines along which the laplacian changes sign : the so called "zero crossings of the Laplacian". According to David Marr, those zero crossings represent the "raw primal sketch" of the image, that is the information on which further vision algorithms should be based

2.2 Geometric and morphological invariance axioms

We begin by considering the morphological assumption, that the scale space should be independent from the (arbitrary) grey-level scale.

Definition 6. *We shall say that a scale space is morphological if*

$$g \circ T_{t+h,t} = T_{t+h,t} \circ g, \tag{2.7}$$

for any nondecreasing function g from $\mathbb{R}$ into $\mathbb{R}$.

We shall now list a series of axioms which state the invariance of scale space with respect to the respective positions of the *percipiens* and *perceptum* in the image generation process. This leads us to define more and more precisely which functions F are admissible for a general purpose, and therefore invariant, scale space. Let R be an isometry of $\mathbb{R}^N$ and denote by Ru the function $Ru(\mathbf{x}) = u(R\mathbf{x})$.

Definition 7. *We shall say that a scale space $T_{s,t}$ is* **euclidean invariant, or isotropic** *if for every isometry R of $\mathbb{R}^N$ into $\mathbb{R}^N$,*

$$RT_{t+h,t} = T_{t+h,t}R \tag{2.8}$$

The next optional geometrical axiom is the scale invariance of image analysis. Set $H_\lambda f(\mathbf{x}) = f(\lambda \mathbf{x})$.

Definition 8. *We say that a scale space is* **scale invariant** *if there exists a rescaling function $t'(t, \lambda)$, defined for for any $\lambda > 0$ and $t \geq 0$, such that*

$$H_\lambda T_{t'} = T_t H_\lambda, \tag{2.9}$$

and, more generally :

$$H_\lambda T_{t',s'} = T_{t,s} H_\lambda. \tag{2.10}$$

where we note $t' = t'(t, \lambda)$ and $s' = t'(s, \lambda)$.

In fact, (2.10) implies (2.9) and must be considered as a slightly stronger form. Now, as proven in [3], (2.9) is enough to ensure a normalization of Function $t'(t, \lambda)$. We cannot fix a priori the form of t' because it can vary in concrete examples of scale spaces. One can prove ([3]) that it can be "normalized", which means that we always can, by an adequate rescaling of any scale space, set $t' = \lambda t$.

The scale invariance means that the result of the scale space T_t is independent of the size of the analyzed features : this is very important in "natural images", since the same object can be photographed at very different distances and therefore at very different scales. Thus, it is essential for the stability of shape analysis that the result of an analysis of this object should not yield a different "shapes" at different distances. Thus, the sequence of the shapes obtained by scale space must be independent of the (a priori unknown) size of the object in the picture. Of course, we cannot impose $t' = t$ because scale of smoothing and scale of the image are covariant, as can be appreciated by considering the examples of linear filtering and the heat equation.

Finally, we state an axiom which implies the euclidean and the scale invariances, and also the invariance of the scale space under any affine projection of a planar shape. Combining those transformations leads to an arbitrary linear transform A of the plane. Set for any such transform $Af(\mathbf{x}) = f(A\mathbf{x})$. With the same formalism as for scale invariance, we get :

Definition 9. *We shall say that a scale space is* **affine invariant** *if there is a function $t'(t, A)$, defined for every $t > 0$ and any linear application A of $\mathbb{R}^N$ with $det(A) \neq 0$, such that*

$$AT_{t'} = T_t A. \tag{2.11}$$

and if the stronger version also holds :

$$AT_{t',s'} = T_{t,s} A, \tag{2.12}$$

where again $t' = t'(t, A)$ and $s' = t'(t, A)$.

This relation means that the result of the scale space T_t is independent of the size and position in space of the analyzed features : An affine map corresponds to the anamorphosis of a plane image when it is presented to the eye at any distance large enough with respect to its size and with an arbitrary orientation in space.

2.3 Morphological scale spaces

By the Fundamental Scale Space Theorem 6, all scale space models are defined by a partial differential equation

$$\frac{\partial u}{\partial t} = F(D^2 u, Du, u, x, t), \quad u(0) = u_0$$

where u_0 is the image to analyze (the datum), $u(t)$ is the image analyzed at scale t and $F(A, p, c, x, t)$ depends on a symmetric 2*2 matrix A, a two-dimensional vector p, a constant c, a point of the plane x and a real positive scale t. If we impose the euclidean and contrast invariances, then $T_t u_0$ obeys an equation of the kind

$$\frac{\partial u}{\partial t} = |Du| G(div(\frac{Du}{|Du|}), t) = |Du| G(curv(u), t),$$

where G is a nondecreasing function with respect to its first argument. How can we identify further G ? The next theorem summarizes the answer.

Theorem 8. *(i) Let T_t be a euclidean and contrast invariant scale space, therefore defined by an equation $\partial u/\partial t = |Du|G(curv(u), t)$. If it is scale invariant, then, with a possible rescaling $t = h(t)$, the equation of the scale space is*

$$\frac{\partial u}{\partial t} = |Du| \beta(t \, curv(u)), \tag{2.13}$$

where β is a continuous nondecreasing real function.

(ii) If the scale space is, in addition, affine invariant, then the only possible equation is, up to a rescaling,

$$\frac{\partial u}{\partial t} = |Du| (t \, curv(u))^{1/3}. \tag{2.14}$$

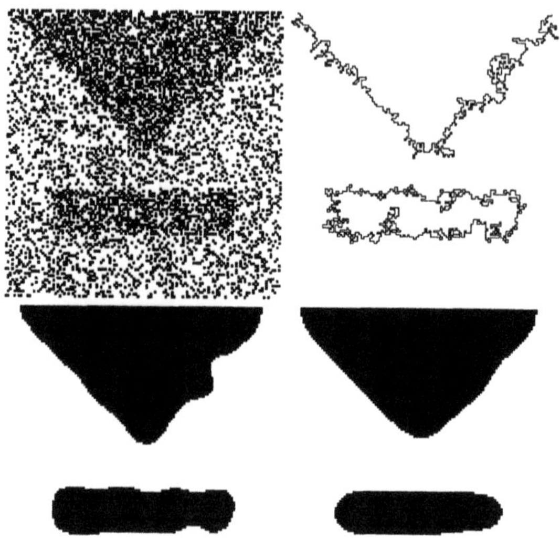

Figure 12. This experiment displays the affine and morphological scale-space applied on a noisy binary image. Up-left : Noisy Triangle and rectangle, up-right : level-line corresponding to the boundary of the two shapes, down-left : filtered version at scale 5, down-right : filtered version at scale 10

3 Convergence of iterated morphological filters to the solution of a partial differential equation

3.1 Erosions and dilations

We now return to the problem of the asymptotic behaviour of iterated filters. Before giving a general statement for iterated morphological filters, let us recall a well known case, elementary to analyse, the case of dilation (or erosion) by a convex set. We define the dilation at scale t of function u_0 with the structuring element B by

$$D_t u_0(\mathbf{x}) = \sup\{u_0(\mathbf{y}), \mathbf{y} - \mathbf{x} \text{ in } tB\}$$

Similarly, the erosion at scale t of function f with structuring element B is defined by

$$E_t u_0(\mathbf{x}) = \inf\{u_0(\mathbf{y}), \mathbf{y} - \mathbf{x} \text{ in } -tB\}.$$

We denote by $||\mathbf{x}||_B$ the seminorm associated with the convex set B, that is,

$$||\mathbf{x}||_B = \sup_{\mathbf{z} \text{ in } B} \prec \mathbf{x}, \mathbf{z} \succ .$$

Proposition 1. *(Lax formula). Set $u(t,\mathbf{x}) = D_t u_0(\mathbf{x})$ (resp. $E_t u_0(\mathbf{x})$). Then $u(t,\mathbf{x})$ is a solution of*

$$\partial u/\partial t = ||Du||_B.$$

(resp. $\partial u/\partial t = -||Du||_B$.) at each point $(t,\mathbf{x})$ where u is C^2 with respect to $\mathbf{x}$.

3.2 Iterated median filters and mean curvature motion

We now examine the convergence of iterated median filters.

Theorem 9. *[10, 11, 22] Let u_0 be a lipschitz function on $I\!R^2$ and set*

$$u_h(\mathbf{x},t) = \text{med}_{D(\mathbf{x},h^{\frac{1}{2}})} u_h(\mathbf{x},(n-1)h) \qquad \text{for} \quad (n-1)h \leq t \leq nh$$

$$\text{and} \quad u_h(\mathbf{x},0) = u_0(\mathbf{x}).$$

Then $u_h(\mathbf{x},t)$ tends uniformly on every compact set of $I\!R^2 \times I\!R^+$ to a function $u(\mathbf{x},t)$ when $n \to +\infty$, where $u(\mathbf{x},t)$ is the unique viscosity solution of

$$\frac{\partial u}{\partial t} = \frac{1}{6}|Du|curv(u) \tag{3.1}$$

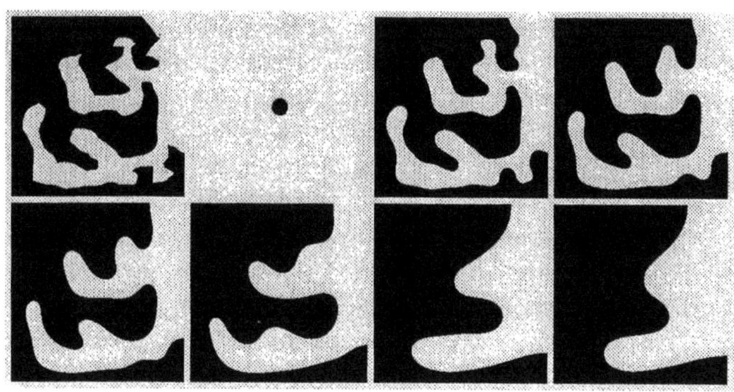

Figure 13. Scale-space based on iterations of the median filter. From left to right and up to down : original shape, the size of the disk used for the median, and the result of the iterated median filter for an increasing number of iterations

Figure 14. Numerically, the iterated median filter and the mean curvature motion are very close, when the curvatures of the level line are not too small. Left : simple shape smoothed by a finite difference scheme of the mean curvature motion, middle : smoothing by a median filter a the same scale, right : difference between left and middle images. The difference is not larger that one pixel width. (Theorem 10)

This equation has recently received a lot of attention because of its geometrical interpretation : indeed, (see S. Osher and J. Sethian [46], L. C. Evans and J. Spruck [22]...) the level lines of the solution move in the normal direction with a speed proportional to their mean curvature (curvature in two dimensions). (This "mean curvature motion" effect will be showed in the experimental results presented above). As we have seen, this "mean curvature motion" property implies that the proposed scale space satisfies the main "morphological" axiom, that is, the invariance of the scale space under any rearrangement of the grey level scale. Thus, the mean curvature motion is a morphological operator. It is quite close to the basic operations of mathematical morphology. Indeed, dilation and erosion can be interpreted as a (forward or backward) motion of level curves at constant spead in the direction of the normal. A theory of weak solutions based upon the so-called viscosity solutions theory has been proposed by Y. G. Chen, Y. Giga and S. Goto [17], L.C. Evans and J. Spruck [22], Y. Giga, S. Goto, H. Ishii and M.H. Sato [24].

Theorem 9 gives a differential interpretation of the median filter and somehow explains why it works so well. Indeed, Equation (3.1) means that all level curves of u are smoothed at a speed depending upon their local radius curvature, so that all small level sets are quickly removed and larger ones have their asperities smoothed out. We shall not prove this theorem, but its proof is easily deduced from the proof of the next convergence result for iterated affine morphological operators.

3.3 Affine invariant iterated morphological filters

As a last example of convergence of iterated filters, let us return to the affine invariant morphological filters : We consider a family $\mathbb{B}$ of convex sets,

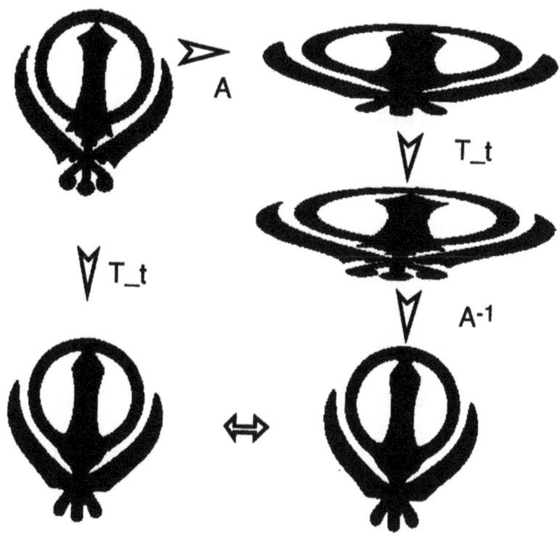

Figure 15. We display in this picture the affine invariance of the affine and morphological scale space. A simple shape (up-left) is smooth using finite differences discretization of the Equation 2.14 (down-left). We apply on the same shape a affine transform, with determinant equal to 1, (up-right), then the smoothing (middle-right), and the inverse of the affine transform (down-right). The results of the two process are equals (see Definition 9)

Figure 16. We display the results of the same process than in the Figure 15 but using the affine "infsup" (see Section 1.4 and Theorem 11). We choose for $\mathbb{B}$ a set of 49 ellipses (see Section 4). The difference between the smoothed shape is due to the lack of affine invariance on a grid

symmetric with respect to zero and invariant under the special linear group (all linear maps with determinant 1). All sets in $\mathbb{B}$ have area s. We set

$$SI_h u(\mathbf{x}) = \sup_{B\in\mathbb{B}} \inf_{\mathbf{y}\in\mathbf{x}+B} u(\mathbf{y})$$

$$IS_h u(\mathbf{x}) = \inf_{B\in\mathbb{B}} \sup_{\mathbf{y}\in\mathbf{x}+B} u(\mathbf{y})$$

where the relation $s = h^{\frac{3}{2}}$ is imposed.

Denoting by T_h one of the schemes IS_h, SI_h or $SI_h IS_h$, we define the result of their iteration n-times by

$$u_h(\mathbf{x},(n+1)h) = T_h u_h(\mathbf{x}, nh),$$

$$\text{and} \quad u_h(\mathbf{x},0) = u_0(\mathbf{x}).$$

As the last result to be stated in this text, on can prove the convergence of every iterated affine filter to the Affine Morphological Scale Space (AMSS) :

Theorem 10. *There exists a constant $c_{\mathbb{B}} > 0$ such that if $u_0(\mathbf{x})$ is any Lipschitz function on $\mathbb{R}^2$, then $u_h(\mathbf{x}, nh)$ tends uniformly on every compact set to $u(\mathbf{x}, t)$, where $u(\mathbf{x}, t)$ is the unique viscosity solution of*

$$\frac{\partial u}{\partial t} = |Du| g(curv(u)), \tag{3.2}$$

where :

$$\lim_{s\to 0} \frac{(T_s u)(\mathbf{x}_0) - u(\mathbf{x}_0)}{s^{2/3}} = c_{\mathbb{B}} |Du| g(curv(u))(\mathbf{x}_0)$$

$$\begin{aligned} \textit{where } g(r) &= c_{\mathbb{B}}(r^-)^{\frac{1}{3}} \quad \textit{if} \quad T_h = SI_h \\ &= c_{\mathbb{B}}(r^+)^{\frac{1}{3}} \quad \textit{if} \quad T_h = SI_h \\ &= c_{\mathbb{B}}(r)^{\frac{1}{3}} \quad \textit{if} \quad T_h = SI_h\, IS_h. \end{aligned}$$

As a consequence, there is a single possibility to smooth an image in a morphological and affine invariant way. All scale space equations considered here have a unique viscosity solution. We have no space to give a generic proof, but the use of iterated filters also is a mathematical tool to prove existence of a solution to any PDE model associated with an iterated morphological filtering. It applies, in particular, to the mean curvature motion (the Osher-Sethian equation).

4 Numerical aspects of morphological filters

In the preceding chapters, we have considered different schemes based on inf-sup or sup-inf operators and on given sets of sets : $\mathbb{B}$. And, we have seen in Theorem 2 that such operators have a fixed asymptotic behaviour.

Now, if we deal with computer images, an image is no longer a function of $\mathbb{R}^2$ into $\mathbb{R}$ but a matrix of values. We shall only consider 8 bits images. An image u is now considered as a matrix $u_{i,j}$, where $u_{i,j} \in \{0, ..., 255\}$. Of course, there are relations between the function u, and the image $u_{i,j}$. From a given computer image $u_{i,j}$, we can define the associated function u by

$$u(\mathbf{x}) = u_{i,j} \text{ if } \mathbf{x} \in [i - \frac{1}{2}, i + \frac{1}{2}[\times[j - \frac{1}{2}, j + \frac{1}{2}[$$

Conversely, from a given function u we can define a computer image $u_{i,j}$ as a sampling of u. This sampling is twofold : a sampling of the space $\mathbb{R}^2$ and a sampling of the grey level values (quantization). We shall consider that the basic sample of the space is the pixel. Conversely, the (i, j) pixel represents the square of $\mathbb{R}^2$ defined as

$$c_{i,j} = [i - \frac{1}{2}, i + \frac{1}{2}[\times[j - \frac{1}{2}, j + \frac{1}{2}[.$$

Thus, with our convention, each pixel has an area equal to 1. Since we do a sampling of the space, we must also define a spatial sampling of the sets of $\mathbb{B}$. Let B be a set of $\mathbb{B}$. As we have seen in the preceding chapters, B can be represented by its characteristic function $\mathbb{1}_B$. We call characteristic mask M_B the sampling of the characteristic function $\mathbb{1}_B$. We define this sampling by

$$M_B(i, j) = \int_{\mathbf{x} \in c_{i,j}} \mathbb{1}_B(\mathbf{x}) dx$$

This corresponds in fact to the area of the intersection of the set B and the pixel $c_{i,j}$. M_B is then a matrix, with real values between 0 and 1. We shall say that the pixel (i, j) belongs to B if $M_B(i, j)$ is positive.

We also define the histogram of u at the point (i_0, j_0) on B by the function $H_{u,B}^{(i_0,j_0)}$ from $\{0, ..., 255\}$ to $\mathbb{R}$,

$$H_{u,B}^{(i_0,j_0)}(k) = \sum_{(i,j), u(i,j)=k} M_B(i - i_0, j - j_0)$$

With this definition, our histogram is not a classical one. First, the histogram is defined on each pixel (i_0, j_0) of the image. Second, it depends upon the set B. Third, $H(k)$ is not defined as the number of the pixels having the grey level value k, but as the sum of the areas defined by the intersection of such pixels with the set $B + (i_0, j_0)$.

Associating with the point $\mathbf{x}_0$, the pixel c_{i_0,j_0} such that $\mathbf{x} \in c_{i_0,j_0}$, we define,

$$\inf_{\mathbf{x} \in B + \mathbf{x}_0} u(\mathbf{x}) = \inf\{k, \sum_{n=0}^{k} H_{u,B}^{(i_0,j_0)}(n) \geq 1\}$$

$$\sup_{\mathbf{x} \in B + \mathbf{x}_0} u(\mathbf{x}) = \sup\{k, \sum_{n=k}^{255} H_{u,B}^{(i_0,j_0)}(n) \geq 1\}$$

and

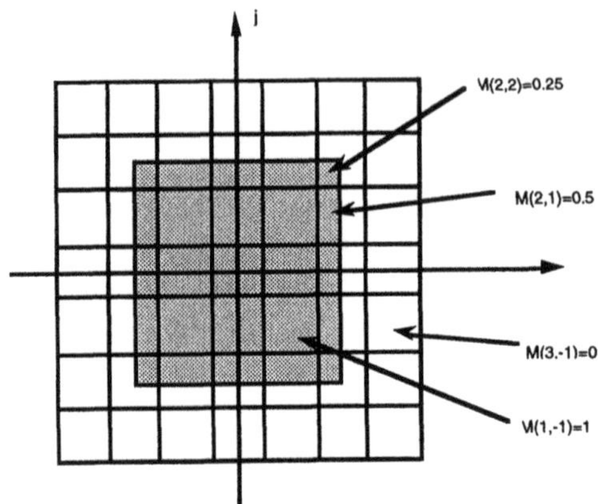

Figure 17. The mask associated to the 4×4 square centered in 0

$$\mathrm{med}^{\alpha,-}_{\mathbf{x}\in B+\mathbf{x}_0} u(\mathbf{x}) = \inf\{k,\ (1-\alpha)(\sum_{n=0}^{k} H^{(i_0,j_0)}_{u,B}(n) - 1) = \alpha(\sum_{n=k}^{255} H^{(i_0,j_0)}_{u,B}(n) - 1)\}$$

$$\mathrm{med}^{\alpha,+}_{\mathbf{x}\in B+\mathbf{x}_0} u(\mathbf{x}) = \sup\{k,\ (1-\alpha)(\sum_{n=0}^{k} H^{(i_0,j_0)}_{u,B}(n) - 1) = \alpha(\sum_{n=k}^{255} H^{(i_0,j_0)}_{u,B}(n) - 1)\}$$

Notice that $\mathrm{med}^{1,-} = \mathrm{med}^{1,+} = \sup$ and that $\mathrm{med}^{0,-} = \mathrm{med}^{0,+} = \inf$. We have also $\mathrm{med}_{\alpha,-} u = -\mathrm{med}_{1-\alpha,+} u$. Moreover, when the histogram H is such that $\forall k, H(k) > 0$, $\mathrm{med}^{\alpha,-}$ and $\mathrm{med}^{\alpha,+}$ are identical. If in addition $\alpha = 1/2$, they correspond to the usual median value.

In our example, we have

$$\inf_{\mathbf{x}\in B} u(\mathbf{x}) = 2$$

$$\sup_{\mathbf{x}\in B} u(\mathbf{x}) = 6$$

$$\mathrm{med}^{1/2,-}_{\mathbf{x}\in B} u(\mathbf{x}) = \mathrm{med}^{1/2,+}_{\mathbf{x}\in B} u(\mathbf{x}) = 4$$

Given a set of sets $\mathbb{B}$, we now consider how to discretize $\inf_{B\in\mathbb{B}} \sup_{\mathbf{y}\in\mathbf{x}+B}$. If $\mathbb{B}$ contains an infinite number of sets, (for example if $\mathbb{B}$ is the set of the ellipses having their area equal to a constant), we see that the computation of that inf is not numerically easy. We then choose to do also a sampling of the set $\mathbb{B}$.

In the experiments, we tried different possibilities for $\mathbb{B}$.

In Figure 9, we process the third row which corresponds to inf-sup on segments, by defining numerically a segment of $\mathbb{R}^2$ as a rectangle a one pixel width.

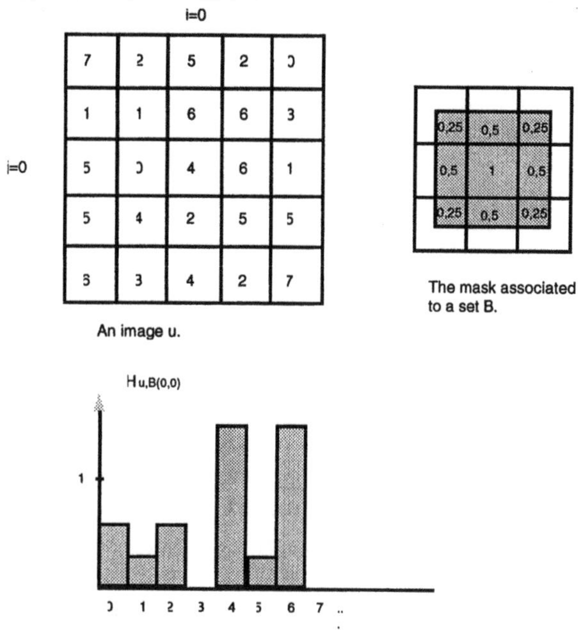

Figure 18. Example of histogram

We also need a sampling with respect to the orientation of the segments. We consider only angles of the type $k\pi/N$ for $0 \leq k < N$. In practice, we have taken $N = 8$, and the length of the segments is $c = 9$ pixels.

In Figure 16, $\mathbb{B}$ is the set of ellipses having their area equal to 50 and centered at 0 : For this set, we use two samplings, one with respect to the orientation of the rectangle, and one with respect to the ratio length/width. Of course, this ratio belongs to $[1, \infty[$. We decide to choose a finite number of values in $[1, \infty[$, such that the maximum of these values reduces the width of the rectangle to the size of one pixel. In practice, in the experiment, we choose 16 angles and 4 ratio length/width, so that our drastic sampling reduces $\mathbb{B}$ to 48 ellipses and one disq.

Acknowledgement

We gratefully acknowledge partial support by EC project ERBCHRXCT930095 (Human capital and mobility). A preliminary version of this work has been published for the tutorial ICIP95, we thank Michael Unser for his support and encouragement. Part of this work has been done by using the MegaWave image processing environment (authors : Jacques Froment and Sylvain Parrino), and Cognitech, Inc. image processing facilities. We also thank Luis Alvarez for much interaction on the subject of this work and Jean-Michel Lasry for a key idea which he confided to us and never published.

References

1. L. Alvarez, F. Guichard, P.L. Lions and J.M. Morel (1992), "Axioms and fundamental equations of image processing" *Report 9216, CEREMADE. Université Paris Dauphine* . Arch. for Rat. Mech. 123, 3, 199-257 (1993).
2. L. Alvarez and Freya Morales (1994) "Affine Morphological Multiscale Analysis of Corners and Multiple Junctions", Ref 9402, August 1994, Dept. Informatica y Systemas, Universidad de Las Palmas de Gran Canaria, to appear in *International Journal of Computer Vision.*
3. L. Alvarez and J.M. Morel (1994) "Formalization and Computational Aspects of Image Analysis", *Acta Numerica*, Cambridge University Press.
4. S. Angenent (1991), "On the formation of singularities in the curve shortening flow" *Journal of Differential Geometry*, 33, 601-633.
5. S. Angenent, G. Sapiro, A. Tannenbaum (1994), "On the affine heat equation for nonconvex curves" Technical report.
6. A. Arehart, L. Vincent and B.B. Kimia. (1993) "Mathematical Morphology : the Hamilton-Jacobi connection." *Int. J. of Comp. Vis.*, 215-219.
7. H. Asada and M.Brady (1986), "The curvature primal sketch", *IEEE Transactions on Pattern analysis and machine intelligence* **8** No. 1.
8. J. Babaud, A. Witkin, A. Baudin and R. Duda. (1986) " Uniqueness of the gaussian kernel for scale-space filtering." *IEEE Trans. PAMI 8.*
9. G. Barles (1985), "Remarks on a flame propagation model", *Technical Report No 464 INRIA Rapports de Recherche.*
10. G. Barles and C. Georgelin (1992), "A simple proof of convergence for an approximation scheme for computing motions by mean curvature", preprint.
11. G. Barles and P.E. Souganidis (1989), "Convergence of approximation schemes for fully nonlinear second order equation", *Asymp. Anal.*, 4, 271-283.
12. Bart M. ter Haar Romey (Ed.) (1994) *Geometry-Driven Diffusion in Computer Vision* Kluwer Academic Publishers.
13. W. Brockett and P. Maragos (1992), "Evolution Equations for continuous-scale morphology", *ICASSP, San Francisco* 23–26.
14. P.J. Burt and E.H. Adelson. (1983) " The Laplacian pyramid as a compact image code", *IEEE Transactions Communications*, 9, 4, 532-540.
15. J. Canny (1986). "A computational approach to edge detection", *IEEE Trans. PAMI*, 8, 6, 679-698.
16. F.Catté, F.Dibos and G.Koepfler (1993) "A morphological approach of mean curvature motion" *Report 9310, CEREMADE. Université Paris Dauphine.*
17. Y-G. Chen, Y. Giga and S. Goto (1991), "Uniqueness and existence of viscosity solutions of generalized mean curvature flow equations", J. Diff. Geometry, 33, 749-786.
18. T. Cohignac, F. Eve, F. Guichard, C. Lopez, J.M. Morel (1993), "Numerical Analysis of the fundamental equation of image processing", preprint, Ceremade, unpublished.

19. T. Cohignac, C. Lopez, and J.M. Morel (1993), "Multiscale analysis of shapes, images and textures", *Proceedings of the eighth workshop on image and multidimensional signal processing, IEEE* , September 8-10, Cannes. 142–143.

20. M.G. Crandall, H. Ishii and P.L. Lions (1992), *User's guide to viscosity solutions of second order partial differential equation*, Bull. Amer. Math. Soc., 27, 1-67, 1992.

21. L.C. Evans (1993) "Convergence of an algorithm for mean curvature motion", *Indiana Univ. Math. J., 42, 533-557*

22. L.C Evans. and J. Spruck (1991) "Motion of level sets by mean curvature 1", *J. Diff. Geometry* 33, 635-681.

23. M. Gage and R.S. Hamilton (1986), "The heat equation shrinking convex plane curves", *J. Differential Geometry* **23**, 69–96.

24. Y. Giga, S. Goto, H. Ishii and M.M. Sato (1991) "Comparison principle and convexity preserving properties for singular degenerate parabolic equations on unbounded domains" *Indiana Univ. Math. J.*, 40, 443-470.

25. F. Guichard (1993), "Multiscale Analysis of Movies",*Proceedings of the eighth workshop on image and multidimensional signal processing, IEEE*, September 8-10, Cannes. 236–237.

26. M. Grayson (1987), "The heat equation shrinks embedded plane curves to round points", *J. Differential Geometry* **26**, 285–314.

27. G. Huisken (1984), "Flow by mean curvature of convex surfaces into spheres", *J. Differential Geometry* 20, 237-266.

28. R. Hummel (1986), "Representations based on zero-crossing in scale-space", *Proc. IEEE Computer Vision and Pattern Recognition Conference*, 204–209.

29. H. Ishii (1995), "A generalization of the Bence, Merriman and Osher algorithm for motion by mean curvature", preprint, Dept. of Math., Chuo Univ., Tokyo 112, Japan.

30. H. Ishii and P.E. Souganidis (1993), "Generalized motion of noncompact hypersurfaces with velocity having arbitray growth on the curvature tensor", preprint.

31. B.B. Kimia, A. Tannenbaum, and S.W. Zucker (1992), "On the evolution of curves via a function of curvature, 1: the classical case", *J. of Math. Analysis and Applications* **163**, No 2.

32. J.J. Koenderink (1984), "The structure of images", *Biol. Cybern.***50**, 363–370.

33. J.J. Koenderink and A.J. van Doorn (1986), "Dynamic shape", *Biol. Cyber.* **53**, 383–396.

34. J.M. Lasry (1992) Unpublished notes, December 1992.

35. Y. Lamdan, J.T.Schwartz and H.J.Wolfson (1988) "Object recognition by affine invariant matching" *In Proc. CVPR 88*

36. C. Lopez, J.M. Morel (1992) "Axiomatisation of shape analysis and application to texture hyperdiscrimination" *Proceedings of the Trento Conference on Surface Tension and Movement by Mean Curvature* De Gruyter Publishers, Berlin.

37. A. Mackworth and F.Mokhtarian (1992) "A theory of multiscale, curvature-based shape representation for planar curves", *IEEE Trans. Pattern Anal. Machine Intell* **14** pp 789–805.

38. P. Maragos (1987) "Tutorial on advances in morphological image processing and analysis" *optical engineering* **26** No. 7.

39. D. Marr (1982) *Vision* Freeman and Co.

40. P. Mascarenhas (1992), "Diffusion generated motion by mean curvature", preprint.

41. G. Matheron (1975), *Random Sets and Integral Geometry*, John Wiley, N.Y.

42. B. Merriman, J. Bence and S.Osher (1992), "Diffusion Generated motion by Mean Curvature", *CAM Report 92-18. Depart. of Mathematics. University of California, Los Angeles CA* 90024.1555.

43. J.M. Morel, S. Solimini (1994) *Variational methods in image segmentation* Birkhäuser.

44. M. Nitzberg, D.Mumford and T. Shiota (1993) *Filtering, Segmentation and Depth.* Lecture Notes in Computer Science, 662, Springer-Verlag.

45. S. Osher and L. Rudin. (1990) " Feature-oriented image enhancement using shock filters." *SIAM J. on Numerical Analysis*, 27, 919-940.

46. S. Osher and J. Sethian(1988), "Fronts propagating with curvature dependent speed : algorithms based on the Hamilton-Jacobi formulation", *J. Comp. Physics* **79**, 12–49.

47. E.J. Pauwels, P. Fiddelaers, T. Moons and L.J. Van Gool. (1993) " An extended class of scale-invariant and recursive scale space filters" *preprint, submitted to PAMI.*

48. P. Perona and J. Malik (1987) "A scale space and edge detection using anisotropic diffusion", *Proc. IEEE Computer Soc. Workshop on Computer Vision.*

49. L. Rudin, S.Osher and E.Fatemi (1992) "Nonlinear Total Variation Based Noise Removal Algorithms" *Proc. Modélisations Matématiques pour le traitement d'images*, INRIA, 149–179.

50. L.I. Rudin (1987) *Images, numerical analysis of singularities and shock filters.* PhD dissertation, Caltech, Pasadena , California No 5250:TR, 1987.

51. G. Sapiro and A. Tannenbaum (1994) "On affine plane curve evolution" *Journal of Functional Analysis*, 119, 1, 79-120.

52. G. Sapiro and A. Tannenbaum (1993) "On invariant curve evolution and image analysis" *Indiana Journal of Mathematics*, 42(3), 985-1009.

53. J. Serra (1982), *Image analysis and mathematical morphology, Vol 1* , Academic Press.

54. J. Sethian (1982) *An analysis of flow propagation*, PhD thesis, University of California.

55. J. Sethian (1985) "Curvature and the evolution of fronts" *Comm. Math. Phys.*, 101, 487-499.

56. M. Soner (1989) " Motion of a set by curvature of its mean boundary." *Preprint*

57. A.P. Witkin (1983), "Scale-space filtering", *Proc. of IJCAI, Karlsruhe* 1019–1021.

The manufacturer's authorised representative in the EU for product safety is Oxford University Press España S.A. of el Parque Empresarial San Fernando de Henares, Avenida de Castilla, 2 – 28830 Madrid (www.oup.es/en or product.safety@oup.com). OUP España S.A. also acts as importer into Spain of products made by the manufacturer.

www.ingramcontent.com/pod-product-compliance
Ingram Content Group UK Ltd.
Pitfield, Milton Keynes, MK11 3LW, UK
UKHW021436280726
14060UKWH00001BA/105

* 9 7 8 0 1 9 8 5 0 0 1 4 8 *